*Micro- and Opto-Electronic Materials and Structures:
Physics, Mechanics, Design, Reliability, Packaging*

*Volume II
Physics Design—Reliability and Packaging*

Micro- and Opto-Electronic Materials and Structures: Physics, Mechanics, Design, Reliability, Packaging

Volume II
Physics Design—Reliability and Packaging

Edited by:

E. Suhir
University of California, Santa Cruz
Santa Cruz, California, USA

University of Maryland
College Park, Maryland, USA

Y.C. Lee
University of Colorado
Boulder, Colorado, USA

C.P. Wong
Georgia Tech
Atlanta, Georgia, USA

E. Suhir
University of California, Santa Cruz
Santa Cruz, California
and
University of Maryland
College Park, Maryland

Y.C. Lee
University of Colorado
Boulder, Colorado

C.P. Wong
Georgia Institute of Technology
Atlanta, Georgia

Micro- and Opto-Electronic Materials and Structures:
Physics, Mechanics, Design, Reliability, Packaging

Library of Congress Control Number: 2006922729

ISBN 0-387-27974-1   e-ISBN 0-387-32989-7
ISBN 978-0-387-27974-9

Printed on acid-free paper.

© 2007 Springer Science+Business Media, Inc.
All rights reserved. This work may not be translated or copied in whole or in part without the written permission of the publisher (Springer Science + Business Media, Inc., 233 Spring Street, New York, NY 10013, USA), except for brief excerpts in connection with reviews or scholarly analysis. Use in connection with any form of information storage and retrieval, electronic adaptation, computer software, or by similar or dissimilar methodology now know or hereafter developed is forbidden.
The use in this publication of trade names, trademarks, service marks and similar terms, even if the are not identified as such, is not to be taken as an expression of opinion as to whether or not they are subject to proprietary rights.

Printed in the United States of America.

9 8 7 6 5 4 3 2 1   SPIN 11055464

springer.com

Contents

Volume I

List of Contributors   xxvii

Preface   xxxi

Materials Physics

Chapter 1
Polymer Materials Characterization, Modeling and Application
L.J. Ernst, K.M.B. Jansen, D.G. Yang, C. van 't Hof, H.J.L. Bressers, J.H.J. Janssen
and G.Q. Zhang   3

1.1. Introduction	3
1.2. Polymers in Microelectronics	4
1.3. Basics of Visco-Elastic Modeling	6
1.3.1. Preliminary: State Dependent Viscoelasticity	6
1.3.2. Incremental Relationship	10
1.3.3. Linear State Dependent Viscoelasticity	13
1.3.4. Isotropic Material Behavior	14
1.3.5. Interrelations between Property Functions	15
1.3.6. Elastic Approximations	17
1.4. Linear Visco-Elastic Modeling (Fully Cured Polymers)	18
1.4.1. Introduction	18
1.4.2. Static Testing of Relaxation Moduli	18
1.4.3. Time-Temperature Superposition Principle	23
1.4.4. Static Testing of Creep Compliances	24
1.4.5. Dynamic Testing	27
1.5. Modeling of Curing Polymers	34
1.5.1. "Partly State Dependent" Modeling (Curing Polymers)	35
1.5.2. "Fully State Dependent" Modeling (Curing Polymers)	49
1.6. Parameterized Polymer Modeling (PPM)	53
1.6.1. PPM Hypotheses	54
1.6.2. Experimental Characterizations	55
1.6.3. PPM Modeling in Virtual Prototyping	62
Acknowledgments	62
References	62

Chapter 2
Thermo-Optic Effects in Polymer Bragg Gratings
Avram Bar-Cohen, Bongtae Han and Kyoung Joon Kim   65

- 2.1. Introduction   65
- 2.2. Fundamentals of Bragg Gratings   67
  - 2.2.1. Physical Descriptions   67
  - 2.2.2. Basic Optical Principles   68
- 2.3. Thermo-Optical Modeling of Polymer Fiber Bragg Grating   70
  - 2.3.1. Heat Generation by Intrinsic Absorption   70
  - 2.3.2. Analytical Thermal Model of PFBG   78
  - 2.3.3. FEA Thermal Model of PFBG   80
  - 2.3.4. Thermo-Optical Model of PFBG   80
- 2.4. Thermo-Optical Behavior of PMMA-Based PFBG   84
  - 2.4.1. Description of a PMMA-Based PFBG and Light Sources   85
  - 2.4.2. Power Variation Along the PFBG   86
  - 2.4.3. Thermo-Optical Behavior of the PFBG–LED Illumination   87
  - 2.4.4. Thermo-Optical Behavior of the PFBG–SM LD Illumination   92
  - 2.4.5. Thermo-Optical Behavior of the PFBG Associated with Other Light Sources   101
- 2.5. Concluding Remarks   102
- References   102
- Appendix 2.A: Solution Procedure to Obtain the Optical Power Along the PFBG   104
- Appendix 2.B: Solution Procedure to Determine the Temperature Profile Along the PFBG   106
  - 2.B.1. Solution Procedure of the Temperature Profile Along the PFBG with the LED   106
  - 2.B.2. Solution Procedure of the Temperature Profile Along the PFBG with the SM LD   106

Chapter 3
Photorefractive Materials and Devices for Passive Components in WDM Systems
Claire Gu, Yisi Liu, Yuan Xu, J.J. Pan, Fengqing Zhou, Liang Dong and Henry He   111

- 3.1. Introduction   111
- 3.2. Tunable Flat-Topped Filter   114
  - 3.2.1. Principle of Operation   114
  - 3.2.2. Device Simulation   116
  - 3.2.3. Design for Implementation   117
- 3.3. Wavelength Selective 2×2 Switch   117
  - 3.3.1. Principle of Operation   118
  - 3.3.2. Experimental Demonstration   119
  - 3.3.3. Theoretical Analysis   121
  - 3.3.4. Optimized Switch Design   123
  - 3.3.5. Discussion   125
- 3.4. High Performance Dispersion Compensators   126
  - 3.4.1. Multi-Channel Dispersion-Slope Compensator   126
  - 3.4.2. High Precision FBG Fabrication Method and Dispersion Management Filters   129
- 3.5. Conclusions   133
- References   133

Chapter 4
Thin Films for Microelectronics and Photonics: Physics, Mechanics, Characterization, and Reliability
David T. Read and Alex A. Volinsky — 135

- 4.1. Terminology and Scope — 135
  - 4.1.1. Thin Films — 135
  - 4.1.2. Motivation — 136
  - 4.1.3. Chapter Outline — 136
- 4.2. Thin Film Structures and Materials — 137
  - 4.2.1. Substrates — 137
  - 4.2.2. Epitaxial Films — 137
  - 4.2.3. Dielectric Films — 140
  - 4.2.4. Metal Films — 141
  - 4.2.5. Organic and Polymer Films — 142
  - 4.2.6. MEMS Structures — 142
  - 4.2.7. Intermediate Layers: Adhesion, Barrier, Buffer, and Seed Layers — 142
- 4.3. Manufacturability/Reliability Challenges — 143
  - 4.3.1. Film Deposition and Stress — 144
  - 4.3.2. Grain Structure and Texture — 147
  - 4.3.3. Impurities — 151
  - 4.3.4. Dislocations — 152
  - 4.3.5. Electromigration and Voiding — 153
  - 4.3.6. Structural Considerations — 155
  - 4.3.7. Need for Mechanical Characterization — 155
  - 4.3.8. Properties of Interest — 156
- 4.4. Methods for mechanical characterization of thin films — 157
  - 4.4.1. Microtensile Testing — 157
  - 4.4.2. Instrumented Indentation — 159
  - 4.4.3. Other Techniques — 164
  - 4.4.4. Adhesion Tests — 165
- 4.5. Materials and Properties — 172
  - 4.5.1. Grain Size and Structure Size Effects — 172
- 4.6. Properties of Specific Materials — 173
- 4.7. Future Research — 175
  - 4.7.1. Techniques — 175
  - 4.7.2. Properties — 175
  - 4.7.3. Length Scale — 175
- References — 176

Chapter 5
Carbon Nanotube Based Interconnect Technology: Opportunities and Challenges
Alan M. Cassell and Jun Li — 181

- 5.1. Introduction: Physical Characteristics of Carbon Nanotubes — 181
  - 5.1.1. Structural — 181
  - 5.1.2. Electrical — 182
  - 5.1.3. Mechanical — 185
  - 5.1.4. Thermal — 186
- 5.2. CNT Fabrication Technologies — 186

5.2.1. Chemical Vapor Deposition of Carbon Nanotubes	187
5.2.2. Process Integration and Development	189
5.3. Carbon Nanotubes as Interconnects	191
5.3.1. Limitations of the Current Technology	191
5.3.2. Architecture, Geometry and Performance Potential Using Carbon Nanotubes	191
5.4. Design, Manufacture and Reliability	194
5.4.1. Microstructural Attributes and Effects on Electrical Characteristics	194
5.4.2. Interfacial Contact Materials	196
5.4.3. End-contacted Metal–CNT Junction	198
5.4.4. Thermal Stress Characteristics	198
5.4.5. Reliability Test	199
5.5. Summary	200
References	200

Chapter 6
Virtual Thermo-Mechanical Prototyping of Microelectronics and Microsystems
A. Wymysłowski, G.Q. Zhang, W.D. van Driel and L.J. Ernst — 205

6.1. Introduction	205
6.2. Physical Aspects for Numerical Simulations	206
6.2.1. Numerical Modeling	208
6.2.2. Material Properties and Models	211
6.2.3. Thermo-Mechanical Related Failures	215
6.2.4. Designing for Reliability	219
6.3. Mathematical Aspects of Optimization	225
6.3.1. Design of Experiments	226
6.3.2. Response Surface Modeling	236
6.3.3. Advanced Approach to Virtual Prototyping	242
6.3.4. Designing for Quality	249
6.4. Application Case	252
6.4.1. Problem Description	252
6.4.2. Numerical Approach to QFN Package Design	253
6.5. Conclusion and Challenges	259
6.6. List of Acronyms	264
Acknowledgments	264
References	264

Materials Mechanics

Chapter 7
Fiber Optics Structural Mechanics and Nano-Technology Based New Generation of Fiber Coatings: Review and Extension
E. Suhir — 269

7.1. Introduction	269
7.2. Fiber Optics Structural Mechanics	270
7.2.1. Review	270
7.3. New Nano-Particle Material (NPM) for Micro- and Opto-Electronic Applications	273
7.3.1. New Nano-Particle Material (NPM)	273
7.3.2. NPM-Based Optical Silica Fibers	274

7.4. Conclusions	277
Acknowledgment	277
References	277

Chapter 8
Area Array Technology for High Reliability Applications
Reza Ghaffarian — 283

8.1. Introduction	283
8.2. Area Array Packages (AAPs)	284
8.2.1. Advantages of Area Array Packages	285
8.2.2. Disadvantages of Area Arrays	285
8.2.3. Area Array Types	286
8.3. Chip Scale Packages (CSPs)	286
8.4. Plastic Packages	288
8.4.1. Background	288
8.4.2. Plastic Area Array Packages	288
8.4.3. Plastic Package Assembly Reliability	289
8.4.4. Reliability Data for BGA, Flip Chip BGA, and CSP	291
8.5. Ceramic Packages	293
8.5.1. Background	293
8.5.2. Ceramic Package Assembly Reliability	294
8.5.3. Literature Survey on CBGA/CCGA Assembly Reliability	295
8.5.4. CBGA Thermal Cycle Test	297
8.5.5. Comparison of 560 I/O PBGA and CCGA assembly reliability	302
8.5.6. Designed Experiment for Assembly	305
8.6. Summary	309
8.7. List of Acronyms and Symbols	310
Acknowledgments	311
References	311

Chapter 9
Metallurgical Factors Behind the Reliability of High-Density Lead-Free Interconnections
Toni T. Mattila, Tomi T. Laurila and Jorma K. Kivilahti — 313

9.1. Introduction	313
9.2. Approaches and Methods	315
9.2.1. The Four Steps of The Iterative Approach	315
9.2.2. The Role of Different Simulation Tools in Reliability Engineering	321
9.3. Interconnection Microstructures and Their Evolution	324
9.3.1. Solidification	324
9.3.2. Solidification Structure and the Effect of Contact Metalization Dissolution	325
9.3.3. Interfacial Reactions Products	330
9.3.4. Deformation Structures (Due to Slip and Twinning)	333
9.3.5. Recovery, Recrystallization and Grain Growth	335
9.4. Two Case Studies on Reliability Testing	335
9.4.1. Case 1: Reliability of Lead-Free CSPs in Thermal cycling	337
9.4.2. Case 2: Reliability of Lead-Free CSPs in Drop Testing	341
9.5. Summary	347

Acknowledgments 348
References 348

Chapter 10
Metallurgy, Processing and Reliability of Lead-Free Solder Joint Interconnections
Jin Liang, Nader Dariavach and Dongkai Shangguan 351

- 10.1. Introduction 351
- 10.2. Physical Metallurgy of Lead-Free Solder Alloys 352
  - 10.2.1. Tin-Lead Solders 352
  - 10.2.2. Lead-Free Solder Alloys 353
  - 10.2.3. Interfacial Reaction: Wetting and Spreading 357
  - 10.2.4. Interfacial Intermetallic Formation and Growth at Liquid–Solid Interfaces 363
- 10.3. Lead-Free Soldering Processes and Compatibility 377
  - 10.3.1. Lead-Free Soldering Materials 378
  - 10.3.2. PCB Substrates and Metalization Finishes 380
  - 10.3.3. Lead-Free Soldering Processes 381
  - 10.3.4. Components for Lead-Free Soldering 384
  - 10.3.5. Design, Equipment and Cost Considerations 387
- 10.4. Reliability of Pb-Free Solder Interconnects 388
  - 10.4.1. Reliability and Failure Distribution of Pb-Free Solder Joints 388
  - 10.4.2. Effects of Loading and Thermal Conditions on Reliability of Solder Interconnection 389
  - 10.4.3. Reliability of Pb-Free Solder Joints in Comparison to Sn-Pb Eutectic Solder Joints 395
- 10.5. Guidelines for Pb-free Soldering and Improvement in Reliability 406
- References 406

Chapter 11
Fatigue Life Assessment for Lead-Free Solder Joints
Masaki Shiratori and Qiang Yu 411

- 11.1. Introduction 411
- 11.2. The Intermetallic Compound Formed at the Interface of the Solder Joints and the Cu-pad 412
- 11.3. Mechanical Fatigue Testing Equipment and Load Condition in the Lead Free Solder 413
- 11.4. Results of Mechanical Fatigue Test 414
- 11.5. Critical Fatigue Stress Limit for the Intermetallic Compound Layer 417
- 11.6. Influence of the Plating Material on the Fatigue Life of Sn-Zn (Sn-9Zn and Sn-8Zn-3Bi) Solder Joints 424
- 11.7. Conclusion 426
- References 426

Chapter 12
Lead-Free Solder Materials: Design For Reliability
John H.L. Pang 429

- 12.1. Introduction 429
- 12.2. Mechanics of Solder Materials 430
  - 12.2.1. Fatigue Behavior of Solder Materials 431
- 12.3. Design For Reliability (DFR) 433

CONTENTS

12.4. Constitutive Models For Lead Free Solders — 435
    12.4.1. Tensile Test Results — 435
    12.4.2. Creep Test Results — 440
12.5. Low Cycle Fatigue Models — 443
12.6. FEA Modeling and Simulation — 448
12.7. Reliability Test and Analysis — 454
12.8. Conclusions — 456
Acknowledgments — 456
References — 456

Chapter 13
Application of Moire Interferometry to Strain Analysis of PCB Deformations at Low Temperatures — 459
Arkady Voloshin

13.1. Introduction — 459
13.2. Optical Method and Recording of Fringe Patterns — 460
    13.2.1. Fractional Fringe Approach — 461
    13.2.2. Grating Frequency Increase — 461
    13.2.3. Creation of a High-Frequency Master Grating — 462
    13.2.4. Combination of the High Grating Frequency and Fractional Fringe Approach — 463
13.3. Data Processing — 463
13.4. Test Boards and Specimen Grating — 463
13.5. Elevated Temperature Test — 465
13.6. Low Temperature Test — 468
13.7. Conclusions — 470
Acknowledgment — 472
References — 473

Chapter 14
Characterization of Stresses and Strains in Microelectronics and Photonics Devices Using Photomechanics Methods — 475
Bongtae Han

14.1. Introduction — 475
14.2. Stress/Strain analysis — 476
    14.2.1. Moiré Interferometry — 476
    14.2.2. Extension: Microscopic Moiré Interferometry — 477
    14.2.3. Specimen Gratings — 479
    14.2.4. Strain Analysis — 480
    14.2.5. Thermal Deformation Measured at Room Temperature — 481
    14.2.6. Deformation as a Function of Temperature — 485
    14.2.7. Hygroscopic Deformation — 494
    14.2.8. Micromechanics — 501
14.3. Warpage Analysis — 505
    14.3.1. Twyman/Green Interferometry — 505
    14.3.2. Shadow Moiré — 509
    14.3.3. Far Infrared Fizeau Interferometry — 514
Acknowledgment — 520
References — 520

Chapter 15
Analysis of Reliability of IC Packages Using the Fracture Mechanics Approach
Andrew A.O. Tay — 523

- 15.1. Introduction — 523
- 15.2. Heat Transfer and Moisture Diffusion in IC Packages — 525
- 15.3. Fundamentals of Interfacial Fracture Mechanics — 527
- 15.4. Criterion for Crack Propagation — 529
- 15.5. Interface Fracture Toughness — 529
- 15.6. Total Stress Intensity Factor — 530
- 15.7. Calculation of SERR and Mode Mixity — 531
  - 15.7.1. Crack Surface Displacement Extrapolation Method — 531
  - 15.7.2. Modified J-integral Method — 532
  - 15.7.3. Modified Virtual Crack Closure Method — 533
  - 15.7.4. Variable Order Boundary Element Method — 536
  - 15.7.5. Interaction Integral Method — 536
- 15.8. Experimental Verification — 538
- 15.9. Case Studies — 542
  - 15.9.1. Delamination Along Pad-Encapsulant Interface — 542
  - 15.9.2. Delamination Along Die-Attach/Pad Interface — 544
  - 15.9.3. Analysis Using Variable Order Boundary Element Method — 546
- 15.10. Discussion of the Various Numerical Methods for Calculating G and ψ — 549
- 15.11. Conclusion — 551
- References — 551

Chapter 16
Dynamic Response of Micro- and Opto-Electronic Systems to Shocks and Vibrations: Review and Extension
E. Suhir — 555

- 16.1. Introduction — 555
- 16.2. Review — 556
- 16.3. Extension: Quality of Shock Protection with a Flexible Wire Elements — 557
- 16.4. Analysis — 558
  - 16.4.1. Pre-Buckling Mode: Small Displacements — 558
  - 16.4.2. Post-Buckling Mode: Large Displacements — 564
- 16.5. Conclusions — 567
- References — 568

Chapter 17
Dynamic Physical Reliability in Application to Photonic Materials
Dov Ingman, Tatiana Mirer and Ephraim Suhir — 571

- 17.1. Introduction: Dynamic Reliability Approach to the Evolution of Silica Fiber Performance — 571
  - 17.1.1. Dynamic Physical Model of Damage Accumulation — 572
  - 17.1.2. Impact of the Three-Dimensional Mechanical-Temperature-Humidity Load on the Optical Fiber Reliability — 575
  - 17.1.3. Effect of Bimodality and Its Explanation Based on the Suggested Model — 576
- 17.2. Reliability Improvement through NPM-Based Fiber Structures — 585

	17.2.1. Environmental Protection by NPM-Based Coating and Overall Self-Curing Effect of NPM Layers	585
	17.2.2. Improvement in the Reliability Characteristics by Employing NPM Structures in Optical Fibers	587
17.3.	Conclusions	593
References		593

Chapter 18
High-Speed Tensile Testing of Optical Fibers—New Understanding for Reliability Prediction
Sergey Semjonov and G. Scott Glaesemann 595

18.1.	INTRODUCTION	595
18.2.	Theory	596
	18.2.1. Single-Region Power-Law Model	596
	18.2.2. Two-Region Power-Law Model	598
	18.2.3. Universal Static and Dynamic Fatigue Curves	599
18.3.	Experimental	602
	18.3.1. Sample Preparation	602
	18.3.2. Dynamic Fatigue Tests	604
	18.3.3. Static Fatigue Tests	605
18.4.	Results and Discussion	606
	18.4.1. High-Speed Testing	606
	18.4.2. Static Fatigue	610
	18.4.3. Influence of Multiregion Model on Lifetime Prediction	613
18.5.	Conclusion	613
References		614
Appendix 18.A: High Speed Axial Strength Testing: Measurement Limits		616
Appendix 18.B: Incorporating Static Fatigue Results into Dynamic Fatigue Curves		620
	18.B.1. Static Fatigue Test	620
	18.B.2. Dynamic Fatigue Test	621
	18.B.3. Discussion	622

Chapter 19
The Effect of Temperature on the Microstructure Nonlinear Dynamics Behavior
Xiaoling He 627

19.1.	Introduction	627
19.2.	Theoretical Development	630
	19.2.1. Background on Nonlinear Dynamics and Nonlinear Thermo-Elasticity Theories	630
	19.2.2. Nonlinear Thermo-Elasticity Development for an Isotropic Laminate Subject to Thermal and Mechanical and Load	631
19.3.	Thin Laminate Deflection Response Subject to Thermal Effect and Mechanical Load	633
	19.3.1. Steady State Temperature Effect	633
	19.3.2. Transient Thermal Field Effect	638
19.4.	Stress Field in Nonlinear Dynamics Response	653
	19.4.1. Stress Field Formulation	653
	19.4.2. Stress Distribution	654
	19.4.3. Failure Analysis	654

19.5. Discussions	660
19.6. Summary	661
Nomenclature	662
Acknowledgment	663
References	663

Chapter 20
Effect of Material's Nonlinearity on the Mechanical Response of some Piezoelectric and Photonic Systems
Victor Birman and Ephraim Suhir — 667

20.1. Introduction	667
20.2. Effect of Physical Nonlinearity on Vibrations of Piezoelectric Rods Driven by Alternating Electric Field	668
20.2.1. Physically Nonlinear Constitutive Relationships for an Orthotropic Cylindrical Piezoelectric Rod Subject to an Electric Field in the Axial Direction	670
20.2.2. Analysis of Uncoupled Axial Vibrations	673
20.2.3. Solution for Coupled Axial-Radial Axisymmetric Vibrations by the Generalized Galerkin Procedure	677
20.2.4. Numerical Results and Discussion	678
20.3. The Effect of the Nonlinear Stress–Strain Relationship on the Response of Optical Fibers	683
20.3.1. Stability of Optical Fibers	684
20.3.2. Stresses and Strains in a Lightwave Coupler Subjected to Tension	686
20.3.3. Free Vibrations	690
20.3.4. Bending of an Optical Fiber	692
20.4. Conclusions	695
Acknowledgment	696
References	697
Index	701

Volume II

List of Contributors	xxvii
Preface	xxxi

Physical Design

Chapter 1
Analytical Thermal Stress Modeling in Physical Design for Reliability of Micro- and Opto-Electronic Systems: Role, Attributes, Challenges, Results
E. Suhir — 3

1.1. Thermal Loading and Thermal Stress Failures	3
1.2. Thermal Stress Modeling	4
1.3. Bi-Metal Thermostats and other Bi-Material Assemblies	5
1.4. Finite-Element Analysis	5

1.5. Die-Substrate and other Bi-Material Assemblies	6
1.6. Solder Joints	8
1.7. Design Recommendations	9
1.8. "Global" and "Local" Mismatch and Assemblies Bonded at the Ends	10
1.9. Assemblies with Low Modulus Adhesive Layer at the Ends	11
1.10. thermally Matched Assemblies	11
1.11. Thin Films	12
1.12. Polymeric Materials And Plastic Packages	13
1.13. Thermal Stress Induced Bowing and Bow-Free Assemblies	14
1.14. Probabilistic Approach	15
1.15. Optical Fibers and other Photonic Structures	15
1.16. Conclusion	16
References	17

Chapter 2
Probabilistic Physical Design of Fiber-Optic Structures
Satish Radhakrishnan, Ganesh Subbarayan and Luu Nguyen   23

2.1. Introduction	23
2.1.1. Demonstration Vehicle	24
2.2. Optical Model	25
2.2.1. Mode Field Diameter	26
2.2.2. Refraction and Reflection Losses	27
2.2.3. Calculations for Coupling Losses	27
2.2.4. Coupling Efficiency	28
2.3. Interactions in System and Identification of Critical Variables	30
2.3.1. Function Variable Incidence Matrix	30
2.3.2. Function Variable Incidence Matrix to Graph Conversion	31
2.3.3. Graph Partitioning Techniques	34
2.3.4. System Decomposition using Simulated Annealing	34
2.4. Deterministic Design Procedures	37
2.4.1. Optimal and Robust Design	40
2.4.2. A Brief Review of Multi-Objective Optimization	42
2.4.3. Implementation	43
2.4.4. Results	43
2.5. Stochastic Analysis	44
2.5.1. The First and Second Order Second Moment Methods	44
2.6. Probabilistic Design for Maximum Reliability	46
2.6.1. Results	49
2.7. Stochastic Characterization of Epoxy Behavior	51
2.7.1. Viscoelastic Models	52
2.7.2. Modeling the Creep Test	53
2.7.3. Dynamic Mechanical Analysis	54
2.7.4. Experimental Results	55
2.8. Analytical Model to Determine VCSEL Displacement	57
2.8.1. Results	63
2.9. Summary	67
References	67

Chapter 3
The Wirebonded Interconnect: A Mainstay for Electronics
Harry K. Charles, Jr. — 71

- 3.1. Introduction — 71
  - 3.1.1. Integrated Circuit Revolution — 71
  - 3.1.2. Interconnection Types — 72
  - 3.1.3. Wirebond Importance — 80
- 3.2. Wirebonding Basics — 81
  - 3.2.1. Thermocompression Bonding — 81
  - 3.2.2. Ultrasonic Bonding — 83
  - 3.2.3. Thermosonic Bonding — 85
  - 3.2.4. Wirebond Reliability — 87
  - 3.2.5. Wirebond Testing — 89
  - 3.2.6. Bonding Automation and Optimization — 93
- 3.3. Materials — 95
  - 3.3.1. Bonding Wire — 95
  - 3.3.2. Bond Pad Metallurgy — 100
  - 3.3.3. Gold Plating — 102
  - 3.3.4. Pad Cleaning — 104
- 3.4. Advanced Bonding Methods — 105
  - 3.4.1. Fine Pitch Bonding — 105
  - 3.4.2. Soft Substrates — 108
  - 3.4.3. Machine Improvements — 110
  - 3.4.4. Higher Frequency Wirebonding — 110
  - 3.4.5. Stud Bumping — 115
- 3.5. Summary — 116
- Acknowledgments — 116
- References — 116

Chapter 4
Metallurgical Interconnections for Extreme High and Low Temperature Environments
George G. Harman — 121

- 4.1. Introduction — 121
- 4.2. High Temperature Interconnections Requirements — 122
  - 4.2.1. Wire Bonding — 122
  - 4.2.2. The Use of Flip Chips in HTE — 127
  - 4.2.3. General Overview of Metallurgical Interfaces for Both HTE and LTE — 129
- 4.3. Low Temperature Environment Interconnection Requirements — 129
- 4.4. Corrosion and Other Problems in Both *HTE*, and *LTE* — 130
- 4.5. The Potential Use of High Temperature Polymers in HTE — 131
- 4.6. Conclusions — 132
- Acknowledgments — 132
- References — 132

Chapter 5
Design, Process, and Reliability of Wafer Level Packaging
Zhuqing Zhang and C.P. Wong — 135

- 5.1. Introduction — 135

5.2. WLCSP	137
5.2.1. Thin Film Redistribution	137
5.2.2. Encapsulated Package	139
5.2.3. Compliant Interconnect	139
5.3. Wafer Level Underfill	141
5.3.1. Challenges of Wafer Level Underfill	142
5.3.2. Examples of Wafer Level Underfill Process	143
5.4. Comparison of Flip-Chip and WLCSP	145
5.5. Wafer Level Test and Burn-In	145
5.6. Summary	149
References	149

Chapter 6
Passive Alignment of Optical Fibers in V-grooves with Low Viscosity Epoxy Flow
S.W. Ricky Lee and C.C. Lo   151

6.1. Introduction	151
6.2. Design and Fabrication of Silicon Optical Bench with V-grooves	152
6.3. Issues of Conventional Passive Alignment Methods	158
6.3.1. V-grooves with Cover Plate	158
6.3.2. Edge Dispensing of Epoxy	161
6.4. Modified Passive Alignment Method	162
6.4.1. Working Principle	162
6.4.2. Alignment Mechanism	163
6.4.3. Design of Experiment	164
6.4.4. Experimental Procedures	164
6.4.5. Experimental Results	165
6.5. Effects of Epoxy Viscosity and Dispensing Volume	168
6.6. Application to Fiber Array Passive Alignment	170
6.7. Conclusions and Discussion	172
References	172

Reliability and Packaging

Chapter 7
Fundamentals of Reliability and Stress Testing
H. Anthony Chan   177

7.1. More Performance at Lower Cost in Shorter Time-to-market	178
7.1.1. Rapid Technological Developments	178
7.1.2. Integration of More Products into Human Life	178
7.1.3. Diverse Environmental Stresses	178
7.1.4. Competitive Market	179
7.1.5. Short Product Cycles	179
7.1.6. The Bottom Line	179
7.2. Measure of Reliability	180
7.2.1. Failure Rate	180
7.2.2. Systems with Multiple Independent Failure Modes	181
7.2.3. Failure Rate Distribution	182
7.3. Failure Mechanisms in Electronics and Packaging	184

       7.3.1. Failure Mechanisms at Chip Level Include   184
       7.3.2. Failure Mechanisms at Bonding Include   184
       7.3.3. Failure Mechanisms in Device Packages Include   185
       7.3.4. Failure Mechanisms in Epoxy Compounds Include   185
       7.3.5. Failure Mechanisms at Shelf Level Include   185
       7.3.6. Failure Mechanisms in Material Handling Include   185
       7.3.7. Failure Mechanisms in Fiber Optics Include   185
       7.3.8. Failure Mechanisms in Flat Panel Displays Include   186
  7.4. Reliability Programs and Strategies   186
  7.5. Product Weaknesses and Stress Testing   187
       7.5.1. Why do Products Fail?   187
       7.5.2. Stress Testing Principle   189
  7.6. Stress Testing Formulation   191
       7.6.1. Threshold and Cumulative Stress Failures   191
       7.6.2. Stress Stimuli and Flaws   192
       7.6.3. Modes of Stress Testing   193
       7.6.4. Lifetime Failure Fraction   194
       7.6.5. Robustness Against Maximum Service Life Stress   195
       7.6.6. Stress–Strength Contour   197
       7.6.7. Common Issues   198
  7.7. Further Reading   201

Chapter 8
How to Make a Device into a Product: *Accelerated Life Testing (ALT), Its Role, Attributes, Challenges, Pitfalls, and Interaction with Qualification Tests*
E. Suhir   203

  8.1. Introduction   203
  8.2. Some Major Definitions   204
  8.3. Engineering Reliability   204
  8.4. Field Failures   205
  8.5. Reliability is a Complex Property   206
  8.6. Three Major Classes of Engineering Products and Market Demands   206
  8.7. Reliability, Cost and Time-to-Market   208
  8.8. Reliability Costs Money   208
  8.9. Reliability Should Be Taken Care of on a Permanent Basis   209
  8.10. Ways to Prevent and Accommodate Failures   210
  8.11. Redundancy   211
  8.12. Maintenance and Warranty   211
  8.13. Test Types   212
  8.14. Accelerated Tests   212
  8.15. Accelerated Test Levels   213
  8.16. Qualification Standards   213
  8.17. Accelerated Life Tests (ALTs)   214
  8.18. Accelerated Test Conditions   215
  8.19. Acceleration Factor   216
  8.20. Accelerated Stress Categories   217
  8.21. Accelerated Life Tests (ALTs) and Highly Accelerated Life Tests (HALTs)   218
  8.22. Failure Mechanisms and Accelerated Stresses   219

8.23.	ALTs: Pitfalls and Challenges	219
8.24.	Burn-ins	220
8.25.	Wear-Out Failures	221
8.26.	Non-Destructive Evaluations (NDE's)	222
8.27.	Predictive Modeling	222
8.28.	Some Accelerated Life Test (ALT) Models	223
	8.28.1. Power Law	224
	8.28.2. Boltzmann-Arrhenius Equation	224
	8.28.3. Coffin-Manson Equation (Inverse Power Law)	225
	8.28.4. Paris-Erdogan Equation	226
	8.28.5. Bueche-Zhurkov Equation	227
	8.28.6. Eyring Equation	227
	8.28.7. Peck and Black Equations	227
	8.28.8. Fatigue Damage Model (Miner's Rule)	228
	8.28.9. Creep Rate Equations	228
	8.28.10. Weakest Link Models	228
	8.28.11. Stress–Strength Models	229
8.29.	Probability of Failure	229
8.30.	Conclusions	230
References		230

Chapter 9
Micro-Deformation Analysis and Reliability Estimation of Micro-Components by Means of NanoDAC Technique
Bernd Michel and Jürgen Keller — 233

9.1.	Introduction	233
9.2.	Basics of Digital Image Correlation	234
	9.2.1. Cross Correlation Algorithms on Gray Scale Images	234
	9.2.2. Subpixel Analysis for Enhanced Resolution	236
	9.2.3. Results of Digital Image Correlation	238
9.3.	Displacement and Strain Measurements on SFM Images	239
	9.3.1. Digital Image Correlation under SPM Conditions	239
	9.3.2. Technical Requirements for the Application of the Correlation Technique	241
9.4.	Deformation Analysis on Thermally and Mechanically Loaded Objects under the SFM	241
	9.4.1. Reliability Aspects of Sensors and Micro Electro-Mechanical Systems (MEMS)	241
	9.4.2. Thermally Loaded Gas Sensor under SFM	242
	9.4.3. Crack Detection and Evaluation by SFM	243
9.5.	Conclusion and Outlook	250
References		250

Chapter 10
Interconnect Reliability Considerations in Portable Consumer Electronic Products
Sridhar Canumalla and Puligandla Viswanadham — 253

10.1.	Introduction	253
10.2.	Reliability—Thermal, Mechanical and Electrochemical	255
	10.2.1. Accelerated Life Testing	255
	10.2.2. Thermal Environment	257

	10.2.3.	Mechanical Environment	257
	10.2.4.	Electrochemical Environment	264
	10.2.5.	Tin Whiskers	267
10.3.	Reliability Comparisons in Literature	267
	10.3.1.	Thermomechanical Reliability	268
	10.3.2.	Mechanical Reliability	270
10.4.	Influence of Material Properties on Reliability	271
	10.4.1.	Printed Wiring Board	271
	10.4.2.	Package	272
	10.4.3.	Surface Finish	272
10.5.	Failure Mechanisms	273
	10.5.1.	Thermal Environment	273
	10.5.2.	Mechanical Environment	276
	10.5.3.	Electrochemical Environment	286
10.6.	reliability test Practices	291
10.7.	Summary	294
Acknowledgments	295
References	295

Chapter 11
MEMS Packaging and Reliability
Y.C. Lee — 299

11.1. Introduction	299
11.2. Flip-Chip Assembly for Hybrid Integration	304
11.3. Soldered Assembly for Three-Dimensional MEMS	309
11.4. Flexible Circuit Boards for MEMS	313
11.5. Atomic Layer Deposition for Reliable MEMS	316
11.6. Conclusions	320
Acknowledgments	320
References	320

Chapter 12
Advances in Optoelectronic Methodology for MOEMS Testing
Ryszard J. Pryputniewicz — 323

12.1. Introduction	323
12.2. MOEMS Samples	324
12.3. Analysis	328
12.4. Optoelectronic Methodology	330
12.5. Representative Applications	334
12.6. Conclusions and Recommendations	338
Acknowledgments	339
References	339

Chapter 13
Durability of Optical Nanostructures: Laser Diode Structures and Packages, A Case Study
Ajay P. Malshe and Jay Narayan — 341

13.1. High Efficiency Quantum Confined (Nanostructured) III-Nitride Based Light Emitting Diodes And Lasers	342
13.1.1. Introduction	342
13.2. Investigation of Reliability Issues in High Power Laser Diode Bar Packages	348
13.2.1. Introduction	348
13.2.2. Preparation of Packaged Samples for Reliability Testing	349
13.2.3. Finding and Model of Reliability Results	350
13.3. Conclusions	357
Acknowledgments	358
References	358

Chapter 14
Review of the Technology and Reliability Issues Arising as Optical Interconnects Migrate onto the Circuit Board
P. Misselbrook, D. Gwyer, C. Bailey, D. Gwyer, C. Bailey, P.P. Conway and K. Williams   361

14.1. Background to Optical Interconnects	362
14.2. Transmission Equipment for Optical Interconnects	362
14.3. Very Short Reach Optical Interconnects	365
14.4. Free Space USR Optical Interconnects	366
14.5. Guided Wave USR Interconnects	367
14.6. Component Assembly of OECB's	370
14.7. Computational Modeling of Optical Interconnects	373
14.8. Conclusions	380
Acknowledgments	380
References	381

Chapter 15
Adhesives for Micro- and Opto-Electronics Application: Chemistry, Reliability and Mechanics
D.W. Dahringer   383

15.1. Introduction	383
15.1.1. Use of Adhesives in Micro and Opto-Electronic Assemblies	383
15.1.2. Specific Applications	384
15.2. Adhesive Characteristics	385
15.2.1. General Properties of Adhesives	385
15.2.2. Adhesive Chemistry	390
15.3. Design Objective	393
15.3.1. Adhesive Joint Design	393
15.3.2. Manufacturing Issues	397
15.4. Failure Mechanism	401
15.4.1. General	401
15.4.2. Adhesive Changes	401
15.4.3. Interfacial Changes	401
15.4.4. Interfacial Stress	401
15.4.5. External Stress	402
References	402

Chapter 16
Multi-Stage Peel Tests and Evaluation of Interfacial Adhesion Strength for Micro- and Opto-Electronic Materials
Masaki Omiya, Kikuo Kishimoto and Wei Yang — 403

16.1. Introduction	403
16.2. Multi-Stage Peel Test (MPT)	407
16.2.1. Testing Setup	407
16.2.2. Multi-Stage Peel Test	408
16.2.3. Energy Variation in Steady State Peeling	409
16.3. Interfacial Adhesion Strength of Copper Thin Film	413
16.3.1. Preparation of Specimen	413
16.3.2. Measurement of Adhesion Strength by the MPT	414
16.3.3. Discussions	415
16.4. UV-Irradiation Effect on Ceramic/Polymer Interfacial Strength	419
16.4.1. Preparation of PET/ITO Specimen	419
16.4.2. Measurement of Interfacial Strength by MPT	422
16.4.3. Surface Crack Formation on ITO Layer under Tensile Loading	424
16.5. Concluding Remarks	426
Acknowledgment	427
References	427

Chapter 17
The Effect of Moisture on the Adhesion and Fracture of Interfaces in Microelectronic Packaging
Timothy P. Ferguson and Jianmin Qu — 431

17.1. Introduction	432
17.2. Moisture Transport Behavior	433
17.2.1. Background	433
17.2.2. Diffusion Theory	434
17.2.3. Underfill Moisture Absorption Characteristics	435
17.2.4. Moisture Absorption Modeling	438
17.3. Elastic Modulus Variation Due to Moisture Absorption	442
17.3.1. Background	442
17.3.2. Effect of Moisture Preconditioning	444
17.3.3. Elastic Modulus Recovery from Moisture Uptake	447
17.4. Effect of Moisture on Interfacial Adhesion	449
17.4.1. Background	449
17.4.2. Interfacial Fracture Testing	451
17.4.3. Effect of Moisture Preconditioning on Adhesion	452
17.4.4. Interfacial Fracture Toughness Recovery from Moisture Uptake	461
17.4.5. Interfacial Fracture Toughness Moisture Degradation Model	462
References	469

Chapter 18
Highly Compliant Bonding Material for Micro- and Opto-Electronic Applications
E. Suhir and D. Ingman — 473

18.1. Introduction	473

CONTENTS

18.2. Effect of the Interfacial Compliance on the interfacial Shearing Stress	474
18.3. Internal Compressive Forces	476
18.4. Advanced Nano-Particle Material (NPM)	476
18.5. Highly-Compliant Nano-Systems	478
18.6. Conclusions	479
References	480
Appendix 18.A: Bimaterial Assembly Subjected to an External Shearing Load and Change in Temperature: Expected Stress Relief due to the Elevated Interfacial Compliance	480
Appendix 18.B: Cantilever Wire ("Beam") Subjected at its Free End to a Lateral (Bending) and an Axial (Compressive) Force	483
Appendix 18.C: Compressive Forces in the NPM-Based Compound Structure	485

Chapter 19
Adhesive Bonding of Passive Optical Components
Anne-Claire Pliska and Christian Bosshard 487

19.1. Introduction	487
19.2. Optical Devices and Assemblies	489
19.2.1. Optical Components	489
19.2.2. Opto-electronics Assemblies: Specific Requirements	489
19.3. Adhesive Bonding in Optical Assemblies	503
19.3.1. Origin of Adhesion	503
19.3.2. Adhesive Selection and Dispensing	508
19.3.3. Dispensing Technologies	515
19.4. Some Applications	518
19.4.1. Laser to Fiber Assembly	518
19.4.2. Planar Lightwave Circuit (PLC) Pigtailing	520
19.5. Summary and Recommendations	522
Acknowledgments	523
References	523

Chapter 20
Electrically Conductive Adhesives: A Research Status Review
James E. Morris and Johan Liu 527

20.1. Introduction	527
20.1.1. Technology Drivers	527
20.1.2. Isotropic Conductive Adhesives (ICAs)	529
20.1.3. Anisotropic Conductive Adhesives (ACAs)	529
20.1.4. Non-Conductive Adhesive (NCA)	529
20.2. Structure	529
20.2.1. ICA	529
20.2.2. ACA	532
20.2.3. Modeling	534
20.3. Materials and Processing	534
20.3.1. Polymers	534
20.3.2. ICA Filler	536
20.3.3. ACA Processing	536
20.4. Electrical Properties	538

20.4.1. ICA	538
20.4.2. Electrical Measurements	544
20.4.3. ACA	544
20.5. Mechanical Properties	546
20.5.1. ICA	546
20.5.2. ACA	547
20.6. Thermal Properties	553
20.6.1. Thermal Characteristics	553
20.6.2. Maximum Current Carrying Capacity	553
20.7. Reliability	554
20.7.1. ICA	554
20.7.2. ACA	557
20.7.3. General Comments	565
20.8. Environmental Impact	565
20.9. Further Study	565
References	565

Chapter 21
Electrically Conductive Adhesives
Johann Nicolics and Martin Mündlein

	571
21.1. Introduction and Historical Background	571
21.2. Contact Formation	574
21.2.1. Percolation and Critical Filler Content	574
21.2.2. ICA Contact Model	575
21.2.3. Results	578
21.3. Aging Behavior and Quality Assessment	595
21.3.1. Introduction	595
21.3.2. Material Selection and Experimental Parameters	595
21.3.3. Curing Parameters and Definition of Curing Time	597
21.3.4. Testing Conditions, Typical Results, and Conclusions	598
21.4. About Typical Applications	602
21.4.1. ICA for Attachment of Power Devices	602
21.4.2. ICA for Interconnecting Parts with Dissimilar Thermal Expansion Coefficient	604
21.4.3. ICA for Cost-Effective Assembling of Multichip Modules	606
21.5. Summary	607
Notations and Definitions	607
References	608

Chapter 22
Recent Advances of Conductive Adhesives: A Lead-Free Alternative in Electronic Packaging
Grace Y. Li and C.P. Wong

	611
22.1. Introduction	611
22.2. Isotropic Conductive Adhesives (ICAs)	613
22.2.1. Improvement of Electrical Conductivity of ICAs	614
22.2.2. Stabilization of Contact Resistance on Non-Noble Metal Finishes	615
22.2.3. Silver Migration Control of ICA	618

22.2.4. Improvement of Reliability in Thermal Shock Environment	618
22.2.5. Improvement of Impact Performance of ICA	619
22.3. Anisotropic Conductive Adhesives (ACAs)/Anisotropic Conductive Film (ACF)	619
22.3.1. Materials	620
22.3.2. Application of ACA/ACF in Flip Chip	621
22.3.3. Improvement of Electrical Properties of ACAs	621
22.3.4. Thermal Conductivity of ACA	623
22.4. Future Advances of ECAs	623
22.4.1. Electrical Characteristics	623
22.4.2. High Frequency Compatibility	623
22.4.3. Reliability	623
22.4.4. ECAs with Nano-filler for Wafer Level Application	625
References	625

Chapter 23
Die Attach Quality Testing by Structure Function Evaluation
Márta Rencz, Vladimir Székely and Bernard Courtois

	629
Nomenclature	629
Greek symbols	629
Subscripts	630
23.1. Introduction	630
23.2. Theoretical Background	630
23.3. Detecting Voids in the Die Attach of Single Die Packages	634
23.4. Simulation Experiments for Locating the Die Attach Failure on Stacked Die Packages	636
23.4.1. Simulation Tests Considering Stacked Dies of the Same Size	637
23.4.2. Simulation Experiments on a Pyramidal Structure	639
23.5. Verification of the Methodology by Measurements	642
23.5.1. Comparison of the Transient Behavior of Stacked Die Packages Containing Test Dies, Prior Subjected to Accelerated Moisture and Temperature Testing	642
23.5.2. Comparison of the Transient Behavior of Stacked Die Packages Containing Real Functional Dies, Subjected Prior to Accelerated Moisture and Temperature Testing	644
23.6. Conclusions	649
Acknowledgments	649
References	650

Chapter 24
Mechanical Behavior of Flip Chip Packages under Thermal Loading
Enboa Wu, Shoulung Chen, C.Z. Tsai and Nicholas Kao

	651
24.1. Introduction	651
24.2. Flip Chip Packages	652
24.3. Measurement Methods	654
24.3.1. Phase Shifted Shadow Moiré Method	654
24.3.2. Electronic Speckle Pattern Interferometry (ESPI) Method	655
24.4. Substrate CTE Measurement	656
24.5. Behavior of Flip Chip Packages under Thermal Loading	661
24.5.1. Warpage at Room Temperature	661

24.5.2. Warpage at Elevated Temperatures ... 662
24.5.3. Effect of Underfill on Warpage ... 666
24.6. Finite Element Analysis of Flip Chip Packages under Thermal Loading ... 668
24.7. Parametric Study of Warpage for Flip Chip Packages ... 669
24.7.1. Change of the Chip Thickness ... 670
24.7.2. Change of the Substrate Thickness ... 670
24.7.3. Change of the Young's Modulus of the Underfill ... 671
24.7.4. Change of the CTE of the Underfill ... 672
24.7.5. Effect of the Geometry of the Underfill Fillet ... 672
24.8. Summary ... 674
References ... 674

Chapter 25
Stress Analysis for Processed Silicon Wafers and Packaged Micro-devices
Li Li, Yifan Guo and Dawei Zheng ... 677
25.1. Intrinsic Stress Due to Semiconductor Wafer Processing ... 677
25.1.1. Testing Device Structure ... 678
25.1.2. Membrane Deformations ... 679
25.1.3. Intrinsic Stress ... 681
25.1.4. Intrinsic Stress in Processed Wafer: Summary ... 683
25.2. Die Stress Result from Flip-chip Assembly ... 685
25.2.1. Consistent Composite Plate Model ... 685
25.2.2. Free Thermal Deformation ... 687
25.2.3. Bimaterial Plate (BMP) Case ... 688
25.2.4. Validation of the Bimaterial Model ... 691
25.2.5. Flip-Chip Package Design ... 695
25.2.6. Die Stress in Flip Chip Assembly: Summary ... 697
25.3. Thermal Stress Due to Temperature Cycling ... 698
25.3.1. Finite Element Analysis ... 698
25.3.2. Constitutive Equation for Solder ... 699
25.3.3. Time-Dependent Thermal Stresses of Solder Joint ... 700
25.3.4. Solder Joint Reliability Estimation ... 701
25.3.5. Thermal Stress Due to Temperature Cycling: Summary ... 703
25.4. Residual Stress in Polymer-based Low Dielectric Constant (low-k) Materials ... 703
References ... 708

Index ... 711

List of Contributors

VOLUME I

Avram Bar-Cohen
University of Maryland
College Park, Maryland, USA

Victor Birman
University of Missouri-Rolla
St. Louis, Missouri, USA

H.J.L. Bressers
Philips Semiconductors
Nijmegen, The Netherlands

Alan M. Cassell
NASA Ames Research Center
Moffett Field, California, USA

N. Dariavach
EMC Corp
Hopkinton, Massachusetts, USA

Liang Dong
Lightwaves 2020 Inc.
Milpitas, California, USA

L.J. Ernst
Delft University of Technology
Delft, The Netherlands

Reza Ghaffarian
Jet Propulsion Laboratory
California Institute of Technology
Pasadena, California, USA

G. Scott Glaesemann
Corning Incorporated
Corning, New York, USA

Claire Gu
University of California, Santa Cruz
Santa Cruz, California, USA

Bongtae Han
University of Maryland
College Park, Maryland, USA

Henry He
Lightwaves 2020 Inc.
Milpitas, California, USA

Xiaoling He
University of Wisconsin
Milwaukee, Wisconsin, USA

Dov Ingman
Technion, Israel Institute of Technology
Haifa, Israel

K.M.B. Jansen
Delft University of Technology
Delft, The Netherlands

J.H.J. Janssen
Philips Semiconductors
Nijmegen, The Netherlands

Kyoung Joon Kim
University of Maryland
College Park, Maryland, USA

Jorma K. Kivilahti
Helsinki University of Technology
Helsinki, Finland

Tomi T. Laurila
Helsinki University of Technology
Helsinki, Finland

J. Liang
EMC Corp
Hopkinton, Massachusetts, USA

Yisi Liu
University of California, Santa Cruz
Santa Cruz, California, USA

Jun Li
NASA Ames Research Center
Moffett Field, California, USA

Toni T. Mattila
Helsinki University of Technology
Helsinki, Finland

Tatiana Mirer
Technion, Israel Institute of Technology
Haifa, Israel

J.J. Pan
Lightwaves 2020 Inc.
Milpitas, California, USA

John H.L. Pang
Nanyang Technological University
Nanyang, Singapore

David T. Read
National Institute of Standards and Technology
Boulder, Colorado, USA

Sergey Semjonov
Fiber Optics Research Center
Moscow, Russia

D. Shangguan
FLEXTRONICS
San Jose, California, USA

Masaki Shiratori
Yokohama National University
Yokohama, Japan

Andrew A.O. Tay
National University of Singapore
Republic of Singapore

W.D. van Driel
Delft University of Technology
Delft, The Netherlands

C. van't Hof
Delft University of Technology
Delft, The Netherlands

Alex A. Volinsky
University of South Florida
Tampa, Florida, USA

Arkady Voloshin
Lehigh University
Bethlehem, Pennsylvania, USA

A. Wymyslowski
Wroclaw University of Technology
Wroclaw, Poland

Yuan Xu
University of California, Santa Cruz
Santa Cruz, California, USA

D.G. Yang
Delft University of Technology
Delft, The Netherlands

Qiang Yu
Yokohama National University
Yokohama, Japan

G.Q. Zhang
Delft University of Technology
Delft, The Netherlands and
Philips Semiconductors
Eindhoven, The Netherlands

Fengqing Zhou
Lightwaves 2020 Inc.
Milpitas, California, USA

VOLUME II

C. Bailey
University of Greenwich
London, United Kingdom

LIST OF CONTRIBUTORS

Christian Bosshard
CSEM SA, Untere Grundlistrasse 1
Alpnach Dorf, Switzerland

Sridhar Canumalla
Nokia
Irving, Texas, USA

H. Anthony Chan
University of Cape Town
Rondebosch, South Africa

Harry K. Charles, Jr.
The Johns Hopkins University
Laurel, Maryland, USA

Shoulung Chen
National Taiwan University
Taiwan

Bernard Courtois
TIMA-CMP
Grenoble Cedex, France

D.W. Dahringer
D.W. Dahringer Consultants
Glen Ridge, New Jersey, USA

Timothy P. Ferguson
Southern Research Institute
Birmingham, Alabama, USA

Yifan Guo
Skyworks Solutions, Inc.
Irvine, California, USA

D. Gwyer
University of Greenwich
London, United Kingdom

George G. Harman
National Institute of Standards and Technology
Gaithersburg, Maryland, USA

D. Ingman
Technion, Israel Institute of Technology
Haifa, Israel

Nicholas Kao
National Taiwan University
Taiwan

J. Keller
Fraunhofer Institute for Reliability and Micro Integration (IZM)
Berlin, Germany

Kikuo Kishimoto
Tokyo Institute of Technology
Tokyo, Japan

S.W.R. Lee
Hong Kong University of Science and Technology
Clear Water Bay, Kowloon, Hong Kong

Grace Y. Li
Georgia Institute of Technology
Atlanta, Georgia, USA

Li Li
Cisco Systems, Inc.
San Jose, California, USA

Johan Liu
Chalmers University of Technology
Goteborg, Sweden

C.C. Lo
Hong Kong University of Science and Technology
Clear Water Bay, Kowloon, Hong Kong

Ajay P. Malshe
University of Arkansas
Fayetteville, Arkansas, USA

Bernd Michel
Fraunhofer MicroMaterials Center
Berlin, Germany

P. Misselbrook
Celestica
Kidsgrove, Stoke-on-Trent, United Kingdom
and
University of Greenwich
London, United Kingdom

James E. Morris
Portland State University
Portland, Oregon, USA

Martin Mündlein
Vienna Institute of Technology
Vienna, Austria

Jay Narayan
North Carolina State University
Raleigh, North Carolina, USA

Johann Nicolics
Vienna Institute of Technology
Vienna, Austria

Luu Nguyen
National Semiconductor Corporation
Santa Clara, California, USA

Masaki Omiya
Tokyo Institute of Technology
Tokyo, Japan

Anne-Claire Pliska
CSEM SA, Untere Grundlistrasse 1
Alpnach Dorf, Switzerland

Jianmin Qu
Georgia Institute of Technology
Atlanta, Georgia, USA

Satish Radhakrishanan
Purdue University
West Lafayette, Indiana, USA

Marta Rencz
MicReD Ltd.
Budapest, Hungary

Ganesh Subbarayan
Purdue University
West Lafayette, Indiana, USA

Vladimir Szekely
Budapest University of Technology and Economics
Budapest, Hungary

C.Z. Tsai
National Taiwan University
Taiwan

Puligandla Viswanadham
Nokia Research Center
Irving, Texas, USA

K. Williams
Loughborough University
Loughborough, United Kingdom

Enboa Wu
National Taiwan University
Taipei, Taiwan

Wei Yang
Tsinghua University
Beijing, P.R. China

Z. Zhang
Georgia Institute of Technology
Atlanta, Georgia, USA

Dawei Zheng
Kotura, Inc.
Monterey Park, California, USA

Preface

This book encompasses a broad area of micro- and opto-electronic engineering materials: their physics, mechanics, reliability, and packaging, with an emphasis on physical design issues and problems. The editors tried to bring in the most eminent engineers and scientists as chapter authors and put together the most comprehensive book ever written on the subjects of materials, mechanics, physics, packaging, functional performance, mechanical reliability, environmental durability and other aspects of reliability of micro- and opto-electronic assemblies, components, devices, and systems. University professors and leading industrial engineers contributed to the book. The contents of the book reflect the state-of-the-art in the above listed fields of applied science and engineering.

The intended audience are all those who work in micro- and opto-electronics, and photonics; electronic and optical materials; applied and industrial physics; mechanical and reliability engineering; electron and optical devices and systems. The expected and targeted readers are practitioners and professionals, scientists and researchers, lecturers and continuing education course directors, graduate and undergraduate students, technical supervisors and entrepreneurs. The book can serve, to a great extent, as an encyclopedia in the field of physics and mechanics of micro- and opto-electronic materials and structures. In the editors' opinion, it can serve also as a textbook, as a reference book, and as a guidance for self- and continuing education, i.e., as a source of comprehensive and in-depth information in its areas. The book's chapters contain both the description of the state-of-the-art in a particular field, as well as new results obtained by the chapter authors and their colleagues.

We would like to point out that many methods and approaches addressed in this book extend far beyond microelectronics and photonics. Although these methods and approaches were developed, advanced and reported primarily in application to micro- and opto-electronic systems, they are applicable also in many related areas of engineering and physics.

The editors are proud of the broad scope of the book, and of the quality of the contributed chapters, and would like to take this opportunity to deeply acknowledge, with thanks, the conscientious effort of the numerous contributors.

February 2006
E. Suhir
C.-P. Wong
Y.-C. Lee

PHYSICAL DESIGN

1

Analytical Thermal Stress Modeling in Physical Design for Reliability of Micro- and Opto-Electronic Systems: Role, Attributes, Challenges, Results

E. Suhir

University of California, University of Maryland, and ERS/Siloptix Co., Los Altos, CA 94024, USA

> "*Mathematical formulas have their own life, they are smarter than we, even smarter than their authors, and provide more than what has been put into them*"
> Heinrich Hertz, German Physicist

> "*If my theory is in conflict with the experiment, I pity the experiment*"
> Friedrich Hegel, German Philosopher

> "*A formula longer than three inches is most likely wrong*"
> Unknown Reliability Engineer

1.1. THERMAL LOADING AND THERMAL STRESS FAILURES

Various areas of engineering differ, from the Structural Analysis and Structural Reliability point of view, by the employed materials, typical structures used, and the nature of the applied loads. The most typical microelectronic (ME) and optoelectronic (OE) structures are bodies made of a large variety of dissimilar materials. The most typical loads are thermal loads. These are caused by CTE mismatch and/or by temperature gradients [1–8].

Thermal loading takes place during the normal operation of the system, as well as during its fabrication, testing, or storage. Thermal stresses, strains and displacements are the major contributor to the finite service life and elevated failure rate of ME and OE equipment. Examples are ductile rupture, brittle fracture, thermal fatigue, creep, excessive deformation or displacement, stress relaxation (that might lead to excessive displacements), thermal shock, stress corrosion.

Elevated thermal stresses and strains can lead not only to structural ("physical") failure, but also to functional (electrical or optical) failure. If the heat, produced by the chip, cannot readily escape, then the high thermal stress in the IC can result in failure of the p-n junction [9]. Low temperature microbending (buckling of the glass fiber within the low modulus primary coating) in dual-coated optical fibers, although might be too small to lead to appreciable bending stresses and delayed fracture ("static fatigue"), can result in appreciable added transmission losses. Loss in optical coupling efficiency can occur, when the displacement in the lateral (often less than 0.2 micrometers) or angular (often less than a split of one percent of a degree) misalignment in the gap between two light-guides or between a light source and a light-guide becomes too large, because of thermally induced deformations or because of thermal stress relaxation in a laser weld. Small lateral or angular displacements in MEMS-based photonic systems (such as, say, some types of tunable lasers) can lead to a complete optical failure of the device. Tiny temperature-change-induced changes in the distance between Bragg gratings "written" on an optical fiber can be detrimental to its functional performance. For this reason thermal control of the ambient temperature is often needed to ensure sufficient protection provided to an optical device, whose performance is sensitive to the change in temperature.

As a matter of fact, the requirements for the mechanical behavior of the materials and structures in OE are often based on the functional (optical) performance of the device/system, rather than on its mechanical (structural) reliability. The requirements for the structural reliability might be much less stringent.

The thermally induced stresses and displacements in ME and OE systems can be linearly or nonlinearly elastic (reversible) or plastic (residual, irreversible), or can be caused by time dependent effects, such as creep, stress relaxation, visco-elastic or visco-plastic phenomena, aging, etc.

The ability to understand the sources of the thermal stresses and strains in ME and OE structures is of significant practical importance, and so is the ability to predict/model/simulate and possibly minimize, if necessary, the induced stresses and displacements.

1.2. THERMAL STRESS MODELING

Thermally induced failures in ME and OE equipment can be prevented only if predictive modeling is consistently used in addition (and, desirably, prior) to experimental investigations and reliability testing [10,11]. Such testing could be carried out on the design (product development) stage, during qualification and manufacturing of the product (qualification testing), or during accelerated or highly accelerated life testing (ALT and HALT).

Accelerated testing, which is the major experimental approach in ME and OE, cannot do without simple and meaningful predictive models. It is on the basis of these models that a reliability engineer decides which parameter should be accelerated, how to process/interpret the experimental data and how to bridge the gap between what one "sees" as a result of accelerated testing and what he/she will supposedly "get" in the actual use condition.

Modeling is the basic approach of any science, whether "pure" or applied [12]. Research and engineering models can be experimental or theoretical. Experimental models are typically of the same physical nature as the actual phenomenon or the object. Theoretical models represent real phenomena and objects by using abstract notions. Such models typically employ more or less sophisticated mathematical methods of analysis, and can be either analytical ("mathematical") or numerical (computational). The today's numerical models are, as a rule, computer-aided, and finite-element analyses (FEA) are widely used in the stress–strain evaluations and physical design of ME and OE structures [13]. Experimental and theoretical models should be viewed, of course, as equally important and indispensable tools for the design of a viable, reliable and cost-effective product [10,11,14,15].

1.3. BI-METAL THERMOSTATS AND OTHER BI-MATERIAL ASSEMBLIES

Pioneering work in modeling of thermal stress in bodies comprised of dissimilar materials was carried out by Timoshenko [16] and Aleck [17]. Timoshenko based his treatment of the problem on a structural analysis (strength-of-materials) approach. Aleck applied theory-of-elasticity method. Both approaches were later extended in application to structures employed in various fields of engineering, including ME and OE [19–45]. Chen and Nelson [18], Chang [19], Suhir [20–22] used structural analysis approach, although the interfacial compliance introduced by Suhir was evaluated based on a theory-of-elasticity method [20,21]. Zeyfang [23], Eischen et al. [24], Kuo [25], Yamada [26] and others used the theory-of-elasticity treatment of the problem.

The application of the structural analysis approach [27] enables one to determine, often with sufficient accuracy and always with extraordinary simplicity, the stresses acting in the constituent materials, as well as the interfacial shearing and through-thickness ("peeling") stresses. This approach results in closed form solutions and in easy-to-use formulas (see, for instance [20]). It can be (and, actually, has been) effectively employed as a part of the physical design process to select the appropriate materials, establish the feasible dimensions of the structural elements, compare different designs from the standpoint of the induced stresses and deformations, etc. On the other hand, the theory-of-elasticity method is based on rather general hypotheses and equations, and provides a rigorous treatment of the problem. Typically, it requires, however, additional use of computers to obtain the final solution to the given problem. The theory-of-elasticity approach is advisable, when there is a need for the most accurate evaluation of the induced stresses. Applied within the framework of linear elasticity, this approach leads, in the majority of cases, to a singularity at the assembly edges or at the corner of a structural element. For this reason its application has been found particularly useful when there is intent to further proceed with fracture analysis of interfacial delaminations, crack initiation and propagation at the corners, etc.

The structural analysis (strength-of-materials) and theory-of-elasticity approaches should not be viewed as "competitors," but rather as different tools, which have their merits and shortcomings, and their areas of application. These two analytical approaches should complement each other in any comprehensive engineering analysis and physical design effort.

1.4. FINITE-ELEMENT ANALYSIS

Finite-element analysis (FEA) has become, since the mid-1950s, the major resource for computational modeling in engineering, including the area of ME and OE (see, for instance, Lau [28], Glaser [29], Akay and Tong [30]). The today's powerful and flexible FEA computer programs enable one to obtain, within a reasonable time, a solution to almost any stress–strain-related problem.

Broad application of computers has, however, by no means, made analytical solutions unnecessary or even less important, whether exact, approximate, or asymptotic. Simple analytical relationships have invaluable advantages, because of the clarity and "compactness" of the obtained information and clear indication of the role of various factors affecting the given phenomenon or the behavior of the given system. These advantages are especially significant when the parameter under investigation depends on more than one variable. As to the asymptotic techniques and formulas, analytical modeling can be successful in those cases, in which there are difficulties in the application of computational methods, e.g., in problems containing singularities.

But, even when the application of numerical methods encounters no significant difficulties, it is always advisable to investigate the problem analytically before carrying out computer-aided analyses. Such a preliminary investigation helps to reduce computer time and expense, develop the most feasible and effective preprocessing model and, in many cases, avoid fundamental errors. Those that have a hands-on experience in using FEA, know very well that it is easy to obtain *a* solution based on the FEA software, but it might be not that easy to obtain *the* right solution.

Preliminary analytical modeling can be very helpful in creating a meaningful and economic preprocessing simulation model. This is particularly true in OE and photonics, where high accuracy is usually required. Special attention should be paid and special effort should be taken to make the existing FEA programs accurate enough to be suitable for the evaluation of the tiny thermo-mechanical displacements in an OE or a photonic system. Another challenge has to do with the necessity to consider visco-elastic and time-dependent behavior of photonic materials, so that the long-term reliability of the device is not compromised.

It is noteworthy that FEA was originally developed for structures with complicated geometry and/or with complicated boundary conditions, when it might be difficult to apply analytical approaches. Consequently, FEA is especially widely used in those areas of engineering, in which structures of complex configuration are typical: aerospace, marine and offshore structures, some complicated civil engineering structures, etc. In contrast, a relatively simple geometry and simple configurations usually characterize ME and OE assemblies and structures. Owing to that, such structures can be easily idealized as beams, flexible rods, circular or rectangular plates, frames, or composite structures of relatively simple geometry, thereby lending themselves to analytical modeling.

1.5. DIE-SUBSTRATE AND OTHER BI-MATERIAL ASSEMBLIES

The mechanical behavior of bonded bi-material assemblies, and particularly die-substrate assemblies, was addressed in numerous studies [31–40]. Typical failure modes in die-substrate assemblies are [31]:

(1) adherend (die or substrate) failure: a silicon die can fracture in its midportion or at its corner located at the interface;
(2) cohesive failure of the bonding material (i.e., failure in the bulk of the die-attach material); and
(3) adhesive failure of the bonding material (i.e., failure at the adherend/adhesive interface).

An adhesive failure is not expected to occur in a properly fabricated joint. If such a failure takes place, it usually occurs at a very low load level, at the product development stage, and should be regarded as a manufacturing or a quality control problem, rather than a material's or structural one.

A crack on the upper ("free") surface of the die is typically due to the normal stress acting in the die cross-sections. This stress is more or less uniformly distributed throughout the die and drops to zero at its ends. The crack at the die's corner at its interface with the substrate should be attributed to the interfacial (shearing and "peeling") stresses. These stresses concentrate, for sufficiently long assemblies with stiff enough interfaces, at the assembly ends and are next-to-zero in its midportion.

Measures that could be taken to bring the induced stresses down depend on the type/category of the stress responsible for the particular failure mode. In the case of a crack in the midportion of the die, it is the improved thermal match between the die and the substrate materials and/or a lower bow of the assembly that can improve the situation. In the case of a crack at the die's corner, the employment of a thicker and/or lower modulus adhesive can be helpful.

Die-substrate assemblies, as well as many other bi-material assemblies, are characterized by a substantial thermal expansion (contraction) mismatch of the adherend (silicon and the substrate) materials, as well as by thin and low modulus adhesive (die-attach) layers, compared to the thickness and Young's moduli of the adherends. Such a situation results in the fact that the attachment (die-attach) material experiences shear only, and also in the fact that only the interfacial compliance of the adhesive (die-attach) layer, and not its coefficient of expansion, is important [20]. It has been shown also [20,22,31] that thermally induced elastic stresses in sufficiently large bi-material assemblies do not increase with a further increase in the assembly size, and that a substantial relief in the interfacial stresses can be achieved by using thick and low modulus adhesives. This can result also in an improved adhesive and cohesive strength of the adhesive (die-attach) material, and reduce the likelihood of occurrence of the brittle crack at the chip's corner.

In the case of small-size assemblies (e.g., those with chips, not exceeding, say, 5 mm), thick and low modulus adhesives can lead to the decrease in the stress in the chip itself as well. For large chips, however (larger than, say, 10 mm), other measures should be taken, if there is a need to bring down the stresses in the midportion of the die: a substrate material with a better match with silicon should be used; a die-attach material with a lower curing temperature (and/or lower glass transition temperature) could be employed; application of a flexible substrate could be considered, etc.

As to the thermal stress modeling in bi-material assemblies, it has been found [32] that the approach, previously used in application to a thin film structure [33], can be successfully employed to further simplify thermal stress prediction in a bi-material assembly as well. This approach suggests that the interfacial shearing stress can be evaluated using an assumption that this stress is not affected by (not coupled with) the "peeling" stress. Such an assumption is conservative, i.e., results in a reasonable overestimation of the maximum shearing stress compared to the stress level obtained from the coupled equations [21]. After the shearing stress function is determined, the "peeling" stress can be computed from an equation that is similar to the equation of bending of a beam lying on a continuous elastic foundation [27].

The developed models were applied to many problems in ME and OE and beyond. The model suggested in [20] was applied by Hall et al. [34] and other researchers [35] to tri-material assemblies. An assembly should be treated as a tri-material one (as opposite to an adhesively bonded assembly), if the adhesive layer in it is not thin and/or if its Young's modulus is not significantly lower than the Young's modulus of the adherend materials. In such a situation the CTE of the bonding material has to be accounted for. An analytical model for a tri-material assembly, in which all the materials are treated as "equal constituents/members" of the assembly, i.e., in which the geometries (thicknesses) of all the assembly components and material's properties (elastic constants and CTEs) of all the materials are important, was developed in [36].

Luryi and Suhir [33] applied the model developed in [20] to the case of semiconductor crystal growth. They suggested a new approach to the high quality (dislocation free) epitaxial growth of lattice-mismatched materials. The authors have shown that lattice-mismatch strain can be simply added to the thermal-mismatch strain and, owing to that, can be easily and naturally incorporated into stress analysis models developed earlier for thermally induced strains. This paper has triggered a substantial experimental effort (numerous citations of it could be found in the literature) and is still widely referenced in the physical literature (see, for instance, [38]).

Suhir and Sullivan [39] have developed an axisymmetric version of the model suggested in [20] and applied it for the evaluation of the adhesive strength of epoxy molding compounds used in plastic packaging of IC devices.

Suhir [22] and Cifuentes [40] addressed plastic and elastoplastic deformations in the solder layer and in the beams experiencing thermal loading.

1.6. SOLDER JOINTS

Numerous models have been developed for the evaluation of thermal stresses in, and prediction of the lifetime of, solder joint interconnections (see, for instance, [41–46]). Typically, the stresses in the solder joints are caused by the thermal expansion (contraction) mismatch of the chip and the substrate materials, or, in assemblies of ball-grid-array (BGA) type, by the mismatch of the package structure and the PCB (system's substrate).

The majority of the suggested models are based on the prediction and improving of the solder joint fatigue, which is caused by the accumulated cyclic strain in the solder material. This strain is due to the temperature fluctuations resulting from either the changes in the ambient temperature (temperature cycling) or from heat dissipation in the package (power cycling). Various (viscoplastic) models for the prediction of the fatigue life-time of the solder material were suggested by Akay and Tong [30], Morgan [41], Hwang [42], Ianuzzelli, Pittaresi and Prakash [43] and many others.

The ultimate strength of solder joint interconnections is typically measured by using shear-off tests. A new, "twist-off," technique for testing of solder joint interconnections was suggested in [44]. It enables one to mimic best the actual state of stress of such interconnections. The technique was developed in application to flip-chip (FC) and ball-grid-array (BGA) assemblies.

One effective way to bring down the thermal stresses in solder joints is by employing a flex circuit [45,46]. The developed models can be used to assess the incentive in the application of such circuits, as well as the expected stress relief. Juskey and Carson [47] suggested that flex circuits be used as carriers for the direct chip attachment (DCA) technology. Flex circuitry offers a low cost and reliable system with a low thermal stress level. The flexible material of choice for today's manufacturing environment is polyamide. This material is able to withstand high temperatures during reflow soldering, and possesses good electrical and mechanical characteristics.

Solder materials and solder joints are as important in photonics, as they are in microelectronics. There are, however, a number of specific requirements for the photonics solder materials and joints: ability to achieve high alignment, requirement for a low creep, etc. [48]. "Hard" (high modulus) solder materials (such as, say, gold-tin eutectics) are thought to have better creep characteristics than "soft" (such as, say, silver-tin) solders. It should be pointed out, however, that "hard" solders can result in significantly higher thermally induced stresses than "soft" solders [49], and therefore their ability to withstand creep might be not as good as expected, not to mention the short-term reliability of the material.

Thermally induced stresses in optical fibers soldered into ferrules were modeled in [49]. Modeling was based on the solution to the axisymmetric theory-of-elasticity problem for an annular composite structure comprised of the metalized silica fiber, the solder ring, and the ferrule. The obtained relationships enable one to design the joint in such a way that the solder ring is subjected to relatively low compressive stresses. It has been shown that neither low expansion ferrules, nor high expansion ones, might be suitable for a particular solder material and particular thickness of the solder ring.

Solders are often used as continuous attachment layers in ME and OE assemblies. Stress concentration at the ends of such attachments can lead to plastic deformations of the solder material. These deformations were addressed by Suhir [22] and Cifuentes [40]. This problem has become recently of significant importance in connection with using Indium or Indium-based alloys as suitable attachments of the quantum wells of GaAs lasers to metal substrates. The developed models enable one to assess the size of the zone, in which the plastic deformations are possible, and to assess the effect on the state of the inelastic strain in the device. The plastic stresses will not propagate inwards the assembly, if its length exceeds appreciably the total length of the areas occupied by the elevated stresses. This consideration provides a practically useful criterion for the selection, if possible, of the length of the continuous solder layer in the application in question.

1.7. DESIGN RECOMMENDATIONS

Based on the modeling of thermal stresses in typical adhesively bonded or soldered assemblies, i.e., in assemblies with appreciable CTE mismatch of the adherends and a homogeneous adhesive or solder layer, the following general recommendations, aimed at the improvement of the ultimate and fatigue strength, of the assemblies, have been developed:

- equalize the in-plane and bending stiffness of the adherends, and use identical adherends, if possible;
- use low modulus adhesives; as an alternative to using a low modulus adhesive throughout the joint, use such an adhesive only at the ends of the joint, i.e., in the region of high interfacial stresses, while a higher modulus adhesive could be used in the midportion of the assembly;
- vary, if possible, the adherend thickness along the assembly in a proper way and/or slant the adherends edges for a thicker adhesive layer at the assembly ends;
- keep the stresses within the elastic range, if possible;
- minimize "peeling" (in the case of multimaterial and thin film structures) and axial (in the case of solder joints) stresses.

1.8. "GLOBAL" AND "LOCAL" MISMATCH AND ASSEMBLIES BONDED AT THE ENDS

In those cases when the adhesive (solder) layer is not homogeneous, or when the components are just partially bonded or soldered to each other, both "global" and "local" mismatch loading takes place [50–53]. The "local" mismatch loading is due to the mismatch of the dissimilar materials within the bonded or soldered region, while the "global" mismatch loading is caused by the mismatch in the unbonded region. Examples are: solder joint interconnections, optical glass fiber interconnects adhesively bonded or soldered at their ends into a ferrule or a capillary; optical glass fibers in micromachined (MEMS) optical switches packaged into a dual-in-line package, etc.

The interaction of the interfacial shearing stresses caused by the "global" and "local" mismatches in a typical bi-material assembly adhesively bonded or soldered at the ends [51–53] can be qualitatively summarized as follows:

- The interfacial shearing stresses caused by the "local" mismatch are antisymmetric with respect to the mid-cross-section of the bonded area: these stresses are equal in magnitude and opposite in directions (signs).
- The "local" shearing stresses concentrate at the ends of the bonded area and, for sufficiently long bonded joints and stiff interfaces (thin and high-modulus adhesive layer), are next-to-zero in the midportion of the bonded area.
- For short-and-compliant bonded areas, the "local" shearing stresses are more-or-less linearly distributed over the length of the bonded area, and their maxima at the assembly ends can be significantly lower than the maximum stresses in long-and-stiff bonded joints.
- The shearing stresses caused by the "global" mismatch act in the same direction over the entire length of the bonded joint. This direction is such, that in the inner portions of the joints (i.e., in the portions located closer to the mid-cross-section of the unbonded region, which is also the mid-cross-section of the assembly as a whole), the total interfacial stress should be computed as the difference between the "local" and the "global" stress. In the outer portions of the bonded joints, the total stress should be computed as the sum of the "local" and the "global" stress.
- In the case of short-and-compliant joints, when both the "local" and the "global" stresses are more or less uniformly distributed over the joint's length, the total stress is indeed larger than each of these stress categories. Since, however, both the "local" and the "global" stresses in short-and-compliant joints can be very low compared to the stresses in long-and-stiff assemblies, the total stress can be very low as well, despite the fact that, for the outer (peripheral) portions of the bonded joints, this stress is obtained as a sum of the "local" and the "global" stresses.
- In the case of long-and-stiff joints, the "global" stresses concentrate at the inner edges of the bonded joints and rapidly decrease with an increase in the distance of the given cross-section from these edges. In such a situation, the interaction of the "local" and "global" stresses is always favorable, i.e., at the inner edge, results in the total stress, obtained as a difference between the "local" and the "global" stress, and, at the outer edge, is due to the "local" stress only.
- For sufficiently long-and-stiff bonded joints, the magnitude of the "global" stress at the inner boundary is equal to the magnitude of the maximum "local" stress, so that the total shearing stress is zero.

Although substantial relief in the total stress can be achieved by employing bonded joints with short-and-compliant attachments, this approach usually cannot be recommended, because insufficient "real estate" of the bonded areas does not allow one to produce reliable enough joints. However, in the case of solder joint interconnections, both the favorable effect of the short-and-compliant joint and the unfavorable effect of the summation of the "local" and the "global" thermal stresses/strains at the assembly ends should be considered.

The interaction of the "local" and the "global" stresses, with consideration of the effect of the coefficient of thermal expansion (contraction) of the epoxy material itself, was studied in [51] for a glass fiber interconnect whose ends are epoxy bonded into capillaries. The necessity of taking into account the CTE of the adhesive material was due to the fact that the cross-sectional area of the adhesive ring was considerably larger than the cross-sectional area of the glass fiber. Therefore the longitudinal compliance of the adhesive ring was comparable with the compliance of the fiber and could not be neglected.

Understanding of the interaction of the "global" and "local" stresses is particularly important in connection with the ME and OE assemblies bonded at the ends. In some ME assemblies of the flip-chip type the solder joint stand-off is so low that it is practically impossible to bring in the underfill material underneath the chip, especially if the chip is large. On the other hand, there might be no need for that, since the underfill material works effectively only at its peripheral portions [51,52]. Modeling of the mechanical behavior of such an assembly is crucial in order to establish the adequate width of the adhesive layer: this width should be large enough to provide sufficient bonding strength of the assembly. The stresses in such an assembly will not be higher than in an assembly with a continuous underfill.

1.9. ASSEMBLIES WITH LOW MODULUS ADHESIVE LAYER AT THE ENDS

Interfacial shearing and peeling stresses in adhesively bonded or soldered assemblies concentrate at the assembly ends. These stresses can be reduced by employing a low modulus material at the assembly ends [54] and/or by slanting the edges of the assembly components [55], thereby increasing the thickness of the adhesive layer at the assembly ends. The stresses at the ends of polymer-coated optical fibers can be reduced by using a low modulus coating at the fiber ends [56,57]. The mechanical behavior of such ME and OE structures is, in a sense, opposite to the situation that takes place in an assembly adhesively bonded at the ends. Indeed, in an assembly with a low modulus adhesive/coating at the end, it is the midportion of the assembly that is characterized by an elevated Young's modulus of the adhesive (coating), while in the case of an assembly bonded at the ends, it is its midportion that is characterized by a "low" (actually, zero) Young's modulus of the "attachment."

1.10. THERMALLY MATCHED ASSEMBLIES

There is an obvious incentive to employ thermally matched materials in ME and OE assemblies. It is this assembly, which is used in a Si-on-Si flip-chip (FC) design [58,59], in a ceramic Cerdip/Cerquad ME package [60]. On the other hand, some photonics structures (say, holographic memory assemblies) are made of identical adherends and a compliant (thick and low modulus) adhesive [61–63].

There is substantial difference in the mechanical behavior of the assemblies with an appreciable mismatch in the CTE of the adherends and the thermally matched assemblies, particularly those with identical adherends. While in assemblies with an appreciable thermal mismatch of the adherends, the CTE of the adhesive material (as long as this material is thin and/or has a low Young modulus) does not affect the mechanical behavior of the assembly, in assemblies with identical adherends the mismatch between the adhesive and the adherends' materials the CTE of the adhesive is definitely important. In addition, the mechanical behavior and reliability of the adhesive material is quite different. In assemblies with mismatched adherends, and thin and low modulus adhesives, the adhesive layer is primarily subjected to the interfacial shear, while in the case of matched assemblies (identical adherends) the adhesive layer elevated experiences both shearing and performs similarly a thin film fabricated on a thick substrate and tensile (compressive) stresses.

Thermal stresses in solder joints in thermally matched silicon-on-silicon flip-chip assemblies achieve their maximum values at the interfaces and concentrate at the joints' corners [58,59]. The stresses, acting in the axial direction (these stresses are analogous to the "peeling" stresses in thin film structures), are the highest, and it is these stresses and strains that are primarily responsible for the joint's reliability.

The case of identical ceramic adherends was considered in connection with choosing an adequate coefficient of thermal expansion for a solder (seal) glass in a ceramic package design [60]. It has been found that the best result can be achieved by using a probabilistic approach, in which the coefficient of thermal expansion of the solder glass is treated as a random variable. The package manufactured in accordance with the developed recommendations exhibited no failures. Based on the performed analysis, it has been concluded that in order to successfully apply a probabilistic approach (see, for instance, [64]) customers should require that vendors provide information concerning both the mean value and the standard deviation of the parameter of interest.

Several thermoelastic models [61–63] were developed for the prediction of the mechanical behavior of the adhesive material in adhesively bonded assemblies with identical nondeformable adherends of different configurations. The analyses were carried out in application to assemblies used in advanced holographic memory devices. It has been shown, particularly, that the interfacial compliance of the adhesive layer, in the case of sufficiently large-and-thin assemblies with thermally matched adherends, is half the magnitude of the interfacial compliance in the case of assemblies with mismatched adherends. It has been shown also that the elevated interfacial shearing stresses are somewhat higher for a circular assembly than for a rectangular one. These stresses also occupy a narrower zone around the assembly edge. An inhomogeneous adhesive layer, which is important for the considered application, was examined. The developed models enable one to establish the conditions, at which the requirement for the undistorted boundaries of the inner "pieces" of the adhesive is fulfilled. This requirement is important from the stand point of the satisfactory optical performance of the assembly.

1.11. THIN FILMS

Typical thermal stress failures in thin films fabricated on thick substrates are interfacial delaminations (including delamination buckling), and film cracking and blistering. Numerous investigators [65–71] analyzed thermal stresses in thin films. Based on the obtained results, practical recommendations for a physical design of a reliable thin film structure have been formulated. Particularly, it has been found that

- the thermal stress in the given film layer of a multilayer film structure is due to the thermal expansion mismatch of this layer with the substrate, and not with the adjacent film layers [68];

- the edge stresses in the film are affected by the edge configuration [70]: circular assemblies are somewhat "stiffer" than the rectangular ones, i.e., result in higher stresses that concentrate at a narrower peripheral ring;
- stress in a thin film, which does not experience bending stresses, is not affected by the assembly bow, while the assembly bow and the stresses in the substrate are strongly affected by the stresses in the film [71].

The effect of lattice mismatch of semiconductor materials during crystal growth of thin Germanium films on a thick Silicon substrate was addressed, along with the effect of thermal mismatch, by Luryi and Suhir [37]. It has been shown that by using a "tower-like" surface of the substrate (such a surface can be achieved by high-resolution lithography, by employment of porous silicon, etc.) one can grow dislocation free semiconductor films.

1.12. POLYMERIC MATERIALS AND PLASTIC PACKAGES

Polymeric materials are widely used in ME and OE engineering (see, for instance, [72–78]). Examples are: plastic packages of integrated circuit (IC) devices, adhesives, various enclosures and plastic parts, polymeric coatings of optical silica fibers, and even polymeric lightguides. There are numerous and rapidly growing opportunities for the application of polymers for diverse functions in the "high-technology" field. Polymeric materials are inexpensive and lend themselves easily to processing and mass production techniques. The reliability of these materials, however, is usually not as high as the reliability of inorganic materials and is often insufficient for particular applications, thereby limiting the area of the technical use of polymers. There exists a crucial necessity for the advancement of the experimental and theoretical methods, techniques and approaches, aimed at the prediction and improvement of the short/long-term performance of polymeric materials for various ME and OE applications.

Recent improvements in the mechanical properties of molding compounds, plastic package designs, and manufacturing technologies have resulted in substantial increase in the reliability of plastic packages. There is, however, one major industry-wide concern associated with these packages—their moisture-induced failures ("popcorn" cracking). Such failures typically occur during surface mounting the packages onto printed circuit boards by means of high temperature reflow soldering. "Popcorn" cracking is usually attributed to the elevated pressure of the water vapor, generated due to a sudden evaporation of the absorbed moisture [73]. It is believed that thermal stresses also play an important role, both directly, due to their interaction with mechanical vapor-pressure-induced stresses in the underchip portion of the molding compound, and indirectly, by triggering the initiation and facilitating the propagation of the interfacial delaminations.

It has been suggested [74] that constitutive equations, obtained as a generalization of von-Karman's equations for large deflections of plates, be used as a suitable analytical stress model for the prediction and prevention of structural failures in moisture-sensitive plastic packages. Such a generalization accounts for the combined action of the lateral pressure, caused by the generated water vapor, and the thermally induced loading. The developed model can be used for the selection of the low stress molding compounds, for comparing different package design from the standpoint of their propensity to "popcorn"-cracking, in the development of "figures-of-merit" [75], which would enable one to separate packages that need to be "baked" and "bagged" from those that do not, etc. This loading is due to both the temperature gradients and the thermal expansion (contraction) mismatch of the dissimilar materials in the package. Since the coefficient of thermal expansion and Young's modulus of the molding compound are temperature dependent, the constitutive equations account for this dependence. The developed equations were applied to the delaminated underchip layer of the molding compound. This layer is treated as a thin rectangular plate clamped at the support contour. It has been shown, in particular, that, from the standpoint of structural analysis, the distinction between "thick" and "thin" packages should be attributed primarily to the level of the in-plane ("membrane"), normal stresses in the underchip portion of the compound: in "thick" packages this portion exhibits bending only, while in "thin" packages it is subjected to both bending and in-plane loading. The obtained data, which are in good agreement with experimental observations, have indicated that the geometric characteristics of the package (the underchip layer thickness, chip and paddle size, etc.) have a strong effect on the package propensity to failure.

The obtained results have been used to develop guidelines ("figures-of-merit"), which enable one to separate packages that need to be "baked" and "bagged" from those that do not, as well as for guidelines aimed at the preliminary selection of the feasible molding compound for the given package design.

1.13. THERMAL STRESS INDUCED BOWING AND BOW-FREE ASSEMBLIES

Thermal stress induced bowing can prevent further processing of BGA packages or of thin (TSOP) plastic packages [79], can lead to cracking of ceramic substrates in thin overmolded packages [80,81], or can have another adverse effect on the design or processing of plastic packages of IC devices. It has been shown [79–81] that employment of additional (surrogate) layers can dramatically improve the situation.

There is an obvious incentive for the use of bow-free (temperature change insensitive) assemblies in ME and OE packaging. It has been shown [83,84] that this can be achieved if a thick enough bonding layer is introduced to produce an appreciable axial (in-plane) force. This force is necessary to create a bending moment that would be able to equilibrate the thermally induced moment produced by the dissimilar adherends. A statically determinate bi-material assembly (i.e., an assembly with a very thin and/or a very low-modulus bonding layer) cannot be made bow free. The thermally induced forces acting in the components of a bi-material assembly are equal in magnitude and opposite in sign, and create a bending moment that can be equilibrated by the elastic moment only. This inevitably leads to nonzero deflections, whether large or small. To be bow-free, a multi-material assembly should be made statically indeterminate. It should contain, therefore, at least three dissimilar materials, so that the resulting bending moment, caused by the induced forces in all the three materials, is zero.

A sufficiently large axial force in the bonding material can be created by one or a combination of two or more of the following measures:

- by using a bonding material with a high elastic modulus;
- by using a bonding materials with a significant thermal mismatch with the adherends;
- by using a bonding material with a high curing temperature, and/or
- by making the bonding material thick.

It is only the last measure, however, that, while resulting in a desirable elevated thermally induced force in the bonding material, does not necessarily lead to an elevated axial stress in it. Computations based on the developed analytical models [83,84] have indicated that the "thick" bonding layer in a bow free assembly can still be made thin enough (about 4 mils or so) to be effective, provided that the material and/or the thickness of at least one of the adherends is adequately chosen.

1.14. PROBABILISTIC APPROACH

Probabilistic models might be very useful in situations, in which the "fluctuations" from the mean values are significant and in which the variability, change and uncertainty play a vital role (see, for instance, [60,64,85,86]). In the majority of such situations the product will most likely fail, if these uncertainties are ignored. So far, probabilistic (statistical) models are used in "high-tech" engineering primarily for the design and analyses of experiments. They are very seldom used yet as a physical design tool. In this connection we would like to emphasize that wide and consistent use of probabilistic models would not only enable one to establish the scope and the limits of the application of deterministic solutions, but can provide a solid basis for a well-substantiated and goal-oriented accumulation, and effective utilization of empirical data. Probabilistic models enable one to quantitatively assess the degree of uncertainty in various factors, which determine the performance of a product. Then a reliability engineer can design a product with a predictable and low probability of failure. A good illustration to these statements is the success of the design described in [60]).

1.15. OPTICAL FIBERS AND OTHER PHOTONIC STRUCTURES

Various problems of the thermal stress modeling in bare and coated optical silica fibers were addressed in [3,7,87–98].

Low temperature microbending can result in substantial added transmission losses in dual-coated optical fibers. Based on the developed analytical stress models [90], it has been shown that the initial curvatures can play an important role in the low temperature behavior of a dual-coated silica fiber and that certain curvature lengths are less favorable than others from the standpoint of the possible fiber buckling. It has been shown also [91] that the magnitude of the spring constant of the elastic foundation provided by the primary coating layer could have a significant effect on the buckling conditions, and that, in the case of thick and relatively low modulus secondary coatings, both coating layers should be considered when evaluating the spring constant. For thin and high modulus secondary coatings, however, only the primary coating material could be considered when evaluating the spring constant of the elastic foundation for the silica glass fiber.

Application of a mechanical approach to the evaluation of low temperature added transmission losses [92] enables one, based on the developed analytical stress model, to evaluate the threshold of the low temperature added transmission losses from purely mechanical calculations, without resorting to optical calculations or measurements. The model (confirmed by actual optical measurements) presumes that the threshold of the elevated added transmission losses coincides with the threshold of the elevated thermally induced stresses applied by the coating (jacket) to the silica fiber.

Various aspects related to thermal stresses and thermal stress related behavior of solder materials for photonics applications were addressed in [53]. Thermal stresses in, and optimal physical design of, solder joints for metalized optical fibers soldered into ferules were analyzed, based on the developed analytical stress model, in [54]. It has been demonstrated that an adequate modeling of what could be expected in an actual joint is a must: the selection of the right enclosing material and the right thickness of the solder preform (for the given solder material) should be conducted and decided upon prior to manufacturing of the joint.

Mechanical behavior and elastic stability of bare, polymer coated and metalized optical fiber interconnects was modeled in [93–95]. It has been shown that low modulus polymer coatings have significant advantages over high modulus metalizations, as far as the stresses in the coating (metalization) are concerned, and typically should be preferred despite of their sensitivity to moisture penetration. An analytical stress model developed in [94] enables pone to select the appropriate enclosure material for minimizing the thermally induced bending stresses in an optical fiber interconnect experiencing ends-offset and thermally induced compressive loading because of its mismatch with the enclosure material. In those cases when low buckling stresses are a problem, thicker polymer coatings can be used to improve the elastic stability of a polymer-coated fiber [95].

Suhir and Vuillamin [96] have demonstrated, based on the developed analytical and FEA models, that the gradient in the distribution of the CTE along one of the diameters of a glass fiber cross-section can be responsible for the undesirable "curling" phenomenon that often occurs during drawing of optical silica fibers.

Optical silica fiber materials exhibit highly nonlinear (but still elastic) behavior when subjected to tension or compression. This effect was considered, along with the effect of the nonprismaticity, in the model [97] developed for a fused biconical taper (FBT) coupler experiencing thermally induced tension. An effective method for thermostatic compensation of temperature sensitive devices was suggested in [98] in application to Bragg gratings. It has been shown that there is no need to use, for particular applications, mechanically vulnerable ceramic materials with a negative CTE: regular and more mechanically reliable materials can be successfully used for the objective in question.

It has been recently demonstrated [99,100] that a newly developed nano-particle material (NPM) can make a substantial difference in the state-of-the-art of coated optical fibers: this material has all the merits of the polymer coated and metalized optical fibers without having their drawbacks.

1.16. CONCLUSION

- Predictive modeling is an effective tool for the prediction and prevention of mechanical and functional failures in microelectronics and photonics materials, structures, packages and systems, subjected to thermal loading.
- Experimental and theoretical models should be viewed as equally important and equally indispensable to the design of a viable, reliable and cost-effective product. The same is true for analytical and numerical (FEA) models.
- Special effort should be taken to make the existing FEA program accurate enough to be suitable for the evaluation of the thermal stresses and displacements in photonics structures. In such a situation, analytical modeling of a simplified structure of interest can be very useful for the selection and mastering of the preprocessing FEA model.

- Application of the probabilistic approach enables one to quantitatively assess the role of various uncertainties in the materials properties, geometrical characteristics and loading conditions, and, owing to that, to design and manufacture a viable and reliable product.
- A newly developed nanomaterial can make a substantial difference in the state-of-the-art of coated optical fibers: this material has all the merits of the polymer coated and metalized optical fibers without having their drawbacks.

REFERENCES

Thermal loading and thermal stress failures

1. J.H. Lau, Ed., Thermal Stress and Strain in Microelectronics Packaging, Van-Nostrand Reinhold, New York, 1993.
2. E. Suhir, R.C. Cammarata, D.D.L. Chung, and M. Jono, Mechanical behavior of materials and structures in microelectronics, Materials Research Society Symposia Proceedings, Vol. 226, 1991.
3. E. Suhir, M. Fukuda, C.R. Kurkjian, Eds., Reliability of photonic materials and structures, Materials Research Society Symposia Proceedings, Vol. 531, 1998.
4. E. Suhir, M. Shiratori, Y.C. Lee, and G. Subbarayan, Eds., Advances in Electronic Packaging—1997, Vols. 1 and 2, ASME Press, 1997.
5. E. Suhir, Thermal stress failures in microelectronic components—review and extension, in A. Bar-Cohen and A.D. Kraus, Eds., Advances in Thermal Modeling of Electronic Components and Systems, Hemisphere, New York, 1988.
6. E. Suhir, B. Michel, K. Kishimoto, and J. Lu, Eds., Mechanical Reliability of Polymeric Materials and Plastic Packages of IC Devices, ASME Press, 1998.
7. E. Suhir, Thermal stress failures in microelectronics and photonics: prediction and prevention, Future Circuits International, issue 5, 1999.
8. E. Suhir, Microelectronics and photonics—the future, Microelectronics Journal, 31(11–12) (2000).
9. G.A. Lang, et al., Thermal fatigue in silicon power devices, IEEE Transactions on Electron Devices, 17 (1970).

Thermal stress modeling

10. E. Suhir, Accelerated Life Testing (ALT) in microelectronics and photonics: its role, attributes, challenges, pitfalls, and interaction with qualification tests, Keynote address at the SPIE's 7-th Annual International Symposium on Nondestructive Evaluations for Health Monitoring and Diagnostics, 17–21 March, San Diego, CA, 2002.
11. E. Suhir, Reliability and accelerated life testing, Semiconductor International, February 1, 2005.
12. E. Suhir, Modeling of the mechanical behavior of microelectronic and photonic systems: attributes, merits, shortcomings, and interaction with experiment, Proceedings of the 9-th Int. Congress on Experimental Mechanics, Orlando, FL, June 5–8, 2000.
13. E. Suhir, Analytical stress–strain modeling in photonics engineering: its role, attributes and interaction with the finite-element method, Laser Focus World, May 2002.
14. E. Suhir, Thermomechanical stress modeling in microelectronics and photonics, Electronic Cooling, 7(4) (2001).
15. M. Schen, H. Abe, and E. Suhir, Eds., Thermal and Mechanical Behavior and Modeling, ASME, AMD-Vol, 1994.

Bi-metal thermostats and other bi-material assemblies

16. S. Timoshenko, Analysis of bi-metal thermostats, Journal of the Optical Society of America, 11 (1925).
17. B.J. Aleck, Thermal stresses in a rectangular plate clamped along an edge, ASME Journal of Applied Mechanics, 16 (1949).
18. W.T. Chen and C.W. Nelson, Thermal stresses in bonded joints, IBM Journal, Research and Development, 23(2) (1979).

19. F.-V. Chang, Thermal contact stresses of bi-metal strip thermostat, Applied Mathematics and Mechanics, 4(3) Tsing-hua Univ., Beijing, China (1983).
20. E. Suhir, Stresses in bi-metal thermostats, ASME Journal of Applied Mechanics, 53(3) (1986).
21. E. Suhir, Interfacial stresses in bi-metal thermostats, ASME Journal of Applied Mechanics, 56(3) (1989).
22. E. Suhir, Calculated thermally induced stresses in adhesively bonded and soldered assemblies, Proc. of the Int. Symp. on Microelectronics, ISHM, 1986, Atlanta, Georgia, Oct. 1986.
23. R. Zeyfang, Stresses and strains in a plate bonded to a substrate: Semiconductor devices, Solid State Electronics, 14 (1971).
24. J.W. Eischen, C. Chung, and J.H. Kim, Realistic modeling of the edge effect stresses in bimaterial elements, ASME Journal of Electronic Packaging, 112(1) (1990).
25. A.Y. Kuo, Thermal stress at the edge of a bi-metallic thermostat, ASME Journal of Applied Mechanics, 57 (1990).
26. S.E. Yamada, A bonded joint analysis for surface mount components, ASME Journal of Electronic Packaging, 114(1) (1992).
27. E. Suhir, Structural analysis in microelectronic and fiber optic systems, Vol. 1, Basic Principles of Engineering Elasticity and Fundamentals of Structural Analysis, Van Nostrand Reinhold, New York, 1991.
28. J.H. Lau, A note on the calculation of thermal stresses in electronic packaging by finite-element method, ASME Journal of Electronic Packaging, 111(12) (1989).
29. J.C. Glaser, Thermal stresses in compliantly joined materials, ASME Journal of Electronic Packaging, 112(1) (1990).
30. J.U. Akay and Y. Tong, Thermal fatigue analysis of an smt solder joint using FEM approach, Journal of Microcircuits and Electronic Packaging, 116 (1993).
31. E. Suhir, Die attachment design and its influence on the thermally induced stresses in the die and the attachment, Proc. of the 37th Elect. Comp. Conf., IEEE, Boston, MA, May 1987.
32. V. Mishkevich and E. Suhir, Simplified approach to the evaluation of thermally induced stresses in bi-material structures, in E. Suhir, Ed., Structural Analysis in Microelectronics and Fiber Optics, ASME Press, 1993.
33. E. Suhir, Approximate evaluation of the elastic interfacial stresses in thin films with application to high-Tc superconducting ceramics, Int. Journal of Solids and Structures, 27(8) (1991).
34. P.M. Hall, et al., Strains in aluminum-adhesive-ceramic trilayers, ASME Journal of Electronic Packaging, 112(4) (1990).
35. E.K. Buratynski, Analysis of bending and shearing of tri-layer laminations for solder joint reliability, in E. Suhir, et al., Advances in Electronic Packaging 1997, ASME Press, 1997.
36. E. Suhir, Analysis of interfacial thermal stresses in a tri-material assembly, Journal of Applied Physics, 89(7) (2001).
37. S. Luryi and E. Suhir, A new approach to the high-quality epitaxial growth of lattice—mismatched materials, Applied Physics Letters, 49(3) (1986).
38. S.C. Lee et al., Strain-reliweved, dislocation-free $In_xGa_{1-x}As/GaAs$(001) heterostructure by nanoscale-patterned growth, Applied Physics Letters, 85(18) (2004).
39. E. Suhir and T.M. Sullivan, Analysis of interfacial thermal stresses and adhesive strength of bi-annular cylinders, Int. Journal of Solids and Structures, 26(6) (1990).
40. A.O. Cifuentes, Elastoplastic analysis of bimaterial beams subjected to thermal loads, ASME Journal of Electronic Packaging, 113(4) (1991).

Solder Joints

41. H.S. Morgan, Thermal stresses in layered electrical assemblies bonded with solder, ASME Journal of Electronic Packaging, 113(4) (1991).
42. J.S. Hwang, Modern Solder Technology for Competitive Electronics Manufacturing, McGraw-Hill, New York, 1996.
43. R.J. Iannuzzelli, J.M. Pitarresi, and V. Prakash, Solder joint reliability prediction by the integrated matrix creep method, ASME Journal of Electronic Packaging, 118 (1996).
44. E. Suhir, Twist-off testing of solder joint interconnections, ASME Journal of Electronic Packaging, 111(3) (1989).
45. E. Suhir, Stress relief in solder joints due to the application of a flex circuit, ASME Journal of Electronic Packaging, 113(3) (1991).
46. E. Suhir, Flex circuit vs regular substrate: predicted reduction in the shearing stress in solder joints, Proc. of the 3-rd Int. Conf. on Flexible Circuits FLEXCON 96, San-Jose, CA, Oct. 1996.

47. F. Juskey and R. Carson, DCA on flex: A low cost/stress approach, in E. Suhir, et al., Eds., Advances in Electronic Packaging, ASME Press, 1997.
48. E. Suhir, Solder materials and joints in fiber optics: reliability requirements and predicted stresses, Proc. of the Int. Symp. on Design and Reliability of Solders and Solder Interconnections, Orlando, FL, Febr. 1997.
49. E. Suhir, Thermally induced stresses in an optical glass fiber soldered into a ferrule, IEEE/OSA Journal of Lightwave Technology, 12(10) (1994).

"Global" and "Local" Mismatch and Assemblies Bonded at the Ends

50. E. Suhir, "Global" and "local" thermal mismatch stresses in an elongated bi-material assembly bonded at the ends, in E. Suhir, Ed., Structural Analysis in Microelectronic and Fiber-Optic Systems, Symposium Proceedings, ASME Press, 1995.
51. E. Suhir, Thermal stress in a bi-material assembly adhesively bonded at the ends, Journal of Applied Physics, 89(1) (2001).
52. E. Suhir, Bi-material assembly adhesively bonded at the ends and fabrication method, U.S. Patent #6,460,753, 2002.
53. E. Suhir, Predicted thermal mismatch stresses in a cylindrical bi-material assembly adhesively bonded at the ends, ASME Journal of Applied Mechanics, 64(1) (1997).

Assembly with Low Modulus Adhesive Layer at the Ends

54. E. Suhir, Thermal stress in an adhesively bonded joint with a low modulus adhesive layer at the ends, Applied Physics Journal, (April) (2003).
55. E. Suhir, Electronic assembly having improved resistance to delamination, U.S. Patent #6,028,772, 2000.
56. E. Suhir, Thermal stress in a polymer coated optical glass fiber with a low modulus coating at the ends, Journal of Materials Research, 16(10) (2001).
57. E. Suhir, Coated optical fiber, U.S. Patent #6,647,195, 2003.

Thermally Matched Assembliess

58. E. Suhir, Axisymmetric elastic deformations of a finite circular cylinder with application to low temperature strains and stresses in solder joints, ASME Journal of Applied Mechanics, 56(2) (1989).
59. E. Suhir, Mechanical reliability of flip-chip interconnections in silicon-on-silicon multichip modules, IEEE Conference on Multichip Modules, IEEE, Santa Cruz, Calif., March 1993.
60. E. Suhir and B. Poborets, Solder glass attachment in cerdip/cerquad packages: thermally induced stresses and mechanical reliability, Proc. of the 40th Elect. Comp. and Techn. Conf., Las Vegas, Nevada, May 1990, see also: ASME Journal of Electronic Packaging, 112(2) (1990).
61. E. Suhir, Adhesively bonded assemblies with identical nondeformable adherends: predicted thermal stresses in the adhesive layer, Composite Interfaces, 6(2) (1999).
62. E. Suhir, Adhesively bonded assemblies with identical nondeformable adherends and inhomogeneous adhesive layer: predicted thermal stresses in the adhesive, Journal of Reinforced Plastics and Composites, 17(14) (1998).
63. E. Suhir, Adhesively bonded assemblies with identical nondeformable adherends and "piecewise continuous" adhesive layer: predicted thermal stresses and displacements in the adhesive, Int. Journal of Solids and Structures, 37 (2000).
64. E. Suhir, Applied Probability for Engineers and Scientists, McGraw Hill, New York, 1997.

Thin Films

65. K. Roll, Analysis of stress and strain distribution in thin films and substrates, Journal of Applied Physics, 47(7) (1976).
66. G.H. Olsen and M. Ettenberg, Calculated stresses in multilayered heteroepitaxial structures, Journal of Applied Physics, 48(6) (1977).
67. J. Vilms and D. Kerps, Simple stress formula for multilayered thin films on a thick substrate, Journal of Applied Physics, 53(3) (1982).
68. E. Suhir, An approximate analysis of stresses in multilayer elastic thin films, ASME Journal of Applied Mechanics, 55(3) (1988).

69. T.-Y. Pan and Y.-H. Pao, Deformation of multilayer stacked assemblies, ASME Journal of Electronic Packaging, 112(1) (1990).
70. E. Suhir, Approximate evaluation of the elastic thermal stresses in a thin film fabricated on a very thick circular substrate, ASME Journal of Electronic Packaging, 116(3) (1994).
71. E. Suhir, Predicted thermally induced stresses in, and the bow of, a circular substrate/thin-film structure, Journal of Applied Physics, 88(5) (2000).

Polymeric Materials and Plastic IC Packages

72. E. Suhir, Applications of an epoxy cap in a flip-chip package design, ASME Journal of Electronic Packaging, 111(1) (1989).
73. G.S. Ganssan and H. Berg, Model and analysis for reflow cracking phenomenon in SMT plastic packages, 43-rd IEEE ECTC., 1993.
74. E. Suhir, Failure criterion for moisture-sensitive plastic packages of integrated circuit (IC) devices: application of von-Karman equations with consideration of thermoelastic strains, Int. Journal of Solids and Structures, 34(12) (1997).
75. E. Suhir and Q.S.M. Ilyas, "Thick" plastic packages with "small" chips vs "thin" packages with "large" chips: how different is their propensity to moisture induced failures?, in E. Suhir, Ed., Structural Analysis in Micro-electronics and Fiber Optics, Symposium Proceedings, ASME Press, 1996.
76. M. Uschitsky and E. Suhir, Predicted thermally induced stresses in an epoxy molding compound at the chip corner, in E. Suhir, Ed., Structural Analysis in Microelectronics and Fiber Optics, Symposium Proceedings, ASME Press, 1996.
77. M. Ushitsky, E. Suhir, and G.W. Kammlott, Thermoelastic behavior of filled molding compounds: composite mechanics approach, ASME Journal of Electronic Packaging, 123(4) (2001).
78. D.K. Shin and J.J. Lee, A study on the mechanical behavior of epoxy molding compound and thermal stress analysis in plastic packaging, in E. Suhir, et al., Advances in Electronic Packaging 1997, Vol. 1, ASME Press, 1997.

Thermal Stress Induced Bowing and Bow-Free Assemblies

79. E. Suhir, Predicted bow of plastic packages of integrated circuit devices, in J.H. Lau, Ed., Thermal Stress and Strain in Microelectronic Packaging, Van Nostrand Reinhold, New York, 1993.
80. E. Suhir and J. Weld, Electronic package with reduced bending stress, U.S. Patent #5,627,407, 1997.
81. E. Suhir, Arrangement for reducing bending stress in an electronics package, U.S. Patent #6,180,241, 2001.
82. E. Suhir, Device and method of controlling the bowing of a soldered or adhesively bonded assembly, U.S. Patent #6,239,382, 2001.
83. E. Suhir, Bow free adhesively bonded assemblies: predicted stresses, Electrotechnik & Informationtechnik, 120(6) (2003).
84. E. Suhir, Bow-free assemblies: predicted stresses, Therminic'2004, Niece, France, Sept. 29–Oct. 1, 2004.

Probabilistic Approach

85. E. Suhir, Probabilistic approach to evaluate improvements in the reliability of chip-substrate (chip-card) assembly, IEEE CPMT Transactions, Part A, 20(1) (1997).
86. E. Suhir, Thermal stress modeling in microelectronics and photonics packaging, and the application of the probabilistic approach: review and extension, IMAPS International Journal of Microcircuits and Electronic Packaging, 23(2) (2000).

Optical Fibers and Other Photonic Structures

87. E. Suhir, Stresses in dual-coated optical fibers, ASME Journal of Applied Mechanics, 55(10) (1988).
88. E. Suhir, Fiber optic structural mechanics—brief review, editor's note, ASME Journal of Electronic Packaging, September 1998.
89. E. Suhir, Polymer coated optical glass fibers: review and extension, Proceedings of the POLYTRONIK'2003, Montreaux, October 21–24, 2003.
90. E. Suhir, Effect of initial curvature on low temperature microbending in optical fibers, IEEE/OSA Journal of Lightwave Technology, 6(8) (1988).

91. E. Suhir, Spring constant in the buckling of dual-coated optical fibers, IEEE/OSA Journal of Lightwave Technology, 6(7) (1988).
92. E. Suhir, Mechanical approach to the evaluation of the low temperature threshold of added transmission losses in single-coated optical fibers, IEEE/OSA Journal of Lightwave Technology, 8(6) (1990).
93. E. Suhir, Coated optical fiber interconnect subjected to the ends off-set and axial loading, International Workshop on Reliability of Polymeric Materials and Plastic Packages of IC Devices, Paris, Nov. 29–Dec. 2, 1998, ASME Press, 1998.
94. E. Suhir, Optical fiber interconnect with the ends offset and axial loading: what could be done to reduce the tensile stress in the fiber? Journal of Applied Physics, 88(7) (2000).
95. E. Suhir, Critical strain and postbuckling stress in polymer coated optical fiber interconnect: what could be gained by using thicker coating? International Workshop on Reliability of Polymeric Materials and Plastic Packages of IC Devices, Paris, Nov. 29–Dec. 2, 1998, ASME Press, 1998.
96. E. Suhir and J.J. Vuillamin, Jr., Effects of the CTE and Young's modulus lateral gradients on the bowing of an optical fiber: analytical and finite element modeling, Optical Engineering, 39(12) (2000).
97. E. Suhir, Predicted stresses and strains in fused biconical taper couplers subjected to tension, Applied Optics, 32(18) (1993).
98. E. Suhir, Apparatus and method for thermostatic compensation of temperature sensitive devices, U.S. Patent #6,337,932, 2002.
99. E. Suhir, Polymer coated optical glass fiber reliability: could nano-technology make a difference? Polytronic'04, Portland, OR, September 13–15, 2004.
100. D. Ingman and E. Suhir, Nanoparticle material for photonics applications, Patent pending, 2001.

2

Probabilistic Physical Design of Fiber-Optic Structures

Satish Radhakrishnan[a], Ganesh Subbarayan[a], and Luu Nguyen[b]

[a]Purdue University, West Lafayette, IN 47907, USA
[b]National Semiconductor Corporation, Santa Clara, CA 95052, USA

2.1. INTRODUCTION

Performance of a fiber-optic system depends on the coupling efficiency and the alignment retention capability. Fiber-optic systems experience performance degradation due to uncertainties in the alignment of the optical fibers with the laser beam. The laser devices are temperature sensitive, generate large heat fluxes, are prone to mechanical stresses induced and require stringent alignment tolerance due to their spot sizes. The performance of a photonic system is also affected by many other factors such as geometric tolerances, uncertainties in the properties of the materials, optical parameters such as Numerical Aperture etc. In this chapter, we apply systematic, formal procedures for designing the system in the presence of the above mentioned uncertainties. A low-cost generic fiber-optic package is used as a demonstration vehicle for the design procedures that are described in this chapter.

We begin the chapter by describing a representative fiber-optic system design (a low-cost VCSEL transreceiver), which is used as a vehicle in the development of the design and analysis techniques. An optical model describing the coupling efficiency between the laser and the fiber is next developed. We then develop a general mathematical formulation for maximizing the coupling efficiency and robustness of the system. The optimal geometry of the elements of the system that yield the maximum coupling efficiency and robustness is determined using the developed procedures.

Since the number of uncertain parameters that influence a photonic system can be large, an assessment of inter-relationship between the parameters with a view to identifying the critical parameters is essential. Towards this end we develop system representation and formal graph partitioning strategies to decompose the system into different subsystems. Through this process, we identify critical variables that have a larger, system-level influence. We also develop a simple to implement simulated annealing algorithm for carrying out the system partitioning. The results of system decomposition using graph partitioning and simulated annealing techniques are compared.

In general, the uncertainty in the performance of a photonic system arises mainly due to idealizations in geometry, material behavior, and loading history. Uncertainties in geometry can be predicted and controlled using tighter tolerances. However, the models currently used to describe material behavior are mostly deterministic. To predict the coupling efficiency of a photonic system to greater degree of confidence stochastic analysis procedures are necessary. As part of this analysis, the behavior of materials must be stochastically characterized. We present extensive experimental data on thermally and UV-cured type of epoxies typically used in photonic packages to enable stochastic analysis. The test data includes the viscoelastic behavior. We present an analytical model to obtain the stochastic variation in the displacement of the bonded VCSEL devices resulting from the stochastic viscoelastic behavior of the bond epoxies. We utilize the analytical model to predict the uncertainty in the optical coupling efficiency.

We finally describe efficient stochastic modeling techniques to determine the uncertainties in the alignment of the laser beams with the optical fibers. These include First Order Second Moment and Second Order Second Moment method. We then describe a stochastic design procedure where the uncertainty in behavior and geometry are considered during the design stage.

2.1.1. Demonstration Vehicle

The methodologies described in this chapter for the design of fiber-optic systems in the presence of uncertainty are demonstrated through a proposed photonic system (Figure 2.1). The system described in the example consists of a base, to which a 12×1 VCSEL array is bonded using epoxy. The curved reflector is positioned over the VCSEL leaving a small projection on the rear side to connect the anodes. The positioning of the reflector is such that the optical beam emerging out of the VCSEL is reflected into the optical fiber positioned horizontally. As the optical beam emerges out of the VCSEL, it diverges. The shape and size of the curved reflector should be such that the diverging optical beam is reflected into the optical fiber with the least coupling loss. The divergence angle of the

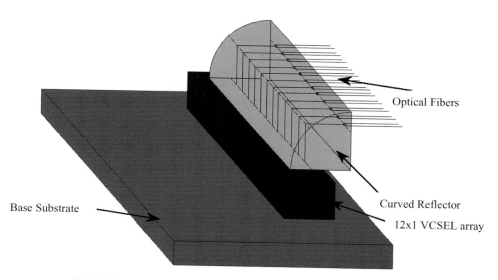

FIGURE 2.1. The fiber-optic system used as a demonstration vehicle in the study.

optical beam after reflection off the curved reflector should be such as to be less than the acceptance angle of the optical fiber to obtain the maximum possible coupling efficiency.

2.2. OPTICAL MODEL

There are many methods such as the Finite Element method, Beam Propagation method, Ray Tracing, etc. that have been applied thus far in the literature for optical modeling. Bierhoff et al. [1,2] describes that methods such as Finite Element Method (FEM), Beam Propagation Method (BPM), etc. can be used very efficiently in modeling single-mode interconnects, but these methods are not applicable for optical multi-mode waveguides guiding more than 1000 propagating modes. Due to the resulting numerical complexity in using FE-BPM, methods based on geometrical optics, called ray tracing, are more effective for modeling multi-mode waveguides. The ray tracing method approximates the output beam of the laser diode by a finite number (N) of rays, each determined by a starting point, a direction, and the time dependent optical power it carries. The propagation of the rays from the laser diode through various optical components as it enters the optical fiber is traced geometrically. This modeling approach leads to an optical model of a laser-diode providing N outputs and an input model of an optical fiber providing M inputs (Figure 2.2).

Kurzweg et al. [3] explains that the ray or geometric optics are the simplest of the optical modeling methods and have the smallest computation time when it comes to modeling and simulation. In the present chapter, the optical beam emerging from the VCSEL is approximated as being Gaussian in character. The laser-fiber coupling analysis of these Gaussian beams is performed using ray tracing analysis. Saleh and Teich, [4] explain that the Gaussian beams are quick to solve, since no explicit integration is necessary to calculate the resulting Gaussian beam at the interface of adjacent components. Ray tracing method has been widely applied in the literature [5–11] to model and to calculate the coupling efficiency in coupling the Gaussian beams with optical fibers and waveguides. The ray tracing method is used for optical modeling in this chapter.

The optical model is described below for the example problem. The equations for the coupling efficiency are symbolically solved in Mathematica [12] using an analytical de-

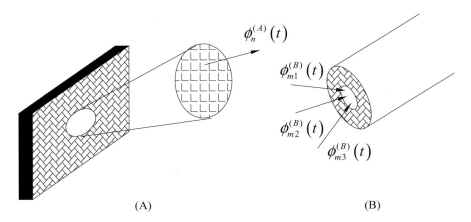

FIGURE 2.2. Equivalent modeling by ray tracing of (A) the output behavior of laser diodes and (B) the input behavior of the optical fibers.

scription of the rays emanating from the VCSEL. The assumptions made in developing the optical model are described below. The optical beam emerging from the VCSEL is assumed to be Gaussian. The curved reflector is assumed to be coated with a highly reflective material with negligible reflective losses. The numerical aperture of the optical fiber is taken as 0.2 for multi-mode operation and 0.14 for single-mode operation. The initial shape of the curved reflector is taken as cylindrical (only a quarter of the cylinder is taken as the curved reflector) with initial radius chosen as 300 microns. The center of the curved reflector is positioned 200 microns away from the center of the opening of the VCSEL. The optical fiber end is placed at the vertical axial plane of the curved reflector. The candidate material for the curved reflector is a cyclic olefin copolymer called Topas [13]. The refractive index of Topas used for the optical model is 1.53 [13].

A 12 × 1 array multi-mode VCSEL operating at 850 nm [14] and single-mode VCSEL operating at 1550 nm [15] was used. Calculations carried out in this study suggest that for single-mode operation, the wavelength of the laser beam has to be greater than 1462 nm, and hence the decision to use a VCSEL operating at 1550 nm. The opening diameter of the optical beams as they emerge out of the VCSEL was 16 microns for multi-mode operation and 8 microns for single-mode operations.

2.2.1. Mode Field Diameter

For a Gaussian beam, mode field diameter is defined as the diameter within which 85% of the power of an optical beam is contained when the light travels in free space. It is also defined as the diameter at which the electric and magnetic fields are reduced to $1/e$ of their maximum values, i.e., the diameter at which the power reduces to $1/e^2$ of the maximum power, because the power is proportional to square of the field strength. Since the optical beam has a Gaussian profile, 100% power would be contained in a beam only when the diameter is infinity. Thus the diameter within which 85% of the power is contained is taken for power calculations.

2.2.1.1. Divergence Angle The ray tracing method is based on geometric analysis and hence the rays propagating from the laser are taken as a set of n rays bounded by two rays tracing the mode-field diameter in two dimensions. The divergence angle corresponding to the mode-field diameter ($1/e^2$ diameter) specified by the VCSEL manufacturers [14,15] are used for ray tracing analysis.

2.2.1.2. Acceptance Angle of the Optical Fiber The optical beam entering the optical fiber propagates along the optical fiber due to total internal reflection. However, only portions of the optical beam directed at the optical fiber will propagate along its length. This is because, for a beam to undergo total internal reflection inside an optical fiber, the incident angle should be less than a critical value known as the acceptance angle. Thus, only the beams incident on the core of the optical fiber at an angle less than the acceptance angle would propagate.

For a fiber to act in single-mode, the numerical aperture (NA), which is a characteristic property of the optical fiber, should be as small as possible. The NA values for commercially available single-mode fibers are currently no lower than 0.14 [16]. Hence a NA value of 0.14 is used for calculating the acceptance angle for single-mode fiber. For multi-mode operation, NA value of 0.2 (a common choice in industry) is used.

The numerical aperture of the optical fiber is directly proportional to the sine of the acceptance angle. Thus, corresponding to the above NA values, the acceptance angle for multi-mode operation is 23° and for single-mode operation it is 16°.

2.2.1.3. V-Number The V-number or the V-parameter is the most important parameter which determines whether an optical fiber would act as a single-mode fiber or a multi-mode fiber. It also indicates the number of modes that propagate along a fiber when operated in multi-mode. For an optical fiber to transmit light in a single-mode, the V-number should be less than 2.405. The optical fiber would transmit multiple modes when the V-number exceeds 2.405. The number of modes N that propagate along the optical fiber is given by $N \approx V^2/2$.

V-number is defined as $\nu = \dfrac{2\pi a}{\lambda}(NA)$.

Thus, for the V-number to be as low as possible the NA value and the core radius (a) of the optical fiber should be as low as possible and the wavelength of the optical beam should be large. The core diameter of the optical fiber is fixed at 8 µm for single-mode fibers and, as stated earlier, 0.14 is the lowest value of NA currently available for single-mode fibers. Thus, the value of the critical wavelength that enables single-mode operation is 1462 nm. Hence, a VCSEL operating at 1550 nm was chosen for the system and used in the calculations of optical behavior.

2.2.2. Refraction and Reflection Losses

The optical beam emerging from the VCSEL passes through the base of the curved reflector and, after reflection, emerges out through the side of the curved reflector. The refractive index of the curved reflector is 1.53 and thus the optical beam entering the curved reflector would undergo a change in the divergence angle and more importantly, it would cause refractive losses. The optical beam would then reflect off the curved section of the reflector. The curved reflector is assumed to have negligible losses on reflection. The reflected beams further undergo refractive losses as they exit out of the reflector. The angle with which the optical beam would exit the reflector would further change due to refraction before entering the optical fiber. A small air gap is assumed between the interfaces of the VCSEL and the curved reflector and between curved reflector and the optical fiber. The magnitude of the refractive loss was estimated at 8.77%.

2.2.2.1. Angle at Entry to the Optical Fiber As explained in the section above, the angle of the optical beam changes as it enters and exits the curved reflector. The divergence angle at the entry to the optical fiber for the initial design was determined using ray tracing to be 30° for both multi-mode and single-mode operations.

2.2.3. Calculations for Coupling Losses

The coupling power of a fiber-optic system can be defined as the ratio of power of the optical beam that enters the optical fiber (when coupled) to the output power of the VCSEL.

$$\text{Coupling Power} = \frac{\text{Power into Fiber}}{\text{Laser Power}}. \tag{2.1}$$

For a Gaussian beam, the power into the fiber is given by $P_0(1 - e^{-2(r/w)^2})$, where, r is the radius of the diverging optical beam corresponding to the acceptance angle of the optical

TABLE 2.1.
Estimated divergence angle and the resulting coupling efficiency at the initial design.

Wavelength	V-number	Divergence angle at VCSEL end	Divergence angle at the optic fiber end	Optical fiber diameter (μm)	Coupling loss (%)	Coupling efficiency (%)
For Multi mode operation (NA = 0.2), Acceptance angle = 23°						
$\lambda = 850$ nm	11.8	27° ($1/e^2$)	30°	50	10	75
For Single mode operation (NA = 0.14), Acceptance angle = 16°						
$\lambda = 1550$ nm	2.27	15° (FWHM)	30°	8	12	73

fiber and w is the mode field radius of the diverging optical beam when it enters the optical fiber. The laser power is taken as the power contained within the mode field diameter of the optical beam $[(P_0(1 - e^{-2})]$ as it exits the VCSEL. The summary of the results for the optical model is given in Table 2.1.

2.2.4. Coupling Efficiency

For the initial design, the coupling efficiency values for both multi-mode and single-mode operation are low (Table 2.1). This is mainly due to the angle of the optical beam at the entry to the optical fiber being much larger than the acceptance angle. One can obtain the maximum possible coupling efficiency when this divergence angle is less than the acceptance angle of the optical fiber. The maximum possible coupling efficiency for both multi-mode and single-mode operations is 83%. The 17% loss is due to refraction as the optical beam traverses through the curved reflector.

2.2.4.1. Minimum Spot Size and Beam Shift The rays tracing the mode filed diameter of the Gaussian beams converge and diverge linearly. However the beams do not converge to a single point. The Gaussian ray profile changes from linear to parabolic shape as the beam converges to result in a minimum spot size (Figure 2.3). The minimum spot size is dependent on the wavelength of the optical beam and is given by $\lambda/\pi\theta$. For the single mode operation, this minimum spot size is 7 microns. The fiber-optic system experiences beam shift due to geometric and assembly tolerances as well due to the material behavior at high temperatures (Figure 2.4). The effect of the beam shift or the misalignment of the optical beam on the coupling efficiency can be seen in Figure 2.5. Here the coupling efficiency is taken as 100% when the beam is aligned without shift (i.e., the divergence angle of the optical beam is equal to the acceptance angle at the fiber). The coupling efficiency value does not change initially for small beam shift since the optical beam usually has a smaller spot size (W_0) as compared to the optical fiber (W) and slowly decreases with increasing beam shift. It can reduce to as low as 30% when the normalized beam shift with respect to the optical fiber is 0.5.

The maximum possible coupling efficiency can be obtained by the optimal design of the shape and size of the curved reflector. The optimized shape and size of the curved reflector should be such that the angle of the diverging beam at the entry to the optical fiber should be within the acceptance angle of the optical fiber. The procedures for achieving such designs are described later in this chapter.

PROBABILISTIC PHYSICAL DESIGN OF FIBER-OPTIC STRUCTURES

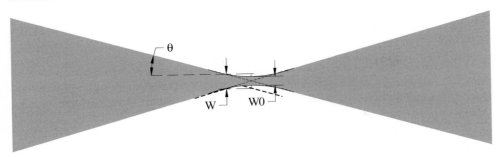

FIGURE 2.3. Gaussian Beams possess a minimum spot size, which is illustrated pictorially here.

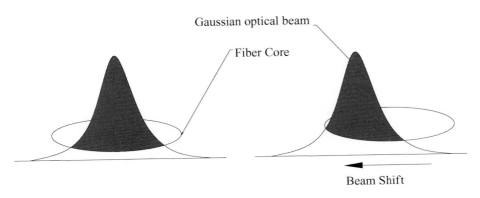

FIGURE 2.4. Shift of Gaussian Beams and the resulting coupling loss.

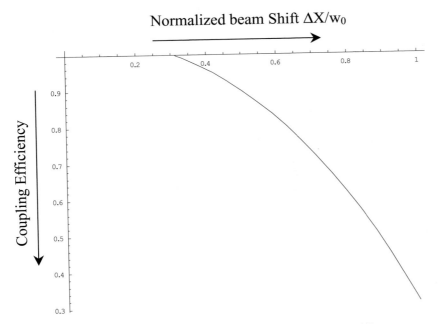

FIGURE 2.5. Coupling efficiency change due to normalized beam shift.

2.3. INTERACTIONS IN SYSTEM AND IDENTIFICATION OF CRITICAL VARIABLES

In general, stochastic analysis of any complex system requires an understanding of the inter-relationships between various elements that together make the complex system. The identification of critical variables to the system would reduce the complexity and the computational expense of the problem. Here, we use the coupling efficiency as the performance measure to evaluate the degradation caused by the misalignment of the laser beam with the optical fiber. To determine the uncertainty in the coupling efficiency of the system, the uncertainties in the variables that influence the coupling efficiency need to be determined first. However, uncertainties of all the variables that affect the coupling efficiency would not contribute equally to the overall uncertainty of the system. A systems analysis procedure based on the function–variable matrix is described here to identify the critical variables, and through their uncertainty determine the uncertainty in the coupling efficiency.

2.3.1. Function Variable Incidence Matrix

System decomposition procedures to identify critical system level parameters have been widely used in the literature. The first step of such a procedure is the formal representation of the functions and the variables that affect them. One such commonly used representation is a table referred to in the literature as the function dependence table or the function–variable incidence matrix [17,18]. The function–variable incidence matrix (referred in the rest of the chapter as the function–variable matrix) is a mechanism for the formal representation of the inter-relationships in the system. The function–variable matrix is partitioned into different subsystems using formal system decomposition techniques to identify the variables that are critical to the system. In other words, variables that strongly tie the functions are the system-level "linking" variables that belong to the whole system as opposed to any one sub-system.

Most systems in general can be characterized as being either hierarchical or non-hierarchical [17]. Starting from a function–variable matrix of the form shown in Figure 2.4, an ideal system would partition into completely independent subsystems with no interaction, but most commonly, engineering systems can only be partitioned into the block angular form shown in Figure 2.6. A system that can be decomposed into independent sub-systems is hierarchical in nature, while a system that partitions into sub-systems that are inter-dependent through a set of system-level "linking" variables is termed a non-hierarchical one. The key feature of non-hierarchical systems is the emergence of two kinds of variables: local variables, which come into play only when describing the functions of specific subsystems, and linking variables, which are shared among subsystems. The linking variables serve to describe the interactions among the function groups.

The systems-level analysis begins with the partitioning of the function–variable matrix into subsystems with local and linking variables. Optimal partitioning into two subsystems will yield the least number of linking variables. However, determining the optimal partition is an NP (Nondeterministic Polynomial time) complete problem [19]. Therefore, commonly, heuristic algorithms are used to carryout the partitioning. These partitioning algorithms and codes based on these algorithms often rely on a graph representation [20] of the system, and therefore there is a need to convert the function–variable matrix into an undirected graph.

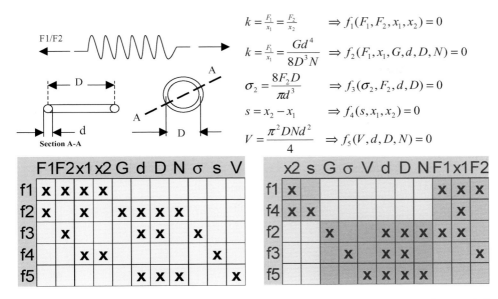

FIGURE 2.6. Function–variable matrix of a simple spring system and its partitioning into two subsystems is illustrated in the figure. Optimal partitioning leads to subsystems that are least interactive. The linking variables F_1, x_1 and F_2 are system-level variables since they determine the nature of the interaction between the subsystems.

System decomposition using graph partitioning techniques is a two step process. The function–variable matrix is first converted to a graph with vertexes and weighted edges. Michelena and Papalamabaros [19] suggested a procedure to convert the function–variable matrix to a hypergraph, which in turn is converted to an undirected graph. In Table 2.2, the function–variable matrix developed for the system described as test vehicle in this chapter is shown. The variables that lead to misalignment of the optical fiber with the laser beam can be characterized as being geometrical, optical, thermo-mechanical, and environmental in nature. The geometric and optical variables affecting the system were obtained from [21]. Not all the variables mentioned above affect the coupling efficiency directly. There are also intermediate functions that in turn affect the final coupling efficiency. These intermediate functions are acceptance angle, divergence angle at laser end, divergence angle at the optical fiber end, fiber offset and ray offset. The intermediate functions become variables for the description of higher level functions. There are thus a total of 7 functions and 23 variables.

The sensitivity values of the geometric variables were calculated based on the optical model described earlier (presented in greater detail in [21]). Some of the variables such as the environmental variables and thermo-mechanical variables do not have any analytical descriptions that enable the easy calculation of the sensitivity values. For these variables, (what was believed to be) a reasonable estimate of the sensitivity value was used.

2.3.2. Function Variable Incidence Matrix to Graph Conversion

Michelena and Papalambros [19] developed techniques to convert the function–variable matrix to graph assuming equal effect of each variable on respective functions and calculating the edge weights of the graph depending on the number of variables on which each function depends. Such an assumption of equal effect of variables on their respective

TABLE 2.2.
Function–variable matrix with sensitivity values for the photonic system used in the present study.

	Environmental effects		Geometrical tolerance						Optical parameters				Thermo-mechanical effects										
	Load on the epoxy between VCSEL and the base	Load on the epoxy between molded block and the base	Thickness of epoxy between VCSEL and base	Thickness of epoxy between molded block and base	Position of optic fiber in X, Y and Z direction	Position of curved reflector (X_0, Y_0)	Dimensions of curved reflector (a, b)	Position of VCSEL with respect to the base	Numerical aperture of fiber	Wavelength of laser beam	Initial diameter of the laser beam	Core diameter of optical fiber	Young's modulus of molded block	Viscoelasticity property of the epoxy between molded block and the base	Viscoelasticity property of the epoxy between the VCSEL and the base	Thermal conductivity of molded block	CTE value of the molded block	Acceptance angle	Divergence angle at laser end	Divergence angle at optic fiber end	Fiber offset	Ray offset	Viscoelasticity behavior
Acceptance angle									0.507									1					
Divergence angle at laser end										1	1								1				
Divergence angle at optical fiber end										1	1									1			
Fiber offset	1	1	1	1	0.0155	0.5	0.5	0.5					0.5	1	1						1		
Ray offset					0.015	0.5	0.5	0.5				0.5	0.2	1	1	0.5	0.5					1	
Viscoelasticity behavior	0.5	0.5	0.5	0.5	0.02																		1
Coupling efficiency	0.4	0.4	0.2	0.2	0.02					0.02	0.02	0.02						0.04	0.0204	0.04	0.02	0.0204	

functions is not always valid, especially when the degree of dependence of the function on different variables varies substantially. The method to calculate the edge weights of the graph is modified in this chapter to include the sensitivity values of the function with respect to each of the variables.

The conversion of the function–variable matrix to graph first consists of representing the function–variable matrix in terms of a hypergraph as shown in Figure 2.7. A hypergraph may be visualized as consisting of solid linkages each representing one variable. The vertexes of the linkage represent the functions that the variables affect and the edges are the hyperedges [19]. The number of edges in a linkage is denoted as p. Assuming equal effect of all variables on a function, one simple estimate of the weight of a hyperedge is $w = 1/(p-1)$. Since the minimum number of edges needed to be cut to partition the vertex set of a p-hyperedge is $(p-1)$, the total weight of the cut edges will be one [19] (Figure 2.7).

The sensitivity values of the variables with respect to the functions can be used to accurately calculate the edge weight. The modified procedure for calculating the edge weight of the graph for the example problem is shown in Figure 2.8. The edges of the hypergraph and graph are exactly as shown in Figure 2.7. However, the weights of hyperedges are calculated differently. The weight of each hyperedge of a variable is different and depends only on the sensitivity value of the functions (vertexes) that the hyperedge connects. The hyperedge weight is calculated as the product of the sensitivity values of the functions connected by the hyperedge with respect to the variable. This can be written mathematically as

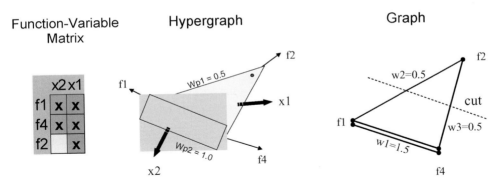

FIGURE 2.7. A schematic illustration of the conversion from function–variable matrix to hypergraph to graph for the application of graph partitioning algorithms [19].

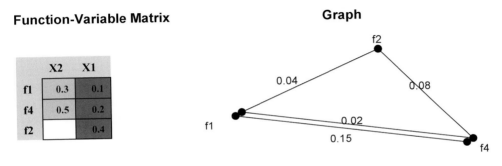

FIGURE 2.8. A schematic illustration of the calculation of edge weights based on sensitivity values.

$$w(i,j)_{i \neq j} = \sum_{k=1}^{n} f(i,k) \cdot f(j,k), \quad i,j = 1, 2, \ldots m, \quad (2.2)$$

where, m = number of functions, n = number of variables. $w(i,j)$ represents the weight of graph edge connecting functions i and j while $f(i,j)$ represents the element of the function–variable matrix. The value of $f(i,j)$ is zero when a variable j does not affect the function i thus resulting in zero edge weight.

The algorithm to convert the function–variable matrix to graph was implemented in Java based on the work in reference [19] using the modified procedure to include sensitivities in calculating the edge weight. The output of the Java code is inline with the input file required by the Chaco program [10] to partition the graph. The graph partitioning algorithms are discussed next.

2.3.3. Graph Partitioning Techniques

The graph partitioning as a means for system decomposition is widely applied in a variety of fields such as power networks [22,23], finite element analysis [24], VLSI design [25], and materials development [26] among others. The algorithms for graph partitioning are well established and largely heuristic in nature owing to the NP completeness of the partitioning problem. The popular Kernighan-Lin algorithm [20] as well as its variant by Fiduccia and Mattheyses [27] are implemented in codes such as Chaco developed by Hendrickson and Leland [28].

The Kernighan-Lin algorithm though popular and widely applied has two major drawbacks. Firstly it needs the system to be represented in the form of a graph. This graph is then partitioned to obtain the optimal system decomposition. Secondly Kernighan-Lin algorithm does not result in optimal partition when the ratio of number of edges to the number of vertexes is low. The algorithm performs poorly when this ratio is less than three and produces nearly optimal solutions when the ratio is higher than five [29].

The graph partitioning algorithms (for example those described in reference [20] and the ones used in this study) enable one to partition (into sub-graphs) any combinatorial problem expressed as a graph. Their most common use has thus far been to divide a finite element domain into sub-domains for efficient parallel computation. Here, such an approach is used for assigning system-level design and material parameters into appropriate sub-systems for the test vehicle described earlier in an automated manner. The common heuristic algorithms such as the Kernighan-Lin algorithm are available in the Chaco program [28] used in the present study. In addition to partitioning a system into two subsystems, simultaneous k-partitioning is possible in Chaco when the number of partitions, k, is specified by the user.

2.3.4. System Decomposition using Simulated Annealing

Another method to partition a graph is to use a globally optimal search strategy such as Simulated Annealing. Graph partitioning using simulated annealing is demonstrated in the literature in references [30–32], but such a partitioning procedure is not common. System decomposition using simulated annealing is a two step process where in the system must be represented in the form of a network or graph and the graph is then partitioned optimally.

Simulated annealing is a generalization procedure similar to Monte Carlo simulation for examining the equation of states and frozen states of n-body systems [33]. The algorithm is inspired by the manner in which liquid freezes or metals recrystallize during the process of annealing. In an annealing process a melt, initially at high temperature and in disordered state, is slowly cooled so that the system at any time is approximately in thermal equilibrium. As cooling proceeds, the system becomes more ordered and approaches a stable ground state at $T = 0$. Hence, the process can be thought as an adiabatic path to the lowest energy state. If the initial temperature of the system is too low or cooling is done fast, the system may become quenched forming defects or freezing out in a metastable state (i.e., trapped in local minimum energy state).

In the original Metropolis scheme [33], an initial state of a thermodynamic system was chosen at energy E and temperature T. Holding T constant, the initial configuration is perturbed and the change in energy ΔE is computed. If the change in energy is negative the new configuration is accepted. If the change in energy is positive it is accepted with a probability given by the Boltzmann factor $\exp(E/kT)$. This process is then repeated sufficient number of times to give good sampling statistics for the current temperature, and then the temperature is decremented and the entire process repeated until a frozen state is achieved at $T = 0$. The flow of control during solution using simulated annealing is described in the flowchart of Figure 2.9.

By analogy, the generalization of this Monte Carlo approach to combinatorial optimization problem is straightforward [32–35]. The current state of the thermodynamic system is analogous to the current solution to the optimization problem, the energy equation for the thermodynamic system is analogous to the objective function, and the ground state is analogous to the global minimum. The major difficulty in implementing the algorithm is the lack of analogy for the temperature T. Whether the final minima obtained is global or local depends on the "annealing schedule," the choice of initial temperature, how many iterations are performed at each temperature, and the magnitude of the temperature decrement as the cooling proceeds.

Many different schemes have been proposed for the length of the Markov chain and for updating the temperature. Aarts [36] suggests that for discrete valued design variables, every possible combination of design variables in the neighborhood of a steady state design variable should be visited at least once with a probability of P. Thus, if there are S neighboring designs then the length of Markov chain is given by

$$M = S \ln\left(\frac{1}{1-P}\right),$$

where $P = 0.99$ for $S > 100$, and $P = 0.995$ for $S < 100$. For discrete valued variables, there are many options for defining the neighborhood of the design. One possibility is to define it as all the designs that can be obtained by changing one design variable to its next higher or lower value [37]. For an n variable design problem, the immediate neighborhood has $S = 3n - 1$ points.

Many different schemes have been proposed for updating the temperature. A frequently used rule is a constant cooling update [36,37].

$$T_{k+1} = \alpha T_k, \quad k = 0, 1, 2, \ldots . \tag{2.3}$$

where, $0.5 \leq \alpha \leq 0.95$.

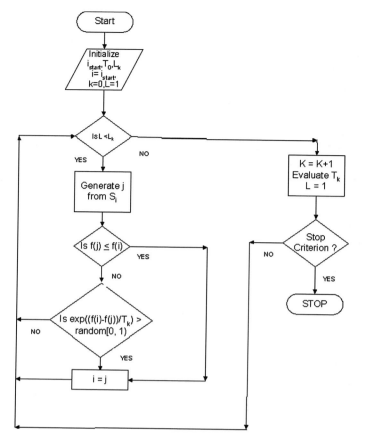

FIGURE 2.9. A flow chart describing the simulated annealing procedure.

In addition to using heuristic graph-partitioning strategies, in the present study, system decomposition is demonstrated by implementing the simulated annealing algorithm. The main reason for using simulated annealing is to reduce the complexity of the partitioning process by converting the two-part operation (conversion of function–variable matrix to graph followed by partitioning of the graph) into one operation. Though the computational time for simulated annealing is large due to the number of iterations required to achieve the convergence criteria, the time required for partitioning does not increase significantly as the number of subsystems increases. Also, as the system size increases, the time taken for partitioning does not increase significantly. Thus for a large system, the time taken for partitioning a system with simulated annealing will compare well with that obtained using the Chaco program. In the present work, the simulated annealing algorithm was implemented using Mathematica [12].

The objective function for this problem can be stated as

$$\min \sum_{j=1}^{n} \sum_{i=1}^{m} L(i, j), \tag{2.4}$$

where m = number of functions, n = number of linking variables. $L(i, j) = n \times m$ is the matrix containing the sensitivity values of all the functions with respect to the linking variables. The goal of the partitioning problem is to rearrange the rows and columns such that the number of linking variables is a minimum. The double summation in the objective function above is defined as the cut-size of the partition. Thus, the cutsize resulting from the partition must be minimized to obtain the optimal partition of the system.

For the photonic system considered in this chapter, the function–variable matrix was first converted to a graph and the edge weights were calculated based on the methodology described earlier. The Chaco program [20] was then used to partition the graph and thereby partition the system into subsystems. A similar partitioning was carried out using the simulated annealing algorithm. The system decomposition results obtained using the simulated annealing method matched exactly with that obtained using the Chaco program.

The result of partitioning the system into two subsystems is shown in Table 2.3. In the table, the seven functions were split into two groups of three and four functions each. We further partitioned the second subsystem into two lower-level sub-systems. The two lower-level sub systems consisted of two functions each. For the first lower-level sub system, there were no local variables. The second partitioning resulted in five linking variables for two lower-level sub systems. The final output of the graph partitioning analysis is shown in Table 2.4. Thus, after partitioning the system into three sub systems, we find that the system is naturally decomposed into largely domain-specific subsystems consisting of optical parameters, material parameters and loading and a few geometric parameters. The partitioning resulted in 15 linking variables for the whole system.

The linking variables were mainly geometric variables: position of optical fiber in the x, y and z direction; the position of the curved reflector in the x and y direction; the dimensions of the curved reflector; the position of the VCSEL with respect to the base; the initial diameter of the laser beam; the core diameter of the optical fiber; the divergence angle at the optical fiber end; the fiber offset and the ray offset; the thickness of the epoxy between base VCSEL and the base; the thickness of the epoxy between the molded block and the base. The only optical parameter among the linking variables was the wavelength of the laser beam, while the load/environmental parameters were the load on the epoxy between VCSEL and the base and the load on epoxy between the molded block and the base. Finally, the parameter defining the viscoelastic behavior of the epoxy between the VCSEL and the base was also a linking variable. These are the most critical variables that describe the interaction between the subsystems identified by the partitioning process. In the following sections, we evaluate the effect of uncertainty in the above identified linking variables on the coupling efficiency of the system.

2.4. DETERMINISTIC DESIGN PROCEDURES

The coupling efficiency for the initial design was determined to be very low as compared to its maximum achievable value (Table 2.1). As mentioned earlier, the coupling efficiency can be maximized through the optimal design of the shape and size of the curved reflector. A system fabricated per the specified design, however, is not guaranteed to achieve the optimal coupling efficiency due to the uncertainties caused by the fabrication/assembly processes as well as the material behavior under the use environment. Ideally, the fiber-optic system should be robust against the uncertainties in the variables. The following sections describe the design procedures for the optimal and robust design of the fiber-optic system.

TABLE 2.3.
The function–variable matrix for the photonic system obtained after the first partition.

Variable	Acceptance angle	Divergence angle at laser end	Coupling efficiency	Divergence angle at optical fiber end	Fiber offset	Ray offset	Viscoelasticity behavior
Numerical aperture of fiber	✓	✓					
Acceptance angle	✓	✓					
Divergence angle at laser end		✓	✓				
Load on the epoxy between VCSEL and the base			✓		✓	✓	
Load on the epoxy between molded block and the base					✓	✓	
Thickness of epoxy between VCSEL and base			✓		✓	✓	
Thickness of epoxy between molded block and base					✓	✓	
Young's modulus of molded block						✓	✓
Viscoelasticity property between molded block and base						✓	✓
Viscoelasticity property of the epoxy between VCSEL and the base						✓	✓
Thermal conductivity of molded block						✓	✓
CTE value of the molded block						✓	✓
Viscoelasticity behavior							✓
Position of optic fiber in X, Y and Z direction					✓	✓	
Position of curved reflector (X_0, Y_0)			✓	✓		✓	
Dimensions of curved reflector (a, b)			✓	✓	✓		
Position of VCSEL with respect to the base			✓	✓	✓		
Initial diameter of the laser beam			✓	✓	✓		
Core diameter of optical fiber					✓	✓	
Divergence angle at optic fiber end			✓	✓	✓		
Ray offset			✓			✓	✓
Fiber offset			✓		✓	✓	
Wavelength of laser beam			✓	✓	✓		

TABLE 2.4.
The function–variable matrix for the photonic system obtained after the second partition.

Function	Optical parameters			Material parameters					Geometry and loading parameters					Geometry tolerances								
	Numerical aperture of fiber	Acceptance angle	Divergence angle at laser end	Young's modulus of molded block	Viscoelasticity property of the epoxy between molded block and the base	Thermal conductivity of molded block	CTE value of the molded block	Viscoelasticity behavior	Load on the epoxy between VCSEL and the base	Load on the epoxy between molded block and the base	Thickness of epoxy between VCSEL and base	Thickness of epoxy between molded block and base	Viscoelasticity property of the epoxy between the VCSEL and the base	Position of optic fiber in X, Y and Z direction	Position of curved reflector (X_0, Y_0)	Dimensions of curved reflector (a, b)	Position of VCSEL with respect to the base	Initial diameter of the laser beam	Core diameter of optical fiber	Divergence angle at optic fiber end	Ray offset	Fiber offset	Wavelength of laser beam
Acceptance angle	✓																						
Divergence angle at laser end	✓	✓																					
Coupling efficiency	✓	✓	✓	✓	✓	✓	✓	✓	✓	✓	✓	✓	✓	✓	✓	✓	✓	✓	✓	✓	✓	✓	✓
Divergence angle at optical fiber end																			✓	✓	✓		✓
Ray offset														✓	✓	✓	✓	✓				✓	✓
Fiber offset														✓	✓	✓	✓					✓	
Viscoelasticity behavior				✓	✓	✓	✓	✓	✓	✓	✓	✓	✓										

The procedures are deterministic since the design problem is formulated through deterministic treatment of functions and variables. In other words, while the design procedure is deterministic, the expectation is that the resulting designs are robust to uncertainties in the variables.

2.4.1. Optimal and Robust Design

In general, to maximize the coupling efficiency of the fiber-optic system, the laser beam shape and size entering the optical fiber needs to be optimized. The objective for the optimization procedure is to ensure that the i) diameter of the beam at the entry to the optical fiber is less than the diameter of the core of the optical fiber, and, ii) divergence angle at the entry to the optical fiber is less than the acceptance angle of the optical fiber.

The specific form of objective function for both single and multi-mode operation chosen in the present study is:

$$\max f \equiv \left(1 - e^{-2(r/w)^2}\right), \qquad (2.5)$$

where r is the radius of the diverging optical beam desired to be less than the acceptance angle of the optical fiber and w is the mode field radius of the optical beam when it enters the fiber. The radius and the divergence angle of the optical beam are calculated using the parameters y_1, y_2, θ_1 and θ_2 as shown in Figure 2.10.

Due to the inherent uncertainties in the design variables, in practice, the function value will vary in the neighborhood of the optimum design. Thus, the optimal design may not necessarily be the most robust one or, the optimal design need not necessarily be least sensitive to perturbations in the design. Thus, we also develop in the present study a mathematical formulation to identify the robust design that will minimize the performance variability while achieving (to the extent possible) the maximum coupling efficiency.

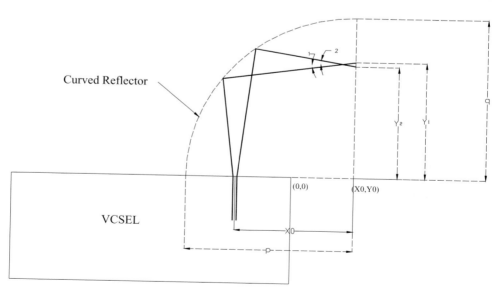

FIGURE 2.10. Parameters used in the optimal design of the reflector.

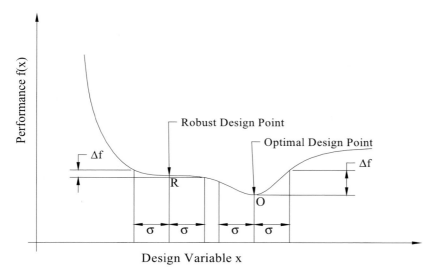

FIGURE 2.11. An illustration of optimal and robust design points.

The difference between the optimal design point and the robust design point can be seen in Figure 2.11. Points R and O are the robust and optimal design points, respectively. For a perturbation of the order of the standard deviation σ in the design variable, it can be seen that the change in the function value Δf for an optimal design point could be high, while for a robust design point, the change in function value Δf is minimal. The extension of these ideas to multiple dimensions is straightforward.

The methodologies developed thus far in the literature [38–43] for robust design are suitable only for smooth functions since these are based on determining sensitivities. Therefore, the methodologies cannot handle steep changes in function values. In this study, we formulate a more general robust optimal design problem using a min-max optimization formulation. The objective function for robust design is as follows:

$$\min_{x} \max_{\Delta x} \left[f - \frac{1}{\Delta x} \int_{x}^{x+\Delta x} f \, dx \right]^2. \tag{2.6}$$

In the above expression, the integral represents the average value of the function over the interval Δx at x. The squared difference between the function value and the average value is first maximized for a particular value of Δx to identify the variation of the function at any x. This function is then minimized with respect to the variable x.

The integral in Equation (2.6) was evaluated using a numerical quadrature due to the complexity of the function. The built-in numerical integration function in Mathematica, while accurate, required a great deal of computational time. Therefore, a simpler trapezoidal rule was used for integration. Since the trapezoidal rule involves the evaluation of the function values at specific intervals, the computation time was reduced by an order of magnitude. The final form of the objective function for solving Equation (2.6) was as follows:

$$\min_{x} \max_{\Delta x} \left[f - \frac{h}{2\Delta x} \left(f_0 + f_n + 2 \sum_{i=1}^{n-1} f_i \right) \right]^2, \quad n = \left(\frac{\Delta x}{h} \right). \tag{2.7}$$

In general, the robust design determined by solving Equation (2.7) could result in a coupling efficiency that is lower than the optimal value. The robust design objective function can be modified to include the optimal design function to yield a robust design with an acceptable coupling efficiency value.

$$\max_x \left\{ c_1 f - c_2 \max_{\Delta x} \left[f - \frac{h}{2\Delta x} \left(f_0 + f_n + 2 \sum_{i=1}^{n-1} f_i \right) \right]^2 \right\}, \quad c_1 + c_2 = 1. \quad (2.8)$$

Through the parameters c_1 and c_2, one can effect a trade-off between the optimal and robust designs. Clearly, when c_2 is unity, the intent is to obtain a robust design and when the parameter is set to zero, the intent is to obtain an optimal design. For other combinations of the parameters c_1 and c_2, one obtains designs with characteristics in between those of optimal and robust designs.

2.4.2. A Brief Review of Multi-Objective Optimization

The design formulation in Equation (2.8) is a constrained multi-objective optimization problem. The general form of bound constrained multi-objective function is:

$$\text{Min: } c_1 f_1(x) + c_2 f_2(x) + \cdots \quad (2.9)$$

Subject to: $\underline{x_l} \leq \underline{x} \leq \underline{x_u} \quad 0 \leq c_1, c_2, c_3, \ldots \leq 1.$

The concept of constrained multi-objective optimization can be explained using Figure 2.12. Let f_1 and f_2 be two objective functions dependent on variables x_1 and x_2. Let x_{m_1} and x_{m_2} be the points that minimize f_1 and f_2, respectively. The constant function value point sets around x_{m_1} and x_{m_2} are also illustrated in the figure. The two curves intersecting the constant value function curves represent the constraints. The curve connecting x_{m_1} and x_{m_2} represents the set of points that minimize $f = c_1 f_1(x) + c_2 f_2(x)$ for

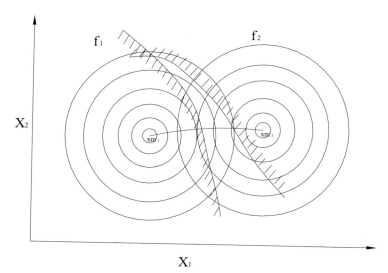

FIGURE 2.12. A schematic illustration of multi-objective optimization.

various combinations of the parameters c_1 and c_2 in the range of [0, 1]. In general, the optimum in a multi-objective problem is an $(n-1)$ dimensional hypersurface of "non-dominated" points from which it is not possible to simultaneously decrease all the function values. For an unconstrained problem, any point in the curve connecting x_{m_1} and x_{m_2} would contribute to a solution set. However, the constraints reduce the solution set size limiting the possible acceptable combinations of c_1 and c_2. Thus for a multi-objective optimization problem, the constraints could result in a design with one or more objective function given greater importance as compared to the other objective functions.

2.4.3. Implementation

Traditional gradient-based search methods such as the quasi-Newton method implemented in the *"find minimum"* function of Mathematica [12] were first used to determine the optimum solution. The *"find minimum"* function in Mathematica, since it uses a quasi-Newton algorithm, is only assured of converging to a local minimum. Even this convergence depends on the satisfaction of "guaranteed descent" conditions such as the Goldstein-Armijo condition [44]. Different starting values for *"find minimum"* yield different solutions indicating multiple local minima for the multi-objective function. For problems with a large number of variables and with potentially large number of local minima, there is no assurance that the global minimum will be identified even if many runs, each with a different starting point, are carried out. Common algorithms that are capable of identifying the global minima use search techniques that randomly sample the design space. Simulated annealing and Genetic algorithms belong to this class of random search algorithms [45]. This procedure of random search, though very effective in identifying the global minimum, becomes computationally expensive for optimization problems with a large number of variables, especially when the cost of analysis is high. In the present study, owing to the relatively low cost of analysis, we chose simulated annealing [46–49] as the optimization algorithm. We used the *"Nminimize"* function implemented within Mathematica to use the simulated annealing optimization algorithm to solve Equation (2.8).

The inner maximization function in Equation (2.8) was solved using the *"Nminimize"* function of Mathematica. However, the output of the inner maximization function could not be used as an input for the outer maximization function within Mathematica. Thus, a simulated annealing code was written to solve the outer maximization function. The computational time to solve Equation (2.8) was about 40 hours.

2.4.4. Results

The procedures described in the previous subsections were applied to the test vehicle. The cross-section of the curved reflector was taken as an ellipse for the purposes of the design. The objective functions in Equation (2.8) was solved using the simulated annealing algorithm to determine the global optimum values of major radius (p), minor radius (q) and the position of the VCSEL with respect to the axis of the curved reflector (x_0) for the respective design objectives. The lower and upper limits on both the major and minor radii of the curved reflector for both single-mode and multi-mode operation were 250 microns and 600 microns respectively. The lower limit of 250 microns was chosen to ensure that it was larger than the outer diameter of the cladding of the optical fibers which is 125 microns. The upper limit of 600 microns was chosen to limit the overall size of the package.

The results for optimal and robust designs for both multi-mode and single-mode operation are tabulated in Tables 2.5 and 2.6 respectively. The values in the last columns in

TABLE 2.5.
Deterministic optimal and robust designs for multi-mode operation.

Method	p (μm)	q (μm)	x_0 (μm)	Coupling efficiency %
Initial design	300	300	200	90
Deterministic optimal design	600	308	530	100
Deterministic robust design	598	331	520	100

TABLE 2.6.
Deterministic optimal and robust designs for single-mode operation.

Method	p (μm)	q (μm)	x_0 (μm)	Coupling efficiency %
Initial design	300	300	200	88
Deterministic optimal design	600	270	545	100
Deterministic robust design	600	275	543	100

Tables 2.5 and 2.6 are the coupling efficiency values for the respective methods. It can be seen that both the methods result in an improved coupling efficiency value with respect to the initial design. Both the designs yielded 100% coupling efficiency for both multi- and single-mode operation.

2.5. STOCHASTIC ANALYSIS

In this section, we develop techniques for the stochastic analysis of the photonic systems. Traditionally, uncertainty analysis has been based on the sampling methods. These methods involve evaluating functional relationships at a set of sample points, and thereby establishing the output uncertainty through exhaustive evaluation of input uncertainties. One such method is the Monte Carlo Simulation. Monte Carlo method involves random sampling from the distribution of inputs and successive model runs until a statistically significant distribution of outputs is obtained. However, its use is limited when complex engineering analyses are required to predict the output function. Added to that, the probability distribution functions may not be well defined for the input variables. The numerous simulations required by Monte Carlo techniques can lead to very high computational time. This high computational time may not be acceptable due to cost and time constraints. Thus the Monte Carlo method, though very accurate, is not suitable for complex physical systems. Thus, the approximation techniques based on the first and second order analysis of numerical models are more suitable due to lower computational cost. These approximate, but computationally efficient methods for stochastic analysis are described in this section.

2.5.1. The First and Second Order Second Moment Methods

The assumption underlying the First and Second Order Second Moment methods is that the important information about the random variables (or functions) of interest can be summarized with the mean representing the expected value of the variable (or function), and the variance representing the second moment about the mean. The First and Second Order Second Moment methods are based on Taylor Series expansion around either mean or critical values of one or more variables. The second moment methods can be used for systems with second moment inputs and parameters with relatively small variance. The third and higher moments are usually ignored since they are relatively small compared to the second moment.

The First Order analysis is the analysis of the mean and the variance of a random function based on its first order Taylor Series expansion. Second order analysis is the analysis of the mean and the variance based on the functions second order Taylor Series expansion.

Let $\mathbf{f(x)}$, be a vector function of random variable $\mathbf{x}$. The Taylor Series expansion of the function about say the mean $\hat{\mathbf{x}}$ is

$$f(\mathbf{x}) = f(\hat{\mathbf{x}}) + \nabla f^T(\hat{\mathbf{x}})(\mathbf{x} - \hat{\mathbf{x}}) + \frac{1}{2}(\mathbf{x} - \hat{\mathbf{x}})^T \mathbf{H}(\mathbf{x} - \hat{\mathbf{x}}) + \cdots, \qquad (2.10)$$

where, $\mathbf{H}$ is the Hessian matrix containing the second partial derivatives of the function. The first order Taylor Series is then,

$$f(\mathbf{x}) = f(\hat{\mathbf{x}}) + \nabla f^T(\hat{\mathbf{x}})(\mathbf{x} - \hat{\mathbf{x}}). \qquad (2.11)$$

Taking the expected value of the above function, the mean of f is estimated to first order by:

$$\hat{f} = E[f(\mathbf{x})]$$
$$= E[f(\hat{\mathbf{x}}) + \nabla f^T(\hat{\mathbf{x}})(\mathbf{x} - \hat{\mathbf{x}})] \quad \mathbf{x} \in R^n$$
$$= f(\hat{\mathbf{x}}). \qquad (2.12)$$

In the above expression, both $f(\hat{\mathbf{x}})$ and $\nabla f(\hat{\mathbf{x}})$ are evaluated at $\hat{\mathbf{x}}$ and therefore are known, deterministic quantities. The first order estimate of the mean is exactly the value obtained through the application of traditional deterministic approach.

Thus, using the expected value, the variance of f may be approximated to first order by:

$$\mathbf{var}[f(\mathbf{x})] = E[(f - \hat{f})^2]$$
$$= \nabla f^T(\hat{\mathbf{x}}) E[(\mathbf{x} - \hat{\mathbf{x}})(\mathbf{x} - \hat{\mathbf{x}})^T] \nabla f(\hat{\mathbf{x}})$$
$$= \nabla f^T(\hat{\mathbf{x}}) \mathbf{cov}(\mathbf{x}) \nabla f(\hat{\mathbf{x}}), \qquad (2.13)$$

where, $\mathbf{cov(x)}$ is the covariance matrix of $\mathbf{x}$. The variance of f is a function of the uncertainty or variability of $\mathbf{x}$, and the sensitivity of f to $\mathbf{x}$ in the neighborhood of $\hat{\mathbf{x}}$.

In a similar manner, using the second order Taylor series expansion of the function, the estimate of the mean can be calculated as:

$$\hat{f} = E\left[f(\hat{\mathbf{x}}) + \nabla f^T(\hat{\mathbf{x}})(\mathbf{x} - \hat{\mathbf{x}}) + \frac{1}{2}(\mathbf{x} - \hat{\mathbf{x}})^T \mathbf{H}(\mathbf{x} - \hat{\mathbf{x}})\right]$$
$$= f(\hat{\mathbf{x}}) + \frac{1}{2} \sum_i \sum_j \mathbf{H}_{ij} \mathbf{cov}(\mathbf{x})_{ij}, \qquad (2.14)$$

where, $cov(x)_{ij}$ is the ijth element of the covariant matrix. Thus, the second order mean includes an additional correction for the difference between the function evaluated at $\hat{x}$ and its true value. It is possible that a first order or deterministic model may yield incorrect results even when the estimate of x is subject to only small errors. However, if the difference between the first order estimate and the second order estimate is large, then the second order estimate may also be inadequate. In such cases, sampling techniques such as the Monte Carlo methods may be required.

The second order estimate of the variance of f may be derived following the procedure used for first order estimate:

$$\begin{aligned}
var[f(x)] &= E\left[\left(\nabla f^T(\hat{x})(x-\hat{x}) + \frac{1}{2}(x-\hat{x})^T H(x-\hat{x})\right.\right. \\
&\qquad \left.\left. - \frac{1}{2}\sum_i\sum_j H_{ij} cov(x)_{ij}\right)^2\right] \\
&= \nabla f^T(\hat{x}) cov(x) \nabla f(\hat{x}) - \left[\frac{1}{2}\sum_i\sum_j H_{ij} cov(x)_{ij}\right]^2 \\
&\quad + \sum_i\sum_j\sum_k \frac{\partial f}{\partial x_i} H_{jk} E[(x_i-\hat{x}_i)(x_j-\hat{x}_j)(x_k-\hat{x}_k)] \\
&\quad + \frac{1}{4}\sum_i\sum_j\sum_k\sum_l H_{ij} H_{kl} E[(x_i-\hat{x}_i)(x_j-\hat{x}_j)(x_k-\hat{x}_k)(x_l-\hat{x}_l)].
\end{aligned}$$

(2.15)

The third and fourth moments of x in the above expression are in general difficult to compute. However, if the variables x_i are independent, then further simplification of the above expressions is possible since for independent variables $x_i, x_j, x_k \ldots$

$$E[(x_i-\hat{x}_i)(x_j-\hat{x}_j)(x_k-\hat{x}_k)\ldots] = 0 \quad \text{if } m \neq n \text{ where } m,n \in \{i,j,k\ldots\}. \quad (2.16)$$

Thus, for independent variables, the above expressions for first order and second order methods reduce to those listed in Table 2.7.

2.6. PROBABILISTIC DESIGN FOR MAXIMUM RELIABILITY

Probabilistic design unlike the deterministic design is aimed at the design of the system in the presence of uncertainty [45,46]. The uncertainties in the design variables are assumed to be known apriori and are included during the design process to account for their effects on the system performance. The quantification of the performance of the system is very critical for fiber-optic systems where the system performance is greatly affected by the uncertainty in the variables. The probabilistic design techniques have found widespread use in the design of structures and mechanical systems [47–52]. Their use for design of electronic packages [53–55] and fiber-optic systems to maximize the system reliability is largely missing.

Deterministic designs may be thought of as specifying a safety factor defined as the ratio of mean strength or allowable limit (μ_S) against the mean system response or

TABLE 2.7.
Expressions for first- and second-order estimates of mean and variance of a general function.

	First order estimate	Second order estimate
Mean (μ_f)	$f(\hat{\mathbf{x}})$	$f(\hat{\mathbf{x}}) + \dfrac{1}{2}\sum_i \dfrac{\partial^2 f}{\partial x_i^2}\sigma_i^2$
Variance (σ_f^2)	$\sum_i \left(\dfrac{\partial f}{\partial x_i}\right)^2 \sigma_i^2$	$\sum_i \left(\dfrac{\partial f}{\partial x_i}\right)^2 \sigma_i^2 - \left(\dfrac{1}{2}\sum_i \dfrac{\partial^2 f}{\partial x_i^2}\sigma_i^2\right)^2$ $+ \sum_i \dfrac{\partial f}{\partial x_i}\dfrac{\partial^2 f}{\partial x_i^2} E\!\left[(x_i - \hat{\mathbf{x}}_i)^3\right]$ $+ \dfrac{1}{4}\sum_i \left(\dfrac{\partial^2 f}{\partial x_i^2}\right)^2 E\!\left[(x_i - \hat{\mathbf{x}}_i)^4\right]$

where, all the derivatives are evaluated at $\hat{\mathbf{x}}_i$, the summations are over the number of variables, and σ_i is the standard deviation of x_i. The third and fourth moment terms in the second order estimate are very difficult to obtain for real systems and therefore are usually ignored. Clearly, the stochastic analysis of any system imposes the need to characterize the mean and variance of all input parameters x_i. In the case of a photonic system, of all the variables, the material behavior expected to play a dominant role in determining the uncertainty in performance. The statistical description of the viscoelastic behavior of common epoxies used to bond the VCSEL to the substrate is the focus of the following section.

applied load (μ_L). By considering only the mean values of allowable limit and system response, it is not possible to estimate the reliability of the system, since it is statistically possible for the design to fail even if the mean strength is larger than the applied load. In the Probabilistic design procedure, on the other hand, the design variable values are determined such that they satisfy a probabilistically stated reliability criterion. The values of design variables for maximum reliability or minimum probability of failure are determined using the design procedure. In Figure 2.13, μ_L and μ_S indicate the mean values of the applied load and the strength for a system design respectively. The two curves $f_L(l)$ and $f_S(s)$ are the probability density functions corresponding to the applied load and the strength. The overlapping region indicates the region of failure. For the reliability of the system to be maximum, this overlap needs to be minimized as discussed below.

Considering normal distributions for both S (μ_S, σ_S) and L (μ_L, σ_L), another random variable $Z = S - L$ is first introduced. Assuming that S and L are independent, Z has a normal distribution with a mean value of $\mu_S - \mu_L$ and a standard deviation of $\sqrt{\sigma_S^2 + \sigma_L^2}$. The system failure can now be defined as when Z is less than zero. Thus, the probability of failure can be written as

$$p_f = 1 - \Phi\left[\frac{\mu_S - \mu_L}{\sqrt{\sigma_S^2 + \sigma_L^2}}\right],$$

where Φ is the cumulative distribution function (CDF) of the standard normal variable. Here,

$$\frac{\mu_S - \mu_L}{\sqrt{\sigma_S^2 + \sigma_L^2}}$$

is termed as the safety margin or reliability index and is denoted by β.

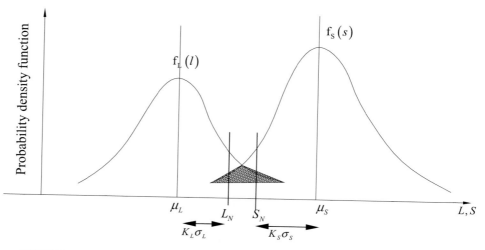

FIGURE 2.13. An illustration of load-strength interference on which the probabilistic design is based.

Thus, the objective function for probabilistic design is formulated as:

$$\max : \beta, \qquad (2.17)$$

where, $\beta = g(f > C)$, where f is the coupling efficiency and C is the specified acceptable value of f. Here, we define the failure of the system to occur when the coupling efficiency drops below 100%; the goal of the optimization is thus to minimize the probability that coupling efficiency drops below 100%. The specific form for β can be written as

$$\beta = \frac{\mu_{CE} - 1}{\sqrt{\sigma_{CE}^2 + 0}}.$$

In maximizing β, μ_{CE} is maximized while σ_{CE} is minimized. We state that Equation (2.17) can be reformulated alternatively as the multi-objective optimization problem shown below:

$$\max : c_1 \mu_{CE} - c_2 \sigma_{CE}. \qquad (2.18)$$

The multi-objective formulation possesses the advantage of allowing the tradeoff between the optimal and probabilistic designs. The mean value of coupling efficiency (μ_{CE}) is given by the function f in Equation (2.5). The standard deviation of coupling efficiency (σ_{CE}) is determined by using the first order approximation given by

$$\sigma_{CE} = \sqrt{\sum_i \left(\frac{\partial f}{\partial x_i}\right) \sigma_{x_i}^2}. \qquad (2.19)$$

For purposes of illustration, the standard deviation of the design variables (major and minor diameter of the curved reflector and position of curved reflector with respect to the VCSEL) was chosen here to be 10% of the value of the core diameter of the optical fiber for multi-mode operation and 20% for the single-mode operation respectively. The optimization of the above function was also performed using the simulated annealing algorithm implemented within Mathematica.

2.6.1. Results

The results of the probabilistic design are listed in Tables 2.8 and 2.9. For the sake of easy comparison, the results of deterministic design listed in Tables 2.5 and 2.6 are repeated here. It is clear from the tables that probabilistic optimal design yields a solution that is identical to the deterministic optimal design. Monte Carlo simulation was carried out at the final design to determine the coupling efficiency uncertainty given the uncertainty in the design variables (Figures 2.14 and 2.15). Figures 2.14 and 2.15 consider only the effect of geometric uncertainty on the coupling efficiency. The optimization of the curved reflector is performed in 2-dimensions and accounts for the effect of coupling loss due to misalignment only in the lateral direction. The material uncertainty, however, results in misalignment of the optical fibers along the longitudinal direction of the curved reflector and is thus not included in the optimization of the shape and size of the curved reflector.

The deterministic and probabilistic optimal and robust design all produce a much improved coupling efficiency performance as compared to the initial design. The coupling efficiency varies in the range of 40–100% (considering three-sigma rule) for the initial design. This variation is reduced to 75–100% and 80–100% for the optimal and robust designs respectively. The robust design procedure performs better than optimal design during the early design stage when the uncertainty in the design variables is usually unknown. The probabilistic design can be used to modify the robust design when experimental data are available to quantify the uncertainty in the design variables. The probabilistic design is however a natural choice when the uncertainties in the design variables are known apriori.

Similar comparison plots are shown for c_1 and c_2 values of 0.5 for robust and probabilistic design (Figures 2.16 and 2.17). The increase in standard deviation values for robust

TABLE 2.8.
Results of deterministic/probabilistic optimal, deterministic robust and probabilistic robust designs for multi-mode operation.

Method	p (μm)	q (μm)	x_0 (μm)	Coupling efficiency %
Initial design	300	300	200	90
Deterministic/probabilistic optimal design	600	308	530	100
Deterministic robust design	598	331	520	100
Probabilistic robust design	600	388	501	100

TABLE 2.9.
Results of deterministic/probabilistic optimal, deterministic robust and probabilistic robust designs for single-mode operation.

Method	p (μm)	q (μm)	x_0 (μm)	Coupling efficiency %
Initial design	300	300	200	88
Deterministic/probabilistic optimal design	600	270	545	100
Deterministic robust design	600	275	543	100
Probabilistic robust design	600	300	534	100

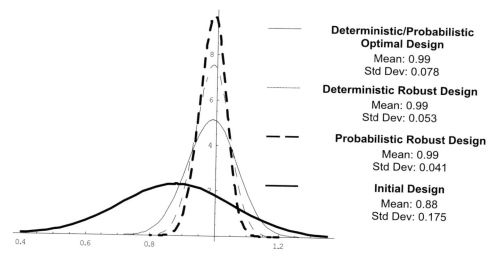

FIGURE 2.14. A comparison of coupling efficiency variation between deterministic/probabilistic optimal, deterministic robust, probabilistic robust and initial design for multi-mode operation.

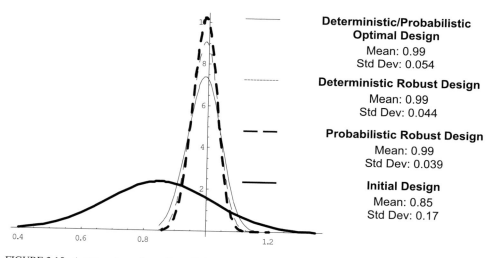

FIGURE 2.15. A comparison of coupling efficiency variation between deterministic/probabilistic optimal, deterministic robust, probabilistic robust and initial design for single-mode operation.

and probabilistic values is a result of trade off with respect to the optimal design. The results indicate that the standard deviation values for all the methods are within a small range. It can be seen that the robust design performs marginally better than the probabilistic design for both the multi-mode and single mode operation. Thus for equal weightage of optimal and robust design, the deterministic design techniques are more desirable as they result in comparable results with respect to probabilistic design and saves time and cost in computing the uncertainty in design variables.

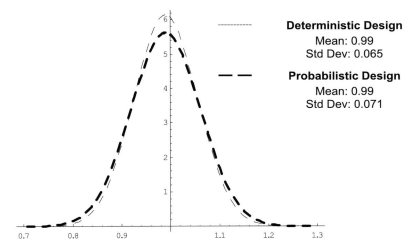

FIGURE 2.16. A comparison of coupling efficiency variation between deterministic and probabilistic designs for multi-mode operation with trade off.

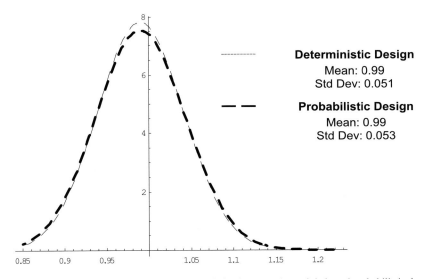

FIGURE 2.17. A comparison of coupling efficiency variation between deterministic and probabilistic designs for single-mode operation with trade off.

2.7. STOCHASTIC CHARACTERIZATION OF EPOXY BEHAVIOR

A typical fiber-optic package experiences high operating temperatures in the range of 70–100°C. The epoxies used to bond the laser to the substrate undergo temperature induced creep behavior due to the operating temperature. The displacement of the laser with respect to the optical fiber induced due to the temperature induced creep behavior of the epoxy could result in significant coupling efficiency loss. A displacement of one radius value of the optical fiber could result in coupling efficiency drop of up to 70% (Figure 2.5). The uncertainty in the material behavior combined with the uncertainty in the geometric parameters would further affect the coupling loss. A stochastic characterization of material uncertainty is thus required to determine its effect on system behavior and thus to enable design decisions. In the following, a representative epoxy material EMI Emcast 501 is stochastically characterized.

2.7.1. Viscoelastic Models

Material models are necessary to parametrically describe material behavior. Towards this end, we begin by describing popular one-dimensional viscoelastic models. The time dependent (viscoelastic) behavior of the epoxies can be simulated through elastic springs and viscous dashpots. The most common models used to describe the viscoelastic behavior are the Maxwell, Kelvin and the Standard solid model. The spring and the dashpot are set in series configuration for Maxwell model and in parallel configuration for Kelvin model. The standard model consists of either the Kelvin model in series with a spring or a Maxwell model in parallel with a spring. The standard solid model is used in this study since the behavior of the epoxy is more similar to solid than liquid.

The subscripts 2 and 1 used in Figure 2.18 represent the parallel Kelvin arrangement and the single spring element respectively.

For the arrangement in series as shown in Figure 2.18,

$$\sigma = \sigma_1 = \sigma_2,$$
$$\varepsilon = \varepsilon_1 + \varepsilon_2. \tag{2.20}$$

The strains in the spring 1 and in the Kelvin arrangement are given by

$$\varepsilon_1 = \frac{\sigma_1}{E_1}, \quad \varepsilon_2 = \frac{\sigma_2}{E_2 + \eta_2 D}. \tag{2.21}$$

Substituting the above equations into the equation for total strain we get the equation of strain and strain rate related to the stress and stress rate for the standard model shown in Figure 2.18 as

$$\dot{\varepsilon}\eta_2 + \varepsilon E_2 = \left(\frac{E_2 + E_1}{E_1}\right)\sigma + \frac{\eta_2}{E_1}\dot{\sigma}. \tag{2.22}$$

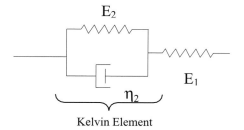

FIGURE 2.18. Standard solid model used to describe EMI Emcast 501 epoxy.

2.7.2. Modeling the Creep Test

The epoxy between the VCSEL and the substrate is subjected to creep induced by high temperatures. Thus the standard solid model described above and governed by Equation (2.22) will have to be used to determine its creep behavior. Creep test is a standard procedure wherein a stress τ_0 is applied at time $t = 0$ on the sample and then maintained constant thereafter. This test can be modeled as a function of time with the aid of the unit step function $[u(t)]$. Thus, during creep test

$$\tau(t) = \tau_0[u(t)]. \tag{2.23}$$

Substituting for $\tau(t)$ in Equation (2.22) and applying Laplace transform, we get

$$\eta_2 s \bar{\varepsilon}(s) + \bar{\varepsilon}(s) E_2 = \left(\frac{E_2 + E_1}{E_1}\right) \bar{\tau}(s) + \frac{\eta_2}{E_1} s \bar{\tau}(s). \tag{2.24}$$

In the case of creep test, $\tau(t) = \tau_0[u(t)]$, thus $\bar{\tau}(s) = \dfrac{\tau_0}{s}$.

Thus, Equation (2.24) simplifies to

$$\eta_2 s \bar{\varepsilon}(s) + \bar{\varepsilon}(s) E_2 = \left(\frac{E_2 + E_1}{E_1}\right) \frac{\tau_0}{s} + \frac{\eta_2}{E_1} \tau_0,$$

$$\bar{\varepsilon}(s)(\eta_2 s + E_2) = \frac{\tau_0}{E_1}\left[(E_2 + E_1)\frac{1}{s} + \eta_2\right], \tag{2.25}$$

$$\bar{\varepsilon}(s) = \frac{\tau_0}{E_1}\left\{\frac{E_2 + E_1}{\eta_2}\left[\frac{1}{s\left(s + \frac{E_2}{\eta_2}\right)}\right] + \frac{1}{\left(s + \frac{E_2}{\eta_2}\right)}\right\}, \tag{2.26}$$

$$\bar{\varepsilon}(s) = \frac{\tau_0}{E_1}\left[\frac{E_2 + E_1}{E_2}\left(\frac{1}{s} - \frac{1}{s + \frac{E_2}{\eta_2}}\right) + \frac{1}{s + \frac{E_2}{\eta_2}}\right]. \tag{2.27}$$

Now, applying inverse Laplace transform, we get

$$\varepsilon(t) = \frac{1}{E_1}\left[\frac{E_2 + E_1}{E_2}\left(1 - e^{-(\frac{E_2}{\eta_2})t}\right) + e^{-(\frac{E_2}{\eta_2})t}\right]\tau_0[u(t)], \tag{2.28}$$

$$\varepsilon(t) = \frac{1}{E_1}\left[\frac{E_2 + E_1}{E_2} - \frac{E_1}{E_2}e^{-(\frac{E_2}{\eta_2})t}\right]\tau_0[u(t)]. \tag{2.29}$$

Equation (2.29) above gives the strain in the epoxy when a constant shear stress τ_0 is applied. Thus, given the applied stress on the epoxy we can calculate the strain and hence the misalignment of the laser beams with the optical fiber as a function of time. The stress experienced by the epoxy is caused by the temperature induced differential expansion between the VCSEL and the substrate. This will be determined later for a bonded three layer system.

2.7.3. Dynamic Mechanical Analysis

The epoxy was characterized using a Dynamic Mechanical Analyzer (DMA). The stress–strain behavior is characterized as a function of temperature as well as frequency. A sinusoidal stress or strain is applied to the material and the output response is measured. Viscoelastic materials exhibit a lag in the output response which is characterized by the phase shift δ as shown in Figure 2.19. The DMA provides two outputs namely the storage modulus which indicates the elasticity of the material and the amount of energy it can store when a stress is applied and a loss modulus which indicates the viscous property of the material and the amount of energy lost to friction and internal motions. The phase lag terms relate the storage and the loss moduli. The tangent of the phase angle δ gives the ratio of loss modulus to storage modulus

$$\tan \delta = \frac{E''}{E'}. \tag{2.30}$$

Equation (2.22) was used to fit the two outputs obtained from the DMA tests to obtain the parameters of the model shown in Figure 2.18. The model parameters E_1, E_2 and η_2 are temperature dependent. The material parameters E' and E'' obtained from the DMA tests however depend on both temperature as well as the frequency at which the stress or strain is applied.

The epoxy samples were tested at four different temperatures of 50°C, 80°C, 110°C, and 150°C. A sinusoidal displacement with maximum amplitude of 5 μm was applied over a frequency range of 0.01 Hz to 10 Hz (0.01 Hz is the minimum value of frequency that could be applied with the equipment that was used in this study).

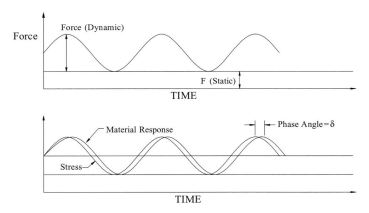

FIGURE 2.19. Principle of DMA.

The equation of modulus for the model in Figure 2.18 can now be obtained by applying a sinusoidal strain to Equation (2.31).

$$\dot{\varepsilon}\eta_2 + \varepsilon E_2 = \left(\frac{E_2 + E_1}{E_1}\right)\sigma + \frac{\eta_2}{E_1}\dot{\sigma}, \qquad (2.31)$$

$$\varepsilon = \varepsilon_0 e^{i\omega t}.$$

The resultant stress is then

$$\sigma = \sigma_0 e^{i(\omega t + \delta)}. \qquad (2.32)$$

Substituting Equation (2.32) in to Equation (2.31), we get

$$\varepsilon_0 E_2 e^{i\omega t} + i\omega\varepsilon_0\eta_2 e^{i\omega t} = \frac{1}{E_1}\left[\sigma_0(E_1 + E_2)e^{i(\omega t + \delta)} + i\omega\sigma_0\eta_2 e^{i(\omega t + \delta)}\right], \qquad (2.33)$$

$$E_c = \frac{\sigma(t)}{\varepsilon(t)} = \frac{E_1(E_2 + i\omega\eta_2)}{(E_1 + E_2 + i\omega\eta_2)},$$

$$E_c = \frac{E_1(E_1 E_2 + E_2^2 + w^2\eta_2^2)}{(E_1 + E_2)^2 + w^2\eta_2^2} + i\frac{E_1^2 w\eta_2^2}{(E_1 + E_2)^2 + w^2\eta_2^2}, \qquad (2.34)$$

where, E_c is the complex modulus of the viscoelastic material dependent on the frequency of the applied stress or strain. The real part of Equation (2.34) is the storage modulus and the imaginary part is the loss modulus. The constants in Equation (2.34) can be found by fitting the equation to the experimental data with respect to frequency.

2.7.4. Experimental Results

Figures 2.20–2.21 show the data fit and the parameters E_1, E_2 and η_2 obtained for one of the 40 samples tested at 50°C, 80°C, 110°C and 150°C respectively. The curve fit was done with respect to the magnitude of the complex modulus of the experimental data.

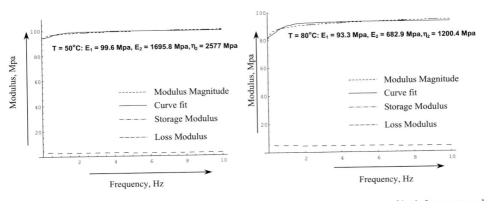

FIGURE 2.20. A fit of the measured modulus value of EMI Emcast epoxy as a function of both frequency and temperature (left plot at 50°C and right plot at 80°C).

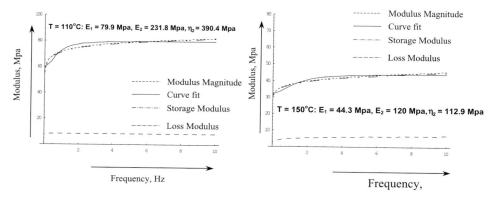

FIGURE 2.21. A fit of the measured modulus value of EMI Emcast epoxy as a function of both frequency and temperature (left plot at 110°C and right plot at 150°C).

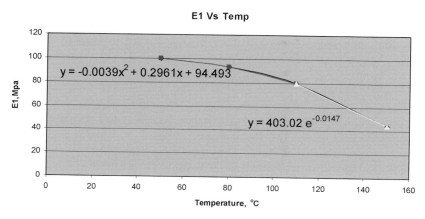

FIGURE 2.22. A fit of the measured value of parameter E_1 as a function of temperature.

The storage modulus was very close to the magnitude of complex modulus as can be seen in Figures 2.20–2.21. The loss modulus values are also plotted in Figures 2.20–2.21. The loss modulus value was in the range of 2.5%–10% of the storage modulus for the temperature range from 50°C to 150°C. The standard solid parameters E_1, E_2 and η_2 were next fitted with respect to temperature to describe the model parameters over the entire temperature range of 50°C to 150°C. Quadratic and an exponential description provided fits with least error for E_1 between 50°C to 110°C and between 80°C and 150°C respectively. A cubic description was the best fit for parameters E_2 and η_2 over the entire temperature range. This data fit with respect to temperature for one of the samples is shown in Figures 2.22–2.24. The data fits shown in Figures 2.20–2.21 are however for only one of the 40 tests we conducted. The stochastic variations of the parameters over the 40 tests are shown in Figures 2.25–2.27.

Similar plots of variation of the material parameters were determined at temperatures 50°C, 110°C and 150°C. The variation in material parameters at intermediate temperatures can be determined using the data fit curves similar to those in Figures 2.22–2.24.

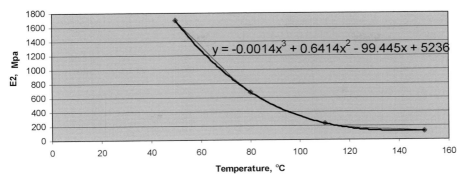

FIGURE 2.23. A fit of the measured value of parameter E_2 as a function of temperature.

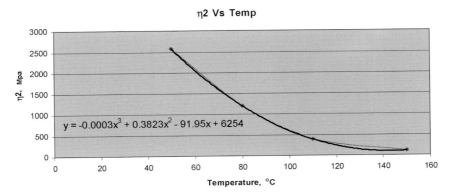

FIGURE 2.24. A fit of the measured value of parameter η_2 as a function of temperature.

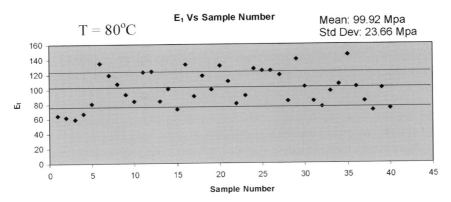

FIGURE 2.25. Variation of E_1 at $80°C$.

2.8. ANALYTICAL MODEL TO DETERMINE VCSEL DISPLACEMENT

The creep model developed in the previous section can be used to determine the magnitude of the shear displacement in the epoxy. The calculation of displacement of the epoxy

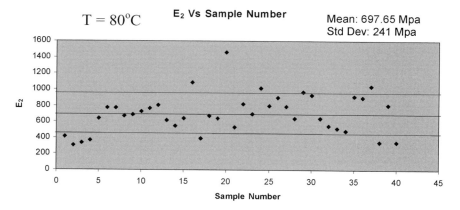

FIGURE 2.26. Variation of E_2 at 80°C.

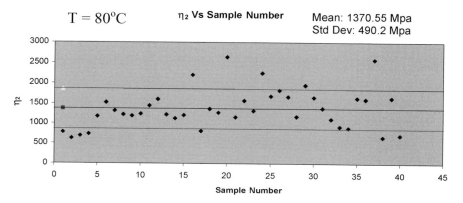

FIGURE 2.27. Variation of η_2 at 80°C.

requires the determination of its stress state. The stresses that are acceptable must satisfy static equilibrium. The simple one-dimensional derivation below determines the displacement by balancing the forces on each layer of the assembly. The derivation is an extension of the classical Timoshenko theory of bi-metal thermostats. The stress distribution analysis of bonded assemblies for electronic packages was first considered by Chen and Nelson [56]. The following derivation was obtained by Suhir [57] for bi-metal thermostats and later extended for tri-material assemblies [58,59].

In Figure 2.28, the 2D analytical model of the VCSEL, epoxy and the substrate are shown. Assuming that the bond epoxy is "soft" relative to the component and the substrate, the forces acting on the component, epoxy and on the substrate can be modeled as shown in Figure 2.29.

The displacement compatibility for the epoxy is given by

$$u_c(x) = u_s(x) - 2\kappa_e \tau(x). \tag{2.35}$$

The last term in Equation (2.35) accounts for the non-uniform distribution of forces and are calculated under the assumption that the corresponding corrections are directly

PROBABILISTIC PHYSICAL DESIGN OF FIBER-OPTIC STRUCTURES

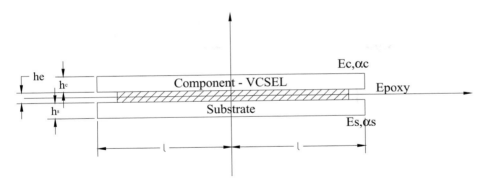

FIGURE 2.28. Two-dimensional model of (the cross-section of) VCSEL, epoxy and substrate.

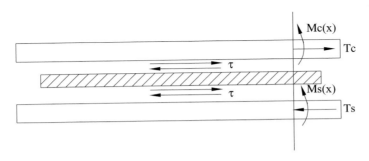

FIGURE 2.29. A free-body diagram of the VCSEL-epoxy-substrate showing the stresses resulting from differential thermal expansion.

proportional to the shearing force in the given cross-section; a further assumption is that the corrections are not affected by the shearing forces in the other cross-sections. Constant κ_e is the interfacial compliance of the epoxy layer.

The displacements $u_c(x)$ and $u_s(x)$ of the component and the substrate are given by

$$u_c(x) = \alpha_c \Delta T x - \lambda_c \int_0^x T_c(\xi)d\xi + \kappa_c \tau(x) + \frac{h_c}{2} \int_0^x \frac{d\xi}{\rho(\xi)}, \qquad (2.36)$$

$$u_s(x) = \alpha_s \Delta T x - \lambda_s \int_0^x T_s(\xi)d\xi + \kappa_s \tau(x) + \frac{h_s}{2} \int_0^x \frac{d\xi}{\rho(\xi)}. \qquad (2.37)$$

The first terms in Equations (2.36) and (2.37) are the unrestricted thermal expansions of the component and the substrate. The second terms are due to the normal forces and are calculated under the assumption that these forces are uniformly distributed over the thickness. The third terms were motivated earlier in Equation (2.35) and the last terms arise out of the curvature (ρ s the local radius) caused by the bending. Also, due to force balance,

$$T_c(x) = T_s(x) = \int_{-l}^{x} \tau(\xi)d\xi, \qquad (2.38)$$

where, $T_c(x)$ and $T_s(x)$ are the normal force (per unit assembly width) along the length direction and α_s and α_c are the thermal expansion coefficients. Further,

$$\lambda_c = \frac{1-v_c}{E_c h_c}, \quad \lambda_s = \frac{1-v_s}{E_s h_s} \qquad (2.39)$$

are the axial compliances of the component and the substrate and

$$\kappa_c = \frac{h_c}{3G_c}, \quad \kappa_s = \frac{h_s}{3G_s} \qquad (2.40)$$

are their interfacial compliances,

$$\kappa_e = \frac{h_e}{3G_e} \qquad (2.41)$$

is the interfacial compliance of the bond epoxy, h_i are the thicknesses of the three elements of the assembly.

For the earlier described standard viscoelastic solid, the shear modulus G_e of the epoxy is given by Equation (2.29), which is repeated here:

$$\frac{1}{G_e} = \frac{1}{E_2}\left[\frac{E_2+E_1}{E_1} - e^{-(\frac{E_2}{\eta_2})t}\right]. \qquad (2.42)$$

The moment balance in the mid-layer of the assembly, at any location x along the length of the assembly yields the following equation:

$$\frac{h_c+h_e}{2}T_c(x) + \frac{h_s+h_e}{2}T_s(x) = M_c(x) + M_s(x), \qquad (2.43)$$

where,

$$M_i(x) = \frac{D_i}{\rho(x)}, \quad D_i = \frac{E_i h_i^3}{12(1-v_i^2)},$$

where, D_i are the flexural rigidity.

If $h = h_c + h_s$, if $h_e \ll h_c, h_s$, $D = D_c + D_s$:

$$\therefore \frac{1}{\rho(x)} = -\frac{h}{2D}T(x). \qquad (2.44)$$

Substituting Equations (2.36, 2.37 and 2.44) in Equation (2.35), we get

$$\tau(x) - \frac{1}{L^2}\int_0^x T(\xi)d\xi = \frac{\Delta\alpha\Delta T x}{\kappa}, \qquad (2.45)$$

where,

$$\frac{1}{L^2} = \frac{\lambda}{\kappa}, \quad \lambda = \lambda_c + \lambda_s + \frac{h^2}{4D}, \quad \kappa = \kappa_c + \kappa_s + 2\kappa_e.$$

Solving Equation (2.45) and applying the boundary condition, $\tau(0) = 0$, $T(l) = 0$, we get

$$\tau(x) = \frac{L}{\kappa}(\alpha_c - \alpha_s)\Delta T \frac{\sinh\left(\frac{x}{L}\right)}{\cosh\left(\frac{l}{L}\right)}. \tag{2.46}$$

Substituting Equation (2.46) into Equation (2.29), we get

$$\frac{u_c - u_s}{h_e} = \frac{1}{E_2}\left[\frac{E_2 + E_1}{E_1} - e^{-(\frac{E_2}{\eta_2})t}\right]\frac{(\alpha_c - \alpha_s)\Delta T L}{\kappa}\frac{\sinh\left(\frac{x}{L}\right)}{\cosh\left(\frac{l}{L}\right)}. \tag{2.47}$$

Substituting Equations (2.41) and (2.42) into Equation (2.47) and rearranging, we get:

$$\frac{u_c}{L} = \frac{u_s}{L} + \frac{3\kappa_e}{\kappa}(\alpha_c - \alpha_s)\Delta T \frac{\sinh\left(\frac{x}{L}\right)}{\cosh\left(\frac{l}{L}\right)}. \tag{2.48}$$

Equation (2.48) gives the normalized displacement of the VCSEL due to the viscoelastic deformation of the epoxy. This normalized displacement is used to determine the misalignment of the optical beams with respect to the optical fibers. This enables the calculation of the coupling loss along the length of the VCSEL array.

It is of interest to compare the above result against the pioneering derivation for stresses in bonded layers by Chen and Nelson [56]. Chen and Nelson's model considered only the normal stresses on the adherents, only the shear stresses on the adhesives and ignored the peeling stresses and the bending moments on the adherents. The equilibrium equations were solved along with the stress–strain constitutive equations to determine the displacements and stresses on the bonded layers. The shear stress and displacement of VCSEL derived using Chen and Nelson model in a manner analogous to the above derivation are:

$$\tau = \frac{\tilde{L}}{3\kappa_e}(\alpha_c - \alpha_s)\Delta T \frac{\sinh\left(\frac{x}{\tilde{L}}\right)}{\cosh\left(\frac{l}{\tilde{L}}\right)}, \tag{2.49}$$

and

$$\frac{u_c}{\tilde{L}} = \frac{u_s}{\tilde{L}} + (\alpha_c - \alpha_s)\Delta T \frac{\sinh\left(\frac{x}{\tilde{L}}\right)}{\cosh\left(\frac{l}{\tilde{L}}\right)}, \tag{2.50}$$

where,

$$\frac{1}{\tilde{L}^2} = \frac{G_e}{h_e}\left(\frac{1}{E_s h_s} + \frac{1}{E_c h_c}\right). \tag{2.51}$$

A comparison of the results for the maximum shear stress value for an elastic system using Chen and Nelson [56] model, Suhir's [57–59] model and Finite Element Analysis (FEA) is given below. The comparison was made for varying thickness and modulus values of the epoxy layer.

It can be seen from Figures 2.30 and 2.31 that the maximum shear stress value predicted by the Suhir model agrees more closely with the FEA results as compared to Chen and Nelson's model for a wide range thickness and modulus values of the epoxy. Chen

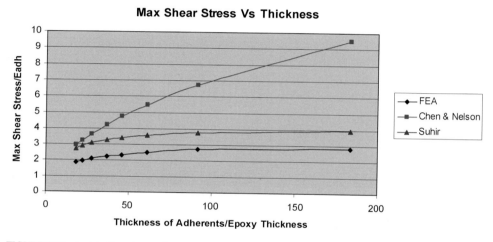

FIGURE 2.30. A plot of maximum shear stress value with respect to the thickness of the epoxy layer determined using Chen and Nelson's model [56], Suhir's model [57–59] and FEA.

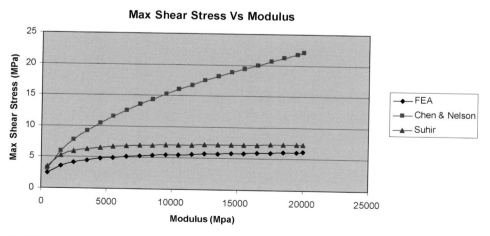

FIGURE 2.31. A plot of maximum shear stress value with respect to the modulus of the epoxy determined using Chen and Nelson's model [56], Suhir's model [57–59] and FEA.

and Nelson model, however, does yield results that are close to those predicted through FEA when the modulus of the epoxy is low and when the thickness of the epoxy layer is relatively large (but smaller than the thickness of the adherents). This is because Chen and Nelson model is applicable only when the ratio of thickness of adherents to the thickness of the epoxy is greater than 10 and less than 30. This validity of Chen and Nelson model over a very small range of epoxy modulus can be explained by the fact that the Chen and Nelson model considers only the normal stresses on the adherent layers and ignores the bending moment caused by the differential thermal expansion between the layers.

2.8.1. Results

The plot for normalized displacement of VCSEL is shown in Figure 2.32. It can be seen that a VCSEL located at the end of the array will undergo exponentially large displacement relative to the device in the center of the array. This displacement of the VCSEL is next used to calculate the coupling efficiency loss [60]. The optical model developed earlier was used to determine the coupling efficiency loss. The extent of coupling efficiency loss when a beam shifts with respect to the optical fiber is shown in Figure 2.33. The coupling efficiency loss is plotted with respect to the normalized beam shift in Figure 2.33. It can be seen that the coupling efficiency does not change initially; however, it drops sharply beyond a certain value of normalized beam shift. This is because the optical beam when it enters the optical fiber has a minimum spot size. The minimum spot size for single-mode operation is ~ 7 µm. Thus the coupling efficiency drops only when the displacement of the VCSEL causes the minimum spot size to shift beyond the diameter of the optical fiber. The coupling loss due to the displacement along the VCSEL array is shown in Figure 2.34. The coupling loss is plotted at four different temperatures. The multiple curves at each temperature represent the increase in loss with increase in time at that temperature. Thus, it can be seen that the coupling loss would be maximum for a VCSEL placed at the end of an array and would further increase with time. The coupling loss would also increase with increase in temperature. Figure 2.34 can be used as a design chart to determine the maxi-

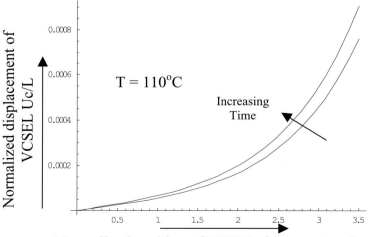

FIGURE 2.32. Normalized displacement as a function of position along the VCSEL array.

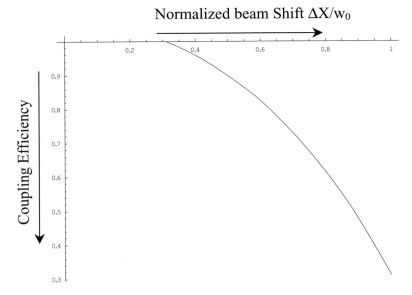

FIGURE 2.33. Coupling efficiency change due to normalized beam shift.

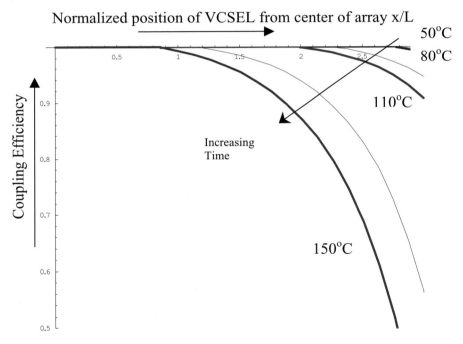

FIGURE 2.34. Coupling efficiency for VCSELs located at various positions along the length of the array at different temperatures.

mum allowable length of a VCSEL array operating at a particular temperature to achieve an acceptable coupling loss.

PROBABILISTIC PHYSICAL DESIGN OF FIBER-OPTIC STRUCTURES

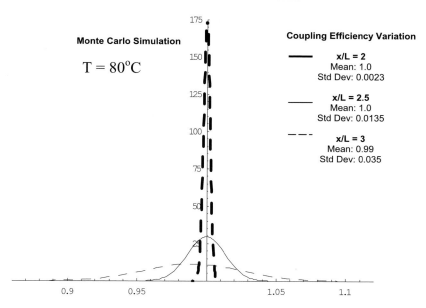

FIGURE 2.35. Stochastic variation of coupling efficiency as a result of material uncertainty for VCSELs located at various positions along the length of the array.

The stochastic variation of the material parameters E_1, E_2 and η_2 can be used to determine the stochastic variation in the coupling efficiency. We performed Monte Carlo simulation to determine the stochastic variation in coupling efficiency caused by the stochastic variation in material behavior. A total of 2000 simulations were conducted within Mathematica [12]. The input variation in material parameters E_1, E_2 and η_2 were propagated through Equation (2.48) to determine the stochastic variation in the VCSEL displacement. The stochastic variation in displacement was then used as input to the optical model to determine the variation in the coupling efficiency. The variation in the coupling efficiency is plotted for three values of normalized position (Figure 2.35) along the VCSEL array. It can be seen from Figure 2.35 that for the end VCSEL of a longer array, the mean coupling efficiency decreases marginally and the standard deviation increases exponentially relative to that for a shorter array. Thus the uncertainty in coupling efficiency is larger at the end of the VCSEL as compared to the center of the array. This again enables design decisions since the probability that a certain prescribed coupling efficiency will be achieved can now be determined.

As explained earlier, the uncertainty in the material behavior is not included in the optimization procedure. The uncertainty in coupling efficiency resulting from material behavior is combined with the geometric uncertainty and evaluated for initial and probabilistic design (Figure 2.36). The effect of material behavior and its uncertainty is a function of temperature and has significant effect on the coupling efficiency value and its uncertainty at temperature above 80°C and when the VCSEL array effective length x/L is larger than 2.5. Thus the material processing parameters of the bonding epoxies must be tightly controlled to reduce the resulting uncertainty in its behavior.

The Monte Carlo Simulation results are compared with the results from the approximation techniques FOSM and SOSM (Figure 2.37). It can be seen that the mean values of coupling efficiency for both FOSM and SOSM are very close with respect to the results

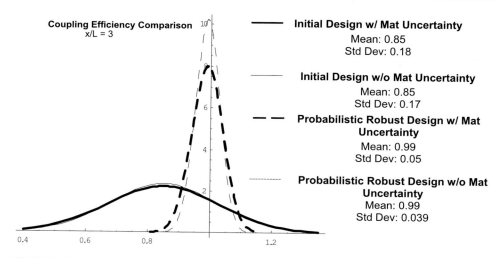

FIGURE 2.36. A comparison of uncertainty in coupling efficiency arising out of material and geometric variations.

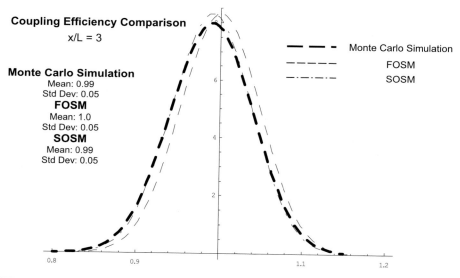

FIGURE 2.37. Comparison of Mote Carlo Simulation, FOSM and SOSM analysis for the mean and variance values of coupling efficiency resulting from material and geometric variations for single-mode operation.

from Monte Carlo Simulation. The standard deviation value for FOSM and SOSM agree reasonably with respect to the values obtained through Monte Carlo Simulation. Thus the first order and second order second moment methods can be used as reasonable approximations to determine the system uncertainty emerging from the variables affecting the system. The use of the approximation methods greatly reduces the computational expense as compared to the more expensive Monte Carlo simulation method.

The probabilistic design procedure demonstrated above reflects the physics and the variability of the behavior and performance of the photonic system much better than the deterministic procedure. The variability and uncertainty are inherent in nature of the material behavior, manufacturing and application conditions and cannot be ignored in the design procedure when their effect on the performance is large as can be seen in Figure 2.36. The probabilistic models provide a goal oriented utilization of the experimental data in the optimizing the performance of the system. It should however be noted that the probabilistic methods though powerful are limited to the availability of the uncertainty quantification through experiments. In cases where the experimental data is not available, the deterministic robust design procedures must be used to optimize the performance of the system.

2.9. SUMMARY

The procedures described in this chapter enable deterministic and probabilistic design of fiber-optic systems. The developed techniques are demonstrated for a 12×1 fiber-optic package using a curved reflector concept. The systems approaches such as the graph partitioning procedure and system decomposition through simulated annealing are very useful in identifying the critical variables that influence the uncertainty in fiber-optic systems. The novel deterministic robust design formulation developed in this chapter minimizes the sensitivity of the objective function to changes in the design variables. The probabilistic robust design technique includes the uncertainty in design variables in determining the design with least probability of failure.

The effect of material behavior on the system performance was demonstrated in the chapter. The analytical model used to determine the displacement of the VCSEL and shear stresses on the interface are applicable for any tri-layered bonded system. The operating temperature of the VCSEL has a strong effect on the coupling loss. The displacement of the VCSEL caused by the viscoelastic behavior of the epoxy increases with operating temperature of the VCSEL and thus results in increased coupling loss. The effect of material uncertainty on the coupling efficiency is dependent on the length of the VCSEL array. The effect of geometric uncertainty on the coupling efficiency is however, dependent on the dimensions and position of the curved reflector and the laser array with respect to the optical fiber. The effect of geometric uncertainty is therefore the same for each VCSEL in the array. For the example system considered in this chapter, the geometric uncertainty initially dominates the coupling loss, however, as the length of the array increases the material uncertainty dominates the coupling loss and this effect is dependent on time due to the viscous behavior of bond epoxy. The material behavior would therefore result in increased coupling loss with time. Overall, the procedures described in this chapter enable stochastic design of photonic systems in general and fiber-optic systems in particular. Such analyses are critical in systems where extreme accuracy in assessing performance is needed or where wide uncertainty is the norm.

REFERENCES

1. T. Bierhoff, A. Wallarabenstein, A. Himmler, E. Griese, and G. Mrozynski, An approach to model wave propagation in highly multimodal optical waveguides with rough surfaces, Proceedings of Xth-International Symposium on Theoretical Electrical Engineering (ISTET), Magdeburg, Germany, 1999, pp. 515–520.
2. T. Bierhoff, A. Wallarabenstein, A. Himmler, E. Griese, and G. Mrozynski, Ray tracing technique and its verification for the analysis of highly multimode optical waveguides with rough surfaces, IEEE Transaction on Magnetics, 37(5), Part 1, pp. 3307–3310 (2001).

3. T.P. Kurzweg, J.A. Martinez, S.P. Levitan, P.J. Marchand, and D.M. Chiarulli, New models for optical MEMS, Proceedings of the SPIE—The International Society for Optical Engineering, 4198, pp. 63–74 (2001).
4. B.E.A. Saleh and M.C. Teich, Fundamentals of Photonics, Wiley-Interscience, New York, 1991.
5. P. Accordi, C.P. Basola, G.P. Bava, G. Chiaretti, and I. Montroseet, Coupling efficiency evaluation of multi-mode fiber devices using GRIN rod lenses, Applied Optics, 29(1), pp. 37–46 (1990).
6. K.Y. Lee and W.S. Wang, Ray optics analysis of the coupling efficiency from a gaussian beam to a rectangular multimode embedded strip waveguide, Fiber and Integrated Optics, 13(3), pp. 321–330 (1994).
7. K. Xue and Q. Zhang, A novel low-cost high performance micro pseudo-spherical lens optimized for fiber collimators, Proceedings of SPIE—The International Soceity for Optical Engineering, 4906, pp. 249–252 (2002).
8. W.H. Cheng, The optimal power coupling from GaAs lasers into spherical-ended fibers, Proceedings of the IEEE, 69(3), pp. 396–397 (1981).
9. B.V. Hunter, Algorithm determines efficiency of fiber-coupling calculation, Laser Focus World, 39(7), pp. 63–66 (2003).
10. K.J. Garcia, Calculating component coupling coefficients, Laser Focus World, (August), pp. 51–54 (2000).
11. K. Kawano, Coupling of laser light to optical fibers, Review of Laser Engineering, 22(4), pp. 226–234 (1994).
12. MATHEMATICA, Version 5.0, Copyright 1988–2005 Wolfram Research Inc.
13. G. Khanarian and H. Celanese, Optical properties of cyclic olefin copolymers, Optical Engineering, 40(6), pp. 1024–1029 (2001).
14. D. Vez, S. Hunziker, S. Eitel, U. Lott, M. Moser, R. Hoevel, H.-P. Gauggel, A. Hold, and K. Hulden, Packaged 850 nm vertical-cavity surface-emitting lasers as low-cost optical sources for transparent fiber-optic links, 33rd European Microwave Conference, 2002, pp. 611–615.
15. 1550 nm, Vertical Cavity Surface Emitting Laser for Single Mode Operation, Product Data Sheet, Vertilas GmbH, Germany, 2003.
16. Corning Incorporated (Private Communication) (2003).
17. R.T. Haftka and A. Gurdal, Elements of Structural Optimization, Kluwer Academic Publications, Dordtecht, Netherlands, 1992.
18. A. Kusiak and N. Larson, Decomposition and representation methods in mechanical design, Transactions of the ASME, Special 50th Anniversary Design Issue, 117, pp. 17–24 (1995).
19. N.F. Michelena and P.Y. Papalambros, Optimal model-based partitioning for power train system design, ASME Design Automation Conference, Boston, 1995.
20. B.W. Kernighan and S. Lin, An efficient heuristic procedure for partitioning graphs, The Bell System Technical Journal, 49, pp. 291–307 (1970).
21. S. Radhakrishnan, G. Subbarayan, L. Nguyen, and W. Mazotti, Optimization and stochastic procedures for robust design of photonic packages with applications to a generic package, Proceedings of the 53rd Electronics Components and Technology Conference, IEEE, 2003, pp. 720–726.
22. T. Bi, Y. Ni, C.M. Shen, and F.F. Wu, An on-line distributed intelligent fault section estimation system for large-scale power networks, Electrical Power Systems Research, 62, pp. 173–182 (2002).
23. C.U. Saraydar and A. Yener, Adaptive cell sectorization for CDMA systems, IEEE Journal of Selected Areas in Communications, 19(6), pp. 1041–1051 (2001).
24. C. Walshaw and M. Cross, Mesh partitioning: a multilevel balancing and refinement algorithm, Society of Industrial and Applied Mathematics, 22(1), pp. 63–80 (2000).
25. M. Ouyang, M. Toulouse, K. Thulasiraman, F. Glover, and J.S. Deogun, Multilevel cooperative search for the circuit/hypergraph partitioning problem, IEEE Transactions on Computer-Aided Design of Integrated Circuits and Systems, 21(6), pp. 685–693 (2002).
26. G. Subbarayan and R. Raj, A methodology for integrating materials science with system engineering, Materials and Design, 20, pp. 1–12 (1999).
27. C.M. Fiduccia and R.M. Mattheyses, A linear time heuristic for improving network partitions, Proceedings of 19th IEEE Design Automation Conference, IEEE, 1982, pp. 175–181.
28. B. Hendrickson and R. Leland, The Chaco User's Guide, Version 2.0, Technical Report SAND95-2344, Sandia National Laboratories, Albuquerque, NM, 1995.
29. M.K. Goldberg and M. Burstein, Heuristic improvement technique for bisection of VLSI networks, Proceedings IEEE International conference on Computer Design: VLSI in Computers, 1995, pp. 122–125.
30. D.S. Johnson, C.R. Aragon, L.A. McGeouch, and C. Schevon, Optimization by simulated annealing: an experimental evaluation; Part I, Graph partitioning, Operations Research, 37(6), pp. 865–892 (1989).
31. J. Sheild, Partitioning concurrent VLSI simulation programs onto a multiprocessor by simulated annealing, IEEE Proceedings—E Computers and Digital Techniques, 134(1), pp. 24–30 (1987).

32. S. Kirkpatrick, Optimization by simulated annealing: quantitative studies, Journal of Statistical Physics, 34(5-6), pp. 975–86 (1984).
33. N. Metropolis, A. Rosenbluth, M. Rosenbluth, A. Teller, and E. Teller, Equation of state calculations by fast computing machines, Journal of Chemical Physics, 21(6), pp. 1087–1092 (1953).
34. S. Kirkpatrick, C.D. Gelatt, Jr., and M.P. Vecchi, Optimization by simulated annealing, Science, 220(4598), pp. 671–380 (1983).
35. V. Cerny, Thermodynamical approach to the traveling Salesman problem: an efficient simulation algorithm, Journal of Optimization Theory & Applications, 45(1), pp. 41–51 (1985).
36. E. Aarts and L. Krost, Simulated Annealing and Boltzmann Machines, A Stochastic Approach to Combinatorial Optimization and Neural Computing, John Wiley and Sons, 1989.
37. R.T. Haftka and A. Gurdal, Elements of Structural Optimization, Kluwer Academic Publications, Dordtecht, Netherlands, 1992.
38. G. Taguchi, Introduction to Quality Engineering, Krauss International Publications, White Plains, NY, 1986.
39. G. Taguchi, System of Experimental Designs, Krauss International Publications, Vol. 1 and 2, White Plains, NY, 1986.
40. J.N. Otto and E. Antonsson, Extensions to the Taguchi method of product design, Design Theory and Methodology, ASME, DE-31, pp. 21–30 (1991).
41. B. Ramakrishna and S.S. Rao, A robust optimization approach using Taguchi's loss function for solving nonlinear optimization problems, Advances in Design Automation, 32, pp. 241–248 (1991).
42. T. Chang, A.C. Ward, J. Lee, and E.H. Jacox, Distributed design with conceptual robustness, a procedure based on Taguchi's parameter design, Concurrent Product Design, ASME, DE-74, pp. 19–29 (1994).
43. S. Sunderasan, K. Ishi, and D.R. Houser, A robust optimization procedure with variaton on design variables and constraints, Engineering Optimization, 20, pp. 163–179 (1992).
44. J.J. Dennis and R. Schnabel, Numerical Methods for Unconstrained Optimization and Nonlinear Equations, Prentice-Hall Inc., Englewood Cliffs, New Jersey, 1983.
45. S. Mahadevan and A. Haldar, Reliability-based optimization using SFEM, Lecture Notes in Engineering, 61, pp. 241–250 (1991).
46. E. Suhir, Applied Probability for Engineers and Scientists, McGraw Hill, New York, 1997.
47. S.V.L. Chandu and R.V. Grandhi, General purpose procedure for reliability based structural optimization under parametric uncertainties, Advances in Engineering Software, 23(1), pp. 7–14 (1995).
48. J.O. Lee, Y.-S. Yang, and W.-S. Ruy, A comparative study on reliability-index and target-performance based probabilistic structural design optimization, Computers and Structures, 80(3-4), pp. 257–269 (2002).
49. H.A. Jensen, Reliability-based optimization of uncertain systems in structural dynamics, AIAA Journal, 40(4), pp. 731–738 (2002).
50. M. Allen, M. Raulli, K. Maute, and D.M. Frangopol, Reliability-based analysis and design optimization of electrostatically actuated MEMS, Computers and Structures, 82(13–14), pp. 1007–1020 (2004).
51. L.L. Howell, S.S. Rao, and A. Midha, Reliability-based optimal design of a bistable compliant mechanism, Proceedings of the 19th Annual ASME Design Automation Conference, Part I, 1998, pp. 441–448.
52. E. Ponslet, G. Maglaras, R.T. Haftka, E. Nikolaidis, and H.H. Cudney, Comparison of probabilistic and deterministic optimizations using genetic algorithms, Structural Optimization, 10(3-4) (1995).
53. E. Suhir, Thermal stress modeling in microelectronics and photonic structures, and the application of the probabilistic approach: review and extension, The International Journal of Microcircuits and Electronic Packaging, 23(2), pp. 215–223 (2000).
54. E. Suhir, Probabilistic approach to evaluate improvements in the reliability of chip-substrate (chip-card) assembly, IEEE CPMT Transactions, Part A, 20(1), pp. 60–63 (1997).
55. E. Suhir and B. Poborets, Solder glass attachment in cerdip/cerquad packages: thermally induced stresses and mechanical reliability, ASME Journal of Electronic Packaging, 112(3), pp. 204–209 (1990).
56. W.T. Chen and C.W. Nelson, Thermal stress in bonded joints, IBM Journal of Research and Development, 23(2), pp. 179–188 (1979).
57. E. Suhir, Stresses in bi-metal thermostats, Journal of Applied Mechanics, 53, pp. 657–660 (1986).
58. E. Suhir, Calculated thermal induced stresses in adhesively bonded and soldered assemblies, Proceedings of the ISHM International Symposium on Microelectronics, Atlanta, Georgia, 1986, pp. 383–392.
59. E. Suhir, Die attachment design and its influence on thermal stresses in the die and the attachment, Proceedings of the IEEE 37th Electronics Components Conference, 1987, pp. 508–517.
60. N. Barbes and P. Walsh, Loss of Gaussian beams through off-axis circular apertures, Applied Optics, 27(7), pp. 1230–1232 (1988).

3

The Wirebonded Interconnect: A Mainstay for Electronics

Harry K. Charles, Jr.

The Johns Hopkins University, Applied Physics Laboratory, 11100 Johns Hopkins Road, Laurel, Maryland 20723-6099, USA

3.1. INTRODUCTION

3.1.1. Integrated Circuit Revolution

3.1.1.1. Device Trends Semiconductor device technology has had an unparalleled rise in density, functionality, and complexity in its history since the invention of the bipolar transistor in 1947 [4] and the birth of the integrated circuit (IC) in 1958 [46]. In fact, IC technology has followed a path of doubling its complexity (or number of devices per single piece of silicon or chip) every 18 months to two years since its birth [57,71]. Electronic fabrication technology has the ability to put over 100 million transistors on a single piece of silicon (chip) less than 2 cm^2 in area. A billion transistors on the same size chip have already demonstrated with 10^{11} devices (transistors) predicted on a similar size semiconductor slice by 2010.

3.1.1.2. Input/Output Trends With this extremely rapid rise in chip density and functionality, the requirement for increased inputs/outputs (I/O) has also risen dramatically. Transistors required three to four interconnects and were the mainstay semiconductor product during the decade of the fifties. Early ICs required a dozen or so interconnect wires; but as the IC revolution continued, I/O requirements increased rapidly. Today, ICs routinely have several hundred I/O pads with some approaching the 1000 mark (random logic and microprocessors). A few devices even have higher I/O numbers, usually in the 1000 to 1500 range. The complex, increased functionality ICs of the future will have I/O requirements in the thousands. It should be remembered, however, that systems will still contain a wide variety of chip types ranging from memory with I/O counts less than 100 to special-purpose microprocessors and random logic with I/Os in the thousands. Thus, an effective interconnection system must be able to transcend the full range of I/O number and density requirements.

3.1.2. Interconnection Types

3.1.2.1. Overview There are three major forms of electrical interconnection for integrated circuits and related packaging applications: (1) wirebonding, (2) flip chip attachment, and (3) tape automated bonding (TAB) as shown schematically in Figures 3.1–3.4, respectively. Many other forms of interconnect exist to meet special needs or performance

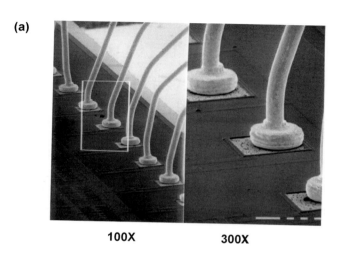

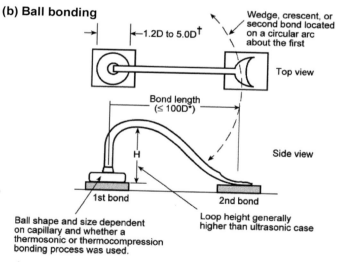

FIGURE 3.1. Ball bonds (thermocompression or thermosonic). (a) Scanning electron microscope photomicrograph of typical ball bonds; (b) schematic representation of ball bonds with important parameters indicated.

requirements. These range from completely deposited multilayer thin film interconnection schemes such as high density interconnect (HDI) [50] to interconnection techniques involving "G-shaped" springs [53], laser written conductors [25,52,79], and elastomeric pressure contacts. Detail description of these techniques is beyond the scope of this work.

Wirebonding, by far, is the most dominant form of electrical chip interconnection. Over four trillion wirebonds are made annually. This staggering number of wirebonds ac-

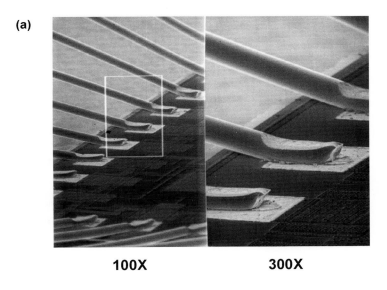

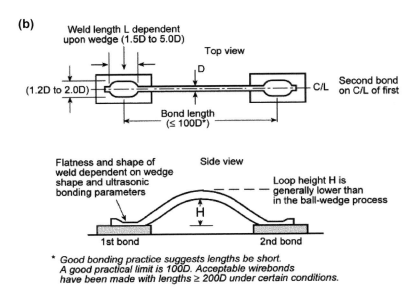

FIGURE 3.2. Ultrasonic bonds (wedge bonds). (a) Scanning electron microscope photomicrographs of typical ultrasonic wedge bonds; (b) schematic representation of ultrasonic bonds with important parameters indicated.

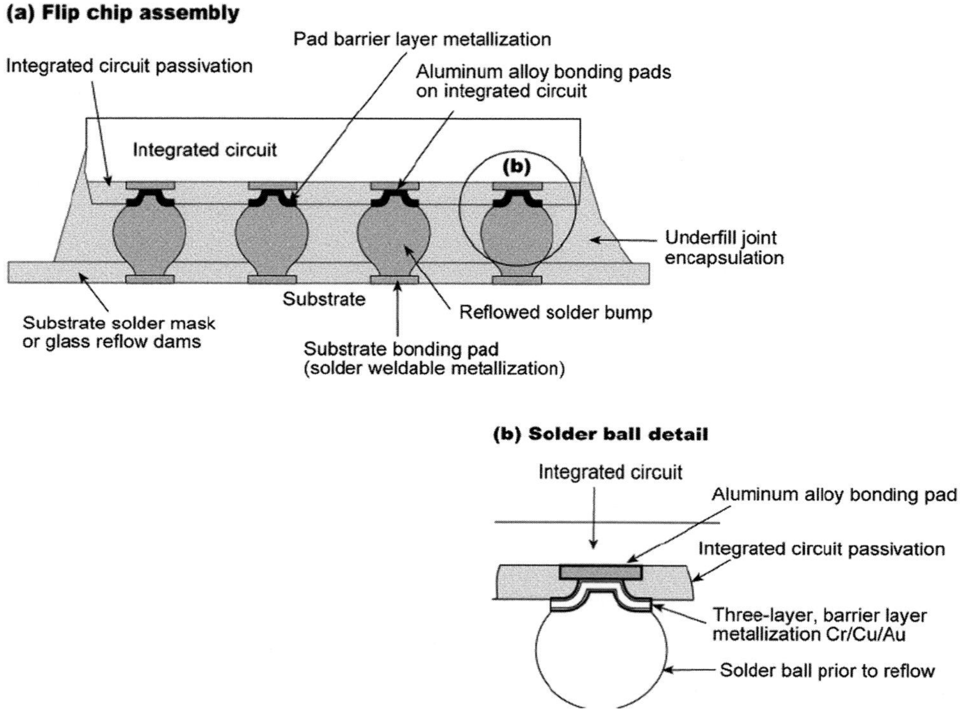

(a) Flip chip assembly

FIGURE 3.3. Schematic representation of the flip chip bonding process. (a) Cross-section of a flip chip assembly; (b) detail of the solder ball and barrier layer metalization prior to reflow.

counts for well over 90 percent of all the first-level interconnects (chip to package or chip to board) produced. The details of the formation, application, and the future of wirebonding in relationship to electronic products are the focus of this article. Prior to describing wirebonding in detail, it is necessary to put wirebonding in perspective with the other forms of chip electrical interconnection.

3.1.2.2. Flip Chip The basic flip chip process was developed by IBM in the early 1960s [56]. In their process called controlled collapse chip connection (C4 for short), solder bumps are formed on wettable chip bonding pads. A mating solder-wettable metalization pattern is created on the package or substrate. The chip of IC is placed upside down (flip chip), and all joints are formed simultaneously by reflow soldering. Figure 3.3 is a schematic representation of an IC attached by the flip chip process.

In the original C4 process, copper spheres were embedded in the solder bumps to keep the edges of unpassivated silicon chips from electrically shorting to the solder-coated substrate metalizations (usually thick films). With the rapid growth of IC technology, including effective die passivations, the copper spheres were removed, and the current flip chip or C4 processes were developed. In the current process, the solder bump is constrained from completely collapsing or flowing out over the entire bonding site by surface tension and the use of solder masks or dams. In thick film circuitry, the dams are glass, while for organic boards, the dams or masks are organic resins. The flow on the chip is constrained by a special bonding pad metallurgy that consists of a pad (circular or hexagonal in shape) of evaporated chromium, copper, and gold. The pad metallurgy is then coated by evaporation with, for example, Sn5 (5 weight percent tin, 95 weight percent lead) or Sn10 (10 weight percent Sn, 90 weight percent Pb) to a thickness of 100–125 µm. Thickness of this plating is, of course, ultimately determined by the pad size (pitch). The high lead content tin-lead solders have excellent strength and fatigue resistance but the reflow temperatures are high (approximately 315°C). This high reflow temperature typically limits these materials to inorganic substrates (ceramics, silicon, etc.) The use of lower melting point alloys such as Sn 63 (63 weight percent Sn, 37 weight percent Pb), and those based on indium have allowed organic boards to be used. Because of the reduced fatigue resistance of the lower melting temperature solder alloys, the larger coefficient of the thermal expansion mismatches for the materials involved (i.e., silicon chip and organic substrate), and the increased size of the IC chips themselves, flip chipping on organic boards requires the use of underfills [41] to achieve the required product lifetimes under most operational scenarios.

Solder pads can be placed over active areas on the ICs, because the bonding process (solder reflow) involves little or no force that could damage sensitive structures. Thus, high density area arrays (of solder bumps) can be formed over the entire IC surface (providing a very high number of I/Os). For example, on a square chip with a side length of 10 mm and 25 µm (in diameter) solder bumps placed on 75 µm centers, an array (133×133) of over 17,000 interconnection sites can be formed. For lower I/O number requirements, both the ball diameter and pitch (center-to-center spacing of solder balls) can be increased, thus, improving overall ease of attachment and increased fatigue resistance. With 100 µm diameter bumps on 250 µm centers, a 40×40 array of solder bumps (1600 I/O) can be formed on a 10 mm square chip.

Besides providing the highest I/O density of the major interconnect types, the flip chip solder joint with its associated low inductance and low capacitance behavior is a very high performance interconnect. A typical flip chip solder joint exhibits low insertion loss even for signals at frequencies greater than 100 GHz. Although wirebonding is the most widely used and least expensive first level interconnection scheme as described below, it is still a relatively slow process (even with automatic wire bonders producing at rates of 10–15 bonds per second) when compared to the mass reflow associated with flip chipping.

3.1.2.3. Tape Automated Bonding (TAB) Tape automated bonding is also a "gang" bonding method in which bonds (on the chip, lead frame, or substrate) are formed simultaneously. Separate processes are necessary, however, for the connection of the tape to the chip (first or inner lead bond) and then for the connection of the chip and its now attached lead structure to the package or substrate (second or outer lead bond). This contrasts to the flip chip case where all bonds (first and second) are made simultaneously.

The initial TAB process involves the bonding of ICs to the tape (prefabricated metallic interconnection pattern, either freestanding or supported on an organic carrier film, usually in a format of one pattern wide by several hundred patterns long) using thermocompression bonding or solder reflow. The choice of thermocompression bonding or solder reflow depends upon the type of interconnection bump used on the chip (bumped chip) or on the tape (bumped tape). Bumping of the chip or the tape finger is necessary not only to effect the interconnect but also to prevent damage of the passivation surrounding the chip bonding pad as the bonding operation is performed. This initial or first bonding operation is called inner lead bonding (ILB). After the completion of the inner lead bonding process, the chip, which is now attached to a lead frame (single layer tape) or a lead pattern on an organic carrier (double layer or multi-layer tape), can be tested, encapsulated, and/or environmentally screened. The strip format of the tape facilitates the use of automated equipment. Subsequently, the individual pre-tested, encapsulated, and/or environmentally

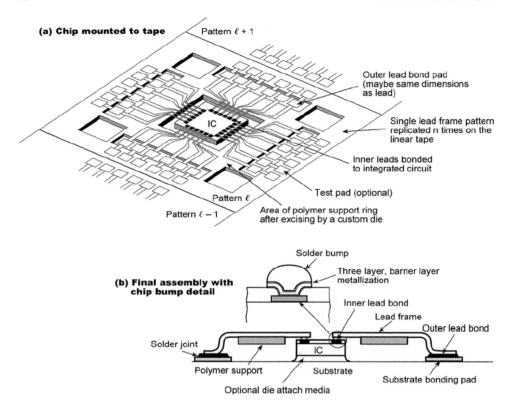

FIGURE 3.4. Schematic representation of the tape automated bonding (TAB) process. (a) Isometric view of chip mounted on carrier tape; (b) cross-sectional view of final chip to package or substrate assembly with inner lead solder bump detail.

screened parts can be excised from the tape and attached to a package, substrates, or board by a process called outer lead bonding (OLB). Basic schematic representations of TAB processes are shown in Figure 3.4.

The pre-patterned tape comes in many forms, widths, and materials depending upon circuit requirements, tape fabrication process, bonding equipment, and the metallurgy(s) involved. The tape can have either single or multiple conductor layers. The multiple conductor layers are separated by intervening dielectric layers, which are typically a form of polyimide. More detailed descriptions of tape construction and materials can be found elsewhere [12].

Bonding the chip to the tape lead frame is usually accomplished with a pre-deposited gold bump (solder bumps are also used). This interconnection bump is either placed on the chip with appropriate interface metallurgy or on the tape (as mentioned above). The bumps are needed to reach the recessed bonding sites (below the top level of the passivation layer) and minimize the TAB lead forces on the passivation surrounding the bonding pad. The interface metallurgy usually consists of several metal layers designed to provide low contact resistance, improved bump adhesion, and a hermetic seal of the pad with its surrounding passivation. Typically, these layers consist of an adhesion layer (chromium, titanium, etc.), a barrier layer (copper, nickel, platinum, or palladium), and, finally, the bump metal (gold, gold/copper, or solder-plated copper). Most mating lead frames consist of gold- or tin-plated copper. Generic interface metallurgies for various tape systems are presented by Tummula et al. ([80], p. II-225).

The ILB process can be accomplished for gold bumps by using thermocompression bonding with a heated thermode [80]. Thermodes exist in many forms, including solid and bladed. Just as in thermocompression wirebonding, the inner lead bond strength is strongly dependent on the temperature, dwell time (length of time the thermode is in contact with the lead), and the loading (applied force) during the bonding process. The bond termination material and the cleanliness of the bond interface also have an effect. If a solder bump is used, then the ILB process is solder reflow. Either a high lead content solder (e.g., Sn5 or Sn10) or a tin-gold eutectic alloy solder (80 weight percent Sn, 20 weight percent Au) is used. In the tin-gold process, tin-plated leads are bonded to gold bumps (or vice versa). The tin-gold eutectic attach process produces low stress, uses a relatively low temperature (280°C) when compared to other ILB process, and is generally applicable to most tapes and bonding situations.

Outer lead bonding is typically performed with a heated blade-type thermode, which forces the TAB leads against the bonding pads on the package, substrate, or board. Typically, the OLB thermodes are larger than their ILB counterparts because of lead fan out and the larger OLB bonding sites. The bonding sites or pads are usually coated with a solder on solder paste, and once the heated thermode causes the solder to reflow, its temperature is reduced and the solder is allowed to solidify prior to the removal of the thermode. Other simultaneous or gang soldering techniques such as vapor phase, infrared, and hot air soldering may be used with appropriate fixturing. Thermocompression bonding between two compatible metallurgies also has been used for OLB.

Prior to OLB, it is necessary to remove any interconnection lead support bars or common plating connections used in the tape fabrication process that may cause lead shorting in the final TAB assembly. Such removal is done by punching or cutting. If further chip testing is required prior to final assembly, it is typically done at this time. In order to test the chips thoroughly it may be necessary to separate the lead frame cells with chips attached and place them in a tape carrier that is compatible with the testing apparatus. With proper design of the tape carrier, testing at speed and full function verification are possible. Once testing is complete, the chip with its TAB lead frame structure is cut from the carrier tape with a metal die. The leads are then bent to shape (e.g., full wing) to provide the proper mechanical compliance.

3.1.2.4. Interconnection Requirements Wirebonded interconnects are usually applied to perimeter bonding pads on ICs. Perimeter bonding pads are located over non-active areas of the chip, thus preventing any damage to the IC, due to forces associated with the bonding process. Using special processes coupled with precision bonding machine control, several researchers and a few manufacturers have wirebonded successfully over active regions; but this is not a widely accepted or recommended practice. Flip chip reflow soldering, on the other hand, can be used over active regions, because it exerts little or no force in the attachment process. TAB, depending upon the tape form (area or perimeter), and the type of inner lead bonding (e.g., solder reflow), can be used in either mode, although perimeter TAB is by far the most common.

Figure 3.5 illustrates current and projected I/O requirements for various types of electronic products. As can be seen, the I/O requirement range from less than 100 to almost 5000 depending upon product type and the time period considered. To gain some understanding of the implications of these large and increasing I/O numbers, let's consider how they might be supported from an interconnection point of view. Figure 3.6 plots

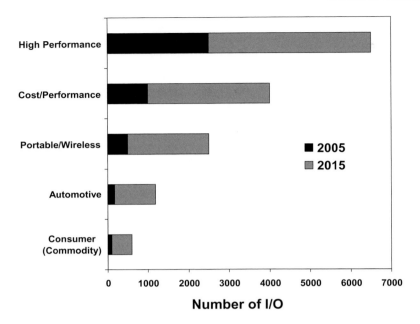

FIGURE 3.5. Maximum expected I/O for different classes of electronic products both now and ten years in the future.

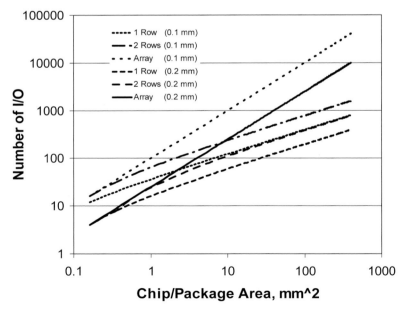

FIGURE 3.6. Number of I/O's as a function of chip package area for both perimeter (1 row and 2 rows) and area array interconnection points (bonding pads).

THE WIREBONDED INTERCONNECT: A MAINSTAY FOR ELECTRONICS

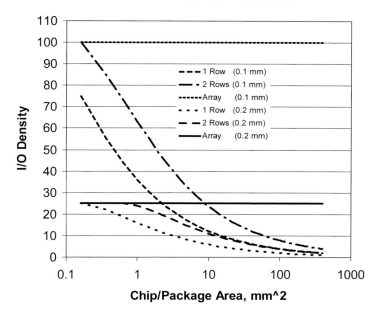

FIGURE 3.7. Package or chip I/O density (I/O per unit area) as a function of the area for both perimeter (1 row and 2 rows) and area array interconnection points (bonding pads).

the number of I/O versus chip area for the two major interconnection types: wirebonding (perimeter attachment) and flip chip (area attachment). Wirebonding, even with two rows of bonds at an extremely fine pitch (e.g., 75 µm), requires a relatively large chip size (350 mm^2) to reach 1000 I/O while a chip of that size can support almost 16,000 I/O using flip chipping. The curves for wirebonding are also applicable for predicting I/O count using perimeter TAB. As mentioned above, area TAB exists but is not widely used. Area array interconnects can easily exceed 1000 I/O even on small-size chips with a relaxed bond-to-bond spacing or pitch. It is common to normalize the I/O numbers with respect to the chip area, thus forming the I/O density (number of I/O per unit area). Figure 3.7 plots I/O density versus chip area for various pitches of the interconnect. For an area array the I/O density is constant for a given pitch regardless of the chip size, while for a perimeter bonded chip, even with double rows, the I/O density falls off exponentially with increasing chip area.

Similarly, many other IC design, process, and material parameters affect IC bondability in addition to increased active device density and rising interconnection requirements. The aluminum-silicon alloy system (Al + 1% Si), which was standard on many early integrated circuits, has been changed by adding copper (up to 4%) to prevent electromigration as the spacing between adjacent lines has decreased. The addition of copper has produced bondability problems. Research has shown [36] that copper content above 2% prevents effective wirebonding. Another manifestation of shrinking line size is that the lines are becoming much more resistive, forcing the replacement of the aluminum-silicon alloy system with a metal having higher electrical conductivity, such as copper. Copper requires trenching encapsulation with either chromium or titanium adhesion layers [29]. The rigid organic dielectric layers on the IC will be replaced by organic materials with lower dielectric constants, such as polyimide, benzocyclobutone, or Teflon®-based materials (polytetrafluoroethylene). The ultimate goal, if interconnect topologies and copper passivation processes can be developed, would be to use air as the dielectric. Using copper as the IC metalization with soft organics as the intervening dielectric layers, present challenges to the first-level (on-chip) interconnection processes, especially wirebonding. Copper metalization pads will necessitate copper wirebonding or a suitable barrier layer metalization cap to allow bonding with gold or aluminum wires. A metal flash on the copper pad to prevent oxidation is also necessary prior to firming flip chip solder balls.

3.1.3. Wirebond Importance

As mentioned above, wirebonding is the dominant form of first-level interconnect because of its flexibility, low interconnect cost, ease of use, and relatively low capitalization costs. Wirebonding is extremely flexible. It can bond various chip types, metallurgies, pad sizes and locations, and configurations, etc., by changing machine parameters, software programs, and, perhaps, bonding tools (capillaries and wedges), bonding wire, wire size, and bond type (e.g., ball bonding to wedge bonding). Such changes are straightforward, can be performed very quickly, and usually at low cost, except in the case of the bond type that will require moving to a different bonding machine with all of its associated acquisition and set-up costs. In addition, the cost of a fully automatic wirebonding machine and a suitable wirebond testing machine along with an organic die attach system can be acquired for less than $250,000 (2005 U.S. dollars). Wirebond interconnections in volume production cost between $0.001 and $0.002 (U.S. dollars) per interconnect.

The other major interconnect types, flip chip and TAB, require major tooling and capital equipment to produce the on-chip solder bumps (flip chip) or the custom pre-patterned tape (TAB). In either technique, a minor change in chip pad geometry will require a costly photo tooling change in addition to a new acquisition cycle for mating substrates or tape. This lack of flexibility and increased complexity (additional processes) produce a per interconnect cost of $0.05–0.10 (U.S. dollars). Given that the cost of interconnect is much greater than wirebonding and the number of I/Os possible is rapidly increasing (fine pitch wirebondings), one might ask, why are the other interconnection methods considered important?

Flip chip has four major advantages: (1) it can produce the highest number of interconnections per unit area (and the interconnection density is constant); (2) all the bonds are contained within the chip area, i.e., there is no second bond or outer lead bond location beyond the chip perimeter; (3) it has extremely low inductance and capacitance per joint, thus allowing operation at very high frequency (up to 100 GHz); and (4) it has the most robust replacement process (of any of the major interconnection schemes) for preserving the under laying board, substrate, or the chip [65]. It is interesting to note that in addition to flip chip's large interconnect potential at fine pitch it can satisfy most practical I/O number requirements at a much larger pitch, making large robust solder joints possible.

While flip chip for performance and wirebonding for cost and flexibility exceed the capabilities of TAB, TAB does have one interesting advantage over the other interconnection types. In the TAB process, the IC is attached to its final lead structure (via inner lead bonding) prior to placement in a package or directly on a substrate. Such attachment allows for testing both at speed and temperature in a lead configuration that is closely representative of its final use state; thus in principle, solving the known good die (KGD) problem [5,7,62]. Die testing techniques for wirebonded interconnected chips (prior to placement at final chip location) have been developed, but they lack the utility of TAB and always involve some form of wirebond lifting (removal) and replacement, which carries with it an inherently greater risk of good die loss. Flip chip test scenarios have also been developed and usually involve reflow to a test substrate or some form of flexible pressure type contact structure or interposer.

Given the discussion above, it is clear that wirebonding will be the dominant form of first-level interconnect (chip to package, substrate, or board) for some time to come in all applications that: (1) can afford the size of the perimeter extension (beyond the chip) required for the second bond; (2) allow wirebonding to affect the required number of I/O interconnections with perimeter bonding pads; and (3) have a frequency of operation low enough (i.e., <10 GHz) so that wire loss is manageable.

If the chip packaging requirements exceed the wirebonding restrictions listed above, then flip chip should be the interconnect of choice. Flip chip usage is increasing rapidly worldwide. It is particularly effective in applications where its special advantages of small size and high performance are required. Flip chipping is clearly the second most important first-level interconnect. Special testing needs may, under certain circumstances, require the use of TAB.

3.2. WIREBONDING BASICS

Figure 3.8 illustrates an example of a modern wirebonded circuits. The wirebonding process begins by firmly attaching the backside of the integrated circuit or wirebondable component to the appropriate substrate location or package bottom by using an organic adhesive, a low melting point glass, the reflow of a metal alloy, or a gold-silicon eutectic alloy process [30]. Once bonded in place (the process is called die or chip attach), wires are attached to the chip bonding pads using special tools (capillaries or wedges) and various combinations of heat, pressure (force), and ultrasonic energy. Depending upon tool type and choice of welding energy (direct heat or ultrasonic heating or both), three major techniques for wirebonding have emerged over the years since microelectronic wirebonding was developed in the mid-1940s to the mid-1950s timeframe [20,36]: thermocompression bonding, ultrasonic bonding, and thermosonic bonding.

Thermocompression bonding and thermosonic bonding methods produce a ball-wedge (first bond-second bond) type bond [Figure 3.1(a)], where the wedge (tail, crescent, or second) bond lies on an arc about the first bond or ball bond as shown in Figure 3.1(b). Ultrasonic bonding produces a symmetric wedge-wedge (first bond-second bond) style bond as shown in Figure 3.2(a). In ultrasonic bonding, the second bond lies along the center line of the first [see Figure 3.2(b)].

3.2.1. Thermocompression Bonding

A thermocompression bond (or weld) is the result of bringing two metal surfaces (bonding wire and the substrate or pad metalization, for example) together in intimate contact during a controlled time, temperature, and pressure (or force) cycle. During this "bonding" cycle, the wire and, to some extent, the underlying metalization undergo plastic deformation and interdiffusion on the atomic scale. This atomic interdiffusion can result in a uniform gold welded interface, if both gold wire and gold pad or substrate metalization are used. Gold-aluminum intermetallics [63] are formed when gold wire and aluminum pads (or vice versa) are used. Regardless, the plastic deformation that occurs at the bonding interface ensures: intimate surface contact between the wire and the pad, provides an

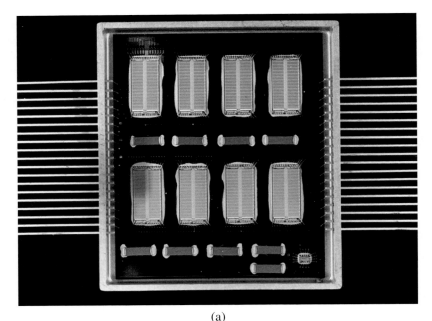

FIGURE 3.8. Examples of wirebonded circuitry. (a) Static RAM module using MCM-D technology. Unit contains 300 gold thermosonic wirebonds. (b) Experimental X-ray detector for use in space. 36-detector chips with bond pads on both sides of the chip. Each chip has over 200 wirebonds per side. Total wirebonds on assembly exceed 18,000. Chips mounted on open frame to allow wirebonder access to both sides.

increase in the interfacial bonding area, and breaks down any interfacial film layer (oxide, contamination, etc.). Surface roughness, voids, oxides, and absorbed chemical species or moisture layers can all impede the intimate metal-to-metal contact and limit the extent and strength of the interfacial weld; thus, causing a poor bond. In some cases, this interfacial contamination (usually on the pad) is so extensive that it prevents bonding altogether.

The interfacial bonding temperatures are typically in the range of 300–400°C [43] for bonds made by thermocompression bonding. The bonding cycle, exclusive of bond positioning, takes a fraction of a second. In thermocompression bonding, the required heat for interface formation is applied by either a heated capillary (the bonding tool through which the wire feeds) or by mounting the substrate and/or package on a heated stage (column). With stage or column heat, the die and package combination must come into thermal equilibrium with the stage, which can take seconds to minutes depending upon mass. Because of the high stage or column temperatures (>200°C) involved in thermocompression bonding, IC or device die attachment is usually limited to the gold-silicon eutectic or certain metal alloy attaches. Also, long times on heated stages can cause reliability problems with previously placed wirebonds, such as uncontrolled intermetallic growth. Most modern thermocompression bonders use a combination of both capillary and column heat. The capillary is made of ceramic, ruby, tungsten carbide, or other refractory material. A typical ball bonding cycle is illustrated in Figure 3.9. There are five major steps in the ball bonding process: (1) ball formation (views a and b, Figure 3.9); (2) ball attachment to IC or substrate pad (first bond) (view c, Figure 3.9); (3) traverse to second bond location (view d, Figure 3.9); (4) wire attachment to package or board pad (second bond) (view e, Figure 3.9); and (5) wire separations (view f, Figure 3.9). The initial ball formation step is accomplished by cutting the wire end as it extends through the capillary with an electronic discharge. This cutting is called flame-off due to the fact that in the early days of wirebonding an open flame hydrogen (or forming gas) torch was used to cut the wire. Once cut, the ends of the wire ball up due to surface tension and capillary action. Figure 3.10 illustrates free air balls produced with gold wire by a negative electronic flame-off system. Heat, time, and pressure or force are the major determining factors in the formation of thermocompression bonds. Typically, the forces used in thermocompression bonding are higher than in other ball bonding methods (i.e., thermosonic ball bonding), resulting in a much more flattened ball. Thus, the first bond is "nail head" shaped rather than just a slightly flattened ball as obtained with standard pitch thermosonic ball bonding [e.g., see Figure 3.1(a)].

Gold wire is used in most thermocompression wirebonding processes because it is easily deformed under pressure at elevated temperature and very resistant to oxide growth that can inhibit proper ball formation. Aluminum wire, because of its rapid oxide growth, has difficulty in forming properly shaped balls on standard bonding machines. Successful aluminum wire ball bonds have been formed using an inert atmosphere around the bonding head to minimize oxide formation [28,60]. Copper and other materials (e.g., palladium and platinum) have also been ball bonded [48] in both thermocompression and thermosonic applications. Also, wedge style thermocompression bonding with many different materials has been performed [6,51]. Wedge style thermocompression bonding forms the basis of the thermode ILB attachment used in TAB.

3.2.2. Ultrasonic Bonding

Ultrasonic bonding (or wedge bonding) is a lower-temperature process in which the source of energy for the metal welding is ultrasonic energy produced by a transducer vibrating the bonding tool (wedge) in the frequency range of 20 to 300 kHz. The most common frequency is 60 kHz ([36], pp. 23–26), although higher frequency ultrasonics are in use or being considered for difficult bonding situations. Thermosonic bonding at higher frequencies will be discussed in Section 3.4.4 below. The ultrasonic wedge bonding process is illustrated in Figure 3.11. In ultrasonic bonding, the wedge tip vibrates parallel to the bonding pad. Ultrasonic bonds are typically formed with aluminum or aluminum alloy wire on either aluminum or gold pads. Gold wire ultrasonic bonding has been performed

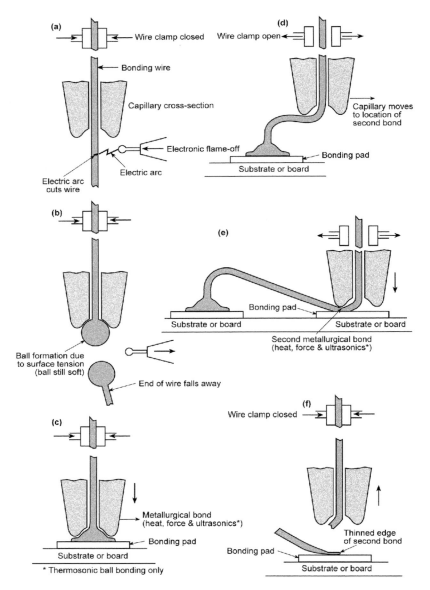

FIGURE 3.9. Schematic representation of the ball bonding cycle: (a) flame-off; (b) ball formation; (c) first bond; (d) transition to second bond; (e) second bond; and (f) separation of wire after second bond.

with both round wire and flat ribbon, although it is not widely used because of cost. Gold ribbon, because of its rectangular cross-section, provides a lower inductance interconnect (compared to a round wire of equivalent cross-sectional area) useful in radio frequency and microwave chip interconnect. In special applications, copper and palladium have been bonded by the ultrasonic process ([30], pp. 409–410). The major advantages of ultrasonic bonding include the ability to effect strong bonds with little or no applied substrate heat (implying the use of low temperature die attachment methods); and it typically can be per-

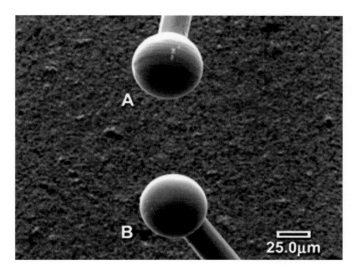

FIGURE 3.10. Scanning electron photomicrograph of free air balls produced by negative electronic flame-off. The gold wire diameter is 25.4 µm. (A) Free air ball made on 100 kHz bonder (62.2 µm diameter); (B) free air ball made on 60 kHz bonder (59.7 µm diameter). Magnification approximately 350×.

formed at finer pitches (because of the elongated, narrow shape of the bond compared to the round ball diameter) than ball bonding methods. Automated wedge or ultrasonic bonders are typically slower than ball bonders due to the requirement that the second bond must be in line with the first bond; i.e., follow the centerline of the wedge. Thus, either the entire package (substrate) or the bonding head must be rotated to bond in different directions. This slows down the bonding process when compared to ball bonding, which can place the second bond anywhere on a circle surrounding the first bond with only transversal movement of the head (or stage). [See Figure 3.1(b).]

3.2.3. Thermosonic Bonding

In thermosonic wirebonding, ultrasonic energy is combined with the ball bonding capillary technique employed in thermocompression bonding. Typically, the thermosonic bonding process is performed in a manner analogous to the thermocompression bonding process, except the capillary is not heated (or held at a lower temperature when compared to the capillary temperature in thermocompression bonding); and the stage or column temperatures are typically 150°C or less. To generate the required interfacial heat for welding at the interface of the wire and the pad, short bursts (tens of milliseconds) of ultrasonic energy are applied to the capillary when the wire and the pad are in contact. Because of the addition of ultrasonic energy (causing localized heat generation at the wire–pad interface), the requirements on stage and capillary heat (as mentioned above) and pressure (force) can be relaxed. The applied forces in thermosonic bonding are typically much less than those encountered in thermocompression bonding, thus allowing bonding over delicate or force sensitive chip or substrate regions. Since interconnections are made with the ICs (and substrates) held at temperatures of 150°C or less, they can be attached with epoxy or other organic adhesives without fear of degradation (i.e., prolonged exposures at temperatures above their glass transition temperature) due to excessive bonder stage or column temperature. Because the temperatures are lower, there is also significantly less risk of uncontrolled

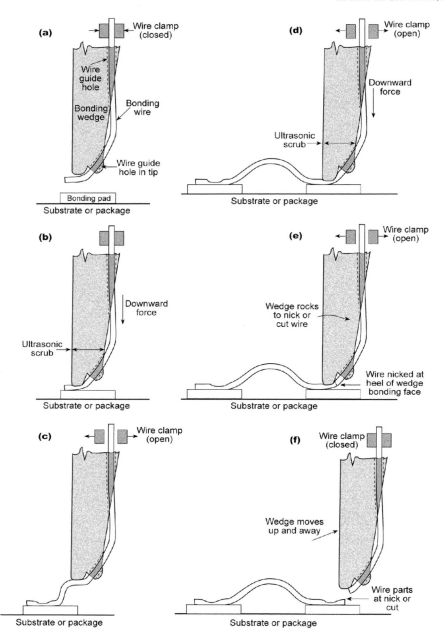

FIGURE 3.11. Schematic representation of the ultrasonic (wedge) bonding cycle: (a) initial wire-wedge configuration; (b) first bond; (c) transition to second bond; (d) second bond; (e) wire nicking or cutting operation; and (f) wire separation after second bond.

intermetallic growth. Thermosonic wirebonding is conducted primarily with gold wire, but aluminum [60], copper [48], and palladium [6] wires have been bonded successfully by the thermosonic process. As the metalization on high performance ICs migrates from aluminum alloys to copper [29], new pad stack configurations (e.g., copper-nickel-gold or, perhaps, just copper) will emerge. These new pad stacks will require the reevaluation of the thermosonic bonding process and, perhaps, the full consideration of the use of copper wire. Copper thermosonic wirebonding has been successfully used in the connection of the ICs to copper alloy lead frames in dual-in-line packages [39].

Such thermosonic bonding evaluations are already underway for the new substrate and pad structures encountered in the development of multichip modules (MCMs) [14]. Some of the results from these new structures and material evaluations are discussed in Section 3.4.2 below.

3.2.4. Wirebond Reliability

Wirebonding has evolved into an extremely reliable first-level microelectronic interconnection technique due to the introduction of fully automated wirebonding in the late 1970s. Coupled with the precision control of automatic wirebonders, such things as improved bonding, pad metallurgy, controlled bonding wire impurity content, effective pad cleaning processes, high purity and stable die attach adhesives, and reduced temperature bonding processes (ultrasonic and thermosonic) have all contributed to the widespread use and reliability of wirebonds. In fact, wirebond defect rates for chips bonded in single chip packages have reached the low parts per million level. Despite these improvements and high wire yields (defects less than 30 parts per million and in several instances as low as 3 parts per million (6σ) [34]), many problems still can occur in wirebonded systems.

These problems can include: mechanical fatigue due to conditions of thermal or power cycling; interactions both chemical and mechanical with encapsulation materials during molding and after cure; corrosion induced by die attach media, the atmosphere, and other process-related conditions; and wire structural changes due to the bonding parameters, such as the uncontrolled grain growth associated with the heat-affected zone. The most widely studied and publicized wirebond reliability probability is associated with the alloying reactions that occur at the gold wire–aluminum alloy bonding pad interface (and, to a much lesser degree, aluminum wire-gold bonding pad interface). Aluminum-gold intermetallic formation occurs naturally during the bonding process and contributes significantly to the integrity of the gold–aluminum interface. Intermetallics (in particular, $AuAl_2$ or purple plague and Au_5Al_2 or white plague) are generally brittle; and, under conditions of vibration or flexing (either mechanically or thermally induced due to coefficient of thermal expansion mismatches), may break due to metal fatigue or stress cracking, resulting in bond failure [63].

At elevated temperatures, aluminum rapidly diffuses into the gold forming the $AuAl_2$ phase, leaving behind Kirkendall voids [63] at the aluminum–$AuAl_2$ interface. Figure 3.12 shows views of extensive intermetallic growth around and under various thermosonic wirebonds (both ball and tail bonds). Kirkendall voiding has also been observed at gold–Au_5Al_2 interfaces. Excessive intermetallic growth can lead to the coalescence of voids, which can lead to a bond crack or lift and an open circuit. Impurities in the bonding wire, on the pad metalization, or at the wirebond–pad interface have been shown to cause rapid intermetallic growth and Kirkendall voiding at temperatures below those associated with normal intermetallic formation [8]. Table 3.1 gives the formation temperature, activation energies, and some notes for the five aluminum-gold intermetallics. The deleterious effects of intermetallics can be controlled if the time of exposure to high temperature is minimized and if proper materials and cleaning procedures are used [82]. Design rules have been developed

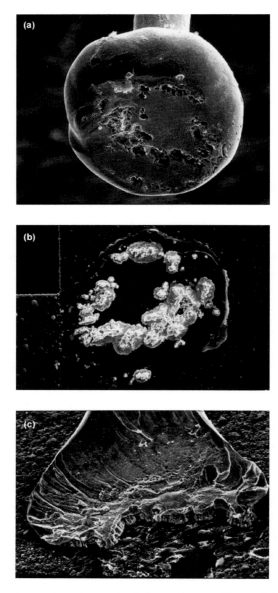

FIGURE 3.12. Scanning electron photomicrographs of advanced intermetallic growth: (a) underside of ball bond with regions of intermetallic voiding (Kirkendall); (b) residual intermetallic left on bonding pad corresponding to the voided regions of the ball in view (a); and (c) tail bond with extensive intermetallic formation under the bond edge and consuming part of the flattened bond region. Magnification approximately 75×.

for minimizing intermetallic void failures by controlling film layer composition and thickness [22]. In addition, proper optimization of the wirebonding process has a significant influence on intermetallic growth.

TABLE 3.1.
Aluminum-gold intermetallic alloy properties.

Alloy[a]	Formation temperature, °C	Activation energy[b]			Comments
		eV	kJ/mol^{-1}	kcal/mol^{-1}	
Au$_5$Al$_2$	23–100	0.62	59.4	14.3	Tan in color
Au$_2$Al	50–80	1.02	98.3	23.5	Metallic gray in color (orthorhombic, randomly oriented monocrystals)
AuAl$_2$	150	1.20	115.8	27.7	Deep purple in color (purple plague-resistivity 8 μΩ cm)
Au$_4$Al	~150				Tan in color
AuAl	~250				White in color

[a] The intermetallic alloys typically form in the order listed (Au$_5$Al$_2$, ..., AuAl) consistent with their temperature of formation.
[b] A range of activation energies from 0.2 eV to 1.2 eV, have been observed for the aluminum-gold system depending upon growth, testing, and contamination conditions.

3.2.5. Wirebond Testing

Since its introduction in the 1970s, the destructive wirebond pull test [37] is the most widely accepted technique for the evaluation and control of both wirebond quality and the associated setup of bonding machine parameters. Despite its widespread use, due to low cost and ease of use, the destructive wirebond pull test has some significant disadvantages. First, since it is destructive, it can only provide information on a lot sample basis for production product. It can be used for pre- and post-lot qualifiers to help setup the bonding machine and, of course, as a post mortem diagnostic tool in failure analysis or as part of routine destructive physical analysis. Thus, it does not provide a measure of quality for each bond. Second, in fine pitch wirebonded circuitry, it is difficult to insert the hook between adjacent wires without touching bonds (wires) other than the one of interest. Third, the destructive pull test provides very little information on the strength or overall quality of the bond interfaces as long as the chief failure mode is a wire break.

Only in the case of catastrophic interface failure, such as those encountered with impurity-driven intermetallic growth [8], will the destructive wirebond pull test yield information other than the relative breaking strength of the wire assuming appropriate correction is made for both the wire and test geometries [9]. This phenomenon is especially true in standard ball bonding situations where the ball of relatively large diameter (nominally 2.5–5.0 times the wire diameter) forms an effective bonding pad attachment that is many times stronger than the breaking strength of the wire. Although usually much stronger than the nominal wire breaking strength, except in the case of very fine pitch ball bonding where the diameter of the ball is approaching that of the wire (e.g., $1.2D$ where D is the diameter of the wire) ([36], pp. 255–260), the strength of the ball-to-bonding pad attachment can vary significantly owing to the influence of bonding machining parameters, composition of the interfacial material, and environmental stresses.

These factors have led to the development of two complementary tests: (1) the 100% nondestructive pull test (NDPT) [1], and (2) the ball shear test [2]. The 100% NDPT provides a degree of confidence that each bond is strong (at least to the nondestructive preset force limit [1]. The ball shear test can be used to investigate not only the interface between the wirebond ball and the bonding pad, but also the influence of both pre- and post-bonding factors. Table 3.2 summarizes the areas of application for both the wirebond pull test and the ball bond shear test. A careful review of Table 3.2 illustrates the complementary nature

TABLE 3.2.
A comparison of areas of applicability between the wirebond pull test (ASTM Standard Test Method F458-84), and the ball bond shear test (ASTM Standard Test Method F1269-89).

Area of applicability	Wirebond pull test	Ball bond shear test
Module geometry	Yes	No
Wirebond geometry	Yes	No
Wire quality, defects, etc.	Yes	No[a]
Second bond	Yes[b]	No
Bonding machine set-up, optimization, etc.	No[c]	Yes
Process development	No[c]	Yes
Substrate, bonding pad quality	No[c]	Yes

[a] Sensitivity to contamination, insensitive to mechanical defects.
[b] Extremely dependent on geometry.
[c] Insensitive unless the effect is catastrophic.

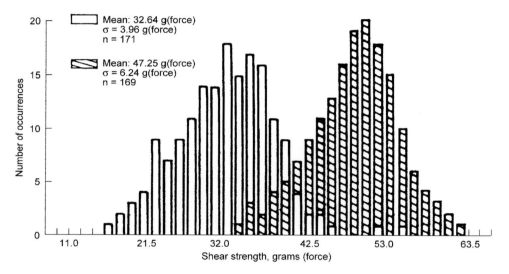

FIGURE 3.13. Histograms of gold thermosonic ball bond shear strengths for bonds placed on aluminum metalization (over silicon). Histogram A (open bars) are the shear test results after the bonding machine was set up using the wirebond pull test ($n = 171$, $\mu = 32.64$ grams (force), $\sigma = 6.24$ grams (force)), Histogram B (shaded bars) are the shear test results after the bonding machine was optimized using the ball shear test ($n = 169$, $\mu = 47.25$ grams (force), $\sigma = 3.96$ grams (force)).

of the destructive wirebond pull test and the ball-bond shear test. Figure 3.13 illustrates the improvement that can be achieved in the strength of the interface between the wirebond ball and the bonding pad by using the ball shear test (instead of the wirebond pull test) to optimize the bonding machine parameters [10].

As mentioned above the most common gauge of wirebond quality has been mechanical testing, i.e., the wirebond pull test and the ball bond shear test. Improvements in wirebond technology have caused both tests to have limitations. The pull test requires a book to be placed under a wire, which is very difficult in situations where the wires are closely spaced without damaging adjacent wires. There is also the difficulty of applying a

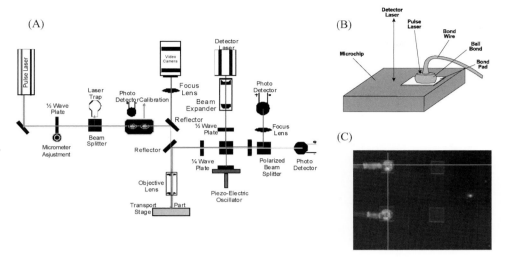

FIGURE 3.14. Laser-induced ultrasonic energy wirebond evaluation system. (A) Optical system schematic; (B) schematic representation of placement of the excitation and detection laser beams relative to the wirebond; (C) a photomicrograph showing the location of the excitation laser (cross-hairs on top of ball bond) and the detection laser (white dot on right).

consistent force to the bond interface, since the tensile and shear forces on the bond vary with the wire length and the hook position along the wire [36]. The ball shear test requires that a ram (wedge-shaped tool with a flat or slightly curved face) be placed on the major diameter of the ball. If the ball is low profile or flat such as those encountered in fine pitch wirebonding (Section 3.4.1), or thermocompression wirebonding, the ram can easily ride up over the ball. With closely spaced bonds (50–60 µm or less separation) the ram can run into adjacent bonds causing damage. Mechanical testing also tends to be time consuming and more importantly destructive. Even in non-destructive modes (see above) wires are deformed and ball edges flatten, thus giving rise to concerns about future product reliability. Hence most people recommend the mechanical testing of product on a lot sample basis only and, of course for the set-up of wirebonding machines.

A new method for wirebond testing is being developed to address the mechanical test limitations [70]. The technique uses a laser to generate an ultrasonic pulse which is passed through the bond interface and detected nearby. The test is non-destructive, fast, and appears to detect bond interface anomalies. The ultrasonic wave train is thermoelastically generated by a sub-nanosecond laser pulse hitting the top of the ball or wedge bond. It next travels through the ball or wedge and the bond interface is then detected on the surface of the integrated circuit by a laser interferometer that measures changes in the surface height. This surface displacement versus time data is then numerically converted to power versus frequency data, or Power Spectral Density (PSD). The laser ultrasonic bond testing has several potential advantages over the standard mechanical tests: (1) it is non-contact and (2) it is non-destructive. All devices produced can be tested, so quality data does not have to be inferred from a lot sample. In addition, the equipment is controlled by computer so the potential exists to fully implement the test for high production rates when attached to a wirebonder for real-time bond assessment.

A schematic representation of the test configuration is shown in Figure 3.14. Figure 3.15 presents displacement versus time curves recorded by the interferometric detection

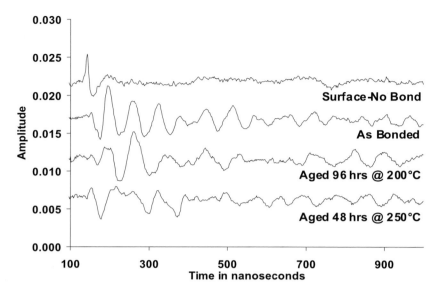

FIGURE 3.15. Displacement amplitude vs. time for bonds with different aging conditions. "No bond" illustrates noise level after bond pad surface is pulsed with a laser. Traces represent averages of at least seven individual trials and have been offset in amplitude for clarity.

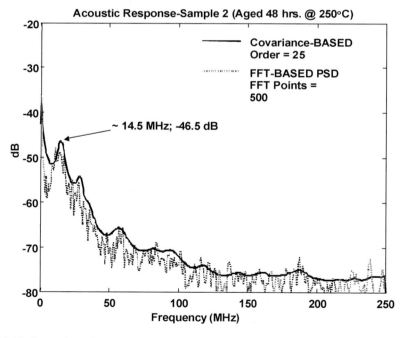

FIGURE 3.16. Comparison of the power spectral density (PSD) resulting from the fast Fourier transform (FFT) and the auto regressive (covariance based) numerical methods.

TABLE 3.3.
The effect of thermal aging on the power spectral density (PSD) behavior of typical thermosonic ball bonds made to various metalizations on silicon.

Sample	Frequency, MHz	Power, dB	Shear strength,[a] grams (force)
Sample 1: Al-1%Si			
Substrate	18.5	−56.0	
As bonded	16.5	−39.0	51.7 ± 1.8
Aged: 96 hrs @ 200°C	13.5	−44.5	60.7 ± 2.6
Sample 2: Al-1% + 0.5%Cu			
Substrate	19.5	−57.0	
As bonded	16.5	−45.5	54.3 ± 2.5
Aged: 48 hrs @ 250°C	14.5	−46.5	57.6 ± 2.2

[a] Shear strength obtained from other samples in the same sample population.

system. The numerical analysis results for a representative sample (bond aged 48 hours at 250°C) are shown in Figure 3.16. The dotted spectrum is the result of applying standard Fast Fourier Transform analysis methods to the displacement versus time curve to extract the PSD. Further analysis using an autoregressive covariance-based technique produced the sold line shown in Figure 3.16. The covariance method clearly shows a resonance response at 14.5 MHz. Applying this method to the other samples produced the data shown in Table 3.3. Table 3.3 presents the fundamental peak frequency and power levels for the aged samples along with shear strength data from bonds of the same population. Details of these results along with complete description of the method can be found in the papers by Romenesko et al. [70].

The laser ultrasonic bond evaluation has correlated a shift in the ultrasonic frequency spectrum with both bond aging and intermetallic growth. The ultrasonic wave detected was shown to be a true surface wave and thus, non-dispersive in nature. Results proving the ultrasonic wave is a surface wave are given in Figure 3.17. This means that the detected frequency shifts cannot be attributable to spectral changes due to dispersion as the detection point is moved farther away from the bond pad. In addition, no significant directional dependence of the spectrum was found—again indicating that the measurements are insensitive to the detector location relative the crystal axes of the semiconductor.

3.2.6. Bonding Automation and Optimization

Originally, wirebonding was done manually where the operator controlled every step of the bonding operation from flame-off to wire clamping and breaking (on large diameter wire even manual cutters were used). In manual bonding, operator skill was paramount to the fabrication of high-quality, reliable wire bonds. Even as the technology evolved and semi-automatic wirebonders appeared (flame-off and bonding cycle under machine control, but positioning or bond alignment was left to the operator), operator skill was key to producing highly successful (reliable) wirebonds [34]. Today, fully automated wirebonders dominate the scene. Both automatic thermosonic and ultrasonic wirebonders are in widespread use. Automatic wirebonders use pattern recognition to locate the bonding pads on both the chip and the package or substrate; and then, under complete computer control, the machines automatically bond all connections at rates exceeding 15 wirebonds (30 welds) per second. Position accuracies at those bonding rates are typically ±2.5 to ±3 μm. Us-

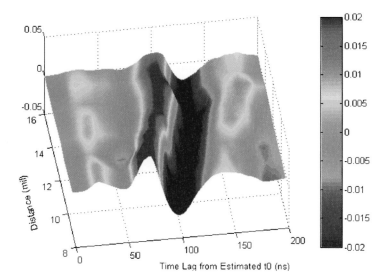

FIGURE 3.17. Results of arrival time measurement with distance. Waveforms are arranged on edge and spaced by the distance to the detector, showing arrival time to be linear with distance. Vertical axis is displacement amplitude.

ing automatic component handlers, automatic bonding machines can sustain such rates for hours. Such automation, with its concomitant accuracy and improved process control has dropped wirebond failure rates for individually packaged parts (single chip packages) into the low part per million range [34].

Failure rates associated with multichip modules, chip-on-board (or COB) and chip-on-flex are significantly higher as a result of the complex structures and new materials present in these advanced packaging structures. Some of the bonding issues for these complex circuits and structures are described in Section 3.4.2 below.

Bonding machine optimization can be accomplished in several ways depending upon the availability of test samples and trained personnel. The most straight forward way is to do a fractional factorial experimental design [10] which minimizes trials and eliminates inherent operator bias. Typically, the machine set up parameters of interest include the ultrasonic energy (P), the substrate temperature (T), and the duration of the ultrasonic energy or dwell time (D). The bonding force is usually not considered (once an initial set up has been done) since it is typically held constant for a given substrate, hybrid, or module configuration. The force is usually set to a level that promotes long capillary lifetime, thus eliminating the need to change capillaries during an experimental set (which helps minimize bond variations and improves reproducibility). For the three variables mentioned above, the bonding parameter experiments would involve a simple 2^3 factorial design with each of the variables in turn being set to expected low (-1) and high ($+1$) range limits as shown in Table 3.4. The experimental design can be unreplicated provided sufficient number of samples (>35) exist for each treatment. Random execution order should be established for all the experimental treatments to eliminate any potential memory effects. The responses denoted as S_i can be the mean shear strengths for first bond analysis (recommended) or the wirebond pull strength for each treatment. The second and third order effects are also shown in Table 3.4. The calculation of any one of these effects is simply the

TABLE 3.4.
2^3 factorial experimental design (unreplicated).

P^a	T^b	D^c	$P \times T$	$P \times D$	$T \times D$	$P \times T \times D$	Response^d
−1	−1	−1	+1	+1	+1	−1	S_1
−1	−1	+1	+1	−1	−1	+1	S_2
−1	+1	−1	−1	+1	−1	+1	S_3
−1	+1	+1	−1	−1	+1	−1	S_4
+1	−1	−1	−1	−1	+1	+1	S_5
+1	−1	+1	−1	+1	−1	−1	S_6
+1	+1	−1	+1	−1	−1	−1	S_7
+1	+1	+1	+1	+1	+1	+1	S_8

$^a P$ = bond power (e.g., first bond power setting), where −1 represents the low power value and +1 represents the high power value.
$^b T$ = temperature (substrate), °C. Again, −1 represents the low temperature setting and +1 represents the high temperature value.
$^c D$ = dwell time, ms. As above, −1 represents the shortest dwell time and +1 the longest.
$^d S_i$ = response function, typically the shear strength.

sum of the products for each level with the corresponding response all divided by $2^{(n-1)}$ where $n = 3$. For example, the effect of bond power is

$$P = (-S_1 - S_2 - S_3 - S_4 + S_5 + S_6 + S_7 + S_8)/4.$$

In order to determine the statistical significance of a particular effect with an unreplicated experimental design, an estimate of the sample variance is needed. A method for estimating the variance and confidence intervals at various significance levels has been described previously [18].

Using the same 2^3 factorial design concept with replicated center points, a linear model for ball bond shear strength in terms of P, T, and D can be constructed. The resultant ball shear equation simplifies the understanding of how the bonding parameters influence bond strength without the need for complex three dimensional plots, although with widespread availability of high performance computers, even on the shop floor, three dimensional contour plots may be preferred. In addition, the linear factorial design provides an efficient means for generating new models should different substrates and substrate metalization be required.

3.3. MATERIALS

3.3.1. Bonding Wire

Microelectronic bonding wire comes in a variety of pure and alloy materials. In addition to round wire, flat-ribbon material is available in some materials for special applications such as radio frequency and microwave circuits. Round wire is by far the most common, and fine round wires with diameters as small as 5 μm are produced commercially. Large diameter round wires up to 500 μm in diameter are used for power applications. Ribbons range from 50 μm to 1200 μm in width and come in various thicknesses.

The major materials used for these wires (and ribbons) are gold (pure and alloys), aluminum (pure), aluminum with 1% silicon, aluminum with magnesium, and, more recently, copper. Typical properties for these wires are given in Tables 3.5 and 3.6. Other wires, such

TABLE 3.5.
Mechanical properties of bonding wire.

Material	Wire diameter,[a] μm	Temper[b]	Elongation, %	Tensile strength, MPa	Comments
Aluminum (99.99% pure)	18–75 (small diameter)	H	2–6	1.9–2.5	Softer than other wire. Sags more than other wires for equivalent diameters. Difficult to handle in small diameters.
		M	6–12	1.7–1.9	
		S	12–18	1.5–1.9	
	75–500 (large diameter)		5–10	1.4–1.5	
			10–20	1.0–1.4	
Aluminum + 1% silicon	25–250	H	1–5	2.9–3.5	Standard integrated circuit bonding wire (wedge bonding). Since 1% silicon greatly exceeds the room temperature solubility of silicon in aluminum, there is a tendency for Si to precipitate at bonding temperatures—unless the alloy is homogeneous at the nanometer level.
		M	5–10	2.2–2.6	
		S	10–20	1.5–1.9	
Aluminum + 0.5–1% magnesium	25–250	H	1–5	2.9–3.5	Does not form a precipitative phase since room temperature solubility in silicon is 2%. Excellent fatigue resistance—mitigates low cycle fatigue in power devices. Sometimes small amounts of palladium (0.1–0.15%) are added.
		M	5–15	2.2–2.6	
		S	10–20	1.5–1.9	
Gold (99.99% pure)	18–50	H	1–3	3.0–4.7	Mainstay ball bonding wire. Sometimes very hard gold wire (>7 MPa tensile strength, <1% elongation) is used for wedge bonding.
		SR	3–6	3.6–4.1	
		A	4–8	3.2–3.8	
Gold (98.5% pure) + 1% palladium	18–37		0.5–3	8.7–10.4	Formulated for stud bumping. Produces consistent uniform sized balls.

[a] Typical wire sizes available from various manufacturers.
[b] Temper: H = hard, M = medium, S = soft, SR = stress relieved, and A = annealed.

as palladium and silver have been bonded in the past as described above. Gold has been the dominant material used for the ball bonding process, while aluminum and its alloys predominate in the wedge (ultrasonic) bonding process. The gold used is extremely pure (99.99%) with total impurities typically less than 100 ppm. Beryllium is the key impurity used to stabilize the wire and control some of its mechanical properties. The gold wire used for stud bumping (single ended ball bonds) is not as pure, with a significant amount of palladium (~1%) added to ensure the formation of uniform balls with minimum tails (residual wire remaining on the ball after wire is broken). Aluminum with 1% silicon matches the common alloy used for semiconductor device metalization and offers improved strength and stiffness over pure aluminum in small diameter applications. Pure aluminum is used in most large-wire applications, while aluminum and magnesium are used in cases where the interconnect is subject to conditions of low-cycle fatigue or on-off power cycling [67].

Because microelectronic bond wires are drawn through a series of dies, the as-drawn wire has significant residual strain and, while strong, is often brittle (low elongation). To

TABLE 3.6.
Thermal and electrical properties of bonding wire materials.

Material	Melting point, °C	Thermal conductivity, w/m K	Coefficient of thermal expansion $\times 10^{-6}/°C$	Electrical resistivity $\times 10^{-6}\,\Omega\,cm$	Electrical conductivity % IACS[a]
Aluminum (99.99% pure)	660	230	23–24	2.49–2.77	69–62
Aluminum + 1% silicon	600–630	195	22–23	2.96–3.18	58–54
Aluminum + 0.5–1% magnesium	654	180–195	22–24	3.01	57
Gold (99.99% pure)	1063	312	14–15	2.20–2.29	78–75
Copper (99.99% pure)	1083	395	16–17	1.72–1.81	100–95
Palladium (99.99%)	1552	75	10–12	10.75–15.63	16–11

[a] IACS = International Annealed Copper Standard. 100% IACS = $5.81 \times 10^5/\Omega\,cm$.

overcome these factors, the wire is typically strain relieved and sometimes annealed to achieve more desirable properties for the bonding process. Some of the effects of these post drawing processes can be seen in Tables 3.5 and 3.6. Figures 3.18 to 3.20 show the effects of time after manufacture (in controlled storage) on bonding wire properties for a few wire types. It is clear that depending upon the temper of the wire, storage time can have a significant affect on wire properties and hence on the quality of the bonds themselves.

There is great interest in replacing gold bonding wire with copper wire both for reduced cost, and, as the integrated circuit metalizations migrate to copper, for direct bonding to the copper pads. Thus, precluding the need for the copper pads to have barrier layer metalization (nickel-gold or titanium-tungsten gold), necessary with gold wire. Copper wire also has a high electrical conductivity and because of its strength, it resists wire sweep during the injection molding integrated circuit encapsulation processes. Since copper rapidly oxidizes in air, the ball formation process must be done in an inert atmosphere requiring significant bonding machine modifications. Copper has higher shear modules than gold (48 GPa versus 26 GPa) and Cu balls are significantly harder than gold balls (e.g., 50 compared to 35 on the Knoop Hardness Scale), thus, creating the potential for damage to delicate chips and substrates in the bonding process. Copper ball bonding produces a significant increase in cratering [21]. Several changes to bonding machine operation have been proposed as possible solutions to the copper hardness problem including increased substrate and capillary heat, reduced ultrasonic energy and a rapid first bond touchdown (to keep the ball hot and hence softer).

Bonding to copper pads, unless barriered as described above, could require significantly more ultrasonic energy due to the formation of copper oxides. In a similar vein, copper ball bonds made to conventional aluminum alloy pads seems to be viable. Copper-aluminum intermetallics exist ($CuAl_2$ and $CuAl$) and some studies have indicated rapid increases in joint resistance during thermal aging [45]. Most studies report the reliability of the copper-aluminum system to be equal to that of the gold-aluminum system. The bondability is probably more of an issue than the reliability, even with the mitigating measures described above, because the hard copper ball is likely to "push" the soft aluminum metalization aside during the bonding process, especially with today's thin IC metalizations (~0.5 µm), resulting in a weak bond or a no stick situation. Similarly, cratering and the

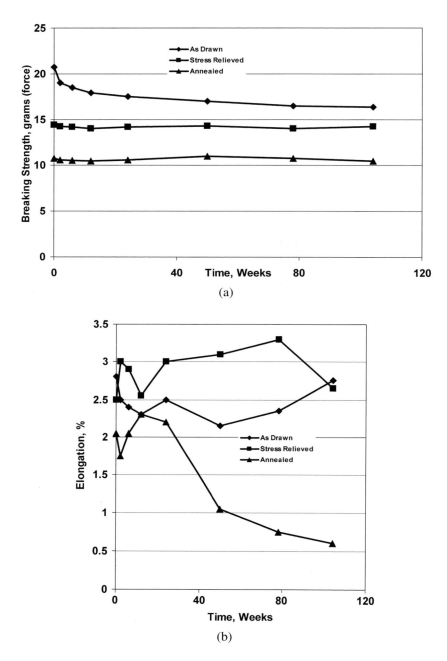

FIGURE 3.18. Breaking strength and elongation of aluminum bonding wire (Al + 1% Si) as a function of storage time for various wire tempers: (a) breaking strength grams (force); (b) elongation in %.

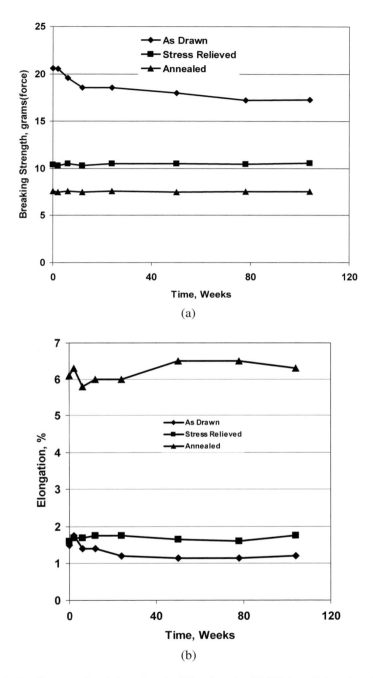

FIGURE 3.19. Breaking strength and elongation of gold bonding wire (99.99% Au + Be) as a function of storage time for various wire tempers: (a) breaking strength in grams (force); (b) elongation in %.

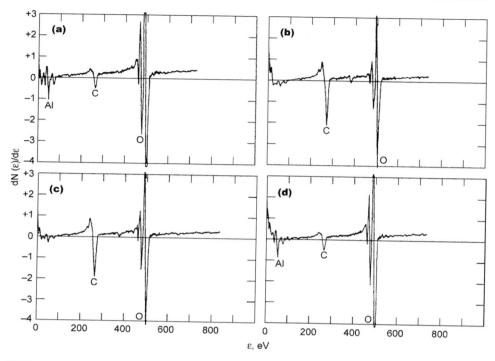

FIGURE 3.20. Auger electron spectra of aluminum metalized silicon substrates both pre- and post-cleaning with solvents and UV-ozone: (a) as processed substrate (uncontaminated); (b) substrate contaminated with photo resist; (c) substrate cleaned with solvent; and (d) substrate cleaned with UV-ozone.

susceptibility of Cu to corrosion (sulfur, halogens, etc.) could inhibit the widespread use of copper ball bonding. It is also very difficult to make small balls such as required for fine pitch wirebonding (see Section 3.4.1 below).

3.3.2. Bond Pad Metallurgy

Various metalization schemes have been used to ensure the bondability of chips, packages, and substrates. Unfortunately, most chip metalizations are selected for reasons other than the ability to form wirebonds. Historically, the typical chip metalization is aluminum containing a small percentage of silicon (typically 1%). The presence of silicon prevents the rapid diffusion of the underlying silicon (in the contact window) into the aluminum and thus reducing the pit formation in the silicon. Such pitting allows aluminum to migrate into the pits, creating aluminum conductive spikes which can damage performance or destroy device operation. Too much silicon in the aluminum can cause silicon precipitation during heat treatment and form silicon crystallites or nodules on the bonding pad surface and in contact with the underlying silicon. Such effects can cause both bonding and electrical problems.

To ensure adequate electromigration resistance as device geometrics shrink [59], alloying elements are also incorporated into or sometimes placed under the standard chip metalization. For example, copper is often added to aluminum and aluminum with silicon in concentrations between 0.5 and over 5% by weight to prevent electromigration. Above about 2 weight percent copper, the wire bondability of the aluminum-copper alloy has been shown to decrease, while lower amounts have exhibited excellent bondability [77]. Aluminum with small amounts of copper, however, is subject to Al_2Cu hillock formation during thermal processing. These hillocks can cause interlayer shorts, etc. Thus, to prevent Al_2Cu hillocks, process engineers add more copper (>4 weight percent), which causes widespread hillock growth; but the hillocks are very limited in height, thus reducing the shorting potential at the expense of bondability. Higher copper levels also increases the susceptibility of aluminum to corrosion and may lead to surface oxide formation, which can further reduce bondability.

Titanium-tungsten or titanium nitride layers are sometimes added under pads to improve adhesion and to stiffen the pads on soft or flexible substrates. If process conditions are improperly controlled, these under layers can reduce bondability. Titanium also has been alloyed with aluminum metalization on chips to reduce electromigration. Again, potential titanium migration to the surface can cause bonding problems. The titanium also increases the hardness of metalizations, which in general requires more aggressive bonding parameters to effect high quality bonds. To achieve the highest bondability in the presence of titanium, bonding temperatures must be substantially increased ($\geq 180°C$), which requires the use of high-temperature die attach (e.g., the gold-silicon eutectic). As recommended by Harman ([36], pp. 243–246), capping with a thin layer of pure aluminum (0.25–0.5 μm in thickness) would allow various metalizations to be used and still provide the best metallurgy for high-yield bonding. Care must be exercised to keep the pure aluminum cap metalization thin, because it has been shown that bond strength decreases with increasing aluminum layer thickness [82].

Gold metalization also can be an effective cap to ensure bondability. Gold was originally used on some semiconductor devices. The pad stack typically was titanium-palladium-gold. Such pad stacks produced excellent bondability providing the gold thickness hardness and morphology were carefully controlled. Today, gold is rarely used on integrated circuits, but is widely used on package bonding pads and substrates to provide a wire bondable surface. The search for bondable gold has been the subject of many articles over the last decade or two. Gold deposited by thin film deposition is inherently bondable due to its purity and fine grain structure. Most gold bonding problems have been associated with either screen printed inks used in thick film or low temperature cofired processes or with plated gold.

The bondability of thick-film metalizations, particularly gold-based films, has been of concern for many years in the microelectronics industry. Statements such as "bondable" gold still appear in various forms in the commercial advertising literature without any quantification. The implication is that if you use the particular company's bondable gold that wirebound performance should approach the ideal, i.e., wirebond pull and ball bond shear strengths close to those obtainable with thin films. Historically, authors such as Jellison and Wagner [44], have shown that with clean substrates and thermocompression bonding, thick film gold substrates yielded similar ball bond shear strengths as comparable bonds made to thin-film gold. Some studies actually showed that bonding to thick film gold was less sensitive than bonding to thin films in the presence of surface contamination. The role of surface cleanliness prior to bonding on both thick and thin films cannot be over emphasized and it has been studied in great detail by several authors including Jellison [43], and Weiner et al. [82].

In the past, the role of surface composition, surface morphology and actual conductor or bonding pad geometry has not been addressed in detail to the same levels as the cleanliness problem. From the studies that have been performed, Spencer [73], Golfarb [31],

Prather et al. [64], it is clear that conductor composition, morphology, and geometry are extremely important factors in thick film bondability. Certain manufacturers over the years have "flattened" or "coined" the thick film at the bonding site by using special tools placed in bonding machines. Such processes are very expensive and time consuming.

Our studies (e.g., [69]), have shown slightly different but not necessarily conflicting results. In our studies we compared different metalization ink types: pure gold, lightly alloyed gold, and heavily alloyed gold. The pure gold was an oxide bonded gold made for wirebonding using gold wire. The lightly alloyed gold was oxide bonded and especially formulated to retard strength loss (due to intermetallic diffusion/formation) that occurred when making aluminum wirebonds. The heavily alloyed gold (which contained significant amounts of platinum and palladium) was primarily made for solder reflow operations. These metalizations were screen printed and fired using a test pattern consisting of various line and bonding pad sizes, ranging from 125 μm to 500 μm. Thin film vacuum deposited pure gold (3 μm in thickness) was also patterned and used as a reference in these bondability studies. Pad surface and line morphology and shape were measured using a scanning electronic microscope and a stylus profilometer, respectively. Surface impurities were analyzed by Auger electron spectroscopy and wirebond quality was assessed by both the ball bond shear test and the wirebond pull test (See Section 3.2.5).

The metalization type had the greatest effect on both the ball bond shear strength and wirebond pull strength. Pure thin film gold demonstrated the best bondability and had the highest average shear strength. The lightly alloyed thick film gold (made for aluminum wirebonding) gave results similar to the thin film gold. The pure thick film gold and the heavily alloyed gold produced significantly poorer results (e.g., 35 grams (force) shear strength compared to 48 grams (force) shear strength on average) for comparably sized and placed bonds. Surface morphologies were different between all four metalizations with the thin film surface being extremely smooth, small grained with no pores. The heavily alloyed gold surface was extremely porous and very rough compared to the other metalizations. The pure thick film gold and the lightly alloyed gold had similar morphology, although the lightly alloyed gold was slightly rougher and more porous.

In a design of experiments study, parameters such as surface porosity, surface curvature and pad or line width size were determined to be secondary effects. Ball location on bonding pads or lines seemed to have little effect on the thin film and pure thick film bonding results. As surface porosity and roughness increased effects associated with ball location became slightly more dominant. Tail bonds seemed to be more affected than the ball bonds. Mechanical operations such as burnishing (scrubbing with an abrasive) or coining appeared to have little effect and in the case of the heavily alloyed gold burnishing significantly reduced the bondability.

Results of the study indicated that surface composition was the key factor in bondability. This result is consistent with findings of Harman [36] in his Chapter 6 on plated golds. He further correlates bondability or lack there of with film hardness, i.e., soft gold is preferred. In our studies the hardness of the thick film layers increased with increasing impurity concentration, based on gold ball deformation, at given force level. No quantitative measurements of hardness were made.

3.3.3. Gold Plating

3.3.3.1. Electroplated Gold Impurities in electroplated gold layers have long been a source of bonding problems. Impurities have caused both low bonding yields and premature failures during accelerated testing or real life operational use. Horsting [40] presented fundamental studies that related gold purity to the formation of "purple plague" and hence bond failures. Horsting believed that the accelerated diffusion of the impurities into bond intermetallic regions caused precipitates to form which acted as nucleation points for vacancies causing more rapid void formation during the normal interdiffusion of gold and aluminum. The actual impurities in the gold were not precisely determined by Horsting due to equipment limitations, but qualitatively he principally found nickel, iron, cobalt, and boron. Later, researchers confirmed Horsting's rapid impurity diffusion theories Newsome et al. [58].

Gold electroplating bathes typically consist of potassium-gold-cyanide solutions plus additives such as buffers, citrates, phosphates, carbonates, and lactates. Impurities such as thallium, lead, and arsenic are added to improve plating deposition rate and as modifiers to reduce grain size—hence changing surface morphology. Thallium has been the impurity most often linked to wirebonding problems [26,27], but work by Wakabayashi [81] identified lead as another significant cause. He also indicated that under certain plating conditions, arsenic could improve bond strength. Impurities such as lead and thallium can cause the gold crystal structure to change on the bonding surface. Surface morphology can also be changed by varying the plating parameters. To date there is not conclusive proof that the subtle changes in surface morphology in plated gold layers have a correlatable effect on bondability and bond strength, unlike the experiences with thick films above.

Other plated gold phenomena such as hydrogen entrapment and film hardness can also cause bonding problems. Hydrogen entrapment can be mitigated by annealing, providing the assembly can withstand the annealing environment (minimum of 2 days at 150°C). Such annealing, while removing hydrogen, also reduces its hardness. Hardness thus becomes a key bonding indicator, if not the root cause, of bondability problems.

3.3.3.2. Electroless Autocatalytic Gold The key to wirebonding on laminate technology for MCMs and COB implementations is the ability to do electroless gold plating on the pre-patterned copper metalization. In working with commercial plating vendors, electroless gold (autocatalytic) plating solutions can be found or developed with standard or modified chemistry that meet the deposition needs (99.99% pure gold up to 1 μm in thickness) for a variety of substrates and applications. Typical laminate processes require a nickel barrier layer over the copper. It is necessary that these autocatalytic gold processes be able to plate on nickel as well as on copper. Two major types of autocatalytic gold plating chemistries exist: (1) high deposition rate strongly basic systems containing cyanide; and (2) neutral pH systems without cyanide. The high deposition rate systems have a pH of about 12 and can erode certain circuit board materials such as polyimide during long plating runs. Several variants of these high deposition rate systems exist including ones which plate gold directly on copper and others which will plate gold onto nickel coated surfaces. Typical plating bath temperatures range from 70°C to 100°C. Such systems have been used to produce bondable gold, but the high bath temperatures, the difficulty in plating on nickel (requires exacting bath chemistry at all times), and the erosion of the substrate material has made these chemistries unsuitable for most organic-based MCMs and COB assemblies. Such chemistries are useful for plating circuits built on ceramic substrates.

The issues associated with the high deposition rate systems caused the development of neutral pH (nominally 7.5) autocatalytic gold processes. These baths contain no cyanide and can operate at 70°C or less and do not erode polyimide. With these systems bondable gold up to 1 μm in thickness can be deposited over nickel barrier layers. Compatible electroless nickel plating solutions exist for copper metalizations. The copper metalization

TABLE 3.7.
Wirebond pull strength for various thicknesses of autocatalytic gold plating over a nickel barrier
(2.5 μm thick) on a copper metalized printed wiring board.

Gold plating thickness, μm	Number of bonds	NDPT[a] failures	Pull strength[b], grams (force)	
			As bonded	After 150°C aging[c]
0.40	129	1	10.6	9.8
0.65	149	0	10.0	10.1
0.90	138	0	9.4	10.6

[a] NDPT = non destructive pull test (at a 2.5 grams (force) limit).
[b] Sample sizes approximately 70 bonds. Standard deviations within ±10%.
[c] 160 hours (polyimide-glass board material).

must first be sensitized with a palladium-based activator. Table 3.7 presents some wirebond reliability data for gold bonds made to various thicknesses of autocatalytically plated gold (neutral pH). The data indicates that bonds remain strong even after extensive thermal aging at 150°C provided the gold is at least 0.65 μm in thickness. Other experiments have shown that a minimum of 0.5 μm is necessary to achieve uniform bonding and reliability after thermal testing.

3.3.4. Pad Cleaning

In order to make high quality, reliable wirebonds, the bonding pads must be clean. Many techniques have been tried over the years, but of all the methods, UV-ozone [82] and oxygen plasma [47] have proved to be the most effective in removing organic contamination. They are also effective against certain inorganic materials that form either a volatile oxide or, if not volatile, one that can be easily removed. While these techniques have been shown to remove a wide variety of contamination types, care must be exercised in their use. Because of the strong oxidizing environments present in O_2 plasma and UV-ozone reactors, metals such as silver, copper, and nickel may oxidize, and thus reduce their bondability. To reduce such effects in plasma reactors, argon is sometimes mixed with the oxygen. These oxygen-argon plasma cleaners are quite effective, combining reactive ion cleaning with physical sputter etching. With any kind of plasma environment, there is a possibility of active circuit radiation damage. Based on this author's experience, this probability is extremely low for oxygen-argon plasma cleaners and should not be viewed as a deterrent to their use. Similarly, because UV radiation can excite impurity states (color centers) in alumina-based ceramics, there is a tendency for white alumina ceramic substrates to appear yellow after UV-ozone treatment. The induced color change can be reversed by a subsequent thermal treatment. Table 3.8 and Figure 3.20 show the effectiveness of UV-ozone cleaning (over solvent cleaning) in removing intentional surface contamination.

Before leaving cleaning, a few comments about ultrasonic cleaning should be made. Historically, there have been several published reports (e.g., [68]), and much anecdotal conversation describing wirebond degradation or failure due to ultrasonic cleaning. Most of the reported incidents center on wirebonds in cavity type packages, such as those used for hybrids or hermetic single chip applications.

As with all mechanical structures, a wirebond has a resonant frequency which if excited will cause the wire to vibrate and in turn may cause fatigue and ultimate failure. The

TABLE 3.8.
Average ball bond shear strength (grams (force)) for various cleaning treatments and thermal aging conditions for thermosonically bonded 25.4 μm gold wire on 1 μm thickness aluminum (on silicon). Average ball diameter was 90 μm (±3 μm).

Sample set	Cleaning conditions	As bonded	Thermally aged
A	No clean[a]	50.9 (±7.1)	47.8 (±7.9)[b]
	Plasma clean[c]	52.2 (±6.5)	52.1 (±6.7)
B	No clean	50.0 (±6.2)	48.6 (±7.1)[d]
	Contaminated[e]	38.9 (±4.1)	40.3 (±5.8)
	Solvent clean	37.3 (±6.1)	37.9 (±7.3)
	Plasma clean[f]	47.5 (±6.0)	47.9 (±6.7)
	UV-Ozone clean	53.0 (±5.1)	54.2 (±5.8)

[a] No clean as received from substrate fabrication.
[b] Sample set A aged for 96 hours at 150°C.
[c] Argon-oxygen plasma (90% Ar, 10% O_2).
[d] Sample set B aged for 168 hours at 125°C.
[e] Contamination agents were photoresist and outgassing products of epoxy cure.
[f] Argon-oxygen plasma (50% Ar, 50% O_2).

resonant frequency of a given diameter bonded wire is dependent on the length and height of the loop. For reasonable geometries and relatively short lengths (<2.5 mm) the resonant frequency of a typical wirebond is quite high (>30 kHz). Historically ultrasonic cleaners operated in the 20 kHz regime, and most of the reported damage occurred with long wire bonds (>2.5 mm) placed in large industrial cleaners (high energy). Thus, the ultrasonic cleaning of cavity type devices with short wires should be safe. Today, ultrasonic cleaners span a broad frequency range from 20 to over 100 kHz. According to Harman ([36], p. 230), it is unlikely that high frequency ultrasonic cleaners (>50–60 kHz) will damage wirebonds.

With pin or ribbon leaded packages in which the pin or ribbon feeds directly inside the package to form the wirebond attachment point, special care needs to be taken to ensure that the external lead structure does not resonant. Resonance in these external leads can set up vibration on the pin or ribbon end inside the package and can cause wire or wirebond failure, especially if the wire is relatively stiff. This would be especially important when parts in quad flat packages are cleaned prior to board attachment.

With today's fully encapsulated microcircuits, the cleaning of parts ultrasonically poses little risk, especially for leadless or short leaded components. The potential danger occurs when cleaning exposed wirebonds in open packages or in COB or flex applications. Another potential danger could be associated with microelectromechanical systems (MEMS) where ultrasonic resonance could cause mechanical failure of the MEMS structures themselves in addition to the potential damage to wirebonds. Again, it is a question of the resonance frequency of the structure compared to the ultrasonic agitation frequency. In all cases with exposed wires and structures, if ultrasonic cleaning methods are employed, cavitation should be avoided [36].

3.4. ADVANCED BONDING METHODS

3.4.1. Fine Pitch Bonding

Fine pitch ball and wedge bonding is continuing to evolve rapidly. While most ball-bonded products are still in a pitch range of 100 μm and above, production quantities of

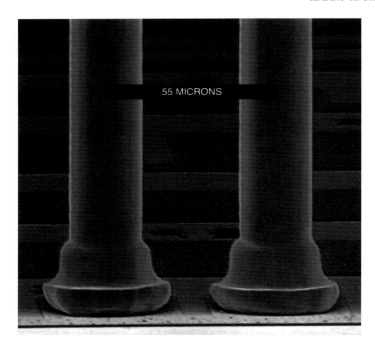

FIGURE 3.21. Scanning electron photomicrograph of ultrafine pitch (55 μm) thermosonic ball bonding. The bonds were made on a K&S Model 8020 automatic ball bonder using 23 μm (0.9 mil) diameter gold alloy wire. Pad metalization was Al + 1% Si + 2% Cu on SiO_2 with a nominal 1 μm thickness (photomicrograph courtesy of L. Levine, K&S).

90 μm pitch are being manufactured. Pitches in the 60 to 90 μm range have begun the transition to volume production, while pitches of 60 μm and below have been demonstrated on a limited scale (see Figure 3.20). Such bonds must be made with bottleneck or stepped-neck capillaries [24]. Today most bonding machines are limited to minimum pitches between 35 μm and 70 μm. An example of fine pitch ball bonding is shown in Figure 3.21. The bond is quite different from a traditional ball bond. It is quite low, almost nail head-like with a "ball" diameter in the range of 1.2–1.5 times the wire diameter. The low height of the nail head (typically 5 to 15 μm) makes the fine pitch ball bond difficult to shear. Most fine pitch ball bonds are still done with 25 μm diameter gold wire, although 18 to 20 μm wire is gaining popularity. Very fine pitches (≤70 μm) and wires smaller than 25 μm in diameter are subject to greater damage in handling and molding operations than their larger more robust counterparts.

Wedge bonding leads the fine pitch parade. Wedge bonds at pitches of 40 μm have been demonstrated using 10 μm diameter gold wire. Wedge bonds at 60 μm and above are made in high volume production using 25 μm diameter gold or aluminum wires. To achieve such fine pitches, the wedge bonds typically have low deformation (1.2 wire diameters). Narrow, cutaway wedge tools are necessary to prevent adjacent wire damage during bonding. An example of 40 μm pitch wedge bonding is shown in Figure 3.22.

Fine pitch bonding is limited by the lack of chips with appropriately sized and spaced bonding pads that can take full advantage of the reduced size, high density wirebonding technology. Shrinking bonding pad size and pitch on chips is further hampered by limitations in test probe placement and movement. High frequency bonding (>60 kHz) has

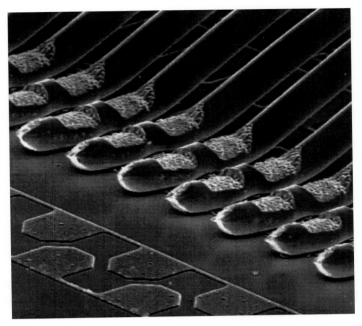

FIGURE 3.22. Scanning electron photomicrograph of ultrafine pitch (40 μm) wedge bonds. The bonds were made on a K&S Model 8060 automatic wedge bonder using 20 μm (0.8 mil) diameter gold alloy wire. Pad metalization was Al + 1% Si + 2% Cu on SiO_2 with nominal 1 μm thickness (photomicrograph courtesy of L. Levine, K&S).

been shown to be beneficial in bonding fine pitch circuitry [32]. More details on higher frequency wirebonding will be given in Section 3.4.4 below.

There are many issues associated with the implementation or use of fine pitch wirebonding. Fine pitch bonding can only be accomplished successfully if the entire process (chip, package or substrate, bonding machine, and bonding practice) is designed from the beginning with fine pitch in mind. The size, placement, and shape of the bonding pads must be coordinated with the selection of the wirebonding machine, the die attach machine and process, and the package or substrate (board) layout. Square bonding pads (hexagonal or round, also) are optimal for ball bonding but pose some limitations for wedge bonding. Ideal wedge bonding pads would be long and narrow [61]; but these are seldom used because of the need to be flexible in bonding method choice and that automatic wedge bonders are, at best, a factor of two slower than automatic ball bonders (due to the need to index either the bonding head or the sample table to maintain wire alignment under the wedge). Thus, a high volume wedge bonded product will cost more than a product interconnected by ball bonding methods, even given the difference in wire cost (aluminum vs. gold, respectively).

It also should be recognized that extremely fine pitch, with any bonding technology, can result in higher costs due to added constraints, reduced throughput (generally lower bonding speeds), and typically a more fragile product. Both equipment and workers associated with the fine pitch process are typically more expensive than those associated with a conventional (low pitch) process. Automatic bonders need the latest in precision pattern recognition coupled with the most accurate placement control. Programming time is greater, and workers must be better trained to master the art of fine pitch. Die attachment machines also must have greater accuracy in the placement process than machines used in conventional pitch processes. Packages, as the wirebonding pitch declines (especially with multiple tiers of bonding pads), must be carefully designed to give maximum bonding tool access, while minimizing chances of wires touching or wire misplacement causing shorting. The ultimate design practice will force package and substrate pitches to those of the chip, thus minimizing fan out and keeping wire lengths short, which will reduce lead inductance and minimize injection molding wire sweep [76]. Copper wire could have an advantage in both electrical performance and mechanical integrity, but the ability to form minimal size balls (necessary for fine pitch) is still in question.

The solution to wires touching and shorting in fine pitch wirebonding could be the use of insulated wire. Insulated wire with appropriate bonding pad capping metalization could also allow COB assemblies to be made without insulating glob tops on overcoats. Insulated bonding wire has been around for over 20 years, but has never received widespread attention, mainly due to a host of implementation/reliability problems including wire coating contamination of the capillary, flame off inconsistencies, and low second bond strength. Recent advances [55] in wire coating technology appear to have made the specter of coated wire viable. Coated bonding wire has obvious advantages including allowing wires to be close together, cross, and even touch. Such ability could solve wire sweep issues and die/wire shorting problems encountered in stacked die or in high density wirebonding in general. The newer coatings appear to be about 0.5 µm in the thickness on 25 µm gold wire with breakdown strengths approaching 200 volts and the ability to survive baking temperatures of 300°C. Wire strength and bonding ability appear not to be reduced by the coating.

3.4.2. Soft Substrates

Deformable or soft substrates in modern wirebonding applications are usually associated with organic-based boards or layers as follows: thin-film, multilayer structures on inorganic carriers such as encountered in multichip modules (MCM-Ds); laminate-type organic constructs such as encountered in printed wiring boards (PWBs), MCM-Ls and COB structures [16]; and chips mounted to unreinforced laminates and/or flexible film layers.

MCM-D modules are made using deposited dielectric and thin-film metal layers. The carrier for these deposited films is usually silicon, although polished ceramics have been used in the past [12]. The dielectric materials are typically spun-on layers of polyimide. Benzocyclobutene (BCB), and several lesser-known polymers [74] also have been used. These dielectric layers usually range in thickness from 5 to 25 µm (or more), with as many as six layers being reported. Metalization schemes have been gold (with suitable adhesion layers such as chromium and tungsten), copper (again with suitable adhesion layers), and aluminum. In addition to organic dielectric layer softness, metal adhesion has been a challenge and requires careful processing to ensure metal layer integrity and inner layer adhesion.

In bonding to MCM-D structures, both thermosonic ball bonding and ultrasonic wedge bonding have been used [54]. In bonding to MCM-Ds, two major issues arise: (1) the size of the bonding pad and (2) the number and thickness of the soft layers (polyimide, BCB, etc.) under the pad. It has been shown [15] that the pad bends or cups under the application of the bonding force. This cupping is due to the compliant nature of the organic material. Elevated temperatures exacerbate the issue, effectively softening the polymer even more. Small bonding pads have less area over which to distribute the load and are thus more susceptible to this cupping or bending phenomenon. Pad deformations under bonding forces and the application of ultrasonic energy have been studied by Takeda et al. [75].

Their results show that normal sized gold pads on copper traces (on polyimide flex boards) can deform as much as 20 μm under normal (but high end of the range) force and ultrasonic energy bonding conditions. They also verified that the use of a nickel under layer (under the gold pad) can significantly reduce the deformation below 10 μm for all bonding conditions. Others have noted similar deformations but the amount of deformation was smaller. In our work, for example, we have observed that for a given bonding force, the deformation increases with organic layer thickness. Pad reinforcement structures and interlayer metalization tend to mitigate deformation. Similarly, a marked decrease in deformation was observed as the bonding force was reduced in all samples, with little or no correlation to changes in sample thickness.

In addition to unreinforced substrate materials, MCM-L and COB implementations can use fiber reinforced organic matrix material such as polyimide or epoxy. The reinforcing fibers are typically glass, although materials such as Kevlar®, quartz, and Aramid® have been used. Sometimes high-frequency circuitry is built on non-fiber reinforced substrates with very low dielectric constants such as Teflon® (polytetrafluoroethylene). Most of these "laminate" technologies use copper metalization protected by thin layers of plated gold (usually with a nickel barrier layer under the gold). The thicknesses of both the metal and dielectric layers are larger than those of the MCM-D technology by factors of 5 for the metals and at least an order of magnitude or more for the dielectrics.

Other MCM-L implementations use fiber reinforced cores with non-reinforced resin layers on their surfaces [35]. Such structures can employ a variety of metalization schemes put in place and patterned by a combination of thin-film deposition (MCM-D) and PWB techniques. Via fills can be plated or actually filled with conductive organic resins [33].

Wirebonding to most MCM-L substrates including those in ball-grid arrays (BGA) and chip-scale packages (CSP) is similar to bonding to PWBs provided the substrates are made with fiber-reinforced resin laminates (e.g., polyimide-glass, epoxy-glass). Direct bonding to PWBs has been done for some time in COB applications. Many problems still exist with bonding to standard PWB fiber reinforced laminates, let alone the new problems associated with reduced pad sizes, unreinforced organic layers, and different via construction techniques found in today's MCM-Ls, BGA and CSP substrates, and integrated circuit redistribution layers. Both aluminum wedge bonds and thermosonic gold ball bonds have been used in COB applications. Wedge bonding is often preferred because it can be done without added substrate heat. Large COB assemblies will tend to warp and possibly soften if heated to or near their glass transition temperature (T_g). FR-4 (epoxy-glass) circuit boards have a T_g around 120°C, while T_g of various polyimide boards exceeds 200°C. Such high-temperature resins can be thermosonically bonded provided proper substrate clamping and backside support is available for large area assemblies. Successful thermosonic bonds have been made at temperatures below 100–110°C so that even FR-4 can be bonded. Even with the thick metalizations typically encountered in the COB arena (e.g., nominally 17–35 μm), anomalies can exist in wirebonding, especially as pads shrink in size. Bonding to BGA and CSP flexible substrates is typically done with gold-ball bonding because of the need for controlled shape bonds and bonds that are very close to the chip edges to keep the package footprint as small as possible. Because of the small area and reduced thickness of the substrate, special care has to be exercised in the bonding process.

In addition to flexible and software substrates, two other difficult bonding situations exist in both: thinned-die and stacked-die (either thinned or not). Thinned die have been around for some time, especially in microwave applications where gallium arsenide (GaAs) microwave devices have been thinned to 100 µm or less to provide better thermal performance. Gallium arsenide is more susceptible to bond cratering and to mechanically induced electrical defects than silicon. For a detailed study of cratering on silicon die, see the paper by Clatterbaugh and Charles [23]. GaAs is weaker than silicon by a factor of 2. The two major material characteristics or parameters that are most relevant to cratering have been shown to be hardness and fracture toughness. Hardness is a measure of the material resistance to deformation while fracture toughness is a measure of the energy (or stress) required to propagate an existing microcrack. The Vicker's hardness for GaAs is 6.9 (± 0.6) GPa while silicon is 11.7 (± 1.5) GPa. In a similar vein, the fracture toughness of GaAs and silicon are 1.0 J/m^2 and 2.1 J/m^2, respectively. Thinned silicon die are now being mounted to flexible circuit boards. Silicon die as thin as 25 µm have demonstrated electrical integrity. Wirebonding, because of the thinness of the die and the softness of the flexible substrate, has proven difficult and most of these assemblies have been flip chipped (i.e., attachment by solder reflow [3]).

Stacked die present their own set of issues, but in general, the problems involve multiple geometries in a given component package with closely spaced wirebonds that can overlap. In addition, sometimes the bonding must be done to chips that are cantilevered over another chip without a means of mechanical support under the bonding pad areas. Fixturing and very careful control of bonding parameters (reduced force and power, higher frequency, and temperature) has allowed successful wirebonding to stacked geometries with as many as six chips. A full discussion of the details of wirebonding to stacked chips is not possible in this work, but some insight can be gained by reading Yao et al. [84].

3.4.3. Machine Improvements

Many bonding machine improvements have been made. While some are manufacturer specific, most are commonly available throughout the industry. These improvements are typically aimed at improving the speed of bonding; increasing the accuracy of the bond placement for fine pitch; improving the bondability of difficult-to-bond metalizations and substrate structures; and controlling the complete bond geometry for greater repeatability, reliability, and, of course, electrical impedance control for high-frequency applications. Such improvements include air bearings to increase bonding speed and reduce machine down-time owing to wear, laser interferometry for precise head positioning, and improved pattern recognition software to enhance learning and bonding speeds as well as encompassing larger chip libraries.

Other software improvements allow complete control of bond shape and length. Such control allows the repeatable fabrication of bonds with a given impedance for microwave and wireless circuitry. Another ramification of bonding machine improvement is the potential use of higher frequency ultrasonics (up to 300 kHz). Research suggests that higher frequencies can reduce bond dwell time and still achieve high quality bonds. Similarly, the application of higher frequency ultrasonics has been reported to enhance the bondability of difficult substrates such as soft ones encountered in MCM-D and COB applications. See Section 3.4.4 below. The introduction of a delay (after force application) prior to the onset of the ultrasonic energy burst has also been shown to be effective in difficult bonding situations.

3.4.4. Higher Frequency Wirebonding

Most of the world's current wirebonding machines have ultrasonic generators and transducers that operate at nominally 60 kHz. The choice of 60 kHz was made several decades ago based on transducer (bonding head) dimensions for microelectronic assemblies and stability during the bonding (transducer loading) operation [36]. Other frequencies from 25 to 300 kHz have been used to attach wires. Ultrasonic welding and material softening have been reported in the range between 0.1 Hz [83] and 1 MHz [49]. Today's interest in higher frequency bonding stems from reports by various authors [32,38, 42,66,72,78] that using higher ultrasonic frequencies produces better welding at lower temperatures in shorter bonding times (dwell times). It has also been indicated that higher frequency wirebonding improves bonding to pads on soft polymer layers such as Teflon® or unreinforced polyimide. While all these improvements were real for the particular situations in hand, few if any controlled studies (systematic, side-by-side experiments on the same substrates with an attempt to control all variables except frequency) have been performed. The following material presents excerpts from the one such study [18,19].

Three metalization schemes were used in this study: (1) aluminum (99.99% pure) with a titanium-titanium nitride (Ti/TiN) adhesion layer; (2) aluminum plus one percent silicon alloy (Al + 1% Si) again with a Ti/TiN adhesion layer; and (3) gold metalization with a titanium-tungsten adhesion layer (TiW). The metal bonding pad formation layers were sputter deposited to thicknesses between 1 μm and 2 μm on silicon base layers. The silicon wafers were p-type with a nominal resistivity of 30–50 Ω·cm. The wafers were thermally oxidized to achieve an SiO_2 thickness of 1 μm prior to metal deposition or spin coating with polyimide. The polyimide layers were between 5 and 20 μm in thickness. The gold metalization was also deposited on highly polished ceramic (99.6% pure alumina) substrates.

Various test structures were photolitographically patterned on each of the metal layers [15,17]. The patterns included: arrays of bonding parts of varying sizes (150 μm to 25 μm square), a daisy chain pattern consisting of almost 650 wirebonds with the resistance of the wirebonds accounting for over 60% of the total resistance of the circuit, and a radially distributed wirebond pattern for shock and vibration testing.

All wirebonding for the study was performed with two semiautomatic thermosonic ball bonders (Marpet Enterprises, Inc., Model 827) equipped with negative electronic flame off (Uthe Technology, Inc., Model 228-1) for uniform control of free air ball size. The flame offs were adjusted to produce 60 ± 2 μm diameter free air balls as shown in Figure 3.10. One of the MEI Model 827 wirebonding machines was equipped with a UTI Model 25ST (64.1 kHz) transducer driven by a standard UTI Model 10G ultrasonic generator. The other Model 827 wirebonding machine was equipped with a UTI Model 4ST (99.5 kHz) transducer which was driven by a UTI 10G generator tuned for 100 kHz. In order to make both transducer waveforms similar since the Model 25ST transducer is much larger than the Model 4ST a short 60 kHz transducer Model 17STL (63.1 kHz) was also used. A comparison of the transducer dimensions is given by Charles et al. [17]. The uniformity of the as-bonded product (both 60 kHz and 100 kHz) is shown in Figure 3.23.

This study has yielded a large amount of data. Key observations and findings include the following. It is clear that significant differences exist between bonding at nominally 60 kHz and bonding at 100 kHz. In addition to differences in transducer electronic waveforms between the standard 60 kHz (long) and the 100 kHz transducer, there exist differences in bonding machine optimization behavior. The 60 kHz system appeared to have a larger bonding window (i.e., for a given force and substrate temperature), and a wider range of ultrasonic power and dwell times produced acceptable bonds (strong, yet not over bonded or with wire damage) when compared to the bonds produced by the 100 kHz system. The 100 kHz bonding window, in addition to being smaller than the 60 kHz window,

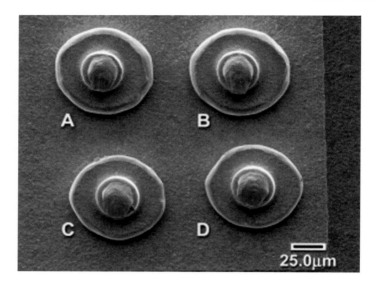

FIGURE 3.23. Scanning electron photomicrograph of single ended ball bonds bonded to substrate pad metalization. Balls A and B were bonded at 100 kHz. Balls C and D were bonded at 60 kHz. The average diameter for balls A and B is 80.5 µm. The average diameter for balls C and D is 77.2 µm. The magnification is approximately 300×.

TABLE 3.9.
Gold thermosonic ball bond shear strength (grams (force)) on gold and aluminum (1% Si) metalizations at both 60 and 100 kHz[a].

Metal	60-kHz	100-kHz	$|\Delta$ means$|$	Significant[b]
Au (on ceramic)	68.4 ± 3.7	84.8 ± 6.5	16.4	Yes (highly)
Al + 1% Si (on silicon)	54.0 ± 3.2	50.6 ± 2.9	3.4	Yes
$|\Delta$ means$|$	14	34.2		
Significant[b]	Yes (highly)	Yes (highly)		

[a] Nominal sample size at each frequency was 100.
[b] 99% confidence that the difference in the means are significant using analysis of variance with the F-test.

was also sharper (i.e., a smaller change in ultrasonic power and/or dwell in relationship to the window edge was required to go from either a no-bond condition or to an over-bonded condition when compared to the 60 kHz system). Despite the smaller, sharper fall-off of the bonding window, the 100 kHz system has one obvious advantage. It formed strong bonds in times that are 30 to 60% shorter than comparable dwells for the 60 kHz system. Comparison of both bonding systems and their transducer waveforms indicate that the 100 kHz system has much faster bonding pulse rise and fall times, along with a more stable voltage (or current) amplitude envelope that the 60 kHz system. Switching to a short 60 kHz transducer with dimensions comparable to those of the 100 kHz transducer produced ultrasonic drive parameters (voltage and current) similar to those of the 100 kHz transducer.

Shear test data on gold substrate metalizations showed that an optimized 100 kHz system produced much stronger bonds than the 60 kHz system (see Table 3.9). As can be seen from Figures 3.10 and 3.23 and Table 3.10, this difference cannot be accounted for by

TABLE 3.10.
Gold thermosonic ball bond average diameters[a] (μm) on gold and aluminum (1% Si) metalizations at both 60 and 100 kHz.[b]

Metal	60 kHz	100 kHz	|Δ means|	Significant[c]
Gold	89.1 ± 4.0	88.3 ± 2.9	0.8	No
Al + 1% Si (on silicon)	91.3 ± 2.3	92.0 ± 2.0	0.7	No
|Δ means|	2.2	3.7		
Significant[c]	Yes	Yes		

[a] Average diameter = $\frac{1}{n}\sum_i^n [(X_i + Y_i)/2]$.
[b] Nominal sample size at each frequency was 100.
[c] 99% confidence that the difference in the means are significant using analysis of variance with the F-test.

TABLE 3.11.
Gold thermosonic ball bond diameters (in μm) in directions perpendicular (X-direction) and parallel (Y-direction) to the direction of the ultrasonic scrub on gold and aluminum (1% Si) metalizations at both 60 and 100 kHz.[a]

Metal	Frequency	X-direction	Y-direction	|Δ means|	Significant[b]
Gold (on ceramic)	60 kHz	84.9 ± 4.8	93.2 ± 5.2	8.3	Yes (highly)
	100 kHz	82.8 ± 3.4	93.8 ± 4.0	11.0	Yes (highly)
Al + 1% Si (on silicon)	60 kHz	93.9 ± 2.9	88.7 ± 3.4	5.2	Yes (highly)
	100 kHz	98.7 ± 2.8	85.4 ± 2.6	13.3	Yes (highly)

[a] Nominal sample size at each frequency was 100.
[b] 99% confidence that the difference in the means are significant using analysis of variance with the F-test.

ball diameters (either pre- or post-bonding), which were essentially the same for both the 60 and 100 kHz systems. When the data was analyzed for the Al + 1% Si metalization (on oxidized silicon), the 60 kHz bonds appeared stronger. Although the difference between the 60 kHz and 100 kHz test results was relatively small (less than 7%). However, when analysis of variance techniques were applied, the difference was significant at the 99% confidence level. Similar results were observed on full thermosonic ball bonds attached to an integrated circuit chip (Al + 1% Si metalization), on which both the ball shear test and the wirebond pull test gave a small edge to the 60 kHz system. Although this data set was relatively small, the student's t-test indicated that the results were significant at the 99% confidence level. Independent of frequency, the differences in ball bond shear strengths between metalization types, were relatively large and highly significant. Bonds on gold were always stronger than bonds on Al + 1% Si metalization consistent with the results shown in many previous studies [11–19].

Other differences were observed such as asymmetry of ball shape with metalization type. No differences in average ball diameters [(X-diameter + Y-diameter)/2] were observed with frequency. Any variations in average ball diameters even those between metalizations (Table 3.10) could be accounted for by differences in the free-air ball size. On the other hand the differences in the X and Y diameter measurements are highly significant and appear to depend on metalization type (Table 3.11). On gold metalization, the as bonded ball diameter in the Y-direction or the direction of the ultrasonic scrub is larger than the orthogonal non-scrub diameter (X-direction) with consistent measurements for

TABLE 3.12.
Gold thermosonic ball bond shear strength (grams (force)) on gold and aluminum (1% Si) metalizations at both 60 and 100 kHz under conditions of thermal aging.[a]

Metal	Aged[b]	60 kHz	100 kHz	|Δ means|	Significant[c]
Gold (on silicon)	No	81.4 ± 4.6	97.4 ± 3.7	16.0	Yes (highly)
	Yes	82.1 ± 3.3	96.4 ± 4.6	14.3	Yes (highly)
|Δ means|		0.7	1.0		
Significant[c]		No	No		
Al + 1% Si (on silicon)	No	47.0 ± 3.7	46.5 ± 4.3	0.5	No
	Yes	57.8 ± 3.3	56.1 ± 4.1	1.7	Yes (slightly)
|Δ means|		10.8	9.6		
Significant[c]		Yes (highly)	Yes (highly)		

[a] Nominal sample size at each frequency was 100.
[b] 120 hours at 150°C.
[c] 99% confidence that the difference in the means are significant using analysis of variance with the F-test.

both 60 and 100 kHz. On Al + 1% Si, the non-scrub direction (X-direction) is larger than the Y-direction by a significant amount for both the 60 kHz and 100 kHz bonding systems. Similar behavior was also observed for pure aluminum metalization. The cause of these phenomena is not well understood but is believed to be associated with the dynamics of the weld formation process. On gold there is the single interdiffusion of the gold wire and gold pad materials. On aluminum and aluminum alloys the formation of gold-aluminum intermetallics is key to the bonding process. The formation of the relatively hard intermetallics may tend to lock the developing bond in the direction of the scrub on the aluminum and aluminum alloy metalizations while on the gold (being relatively ductile) the bond may be able to fully expand in the scrub direction.

Table 3.12 shows results for both 60 and 100 kHz bonded samples under conditions of thermal aging (120 hrs at 150°C). Aging at 150°C has been shown to be very effective [15] for assessing wirebond (ball bond) quality and reliability without introducing unwanted effects caused by substrate interactions and other heat-related phenomena. Table 3.12 again illustrates the significant improvement in shear strength using 100 kHz bonding on gold metalization, this time for gold on a silicon substrate as compared to the gold on ceramic data given in Table 3.9. The small observed differences on the Al + 1% Si metalization for 60 kHz versus 100 kHz is also consistent with the results in Table 3.9, although in this case the difference is statistically insignificant at the 99% confidence level. Again, large and significant differences were observed in the shear strengths between the two metalizations with bonds to gold being much stronger than bonds on Al + 1% Si metalization. These results are consistent regardless of the bonding frequency. After aging, the shear strength of the bonds on gold, at both frequencies, remained essentially unchanged. On the Al + 1% Si metalization the strength of the bonds increased significantly for both frequencies. Again, 100 kHz bonding produced stronger bonds on gold metalization, while 60 kHz bonding appeared to have a slight edge on Al + 1% Si. The increase strength for the aged bonds on the Al + 1% Si metalization is consistent with similar increases reported previously under aging [13], but the timeframe for the existence of the increased strength above the as-bonded condition appears to be longer in this particular experimental series.

3.4.5. Stud Bumping

Alternative forms of flip chip technology, make use of wirebonder-produced bumps. These alternatives include the "stud bump and glue technique" using standard or anisotropically conductive adhesive as shown in Figure 3.24.

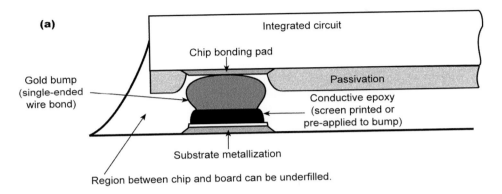

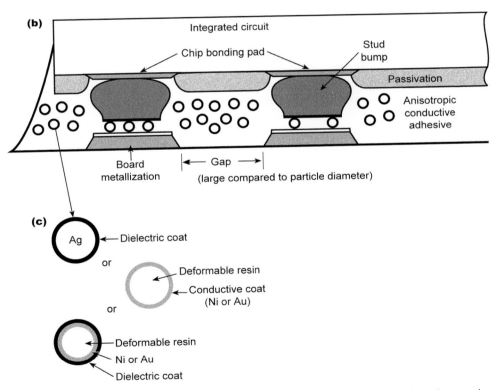

FIGURE 3.24. Schematic representation of attachment of a gold stud-bumped chip to a mating substrate using adhesive processes: (a) flip chip attachment using conventional conductive epoxy (screen printed or pre applied to the bump). Following adhesive cure the region between the chip and the board could be underfilled; (b) flip chip attachment using anisotropic conducting film. The anisotropic film not only makes the electrical contact but also acts as an underfill; (c) anisotropic adhesive conductive particle detail.

In the stud bump and glue technique, single-ended, thermosonic wirebonds are placed on the chip bonding pads by means of an automatic wirebonder using special bonding wire (Table 3.4). The balls are then coined or tamped to a uniform height using a special tool placed in the wirebonder. The stud-bumped chip is then pressed on a plate containing a thin layer of conductive adhesive (epoxy). As the chip is lifted from the plate, a small amount of conductive adhesive adheres to each bump. The chip is then placed on the corresponding substrate pads and the adhesive is cured, resulting in the geometry shown in Figure 3.24. In an alternative method, the epoxy can be preapplied to the substrate pads by screen printing or automated dispensing.

An anisotropic adhesive is an adhesive that has small conductive particles embedded in its non-conducting organic matrix. A bumped chip is then pushed down into the adhesive, capturing a few conducting particles between the bump and the mating bonding pad on the package or substrate. When the adhesive is cured, an electrical interconnect is made. In addition, the region between the chip and the board becomes rigid, mechanically holding the chip to the board. Thus, the adhesive also serves as an underfill [41].

3.5. SUMMARY

Wirebonding continues to be the dominant form of first-level chip connection. Approximately 90% of the world's chip production is wirebonded. Because of its sheer volume, flexibility, and low cost, it will continue to dominate chip interconnect for decades to come. Wirebonding is accomplished by three basic techniques using a variety of wire and pad metallurgies. Wirebonding is robust and, on rigid substrates, has been shown to be extremely reliable (defects in the low part per million range). Bonding to softer substrates, small pads, unconventional metallurgies and stacked components has presented challenges—challenges that wirebonding has been successful in meeting. With appropriate care and understanding of the processes, wirebonding, even under these challenging conditions, can be performed reliably with high yield. Wirebonding is continually improving through advancements in automation, refinement of welding kinetics, improvements in wire and pad metallurgies, improved cleaning methods, and a better and widespread understanding of wirebonding science.

ACKNOWLEDGMENTS

The author greatly acknowledges the support of JHU/APL's Electronic and Mechanical Services Groups on sample preparation and testing. Special thanks is given to Ms. Nancy L. Pickett for manuscript preparation.

REFERENCES

1. ASTM Standard Test Method: F458-84(1995)el, Standard non-destructive pull testing of wire bonds, Annual Book of ASTM Standards, West Conshohocken, Pennsylvania, USA, 1995.
2. ASTM Standard Test Method: F1269-89(1995)el, Test method for destructive shear testing of ball bonds, Annual Book of ASTM Standards, West Conshohocken, Pennsylvania, USA, 1995.
3. C.V. Banda, D.J. Mountain, H.K. Charles, Jr., J.S. Lehtonen, A.C. Keeney, R.W. Johnson, T. Zhang, and Z. Hou, Development of ultra-thin flip chip assemblies for low profile SiP applications, Proc. 37th Int. Microelectronics Symposium, Long Beach, California, 2004, pp. 551–555.

4. J. Bardeen and W.H. Brattain, The transistor, a semiconductor triode, Physical Review, 74, p. 230 (1948).
5. W.D. Barnhart, J. Van Rij, J.M. Petek, and H.K. Charles, Jr., The impact of KGD and module repair on multichip module costs, Proc. 32nd International Microelectronics Symposium, Chicago, Illinois, October 26–28, 1999, pp. 373–387.
6. A. Bischoff, F. Aldinger, and W. Heraeus, Reliability criteria of new low cost materials for bonding wires and substrates, Proc. 34th Electronic Components Conference, New Orleans, Louisiana, USA, 1984, pp. 411–417.
7. H.K. Charles, Jr., Tradeoffs in multichip module yield and cost with KGD probability and repair, Microelectronics Reliability, 41(5), pp. 715–733 (2001).
8. H.K. Charles, Jr., B.M. Romenesko, O.M. Uy, A.G. Bush, and R. Von Briesen, Hybrid wirebond testing—variables influencing bond strength and reliability, The International Journal for Hybrid Microelectronics, 5(1), pp. 260–269 (1982).
9. H.K. Charles, Jr., B.M. Romenesko, G.D. Wagner, R.C. Benson, and O.M. Uy, The influence of contamination on aluminum-gold intermetallics, Proc. Int. Reliability Physics Symposium, San Diego, California, USA, 1982, pp. 128–139.
10. H.K. Charles, Jr., G.V. Clatterbaugh, and J.A. Weiner, The ball bond shear test: its methodology and application, in D.C. Gupta, Ed., Semiconductor Processing, ASTM STP 850, 1984, pp. 429–457.
11. H.K. Charles, Jr., Ball bond shearing: an interlaboratory comparison, Proc. International Microelectronics Symposium, Atlanta, Georgia, 1986, pp. 265–274.
12. H.K. Charles, Jr. and G.V. Clatterbaugh, Thin film hybrids, in M.L. Minges, Ed., Electronic Materials Handbook, Vol. 1, Packaging, ASM International, Materials Park, Ohio, USA, 1989, pp. 313–331.
13. H.K. Charles, Jr., K.J. Mach, and R.L. Edwards, Multichip module (MCM) wirebonding, Proc. International Symposium on Electronic Packaging Technology (ISEPT '96), Shanghai, Peoples Republic of China, 1996, pp. 336–341.
14. H.K. Charles, Jr., K.J. Mach, R.L. Edwards, S.J. Lehtonen, and D.M. Lee, Wirebonding on various multichip module substrates and metallurgies, Proc. 47th Electronic Components and Technology Conference, San Jose, California, USA, 1997, pp. 670–675.
15. H.K. Charles, Jr., K.J. Mach, R.L. Edwards, A.S. Francomacaro, S.J. Lehtonen, and J.S. DeBoy, Wirebonding: reinventing the process for MCMs, Proc. International Symposium on Microelectronics, San Diego, California, USA, 1998, pp. 645–655.
16. H.K. Charles, Jr., K.J. Mach, R.L. Edwards, A.S. Francomacaro, S.J. Lehtonen, and J.S. DeBoy, Multichip module and chip-on-board wirebonding, Proc. 12th European Microelectronics Conf., Harrogate, Yorkshire, England, 1999, pp. 525–532.
17. H.K. Charles, Jr., K.J. Mach, R.L. Edwards, A.S. Francomacaro, J.S. DeBoy, and S.J. Lehtonen, High frequency wirebonding: its impact on bonding machine parameters and MCM substrate bondability, Proc. 34th International Microelectronics Symposium, Baltimore, Maryland, 2001, pp. 350–360.
18. H.K. Charles, Jr., K.J. Mach, S.J. Lehtonen, A.S. Francomacaro, J.S. DeBoy, and R.L. Edwards, High-frequency wirebonding: process and reliability implications, Proc. 52nd IEEE Electronic Components and Technology Conference, San Diego, California, 2002, pp. 881–890.
19. H.K. Charles, Jr., K.J. Mach, S.J. Lehtonen, A.S. Francomacaro, J.S. DeBoy, and R.L. Edwards, Wirebonding at high ultrasonic frequencies: reliability and process implications, Microelectronics Reliability, 43, pp. 141–153 (2003).
20. G.K.C. Chen, The role of micro-slip in ultrasonic bonding of microelectronic dimensions, Proc. 1972 International Microelectronic Symposium, Washington, DC, October 30–November 1, 1972, pp. 5-A-1-1–5-A-1-9.
21. T.B. Ching and W.H. Schroen, Bond pad structure reliability, 24th Annual Proc. Reliability Physics Symposium, Monterey, California, 1988, pp. 64–70.
22. G.V. Clatterbaugh, J.A. Weiner, and H.K. Charles, Jr., Gold-aluminum intermetallics, ball bond shear testing and thin film reaction couples, IEEE Trans. Components, Hybrids Manufacturing Technology, CHMT-7(4), pp. 349–356 (1984).
23. G.V. Clatterbaugh and H.K. Charles, Jr., The effect of high temperature intermetallic growth on ball shear induced cratering, IEEE Trans. Components, Hybrids and Manufacturing Technology, CHMT-13(4), pp. 167–175 (1990).
24. J.C. Demmin, Ultrasonic bonding tools for fine pitch, high reliability interconnects, Proc. Int. Conference on Multichip Modules, Denver, Colorado, USA, 1996, pp. 397–402.
25. V.J. Ehrlich and J.Y. Tsao, Laster direct writing for VLSI, VLSI Electronics: Microstructure Science, Vol. 7, Academic Press, 1983, pp. 129–164.
26. H.W. Endicott, H.K. James, and F. Nobel, Effects of gold-plating additives on semiconducting wire bonding, Plating and Surface Finishing, V, pp. 58–61 (1981).

27. K.L. Evans, T.T. Guthrie, and R.G. Hayes, Investigations of the effect of thallium on gold/aluminum wire bond reliability, Proc. ISTFA, Los Angeles, California, 1984, pp. 1–10.
28. B.L. Gehman, Bonding wire for microelectronic interconnections, IEEE Trans. Components Hybrids and Manufacturing Technology, CHMT-3(8), pp. 375–380 (1980).
29. L. Geppert, Solid state, IEEE Spectrum, 35(1), pp. 23–28 (1998).
30. A.B. Glaser and G.E. Subak-Sharpe, Integrated Engineering: Design Fabrication and Applications, Addison-Wesley, Reading, West Virginia, USA, 1979.
31. S. Goldfarb, Wirebonds on thick film conductors, Proc. 21st IEEE Electronics Components Conference, 1971, pp. 295–299.
32. B. Gonzalez, S. Knecht, and H. Handy, The effect of ultrasonic frequency on fine pitch Al wedge wirebonds, Proc. 46th Electronic Components and Technology Conference, Orlando, Florida, USA, 1996, pp. 1078–1087.
33. C.G. Gonzalez, R.A. Wessel, and S.A. Padlewski, Epoxy-based aqueous-processable photodielectric dry film and conductive via plug for PCB build-up and IC packaging, Proc. 48th Electronic Components and Technology Conference, Seattle, Washington, USA, 1998, pp. 138–143.
34. G.G. Harman, Wirebonding—towards 6σ yield and fine pitch, Proc. 42nd Electronic Components and Technology Conference, San Diego, California, USA, 1992, pp. 903–910.
35. G.G. Harman, Wire bonding to multichip modules and other soft substrates, Proc 1999 International Conference and Exhibition on Multichip Modules, Denver, Colorado, USA, 1995, pp. 292–301.
36. G.G. Harman, Wire Bonding in Microelectronics: Materials Processes, Reliability and Yield, McGraw-Hill, New York, USA, 1997.
37. G.G. Harman and C.A. Canon, The microelectronic wire bond pull test, how to use it, how to abuse it, IEEE Trans. Components, Hybrids and Manufacturing Technology, CHMT-1(3), pp. 203–210 (1978).
38. G. Heinen, R.J. Stierman, D. Edwards, and L. Nye, Wire bond over active circuits, Proc. 44th Electronic Components and Technology Conference (ECTC), Washington, DC, 1994, pp. 922–928.
39. J. Hirota, K. Machinda, T. Okuda, M. Shimotomai, and R. Kawanaka, The development of copper wire-bonding for plastic molded semiconductor packages, Proc. 35th IEEE Electronics Component Conference, Washington, DC, 1985, pp. 116–121.
40. C. Horsting, Purple plaque and gold purity, 10th Annual Proc. IRPS, Las Vegas, Nevada, 5–7 April 1972, pp. 155–158.
41. S. Ito, M. Kuwamura, S. Akizuki, K. Ikemura, T. Fukushima, and S. Sudo, Solid type cavity fill and underfill materials for new IC packaging applications, Proc. 45th IEEE Electronic Components and Technology Conference, Las Vegas, Nevada, USA, 1995.
42. V.P. Jaecklin, Room temperature ball bonding using high ultrasonic frequencies, Proc. Semicon: Test, Assembly and Packaging, Singapore, 1995, pp. 208–214.
43. J.L. Jellison, Effect of surface contamination on the thermocompression bondability of gold, IEEE Trans. Parts, Hybrids and Packaging, PHP-11, pp. 206–211 (1975).
44. J.L. Jellison and J.A. Wagner, Role of surface contaminates in the deformation welding of gold to thick and thin films, Proc. 29th IEEE Electronic Components Conferences, 1979, pp. 336–345.
45. C.N. Johnston, R.A. Susko, J.V. Siciliano, and R.J. Murcko, Temperature dependent wear-out mechanism for aluminum/copper wire bonds, Proc. International Microelectronics Symposium, Orlando, Florida, 1991, pp. 292–296.
46. J.S. Kilby, Invention of the integrated circuit, IEEE Trans. Electronic Devices, ED-23, pp. 648–654 (1976).
47. H.P. Klein, U. Durmutz, H. Pauthner, and H. Rohrich, Aluminum bond pad requirements for reliable wire bonds, Proc. IEEE Int. Symposium on Physics and Failure Analysis of ICs, Singapore, 1989, pp. 44–49.
48. J. Kurtz, D. Cousens, and M. Defour, Copper wire ball bonding, Proc. Int. Electronic Packaging Society Conference, New Orleans, Louisiana, USA, 1984, pp. 1–5.
49. B. Langenecker, Effects of ultrasound on deformation characteristics of metals, IEEE Transactions on Sonics and Ultrasonics, SU-13, pp. 1–8 (1966).
50. L.M. Levinson, C. Eichelberger, W. Wognarowski, and R.O. Carlson, High-density interconnect using laser lithography, Proc. International Symposium on Microelectronics, Seattle, Washington, October 17–19, 1988, pp. 301-306.
51. J. Ling and C.E. Albright, The influence of atmospheric contamination in copper to copper ultrasonic welding, Proc. 34th Electronic Components Conference, New Orleans, Louisiana, USA, 1984, pp. 209–218.
52. D. Liu, C. Zhang, J. Graves, and T. Kegresse, Laser direct-write (LDW) technology and its applications in low temperature co-fired ceramic (LTTC) electronics, Proc. 2003 International Symposium on Microelectronics, Boston, Massachusetts, Nov. 18–20, 2003, pp. 298–303.

53. G. Lo and Sitaraman, G-Helix: Lithography-based, wafer-level compliant chip-to-substrate interconnect, Proc. 54th Electronic Components and Technology Conference, Las Vegas, Nevada, June 1–4, 2004, pp. 320–325.
54. C. Meisser, Bonding techniques for plastic MCMs, Semiconductor International 14, pp. 120–124 (1991).
55. Microbonds, Inc., 151 Amber Street, Unit 1 Markham, Ontario, Canada L3R3B3, www.microbonds.com.
56. L.F. Miller, Controlled collapse reflow chip joining, IBM J. Res. Dev., 13, pp. 239–250 (1969).
57. G.E. Moore, VLSI: some fundamental challenges, IEEE Spectrum, 16(4), pp. 30–37 (1979).
58. J.L. Newsome, R.G. Oswald, and W.R. Rodregues de Miranda, Metallurgical aspects of aluminum wirebonds to gold metallization, 14th Annual Proceedings Reliability Physics, 1976, pp. 63–74.
59. H. Onoda, K. Itashimoto, and K. Touchi, Analysis of electromigration-induced failures on high temperature sputtered Al-alloy metallization, J. Vacuum Science Technology, A(13), pp. 1546–1555 (1995).
60. J. Onuki, M. Suwa, T. Iizuka, and S. Okikawa, Study of aluminum ball bonding for semiconductors, Proc. 34th Electronic Components Conference, New Orleans, Louisiana, USA, 1984, pp. 7–12.
61. K. Otsuka and T. Tamutsa, Ultrasonic wire bonding technology for custom LSIC with large number of pins, Proc. 31st IEEE Electronic Components Conference, Atlanta, Georgia, USA, 1981, pp. 350–355.
62. J.M. Petek and H.K. Charles, Jr., Known good die, die replacement (rework) and their influences on multichip module costs, Proc. IEEE 48th Electronic Components and Technology Conference, Seattle, Washington, USA, 1998, pp. 909–915.
63. E. Philofsky, Intermetallic formation in gold-aluminum systems, Solid State Electronics, 13(10), pp. 1391–1399 (1970).
64. J.B. Prather, S.D. Robertson, and J.W. Slemmons, Aluminum wire bonding to gold thick-film conductors, Electronic Packaging and Production, (May), pp. 68–71 (1974).
65. K.J. Puttlitz, An overview of flip chip replacement technology on MLC multichip modules, Proc. International Electronic Packaging Conference, 1991, pp. 909–928.
66. T.H. Ramsey and C. Alfaro, The effect of ultrasonic frequency on intermetallic reactivity of Au-Al bonds, Solid State Technology, 34, pp. 37–38 (1991).
67. K.V. Raui and E.M. Philofsky, Reliability improvement of wire bonds subjected to fatigue stresses, Proc. 10th IEEE Reliability Physics Symposium, Las Vegas, Nevada, USA, 1972, pp. 143–149.
68. J. Riddle, High cycle fatigue (ultrasonic) not corrosion in fine microelectronic bonding wire, Proc. 3rd ASM Conference on Electronics Packaging, Materials, Processes, and Corrosion in Microelectronics, Minneapolis, Minnesota, 1987, pp. 185–191.
69. B.M. Romensko, H.K. Charles, Jr., G.V. Clatterbaugh, and J.A. Weiner, Thick-film bondability: geometrical and morphological influences, The Int. J. for Hybrid Microelectronics, 8, pp. 408–419 (1985).
70. B.M. Romenesko, H.K. Charles, Jr., J.A. Cristion, and B.K. Sui, Gold-aluminum wirebond inteface testing using laser-induced ultrasonic energy, Proc. 50th Electronic Components and Technology Conference, Las Vegas, Nevada, 2000, pp. 706–710.
71. R.R. Schaller, Moore's law: past, present, and future, IEEE Spectrum, 34(6), pp. 53–59 (1997).
72. Y. Shirai, K. Otsuka, T. Araki, I. Seki, K. Kikuchi, N. Fujita, and T. Miwa, High reliability wire bonding technology by the 120 kHz frequency of ultrasonic, Proc. 1993 International Conference on Multichip Modules, Denver, Colorado, 1993, pp. 366–375.
73. T.H. Spencer, Thermocompression bond kinetics—the four principle variables, International Journal for Hybrid Microelectronics, 5(2), pp. 404–409 (1982).
74. T. Takahashi, E.W. Rutter, Jr., E.S. Moyer, R.F. Harris, D.C. Frye, V.L. St. Joor, and F.L. Oakes, A photodefinable benzocyclobutene resin for thin-film microelectronic applications, Proc. Int. Microelectronics Conference, Yokohama, Japan, 1992, pp. 64–70.
75. K. Takeda, M. Ohmasa, N. Kurosu, and J. Hosaka, Ultrasonic wirebonding using gold plated wire onto flexible printed circuit board, Proc. 1994 International Microelectronics Conference, Oamya, Japan, 1994, pp. 173–177.
76. A.A.O. Tay, K.S. Yeo, and J.H. Wu, The effect of wirebond geometry and die setting on wire sweep, IEEE Trans. on Components, Packaging and Manufacturing Technology, Part B, 18(1), pp. 201–209 (1995).
77. A. Thomas and H.M. Berg, Micro-corrosion of Al-Cu bonding pads, Proc. 23rd IEEE Reliability Physics Symposium, Orlando, Florida, USA, 1985, pp. 153–158.
78. J. Tsujino, T. Mori, and K. Hasegawa, Characteristics of ultrasonic wire bonding using high frequency and complex vibration systems, Proc. 25th Annual Ultrasonic Industry Association Meeting, Columbus, Ohio, 1994, pp. 17–18.
79. D.B. Tuckerman, D.J. Ashkenas, E. Schmidt, and C. Smith, Die attach and interconnection technology for hybrid WSI, 1986 Laser Pantography States Report UCAR-10195, Lawrence Livermore Laboratories, 1986.

80. R.R. Tummula, E.J. Rymazewski, and A.G. Klopfenstein, Microelectronics Packaging Handbook, Vols. I, II, & III, Chapman Hall, NY, USA, 1997.
81. S. Wakabayashi, A. Murata, and N. Wakobauashi, Effects of grain refinement in gold deposits on aluminum wire-bond reliability, Plating and Surface Finishing, V, pp. 63–68 (1981).
82. J.A. Weiner, G.V. Clatterbaugh, H.K. Charles, Jr., and B.M. Romenesko, Gold ball bond shear strengths effects of cleaning, metallization and bonding parameters, Proc. IEEE 33rd Electronic Components Conference, Orlando, Florida, USA, 1983, pp. 208–220.
83. The Welding Handbook, Vol. 2, eighth edition, Ultrasonic Welding, 1991, pp. 784–812.
84. Y.F. Yao, T.Y. Lin, and K.H. Chua, Improving the deflection of wirebonds in stacked chip scale packages CSP, Proc. 53rd Electronic Components and Technology Conference, New Orleans, LA, 2003, pp. 1359–1363.

4

Metallurgical Interconnections for Extreme High and Low Temperature Environments[*]

George G. Harman

Semiconductor Electronics Division, National Institute of Standards and Technology, Gaithersburg, MD 20899-8120, USA

Abstract The material properties and requirements for wire bond and flip chip interconnections that can be used in packaging chips for extreme high and low temperature environments [from +460°C (*HTE*) down to −200°C (*LTE*)] are described. The most commonly used Au–Al wire bonds should be avoided in the *HTE* range, along with any other metallurgical interfaces that form brittle intermetallics and/or Kirkendall voids. Gold–gold bonds improve with time and temperature. Thus, a clear preference is given for gold (or other noble metals) in the *HTE* environment for both wire and flip–chip bonds. For *LTE* and intermediate temperature ranges, such as on Mars and most earth satellites, conventional interconnections (Au and Al wire bonds) to Al chip metalization (bond pads) are acceptable. Also, normal flip-chip solder bumps are acceptable, but without plastic underfill. Information and techniques for using extreme temperature range materials, such as coefficient of thermal expansion (CTE) matching between chip and substrate, high temperature polymers, etc., are presented. Unusual failure mechanisms, such as possible electromigration of wire interconnections in *HTE*, are described. It is concluded that, with proper selection of materials, interconnections can be reliable in both extreme environments.

4.1. INTRODUCTION

The materials and requirements for wire bond and flip chip interconnections that can be used in packaging chips for extreme high and low temperature environments [from +460°C (*HTE*) down to −200°C (*LTE*)] are described. Devices capable of operating in these environments are needed for future space probes to other solar system planets, well-logging, geothermal measurements, sensors near rocket and jet engines, and, to a lesser

---

[*]This work is expanded from an invited presentation at the "Workshop on Extreme Environments Technologies for Space Exploration," Sponsored by the Jet Propulsion Laboratory in May, 2003. Pasadena, CA.

FIGURE 4.1. Pictorial representation of the Solar system planetary temperature ranges for which space probes must be designed in order to survive. [Note: The average nighttime temperature of I_o is about $-170°C$, but it has hot spots over $1000°C$.] (Picture, courtesy J. Patel, NASA/JPL, as modified by R. Kirschman.)

extent, on and in internal combustion engines. The normal metallurgy used for interconnections as well as that on chips (wire bonds, flip chips, aluminum metalization) fails in *HTE*, and may have special requirements to be reliable in *LTE*. The required material changes, such as gold–gold interfaces, substrate/chip CTE matching, avoiding underfill, etc., are necessary for systems to survive in these thermal and changing temperature environments. Some of the concepts and materials below have been described in the literature [1]. For an overview of the Solar system planetary temperature ranges that may be encountered by future space probes, see Figure 4.1.

4.2. HIGH TEMPERATURE INTERCONNECTIONS REQUIREMENTS

4.2.1. Wire Bonding

The most commonly used Au–Al wire bonds have been investigated many times for possible high temperature use, with limited success. Such temperature exposure results in the formation of brittle intermetallic compounds and Kirkendall voids, both resulting in interconnection failure. The most successful approach to limiting such failure was to introduce hydrogen into the sealed package environment [2]. The result of this work is illustrated in Figure 4.2.

This approach, however, has not been used because of possible hermetic leakage of the hydrogen and occasional problems of H_2 effecting device performance. The quoted tests were only run for 1 h at 400°C. Therefore, more work should be performed to ascertain any advantages of H_2 in long term *HTE*. Teverovsky [3] studied the effects of high temperature degradation (in a vacuum for space application) of Au–Al bonds in plastic encapsulated devices. A mean lifetime of about 700 h at 225°C was obtained. This was better than in air, but not acceptable for most *HTE* usage. Wire manufacturers are currently developing doped Au bonding wires with ≥100 ppm of added (proprietary) stabilizing impurities (vs normal <10 ppm). Some have as much as 1%, usually Pd, added. Such wires demonstrate increased bond intermetallic lifetimes [4]. However, as above, preliminary data suggest that this will not be adequate for *HTE*. Thus, for the present, Au–Al interfaces should be avoided in the *HTE* range, along with any other metallurgical interfaces that

METALLURGICAL INTERCONNECTIONS FOR EXTREME TEMPERATURE ENVIRONMENTS

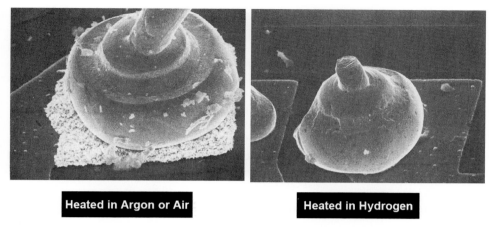

FIGURE 4.2. An example of using H_2 atmosphere to inhibit Au–Al intermetallic compound formation for Au–ball bonds on Al metalization (Heated 400°C for 1 h.) Left photograph bond(s), tested in air, N_2, or Ar (they are equivalent). Note the extensive intermetallic compound surrounding the ball bond. The right figure was tested in an H_2 ambient at 400°C for 1 h also. No intermetallic is evident. Both wires are 25 μm diameter [2].

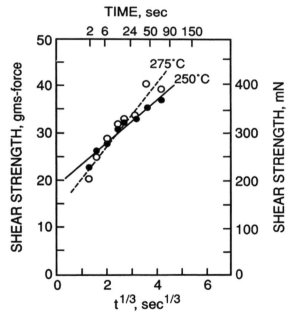

FIGURE 4.3. Improvement in the ball shear strength of organic-contaminated Au–Au wire bonded interfaces with time and temperature. Au–Au bonds are not degraded by high temperatures, as are Au–Al interfaces. After Jellison [5].

form brittle intermetallics and/or Kirkendall voids. In contrast to Au–Al interfaces, Au–Au bonds have long been known to improve with time and temperature [5], Figure 4.3.

Recently, Benoit et al. [6] have exposed Au–Au interfaces to 350°C for up to 300 h, without interface failure. Therefore, a clear preference is given for high melting point monometallic welds of metals or ones that form solid solutions (e.g., gold to other noble metals, etc.) in the *HTE* environment for both wire and flip-chip bonds. Also consider the Al–Ni interface which has withstood equivalent high temperatures (Section 4.2.1.1) and should be useful, especially for large diameter Al wire bonds on Ni pads, as used on power devices. Tri-metallic diffusion barriers have long been used to prevent interdiffusion on hybrid packages. Examples might consist of a Mo, or Kovar base, which has a top layer of deposited Al and on the other side, a layer of Au. An Al wire would be bonded to the top (Al side), and the bottom would be thermocompression (TC) bonded (or welded) to thick film Au on the substrate. The Mo/Kovar serves as an effective diffusion barrier, preventing Au–Al intermetallic and voiding from forming. Traditionally, these have been called molytabs, but other names have been used, as well.

4.2.1.1. Effect of Wire Annealing in HTE   The primary concern in *HTE* is that the bonded/welded interfaces remain strong, even if the wire is annealed and mechanically weaker. *HTE* will result in fully annealing small diameter Al wires, causing their breaking load, and to an extent the bond pull strength, to decrease to less than 30% of initial values. Such annealing does not cause a reliability problem in hermetic (open cavity) packages unless the heel of a wedge bond is overly thinned and subject to temperature cycling (see Section 4.2.1.3). Similar complete annealing occurred in billions of Al wire bonds to Al pads in CERDIP (ceramic-glass) sealed packages (processed at $\geq 400°C$ for 30 min) without reliability problems. Long periods at high temperatures may possibly change the structure of Al wire, weakening it. However, the weakest, the bamboo structure, has only been reported for wire bonds once in the literature [7] and may not occur in current production wires. (That structure is more associated with electromigration failure, see Section 4.2.1.4.)

Unless bond-lifts are experienced, the pull test is not a good indication of wire strength or reliability in *HTE*, since annealing increases the wire elongation during the test, resulting in some compensation for the breaking load decrease. (The pull test is described by a resolution of forces equation [1].) As the loop height increases due to elongation during the test, the force on the bond heel decreases. This is especially apparent in annealed Al wires which can elongate from 10% to 30%. Aluminum–aluminum or Al–Ni *bond interfaces* [6] will remain strong at high temperatures, (350°C for 300 h). Thus, any annealing of the wire (tensile strength decrease) is not considered a reliability problem (see CERDIP above). Gold wire for ball bonding is stabilized and annealed, and its strength and elongation will change less at high temperature than for Al. However, the strength of small diameter gold wires used for wedge bonding will anneal (soften), but less than that of equivalent Al wire. The many different proprietary dopants and quantities used in Au wire (e.g., Be, Ca, Pd, Pb, rare earths, etc. and in amounts ranging from 10 ppm up to 1000 ppm) make it difficult to predict the exact degree of softening that will occur at high temperatures, and such data will vary among manufacturers.

Large diameter Al wire (≥ 100 μm) is normally annealed (to varying degrees) when produced, so its breaking load will decrease minimally in *HTE*. The interface reliability is the same as for small diameter wires/pads.

4.2.1.2. Other Noble Metal Wire Bond Characteristics for HTE   Platinum [8] and Pd wires are occasionally used in *HTE* and *LTE* as well. They can be used in long term *HTE* (annealing has minimal effect on their strength, compared to Al). Both have high resistivity ($\sim 5\times$ Au and Al) and low thermal conductivity ($\sim 25\%$ of Au and Al). As such, both wires have been used to thermally isolate chips that may run much colder or hotter than their environment (e.g., superconducting devices—SQUIDS, etc.). However, both Pt and

Pd are harder than Au and Al. Thus, they require more ultrasonic energy (US) and force to bond according to the simplified formula [9], $E \propto (H)^{3/2}$, where E is the ultrasonic energy and H is the metal hardness. The nanohardness of these three metals is, Au = 1.77 GPa, Pt = 3.55 GPa, and Pd = 2.87 GPa. Thus, Pt takes ~2.8 times and Pd ~2 times as much ultrasonic power as Au to wedge bond (cold US welding), everything else being constant. Since US energy is a major cause of chip fracturing damage [1], both metal wires will tend to crater semiconductors during bonding, but are acceptable for bonding metalization on ceramic substrates. (SiC chips are as hard and have as high a fracture-toughness as most ceramics, and therefore should be safely bondable with these metals, see Table 4.1, below.) Also, high temperature thermocompression bonding (>300°C) to Si, GaAs, etc. can be achieved, without damage, in most cases. However, the bonding process may weaken the thin-film bond-pad adhesion to the chip, which may then be subject to delamination at high temperatures. This must be qualified for the anticipated temperature/bonding stress.

4.2.1.3. Wire Bond Fatigue in Temperature Cycling Environments Metallurgical fatigue damage to wires can occur during large ΔT temperature/power-cycling in both *HTE* and *LTE* (e.g., system [chip] temperatures on Mars' surface can range from −120°C to 85°C and on Jupiter, possibly from −140°C to >380°C). On Earth, well logging and various sensor applications can be cycled through wide temperature ranges as well. Only minimal and limited work has been done to determine the effects of high temperature cycling on 25 μm diameter Au and Al (1% Si) wire fatigue. Deyhim [10] studied fatigue (cycles-to-failure: C-F) on wires of both materials at temperatures up to 125°C, using various strain rates (0.7%, 5% and 10%). The first is considered more similar to temperature cycling of wires in hermetic packages. In that study, on one type of Au wire (unspecified dopants), the C-F decreased from ~5000 25°C to ~600 at 125°C. However, for Au wire made by a different manufacturer, the C-F was ~30,000 at room temperature, and there was only a minimal decline at the high temperature ranges. No data were obtained at temperatures higher than 125°C in this test, and no test methods are available that are capable of measuring such wire fatigue at *HTE* temperatures. Benoit et al. [6] studied the fatigue of wires (at room temperature and low strain rates) after annealing at 300°C and found that all wires failed more rapidly after annealing for longer times. Thus, if a planetary probe (or other *HTE* application) will experience high temperature cycling, then any gold wires to be used must be qualified for fatigue resistance in the design cycle. This is essential for Au wires, since manufacturers frequently change their proprietary, stabilizing impurities. Fine Al wire has changed minimally in recent years, so older data should suffice.

Small diameter wires (e.g., 25 μm) are flexible and stress is mostly applied at the bond heel, and they typically break there during temperature cycling. A high loop height minimizes the stress on the heel and can prevent or limit such damage, (see Figure 4.4). A variety of factors (wire diameter, shape, loop height, metallurgy, and strain-rate) determine the fatigue susceptibility and life of a wire bond [1]. Also, any cracks or kinks in the heel/neck, or sharp bends in loops, such as can be made in worked-loop formation by modern autobonders, may increase fatigue damage. This has been minimally studied.

Large diameter Al wires are stiff, and stress is concentrated at all bends and at the weld attachments (bond heels); failures can occur at these positions. An example of large diameter Al wire fatigue is shown in Figure 4.5(a). Low bond deformation [see Figure 4.5(b)] and uniform looping gives the best fatigue protection for large diameter wires. Other possible solutions are to use Mg doped wire (0.5–1%) [7] and to make all bends smooth.

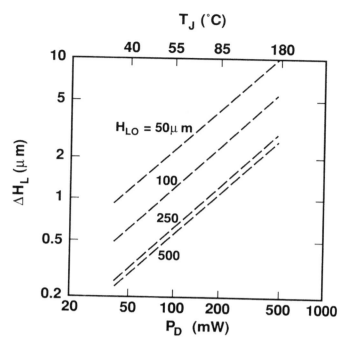

FIGURE 4.4. Calculation of wire bond flexing due to power cycling of a transistor. Top scale is the junction temperature with input power of P_D (mW). The maximum flexing of a semicircular wire bond loop ΔH_L is given on the Y axis. The as-made loop height, H_{LO}, is given by each curve. Minimum flexing occurs with highest loop height. (Analysis was made for a 25 μm diameter Al wire having a 1 mm bond to bond length.)

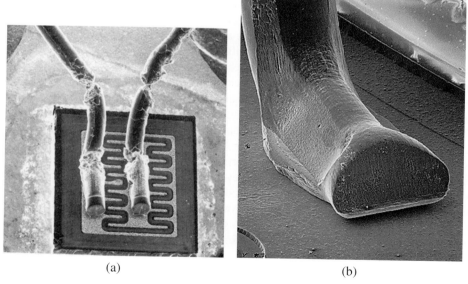

FIGURE 4.5. (a) Fatigue of a large diameter (200 μm), 99.99% pure, Al bonding wire resulting from power cycling [1] from 25°C to 180°C. These underwent 18,606 temperature-power cycles from 25°C to 125°C in a telemetry application. (b) The desired bond shape, to minimize heel fatigue, of a 250 μm diameter Al wire (courtesy of Orthodyne Electronics).

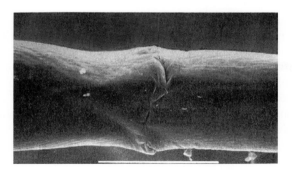

FIGURE 4.6. A SEM photograph of an electromigration-induced bamboo structure in a 25 μm Al wire with a "knuckle" perpendicular to the high current flow [13]. The bar at the bottom is ∼25 μm.

4.2.1.4. Current Carrying Capacity of Conductors at High Temperature The burnout current for various wire diameters and compositions has been discussed in many papers, where measurements/calculations were obtained in room temperature ambients [11,12]. For *HTE*, but not for *LTE*, the current carrying capacity of these wires must be appropriately derated to avoid burnout at low current densities. Aluminum wire, with its 660°C melting point, must be derated more than Au (melting point 1064°C). Tse and Lach [13] found electromigration failures in bonded Al wires exposed to high current density and temperature over several years, and this mechanism could certainly contribute to *HTE* failures. As the ambient increases, this failure mode should increase according to Black's equation [14], ($t_{50} \propto J^{-n} e^{E/T}$, where t_{50} is the median lifetime, J is the current density, T is the absolute temperature, n is ∼2, and E is the activation energy, which varies from 0.5 eV to 0.7 eV for Al, and ∼0.8 eV to 0.9 eV for Au). Aluminum wires will fail more rapidly than Au, Pt, or Pd, by this mechanism. However, values of E for the latter two are not available at this time. Aluminum wires may also develop the weak bamboo structure, as in [13], where knuckle-like joints develop perpendicular to the current flow (see Figure 4.6). These would be very susceptible to ΔT fatigue (Section 4.2.1.3). The Author notes that electromigration is well documented in thin films; however, [13] is the only reported observation of electromigration in bonding wires, and this effect needs to be further studied. It also needs to be verified in contemporary Au bonding wire, since those are the most logical choice for interconnections in extreme *HTE*.

The final wire failure will result from the combination of increasing wire resistance due to the increasing resistivity with ambient (*HTE*) temperature, the wire self-heating due to $I^2 R$ heating, as well as any electromigration that may occur. This affects all of the discussed wires (Au, Pt, Pd, and Al), but Al, with its low melting point, will be the most affected. Thus the potential for wire failure will result in greatly reducing the rated current density of wires in both power and small-wire devices when used in *HTE*.

4.2.2. The Use of Flip Chips in HTE

Efforts have been made to evaluate traditional soft-solder bump flip chips for use in higher temperature environments [15]. Obviously, if the melting point of the solder is approached, this technology will not be viable. *(Hard-solder flip-chip bumps will strain and crack the chips and thus cannot be used.)* If underfill is required, due to chip-substrate CTE mismatch, this will also limit such use, since underfill is not usable above its glass transition temperature, (T_g). The low temperature material properties of underfill are not available and thus may rule out its use in extreme low temperatures, *LTE*. Normal plastic substrates/boards, such as FR-4, BT, etc., cannot be used in *HTE* or *LTE*. New high temperature plastics may be used for boards in the future, but more development is needed (see Section 4.5 below). Gold ball bumps can be used in place of the normal solder-ball flip-chip interconnections. Chips can be ball-bumped or stud-bumped with gold and then thermocompression/thermosonic bonded to the gold metalization on ceramic substrates for reliable *HTE* flip chip interconnections, Figure 4.7.

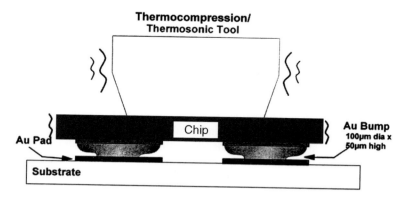

FIGURE 4.7. A gold ball-bumped flip chip that is being TC or TS bonded to gold metalization on a ceramic substrate. Note that these are usually used for low leadcount devices, which is typical of ones used for *HTE*. (Courtesy of Karl Puttlitz.)

TABLE 4.1.
Some material properties of chips and substrates for use in both *HTE* and *LTE*. Components should be chosen to minimize expansion differences in no-underfill flip-chips. Normal letters/numbers are approximate matches for equivalent chip and substrates and large, <u>*bold-underlined*</u> for the others. Generally, for flip chips, the substrate should have slightly higher CTE than the chip (which can heat up). But, for face-up die attach, a more exact match is desirable.

Component	Material	CTE (ppm/°C)	Thermal cond. (W/cm °C)	Fracture toughness MPa m$^{1/2}$
	Si	2.6	1.57	0.83–0.95
	SiC (β)	1–3 (T-dep.)	~5 (>T-dep.)	2.8
	GaN	~3	1.3	0.8
	<u>*GaAs*</u>	<u>*5.7*</u>	<u>*0.48*</u>	<u>*~0.5*</u>
Substrate	SiC	1–3	0.8–2	2.5–4.5
	AlN	4.6	1.75	2.8
	Si$_3$N$_4$	2.8–3.6	0.3–0.4	5.0–6.1
	<u>*Al$_2$O$_3$*</u>	<u>*~6*</u>	<u>*0.35*</u>	<u>*3.1–3.3*</u>
	<u>*BeO*</u>	<u>*~6–7*</u>	<u>*~3*</u>	<u>*3.7*</u>

[Note: Thermal-conductivity values can/will change at low temperatures, and exact data may not be available in the literature. Values can change for ceramic/polycrystalline substrates depending on temperature range, preparation, impurities, etc. and for single crystals, the orientation. Such values decrease in high temperatures.] Data from SRC-CINDAS and NIST Database. Values are rounded and should be used as a guide only. Obtain exact data from manufacturer or other specific measurements.

METALLURGICAL INTERCONNECTIONS FOR EXTREME TEMPERATURE ENVIRONMENTS

If extreme temperature cycling is expected, then a double or triple height ball-bumped structure can be applied at each bond pad to achieve higher chip standoff and minimize stress in the chip and substrate. For both extreme environments, substrate-and-chip CTE matching is essential for flip chips, if temperature cycling is expected (and it usually does occur). See Table 4.1 for some material property choices in designing packages for HTE flip chips. Components should be chosen to minimize the *expansion differences* in no-underfill flip-chips. (Ideally the substrate should have slightly higher CTE than the chip and substrate, since the chip is electrically heated.) However, for face-up chip die attach in *HTE*, one must also consider the thermal conductivity of both chip and substrate. High CTE substrates tend to equalize the temperature difference between the chip; thus a closer match in CTEs' is needed than for flip chip devices to be reliable in large temperature change environments as in Table 4.1.

4.2.3. General Overview of Metallurgical Interfaces for Both HTE and LTE

An overview of metallurgical interface (bonding) reliability for both *HTE* and *LTE* is shown in Figure 4.8. Dark lines indicate the interfaces appropriate for *HTE*. All interfaces, except for Al–Ag, are considered acceptable for *LTE* applications.

4.3. LOW TEMPERATURE ENVIRONMENT INTERCONNECTION REQUIREMENTS

Although a larger portion of this chapter is devoted to problems of interconnections in the *HTE* environment, most of the solar system is cold. This is readily seen in Figure 4.1. The problems of maintaining good chip and package interconnections is relatively simple when compared to those of *HTE*. (Other problems, such as battery performance, operation of some semiconductor devices, resistors, plastic delamination, etc., are of major impor-

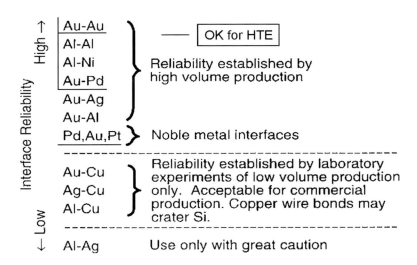

FIGURE 4.8. Metallurgical interfaces that can be used reliably in both extreme environments. All are acceptable for *LTE*, including soft solder based flip chips. However, only the dark underlined, boxed ones are acceptable for *HTE*.

tance but beyond the scope of this work.) Nevertheless, extreme low temperature interconnections do present challenges. Some of these are further discussed along with appropriate HTE problems or in combined sections such as in Section 4.4.

For use in *LTE* on the outer planets as well as for intermediate temperature ranges, such as on Mars, conventional Si or SiGe based devices can be used with normal Al metalization. Interconnections can be made with Au and Al wire bonds. Also, normal flip-chip solder bumps are acceptable on ceramic substrates, but without plastic underfill. The plastic-to-chip CTE mismatch at extreme low temperatures, could fracture the chip. (Future underfill development may improve this.) One can use normal epoxy glass-laminated substrates (e.g., BGAs), but ceramic is preferred. For both extreme environments, but especially for *HTE*, ceramic substrate-and-chip CTE matching is essential for flip chips, if temperature cycling is expected (and it usually does occur). See Table 4.1 for some material property choices for various chips with different ceramic substrates and packages.

4.4. CORROSION AND OTHER PROBLEMS IN BOTH *HTE*, AND *LTE*

There is no *liquid* H$_2$O at the temperatures of either extreme environment (Mars is intermediate, water condenses and materials may corrode); therefore, electro/chemical corrosion of interconnections and metalization is less likely. However, all such devices are built, qualified, and stored in normal Earth environments where water (moisture) and ionic impurities are plentiful and corrosive processes could initiate. Chips must be packaged using the best high-reliability, hermetic procedures. *HTE* chips would normally have gold/noble metalization, and, while electric-field-driven metal migration and high temperature gaseous attack is possible for these metals, it rarely occurs without *liquid* water.

Devices intended for *LTE* operation typically use normal aluminum-metalization on the chips and are subject to normal corrosion. These devices may be cycled through the liquid water (corrosion) range at various times during the device/system life. For Mars, the effective temperature range of semiconductor devices can vary from −120°C to 85°C, which includes self heating. Thus, appropriate low-moisture hermetic precautions are required. [*For comparison, Earth's south-polar-region temperatures range from approximately −80°C to −20°C and, in-situ temperature cycling would result in neither water based corrosion nor significant fatigue.*]

The packages for *HTE* will usually be made of metal/glass/ceramic (classical thick film hermetic hybrid packages). Reliability problems from hysteresis, creep, and/or cracking of normal glass-metal seals (Kovar-glass) in hermetic packages can cause failure during temperature cycling in both *HTE* and especially in *LTE* (see Kohl [16]). The glass-metal seals undergo expansion/contraction hysteresis resulting in cracking or delamination in the range starting about −100°C. Metal leads extending through *HTE*-softened glass seals will yield under stress in the *HTE* environment. Thus ceramic-metal seals should be used for both extreme environments.

Diffusion processes generally follow the Arrhenius equation ($K = Ae^{-E/RT}$, where K is the rate constant, E the activation energy, T is the absolute temperature, and R is the molar gas constant). Thus, diffusion will be greatly accelerated in *HTE*. Possible failure of metal adhesion layers between gold metalization and ceramic, as well as any diffusion barriers under/over the chip metalization must be considered as potential reliability hazards as well. However, the reverse is true in *LTE*, and various detrimental diffusion reactions such as Au–Al intermetallic compounds will not form or measurably increase in thickness.

Chips used for *HTE* will not be conventional Si based. Most likely they will be SiC or possibly GaP, GaAs, etc. Their metalization will most likely consist of complex layers with noble metals and diffusion barriers between the chip and the conductors. Interdiffusion of these layers can lead to adhesion failure [8] and degradation of the device as with the substrate metalization problems, above. The possibility of electromigration of the chip intraconnecting metalization as well as the wire bonds must be considered, as in Section 4.2.1.4. The die attach will have to be metallurgical rather than polymer/epoxy, as currently used on normal earth-bound devices. In some intermediate temperature ranges, silver-glass (max $T \sim 350°C$) [6] my be a satisfactory die attach material. For SiC chips, other possibilities could be Ni/Ti/TaSi$_2$, Ti/TaSi$_2$/Pt, Ni/Ti/Pt/Au, or Au. Currently, neither these nor any other die attach materials have been qualified for long term *HTE* at 460°C. Further investigations must be carried out before such missions take place.

4.5. THE POTENTIAL USE OF HIGH TEMPERATURE POLYMERS IN HTE

Some new high temperature polymers, such as Lo-k materials developed for advanced chips with copper metalization, may be considered for insulators and circuit boards in future extreme environments in cases where non-ceramic boards for interconnections are needed. (See Table 4.2, below.) Some of these materials can be used at temperatures above 400°C. Many of the problems of circuits on polymers, such as metal adhesion at high temperatures, have been solved by the semiconductor industry, but further development is required for specific *HTE* uses. Recently, the high temperature organic materials listed in Table 4.2 [17] have not been acceptable for semiconductor low-k incorporation because of processing or manufacturing compatibility reasons (e.g., SiLK, HOSP, FLARE). If these are available, however, they could still be useful in *HTE*. Recent developments continue in the field of Lo-k, and high temperature materials. For example, high modulus carbon substituted borazine polymers [18] may be appropriate for *HTE* circuit boards. Their temperature characteristics would be similar to the polymers in Table 4.2. Research in other

TABLE 4.2.
Low dielectric constant materials and the maximum operating temperature of ones possibly useful for *HTE* circuit boards or other *HTE* polymer/insulator uses. Weight losses at the indicated temperatures are available from manufacturers, who also supplied the quoted data (right hand column) (Ref. [17]).

High temperature, low dielectric constant, insulator materials						
Material	Max temp (°C)	Modulus (GPa) 25°C	Hardness (GPa)	Fracture-tough (MPa m$^{1/2}$)[c]	CTE[b] (10^{-6}/°C)	References/ sources
DVS-BCB	375	2.9	0.37	0.37	52	Dow Chem.
SiLK-H	450	2.45	0.31	0.6+ to 0.42	62	Dow Chem.
FLARE	>350	2.5	0.35	—	≈60	Honeywell (Allied Signal)
Parylene AF-4	450	2.28	—	—	30–80	Union Carbide

[a] Trade names are used to describe a material when no other identifier is available. This does not imply any endorsement.
[b] CTEs of organic LoK materials generally increase with temperature. Reported values are average and in the range of 25 EC to 100 EC.
[c] Fracture toughness of "Material" interface with SiO$_2$, SiN, Ta, or TaN.
—Development is dynamic and any product above may be improved or discontinued.

high temperature polymers (such as phthalonitriles, cyanate esters, and inorganic-organic hybrid polymers) for operation between 300°C and 510°C in oxidizing atmospheres, also show promise for *HTE* applications [19]. In addition, polyimidebenzoxazole [19] has been studied for such high temperature applications. However, much more development must be accomplished before these can be used.

4.6. CONCLUSIONS

This chapter presented possible materials as well as design considerations for chip interconnections/systems for extreme high temperature and/or low temperature environments. They are needed for future space-craft solar system exploration, well-logging, geothermal measurements, sensors near rocket and jet engines, etc. These cover the range from about −200°C to 460°C. Unusual problems, seldom encountered in normal environments can occur. Examples might be electromigration of interconnection wires and extreme temperature cycling induced fatigue. By using noble metal interconnections and ceramic circuit boards/substrates or possibly new high temperature polymers (still under development), these interconnection needs should be met. Other areas still needing specific development, such as die attach materials for *HTE*, are discussed. Except for these, most requirements are or will be achievable in the near future. For additional overviews of high and low temperature electronics materials, devices, and interconnections, see Kirschman [20] and McCluskey [21].

ACKNOWLEDGMENTS

The author acknowledges valuable discussions/information from R. Kirschman, Extreme-Temperature Electronics, J. Patel, NASA/JPL, and P. McCluskey, UMD. The paper could not have been written without support from NIST Office of Microelectronics Programs and the Semiconductor Electronics Division.

REFERENCES

1. G.G. Harman, Wire Bonding in Microelectronics, Materials, Processes, Reliability, and Yield, Second Edition, McGraw Hill (1997).
2. D.Y. Shih and P.J. Ficalora, The reduction of Au–Al intermetallic formation and electromigration in hydrogen environments, 16th Annual Proc. IEEE Reliability Physics Symposium, San Diego, California, 1978, pp. 268–272 (Figure 4.2, © IEEE, 1978).
3. A. Teverovsky, Effect of vacuum on high-temperature degradation of gold/aluminum wire bonds in pems, 42nd Annual Proc. IEEE International Reliability Physics Symposium, Phoenix, 2004, pp. 547–556.
4. C. Breach, F. Wulff, K. Dittmer, D.R. Calpito, M. Garnier, V. Boillot, and T.C. Wei, Reliability and failure analysis of gold ball bonds in fine and ultra-fine pitch applications, Proc. 2004 Semicon, Singapore, May 4–6, 2004, pp. 1–10.
5. J.L. Jellison, Kinetics of thermocompression bonding to organic contaminated gold surfaces, IEEE Trans. Parts, Hybrids, and Packaging, PHP-13, pp. 132–137 (1977) (Figure 4.3, © IEEE, 1977).
6. J. Benoit, S. Chen, R. Grzybowski, S. Lin, R. Jain, and P. McCluskey, Wire bond metallurgy for high temperature electronics, Proc. 4th Int'l High Temperature Electronics Conference, Albuquerque, NM, June 14–18, 1998, pp. 109–113.
7. K.V. Ravi and E. Philofsky, The structure and mechanical properties of fine diameter Al, 1-pct Si wire, Met. Trans., V2 (March), pp. 711–717 (1972). Also see same authors, Reliability improvement of wire bonds

subjected to fatigue stresses, 10th Annual Proceedings IEEE Reliability Physics Symposium, Las Vegas, Nevada, April 5–7, 1972, pp. 143–149.
8. J.S. Salmon, R.W. Johnson, and M. Palmer, Thick film hybrid packaging techniques for 500°C operation, Proceedings of the 4th International High Temperature Electronics Conference, Albuquerque, NM, June 16–19, 1998, pp. 103–108.
9. Simplified version of an empirical equation derived from US welding, Welding Handbook 8th Edition, Vol. 2, Am. Welding Soc., 1991. The constants of the full equation were developed for the US welding of thick materials, but the hardness relationship is indicative of the values observed in microelectronics.
10. A. Deyhim, B. Yost, M. Lii, and C.-Y. Li, Characterization of the fatigue properties of bonding wires, Proc. 1996 ECTC, Orlando FL, May 28–31, 1996, pp. 836–841.
11. E. Loh, Heat transfer of fine wire fuse, IEEE Trans. CHMT, V-7(Sept.), pp. 264–267 (1984).
12. A. Mertol, Estimation of aluminum and gold bond wire fusing current and fusing time, IEEE Trans. CPMT, Part B, V-18(Feb.), pp. 210–214 (1995).
13. P.K. Tse and T.M. Lach, Aluminum electromigration of 1-mil bond wire in octal inverter integrated circuits, Proc. 45th IEEE ECTC, Las Vegas, NV, 1995, pp. 900–905 (Figure 4.6, © IEEE, 1977).
14. J.R. Black, Electromigration—a brief survey and some recent results, 6th Annual Proc. IEEE Int. Reliability Physics Symposium, Dec. 1968, pp. 338–347.
15. T. Braun, K.F. Becker, M. Koch, V. Bader, R. Aschenbrenner, and H. Reichi, High temperature potential of flip chip assemblies, Intl. High Temperature Electronics Conference, Santa Fe, NM, May 17–20, 2004, pp. TP1-3. See also, Flip chip technology for high temperature automotive applications, 36th International Symposium on Microelectronics, Boston, MA, 2003, pp. 853–858.
16. W.H. Kohl, Materials Technology for Electron Tubes, Reinhold, N.Y., 1951.
17. G.G. Harman and C.E. Johnson, Wire bonding to advanced copper-low-K integrated circuits, the metal/dielectric stacks, and materials considerations, IEEE Trans. CPT, V-25(4), pp. 677–683 (2002).
18. I. Masami, S. Sekiyama, K. Nakamura, S. Shishiguchi, A. Matsuura, T. Takuya, Fukuda, H. Yanazawa, and Y. Uchimaru, Borazine-siloxane organic/inorganic hybrid polymer, Review of Advanced Material Science, V-5, pp. 392–397 (2003).
19. T.M. Keller, D.D. Dominguez, and M. Laskowski, High temperature polymers for geothermal and electronic packaging applications, Intl. High Temperature Electronics Conference, Santa Fe, NM, May 17–20, 2004, pp. TP2. Also see D.A. Dalman and F.F. Hoover, Eds., PBIO film dielectric for advanced microelectronics packaging, ibid, TA2.
20. R. Kirschman, Ed., High Temperature Electronics, IEEE Press, New York, NY, 1999. Also see ibid, Low-Temperature Electronics, 1986, These both present excellent overviews, as well as problems and solutions for extreme environment packaging.
21. F.P. McCluskey, R. Grzybowsky, and T. Podlesak, High Temperature Electronics, CRC Press, New York, 1997. A good recent overview and data on high temperature electronics.

5

Design, Process, and Reliability of Wafer Level Packaging

Zhuqing Zhang[b] and C.P. Wong[a]

[a] *School of Materials Science and Engineering, Georgia Institute of Technology, 771 Ferst Drive, Atlanta, GA 30332-0245, USA*
[b] *Hewlett-Packard Co., 1000 NE Circle Blvd., Corvallis, OR 97330, USA*

Abstract   Wafer level packaging (WLP) has been growing continuously in electronics packaging due to its low cost in batch manufacturing and the potential of enabling wafer test and burn-in. A variety of wafer level packages have been devised, among which four important categories are identified including thin film redistribution and bumping, encapsulated package, compliant interconnect, and wafer level underfill. This chapter reviews the different WLP technologies with an emphasis on challenges and processes of the wafer level underfill. The wafer level packaging integrated with wafer burn-in, test and module assembly shows great attraction due to the dramatic cost reduction. Cost effective ways of building wafer level test and burn-in are under investigation.

5.1. INTRODUCTION

As a result of rapid advances in integrated circuit (IC) fabrication and the growing market for faster, lighter, smaller, yet less expensive electronic products, high performance low cost packaging is needed by the electronics industry. The conventional discrete IC packaging is inefficient, as such, a paradigm shift to wafer level packaging is apparent. Wafer level packaging (WLP) is a packaging technology where most or all of the IC packaging process steps are carried out at the wafer level. In the conventional discrete IC packaging process, the wafers are diced into individual IC chips first and then the chips are redistributed and packaged individually. In the WLP process, redistribution and packaging are performed at the wafer level. After wafer dicing, the individual components are ready to ship and assemble onto the substrates or printed wiring boards (PWBs) by the standard surface mount technology (SMT) process. A comparison of the conventional discrete packaging and the wafer level packaging is illustrated in Figure 5.1. The WLP makes possible 100% silicon efficiency (defined as the ratio of IC area over the entire IC package area) and low packaging cost due to the wafer level batch processing.

There are two major market drivers for wafer level package. In the cost-driven market, the wafer level Chip Scale Packaging (CSP) has the advantage that the cost per device goes down as the wafer size increases and/or the IC size decreases [1]. The wafer level CSPs (WLCSPs) are mainly designed for small dies and low input/output (I/O) devices in consumer product market. Many of the technologies developed for these applications are based on simple peripheral pad redistribution followed by the solder ball attachment. These technologies are finding applications in low I/O counts functions, integrated passives and Rambus™ DRAMs. In this market, the WLP is a technology targeting at lower cost for packaging, and therefore the packaging cost per wafer and the number of chips per wafer (CPW) are the critical measures for success. The transfer to 300 mm wafer fabrication favors the WLP by increasing the number of CPW significantly.

The other market driver of WLP lies in large dies and high I/Os, high performance devices, for which flip-chip is currently the dominant first-level interconnect method. However, the major concern of the flip-chip is the solder joint reliability that is shortened by the thermo-mechanical shear stress due to the coefficient of thermal expansion (CTE) mismatch between the silicon chip and the organic substrate. The use of underfill increases the reliability of the packaging by stress redistribution, but it also increases the cost of flip-chip assembly due to the tedious dispensing and curing process. Wafer level packaging, through its design of stress buffering and/or compliance, promises to improve the reliability of the flip-chip without the additional underfilling steps in the assembly process. In both applications, wafer level packaging may enable wafer test and burn-in, resolving the known good die (KGD) issue, which will further reduce the cost of electronics manufacturing.

A variety of wafer level packages have been devised, among which four major categories are identified as follows:

- Thin film redistribution and bumping.
- Encapsulated package.
- Compliant interconnect.
- Wafer level underfill.

There often exists confusion between the concepts of flip-chip and wafer level packaging. WLP, by definition, requires no more packaging or encapsulation at the board level assembly. This means flip-chip on board (FCOB) can also be considered as a WLP. However, in most cases of FCOB, underfill is needed as mentioned in the above context. Some argued that flip-chip with underfill is not a WLP. Nevertheless, if underfill is moved onto the wafer level and no additional underfilling step is needed at board level assembly, flip-chip

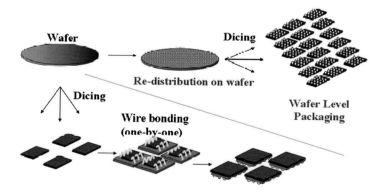

FIGURE 5.1. A comparison between the conventional packaging and the wafer level packaging.

DESIGN, PROCESS, AND RELIABILITY OF WAFER LEVEL PACKAGING

with wafer level underfill is often considered as a type of WLP. In this chapter, different WLP technologies are reviewed and compared with an emphasis on the wafer level underfill. The first three categories of WLP would be discussed in Section 5.2 and wafer level underfill in Section 5.3. The recent development in wafer level test and burn-in is also reviewed.

5.2. WLCSP

5.2.1. Thin Film Redistribution

The thin film redistribution packages provide cost effective wafer level process and standard SMT compatible assembly, and are the major techniques used in the commercial WLCSPs. One example of this type of package is the Ultra CSP™ by K&S Flip Chip Division [2]. The manufacturing process of the Ultra CSP™ is illustrated in Figure 5.2. It utilizes two layers of benzocyclobutene (BCB) dielectric and one redistribution layer of Al/NiV/Cu. After the fabrication of the thin film layers, the solder balls are attached by flux, ball placement and reflow. One advantage of the Ultra CSP™ concept is that it uses standard IC processing technology for the package manufacturing. This makes the Ultra

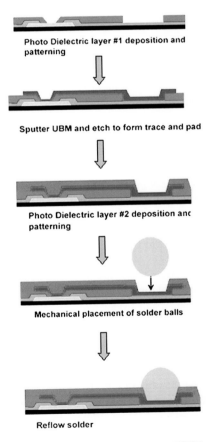

FIGURE 5.2. Process flow of Ultra CSP™.

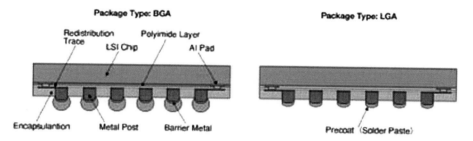

FIGURE 5.3. Structure of the Super CSP™.

FIGURE 5.4. Structure of FIP double bump wafer level package.

CSP™ ideal for both insertion at the end of the wafer fab as well as facilitation of wafer level test and burn-in options.

Typical products using the Ultra CSP™ are small packages with low number of I/Os. In order to increase the solder joint reliability of larger packages, a polymer reinforcement was designed and a technology called Polymer Collar WLP™ was developed by K&S Flip Chip Division [3]. In a standard WLP bumping process, a flux layer is usually applied before the solder ball placement to facilitate the solder wetting on the bond pads during wafer reflow. In the Polymer Collar WLP, a polymeric material is used instead of the flux and remains after the reflow to build reinforcement around the solder joint neck so as to block the shear deformation of the solder. As such, the reliability can be increased. In the Ultra CSP™ package with solders of maximum distant to the neutral point (DNP) being 3.18 mm, a 64% increase in cycle fatigue life-time was observed with the "polymer collar."

Another example of the thin film redistribution WLP is the Super CSP™ developed by Fujitsu Ltd. [4]. Figure 5.3 shows the structure of the Super CSP™ BGA (Ball Grid Array) and LGA (Land Grid Array) type packages. The manufacturing process of the Super CSP™ involves the formation of the redistribution layer by a polyimide film and electrolytic-plated metal trace. After redistribution, the resist is patterned and the copper posts are formed by electrolytic plating. Then the whole wafer is encapsulated with an epoxy molding compound (EMC), and solder balls or solder pastes are applied on top of the copper posts. The board level reliability of the Super CSP™ is good mainly due to high stand-offs of the copper post as well as the low CTE of the EMC encapsulation material which effectively reduces the stress occurring in the solder joint interconnect.

Similar to the Super CSP™, the Fab Integrated Packaging (FIP) invented by Fraunhofer IZM, Berlin uses a stress compensation layer (SCL) which embeds the solder balls before the second solder balls are attached on the top of embedded balls as shown in Figure 5.4. According to the thermal mechanical simulation, the SCL reduces the accumulated equivalent creep strain of the solder balls and also serves as mechanical support for the second solder ball to achieve taller solder heights compared to the standard redistribution

DESIGN, PROCESS, AND RELIABILITY OF WAFER LEVEL PACKAGING

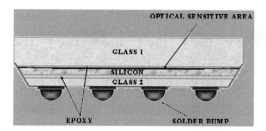

FIGURE 5.5. Structure of ShellOP CSP, an example of encapsulated WLP.

technology [5]. The double bump structure was evaluated by Motorola with different SCL materials [6].

5.2.2. Encapsulated Package

The ShellOP CSP by ShellCase is an example of the encapsulated wafer level packaging [7]. It sandwiches the silicon chip between two glass plates that prevent the silicon from being exposed and ensures mechanical protection as shown in Figure 5.5. The compliant polymer layer under the solder bumps provides board level reliability. This type of package is ideal for optical display.

5.2.3. Compliant Interconnect

For large die applications, many compliant interconnect technologies have been developed to improve the interconnect reliability. Several examples of the compliant interconnect are Microspring Contact on Silicon Technology (MOST) by FormFactor [8], Wide Area Vertical Expansion (WAVE) by Tessera, Sea of Leads (SoL) by Georgia Tech [9], G-Helix Interconnect by Georgia Tech [10], Elastic-Bump on Silicon Technology (ELASTec®) by Infineon [11], On-Wafer Floating Pad Technology by GE Global Research [12], Compliant Bump WLCSP by TI and Fujikura, etc. A common feature of the compliant interconnect is that the interconnect structure is designed to provide movement into x and y directions to accommodate the CTE mismatch during the thermal cycling. In most cases, z direction compliance is also provided to address the substrate coplanarity and wafer testing issues.

The MicroSpring™ technology was first invented for wafer probe cards and LGA production sockets. This technology was recently extended to a wafer level package called MOST™, in which the microspring contacts are fabricated directly on silicon at the wafer level. Figure 5.6 shows a picture of the MOST™ package. The microspring contacts are fabricated by gold wire bonding process and are plated with a Ni alloy, called "spring alloy." These contacts can be attached to the substrate through soldering. They decouple the CTE mismatch between the silicon die and the board. Hence, the reliability far exceeds any solder ball based technology. The microsprings require around one gram of compression force for every 25 micron of displacement and exhibit low contact resistance. They can be used as a fine pitch contact down to 225 microns, a similar pitch size compared with the current flip-chip production. The MOST™ technology integrated with wafer-level test and burn-in has been developed into a "Wafer on Wafer" (WOW) process as discussed later in the text.

FIGURE 5.6. A picture of the MOST™ package.

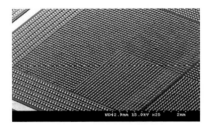

FIGURE 5.7. A picture of the SoL package.

As the semiconductor industry moves toward the development of giga-scale integration, the demand for high packaging density is increasing. Sea of Leads (SoL) by Georgia Tech is an example of the high-density wafer level packaging employing a compliant interconnect technology. SoL extends front end batch processing of the on-chip interconnect on the wafer to include x–y–z compliant chip I/Os through the fabrication of "slippery" leads and embedded air-gaps in the polymer film. A picture of the package is shown in Figure 5.7. There are several methods of allowing the lead movement during the thermal cycling. One method of fabricating the "slippery leads" uses a seed layer plated onto the leads that is selectively etched when the leads are ready to be released from the surface. The embedded air gaps are created through the decomposition of a patterned sacrificial polymer layer on the wafer. The density of the SoL package reaches $12 \times 10^3/\text{cm}^2$. The package supports high frequency signals up to 45 GHz. Similar interconnect structure can be found in the ELASTec by Infineon Technology. In the ELASTec package, the redistribution traces routed from the I/O pads are plated and form a spiral pattern on resilient silicon bumps. where an S-shaped metal layer was plated on a resilient bump made of silicone.

Another example of the high density compliant WLP is the G-Helix also designed by Georgia Tech. The G-Helix is a free-standing compliant interconnect fabricated by photolithography process. Figure 5.8 shows the pictures of the G-Helix Cu interconnect on a 200 μm pitch wafer. The advantage of the compliant interconnect lies in the design flexibility that can be optimized to offer the best mechanical and electrical properties. The drawback may be due to the high cost of the three mask-sets required for the fabrication.

The On-Wafer Floating Pad Technology by GE Global Research and the Compliant Bump WLCSP by TI/Fujikura shared the same characteristic of building the solder bumps on an array of polymer islands. Figure 5.9 shows the structure of Compliant Bump WLCSP. In this case, a resin post (polyimide core) is formed on the wafer to provide a compliant stand-off for the bumps. An encapsulant is molded over the structure to provide a protective

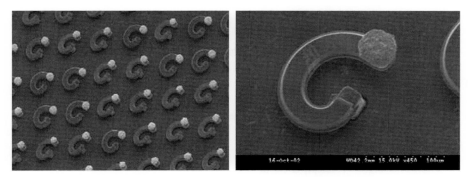

FIGURE 5.8. SEM pictures of G-Helix compliant interconnect WLP.

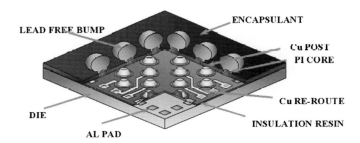

FIGURE 5.9. Structure of compliant bump WLCSP.

layer. The polymer core absorbs the strain between the mounted chip and PWB and can also provide some lateral movement when compressed.

The compliant interconnects usually build onto the thin film redistribution WLCSP, but have showed much superior thermal mechanical reliability, which intrigues enormous research effort from different companies. Nevertheless, the major drawback of the compliant interconnects is their high fabrication cost and the lack of infrastructure. The added inductance of some compliant interconnects also limits their application in high frequency devices. For these reasons, they are still at R&D level and the market is yet to mature.

5.3. WAFER LEVEL UNDERFILL

The wafer level underfill was initially proposed as a SMT compatible flip-chip process to achieve low cost and high reliability [13–16]. It can be used on WLSCPs as well to enhance their board level reliability (MicroFill by National Semiconductor, for instance). The schematic process steps are illustrated in Figure 5.10. In this process, the underfill is applied either onto a bumped wafer or a wafer without solder bumps, using a proper method, such as printing or coating. Then the underfill is B-staged and wafer is diced into single chips. In the case of unbumped wafer, the wafer is bumped before dicing when the underfill can be used as a mask. The individual chips are then placed onto the substrate by standard SMT assembly equipment.

It is noted that in some types of WLCSP, a polymeric layer is also used on the wafer scale to redistribute the I/O and/or to enhance the reliability. However, this polymeric layer

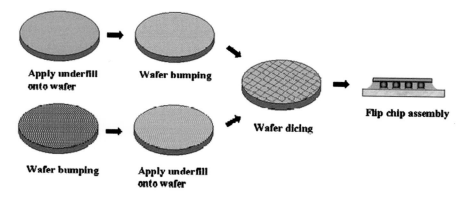

FIGURE 5.10. Process steps of wafer level underfill.

usually does not adhere to the substrate and cannot be considered as underfill. The wafer level underfill discussed here is an adhesive to glue chip and substrate together and functions as a stress-redistribution layer rather than a stress-buffering layer. The attraction of the wafer level underfill lies in the potential low cost (since it does not require a significant change in the wafer back-end process) and high reliability of the assembly enhanced with the underfill. Since this process suggests a convergence of front-end and back-end in package manufacturing, close cooperation between chip manufacturers, packaging companies, and material suppliers are required. Several cooperative research programs have been carried out in this area, including the team between Motorola, Auburn University, and Loctite Electronic Materials sponsored by National Institute of Standards and Technology Advanced Technology Program (NIST-ATP) [17], the team between National Semiconductor, IBM, National Starch and Chemical Company, and Georgia Institute of Technology sponsored also by NIST-ATP [18], and the team between 3M Company and Delpi-Delco Electronic Systems [19].

5.3.1. Challenges of Wafer Level Underfill

The material and process challenges for wafer level underfill have been identified and can be summarized as follows. First, a robust underfill deposition process is required; the resulted underfill layer must be of sufficient uniformity and consistency to enable a high yield in the assembly process, good solder joint formation and acceptable underfill fillet. Different deposition processes have been explored including spin coating, vacuum lamination, screen printing and stencil printing, etc. The underfill needs to be partially cured, or B-staged, if the original form is a liquid to facilitate the later handling including dicing and storage. One method is to use solvent in the deposition process and then drive off the solvent to B-stage the underfill. However, the use of unreactive solvent might leave residue which is likely to cause voiding during the later assembly [20]. B-stage cure can be used with careful control of the curing degree not to interfere with solder joint formation in the solder reflow. Wafer dicing presents another challenge for the underfill since the uncured material would be exposed to water that is used for cooling. If the wafer is to be diced with the underfill, the material also needs good mechanical property to prevent cracking. Unlike liquid underfill that is usually freeze-stored, wafer level underfill requires long shelf-life for packing, shipping and storage of the dies. Fortunately, B-staged material usually has the glass transition temperature above room temperature, at which the mobility of the molecules is low to prevent large-scale reactions.

The issues related to the wafer level underfill in the assembly process start with the vision recognition at the placement machine. Normally, either fiducials or solder bumps on the die are located using the vision system in a pick-and-place equipment for flip-chip bonding alignment. Being covered by the underfill that is often heavily filled with silica fillers and hence translucent, these registration marks are difficult to be recognized by the vision system. Fortunately, many placement machines can adjust illumination angle, light intensity and image acceptance transforms, etc. to optimize imaging [21]. The coating color can also be adjusted to enhance the recognition. Some work has shown that black color provides the best contrast to the coated bumps [22]. If no additional flux is to be dispensed on the board, the wafer level underfill has to provide some tackiness to hold the chip in place. Several methods have been proposed including heating the board, heating the chip in a separate station, and heating the underfill through the pick-up nozzle. Similar to no-flow underfill, self-fluxing capability is required to eliminate the flux dispensing process. However, flux is known to degrade the stability of epoxy-based systems and shorten the shelf-life of the wafer level underfill. Hence, wafer-level underfill usually contains separating materials with different functions to achieve the desired result [23]. The solder wetting process with a wafer level underfill presents challenges to high interconnect yield, because the wetting is constrained by the presence of the partially cured underfill. Numerical simulation has been performed to predict the solder joint formation under constrained boundaries [24]. The solder joint interconnection is highly dependent on the fluxing capability and the viscosity of the underfill. However, it was found that the wetting process could be complicated by underfill outgassing and chip motion driven by forces other than surface tension of the solder [25]. The thickness of the underfill coating was critical for an optimal solder joint formation; deficiency in underfill could result in a gap between the bumps and excess underfill would hinder the solder joint formation. Other issues such as the desire for no post cure and reworkability are being addressed as well in the wafer level underfill process.

5.3.2. Examples of Wafer Level Underfill Process

In order to address the previous challenges, different wafer level underfill processes and the corresponding materials have been developed by various research teams, each providing unique solutions to the issues mentioned above. Illustrated in Figure 5.11 is the wafer scale applied reworkable fluxing underfill process developed by Motorola, Loctite and Auburn University [17]. Since uncured underfill materials are likely to absorb moisture that leads to potential voiding in the assembly, in this process, wafer is diced prior to underfill coating. Two dissimilar materials are applied; the flux layer coating by screen or stencil printing and the bulk underfill coating by a modified screen printing to keep the saw street clean. The separation of the flux from the bulk underfill material preserves the shelf life of the bulk underfill as well as prevents the deposition of fillers on top of the solder bump so as to ensure the solder joint interconnection in the flip-chip assembly. In this process, no additional flux dispensing on board is needed and hence the underfill needs to be tacky in the flip-chip bonding process to ensure the attachment of the chip to the board, as discussed previously.

Underfill deposition on wafer using liquid material via coating or printing requires subsequent B-staging, which is often tricky and problematic. The process developed by 3M and Delphi-Delco circumvents the B-stage step using film lamination [26]. The process

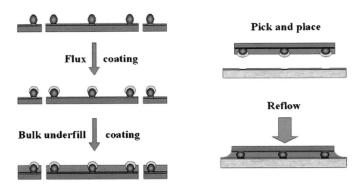

FIGURE 5.11. A wafer scale applied reworkable fluxing underfill process.

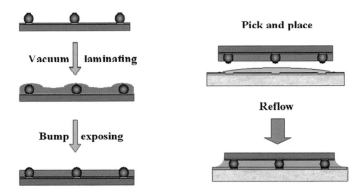

FIGURE 5.12. A wafer-applied underfill film laminating process.

steps are shown in Figure 5.12, in which the solid film comprised of thermoset/ thermoplastic composite is laminated onto the bumped wafer in vacuum. Heat is applied under vacuum to ensure the complete wetting of the film over the whole wafer and to exclude any voids. Then a proprietary process is carried out to expose the solder bump without altering the original solder shape. The subsequent flip-chip assembly is carried out with a curable polymeric flux adhesive pre-applied on the board.

Wafer level underfill can also be applied before the bumping process. Figure 5.13 shows a multi-layer wafer-scale underfill process developed by Aguila Technologies, Inc. [27]. The highly filled wafer level underfill is screen printed onto an unbumped wafer and then cured. Then this material is laser-ablated to form microvias that expose the bond pads. The vias are filled with solder paste and reflowed. Bumps are formed on top of the filled vias. The flip-chip assembly is similar to the previous approach with a polymer flux.

The wafer level underfill process has been successfully implemented in a commercial WLCSP MicroSMD by National Semiconductor [28]. Full area array at 200 micron pitch flip-chip assembly with wafer level underfill was also demonstrated by Georgia Tech [29].

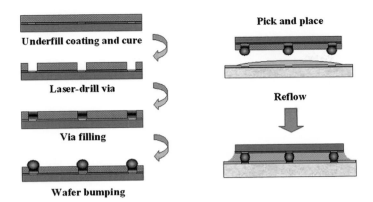

FIGURE 5.13. A multi-layer wafer-scale underfill process.

5.4. COMPARISON OF FLIP-CHIP AND WLCSP

Both flip-chip and WLCSP have the advantage of the actual package size being the same as the chip size. Miniaturization and low profile are the drivers of these two packages. As the interconnect/terminal is concerned, these two packages are similar in that most WLCSPs employ solder technology in an area array. In the assembly process, depending on the reliability requirement, WLCSPs are usually assembled in a standard SMT process without the need of underfill, while for flip-chip on organic substrate, underfill is usually required to ensure the reliability. The use of underfill substantially increases the assembly cost for flip-chip packages. In addition, the repair on the board level becomes difficult due to the fact that most underfill materials are thermosetting resins that are non-reworkable.

A major drawback of the flip-chip technology is the Known Good Die (KGD) issue. Test and burn-in for flip-chip packages are usually conducted at the component level. Wafer level package, on the other hand, has the potential of enabling wafer level test and burn-in, which would substantially reduce the cost and solve the KGD issue. However, due to the complexity of the IC wafer, a low cost and robust wafer level test and burn-in process is still under development.

The current market for WLCSP is low cost, low I/O, and small devices. The high performance devices mainly rely on flip-chip due to its high I/O capability and good electrical performance. However, flip-chip underfill technology is facing challenges as the I/O density increases and the pitch distance decreases. Convergence of the IC fab front-end and back-end of packaging manufacturing and potential cost reduction through wafer level test and burn-in drive the development of high density wafer level package. Compliant interconnects are attempts to bring the high density packaging to the wafer level. On the other hand, wafer level underfill is converging flip-chip technology with wafer level packaging, providing a low cost solution for flip-chip manufacturing.

5.5. WAFER LEVEL TEST AND BURN-IN

One of the attractions of wafer level packaging is the possibility of wafer level test and burn-in. Test (at speed) and burn-in is an expensive step in the semiconductor business. The transfer to full parallel test on wafer level could dramatically reduce the overall cost.

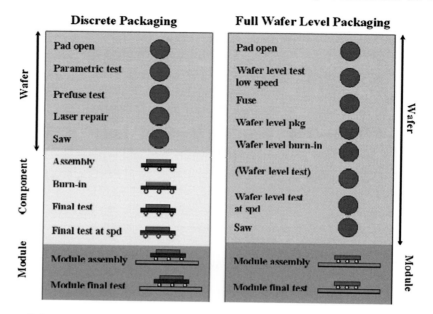

FIGURE 5.14. Assembly and test in discrete packaging and full wafer level packaging.

Estimation indicates that savings up to 50% for the transfer from component to wafer test. Figure 5.14 shows the assembly and test process flow for conventional discrete packaging and full wafer level packaging. The conventional process flow is shown in the left side in which the assembly and tests are done on the component level. On the right side, the integrated assembly and test process on wafer level is illustrated with test and burn-in performed on the wafer level.

However, several critical issues need to be addressed before the implementation of wafer level test and burn-in. The challenges for the probe card include high-density interconnects onto the wafer, CTE matching of the contactor to silicon, coplanar probe tips, high forces to make electrical connection with low resistance, uniform load to all the bumps, etc. In addition, precise alignment of probe to wafer is needed. Thermal management in wafer level burn-in is also critical. All the dies on the wafer should be subjected to a uniform stress; therefore the voltage, temperature, and the ramping rate need to be carefully controlled. Above all, the main barrier to the success of wafer level test and burn-in is the cost/performance ratio.

Several wafer level test and burn-in examples have been demonstrated in industry. Motorola announced the wafer level burn-in technology in 1998 with the partnership of Motorola Semiconductor Products Sector, Tokyo Electron Limited (TEL), and W.L. Gore and Associates, Inc. (GORE) [30]. The developed approach uses TEL wafer-probe technology in a controlled environment and allows each chip on a silicon wafer to be electrically stressed across a range of temperatures from 125°C to 150°C. Using this new technology, a silicon wafer of completed circuits is placed on a thermal chuck with an extremely flat surface. An electrical contact head, with thousands of contacts, is aligned to the wafer and contact is made through a sheet of contact material as shown in Figure 5.15. Critical to the process is the unique full-wafer contact material, called GoreMate™ wafer contactor, placed between the contact head and the test wafer. GORE also developed a thermally

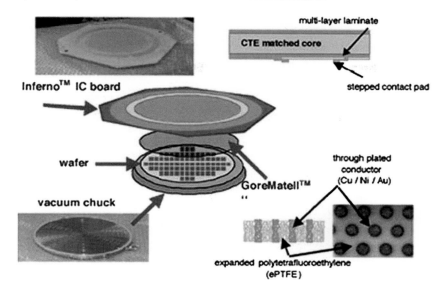

FIGURE 5.15. Motorola wafer level burn-in strategy.

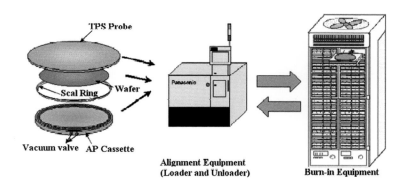

FIGURE 5.16. Matsushita wafer level burn-in overview.

matched (Inferno™) interconnect board, designed to have the same coefficient of expansion as silicon. As estimated from Motorola, through simplification and consolidation of product testing operations, manufacturing cost savings are expected to be as high as 15 percent and improvements in manufacturing cycle time will range up to 25 percent.

Matsushita Electric Industrial Co. Ltd. has also developed a wafer level burn-in strategy as shown in Figure 5.16. A three-part-structure (TPS) probe is used which consists of a glass substrate multilayer wiring board, a compliant z-axis conductor using conductive rubber, and a polyimide membrane with bumps for contacting [31]. The structure of the TPS probe is shown in Figure 5.17. A uniform contact force is provided by the atmosphere when vacuum is applied between the wafer and the TPS probe through the vacuum valve on the AP Cassette as shown in Figure 5.16. The conductive rubber acts to provide the absorption of the bump height differences. Firm contacts have been achieved on 2756 bumps which have remained stable up to 125°C.

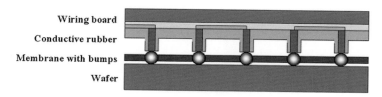

FIGURE 5.17. TPS probe structure of Matsushita wafer level burn-in.

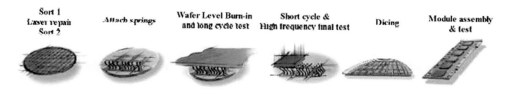

FIGURE 5.18. Process flow of WOW™.

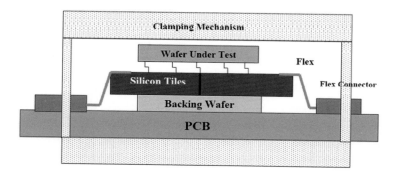

FIGURE 5.19. WOW™ wafer level burn-in structure.

Wafer level packaging, through the design of interconnect on wafer, is enabling full wafer test and burn-in to construct an integrated wafer level packaging, test and burn-in and assembly process to achieve ultimate low cost of electronic packaging. Many compliant interconnect techniques aim at providing flexible bumps that can be pressed down by a low force onto a flat contactor board. ELASTec WLP by Infineon has illustrated the benefit of the resilient bumps. Approximately 2 grams per bump are enough to form a reliable contact, taking into account the height tolerances of bumps and board pads [11]. The compliance of the interconnects also serves to solve the CTE mismatching problem of the wafer and the test board. A good example of WLP enabling wafer level test and burn-in is illustrated by the WOW™ (wafer on wafer) technology by FormFactor. WOW™ is IC industry's first back-end process that provides fully integrated wafer level package, burn-in, test and module assembly.

The microsprings of the MOST™ technology can provide the permanent interconnect onto the final product, as well as temporary connection under pressure during test and burn-in. These microsprings can be located anywhere on the die surface including directly on the bond pads. Figure 5.18 shows the process flow of integrated wafer level package, burn-in, test and assembly in WOW™. The wafer level burn-in structure can be seen in Figure 5.19. Silicon wafer is used for building the contactors due to the matched CTE with the wafer under test and also the well-understood interconnect materials and process. However, it is challenging to build perfect yield wafer larger than 200 mm cost effectively. Therefore silicon tiles with smaller area are placed on and connected to a backing wafer. The test wafer is clamped against the contactors and tested from 25°C to 150°C. The test can also be carried out on a single die level and a multi die (module) level in addition to wafer level for different testing scenario. FormFactor's WOW™ process opens the door for vast business opportunities. However, cost effective wafer alignment and clamping systems, wafer temperature forcing systems, and wafer level test and burn-in electronics are to be sought.

5.6. SUMMARY

Wafer level packaging has been growing continuously in electronics packaging because its low cost in batch manufacturing and the potential of enabling wafer level test and burn-in. A variety of wafer level packages have been devised, among which four important categories are identified including thin film redistribution and bumping, encapsulated package, compliant interconnect, and wafer level underfill. The current WLCSPs mainly use thin film redistribution technology due to its low cost and are found in the consumer electronics market. Many compliant interconnect structures have been developed for large die applications. These compliant interconnects are designed to provide compliance in x, y, and z direction to accommodate the CTE mismatch between the chip and the substrate, and to address the substrate coplanarity as well. Wafer level underfill process suggests the convergence of flip-chip underfill and wafer level CSP, and may provide a low cost solution for high density WLP. However, the unique process of the wafer level underfill presents great challenges for the materials. Several wafer level underfill processes are reviewed and discussed. The wafer level packaging integrated with wafer burn-in, test and module assembly shows great attraction due to the dramatic cost reduction. Cost effective ways of building wafer level test and burn-in are under investigation.

REFERENCES

1. P. Garrou, Wafer level chip scale packaging: an overview, IEEE Transactions on Advanced Packaging, 23(2), pp. 198–205 (2000).
2. P. Elenius, S. Barrett, and T. Goodman, Ultra CSP™—a wafer level package, IEEE Transactions on Advanced Packaging, 23(2), p. 220 (2000).
3. D.H. Kim, P. Elenius, M. Johnson, S. Barrett, and M. Tanaka, Solder joint reliability of a polymer reinforced wafer level package, Proceedings of the 52nd Electronic Components and Technology Conference, 2002, p. 1347.
4. T. Kawahara, Super CSP™, IEEE Transactions on Advanced Packaging, 23(2), p. 215 (2000).
5. M. Topper, J. Auersperg, V. Glaw, K. Kaskoun, E. Prack, B. Beser, P. Coskina, D. Jager, D. Petter, O. Ehrmann, K. Samulewiez, C. Meinherz, S. Fehlberg, C. Karduck, and H. Reichl, Fab integrated packaging (FIP): a new concept for high reliability wafer-level chip size packaging, Proceedings of the 50th Electronic Components and Technology Conference, 2000, pp. 74–80.
6. B. Keser, E.R. Prack, and T. Fang, Evaluation of commercially available, thick, photosensitive films as a stress compensation layer for wafer level packaging, Proceedings of the 51st Electronic Components and Technology Conference, 2001, pp. 304–309.
7. A. Badihi, Ultrathin wafer level chip size package, IEEE Transactions on Advanced Packaging, 23(2), p. 212 (2000).
8. J. Novitsky and D. Pedersen, FormFactor introduces an integrated process for wafer-level packaging, burn-in test, and module level assembly, Proceedings of 1999 International Symposium on Advanced Packaging Materials, 1999, p. 226.

9. M.S. Bakir, H.A. Reed, P.A. Kohl, K.P. Martin, and J.D. Meindl, Sea of leads ultra high-density compliant wafer-level packaging technology, Proceedings of the 52nd Electronic Components and Technology Conference, 2002, p. 1087.
10. Q. Zhu, L. Ma, and S.K. Sitaraman, Design and optimization of a novel compliant off-chip interconnect—one-turn Helix, Proceedings of the 52nd Electronic Components and Technology Conference, 2002, p. 910.
11. H. Hedler, T. Meyer, W. Leiberg, and R. Irsigler, Bump wafer level packaging: a new packaging platform (not only) for memory products, Proceedings of 2003 International Symposium on Microelectronics, 2003, pp. 681–686.
12. R. Fillion, L. Meyer, K. Durocher, S. Rubinsztajin, D. Shaddock, and J. Wrigth, New wafer level structure for stress free area array solder attach, Proceedings of 2003 International Symposium on Microelectronics, 2003, pp. 678–692.
13. S.H. Shi, T. Yamashita, and C.P. Wong, Development of the wafer-level compressive-flow underfill process and its required materials, Proceedings of the 49th Electronic Components and Technology Conference, 1999, p. 961.
14. S.H. Shi, T. Yamashita, and C.P. Wong, Development of the wafer-level compressive-flow underfill encapsulant, IEEE Trans. on Components, Packaging, Manuf. Technol., Part C, 22(4), p. 274 (1999).
15. K. Gilleo and D. Blumel, Transforming flip chip into CSP with reworkable wafer-level underfill, Proceedings of the Pan Pacific Microelectronics Symposium, 1999, p. 159.
16. K. Gilleo, Flip chip with integrated flux, mask and underfill, W.O. Patent 99/56312, Nov. 4, 1999.
17. J. Qi, P. Kulkarni, N. Yala, J. Danvir, M. Chason, R.W. Johnson, R. Zhao, L. Crane, M. Konarski, E. Yaeger, A. Torres, R. Tishkoff, and P. Krug, Assembly of flip chips utilizing wafer applied underfill, Presented at IPC SMEMA Council APEX 2002, Proceedings of APEX, San Diego, CA, 2002, pp. S18-3-1–S18-3-7.
18. Q. Tong, B. Ma, E. Zhang, A. Savoca, L. Nguyen, C. Quentin, S. Lou, H. Li, L. Fan, and C.P. Wong, Recent advances on a wafer-level flip chip packaging process, Proceedings of the 50th Electronic Components and Technology Conference, 2000, pp. 101–106.
19. S. Charles, M. Kropp, R. Kinney, S. Hackett, R. Zenner, F.B. Li, R. Mader, P. Hogerton, A. Chaudhuri, F. Stepniak, and M. Walsh, Pre-applied underfill adhesives for flip chip attachment, IMAPS Proceedings, International Symposium on Microelectronics, Baltimore, MD, 2001, pp. 178–183.
20. B. Ma, Q.K. Tong, E. Zhang, S.H. Hong, and A. Savoca, Materials challenges for wafer-level flip chip packaging, Proceedings of the 50th Electronic Components and Technology Conference, 2000, pp. 171–174.
21. C.D. Johnson and D.F. Baldwin, Wafer scale packaging based on underfill applied at the wafer level for low-cost flip chip processing, Proceedings of the 49th Electronic Components and Technology Conference, 1999, p. 951.
22. M. Chason, J. Danvir, N. Yala, J. Qi, P. Neathway, K. Tojima, W. Johnson, P. Kulkarni, L. Crane, M. Konarski, and E. Yaeger, Development of wafer scale applied reworkable fluxing underfill for direct chip attach, Assembly Process Exhibition and Conference 2001, San Diego, CA, 2001.
23. L. Crane, M. Konarski, E. Yaeger, A. Torres, R. Tishkoff, P. Krug, S. Bauman, W. Johnson, P. Kulkanari, R. Zhao, M. Chason, J. Danvir, N. Yala, and J. Qi, Development of wafer scale applied reworkable fluxing underfill for direct chip attach, Part II, Presented at IPC SMEMA Council APEX 2002, Proceedings of APEX, San Diego, CA, 2002, pp. S36-2-1–S36-2-6.
24. L. Nguyen and H. Hguyen, Solder joint shape formation under constrained boundaries in wafer level underfill, Proceedings of the 50th Electronic Components and Technology Conference, 2000, pp. 1320–1325.
25. J. Lu, S.C. Busch, and D.F. Baldwin, Solder wetting in a wafer-level flip chip assembly, IEEE Transactions on Electronics Packaging Manufacturing, 24(3), pp. 154–159 (2001).
26. R.L.D. Zenner and B.S. Carpenter, Wafer-applied underfill film laminating, Proceedings of the 8th International Symposium on Advanced Packaging Materials, 2002, pp. 317–325.
27. R.V. Burress, M.A. Capote, Y.-J. Lee, H.A. Lenos, and J.F. Zamora, A practical, flip-chip multi-layer pre-encapsulation technology for wafer-scale underfill, Proceedings of the 51st Electronic Components and Technology Conference, 2001, pp. 777–781.
28. L. Nguyen, H. Nguyen, A. Negasi, Q. Tong, and S.H. Hong, Wafer level underfill—processing and reliability, Proc. 27th Int. Electron. Manuf. Tech. Symp., July 17–18, San Jose, CA, 2002.
29. Z. Zhang, Y. Sun, L. Fan, R. Doraiswami, and C.P. Wong, Development of wafer level underfill material and process, Proceedings of 5th Electronic Packaging Technology Conference, Singapore, Dec. 2003, pp. 194–198.
30. G.W. Flynn, Wafer level burn-in (WLBI) at Motorola—outline, Fleck Research Chip Scale International, 1999.
31. Y. Nakata and M. Kawai, A wafer-level burn-in technology, ULSI Process Technology Development Center, Matsushita Electronics Corporation.

6

Passive Alignment of Optical Fibers in V-grooves with Low Viscosity Epoxy Flow

S.W. Ricky Lee and C.C. Lo

Electronic Packaging Laboratory, Center for advanced Microsystems Packaging, Hong Kong University of Science & Technology, Clear Water Bay, Kowloon, Hong Kong

Abstract Optical fibers are one of the most commonly used light transmitting media in optoelectronic systems for telecommunication applications. Because the core diameter of optical fibers is very small, active alignment methods are usually employed for the coupling between optical fibers and other optoelectronic devices. In general, the equipment cost of active alignment is very high and the processing time is relatively long, especially for fiber array alignment. Therefore, the conventional fiber alignment process becomes rather expensive and the throughput is quite low. In recent years, passive alignment using low cost epoxy adhesives and precisely etched V-grooves on silicon optical benches is attracting more attention due to its reduced production cost and short processing time. During the passive alignment process, the optical fiber may be lifted up by the buoyancy of epoxy flow and, hence, an extra cover plate is required to press the fiber against the walls of the V-groove. An effort is made to develop a modified passive alignment method without using the cover plate. Several parameters may affect the yield and need to be optimized. It is found that the amount of epoxy dispensed to the V-groove is critical in the process. Also the viscosity of the epoxy determines the characteristics of the flow in the V-groove and, hence, affects the results of passive alignment. In this chapter, the design and configuration of the modified passive alignment method will be introduced. The effect of the volume and viscosity of epoxy will be presented. The application to multiple fiber alignment will be demonstrated. The newly developed passive alignment method is capable of aligning an array of 8 fibers up to 1 micron accuracy.

6.1. INTRODUCTION

Alignment of optical fibers is very critical for optoelectronic packaging. A slight offset in any direction will affect the performance of the photonic devices. The tolerance of alignment is very tight, especially for single-mode optical fibers of which the core diameter is only 9 μm [1–7]. Active alignment method is commonly used in the industry since the coupling efficiency is optimized. However, the processing time of active alignment is relatively long and the equipment cost is rather high [8–10].

Recently, passive alignment of optical fibers is attracting more attention due to its lower manufacturing cost and shorter processing time, compared with active alignment. Passive alignment is usually implemented on a silicon optical bench (SiOB) with V-grooves [11–17]. The position of optical fibers in passive alignment is defined by the geometry of the V-groove. The conventional method is to dispense the mounting epoxy a glob-top manner. However, the optical fiber may be lifted up due to the buoyancy. In order to avoid this problem, a cover plate is usually required to press the fiber against the wall of the V-groove [18–23]. Although the fiber is well aligned by pressing the cover plate, the applied stress may deform the optical fiber and affect the optical performance. If this pressing process is not well controlled, the optical fiber may be damaged and the reliability of the package will be decreased. Besides, in some applications, there may be not enough space for the mounting of the cover plate [24,25].

In this chapter, a new method of dispensing the epoxy with passive alignment capability will be introduced [26]. It is observed that if a suitable amount of epoxy is flowing to the gap between the optical fiber and the bottom of the V-groove, the optical fiber will not be lifted up by the buoyancy of the epoxy. On the other hand, the optical fiber will be pulled downward and sit against the walls of the V-groove. Based on this self-alignment property, a new design of V-grooves on the SiOB is developed. A "reservoir" is placed right next to the V-groove. The purpose of this reservoir is to let the epoxy flow into the gap between the optical fiber and the bottom of the V-groove. In order to obtain a steady flow with a gentle motion, the epoxy should have relatively low viscosity. The reservoir is patterned and etched together with the V-grooves so that there is no additional cost and time for fabricating the reservoir.

The proposed new passive alignment design has been characterized experimentally. The testing results show that, with the present approach, an optical fiber with an initial offset of 60 μm from its intended position will be aligned to the centre of the V-groove with a deviation less than 0.5 μm. Once the fiber is aligned, more epoxy is dispensed in a glob-top manner to enhance the fiber mounting strength and reliability. In this chapter, the design and fabrication of SiOB, the specifications of the epoxy, and the testing procedures and results will be presented in details.

6.2. DESIGN AND FABRICATION OF SILICON OPTICAL BENCH WITH V-GROOVES

In this section, the design and fabrication of SiOB with V-grooves for the passive alignment will be introduced. Figure 6.1 shows the mask design of SiOB. Two sizes of V-grooves are fabricated on the SiOB. They are used to hold the jacket and the cladding of the optical fibers. In addition to V-grooves, two newly designed features, "reservoir" and "canal," are also fabricated at the same time with the V-grooves. The reservoir and canal are only used in the modified passive alignment method illustrated in the later section. The design shown in Figure 6.1 is suitable for all the other experiments discussed throughout this chapter.

In this experiment, both low and high profile SiOB are fabricated. When the optical fiber is placed on the former SiOB, it is completely underneath the surface as the V-groove is deep enough to hold the cladding. On the other hand, part of the fiber is above the top

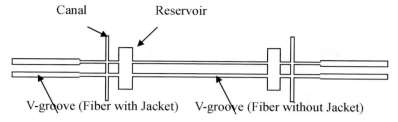

FIGURE 6.1. Mask layout of SiOB for passive alignment of optical fiber.

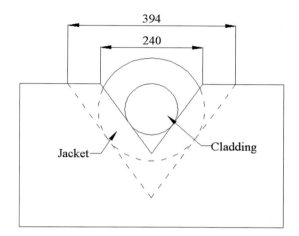

FIGURE 6.2. Low profile SiOB (all units are in micron).

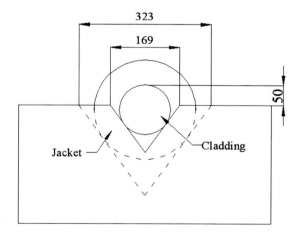

FIGURE 6.3. High profile SiOB (all units are in micron).

surface when it is placed on the latter kind of SiOB. The corresponding dimensions of the low and high profile SiOB are shown in Figures 6.2 and 6.3 respectively.

In the experiment, a 4-inch {1 0 0} silicon wafer with about 500 μm thickness is used. The wafer is cleaned before fabrication. After cleaning, a thin layer of low stress

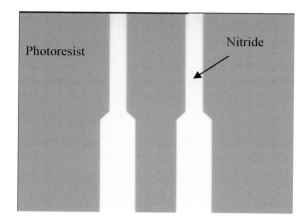

FIGURE 6.4. Patterning of low stress nitride by photoresist.

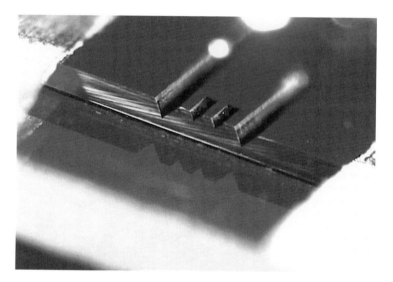

FIGURE 6.5. 3D view of V-grooves on silicon substrate.

nitride with 7000 Å thickness is then deposited on both sides of the wafer. This setup is critical because the wafer is then etched with strong alkaline at high temperature. A thin layer of low stress nitride is needed to act as the passivation layer. The nitride is patterned by photolithography process as shown in Figure 6.4. After hard baking the photoresist, the nitride which is not covered by the photoresist is then dry-etched away. The photoresist is stripped off after the dry etching process and the wafer is now ready for wet etching. The whole process is summarized in Table 6.1.

V-grooves formed by the {1 1 1} planes are obtained after the etching process. During wet etching, {1 1 1} planes are gradually formed [29–36]. When two {1 1 1} planes touch together and form the V-groove, the etching process stops. It is because the etch rate of {1 1 1} planes are the slowest among all crystal planes, and only the {1 1 1} planes are exposed to the solution. The depth of the V-groove depends on dimension of the window

TABLE 6.1.
Process flow of V-groove fabrication.

Process flow	Cross-section view

(1) Wafer cleaning
  – Dip in concentrated sulfuric acid for 10 minutes at 120°C
  – Dip in HF:H20 (1:50) solution for 1 minute at room temperature
	Silicon Wafer

(2) Deposit 7000 Å low stress nitride on both side	Low Stress Nitride

(3) Patterning
  – Spin coat photoresist PR204 on the wafer at 4000 rpm for 30 seconds
  – Soft bake at 110°C for 1 minute
  – Expose to UV light for 5 seconds
  – Develop by FDH-5 for 60 seconds
  – Hard bake at 120°C for 30 minutes
	PR204

(4) Passivation opening
  – Dry etch the nitride layer
  – Remove the photoresist by dipping the MS2001 solution at 70°C for 5 minutes	

(5) Silicon wet etching – Dip in 30% KOH at 85°C for 4 hours	

opening. Because of this, it is possible to have a design which has V-grooves with different depths on the same substrate by wet etching.

Figure 6.5 shows a 3D view of the V-groove obtained. Figure 6.6 shows V-grooves formed by {1 1 1} crystal planes. The cross-sectional view of V-grooves with different sizes and depths is clearly observed.

Figure 6.7 shows the cascaded V-grooves for holding the jacket and the cladding of the optical fiber fabricated on the same SiOB. The newly designed features, reservoir and canal are shown in Figure 6.8. It is observed that the reservoir and the canal are not rectangular in shape on designed the mask due to undercutting at the corner. However, this geometry is useful in the modified passive alignment method and will be discussed in following section.

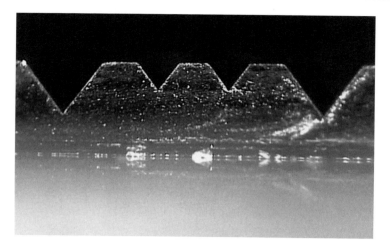

FIGURE 6.6. Cross-section view of V-grooves in different sizes.

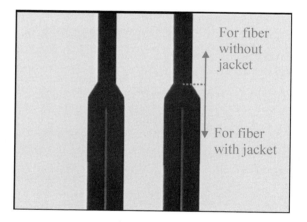

FIGURE 6.7. Cascaded V-grooves.

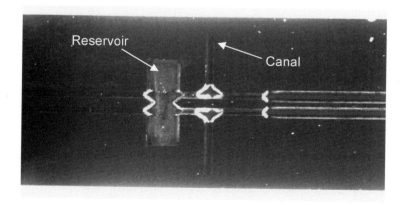

FIGURE 6.8. Reservoir and canal.

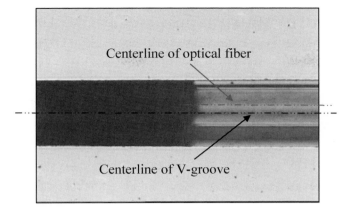

FIGURE 6.9. Initial misalignment.

FIGURE 6.10. Fixing jacket by epoxy.

In the present study, the jacket of the optical fiber is stripped off at the beginning. The stripped length is dependent on the experiment performed. The cladding portion is dipped into the IPA solution for cleaning. Then, the fiber is placed onto the SiOB with V-grooves fabricated as mentioned earlier. In order to test the alignment properties of different passive alignment method, the fiber is intentionally misaligned in the V-groove as shown in Figure 6.9. Placement of the sample fiber is performed under an optical microscope.

Once the fiber is placed in the desired position, epoxy is dispensed onto the jacket in a glob-top manner. After curing the epoxy, the optical fiber can only move in transverse direction but not in axial direction as shown in Figure 6.10.

Two types of epoxies, Epoxy A and Epoxy B, are used in the experiment as listed in Table 6.2. Both are UV curable. Epoxy A is used to hold the jacket, as it has a higher vis-

TABLE 6.2.
Epoxies specifications.

	Epoxy A	Epoxy B
Viscosity	~14000 mPa s	~5000 mPa s
UV Cure	Yes, 365 nm UV light	Yes, 365 nm UV light
Heat Cure	Yes, 30 mins at 121°C	No

cosity. Therefore, it will not flow along with the V-groove rapidly and affect the experiment process at the room temperature.

6.3. ISSUES OF CONVENTIONAL PASSIVE ALIGNMENT METHODS

In this section, some conventional passive alignment methods and their drawbacks will be discussed.

6.3.1. V-grooves with Cover Plate

In order to prevent the buoyancy of the epoxy which may lift up the fiber and cause misalignment, a new procedure is added to the experiment by pressing the fiber with a cover plate. In this experiment, epoxy is first dispensed on top of the fiber. An additional cover plate, which is made by silicon wafer, is then placed on top. A small dead weight, around 2 g, is placed on the cover plate to provide a static force to press the fiber. Figure 6.11 shows the experimental setup. Epoxy A instead of Epoxy B is used, as the epoxy is in-between the SiOB and the cover plate. UV light cannot be used to cure the epoxy. The sample is placed in an oven at 121°C for 30 minutes for the curing process.

Figure 6.12 shows the cross-section inspection obtained by pressing the fiber with a cover plate on a low profile SiOB. The experiment shows that the fibers are not well aligned with the V-grooves even if a cover plate is placed on top. The whole optical fiber is underneath the top surface. This proves that, the cover plate cannot press the fiber effectively. Voids are also found in the cross-section inspection, which add an additional unfavouring result to the experiment.

In order to check whether a low profile SiOB is the factor leading to these unfavouring results, the experiment is repeated with a high profile SiOB. The experimental results are shown in Figure 6.13. From the figure, it is observed that both fibers are well aligned with the V-grooves when a high profile SiOB is used. The cover plate can effectively press the fibers down. However, voids are still found between the cover plate and the SiOB.

This experiment is repeated again by changing one experimental element. This time, only one fiber is placed on the SiOB. The cross-section inspection of this experiment is shown in Figure 6.14.

Figure 6.14 proves that the application of a cover plate is not suitable when there is only one fiber on the SiOB. The cover plate may be tilted when it is placed on the SiOB, because there is only one fiber to support the cover plate.

A die crack is found on the cover plate. It may be resulted from the tilted plate during the curing of epoxy. When cured, the volume of epoxy decreases due to shrinkage. The tilting of the plate leads to asymmetric distribution of epoxy between the gap. During the curing process, a larger epoxy volume change is experienced in the region with more epoxy.

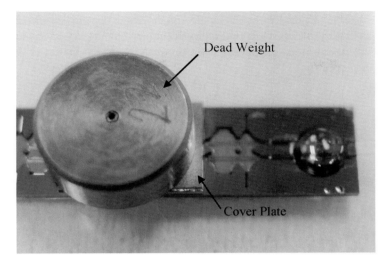

FIGURE 6.11. Application of cover plate.

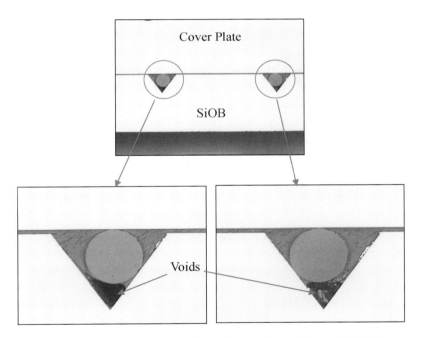

FIGURE 6.12. Cross-section inspection (with cover plate and low profile SiOB).

Therefore, a larger downward force is created on that region. The non-uniformly disturbed force will bend the plate, cracking the die when the force is large enough.

The epoxy shrinks not only during the curing process, but also due to a temperature drop. As the curing temperature of epoxy is 121°C, there is a hundred degree difference when compared with the room temperature. When the sample is removed from the oven, epoxy starts to shrink immediately. The degree of shrinkage depends on the coefficient of thermal expansions of the epoxy. Based on the location of the die crack, it shown that the highest stress is developed at the tip of the optical fiber.

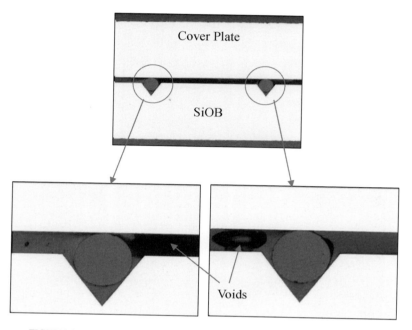

FIGURE 6.13. Cross-section inspection (with cover plate and high profile SiOB).

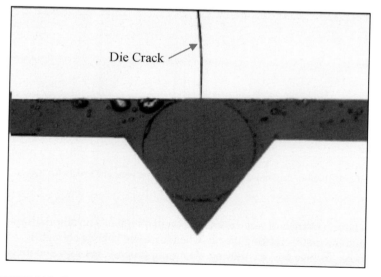

FIGURE 6.14. Cross-section inspection (with cover plate, high profile SiOB, single fiber).

PASSIVE ALIGNMENT OF OPTICAL FIBERS IN V-GROOVES   161

6.3.2. Edge Dispensing of Epoxy

Besides glob top dispensing of epoxy and pressing the fiber with a cover plate, another passive alignment method, edge dispensing of epoxy is studied. A cover plate is used in this method. However, unlike the methods evaluated earlier, this time the cover plate is put on top of the fiber first. Epoxy is then dispensed at the edge of the cover plate. Again, Epoxy A is used because thermal cure is needed. Epoxy will flow into the gap as its viscosity decreases at high temperature.

Figure 6.15 shows the experimental setup and the epoxy dispensing direction. The way epoxy is dispensed is similar to the one used in dispensing underfill in flip chip technology. Figure 6.16 shows the results when the epoxy is dispensed at the edge.

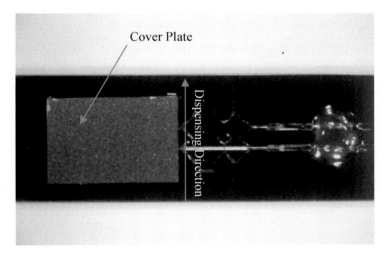

FIGURE 6.15. Edge dispensing method.

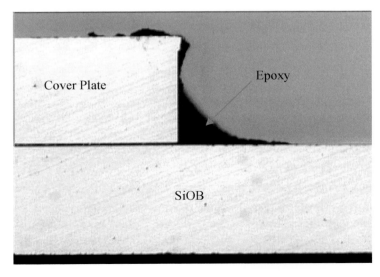

FIGURE 6.16. Epoxy dispensed at the edge.

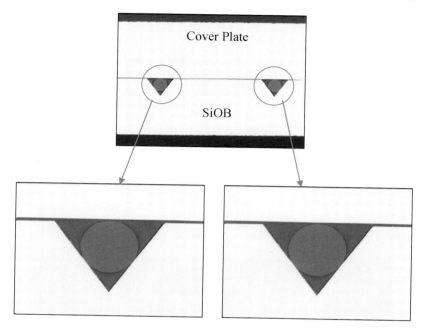

FIGURE 6.17. Cross-section inspection (edge dispensing method, low profile SiOB).

Figure 6.17 shows the cross-section inspection obtained by using edge dispensing method on a low profile SiOB. The figure shows that the optical fibers are well aligned and no void and air bubble is found in the epoxy because the gap is filled up by the epoxy flow. The buoyancy of epoxy is minimized so that both the optical fibers and the cover plate are not lifted up.

The results of applying edge dispensing of epoxy is encouraging. The fibers are well aligned without air bubble. However, the cover plate is still tilted when there is only one fiber on the SiOB. Figure 6.18 shows the tilted cover plate. Although no crack is found from the cross-section inspection, the stress of the cover plate and the fiber may still be very high due to the reasons mentioned before.

6.4. MODIFIED PASSIVE ALIGNMENT METHOD

In earlier sections, some drawbacks of the conventional passive alignment methods are presented. In this section, new modified passive alignment method will be introduced. This method focuses on aligning a single fiber on a SiOB without the help of a cover plate.

6.4.1. Working Principle

The problems found in the conventional passive alignment method are mainly caused by the buoyancy of the epoxy and the application of the cover plate. If the dispensing method is not well controlled, voids and air bubbles may exist in the epoxy. This can be solved by applying the edge dispensing method. However, if only one fiber is placed in the SiOB, the cover plate will be tilted. This leads to non-uniform stress distribution and die cracking.

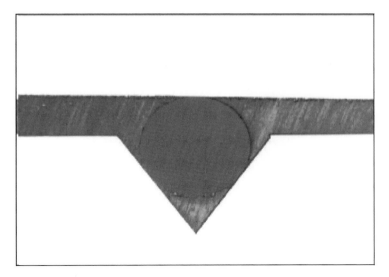

FIGURE 6.18. Cross-section inspection (edge dispensing method, high profile SiOB, single fiber).

From the lessons learned, the edge dispensing method is developed which has a good alignment result with the help of epoxy flow. The steady flow of epoxy from the edge to the gap will prevent any void or air bubble from being trapped. However, if the cover plate is removed, the flow of epoxy will not be well controlled.

To improve the conventional passive alignment method, two new features, reservoir and canal are introduced. Their function is to induce a steady epoxy flow into the gap between the optical fiber and the bottom of the V-groove. This can completely replace the use of the cover plate.

Since the epoxy flow into the gap between the fiber and the bottom of the V-groove, it only wets the bottom part of the fiber. Unbalanced surface tension is acted on the fiber during the epoxy flow. This surface tension will pull down the fiber against the wall of the V-groove. Consequently, the modified passive alignment method has a self-aligning property like the reflow of solder joints in surface mount technology (SMT).

6.4.2. Alignment Mechanism

In order to have epoxy flowing into the gap between the optical fiber and the bottom of the V-groove, epoxy is dispensed and accumulated in the reservoir as shown in Figure 6.19. The epoxy then gathers around the center of the reservoir. It runs along the axial direction of the fiber and finally flows into the V-groove by capillary effect. As presented earlier, undercutting is created during the fabrication process and the reservoir will no longer be rectangular in shape. The feature is similar to a funnel which guides the epoxy to flow steadily into the gap instead of inducing a sharp turn. The gap between the fiber and the bottom of the V-groove is then filled up completely. However, as epoxy flows both toward the fiber tip and the back, if too much epoxy is accumulated at the back, the fiber will be lifted up. At this stage, the flow direction cannot be controlled. Therefore a canal is added to act as a stopper to prevent the epoxy from flowing backward and drive the excess epoxy away. It is redundant to have a canal if the reservoir is placed in the middle of the package. However, this will increase the package size.

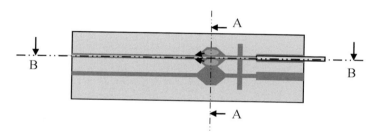

FIGURE 6.19. Alignment mechanism (top view).

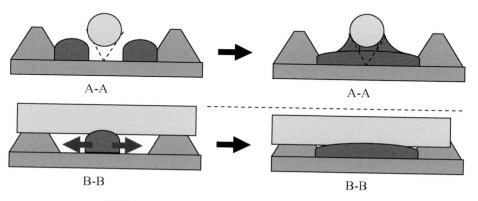

FIGURE 6.20. Alignment mechanism (cross-sectional view).

The cross-section shown in Figure 6.20 explains the alignment mechanism. When the epoxy which gathered in the reservoir touches the fiber, it wets the bottom part of the optical fiber only. This generates an unbalanced surface tension. With the flow of the epoxy along the axial direction by capillary effect, the fiber will be pulled down by surface tension. Thus, the fiber is then aligned against the wall of the V-groove without using an additional cover plate to press it.

6.4.3. Design of Experiment

From the preliminary exercise, it is identified that the amount of epoxy dispensed may play an important role in the quality of passive fiber alignment. Besides, since the overhang length of the stripped fiber (the part with cladding) may affect the gap spacing between the fiber and the walls of the V-groove, this parameter is also considered as a potential factor that may have a certain effect on quality of the passive alignment. Hence, a series of parametric studies with various combinations of epoxy volume and overhang fiber length are performed as listed in Table 6.3. Unfortunately, due to the availability of epoxy materials, it is unable to investigate the effect of various epoxy viscosity on the quality of the passive alignment.

6.4.4. Experimental Procedures

When the fiber is placed in the desired position, Epoxy B is dispensed into the reservoir. Epoxy B is used because its viscosity is lower. Excess epoxy is directed to the canal

TABLE 6.3.
Test matrix for parametric study.

Epoxy weight	Fiber length (w/cladding)		
	8 mm	11 mm	14 mm
0.3 mg	Sample #1	Sample #2	Sample #3
0.4 mg	Sample #4	Sample #5	Sample #6
0.6 mg	Sample #7	Sample #8	Sample #9

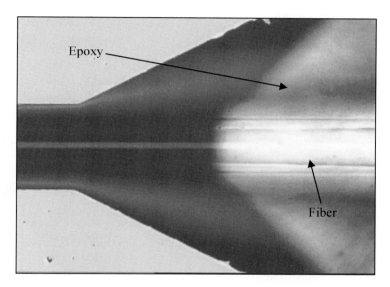

FIGURE 6.21. Epoxy flow into the V-groove from the reservoir.

next to the reservoir. As a result, epoxy flows from the reservoir into the V-groove as shown in Figure 6.21.

During the running of epoxy flow, the fiber is pulled by the surface tension. Once the lateral surface of the fiber touches the walls of the V-groove, the fiber is aligned accordingly. For the current configuration and dimensions, the epoxy flow normally takes about 5 minutes to stop. Then, the epoxy is cured under UV light to fix the fiber position. Here, UV light is used as the fiber is transparent.

6.4.5. Experimental Results

In most optoelectronic applications, the accuracy requirement for optical fiber alignment may reach 0.5 μm. Therefore, in the present study, 0.5 μm offset from the perfectly aligned position is used as a benchmark to evaluate the quality of fiber alignment. The results of the parametric study mentioned earlier are listed in Table 6.4. It is found that some differences exist in the initial misalignment, it is because the fiber is placed by human hand. The fiber aligns with the wall of V-groove even if the initial misalignment is greater than 60 μm. Although there are some differences in the wetted length (the length of the fiber with cladding that is wetted by the epoxy), there is only one case fails to meet the 0.5 μm criterion among all 9 cases.

TABLE 6.4.
Results of parametric study.

	Initial misalignment (μm)	Wetted length (mm)	Within 0.5 μm
Sample #1	48.0	3.3700	Yes
Sample #2	68.0	3.1035	Yes
Sample #3	47.5	3.5670	No
Sample #4	44.8	3.6465	Yes
Sample #5	51.5	3.1865	Yes
Sample #6	39.0	2.7200	Yes
Sample #7	56.0	2.2175	Yes
Sample #8	49.0	3.5300	Yes
Sample #9	51.3	3.1175	Yes

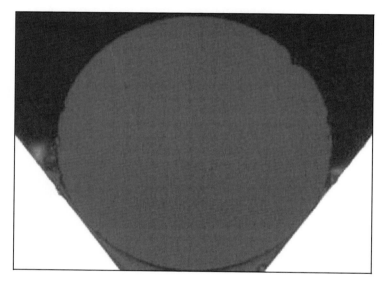

FIGURE 6.22. Well aligned fiber with V-groove.

Figure 6.22 shows the cross-section of a well-aligned fiber using the modified passive alignment method. It is observed that the surface of the fiber is well aligned with the walls of the V-groove. It should be noted that there is little epoxy in the neighborhood of contact points. In fact, most epoxy gathers underneath the fiber, which generates the force to pull down the fiber during the running of epoxy flow. Figure 6.23 shows the cross-section of a fiber with poor alignment. This defect is resulted from epoxy overflow. Several parameters, such as the initial misalignment and the amount of the epoxy added are related to this overflow.

The capability of the newly designed passive alignment method is also tested by monitoring the movement of the optical fiber during the process in real time. The experiment is conducted by coupling optical fibers on one SiOB. By detecting the power received, the movement of the fiber during the passive alignment process can be monitored.

In the experiment, one fiber is initially misaligned and the other one is aligned to the V-groove by the method mentioned above as shown in Figure 6.24. The cladding lengths of

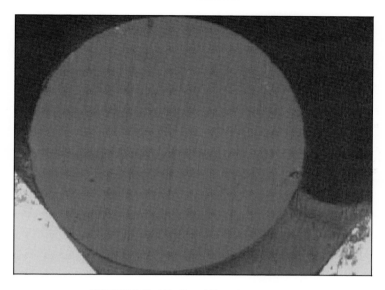

FIGURE 6.23. Misaligned fiber with V-groove.

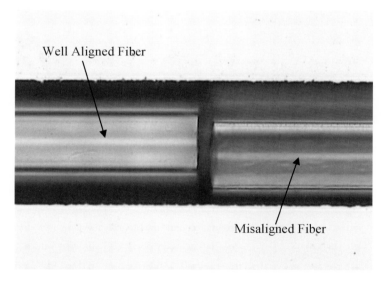

FIGURE 6.24. Initial misalignment.

the fibers are 8 mm. Epoxy B is dispensed in the reservoir next to the misaligned fiber. Light is coupled from the misaligned fiber to the aligned fiber during the alignment process. By simply monitoring the power received from the aligned fiber, the movement of the optical fiber can be evaluated in real time.

Both single-mode and multi-mode optical fiber are used. Figure 6.25 shows the results obtained by coupling a single-mode fiber to another single-mode fiber, and Figure 6.26 shows the results obtained by coupling a multi-mode fiber to another multi-mode fiber. The input power of the single-mode fiber is 2.3 mW where that of the multi-mode fiber

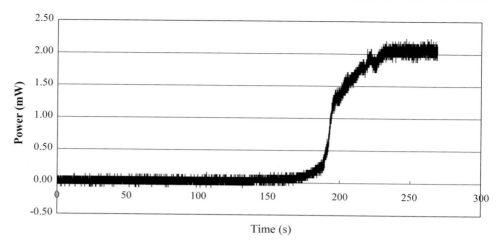

FIGURE 6.25. Real time power monitoring (single-mode to single-mode).

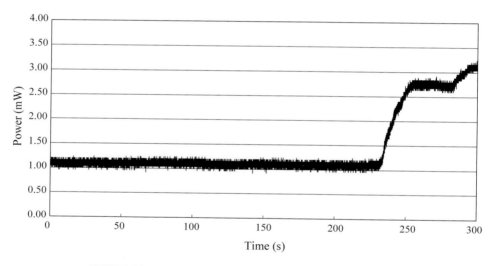

FIGURE 6.26. Real time power monitoring (multi-mode to multi-mode).

is 3.0 mW. The results show that the method achieves higher than 90% coupling efficiency. The whole alignment process takes less than one minute but the flow time is about 5 minutes.

6.5. EFFECTS OF EPOXY VISCOSITY AND DISPENSING VOLUME

From the experimental results, it is observed that the epoxy viscosity is very critical in the alignment process. The viscosity of epoxy affects the flow length, process time and the performance. Besides, the dispensing volume plays a very important role in the yield. If too much epoxy is dispensed, epoxy may overflow and cause misalignment. However, if too little epoxy is dispensed into the reservoir, the epoxy may not flow into the gap between optical fiber and the wall of V-groove. Also, it is difficult to control if the dispensing volume is too small.

In this section, four types of epoxy with different viscosity are used. For each type of epoxy, different epoxy volume is dispensed. In this parametric study, only one fiber is placed on the SiOB. Cross-section inspection is preformed to verify the alignment of the optical fiber. It is claimed to be aligned if the offset of the core centre is less then 1 μm. The theoretical position of the well aligned fiber is shown in Figure 6.27. The test matrix and the results are shown in Table 6.5. The columns of the matrix are arranged by increasing the viscosity and the row of the matrix are arranged by increasing the dispensing volume of epoxy. By completing the matrix, the effect of the epoxy viscosity and dispensing volume are analyzed.

For each combination of epoxy viscosity and dispensing volume, ten samples are tested and analyzed. From the experimental result, it shows both epoxy viscosity and dispensing volume are important factors. The yield of alignment process decrease when the dispensing volume increases. It is because the epoxy may overflow if too much epoxy is dispensed. The canal cannot accommodate that excess epoxy and hence the optical fiber is lifted up can cause the misalignment. Experiment results also show that high viscosity epoxy causes misalignment. If the viscosity of the epoxy is too high, the epoxy cannot completely flow into the gap between the optical fiber and the wall of V-groove. Therefore, the optical fiber is not pulled down by the surface tension of the epoxy. The self alignment capability of the process is further reduced by the epoxy accumulated in the reservoir which may lift up the fiber. In general, dispensing low viscosity epoxy with a right amount of volume provides the best results.

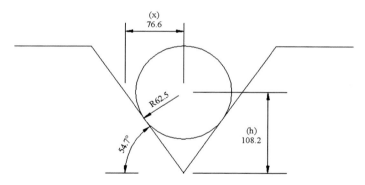

FIGURE 6.27. Theoretical position of well aligned fiber.

TABLE 6.5.
Effect of dispensing volume and viscosity.

Viscosity (Pa s)	Dispensing volume (mm³)			
	0.28	0.56	0.84	1.12
7.5	7/10 aligned	3/9 aligned	2/10 aligned	1/10 aligned
12.2	4/10 aligned	4/8 aligned	1/10 aligned	0/10 aligned
19.5	2/10 aligned	4/10 aligned	4/10 aligned	1/10 aligned
24.8	2/10 aligned	0/10 aligned	0/10 aligned	1/10 aligned

6.6. APPLICATION TO FIBER ARRAY PASSIVE ALIGNMENT

The newly invented method is also applied to aligning fiber arrays. In the present study, fiber arrays with 2, 4 and 8 channels are also tested. There are two pitch sizes, 250 μm and 500 μm, for each array configuration. Figures 6.28 and 6.29 show the mask layout of 4 fiber array with 250 μm pitch and 500 μm pitch, respectively. The fabrication process is the same as mentioned above. Figure 6.30 shows the fiber array SiOB with reservoir and canal.

The experimental procedures are same as aligning single fiber. Fibers are first fixed on the SiOB with high viscosity epoxy. All the fibers on the SiOB are placed with intended initial misalignment. Low viscosity epoxy is dispensed on the reservoir. After the epoxy flow, the epoxy is cured by UV light and cross-section inspection is performed to analyze the alignment performance.

The experimental results are tabulated in Table 6.6. Cross-section inspection is performed to measure the alignment. Figures 6.31 and 6.32 show the cross-section view of 8 fibers array with 250 μm pitch and 500 μm pitch respectively. The samples are clamped to be aligned only all the fibers on the fiber array are aligned within 2 μm.

From the experimental results, it is found that the yield decrease when the number of fibers on the array and the pitch size increase. In this study, the epoxy is only dispensed once on the reservoir for aligning the fiber array. The epoxy is not evenly flow into each gap. Some fibers may be lifted up by the excess epoxy and the outermost fiber may not be aligned due to lack of epoxy. These are the possible reasons to the relatively low yield is obtained when the method is applied to fiber array.

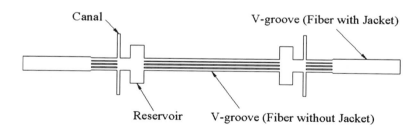

FIGURE 6.28. Mask layout of 4 fibers array (250 μm pitch).

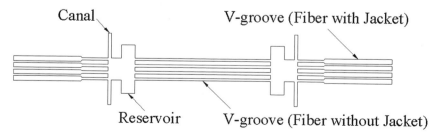

FIGURE 6.29. Mask layout of 4 fibers array (500 μm pitch).

TABLE 6.6.
Experimental results of fiber array.

	Pitch (μm)	Results
2 Fibers array	250	6/10 aligned
	500	8/10 aligned
4 Fibers array	250	6/10 aligned
	500	6/10 aligned
8 Fibers array	250	4/10 aligned
	500	2/10 aligned

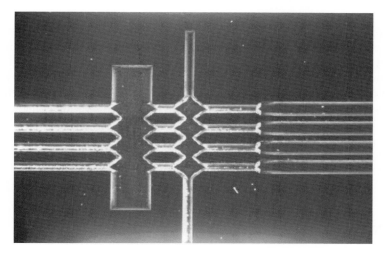

FIGURE 6.30. 4 fibers array (500 μm pitch).

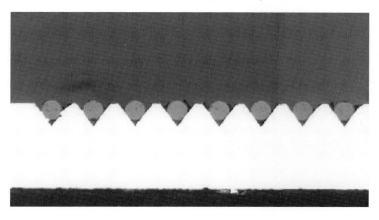

FIGURE 6.31. Cross-section view of 4 fibers array (250 μm pitch).

FIGURE 6.32. Cross-section view of 8 fibers array (500 μm pitch).

6.7. CONCLUSIONS AND DISCUSSION

In this chapter, several conventional passive alignment methods have been discussed. These methods have certain drawbacks which are mainly caused by the buoyancy of epoxy. Though the buoyancy can be overcome by using a cover plate, a high profile SiOB instead of a low profile one must be used. However, voids and air bubbles are easily found in the epoxy. This creates problem in the long-term reliability and position stability.

The situation can be improved by dispensing epoxy at the edge of the cover plate. Experiments prove that in this case, a low profile SiOB can be used and no void and air bubble are trapped in the epoxy. However, the cover plate is tilted when there is only one fiber on the SiOB. The epoxy underneath the tilted cover plate generates non-uniform distributed force. Both the optical fiber and the cover plate suffer from high compressive stresses.

Based on the advantages of the epoxy flow observed in the edge dispensing process, a modified passive alignment method is introduced. The epoxy is dispensed in the reservoir and flow into the gap between the optical fiber and the bottom of the V-groove. Parametric studies and real time monitoring experiment show this method has a self-align property. With this method, the optical fiber will align with the V-grooves without the use of cover plate.

The modified passive alignment method only eliminates the use of cover plate but does not improve the alignment accuracy. The overall alignment depends heavily on the precision of the V-groove fabrication process and the geometry of the optical fiber. This is the major disadvantage of the passive alignment when compared with active alignment.

REFERENCES

1. M.F. Dautartas, J. Fisher, H. Luo, P. Datta, and A. Jeantilus, Hybrid optical packaging, challenges and opportunities, Proc. 52nd ECTC, San Diego, CA, May 2002, pp. 787–793.
2. M.W. Beranek, et al., Passive alignment optical sub-assemblies for military/aerospace fiber-optic transmitter/receiver modules, IEEE Transactions on Advanced Packaging, 23(Aug.), pp. 461–469 (2000).
3. G. Keiser, Optical Fiber Communications, McGraw-Hill, New York, 2000.
4. D.K. Mynbaev and L.L. Scheiner, Fiber-Optic Communications Technology, Prentice Hall, New Jersey, 2001.
5. R.R. Tummala, Fundamentals of Microsystems Packaging, McGraw-Hill, New York, 2001.
6. F.G. Smith and T.A. King, Optics and Photonics, John Wiley & Sons, Chichester, 2000.
7. J.A. Buck, Fundamentals of Optical Fibers, John Wiley & Sons, Chichester, 1995.
8. P. Karioja, et al., Comparison of active and passive fiber alignment techniques for multimode laser pigtailing, Proc. 50th ECTC, Las Vegas, ND, May 2000, pp. 244–248.
9. S.H. Law, T.N. Phan, and L. Poladian, Fibre geometry and pigtailing, Proc. 51st ECTC, Orlando, FL, May 2001, pp. 1447–1450.
10. K. Ishikawa, An integrated micro-optical system for laser-to-fiber active alignment, Proc. IEEE 15th MEMS, Jan. 2002, pp. 491–494.
11. J. Goodrich, A silicon optical bench approach to low cost high speed transceivers, Proc. 51st ECTC, Orlando, FL, May 2001, pp. 238–241.
12. R. Hauffe, U. Siebel, K. Petermann, R. Moosburger, J.-R. Kroop, and F. Arndt, Methods for passive fiber chip coupling of integrated optical devices, IEEE Transactions on Advanced Packaging, 24(Nov.), pp. 450–455 (2001).
13. R.A. Boudreau, Passive alignment in optoelectronic packaging, Optical Fiber Communication, OFC 97, Feb. 1997, pp. 109–110.
14. S.J. Park, et al., A novel method for fabrication of a PLC platform for hybrid integration of an optical module by passive alignment, IEEE Phton. Technol. Lett., 14(Apr.), pp. 486–488 (2002).
15. H. Mori, et al., LD and PD array modules assembled in a new plastic package with auto-alignment projections for silicon optical bench, Pro. OFC/IOOC, 3(Feb.), pp. 198–200 (1999).

16. G. Grand and C. Artigue, Hybridization of optoelectronic components on silicon substrate, Proc. ECOC'94, 1994, pp. 193–200.
17. R. Moosburger, B. Schüppert, U. Fischer, and K. Petermann, Passive alignment technique for all-silicon integrated optics, Proc. Integr. Photon. Res., Boston, MA, IWH3, Apr. 1996, pp. 565–568.
18. R. Moosburger, R. Hauffe, U. Siebel, D. Arndt, J. Kropp, and K. Petermann, Passive alignment of single mode fibers to integrated polymer waveguide structures utilizing a single mask process, IEEE Photon. Technol. Lett., 11, pp. 848–850 (1999).
19. M.W. Beranek, et al., Passive alignment optical sub-assemblies for military/aerospace fiber-optic transmitter/receiver modules, IEEE Trans. Advanced Packaging, 23(Aug.), pp. 461–469 (2000).
20. M.F. Grant, et al., Self-aligned multiple fibre coupling for silica-on-silicon integrated optics, Proc. 9th Annual European Fibre Optic Conference, London, UK, Jun. 1991, pp. 269–272.
21. J.W. Osenbach, et al., Low cost/high volume laser modules using silicon optical bench technology, Proc. IEEE 48th ECTC, May 1998, pp. 581–587.
22. K. Kurata, et al., A surface mount single-mode laser module using passive alignment, IEEE Transactions on Components, Packaging, and Manufacturing Technology, 19(3), pp. 524–531 (1996).
23. K. Yamauchi, et al., Automated mass production line for optical module using passive alignment technique, Proc. 50th ECTC, Las Vegas, ND, May 2000, pp. 15–20.
24. C.B. Probst, A. Bjarklev, and S.B. Andreasen, Experimental verification of microbending theory using mode coupling to discrete cladding modes, 7(Jan.), pp. 55–61 (1989).
25. C. Unger and W. Stocklein, Investigation of the microbending sensitivity of fibers, Journal of Lightwave Theory, 12(Apr.), pp. 591–596 (1994).
26. J.C.C. Lo and S.W.R. Lee, Experimental assessment of passive alignment of optical fibers with V-groove on silicon optical bench, Proc. 6th EPTC, Singapore, December 2004, pp. 375–380.
27. J. Lo, R. Lee, S. Lee, J.S. Wu, and M. Yuen, Modified passive alignment of optical fibers with low viscosity epoxy flow running in V-grooves, Proc. IEEE 54th ECTC, Jun. 2004, pp. 830–834.
28. J. Lo, C.S. Yung, R. Lee, S. Lee, J.S. Wu, and M. Yuen, Passive alignment of optical fiber in V-groove with low viscosity epoxy flow, Proc. ASME IMECE, Nov. 2003, paper IMECE 2003/43902.
29. K.E. Bean, Anisotropic etching of silicon, IEEE Trans Electron Devices, ED-25, pp. 1185–1193 (1978).
30. C.W. Chang and W.F. Hsieh, Micromachined double-side 45° silicon reflectors for dual-wavelength DVD optical pickup heads, Proc. IEEE 54th ECTC, Jun. 2004, pp. 1390–1395.
31. C. Strandman, et al., Fabrication of 45° mirrors together with well-defined v-grooves using wet anisotropic etching of silicon, Journal of Microelectromechanical System, 4(Dec.), pp. 213–219 (1995).
32. S.A. Campbell and H.J. Lewerenz, Semiconductor Micromaching Volume 1 Fundamental Electrochemistry and Physics, John Wiley & Sons, Chichester, 1998.
33. S.A. Campbell and H.J. Lewerenz, Semiconductor Micromaching, Volume 2, Techniques and Industrial Applications, John Wiley & Sons, Chichester, 1998.
34. E. Bassous, Fabrication of novel three-dimensional microstructures by the anisotropic etching of (100) and (110) silicon, IEEE Trans Electron Devices, ED-25, pp. 1178–1185 (1978).
35. M. Sekimura, Anisotropic etching of surfactant-added TMAH solution, Proc. IEEE 12th MEMS, Jan. 1999, pp. 650–655.
36. W. Sonphao and S. Chaisirikul, Silicon anisotropic etching of TMAH solution, Proc. IEEE ISIE, Jun. 2001, pp. 2049–2052.

RELIABILITY AND PACKAGING

7

Fundamentals of Reliability and Stress Testing

H. Anthony Chan

Department of Electrical Engineering, University of Cape Town, Rondebosch, 7701, South Africa

This chapter discusses the concepts which in the author's opinion are fundamental to understand the reliability of electronics and packaging. It also summarized some conventional reliability backgrounds.

Reliability in electronics and packaging is often interpreted differently in different contexts. In some reliability programs, reliability is an interdisciplinary science aimed at predicting, analyzing, preventing and mitigating failures over time. To a manufacturer, reliability may simply be the probability of "not failing" for a "specified" period of time and under "specified" conditions when used in the manner and for the purpose intended. In a seller market where the demand from the customers exceeds the availability of the merchandise, it is tempting for the manufacturers to make the specifications only according to the technical capability of their products. Yet the customers who encounter failures may often be upset. The customers may disagree with the manufacturers on various issues, especially when the specifications are made from the perspectives of the manufacturers alone. Especially in a buyer market, reliability is then the avoidance of failures as experienced by the customers and defined by the customers.

It follows that a good approach in reliability is to understand the causes of failures and then to avoid these causes.

Section 7.1 gives a non-technical discussion of the challenges and trends in reliability, which is a non-technical introduction for management staff, application engineers, and anyone interested in reliability without going into technical depth. Two questions are often addressed. One wants to know how often failures occur. The fundamentals on failure rate and failure distributions are given in Section 7.2. One also wants to understand why failures occur by conducting failure analysis and root cause analysis. Section 7.3 list the failure mechanisms in electronics and packaging. Reliability programs to improve reliability are explained in Section 7.4. The fundaments of product weaknesses and stress testing are given in Section 7.5. Finally, the formulation of stress testing is explained in Section 7.6.

7.1. MORE PERFORMANCE AT LOWER COST IN SHORTER TIME-TO-MARKET

The challenges faced in the electronics industry is not just how to make more reliable products but to do it in the ever more competitive market. Reliability is becoming essential in electronics as the electronics are being used in more products that have become part of human life and culture. Product performance is increasing. Yet product cycles are short, and product cost continues to be cut.

7.1.1. Rapid Technological Developments

The electronics and packaging technology has been rapidly progressing toward higher complexity, system integration, and product miniaturization. With these fast technological improvements, one is always working on the reliability of a new product. Failure modes that did not show up in a previous technology may now be important under factors like increased power, density and speed of operation. New failure modes may also arise with new materials and manufacturing processes under a new technology.

In addition, the three major products of communications, computing and consumer electronics have merged. Before the late 1990's, these three types of products had different reliability requirements. A telecommunication system may not tolerate any failure in over 30 years. Very high reliability is an important consideration for a customer to buy such a system although it may cost more to achieve that reliability level. On the other hand, the sale of a consumer product may be affected primarily by its price and features. A comparatively lower reliability level may be enough. For example, a failure fraction of 2–5% or higher over a product life of 5–10 years is often reported for consumer products. Software products with high failure rates have been penetrating the market of personal computers. Yet, when a low cost product is merged with other systems that need to be highly reliable, this product is contributing to and interacting with the reliability of the overall system. The reliability requirements for these low cost products need be higher now.

7.1.2. Integration of More Products into Human Life

Products from calculators to microwaves and cellular phones are changing human life. As people are becoming more dependent on these products, product failures are affecting people more than before. In the last millennium, vendors often advertised their products in terms of functionality alone. The warranty periods were also short for these products (e.g., 1 year or 90 days). Today, more customers are concerned about possible failures and tend to check with their friends and relatives before they buy a product. Customers also often read reports on the repair history of a commodity under different brand names. As for manufacturers, they now tend to adopt a longer warranty policy to compete in the market. Warranty periods from three to seven years are becoming competitive edges.

7.1.3. Diverse Environmental Stresses

The field stresses seen by customers of electronic devices are diverse and dynamic. Portable and hand held electronics are also vulnerable to outdoor environments. The extreme temperatures (e.g., −20 to 70°C) and temperature cycles differ in different parts of the world. Humidity and the extent of corrosive contamination in the air also differ. Handling by hands and tools often imposes hazardous stresses. These include mechanical shocks when accidentally dropped and electrostatic discharges of several thousand volts when a metallic lead is touched by another conducting body. In addition, users are usually not trained operators and often do not read manuals. Random on and off switching, unexpected modes of operation, attempts to plug into an improper power source and attempts to connect to an incompatible interface are common mistakes. If the products are not robust against the stresses under these user environments, excessive failures may result.

7.1.4. Competitive Market

Today, the world has moved to an Age of Cost Reduction, which is affecting industry, government, commerce and practically everything else. The need to cut cost in highly competitive markets tends to leave reliability efforts to a minimum. Yet, short-sighted cost cutting at the expense of product reliability is expensive in the long run. Early product failures result in warranty repairs, which usually cost much more than would reliability programs to avoid these failures. Early product failures also affect the buying decisions of customers on both current and future products under the same brand name.

7.1.5. Short Product Cycles

Short product cycles make traditional reliability programs difficult to implement owing to the short failure history available and a lack of data for failure analysis for new products. The urgency to bring a product to market in a short product cycle can no longer accommodate reliability programs that are passively only in response to field returns. A pro-active approach is needed to consider reliability, starting before the product design stage.

7.1.6. The Bottom Line

Despite the reliability concerns, revenue growth is the bottom line for investors and corporate owners whose investments govern manufacturers and their research and development programs. Cost-cutting programs have been abundant for many companies since the mid 1980s. The short-term benefits of cost- cutting usually dominate over the long-term health considerations of the business. Few product and process owners realize the long-term importance of reliability to the business. Even for the minority who are willing to put resources into reliability programs, they still need to see the monetary returns of their investment in reliability.

A major selling point of reliability is the avoidance of repair costs, which grow exponentially as a product goes through various stages from early design to maturity. A full stream cost consideration, which includes the repair cost, is needed. In addition, the effects of failures on customers may be serious. Yet, the effects on the vendor may again be judged by how much business will be lost when the investment in proper reliability efforts is not in place.

Some products may be critical to the customer's revenue. For example, a service provider that buy telecommunication systems ought to be cautious that a failure may result in substantial losses to the provider. The cost of providing redundancy to avoid such losses can also be high. For such situations, it is critical for the manufacturer to ensure the highest reliability of the product being sold to the service provider. In other markets, such as consumer products, a customer may buy a product based principally on first cost or features. Yet, the product reliability should still be sufficient for the customer. Excessive failures will damage the brand name of the manufacturer, resulting in loss of sales even for other unrelated products under the same brand name. For either high reliability systems or consumer products, it is important to have a clearly defined reliability objective and understand the economic factors influencing the setting of that objective.

7.2. MEASURE OF RELIABILITY

Two questions are often addressed. One wants to know how often failures occur by measuring the failure rate. One also wants to understand why failures occur by conducting failure analysis and root cause analysis. These data are helpful to the design and manufacturing processes to prevent failures. We summarize the measure of reliability first.

The reliability of different systems may be characterized in different ways.

Failure Rate (Hardware components and systems): The metric for the reliability of hardware components and systems is often expressed in terms of the measure of "unreliability" or the failure rate, which is also known as hazard rate.

Failure Intensity (Software): The metric for the reliability of software is the same as that for hardware systems but is called failure intensity.

Availability (Service): The metric for reliability of service is often called availability. The availability of a system is the probability that the system will be available to perform the intended actions.

Downtime (Computer, Telecommunication): The reliability for computer and telecommunication is often measured as the downtime.

Risk (network): The reliability for a network is also called the risk.

7.2.1. Failure Rate

Cumulative distribution function, or Cumulative fraction failed, $F(t)$ is the probability that a system first fails at or before time t. Denoting the service life of a product by ts Cumulative fraction failed over the service life is $F(ts)$.

Reliability function $R(t)$, or Survivor function $S(t)$ is the probability that a system survives to time t without failure:

$$R(t) = S(t) \equiv 1 - F(t). \tag{7.1}$$

Probability density function $f(t)$ is the probability of failure per unit time (per unit product born at time $t = 0$) occurring at time t. $f(t)$ is related to $F(t)$ by

$$f(t) \equiv \frac{d}{dt} F(t), \tag{7.2}$$

but $f(t)$ is NOT the failure rate.

The (instantaneous) failure rate $\lambda(t)$, or Hazard rate $h(t)$, is defined as the probability of failure per unit product that was working at time $<t$. This failure rate is equal to $f(t)$ over the fraction that survives to time t.

$$h(t) = \lambda(t) \equiv \frac{f(t)}{1 - F(t)}. \tag{7.3}$$

The probability density function $f(t)$ is therefore not identical to the failure rate because some product units born at $t = 0$ would have already failed before time <0, leaving only a the fraction $1 - F(t)$ to survive to time t.

Failure in time FIT is a unit of $\lambda(t)$ often used for electronic components. It is equal to the number of failures per 10^9 hours.

When $\lambda(t)$ is equal to n FITs, it means

$$\lambda(t) = \frac{n}{10^9 \text{ hours}}. \tag{7.4}$$

$$10 \text{ FITs} \cong 1 \text{ in} 10^4 \text{ per year.} \tag{7.5}$$

Failure rate is time dependent.

Instead of counting the number of failures per working product unit per unit time, an alternate measure is the time from time $= 0$ to the time a working unit has just failed. This is called the time to failure for any given particular product unit. The arithmetic average over a large number of product unit of such time to failure is called the mean-time-to-failure, MTTF, of the product. Many electronic components, modules, and sub-systems are non-repairable, so that the only meaningful MTTF for them is the mean-time to the first failure.

Mean Time to Failure MTTF is the average time to first failure:

$$\text{MTTF} \equiv \int_0^\infty dt \, t f(t). \tag{7.6}$$

For a repairable product unit, it may be repaired after the first failure and is then working again. After some time, it may fail again. This time is the time between the first and second failures. In the particular case that the average time (over many units of the same product) between the first and second failures is equal to that between the second and third failures and so on, we can define the mean-time-between-failures, MTBF, for that product. Systems comprising separate components or modules are often repaired by replacing the defective sub-systems with new sub-systems, so that one may talk about the MTBF if we assume that the rest of the system were as good as when they were new.

For repairable systems and assuming the repairs are ideal, the Mean time between failures MTBF is the average cycle time including operation and down times for the system:

$$\text{MTBF} \equiv \text{MOT} + \text{MDT}, \tag{7.7}$$

where MOT(t) is mean operating time, and MDT(t) is mean downtime.

The availability is related to MOT and MDT by

$$\text{Availability} \equiv \frac{\text{MOT}}{\text{MOT} + \text{MDT}}. \tag{7.8}$$

7.2.2. Systems with Multiple Independent Failure Modes

A system usually has multiple failure modes. In the simplest case, let us assume or take the approximation that these failure modes are independent of each other. Such a system may be modeled as a series system or parallel system.

A series system fails when any (one) failure mode occurs. An example of a series system is an IC that may fail caused by electromigration, overvoltage, or corrosion, etc. Another example is a system consisting of different subsystems.

Consider a series system with N independent failure modes. If the reliability functions for each of the individual failure modes are $R_1(t), R_2(t), \ldots, R_N(t)$, the reliability function of the system $R_{12\ldots N}(t)$ is then given by

$$R_{12\ldots N}(t) = R_1(t) R_2(t) \ldots R_N(t). \tag{7.9}$$

When $F_1(t), F_2(t), \ldots, F_N(t)$ are small, i.e., $R_1(t) = 1 - F_1(t)$ is close to unity, one can take the approximation to relate the cumulative functions:

$$F_{12\ldots N}(t) \cong F_1(t) + F_2(t) + \cdots + F_N(t). \tag{7.10}$$

When $F_{12\ldots N}(t)$ is also small, one may take further approximation to relate the failure rates by:

$$\lambda_{12\ldots N}(t) \cong \lambda_1(t) + \lambda_2(t) + \cdots + \lambda_N(t). \tag{7.11}$$

A parallel system with N failure modes fails when all N failure mode occurs. Consider a communication system linking A and B with N channels. If some of these channels are down, communication between A and B will be downgraded. Yet in the case that a low data rate is needed, the communication will be out when all the N channels are out.

If the cumulative distribution functions for each of the individual failure modes are $F_1(t), F_2(t), \ldots, F_N(t)$, the cumulative distribution function of the system $F_{12\ldots N}(t)$ is then given by

$$F_{12\ldots N}(t) = F_1(t) F_2(t) \ldots F_N(t). \tag{7.12}$$

When $F_1(t), F_2(t), \ldots, F_N(t)$ are small, one can take approximation to relate the failure rates by:

$$\lambda_{12\ldots N}(t) \cong \lambda_1(t) \lambda_2(t) \ldots \lambda_N(t). \tag{7.13}$$

The parallel system design provides a way to achieve much smaller failure rate of the overall system than that of the individual subsystems by providing redundant subsystems. Yet there is cost in providing redundancy.

7.2.3. Failure Rate Distribution

The dependency of failure rate on time is typically pictured with a bath-tub curve. At large times, the failure rate rises up as the product wears out. At much shorter time typically within the first year for electronics, the failure rate may also be higher. These early failures are also known as infant mortality and their failure mechanisms are generally different from those of the long-term reliability.

7.2.3.1. Weibull Failure Rate Distribution

One way to model the failure rate distribution is to use the Weibull distribution:

$$\lambda(t) \equiv \lambda_1 t^{-\alpha}, \text{ where } \alpha < 1. \tag{7.14}$$

Weibull distribution may be derived for the failure rate of a system with a large number of independent and competing failure mechanisms, and is therefore applicable to many different systems. This failure rate is quite general in the sense that it remains constant (exponential distribution) for $\alpha = 0$, decreases with time for $0 < \alpha < 1$, and increases with time for $\alpha < 0$. In order to model the bathtub behavior of the failure rate, one may model the different regions separately.

The reliability function for Weibull distribution is given by

$$1 - F(t) = R(t) = \exp\left(-\frac{\lambda_1 t^{1-\alpha}}{1-\alpha}\right) \equiv \exp\left[-\left(\frac{t}{t_e}\right)^{1-\alpha}\right]. \tag{7.15}$$

The term $1 - \alpha$ is called the Weibull shape, and the time when $R(t)$ has dropped to $1/e$ is t_e.

It is customary to plot $\ln F(t)$ against $\ln t$, called the Weibull plot.

Taking natural log twice for the reliability function, one obtains

$$\ln\left[\ln\frac{1}{1-F(t)}\right] = (1-\alpha)\ln t + \ln\frac{\lambda_1}{1-\alpha}. \tag{7.16}$$

In the approximation with small $F(t)$, the left hand side is approximately equal to $\ln F(t)$, so that the relation between $\ln F(t)$ and $\ln t$ is linear.

$$\ln F(t) \cong (1-\alpha)\ln t + \ln\frac{\lambda_1}{1-\alpha}. \tag{7.17}$$

The slope in the Weibull plot is the Weibull shape. When there are 2 different failure modes that dominates in different regions of time, the Weibull plot will change from a straight line segment in one range of time with one value of slope to another straight line segment in another range of time with a different value of slope. This is called the S-shape Weibull plot.

The above Weibull distribution is also called 2-parameter Weibull distribution. One physical difficulty with the 2-parameter distribution is that there is a non-zero probability of occurrence at all values of $t < 0$. A generalization is the 3-parameter Weibull distribution with the reliability function given by:

$$1 - F(t) = R(t) \equiv \exp\left[-\left(\frac{t-t_0}{t_e - t_0}\right)^{1-\alpha}\right], \quad \text{for } t > t_0, \tag{7.18}$$

and

$$1 - F(t) = R(t) \equiv 1, \quad \text{for } t < t_0. \tag{7.19}$$

This distribution includes a threshold time to failure t_0 so that failure may not occur before the time t_0. This is the case when certain processes must proceed to certain extent before an event may occur. An analogy is an egg hatching process. A series of certain metabolism and growth processes must precede taking a minimum number of days. The egg therefore cannot hatch before that minimum amount of time.

7.2.3.2. Lognormal Failure Rate Distribution Another way to model failure rate distribution is to use lognormal distribution for which the probability density function $f(t)$ versus the natural logarithm of time $\ln(t)$ exhibits normal distribution with a mean of $\ln(t_{50})$ and a standard deviation of σ.

$$f(t) \equiv \frac{1}{t\sigma\sqrt{2\pi}} \exp\left[-\frac{1}{2}\left(\frac{\ln t - \ln t_{50}}{\sigma}\right)^2\right]. \tag{7.20}$$

The cumulative distribution function is given by

$$F(t) \equiv \frac{1}{\sigma\sqrt{2\pi}} \int_0^t \frac{dt'}{t'} \exp\left[-\frac{1}{2}\left(\frac{\ln t' - \ln t_{50}}{\sigma}\right)^2\right], \tag{7.21}$$

which can be expressed as a function of $\ln(t)$ as follows:

$$F(t) = \frac{1}{\sqrt{\pi}} \int_{-\infty}^{\frac{\ln t}{\sqrt{2\pi}}} d\left(\frac{\ln t'}{\sqrt{2}\sigma}\right) \exp\left[-\left(\frac{\ln t'}{\sqrt{2}\sigma} - \frac{\ln t_{50}}{\sqrt{2}\sigma}\right)^2\right]. \tag{7.22}$$

The lognormal failure rate distribution is usually applicable when there is a process such as diffusion, electromigration, corrosion, etc. that takes time to proceed to failure.

7.3. FAILURE MECHANISMS IN ELECTRONICS AND PACKAGING

An often asked question is why failures occur. It is customary to conduct failure analysis on failed product units to find out the failure modes. The reliability program will then try to avoid these failures. A list of failure mechanisms in electronics and packaging are given here.

7.3.1. Failure Mechanisms at Chip Level Include

Bulk semiconductor: second breakdown, latch-up, single-event upset, bulk radiation effects, chip fracture.

Semiconductor–dielectric interface: alkali ion migration, slow trapping instability, hot-carrier effects, surface charge spreading, polarization, ionizing radiation effects.

Dielectric: time-dependent dielectric breakdown, dielectric wear-out, fracture, passivation layers, EOS/ESD breakdown.

Conductor and metalization: electromigration, microcracks, corrosion, metal migration, contact spiking, hillock formation.

7.3.2. Failure Mechanisms at Bonding Include

Die bond: voids in bonding alloys or adhesives, disbonding, excessive thermal resistance, contamination from die bonding materials, die cracking by die-bond-induced stress.

Wire bonds: underbonding, overbonding, purple plague (Au-Al intermetallic), contaminant effects, wire break above bond.

7.3.3. Failure Mechanisms in Device Packages Include

Hermetic packages: leak–moisture–corrosion/electrical isolation, movement of bond wires and loose particles, contaminants–halide–Al metal film corrosion, electrolytic etching of Au wire bonds.

Plastic encapsulation: moisture penetration along plastic–lead interface or through plastic—delamination and popcorning during IR solder reflow—corrosion of chip metalization; package stresses from temperature cycling with TCE mismatch (large package)—shear bond wire, dielectric stress cracking, open, short; Ionic impurities, fire retardant—corrosion; radiation from plastic materials.

Electrochemical: solderability, bondability, delamination/adhesion, corrosion, ionic contamination, electromigration.

Thermomechanical: thermal conductivity deficiency, thermal stress due to CTE mismatch, delamination/popcorning, lead bend fatigue, lead break, stress/die attach pad shift.

7.3.4. Failure Mechanisms in Epoxy Compounds Include

Package assembly: moldability (flow and release), wire sweep, void level, mark permanency, wire bond pull strength, line movement (chip metal traces), board adhesion.

Failure mechanisms in soldering include:

Package defect: delamination, popcorn, microcrack.

Joint: intermetallic, non-wetting (oxidation, contamination), no solder—tombstone problem (for small chip discretes); No solder—lead coplanarity problem; no solder—large heat sinking, no flux; Icycle—insufficient flux for wave soldering; re-reflow problem—especially for high lead-count fine-pitch packages; low solder—weak against temperature cycling; excessive voids.

7.3.5. Failure Mechanisms at Shelf Level Include

Mechanical: poor tolerance—misalignment, structural failure, loose interconnection, connector fall off.

Thermal—improper cooling: nonuniform airflow, blocking airflow—some circuits too hot; excessive back pressure; airflow leakage.

7.3.6. Failure Mechanisms in Material Handling Include

Shelf life, electrolytic capacitors (oxide dielectric dissolves); lead solderability (oxidation); moisture absorption—popcorn; connector gold finger contamination; gouged thick film; vacuum pickup, kitting, board flexure; bent lead; mechanical shock—crystals/oscillators; ESD/EOS.

7.3.7. Failure Mechanisms in Fiber Optics Include

Component: output power degradation; facet erosion; dark-line defect—nonradiative recombination due to dislocations; misalignment at fiber-component connection–assembly problem: mounting, soldering, welding, fiber jacket shrinkage—connector misalignment.

Fiber: microcrack growth; fiber connection misalignment; fatigue-stress from fiber bend, stress corrosion; coating problem, moisture.

Fiber cut: digging ground, shark bite, animal.

Static stress fatigue from surface damage: surface cleanliness—fiber manufacture process, OH environment, minimum bending radius.

Passive components (splitters, couplers): failure modes of active components are absent but need high yield, high reliability interconnections.

7.3.8. Failure Mechanisms in Flat Panel Displays Include

Polycrystalline Si defects; rubbing process problem (line mura); cell gap problem (uniformity, chromaticity, flicker, contrast variations); spacer balls/backlight non-uniformity; high concentration of spacer balls (spot mura); backlight variations (color defects); polarizer misalignment; Na ion; open/short in array (pixel, line defect); electrical bias (flicker); close grouping of pixel defects (cluster); poor bonding, defective driver (driver related pixel and line defects); fillport contamination (fillport mura); fillport seal degradation (moisture leak to liquid crystal); uncured epoxy reacting with liquid crystal (edge mura).

7.4. RELIABILITY PROGRAMS AND STRATEGIES

Reliability needs attention in every stage of product development and manufacture. In the design process, one may build in enough design margins to avoid failures or make the design more robust against failures. In the manufacturing process, one may tighten process variations or use more robust processes to avoid failures. In the testing process, one may test whether the products have achieved adequate level of reliability and gather useful information needed to avoid future failures.

Many different reliability methodologies exist. As failure mechanisms are interdisciplinary, caution is needed to extrapolate beyond the bounds of any successful program. In the earlier days of electronics (before 1980's), much emphasis was to achieve long-term reliability. The studies for long-term reliability were primarily using elevated temperature to accelerate a large class of failure mechanisms in electronics that obey Arrhenius equation. As those failure mechanisms were under control during the IC fabrication process, the additional failure mechanisms introduced in the electronic packaging processes are different and do not need to follow Arrhenius equation. In addition, elevated temperature is not the only important acceleration factor. Many other stresses need to be taken into account.

While the field of long-term reliability is generally mature, infant mortality is less well understood. Early failures are now presenting more problems in some products. In addition, many products have a low cost requirement. The additional cost of built-in redundancy may be perceived as incompatible to cost reduction efforts. Some vendors also tend to focus on short-term cost reduction and overlook the effects of reliability on the long-term business.

Although various reliability programs are available for implementation at different stages of development for any given product, resources for reliability programs may be limited. It is therefore a challenge to allocate these resources to implement the most cost effective reliability programs. Reliability engineers generally agree that pro-active reliability measures are more cost effective and deserve higher emphasis. Yet in the commercial world, the measures to fix problems in response to field returns and test data may be more common, even though such measures are more expensive than the pro-active measures. The reason is that the measures needed to improve product robustness may be cost effective only for the full-stream cost in the overall product cycle. The need to invest in pro-active reliability programs at an early stage of product development is therefore often not obvious. Pro-active reliability programs are consequently candidates for cancellation in this Age of Cost Cutting. Even when pro-active reliability programs have been approved, the available resources are limited and one still needs to optimize the effectiveness of the investment in reliability.

Both short term and long-term reliability strategies are needed for every product family. Yet, the current state-of-the-art for the reliability programs of many products is lagging behind the technologies that produce these products.

The field of reliability needs much exploratory work to advance. As the world turns to focusing more on short-term investments, support from companies and the government for more forward looking work are desired. Publications, sharing of information and discussions among different companies are fruitful. While the research efforts of individual companies may be limited, sharing of information among these companies provides a pool of resources to advance this field. If this field advances faster, a better understanding of the basic science and more effective reliability programs will become available to all, creating a win-win situation for these companies.

7.5. PRODUCT WEAKNESSES AND STRESS TESTING

7.5.1. Why do Products Fail?

The causes of failures are product dependent and are numerous for any given product or process. A partial list of failure modes are given in the previous section. Yet, the causes of hardware failures may be categorized as problems in design, materials, manufacture, test and field use, as shown in Figure 7.1. The causes of software failures may be categorized in a similar way.

The following is a systematical understanding of failures and stress testing without being complicated by the specific behavior of different products and their numerous failure behaviors.

Any unit of a product at any stage during development and manufacture may be grouped according to the explicitness of their defects. Thus, these units may belong to the good (solid line), weak (dashed line) or bad (dotted line) groupings, as shown in Figure 7.2.

A product unit can be thought of as going through a process, such as design, manufacture or a test process at every stage of development and production. Each process can be modeled as a process operator. The product unit may either pass or be dropped out in passing through an operator. For each product unit that passes the operator, it may undergo a transition from one group (e.g., good) to another group (e.g., bad).

Figure 7.3 shows three types of operators: process operator, ideal stress operator, and ideal test operator.

Good product units meet all design objectives under all nominal stress conditions. Bad product units have hard defects or parametric deficiencies. These products are, in principle, detectable although they may escape the tests because practical tests are non-ideal and do not have 100% test coverage. It is important to note that an ideal test with 100% test

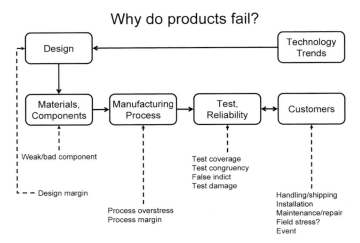

FIGURE 7.1. The causes of hardware failures may be categorized as problems in design, materials, manufacture, test and field use.

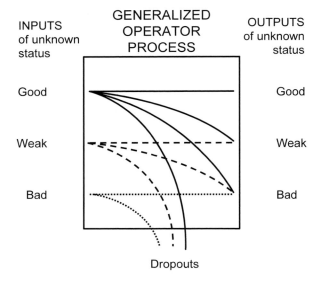

FIGURE 7.2. A model to categorize each product as in the state of good, weak, or bad in different stages during development and manufacture. Each stage is an operator that may change the state of the product.

coverage can only detect all the bad units but as will be explained later cannot detect the weak units.

Weak product units may be degradable or marginal. Marginal products may fail intermittently or only under certain stress conditions. Degradable products have latent defects and may degrade irreversibly into bad or marginal products. The degradation may be stimulated or accelerated by certain stresses.

The ideal stress operator will change the state of all weak units into the bad state during the stressing process.

In software reliability, a hard defect is a fault and a product weakness is a dormant fault.

FUNDAMENTALS OF RELIABILITY AND STRESS TESTING

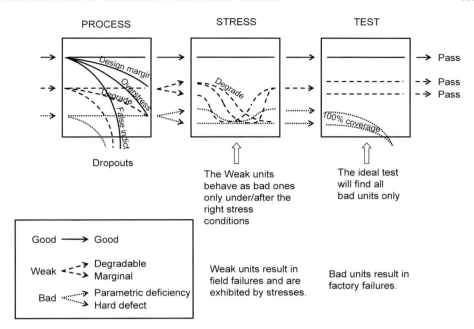

FIGURE 7.3. Three types of operators: process operator, ideal stress operator, and ideal test operator.

The process operator in Figure 7.3 may be any process such as design, development, or manufacture. It shows that both weak and bad products are produced with problems in design margin, manufacture, assembly, or material quality control.

Let us define an ideal fault-detection process as one that does not stress the product and yet achieves 100% test coverage. In the absence of stress, only the bad product units are detected. Weak product units are not detectable. They pass these tests and are shipped to customers, resulting in early failures in the field. Weak product units may degrade or exhibit failures only under certain stress conditions.

We also define an ideal stress process as one that degrades all degradable product units and shows up all marginal units without damaging the good ones. The exit from the ideal stress operator shows what happens to the units upon the removal of stresses. Here, the formerly degradable units remain in the bad state whereas the formerly marginal units return from the bad state to the weak state.

7.5.2. Stress Testing Principle

The principle of stress testing is shown in Figure 7.4. We have conceptually separated test into stressing and fault-detection with 2 separate roles. 100% test coverage refers to successful fault-detection of all bad units, whereas an ideal stressing process turns all hidden faults in the weak units into detectable faults, at least temporarily. The ideal stress-testing operator is simply the cascade of the ideal stressing and the ideal fault-detection operators. This cascaded operator shows that testing with the above stresses will find both weak and bad product units. Corrective actions may then be taken.

Notice that the cascaded process includes testing both during and after the stressing process. The weak product units that are marginal need to be tested while they are subjected

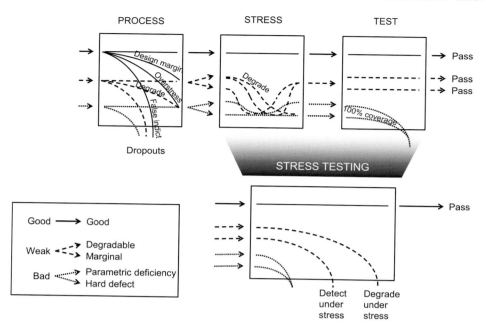

FIGURE 7.4. The principle of stress testing is shown by cascading the ideal stressing and the ideal fault-detection operators to obtain an ideal stress-testing operator.

to the proper stress conditions. One the other hand, the degradable product units may be tested after the stress conditions have been applied to degrade them into bad units.

7.5.2.1. Screening versus Corrective Actions Stress tests include screening without corrective actions and stress testing with corrective actions, shown in Figure 7.5.

Environmental Stress Screening (ESS). ESS applies stresses to stimulate observable failures for weak units. The stress level applied during screening needs to be carefully adjusted to achieve an efficient screening process. A mild test regimen may pass too many weak units whereas too much stress may damage good ones. Note that screening only screens out weak units. It cannot improve the strength of the rest of the units. Although one can reduce field failures by rejecting the weak units, a draw back of screening is that the weak units continue to be manufactured.

Accelerated Stress Testing with Corrective Actions (AST). In AST, stresses are used to identify the weaknesses of the product and the root causes of these weaknesses are investigated. Corrective actions are then taken to avoid producing these weaknesses through achieving robust design, excellent component quality, and robust processes. Once these objectives have been met, product reliability is assured without screening. Yet, AST on a sampling basis (Production-Sampling AST) to monitor and to maintain high quality may be required.

AST with corrective actions is generally most valuable during the design and qualification stage of a product (Design-Qualification AST). ESS may be applicable (Ongoing ESS) when a manufacturing process is new or has not been improved sufficiently. For example, some high performance VLSI IC components are manufactured using the newest technology, which is not yet mature, and tends to have low yield. In this case, ESS at the earliest possible level is useful to screen out the weak units.

Screening versus Stress Testing

- **Environmental Stress Screening (ESS)**
  - Apply stresses to stimulate observable failures for weak units.
  - Weak products are screened out, but may continue to be produced.

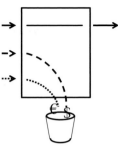

- **Accelerated Stress Testing (AST)**
  - Identify weaknesses of product to withstand stresses.
  - Take corrective actions to achieve product robustness.
  - Use sampling to monitor and maintain quality

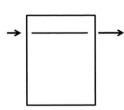

FIGURE 7.5. Comparison between stress screening and stress testing with corrective actions.

The concept of AST in hardware reliability may be extended to software reliability, as well as the reliability of a system where hardware weaknesses and software faults are interacting. Exceptions include the concept of screening, which is not applicable to software faults alone when all the product units have the same copy of software or operating system. The purpose of screening in this case will be to look for hardware and system weaknesses that show up under certain combinations of hardware and software stress conditions.

7.6. STRESS TESTING FORMULATION

7.6.1. Threshold and Cumulative Stress Failures

Products may fail in different ways when they are subjected to stress. The appropriate way to perform AST on a product depends on the nature of the failure modes under consideration. There are many failure modes, but in terms of stress-failure, there are three principal types.

7.6.1.1. Threshold Stress Failure A product typically exhibits a statistically distributed strength to withstand a given stress. In the simplest case, a failure occurs in a specific product sample when the stress level exceeds the threshold strength of the sample. A simple example of the threshold strength is the mechanical strength of a brittle material. Other examples are the threshold power to burn a device and the threshold electric field for avalanche breakdown to occur. A product with low-threshold strength may fail when a high maximum stress occurs during assembly, handling, transportation, installation, or field operation. On the other hand, a product with high-threshold strength is robust to resist these stresses. A screening process at a certain stress level may sort out the weak product units without damaging or wearing out the good ones, which possess threshold strength higher than the screening stress level. Yet, the product yield in screening is an also important consideration.

7.6.1.2. Cumulative-Stress Failure Some failures occur under the cumulative effects of repeated low-level stresses. Although these stresses may be well below the level that will cause immediate product failure, each time they are applied they produce some irreversible changes in the product. Thus, over repeated exposure to these stresses, the cumulative change grows to a level that results in product failure. Well-known examples of cumulative-stress failure are electromigration in a device under high-current density, and fatigue failure of solder joints due to extended thermal cycling. Screening weak products by applying stresses is not always appropriate for these failure modes because the cumulative effect of the damage produced in the units during screening will also shorten the product life of the good units.

7.6.1.3. Combined Threshold—Cumulative Stress Failure A combination of threshold stress and cumulative stresses may also stimulate certain failure modes. Here, a high maximum stress starts an incipient failure site that is later driven to a hard failure by cumulative stresses. For example, a device with a cracked package is more vulnerable to corrosion. Use of high stress to screen out such weak products needs caution. Exceeding the threshold stress alone may not show an immediately detectable failure. If the incipient failure site induced by exceeding the threshold strength is not driven to an observable failure, early field failure may occur.

A product that is robust to a large threshold stress, e.g., a big mechanical shock, may also be weakened by cumulative stresses, e.g., vibrations. After that, it may become vulnerable to a relatively smaller shock.

7.6.2. Stress Stimuli and Flaws

A variety of stress stimuli can be applied to a product as part of an AST regimen. We will review the most commonly used stimuli here, and note some of the types of product deficiencies they are likely to stimulate to failure.

Elevated temperature: testing a product for an extended period at an elevated temperature, or burn-in, is probably the most common form of stress testing. Marginal product designs often exhibit a temperature threshold above which the product will not function satisfactorily, and failure modes that involve chemical or diffusion processes can often be effectively accelerated at elevated temperature. However, for many electronic products there is a tendency to only test the product to the nominally specified upper temperature limit. By testing beyond the nominal temperature limit, one can better assess the robustness of the product.

Power cycling: turning a product on and off is another common form of stress testing; it is commonly done in conjunction with other types of stress testing. The temperature transients that occur during power-up can often stimulate thermal-mechanical defects. For electronic systems, the possibly variable conditions resulting from an abrupt shut-down or from on/off powering may also reveal design deficiencies.

Temperature cycling: a number of interconnection and packaging failure modes may best be stimulated by temperature cycling. The higher transient temperature conditions that occur during temperature cycling can also reveal design deficiencies not normally found during the slow transients normally associated with getting up to burn-in temperatures. Cycling to low temperatures can also reveal temperature threshold problems not covered by traditional burn-in.

Voltage variations: varying the voltage supplied to an electronic system can reveal design margin problems and marginal performance of specific components. This type of stressing is often combined with testing at temperature limits to increase the detection of marginal conditions.

Clock variations: varying the clock rate can reveal timing margin problems. This type of stressing is often difficult to achieve because of the high degree of integration of clocks in the product circuitry, and special design provisions must be made to make it feasible.

Vibration and mechanical shock: vibration or mechanical shock stressing has traditionally been used to reveal structural support problems, problems in the securing of specific (often large and/or heavy) components, and connectorization or cabling problems. In addition, problems with surface mount solder joints are also increasingly being addressed using vibration testing.

Elevated humidity: elevated humidity testing is usually done in conjunction with high temperature testing to reveal problems with corrosion or high voltage isolation breakdown. Extended testing is often required to get results.

Electrostatic discharge (ESD), power surge: specialized testing is often done for ESD and power surges to check the adequacy of isolation and grounding designs. Such testing is generally not performed on product units that are shipped.

Electromagnetic interference (EMI) susceptibility: EMI testing is also often done to verify design robustness, though often only at ambient conditions. However, EMI testing done in conjunction with temperature cycling can reveal additional problems with leakage paths not found at static conditions.

Software stresses: an excess load may be applied to a system while new software is being tested. Servers and database systems may be tested at high network usage situations that may overload the system. Besides extreme values, out-of-bound or invalid data may also be input.

Combination and order of stresses: combinations and interactions of any lists of stresses from above often produce new failure modes not observed with any single type of stress. The order of these stresses can also produce different modes. One way to pick the combinations and order of these stresses is to investigate the "Use Cases". Good questions are how and under what conditions the system will be used, and what can happen.

7.6.3. Modes of Stress Testing

We discuss several modes of doing AST, in the order of descending desirability.

Design-Qualification AST: Product is tested near the end of the design stage to see if it is robust with respect to stress levels in excess of those likely to be encountered in the use environment. If deficiencies are found, the root cause is determined and corrective actions are taken to fix the underlying causes of the problem. This is the most productive mode of doing AST because the benefits are realized over the whole life of the product.

Manufacturing-Qualification AST: A representative sample of product is subjected to AST during manufacturing ramp-up to identify deficiencies in component quality or manufacturing processes. In addition, design margin deficiencies that did not show up in design qualification AST may be found. The emphasis again is on Failure Analysis (FA) and corrective actions, so that deficiencies in the product can be quickly eliminated before production volumes become large.

Production-Sampling AST: For products requiring high reliability, it is useful to continue to perform AST on a sampling basis to monitor the production process for manufacturing or component quality variations, even after satisfactory qualification has been achieved. The emphasis continues to be on determining the root cause of any problems found and taking corrective action to fix them.

Ongoing AST: It sometimes occurs that AST is performed with the intention of doing good FMA and correcting any problems found, but because of a lack of sufficient resources, quality problems persist. In this case, it may still be economically feasible to continue doing AST on an ongoing basis, with some degree of FMA and corrective action to achieve a high-reliability product. Yet, this mode is certainly less desirable than the approaches mentioned above.

Ongoing ESS: Traditionally called environmental stress screening (ESS), this approach is mostly directed at sorting the good product from the bad, which is then repaired, but with little attention to FMA and corrective action. Although this mode may be suitable for limited production volumes where there is little opportunity for product improvement, it is not nearly as productive for higher volume products as the approaches mentioned above.

7.6.3.1. Acronyms of Stress Testing The following names of stress testing are ordered according to the approximate date of their first use in literature. Historically, many names have been given to emphasize its different aspects. Yet, as different companies are learning from each other, their processes will become less and less distinguishable. While these acronyms may appear different and confusing at this point, they may eventually be treated as meaning the same thing.

ESS—Environmental Stress Screening: the emphasis of ESS is in screening in the early developments of theses process in military applications, but people have begun to do corrective actions while still calling it ESS.

HALT—Highly Accelerated Life Testing: step stress to look for the upper/lower operational stress levels, and also the upper/lower destructive stress levels. Perform FMA and corrective actions to raise these levels. Iterate to raise these limits as far as possible. This term is originally used by Greg Hobbs.

HASS—Highly Accelerated Stress Screening: production screening process using HALT to establish elevated stress levels, which are usually established through HALT. This term is originally used by Greg Hobbs.

STRIFE—Stress test and improve design to achieve product robustness with longer service life. This term is originally used in Hewlett-Packard.

EST—Environmental Stress Testing: this term is originally used in Lucent Technologies, formally AT&T. The name tries to emphasize that the process is different from the conventional ESS process.

AST—Accelerated Stress Testing: this non-company specific term is originally used in naming the annually IEEE Workshop on AST. It also emphases that stresses can be general and are not limited to environmental stresses.

7.6.4. Lifetime Failure Fraction

Customers' perspective.

Reliability has different meanings in the literature. We take the perspective of the end customer here.

Knowing only the mean-time-between-failures, MTBF, is not enough for many applications. Customers need to be protected from both excessive infant mortality (time $< T1$) and early wear-out (time $> T2$), shown in Figure 7.6(a).

Customers expect reliability throughout the product's service life. A product is expected to properly function throughout the time it is in service. This time span defines

FUNDAMENTALS OF RELIABILITY AND STRESS TESTING

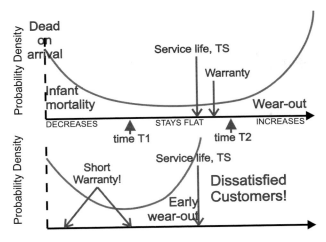

FIGURE 7.6. (a) Upper drawing: Example of a product life longer than service life. (b) Lower drawing: example of a product life shorter than service life.

the service life, *TS*, which may be up to ten years for some consumer electronics. Vendors sometimes only measure T2 only to compare with the warranty period, because of the warranty repair costs incurred by the vendor. However failures that occur after the warranty period continue to upset the customers. If a problematic product with a short T2 [Figure 7.6(b)] is marketed, the excessive failures right after the warranty period will simply turn future customers away. To prevent early wear-out, T2 needs to be large than the expected service life.

Reliability may be defined as the avoidance of failures, which customers see over the life of a product. The ability of a product to properly function just under the conditions specified by the vendor is not enough. If the product cannot withstand the various stresses that it may encounter in a customer's environment at any time during its service life, the customer will experience a product failure.

A convenient measure of reliability is the cumulative fraction failed over the product's service life. A traditional measure of a product's reliability is the (instantaneous) failure rate, which is the (instantaneous) probability density function $f(t)$ at age t divided by the fraction that has not failed. This failure rate is time-dependent and is complicated by infant mortality and early wear-out. For consumer electronics, all the product failures occurring from time $t = 0$ to $t = TS$ are important to the customer. Here, *TS* is counted from when a customer buys a product up to the time the customer replaces or forgets it. A simple measure of a product's reliability is the (time-independent) fraction failed over the product's service life, which is defined as

$$F(TS) \equiv \int_{t=0}^{t=TS} f(t)dt,$$

and is shown in Figure 7.7.

7.6.5. Robustness Against Maximum Service Life Stress

Product strength and lifetime maximum stress.

Examples of product strength distribution and lifetime maximum stress distribution are shown in Figure 7.8.

For threshold stress failures, each unit of a product needs to survive not simply the nominal stress but all the peak stresses during the product life. The highest peak stress encountered over the product life is defined here as the lifetime maximum stress, X. The maximum value of X that a specific product unit can withstand without failing is a measure of the robustness of that unit, and is called the product (yield) strength, Y.

For cumulative stress failures, one picture is to look at the instantaneous strength as being weakened over time by the (time dependent) instantaneous stress. Yet, it is desirable to skip the details of time dependence here and use the same unified formulation for both threshold and cumulative stress failures. This is achieved by defining the lifetime cumula-

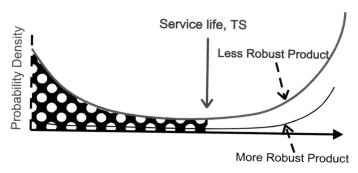

FIGURE 7.7. Fracture failed over service life is the area marked under the curve.

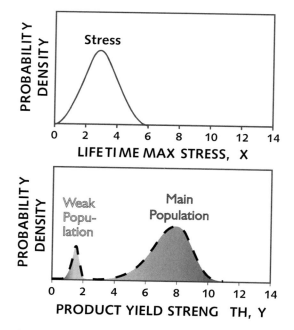

FIGURE 7.8. (a) Upper drawing of the probability density distribution of lifetime maximum stress. (b) Lower drawing of the probability density distribution of product yield strength.

FUNDAMENTALS OF RELIABILITY AND STRESS TESTING 197

tive stress, X, as an effective stress parameter that is proportional to the overall weakening of the instantaneous strength by a physical stress over the product life. The corresponding initial product strength against such an X is again denoted by Y. An example is 85°C, 85% humidity and under 5 V bias for 1000 hours.

The distribution of the lifetime maximum stress is determined by the customer's environment, whereas the product strength is a statistical distribution of the product units.

The product strength distribution usually has one or more weak sub-populations in addition to a main population. The weak sub-population is generally a main contributor to freak failures and infant mortality, which show up in a typical bathtub curve. Most units in the main population possess enough design margins to withstand incurred stresses. Therefore, for the case of threshold stress failure, they generally fail only under extreme stress conditions such as lightning-surge or electrostatic discharge (ESD). For the case of cumulative stress failure, they may fail only under true long-term wear-out. Yet, those falling in the low strength tail of the main population may also contribute to early wear-out.

The presence of weak sub-populations separated from the main population is consistent with the bathtub curve commonly observed in many products, and may model many hardware systems. Yet a wide distribution of the main population can also give rise to higher failure rate in the infant mortality stage. This latter category may be more appropriate for software weaknesses.

7.6.6. Stress–Strength Contour

The occurrence of field failures is determined by the distributions of both the stress X and strength Y. It is therefore convenient to look at the contour map of this joint probability distribution of X and Y (Figure 7.9).

In this contour map, the population lying in the $Y < X$ region will result in field failures whereas those in the $Y > X$ region will not fail in the field. An example of a weak population is one that lies mostly in the $Y < X$ region, while a tiny fraction on its left

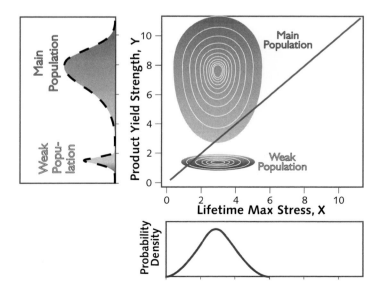

FIGURE 7.9. Joint probability density distribution of lifetime maximum stress and product yield strength.

does survive a benign stress environment. An example of a main population is one that lies mostly in the $Y > X$ region, but its lower-right part still falls into the $Y < X$ region. Then, even though the nominal strength of this main population may seem high enough compared to the nominal maximum stress, a significant fraction may still fail. These failures are owing to the statistical spread in the distributions of both the stress and the strength.

7.6.7. Common Issues

7.6.7.1. How Does Stress Testing Affect the Product? We define a maximum AST stress, X^{ST}, which is analogous to the definition of lifetime maximum stress. For threshold stress failure, it is the maximum stress applied to a unit during ESS or AST. For cumulative stress failure, it is the effective stress parameter proportional to the overall irreversible change that the stress testing process has made on a unit. Then the effects of stress testing at a given maximum stress level, X^{ST}, are shown by a dividing line on the product strength into a region of stress test failure below this line and a region of stress test pass above it.

In the stress-strength contour shown Figure 7.10, the solid horizontal line shows an X^{ST} level that catches all the weak population. Yet, it still does not catch the lower-right part of the main population. This part will fail in the field because it happens to experience a higher level of lifetime maximum stress. We may eliminate this part of the population if we raise X^{ST} to the level shown by the dashed line. Yet we would then also fail a significant portion of the main population in the lower-left part. For ESS, the yield will then be too low so that the screening process is not economical. Thus, the separation between the weak population and the main population must be large enough for a working window for screening to be feasible. For AST during the design stage of a product, the purpose of AST is to find weaknesses in design and manufacturing and to take corrective actions.

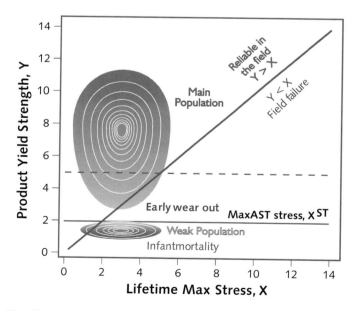

FIGURE 7.10. The effects of stress testing at a given maximum stress level, X^{ST}, are shown by a dividing line on the product strength into the lower region of stress test failure from the upper region of stress test pass.

FUNDAMENTALS OF RELIABILITY AND STRESS TESTING

Some failure modes may have a somewhat low field failure rate but are still not acceptable for a reliable product. The use of moderate stress during AST will not be an effective way to find them. A higher level of such as the one shown as the dashed line in Figure 7.10 is a more effective way to find these weaknesses. This holds if those failure modes in the field will also occur with the higher X^{ST}.

7.6.7.2. Will Stress Testing Damage Good Products? This is usually a concern for ESS only, because the purpose of Design and Manufacturing Qualification AST is to effectively identify and correct potential problems. The corrective actions are essential to achieve the robustness such as the one shown in the contour in Figure 7.11, where the robust main population is safe against stress testing, including Production Sampling AST. The prerequisite of having a robust main population is important. Indeed, incorrect application of screening without first meeting this requirement may damage more weak units to catch the weaker units. The result from such an improper stress-testing program may mislead people and cause them to step back to the use of mild stresses for all stress testing programs. Figure 7.10 shows such a non-robust product, where screening with either a mild stress or an elevated stress cannot improve its robustness. Corrective actions are essential here.

For threshold stress failures, stress testing does not affect the good products, even for Ongoing ESS. When a stress level X^{ST} is used in stress testing, the weak units whose threshold strength is below X^{ST} are detected. The good units whose threshold strengths are above X^{ST} are not weakened by the stress testing process.

For cumulative stress failures, the product must possess a robust main population that is well separated from the weak populations before screening may be applicable. For example, consider a product that has a main population with a robustness of 1000 stress

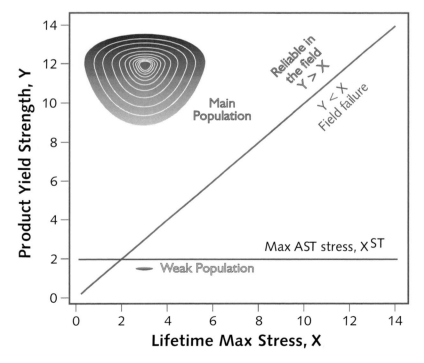

FIGURE 7.11. A more robust product than that represented by Figure 7.10.

units, (Ux), but its weak population is mostly below 10 Ux in strength. An X^{ST} of 10 stress units will screen out most of the weak units but the strengths of all the units are then weakened by 10 Ux after screening (Figure 7.11). Suppose there is a weak unit that has strength in the 100–110 Ux range before screening and will experience an X^{ST} of 100 Ux in a certain customer's environment. The dilemma is that this weak unit will encounter field failure but it could have escaped from it if it had not been screened. The answer to this dilemma is to compare the $F(TS)$ for the units that have been screened to the $F(TS)$ for those that have not been screened. When the main population is very robust, the weak populations may be screened out at the expense of a small decrease in the useful life of the product. Because there are far fewer units in the 100–110 Ux range than in the 0–10 Ux range, screening will decrease the cumulative failure fraction over the product life $F(TS)$ in the field for all the units.

7.6.7.3. Safety Testing AST should precipitate flaws in marginal products before they are shipped from the factory, but should not induce flaws or failure modes that normally would not be present in good products. The useful life of a product also should not be diminished by AST. To prevent this from happening, safety testing should be applied to the candidate regimen.

The preferred safety test method is to repeatedly apply a candidate stress testing regimen X^{ST} to a product until failure occurs at some level, X^{ST}. For example, if a candidate regimen of 10 thermal cycles is proposed and it is observed that on the order of 1000 thermal cycles is needed to eventually break the product, one may reasonably conclude that since $X^{ST} = 1000$ is $\gg$ than $X^{ST} = 10$, the regimen will not significantly reduce the useful life of good products.

In using highly accelerated stress testing, it is also necessary to perform sufficient safety testing on potential combined threshold-cumulative failure modes to be sure that the elevated stress levels are not causing incipient damage that can lead to later failures due to the cumulative effect of a secondary stress. As mentioned earlier, the corrosion

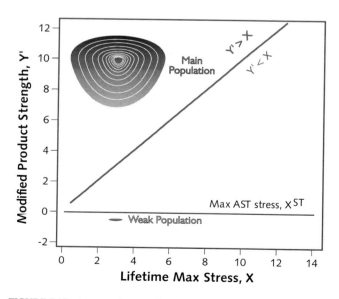

FIGURE 7.12. A more robust product after subjecting to cumulative stress.

of metalization within a device package cracked by an elevated mechanical stress is an example of such a concern.

The burden of doing safety testing is greatly diminished as one gains experience, since the results from safety tests for a technology used on one product may later be applied to other products using a similar technology. It is important to develop a mechanistic understanding of the relationship between applied stress conditions and observed failures so that one can properly judge when it is appropriate to extrapolate from previous safety testing results and when additional testing is required.

Another requirement of using a much higher stress level than those encountered in the nominal operating conditions is that the higher stress level should only increase the probability of the same failures that occur under the lifetime maximum stress conditions. The failures that take place at different values X^{ST} should be of the same failure mode although at different probabilities as evident in the strength distribution. The limit on elevating the stress used is that new physical phenomenon that will not occur under the lifetime maximum stress conditions should not occur under X^{ST}.

7.6.7.4. Are Failures from AST Related to Field Failures? A common question in reliability is whether a suspected product weakness will result in real field failures. Therefore, a stress testing program should choose the types of stresses that are likely to show the same types of failures found in the field environment. Whether these stresses should be thermal, mechanical, thermal-mechanical or chemical do not have to relate to the field stresses. Rather, the considerations are whether these stresses are relevant to the types of field failure modes that are likely in the field environment.

7.7. FURTHER READING

Further readings for Sections 7.1, 7.4, 7.5, and 7.6 may be found in H.A. Chan and P. Englert, "Stress Testing Handbook for Quality Products with Case Studies in Telecommunication and Computer Products," ISBN 0-7803-6025-7, IEEE Press and John Wiley & Sons, 2001. There is a bibliography of over 300 references there. Further reading for Section 7.2 may be found in most books on reliability.

8

How to Make a Device into a Product: *Accelerated Life Testing (ALT), Its Role, Attributes, Challenges, Pitfalls, and Interaction with Qualification Tests*

E. Suhir

University of California, Santa Cruz, CA, University of Maryland, College Park, MD, and ERS/Siloptix Co., Los Altos, CA, USA

> "*You can see a lot by observing*"
> Yogi Berra, American Baseball Player
>
> "*It is easy to see, it is hard to foresee*"
> Benjamin Franklin, American Scientist and Politician

8.1. INTRODUCTION

> "*Vision without action is a daydream. Action without vision is a nightmare*"
> Japanese Saying

Accelerated life tests (ALTs) are aimed at the revealing and understanding the physics of expected or occurred failures. Another objective of the ALTs is to accumulate representative failure statistics. Thus, ALTs are able to both detect the possible failure modes and mechanisms and to quantitatively evaluate the roles of various phenomena and processes that might lead to failures. Adequately designed, carefully conducted, and properly interpreted ALTs provide a consistent basis for obtaining the ultimate information of the reliability of a product—the probability of failure.

ALTs can dramatically facilitate the solutions to the problems of **cost effectiveness and time-to-market**. Because these tests can help a manufacturer to make his device into a product, they should play an important role in the evaluation, prediction and assurance of the reliability of micro- and opto-electronic devices and systems.

In the majority of cases, ALTs should be conducted in addition to the qualification tests required by the existing standards. There might be also situations, when ALTs can be used as an effective substitution for qualification tests, especially for new products, for which qualification requirements have not been developed yet. Whenever possible, ALTs should be used as a consistent basis for the improvement of the existing qualification specifications.

In this chapter, we discuss the role, objectives, attributes, challenges and pitfalls, associated with the use of ALTs, as well as their interaction with qualification tests. The emphasis is on the **role that ALTs** should play in the development, design, qualification and manufacturing of micro-, opto-electronic and photonic components and devices.

8.2. SOME MAJOR DEFINITIONS

"One should always to say 'tables, chairs, glasses of beer,' instead of 'points, straight lines and planes' "
David Gilbert, German Mathematician

The following major definitions are used in engineering reliability.

Failure mode identifies what happened (or might happen), the objective effect by which a failure is observed. Failure mode is what has been detected, observed and/or reported as a failure, whether functional (electrical, optical, thermal, etc.), mechanical (structural, "physical") or environmental (high- or low-temperature induced, high-humidity induced, radiation induced, etc.) failure. The failure mode is the evidence, manifestation, by which the failure is observed. Examples are: shorts and opens; low or distorted output signal; high optical losses; low coupling efficiency; material's failure; loss of structural integrity; brittle fracture, etc.

Failure mechanism identifies what phenomenon or process resulted in a failure. Such a phenomenon or process could be of physical, chemical, mechanical, thermal, or technological nature. Examples are: voltage breakdown, corrosion, fatigue, material's aging, electro-migration, excessive heat, elevated stress, high level of current, initiation and propagation of cracks, excessive displacement, division by zero, etc.

Failure site identifies the location of the failure, i.e., where the failure has occurred.

Load is the mechanical, electrical, thermal, chemical, or physical condition that is able to precipitate failure.

Stress is the level (intensity) of the applied load at the given failure site. Stress does not have to be mechanical, but could be due to an electrical, optical, thermal or other phenomenon.

"Root" cause identifies why a particular failure occurred. Examples are: poor design, selection of an inappropriate material, overstress, use of an inadequate technology, manufacturing deficiencies, misuse or abuse of the employed equipment, human error, etc.

One is supposed to possess a "gut feeling" on what can possibly "go wrong" and should make a preliminary decision on how to detect if something is "going wrong" indeed (in terms of the methodologies and measuring equipment used, qualification of the personnel, available resources, etc.). The factors that are being considered in the experimental design effort can and should change with the changes in the materials and technologies, as well as with the changes in the market demand and competition.

8.3. ENGINEERING RELIABILITY

"Reliability it is when the customer comes back, not the product"
Unknown Reliability Manager

Reliability is the ability of a product to be consistently good in performance, and so to elicit trust of both the manufacturer and the customer. Reliability Engineering deals with failure modes and mechanisms, "root" causes of occurrence of various failures, and methods to estimate and prevent failures. In products, for which a certain failure rate is considered acceptable, Reliability Engineering examines ways of bringing this rate down to an allowable level. A reliable item or a system are able to survive and to satisfactorily perform a required function, without failures or breakdowns, for a specific envisaged period of time under the stated operation and maintenance conditions. In **probability-based reliability engineering**, reliability is defined as the probability of an item or a system to function in the above indicated way.

If reliability is not defined and taken care of, the device will never be able to operate in accordance with the customer expectations and specifications, and the manufacturer will never be able to assess if/which state-of-the-art technologies would enable him to fulfill the customer's reliability and quality requirements.

It goes without saying that, in a sense, reliability related failures are similar to the consequences of a fire: it is easier to determine them in advance and to take measures to prevent them, rather than to eliminate their consequences and fix the design. If the reliability related bottlenecks of a particular design are anticipated and assessed well in advance, then the manufacturer would be able to compare various competing designs, manufacturing technologies and available metrological means from the viewpoint of the product's reliability and cost, and to establish the most feasible trade-offs between the reliability and guaranteed warranty.

8.4. FIELD FAILURES

"Nothing is impossible. It is often merely for an excuse that we say things are impossible"
Francois de la Rochefoucould, French Philosopher

The information of a failure could be obtained at different times, by using different means and at different locations. As far as "when" is concerned, the failures could be detected (observed) during the fulfillment of qualification tests, during screening tests that are carried out in the process of manufacturing, during burn-in tests, etc. As to "where" the information is obtained, it could be carried out at the vendor's site or at the customer's site, in the laboratory or in the field.

Field failures play a special and an important role in reliability engineering. It is "life itself," usually not distorted by accelerated test conditions. Field failure analysis is carried out on products recovered from the field after failure. It is the information from the field failure that is the most valuable in terms of the ability to accurately predict the likelihood of failure in the field, i.e., in actual use conditions.

It would not be an exaggeration to say that while the reliability engineer "supposes" what, when and why might fail, the field "disposes," and therefore a reliability engineer can often learn more from what happened in the field than from what has been observed in the lab or at the manufacturing site. An ironical thing is that although the reliability engineer does not want that field failures happen at all, he/she could learn a lot from an actually occurred field failure.

Field failure analyses require close cooperation between the customer and the vendor, as well as between designers, part suppliers, assemble line, and all those who in one way or another are involved in the product manufacturing and supply. Human error is often a cause of a field failure, and therefore it is often the human ego that is an obstacle to the analysis of a field failure. Field failure analysis requires the creation and availability of a user-friendly flexible and informative data base.

Mechanical failures are the most common type of failure in micro- and opto-electronic and photonic devices and equipment. The overwhelming majority of field failures are often related to a particular failure mechanism (the "weakest link") and to a small number of insufficiently reliable components.

8.5. RELIABILITY IS A COMPLEX PROPERTY

"Truth is rarely pure and never simple"
Oscar Wilde, The Importance of Being Earnest

Reliability is a complex property. It includes the item's (system's) **dependability**, **durability**, **maintainability**, **reparability**, **availability**, **testability**, etc. Each of these qualities can be of a greater or lesser importance depending on the particular function and operation conditions of the item or the system. In this chapter, we include quality, which is typically associated with manufacturing and production, into reliability, which is typically associated with materials and design only, but could be treated in a much broader sense, if one considers also the consistency (predictability, stability, repeatability) of the manufacturing processes, reliability of software, etc. In other words, reliability can be defined as the probability of a certain specified and unacceptable deviation from the pre-established functional, mechanical, environmental or some other type of performance.

8.6. THREE MAJOR CLASSES OF ENGINEERING PRODUCTS AND MARKET DEMANDS

"Plus usus sine doctrina, quam citra usum doctrina valet"
(*"Practice without theory is more valuable than theory without practice"*)
Latin Proverb

The following three major classes of engineering products could be distinguished, as far as their objectives and requirements for their reliability are concerned:

Class I. The product has to be made as reliable as possible, and failure should not be permitted. Such products are being typically manufactured for the military and similar "markets," when cost is not viewed as the most important factor, and the object has to be made reliable by all means. Such "products" (some warfare, military aircraft, battleships, space apparatus, etc.) are seldom manufactured in large quantities, and their failure is viewed as a catastrophe. The consequences of failure of the Class I products are the most severe ones, and can be associated with bad publicity, legal actions and even with the country's security and/or prestige. These "products" usually have a single customer, such as the government or a big firm. Traditionally, the reliability requirements for these products are defined in the form of government standards, such as US Military Standards (MIL-STDs). These standards not only formulate the reliability requirements for the product, but also specify the methods that are to be used to prove (demonstrate) the reliability, and even prescribe how the product must be manufactured, tested and screened. It is the customer, not the manufacturer, who sets such reliability standards.

Class II. The product has to be made as reliable as possible, but only for a certain level of demand (load). If the actual demand happens to be larger than the design demand, the product might fail, although the probability of such a failure should be (made) very small. Examples are civil engineering structures, passenger elevators, ocean-going vessels, offshore structures, commercial aircraft, railroad carriages, some medical equipment, etc. These are typically highly expensive products, which, at the same time, are produced in large quantities, and, therefore, application of Class I requirements to such products, important as these products might be, will lead to unjustifiable and unacceptable expenses, which are, for this reason, deemed to be economically unfeasible. The failure of products of Class II is often associated with loss of human lives, and, like the Class I product failures, is viewed as a catastrophe.

The products of the Class II are typically intended for an industrial market, rather than government or individual consumers. This market is characterized by relatively high volume of production items (buildings, bridges, commercial ships, commercial aircraft, telecommunication networks), but also by fewer and more sophisticated customers than in the commercial market (see Class III below). The reliability standards/specifications for the Class II products come as industrial standards, such as Telcordia, JEDEC, ASTM, etc. These standards often include some MIL-STDs or MIL-STDs requirements as their constitutive parts. The vendor and the customer usually negotiate some form of a reliability-and-quality contract for the Class II products. This contract typically includes both the appropriate industrial specification requirements and, in addition, the requirements of a particular customer.

If the device/component/subsystem passes the required (qualification) functional, mechanical and environmental tests, it becomes a "product," and is "qualified" to be shipped to the customer and to be installed into the customer's equipment. For some types of the Class II products, a low number of field failures might be considered acceptable, and could be even specified beforehand in the contract. For the Class II products, like for the Class I products, it is the customer, rather than the manufacturer who sets the reliability requirements and standards. These, however, could be based on the agreeable and generally acceptable industry standards.

Class III. The reliability does not have to be very high. Failures are permitted, provided that their level is not too high. The demand for the product is driven more by the cost of the product, than by its reliability. The product is relatively inexpensive, produced in massive quantities, and its failure is not viewed as a catastrophe, i.e., a certain level of failures during normal operation is considered acceptable. Examples are various household items, consumer products, agricultural equipment, etc. The typical market for these products is the consumer market. An individual consumer is a very small part of the total consumer base. Consumer intended micro- and opto-electronic products is a typical example of such a market.

Field failures are allowed and are expected to occur, as long as the failure rate is within the anticipated/expected range. The reliability testing is limited, and the improvements are implemented based on the field and market feedback. No special reliability standards are often followed, and it is the customer satisfaction (on the statistical basis), which is the major criterion of the reliability and quality of the product. It is typically the manufacturer, not the consumer, who sets the reliability standards for the product. Relatively simple and innovative Class III consumer products, which have a high degree of customer appeal and are therefore in significant demand, may be able to prosper, at least for a short period of time, even if they are not very reliable.

8.7. RELIABILITY, COST AND TIME-TO-MARKET

"Be grateful for luck, but do not depend on it"
William Feather, American Politician

For many Class II and all the Class III products, **cost and time-to-market** are key issues in competing in the global market place. For Class II products, developing reliable products that cost less is the primary goal of the industry. It is equally important that the product is delivered to the market on time. Reliability, cost and time-to-market considerations play an important role in the design, materials selection and manufacturing decisions. A company cannot be successful, if its products are not cost effective, or do not have a worthwhile lifetime and service reliability to match the expectations of the customer. Product failures have an immediate, and often dramatic, effect on the profitability and even the very existence of a company. This is even true for the Class III products, not to mention the Class II products.

Failure to provide adequate reliability can be costly to a business through the cost of reworking or scrapping of products during manufacturing, as well as through the cost of additional inspections and testing. Warranty repairs after the product is shipped may not only be expensive to the manufacturer, but might be even more costly to the customer in terms of loss of service and/or increased maintenance cost. Profits decrease as the failure rate increases. This is due not only to the increase in the cost of replacing or repairing parts, but, more importantly, to the losses associated with the interruption in service, not to mention the "moral losses." Such "moral losses" make obvious dents in the company's reputation and, as the consequence of that affect its future sales. Too low a reliability can lead to a total loss of business. Many small and large companies that are failing today fail because of insufficient reliability of their products.

On the other hand, there is a permanent "struggle" between the recognition of the industry that high and well-predicted product reliability is a "must" and a strong business pressure that tends to compromise product's reliability in order to shorten the time-to-market and to reduce manufacturing costs. In response to the time and cost pressures of the markets and shareholders (investors), businesses frequently take a much lower approach to reliability than they would have taken otherwise. Businesses attempt to establish a minimum level of product testing or inspection that will provide a level of reliability, which they feel is adequate for the market they serve. Certainly, it is always a challenge to establish, for each particular product and a particular situation, the most reasonable balance between the level of reliability and market demands, in terms of schedule and cost. In the past, it used to be said that of quality, schedule and price, the customer could have any two. Today, none of these items could be compromised for the other two, and the best engineering and business decisions should consider the best trade-off among them.

8.8. RELIABILITY COSTS MONEY

*"Be thankful for problems. If they were less difficult,
someone with less ability might have your job"*
Unknown Reliability Engineer

It is common knowledge that "reliability costs money." Conducting reliability evaluations is not cheap, and the cost of the subsequent failure mode analysis and corrective actions may be even more expensive. It is very undesirable, of course, that a business incurs excessive expenses pursuing high reliability standards with very little payback. The cost of improving or maintaining a certain level of product reliability must always be weighed against the benefits obtained. From the cost and business point of view, there is always an adequate, though less than perfect, level of reliability appropriate for the given product or system. This is always the case for the Class II and Class III products.

Relatively simple and innovative Class III consumer products, which have a high degree of customer appeal and are of significant demand, may be able to prosper, at least for some time, even if they are not very reliable.

A business must understand the cost of reliability, both "direct" cost, i.e., the cost of its own operations, and the "indirect" cost, i.e., the cost to its customers and their willingness to make future purchases and to pay more for more reliable products. Having this in mind, each business, whether small or large, should try to optimize its overall approach to quality and reliability. He/she should also have in mind that the time to develop and time to produce products is rapidly decreasing. This circumstance places a significant pressure on reliability engineers. They are supposed to come up with a reliable product and to confirm its long-term reliability in a short period of time to make their device into a product and to make this product successful in the marketplace.

8.9. RELIABILITY SHOULD BE TAKEN CARE OF ON A PERMANENT BASIS

"The probability of anything happening is in inverse ratio to its desirability"
John W. Hazard, American Writer

There is a story about a young couple who had a newborn baby and asked George Bernard Shaw, who was famous of his wisdom, for an advice. "Our baby is four months old. When should we start bringing it up?" "You are four months late" was the answer. This is true also about when to start being concerned about reliability.

In order that a product is successful in the market place, the manufacturer must understand the physics of failures of his/her product(s). He/she should be able also to design and manufacture a product with the predicted and sufficiently low probability of failure. In other words, he/she should know the ways, in which the useful service life of a material, device, structure, or a system can be predicted and, if necessary, improved, without bringing the product's cost up or postponing its delivery. A reliability engineer should develop effective methods to predict failures, to measure/detect them, to develop reliable methodologies for the prediction of the probability of failure, and, on this basis, to develop methods to minimize and/or to prevent failures at all the stages of the product design, manufacturing, testing and production. The reliability evaluation and assurance cannot be delayed until the device is made (although it is often the case in many actual industries). Reliability of a product should be

- "conceived" at the early stages of its design (a reliability and optical engineers should start working together from the very beginning of the optical device engineering),
- implemented during manufacturing (quality control is certainly an important part of a manufacturing process),
- qualified and evaluated by electrical, optical, environmental and mechanical testing (both the customer requirements and the general qualification requirements are to be considered),
- thoroughly checked (screened) during production, and, if necessary,
- maintained in the field during the product's operation, especially at the early stages of the product's use.

New products present natural and particular reliability concerns, as well as significant challenges at all the stages of their design, manufacture and use. These concerns and challenges have to do with the evaluation and assurance of both the functional (electrical and optical) performance and the structural (mechanical) reliability of the product. One of the major challenges, associated with new product development and reliability, is design and implementation of the adequate accelerated qualification tests and accelerated life test (ALT) approaches, methodologies and procedures [1–8].

A key bottleneck to meet the cost and time-to-market objectives is the product qualification and quality assurance. The required level of reliability is being typically proven based on the standardized qualification tests (QTs) and specifications (acceptance criteria). It is primarily the QTs that make a photonic device into a product. But it is the ALTs that enable a reliability specialist to understand the engineering and science behind the product. It is also the ALTs that enable him/her to create, on the basis of the developed understanding, a viable and a reliable product with the predicted and sufficiently low probability of failure.

8.10. WAYS TO PREVENT AND ACCOMMODATE FAILURES

> *"It is common sense to take a method and try it. If it fails, admit it frankly, and try another. But above all, try something"*
> Franklin D. Roosevelt, American President

The best way to prevent failures is to understand well the physics of failure, to **anticipate the failure modes** that might occur in a particular system, and to design this system in such a way that the likelihood that these failures occur be sufficiently low. In order to achieve this one should be able to

- develop an in-depth understanding of the possible modes and mechanisms of failure in his/her design,
- understand and to distinguish between operational (functional), structural/mechanical (caused by mechanical loading) and environmental (caused by harsh environmental conditions) failures,
- assess the likelihood (the probability) that the anticipated modes and mechanisms might occur in service conditions,
- distinguish between the materials and structural reliability,
- assess the effect of the mechanical and environmental behavior of the materials and structures in his/her design on the functional performance of the product,
- understand the difference between the requirements of the qualification specifications and standards, and the actual operation conditions,
- understand well the qualification test conditions and to design the product not only that it would be able to withstand the operation conditions on the short- and long-term basis, but also to pass the qualification tests,
- control, if necessary, the product's operation and operating environment.

One should have in mind that no failure statistics, nor the most effective ways to accommodate failures, can replace good (robust) physical design. Nonetheless, some proactive measures can be very helpful and can minimize considerably the likelihood of a failure, provided that the best materials are selected and a good design is carried out.

8.11. REDUNDANCY

> *"It is tough to make predictions, especially for the future"*
> Yogi Berra, American Baseball Player

The most effective method to increase the reliability of a system comprised of not-very-reliable components is redundancy. The number of the redundant components does not have to be very large to build a reliable system out of relatively unreliable components. For instance, if one wants to design a system whose reliability (probability of non-failure) is as high as 99%, while the reliability (dependability) of the components that the system is built of is only 80%, one can employ just four redundant components (in parallel) to build such a system. If one, two, three or even four components fail, the system will still operate. Note that if the same components were arranged in series, the overall reliability (the probability of non-failure) of the system would be as low as about 33%. If the system has a good enough reparability, the customer will never know that there was failure in the system, because the system will always be available to him/her.

That is why it is the availability, and not the dependability, which is the appropriate reliability characteristic of the system. High **availability** (i.e., high probability that the system is available to the user when needed) can be achieved even with a not-very-high dependability (i.e., probability of non-failure) of its components, as long as the reparability level (i.e., the probability that the system's workability is restored within the given and short enough period of time) of the system is sufficiently high.

In some cases, a system can be designed in such a way that, if one or more of its parts fail, the system can still operate, with its capabilities impaired to a greater or lesser extent. A two-engine aircraft can still operate, if one of its engines fails. A passenger ship will still not sink, even if two adjacent compartments at her fore- or after-body are flooded.

8.12. MAINTENANCE AND WARRANTY

> *"Only life insurance policy is able to provide a 100% warranty"*
> Unknown Insurance Agent

Maintenance is another failure accommodation method. There are two extreme approaches to accommodate failures, using appropriate maintenance: (1) preventive maintenance and (2) reactive maintenance. When preventive maintenance is used, items are checked and, if necessary, replaced (even if they are still good) in accordance with some more or less well-justified schedule. When reactive maintenance is used, the faulty items are replaced when they fail. In the case of preventive maintenance, one relies on routine procedures for checking and replacing parts. In the case of reactive maintenance, one relies on good technical diagnostics and keeps a highly qualified, highly flexible and highly mobile "rescue squad" to find and to fix the occurred problem. The reactive maintenance approach is more risky, but might be much less costly.

In the Class III systems, a widely used way to mitigate the consequences of failure is to provide a warranty, i.e., a guarantee that the manufacturer will repair or replace, when necessary, the faulty item at no cost to the customer. Still, the manufacturer should bring this practice to the minimum, because nothing can replace the customer's inconvenience, time, irritation, and, hence, dissatisfaction.

8.13. TEST TYPES

"Well done is better than well said"
Benjamin Franklin, American Scientist and Politician

The integrated test program usually includes the following types of tests:

1. Functional (optical, electrical) testing;
2. Mechanical testing;
3. Environmental testing;
4. Safety testing.

A crucial component of these tests is the adequate definition of failure criteria. These can be established, based on the customer requirements, qualification standards, state-of-the-art in the given area of engineering, etc. The peculiarities of a particular test program depend on the resources available, reliability requirements, product application, qualification of the personnel, allocation of facilities and equipment, priorities, etc.

8.14. ACCELERATED TESTS

"The golden rule of an experiment: the duration of the experiment should not exceed the lifetime of the experimentalist"
Unknown Physicist

Shortening of product design and product development time does not allow for time-consuming reliability evaluations. To get maximum information and maximum reliability-and-quality in minimum time and at minimum cost is the major goal of a manufacturer. One certainly wishes to have guidelines/methodologies that would enable him/her to quickly and economically evaluate the reliability of a product, and to afford an opportunity to fix reliability problems long before they lead to major losses. It is impractical and uneconomical to wait for failures, when the mean-time-to-failure for a typical today's micro- or opto-electronic device (equipment) is on the order of hundreds of thousands of hours.

Accelerated tests use elevated stress level and/or higher stress-cycle frequency to precipitate failures over a much shorter time frame. As has been mentioned above, the "stress" does not necessarily have to be a mechanical or a thermo-mechanical one: it can be electrical current or voltage, high (or low) temperature, high humidity, high frequency, high pressure or vacuum, cycling rate, or any other factor responsible for the reliability of the device or the equipment. In order to accelerate the material's (device's) degradation and/or failure, one has to deliberately "distort" one or more parameters (temperature, humidity, load, current, voltage, etc.) affecting the device's functional and/or mechanical performance.

Accelerated tests enable one to gain greater control over the reliability of a product. They have become a powerful means in improving reliability. In accelerated tests one applies a high level of stress over a short period of time to a device/product presuming/assuming that there will be no "shift" in the failure modes and mechanisms. This is true regardless of whether failures will actually occur during the tests (Accelerated Life Tests, which are aimed at "testing to fail") or not (Qualification Tests, which are aimed at "testing to pass"). The accelerated tests must be specifically designed for the product under test. The experimental design should consider the anticipated failure modes and mechanisms, typical use conditions, and the required or available test resources, approaches and techniques.

8.15. ACCELERATED TEST LEVELS

> *"If you do not raise your eyes, you will think that you are at the highest point"*
> Antonio Porchia, Italian Poet

Accelerated tests can be performed at the part level, at the component level, at the module level, at the equipment level and even at the system level. In each particular case, the decision should be made on how to break down the equipment of interest, so that the number of failure modes of the object under testing would not be very large. For this reason, accelerated testing is usually conducted at the part (assembly) or at the component (device) level. If the reliability characteristics of all the components are established, then the reliability characteristics ("indices") of the equipment or the system can be evaluated theoretically, using methods of probabilistic (statistical) analyses. In this connection it should be pointed out that different reliability criteria are (and should be) used depending on whether it is an assembly, a component, a subsystem, a piece of equipment or a large system.

While the probability of failure (dependability) might be the right criterion for a non-reparable component, a piece of equipment should be characterized by its availability, i.e., the probability that this piece of equipment will be available to the user, when it is needed. As to a large and a complex system (say, a switching system or a highly complex communication/transmission system, in which its "end-to-end reliability" is important, including the "reliability" of software), it is the "operational availability" that is of importance. This can be defined as the probability that the system is available "today" and will remain available to the user for the given period of time "tomorrow." What this, actually, means is that the system performs as expected every time the customer accesses it or needs it, whether it is 300 million voice attempts a day or 675 trillion bytes of data a network carries each day. To achieve that one does not have to necessarily keep the dependability of a particular component or even of a subsystem at a very high level. He/she can run a highly available system by achieving high reparability, reasonable redundancy, high-level of trouble shooting, etc.

Because of that, there is a rather wide spectrum of reliability requirements, ranging from very high requirements for large and complex systems, in which a failure is considered a catastrophe, down to simple consumer products, for which the consequences of failure are not as catastrophic as they are for large systems. The reliability (availability) of the contemporary communication networks is as high as 0.999. For consumer products, however, it is the cost and time-to-market that are the major driving forces, and their reliability (typically, dependability) should only be adequate for customer acceptance and reasonable satisfaction. No wonder that in reliability communities one can find a variety of opinions, attitudes and approaches to, and actual practices in, reliability assurance. It depends on the driving market forces and a particular business, whether it is component/device making, equipment manufacturing, or service provision.

8.16. QUALIFICATION STANDARDS

> *"By asking impossible obtain the best possible"*
> Italian Saying

The today's qualification standards and specifications (such as, say, Telcordia requirements for photonics equipment) enable one to "reduce to a common denominator" different products, as well as similar products, but produced by different manufacturers. These standards reflect, to a great extent, the state-of-the-art in a particular field of engineering, as well as more or less typical requirements for the performance of a product intended for a particular application. Industry cannot do without accelerated qualification tests and qualification standards.

However, qualification standards and requirements are only good for what they are intended—to confirm that the given product (provided that it passed the tests) is indeed "qualified" to serve in a particular capacity. In some cases, especially for new products and new technologies, when no experience has been yet accumulated, the general qualification standards, based on the previous generations of the device or on other, "similar," devices and components, might be too stringent. An unreasonable ("torture"/"sledgehammer") qualification test that does not reflect the actual field conditions might result in a rejection of a good product, i.e., of a product that would be able to perform successfully in the field for many years ("supplier's/vendor's risk"). In other cases, the qualification specifications might not be stringent enough for a particular application or particular use conditions, and a product with a not high enough reliability level might be shipped to the customer ("consumer's/customer's risk").

If a product passed the standardized qualification tests, it is not always clear why this product was good, and if the product failed, it is equally unclear what could be done to improve its reliability. Since qualification tests are not supposed to be destructive, they are unable to provide the most important ultimate information about the reliability of the product—the information about the probability of its failure after the given time in service under the given conditions of operation. If a product passed the qualification tests, it does not mean that there will be no failures in the field, and it is unclear how likely or unlikely these failures might be, nor what could be done to improve the product's reliability.

8.17. ACCELERATED LIFE TESTS (ALTS)

"In a long run we are all dead"
John Maynard Keynes, British Economist

The body of knowledge in the accelerated life tests (ALTs) has come a quite long way in a rather short time. ALTs are aimed at the revealing and understanding the physics of the expected or occurred failures. Unlike QTs, ALTs are able to detect the possible failure modes and mechanisms. Another objective of the ALT's is to accumulate sufficiently representative reliability/failure statistics. Thus, ALT's deal with the two major areas of Reliability Engineering—physics and statistics of failure. ALT's should be planned, designed and conducted depending on the projected lifetime of the product, the expected operational and non-operational loading conditions and environment, the frequency and duration of such loading and environmental conditions, etc.

Adequately planned, carefully conducted, and properly interpreted ALTs provide a consistent basis for the prediction of the probability of failure after the given time in service. This information can be extremely helpful in understanding of what and how should be changed in order to design a viable and reliable product. Indeed, any structural, materials and/or technological improvement can be "translated," using the ALTs data, into a reduced probability of failure for the given duration of operation under the given service (environmental) conditions. This is, in effect, the substance of a probabilistic approach to physical (structural) and functional (electrical or optical) design of a component or a device [11,12].

Well-designed and thoroughly implemented ALTs can dramatically facilitate the solutions to many business-related problems, associated with the cost effectiveness and time-to-market. Therefore ALTs, along with the (accelerated) product development/verification tests (PDTs) and qualification tests (QTs), play an important role in understanding and predicting the short- and long-term reliability of microelectronic and photonic equipment and devices.

In the majority of cases, various ALTs should be conducted in addition to, and, preferably, long before (or, at least, concurrently with) the qualification tests. There might be also situations, when accelerated testing can be used as an effective substitution for the qualification tests and standards, especially for very new products, when "reliable" (widely acceptable) qualification standards do not yet exist. This might result in a better understanding of the modes and mechanisms of failure, in the reduced cost of the product and in a shorter time to market, without compromising the product's reliability.

Unfortunately, quite often different manufacturers have to run the same ALTs and quite often learn reliability lessons from their own mistakes. This is because ALTs methodologies, studies, and, especially, test data are generally considered highly proprietary information, which is seldom published.

8.18. ACCELERATED TEST CONDITIONS

"If a man will begin with certainties, he will end with doubts;
but if he will be content to begin with doubts, he shall end in certainties"
Francis Bacon, French Philosopher

The accelerated test conditions are selected based on

- the expected failure modes and mechanisms,
- the most likely use conditions,
- anticipated environmental conditions,
- possible mechanical loadings, and
- qualification test conditions and requirements.

The most common accelerated test conditions (in any type of accelerated tests) are:

- High Temperature (Steady-State) Soaking/Storage/ Baking/Aging/ Dwell,
- Low Temperature Storage,
- Temperature (Thermal) Cycling,
- Power Cycling,
- Power Input and Output,
- Thermal Shock,
- Thermal Gradients,
- Fatigue (Crack Initiation and Propagation) Tests,
- Mechanical Shock,
- Drop Shock (Tests),
- Sinusoidal Vibration Tests (with the given or variable frequency),
- Random Vibration Tests,
- Creep/Stress-Relaxation Tests,
- Electrical Current Extremes,
- Voltage Extremes,
- High Humidity,
- Radiation (UV, cosmic, X-rays),
- Altitude,
- Space Vacuum,

- Industrial Pollution,
- Salt Spray,
- Fungus,
- Dirt,
- High Intensity Noise.

Some of the existing accelerated test equipment enables one to carry out also a combination of these tests.

This is done to detect and evaluate certain types of failure modes. Examples are: temperature/humidity bias (typically, 85°C/85%RH), fatigue tests at elevated temperature conditions, vibration tests at elevated temperature conditions, temperature cycling with voltage variations, etc.

If one cannot define the appropriate test condition with sufficient certainty, it is always advisable to assess this condition in an approximate fashion, probably, with a certain "margin of safety," rather than to ignore a particular test condition at all. If the customer, in the case of Class I or Class II products, does not define a particular test condition, it is the manufacturer who should do that.

8.19. ACCELERATION FACTOR

"A theory without an experiment is dead.
An experiment without a theory is blind"
Unknown Reliability Engineer

Once relevant accelerated stress conditions ("stimuli") are selected, appropriate stress levels must be determined. These levels are product and application specific. For a PCB, for instance, operated in an environment at the temperatures between zero and 50°C, the qualification test design margins of 10°C, 20°C and 30°C above the specified limit are considered "marginally robust," "robust" (acceptable) and very robust ("excellent"), respectively [20]. Another approach [21] suggests that, in order to establish the appropriate stress levels for a particular product, the stress levels should be incrementally increased until a significant percentage (say, larger than 50%) of the sample size no longer functions.

The degree of stress acceleration is described by an acceleration factor. This factor is defined as the ratio of the lifetime (cycles) under normal use (field) conditions to the lifetime (cycles) under the accelerated conditions. The acceleration factor can be interpreted as the number of times the particular failure mechanism has been accelerated during the tests, because of making the conditions more severe than those anticipated in the actual service. The acceleration factor can be established after an appropriate predictive model is agreed upon. It is presumed that such a model holds for both field and test conditions.

The design of the ALTs should consider all the possible failure mechanisms caused by a particular stressing environment. In light emitting diodes (LEDs) and lasers, for instance, the ambient temperatures, the magnitude of the injected current and the light output power level are generally used as acceleration factors. Elevated humidity, temperature cycling and mechanical vibrations can also be used to stimulate failures.

It should be pointed out, however, that high acceleration couldn't always be applied to optical devices. For instance, the internal quantum efficiency of LEDs and lasers (i.e., the efficiency of converting the injected current into light) is very sensitive to the ambient temperature. Most lasers stop lasing at around 100°C. For this reason, extrapolation with a rate of degradation is usually used to estimate the lifetime of an active optical device.

Because no device failure can be typically observed through the time of testing, random failure rate does not occur and therefore failure rate or the probability of failure cannot be used as statistical characteristics of functional failures.

8.20. ACCELERATED STRESS CATEGORIES

> *"Say not 'I have found the truth,' but rather 'I have found a truth'"*
> Kahlil Gibran, Lebanese Poet and Artist

Accelerated tests can be divided, from the standpoint of their objectives, into the following three major types (categories):

- Product development/verification tests (PDTs), or design testing,
- Qualification ("screening") tests (QTs), or production testing,
- Accelerated life tests (ALTs), and highly accelerated life tests (HALTs).

All these tests use harsh environment (elevated stresses) to accelerate the precipitation of dormant defects and potential failures. The tests differ by their objectives, end points, success/failure criteria, and the subsequent action of the human analyst to detect failures (Table 8.1) [7].

The objective of the **product development/verification tests** (PDTs) is to obtain information on the product reliability during design, development, and early manufacturing stages. Many reliability problems are caused by inadequate design margins or variations in manufacturing processes or component quality. To create a reliable product, one must achieve robust (not very stringent) functional design margins and tighten the control of materials and structural variations. The PDTs are supposed to pinpoint the weaknesses and limitations of the design, materials, and the manufacturing technology or process. These tests are used also to evaluate new designs, new processes, the appropriate correction actions, or to compare different products from the standpoint of their reliability. This type of

TABLE 8.1.
Accelerated test types (categories).

Accelerated test type (category)	Product development (verification) tests (PDT)	Qualification ("screening") tests	Accelerated life tests (ALT) and highly accelerated life tests (HALT)
Objective	Technical feedback to make sure that the taken design approach is viable/acceptable	Proof of reliability; demonstration that the product is qualified to serve in the given capacity	Understand the modes and mechanisms of failure and to accumulate failure statistics
End Point	Time, type, level, or number of failures	Predetermined time, or the number of cycles, or the excessive (unexpected) number of failures	Predetermined number (or percent) of failures
Follow-up Activity	Failure analysis; design decision	Pass/fail decision	Failure analysis and statistical analysis of the test results
The Perfect/Ideal Test	Specific definitions	No failure in a long time	Numerous failures in a short time

testing is often limited by time (when almost no failure occurs), and has to be followed by an analysis of the observed failures, or by another in-depth ("independent") investigation. PDTs are, as a rule, destructive. Shear-off tests are a typical example of PDTs aimed at the selection and evaluation of the adequate bonding material.

The objective of the **qualification tests** (QTs) is to prove that the reliability of the product-under-test is above a specified level. This level is usually measured by the percentage of failures per lot and/or by the number of failures per unit time (failure rate). Testing is time limited. The analyst of the test results usually hopes to get as few failures as possible. The pass/fail decision is based on a go/no-go criterion. The typical requirements are no more than a few percent failing parts out of the total lot (population). Although the QTs are unable (and are not supposed) to evaluate the failure rate, their results can be, nonetheless, sometime used to suggest that the actual failure rate is at least not higher than a certain value. This can be done, in a very tentative way, on the basis of the observed (or anticipated) percent defective in the lot. Qualification tests, in the best case scenario, are nondestructive, but some level of failures is acceptable.

8.21. ACCELERATED LIFE TESTS (ALTS) AND HIGHLY ACCELERATED LIFE TESTS (HALTS)

"If you come to a fork, take it"
Yogi Berra, American Baseball Player

The objective of the **accelerated life tests** (ALTs and HALTs) is to reveal the physics of failure, i.e., to establish/reveal the modes and mechanisms of failure, to identify parametric degradation of the materials and structures under test, and the longer-term failure mechanisms. In addition, ALTs are supposed to collect (accumulate) sufficiently representative statistical information about the product-under-test through its failures [13–16]. Qualification tests (QTs) give no indication on the probability of failure. ALTs do.

The ALT and HALT analyst needs to generate as many failures as feasible and as fast as possible. The ALTs are terminated, when the modes and mechanisms of failure are established, and enough failure statistics is collected. The typical acceptable failure ratio is 50%. ALT's and HALT's are destructive tests.

The difference between the accelerated life tests (ALTs) and highly accelerated life tests (HALT's) is that the HALTs are carried out to obtain, as soon as possible, the preliminary information about the reliability of the products, and the principal physics of their failures. The goal of the HALT's is to determine the "weakest links," "bottlenecks" of the design, and to obtain the preliminary information about the major modes and mechanisms of failure. The HALT's are conducted with a smaller number of samples and at higher acceleration factors than the ALTs, so that the duration of tests could be made short enough.

Typically, the duration of HALTs does not exceed two or three months. ALTs, on the other hand, enable one to obtain more realistic information about the product's failure. ALTs are conducted with a larger number of samples and for a longer time than HALT's. It is on the basis of the ALTs (not HALTs) that a reliability engineer can accumulate sufficiently representative failure statistics and to establish the probability of failure in the filed conditions after the given time of operation.

New products should be evaluated, based on both categories of the accelerated life tests, since these products are leading edge technology, are often rather complex and relatively high-cost, and might have long-term failure modes that can be time-consuming to isolate and resolve. Both types of accelerated life testing are usually required to ensure customer satisfaction. The cost of such tests, compared with the likelihood of encountering a field problem months after a large deployment of the product to the field, is usually considered worth the investment. An early warning of a potential failure can easily pay off the test cost, especially when the product ramp rates are steep. The ALTs give both the supplier and the customer an indication of the actual reliability of the product and its components.

One should always have in mind that a field failure might occur even if the product passed all the QTs. QTs are not supposed to reflect the actual use conditions. ALTs are.

8.22. FAILURE MECHANISMS AND ACCELERATED STRESSES

> "*All life is an experiment. The more experiments you make the better*"
> Ralph Waldo Emerson, American Poet and Philosopher

Typically, there is a predominant stress leading to a particular failure mechanism. Some of the failure mechanisms and the corresponding predominant accelerated stresses are summarized in Table 8.2 [7].

8.23. ALTS: PITFALLS AND CHALLENGES

> "*In every big cause one should always leave something to a chance*"
> Napoleon, French Emperor

Sometimes, accelerated test conditions may hasten failure mechanisms that are different from those that could be actually observed in service conditions. Examples are: change in materials properties at high or low temperatures, time-dependent strain due to diffusion, creep at elevated temperatures, occurrence and movement of dislocations caused by an elevated stress, etc. Because of the existence of such a "pitfall," it is always necessary to correctly identify the expected failure modes and mechanisms, and to establish the

TABLE 8.2.
Failure mechanisms and the corresponding accelerated stresses.

Failure mechanisms	Accelerated stresses and parameters
Corrosion (electrochemical, gaseous, galvanic, diffusion-controlled, in the presence of polymer coatings, nonelectrolyte, etc.)	Corrosive atmosphere, temperature, relative humidity
Creep and stress relaxation (static, cyclic)	Mechanical stress, temperature
Delamination	Temperature cycling, relative humidity, frequency
Dendrite growth and/or intermetallics formation	Voltage, humidity
Diffusion	Temperature, concentration gradient
Electromigration and thermomigration (forced diffusion due to electric potential or thermal gradients)	Current density, temperature
Fatigue (high- or low-cycle) crack initiation & propagation	Mechanical stress range, cyclic temperature range, frequency
Interdiffusion	Temperature
Radiation damage (radiation induced embrittlement, charge trapping in oxides, etc.)	Intensity of radiation, total dose of radiation
Stress corrosion cracking	Mechanical stress, temperature, relative humidity
Contacts' wear	Contact force, frequency, relative sliding velocity

appropriate stress/temperature limits, in order to prevent the distortion of ("shift" in) the original (actual) dominant failure mechanism. If, for one reason or another, such a situation cannot be avoided, it should be well understood and adequately interpreted, so that the ALTs do not lead to an erroneous conclusion. In this connection, it should be pointed out that different failure mechanisms are characterized by different activation energies (if, say, Boltzmann-Arrhenius type of equation is used to extrapolate the test data for the use conditions). A simple superposition of the effects of two mechanisms can result in erroneous reliability projections and, as a rule, should not be used.

Another pitfall has to do with the situation, when the accelerated test conditions lead to a bimodal distribution of failures, i.e., to a situation when a dual mechanism of failure takes place. Particularly, infant mortality ("early") failures might occur concurrently with the anticipated ("operational") failures. It is important to make sure that the "early" and "operational" failures are well separated in the tests. Infant mortality failures are usually due to the shortcomings of the manufacturing process and, although should be viewed as "atypical," are, in effect, inevitable. The most common infant mortality failures in microelectronic and photonic structures are: weak boundaries and delaminations, inclusions and voids, imperfections in geometry and materials (leading to elevated stress concentration), uneven coatings and nonuniform adhesive layers, current leakage, etc.

8.24. BURN-INS

"There is no such thing as failed experiment.
There are only experiments with unpredictable outcomes"
Unknown Reliability Engineer

Burn-in ("screening") tests are widely implemented to detect and eliminate infant mortality failures. The rationale behind the burn-in tests is based on a concept that mass production of devices generates two categories of products that pass qualification specifications: robust ("strong") components that are not expected to fail in the field and relatively unreliable ("week") components ("freaks") that most likely will fail in the field in some future time, if shipped to the customer.

Burn-ins are supposed to stimulate failures in defective devices by accelerating the stresses that will cause defective items to fail without damaging good items. Burn-ins are needed to stabilize the performance of the device in use. Burn-ins can be based on high temperatures, thermal cycling, voltage, current density, high humidity, etc. In burn-ins the stress is highly enhanced to generate failure of the "weakest link"/weakest-element in a very short time. These tests strongly accelerate the failure mechanisms' kinetics and cause defective parts to fail, thereby, supposedly, excluding the risk of their failure in the field. In other words, burn-in tests are intended to eliminate the infant mortality portion of the bathtub curve.

For products that will be shipped out to the customer, burn-ins are nondestructive tests. Burn-ins are mandatory on most high-reliability procurement contracts, such as defense, space, and telecommunication systems. In the today's practice burn-ins are often used for consumer products as well. For military applications the burn-ins can last as long as a week (168 hours). For commercial applications burn-ins typically do not last longer than two days (48 hours).

Optimum burn-in conditions can be established by assessment of the main expected failure modes and their activation energies, and from the analysis of the failure statistics during burn-in. Burn-ins are performed by either manufacturer or by an independent test house. Burn-in is a costly process, and therefore its application must be thoroughly monitored. Special investigations are usually required, if one wishes to ensure that cost-effective burn-in of smaller quantities is acceptable. Another cost-effective simplification can be achieved, if burn-in is applied to the complete equipment (assembly or subassembly), rather than to an individual component, unless it is a large system made up of several separately testable assemblies.

Although there is always a possibility that some defects might escape the burn-in tests, it is more likely that burn-in will introduce some damage to the "healthy" structure, i.e., will "consume" a certain portion of the useful service life of the product. This is because burn-ins not only "fight" the infant mortality, but accelerate the very degradation process that takes place in the actual operation conditions, unless the defectives have a much shorter lifetime than the "healthy" product and have a more narrow (more "deterministic," more "delta-like") probability-of-failure distribution density.

Some burn-in tests (high electric fields for dielectric breakdown screening, mechanical stresses below the fatigue limit, and some others) are harmless to the materials and structures under test, and do not lead to an appreciable "consumption" of the useful lifetime (field life loss). Others, although do not trigger any new failure mechanisms, might consume some small portions of the device lifetime. Therefore, when planning, conducting and evaluating the results of the burn-in tests, one should make sure that the stress applied by the burn-in tests is high enough to weed out infant mortality failures, but, at the same time, is low enough not to consume a significant portion of the product's lifetime, nor to introduce a permanent damage.

A natural concern, associated with the burn-in tests, is that there is always a jeopardy that burn-in might trigger some failure mechanisms that would not be possible in the actual use conditions and/or might affect the components that should not be viewed as defective ones.

8.25. WEAR-OUT FAILURES

> *"The problem is not that old age comes.*
> *The problem is that young age passes"*
> Common Wisdom

The bathtub curve of a device that underwent burn-in is supposed to consist of a steady state and wear-out portions only. In lasers, the "steady-state" portion is, in effect, not a horizontal, but a slowly rising curve. Standard production burn-in tests should be combined for laser devices with the long-term life testing. Burn-in for laser devices is typically conducted in dark forced-air ovens at different combinations of constant temperature and current. Periodically parts are removed from the oven and dc tested (at room temperature). Failure can be defined, for instance, as a 2 dB reduction in the output power at the given current.

There is another pitfall associated with the wear-out failures. For a well-designed and adequately manufactured product, the were-out failures should occur at the late stages of operation and testing. If one observes that it is not the case (the steady-state portion of the "bathtub" curve is not long enough or does not exist at all), one should revisit the design and to choose different materials and/or different design solutions, and/or a different (more consistent) manufacturing process, etc. In photonics products the wear-out part of the bathtub curve can occupy a significant portion of the product's lifetime, and should be carefully analyzed.

8.26. NON-DESTRUCTIVE EVALUATIONS (NDE'S)

"It is always better to be approximately right than precisely wrong"
Unknown Reliability Engineer

Many nondestructive means of failure detection and evaluation (NDE) can be very useful: ultrasonic methods, X-raying, Moiré interferometry, IR defectometry, etc. In connection with the use of nondestructive methods, it is noteworthy that some observed defects should not be necessarily viewed as reliability concerns, but should be rather considered as "quality defects." One should have in mind that it is the size and location of a defect, and the loading (stress) conditions that should be considered when deciding if this defect should be tolerated or might cause a reliability problem. For instance, even a large void in the middle of a solder joint might be acceptable and should be viewed as a quality, rather than reliability, defect. However, even a small void (especially a number of "organized," "lined-up," small voids) at the interface (especially at the solder bump corner) can lead to the fatigue (and then brittle) crack initiation and propagation, and should be avoided.

Another aspect, associated with nondestructive evaluations, concerns the resolution (measurement accuracy) of the available/affordable equipment. If it is likely that the level of defectives that might escape inspection exceeds the tolerance limits for the given measuring device, then it is incumbent that burn-in is implemented. For instance, it is well known that the "accuracy" of the operation of a laser diode might very well exceed the accuracy of the equipment, which is used to measure its performance.

8.27. PREDICTIVE MODELING

"Any equation longer than three inches is most likely wrong"
Unknown Physicist

"God created the world such that what is simple is true and what is complicated is false"
Gregory Skovoroda, Ukrainian Philosopher

ALTs cannot do without simple and meaningful predictive models. It is on the basis of such models that a reliability engineer decides which parameter should be accelerated, how to process the experimental data and, most importantly, how to bridge the gap between what one "sees" as a result of the accelerated testing and what he/she will possibly "get" in the actual operation conditions [9,10].

For a manufacturer, the existing qualification standards for the Class I and Class II are "the bible," and, when implementing these standards, he/she can make his/her product qualified without even knowing the actual modes and mechanisms of failure. However, for an engineer who is developing qualification standards, predictive modeling is as important as the actual experimental data are. These models are supposed to provide meaningful relationships that clearly indicate "what affects what and what is responsible for what" and that are able to quantitatively describe these effects. These relationships may or may not include time. If the constitutive relationships do not include time, it usually means that they describe the "steady state" conditions that occur during the mid-portion of the product's lifetime. The relationship could be analytical or based on computer simulations, could be of deterministic or probabilistic nature, could be based on an apriori (probabilistic) analysis (prediction) or on a posteriori (statistical) processing of the obtained experimental data, etc.

Predictive modeling, both functional performance and materials reliability related, should be viewed as an important constituent part of the reliability evaluations. Computation of the expected reliability at conditions other than the actual or accelerated test environment can provide important information about the device performance after a certain time in service or during accelerated testing at the given conditions. By considering the fundamental physics that might constrain the final design, predictive modeling can result in significant savings of time and expense.

Modeling can be very helpful, for instance, in optimizing the performance and lifetime of a device. For instance, the threshold current in an oxide-aperture VCSEL can be brought down by reducing the oxide aperture diameter. This, however, will result in a higher electrical resistance and higher thermal impedance, because the current must pass through a smaller constriction [17]. Since there are size-related trade-offs between the functional performance and structural/materials reliability, an optimal combination of the design possibilities (oxide thickness, vertical placement, aperture diameter, mirror and active region design, etc.) can possibly exist. Clearly, such a design optimization can be achieved only on the basis of predictive modeling.

As far as photonics applications are concerned, high precision in modeling is as important, as high precision in manufacturing. For instance, special effort should be taken to make the existing finite element programs accurate enough to be suitable for the evaluation of the stresses in, and the displacements of, the structural elements in a photonic device. Based on our recent experience, we suggest that analytical ("mathematical") modeling be more widely used to master the preprocessing models in finite element analyses, or, in some cases, even to carry out the entire modeling process [18]. This provides an obvious challenge for reliability and design engineers.

Another challenge, associated with predictive modeling in photonics reliability engineering, is the necessity for considering time-dependent behavior of a material or a structure. Creep and stress relaxation are crucial phenomena to be considered and, if possible, adequately modeled, when a photonic product is designed for a high long-term reliability. In lasers, significant temperature acceleration cannot be applied, since lasers stop lasing at about 100°C. Consequently, extrapolation with a degradation rate is usually employed to estimate the lifetime. The degradation rate is given by the change in the monitored characteristics as a function of aging.

It is the structure of a particular analytical model, and not the numerical values of the parameters, that makes it generic and, therefore, useful. Although in some situations a particular model might be inadequate for the given application or a new situation, it is important that it is amenable to updates and revisions, if necessary, and that it "reduces to the common denominator" the accumulated knowledge to provide continuity. A good predictive reliability model does not need to reflect all the possible situations, but rather should be simple, should clearly indicate "what affects what" in the given phenomenon or structure, and be suitable/flexible for new applications, with new environmental conditions and new technology developments.

8.28. SOME ACCELERATED LIFE TEST (ALT) MODELS

"All the general theories stem from examination of specific problems"
Richard Courant, German Mathematician

It is expected that an accelerated life test model is simple enough, yet meaningful, to be useful for the application in question. It does not have to be comprehensive, but has to be sufficiently generic, and should include all the major variables affecting the phenomenon (failure mode) of interest. As Einstein said, "a good model should be as simple as possible, but not one bit simpler." In other words, a good model should contain all the most important parameters that are needed to describe and to characterize the phenomenon of interest, while parameters of the second rate of importance should not be included into the model. A good life test model should be suitable for the accumulation, on its basis, the reliability statistics and should be flexible enough to account for the role of materials, structures, loading (environmental) conditions, new designs, etc. The scope of the model depends on the type and the amount of information available. ALT models take inputs from various theoretical analyses, test data, field data, customer requirements, qualification spec requirements, state-of-the-art in the given field, consequences of failure for the given failure mode, etc.

Here are some major ALT models (constitutive equations, relationships) used in ALTs of microelectronic and photonic structures. They are all deterministic, and the majority of them apply to the steady state conditions only, i.e., do not consider time related effects. For this reason these relationships are not applicable to the infant-mortality and wear-out portions of the bathtub curve.

8.28.1. Power Law

For some failure mechanisms the analytical models that are used to predict reliability (as represented by the time-to-failure, or cycles-to-failure) have a power law structure:

$$T = C\sigma^n, \tag{8.1}$$

where σ is the stress parameter, and C and n are material parameters.

The power law is used, for instance, to describe degradation in lasers, when the injection current or the light output power are used as acceleration parameters. It is used also to describe the "static fatigue" (delayed fracture) of silica material in optical lightguides. In this case, T in the formula (8.1) is time-to-failure and the exponent n is negative: $n = -18 \to -20$.

8.28.2. Boltzmann-Arrhenius Equation

If Boltzmann-Arrhenius equation is used, the mean time-to-failure can be sought as

$$\tau = \tau_o \exp\left[\frac{U_a}{k(T - T_*)}\right], \tag{8.2}$$

where U_a, eV, is the activation energy, $k = 8.6174 \times 10^{-5}$ eV/K is Boltzmann's constant, T is the absolute temperature, T_* is the temperature sensitivity threshold (if any), and τ_o is the time constant. The equation was first obtained by the German physicist L. Boltzmann in the statistical theory of gases, and then applied by the Swedish chemist S. Arrhenius to describe the inversion of sucrose.

Boltzmann-Arrhenius equation is applicable, when the failure mechanisms are attributed to a combination of physical and chemical processes. Since the rates of many physical processes (such as, say, solid state diffusion, many semiconductor degradation mechanisms) and chemical reactions (such as, say, battery life) are temperature dependent, it is the temperature that is used as an acceleration parameter. The activation energy has been determined for many materials and failure mechanisms used in micro- and opto-electronics. For semiconductor device failure mechanisms the activation energy ranges from 0.3 to 0.6 eV; for intermetallic diffusion it is between 0.9 and 1.1 eV.

Activation energies for some typical failure mechanisms in semiconductor devices are [22]:

- for metal migration 1.8 eV
- for charge injection 1.3 eV
- for ionic contamination 1.1 eV
- for Au-Al intermetallic growth 1.0 eV
- for surface charge accumulation 1.0 eV
- for humidity-induced corrosion 0.8–1.0 eV
- for electromigration of Si in Al 0.9 eV
- for Si junction defects 0.8 eV
- for charge loss 0.6 eV
- for electromigration in Al 0.5 eV
- for metalization defects 0.5 eV

The Boltzmann-Arrhenius equation can be used to model temperature induced degradation in many electronic and photonic products, including lasers. It is presumed that the rate of degradation in lasers is due to diffusion, precipitation, oxidation and other temperature dependent phenomena, so that the degradation, D, rate can be described by the equation

$$dD/dt = A \exp(-U_0/kT). \tag{8.3}$$

Solid-state diffusion can form brittle intermetallic compounds, weaken local areas, cause high electrical impedance.

The effect of the relative humidity (RH) can be accounted for, if the relationship (8.2) is used, by multiplying the right part of this equation by the factor $1/(RH)^n$, where n is an empirical parameter. This relationship can be used, for instance, to describe the results of ALTs for planar lightwave circuit (PLC) devices.

The activation energy and the temperature sensitivity threshold should be established experimentally for a particular application. As to the time constant τ_o, it does not have to be determined, if it is the acceleration factor that is of interest.

8.28.3. Coffin-Manson Equation (Inverse Power Law)

The Coffin-Manson equation (inverse power law) is applicable when the lifetime of the material or a structure is inversely proportional to the applied stress [23,24]. In accordance with this equation, the median number-of-cycles-to-failure in the low-cycle fatigue conditions can be found as

$$N_f = C\sigma_r^{-m}, \tag{8.4}$$

where σ_r is the cyclic mechanical stress range ($\sigma_r = \Delta\sigma = \sigma_{\max} - \sigma_{\min}$) and C and m are material's constants. This formula was applied by many investigators to evaluate the lifetime of solder joints in micro- and opto-electronics. W. Engelmaier suggested the following formula to predict the lifetime of solder joint interconnections

$$N_f = \frac{1}{2}\left(\frac{\varepsilon_r}{2\varepsilon_f}\right)^b, \tag{8.5}$$

where ε_r is the plastic strain, $\varepsilon_f = 0.325$ is the fatigue ductility coefficient [25],

$$b = \left[1.74 \times 10^{-2} \ln(1+f) - 6 \times 10^{-4} T_s - 0.442\right]^{-1} \tag{8.6}$$

is the fatigue ductility exponent, f is the cyclic frequency ($1 \leq f \leq 1000$ cycles per day), and T_s is the mean cyclic temperature.

In random vibration tests, the mean-time-to-failure can be found in accordance with the Steinberg equation

$$\tau = C\sigma_r^{-m/2}, \tag{8.7}$$

where σ_r is the stress at the resonant frequency, and C and m are material's constants. The Equation (8.7) indicates that the mean-time-to-failure is proportional to the square root of the stress, induced at the resonance frequency.

Inverse power law is used also to model aging in lasers in the cases of current or power acceleration:

$$\tau = AI^{-n}, \quad \text{or} \quad \tau = AP^{-n}.$$

Inverse power law is used also to assess the lifetime of a silica material in optical fibers from the measured time-to-failure during accelerated stress.

8.28.4. Paris-Erdogan Equation

This equation establishes the relationship between the fatigue crack growth rate and the variation in the cyclic stress intensity factor:

$$\frac{da}{dN} = A(\Delta K)^{m_p}, \tag{8.8}$$

where da/dN is the crack growth rate, A and m_p are material constants,

$$K = G\sigma\sqrt{2\pi a} \tag{8.9}$$

is the stress intensity factor, a is the crack length, σ is the nominal stress, and the factor G is a function of geometry. The stress intensity factor range, ΔK, in the Equation (8.8) is

$$\Delta K = G\sigma_r\sqrt{2\pi a}, \tag{8.10}$$

where σ_r is the nominal stress range. The Equations (8.8)–(8.10) are applicable when the stress intensity factor range ΔK is larger than a certain threshold for the given material, below which no crack growth can occur, or below which the crack growth rate is very low. Generally, in most electronic and optoelectronic devices under normal use conditions, the initial cracks are very small, and so is the nominal stress range. It could be expected that in normal operating conditions, the ΔK value is smaller than, but not far below, the threshold value. In such a case, the fatigue life is dominated by crack initiation only. However, if the stress range increases, as it takes place in an accelerated test, then the stress intensity factor range ΔK my increase beyond the threshold value, and the failure mechanism might shift from crack initiation to crack propagation.

8.28.5. Bueche-Zhurkov Equation

Bueche-Zhurkov's equation contains not only the absolute temperature, but also the applied stress as an acceleration factor:

$$\tau = \tau_o \exp\left[\frac{U_a - \gamma\sigma}{kT}\right], \tag{8.11}$$

where γ is the stress sensitivity factor, which depends on the structure of the material and the degree of the accumulated damage. The experimentally found stress sensitivity factor for a non-oriented condition of a polyamide is about $\gamma = 1.3 \times 10^{-27} \text{m}^3$. For an oriented condition it can be significantly lower.

The Equation (8.11) underlies the kinetic approach to the evaluation of the strength of materials. In accordance with this approach, it is the random thermal fluctuations of particles (atoms) that are primarily responsible for the materials strength (failure), while the role of the external stress is reduced simply to lowering the activation energy. In many practical applications, it is only the governing relationships of the type (8.11) that is considered, while the numerical values of the parameters are evaluated experimentally for a particular application.

8.28.6. Eyring Equation

In the Equation (8.11) the effect of the external stress is considered indirectly, by reducing the level of the activation energy. This effect is considered directly in the Eyring equation:

$$\tau = A\sigma^{-1} \exp\left(\frac{U_a}{kT}\right). \tag{8.12}$$

Unlike in the Equation (8.11), the stress σ in the Eyring equation does not have to be necessarily a mechanical stress: it could be voltage, humidity, etc.

8.28.7. Peck and Black Equations

Peck's equation is, in effect, Eyring equation expanded and modified for modeling the time-to-failure in the temperature humidity bias conditions:

$$\tau = A(RH)^{-n} \exp\left(\frac{U_a}{kT}\right). \tag{8.13}$$

Here RH is the percent relative humidity. In Black's equation, the RH is substituted with the current density J, the A value is a constant related to the geometry of the conductor, and n is a parameter related to the current density, which accounts for the effects of current flow other than joule heating of the conductor.

8.28.8. Fatigue Damage Model (Miner's Rule)

Fatigue damage rule (the law of linear accumulation of damages) can be formulated as

$$D = \sum_{i=1}^{m} \frac{n_i}{N_i} \leq 1, \qquad (8.14)$$

where D is the cumulative damage, n_i is the actual number of cycles applied at the i-th stress level, N_i is the number of cycles to failure under this stress, and m is the total number of different stress levels. The law of linear accumulation of damages is, generally speaking, applicable only for stresses, not exceeding the yield stress. This linear law is, strictly speaking, not applicable for the assessment of the low-cycle-fatigue lifetime. It is nonetheless often used to estimate the number-of-cycles-to-failure when a wide range of applied stresses, both below and above the yield point, are likely.

8.28.9. Creep Rate Equations

Assuming that the (static) creep rate is constant throughout the test, and that the phenomenon is dominated by the secondary stage, one can evaluate the strain rate due to creep as (Norton creep law)

$$\dot{\varepsilon} = A\sigma^n \exp\left(-\frac{U_a}{kT}\right), \qquad (8.15)$$

where σ is the applied stress, and A and n are material's parameters. It is important that the Equation (8.15) is capable to represent (more or less) the entire creep curve. If the creep phenomenon is heavily dominated by the tertiary stage, the Equation (8.15) might not be adequate. Another widely used relationship for creep rate is Prandtl's law:

$$\dot{\varepsilon} = A[\sinh(B\sigma)]^n \exp\left(-\frac{U_a}{kT}\right).$$

The following Graham-Walles equation was suggested to represent all the creep stages:

$$\dot{\varepsilon} = A\sigma^n \exp\left(-\frac{U_a}{kT}\right)(a + bt^{-2/3} + ct^2), \qquad (8.16)$$

where A, n, a, b and c are experimentally determined constants. It is noteworthy that if the stress σ is due to the thermal expansion mismatch of the dissimilar materials in the structure, it is not an independent variable, but is a function of the temperature T.

Creep tests are much easier to conduct than stress relaxation tests. On the other hand, phenomena associated with stress relaxation (time dependent stress for the given deformation) can be predicted, with sufficient accuracy, theoretically, if creep (time dependent deformation for the given stress) test data are available.

8.28.10. Weakest Link Models

The weakest link model assumes that the material (device) failure originates from the weakest point. This model is applicable, when the physics of the failure phenomenon confirms that this is indeed the case. Failures due to crack generation and propagation, and dielectric breakdown are examples of weakest link failures.

8.28.11. Stress–Strength Models

These models are widely used in various problems of structural (physical) design [11]. In this model the interaction of the probability density functions for the strength and stress distributions is considered. In aerospace, civil, ocean and other structures, the probability density functions are steady state, i.e., do not change with time. In lasers, however, one can assume that the stress distribution function is indeed time independent, but the strength distribution function becomes broader and shifts toward the stress distribution function, when time progresses. At the initial moment of time the two functions are well separated, and the distance between their end points provides an appreciable margin of safety. At a certain moment of the lifetime, the right end of the strength distribution "touches" the left end of the stress distribution (the marginal state). When the time of operation exceeds the moment of time that corresponds to the marginal state, the two curves start to overlap, and the probability of failure is not zero anymore. It does not mean, however, that the device cannot be operated beyond the marginal point of time, provided that the probability of failure can be predicted with sufficient accuracy, and be made low enough for the required (specified) time of operation.

8.29. PROBABILITY OF FAILURE

> "If you bet on a horse, that's gambling. If you bet you can make three spades, that's entertainment. If you bet the device will survive for twenty years, that's engineering. See the difference?"
> Unknown Reliability Engineer

> "Probability is too important to be left to the mathematicians"
> Unknown Reliability Engineer

Based on the accelerated test data, one can predict the probability of failure at the end of the given time of the device operation. Different approaches can be used to evaluate such a probability (see, for instance, [11,12]). The most typical ("parametric") approach, used in engineering practice, is based on an assumption that not only the relationships of the previous section hold for both the accelerated and use conditions, but that the laws of the probability distributions for the parameter of interest do not change either. Recently, there were suggested several ("non-parametric") approaches, based on the extreme value distributions that enable one to successfully process the ALT data, even if the lifetime distribution is stress level dependent [19].

If, for instance, the Bueche-Zhurkov's Equation (8.11) is used, the probability of failure (in a long run) can be found as

$$P = \exp\left(-\frac{U_e - \gamma\sigma}{kT}\right). \tag{8.17}$$

If Engelmaier's Equation (8.5) is applied, then the formula

$$P = 1 - \exp\left[-2N_f\left(\frac{2\varepsilon_f}{\varepsilon_r}\right)^b\right] \tag{8.18}$$

can be used to evaluate the probability of failure after N_f cycles of loading.

It should be pointed out that many manufacturers are not familiar enough with the "mathematical" analyses underlying the "quantitative" part of the "probabilistic/statistical" reliability. For this reason, the processing of the experimental data is usually practiced by statisticians who are typically not very well familiar with and are not directly involved in the design and manufacturing of the product. In addition, their activity is associated only with what happened after (a posteriori assessment), and not prior to (apriori evaluations), the experiment. This is, of course, a significant shortcoming of the to-day's practices. This is also the reason why the probabilistic design, i.e., a design with the predicted probability of failure of the component or a device, is not even present in the to-day's micro- and opto-electronics industry.

8.30. CONCLUSIONS

"Life is the art of drawing sufficient conclusions from insufficient premises"
Samuel Butler

The following conclusions can be drawn from the above discussion:

- Accelerated life tests (ALTs) are aimed at the revealing and understanding the physics of the expected or occurred failures, and are able to detect the possible failure modes and mechanisms. Another objective of the ALTs is to accumulate sufficiently representative failure statistics.
- Adequately designed, carefully conducted, and properly interpreted ALTs provide a consistent basis for the prediction of the probability of failure of the product after the given time of service. Such tests can dramatically facilitate the solution to the cost effectiveness and time-to-market problems.
- ALTs should play an important role in the evaluation, prediction and assurance of the reliability of micro- and opto-electronics devices and systems. ALTs should be conducted in addition to (and, preferably, should start prior to) qualification tests required by the existing standards.

REFERENCES

1. G. Di Giacomo, Reliability of Electronic Packages and Semiconductor Devices, McGraw-Hill, New York, 1997.
2. M. Fukuda, Reliability and Degradation of Semiconductor Lasers and LEDs, Artech House, 1991.
3. O. Svelto and D.C. Hanna, Principles of Lasers, Plenum, 1998.
4. E. Suhir, M. Fukuda, and C.R. Kurkjian, Eds., reliability of photonic materials and structures, Materials Research Society Symposia Proceedings, Vol. 531, 1998.
5. E. Suhir, R.C. Cammarata, D.D.L. Chung, and M. Jono, Mechanical behavior of materials and structures in microelectronics, Materials Research Society Symposia Proceedings, Vol. 226, 1991.
6. A. Katz, M. Pecht, and E. Suhir, Accelerated testing in microelectronics: review, pitfalls and new developments, Proceedings of the International Symposium on Microelectronics and Packaging, IMAPS, Israel, 2000.
7. E. Suhir, Microelectronics and photonics-the future, Microelectronics Journal, 31(11-12) (2000).
8. E. Suhir, Analytical modeling in structural analysis for electronic packaging: its merits, shortcomings and interaction with experimental and numerical techniques, ASME Journal of Electronic Packaging, 111(2) (1989).
9. E. Suhir, Thermo-mechanical stress modeling in microelectronics and photonics, Electronic Cooling, 7(4) (2001).
10. E. Suhir, Applied Probability for Engineers and Scientists, McGraw-Hill, 1997.

11. E. Suhir and B. Poborets, Solder glass attachment in cerdip/cerquad packages: thermally induced stresses and mechanical reliability, Proc. of the 40th Elect. Comp. and Techn. Conf., Las Vegas, Nevada, May 1990; See also: ASME Journal of Electronic Packaging, 112(2) (1990).
12. E. Suhir, Analytical stress-strain modeling in photonics engineering: its role, attributes, challenges, and interaction with the finite-element method, Laser Focus World (May) (2002).
13. E.M. Baskin, Processing of the results of the accelerated life tests for the unspecified time-to-failure distribution function, Proceedings of the Academy of Sciences of the USSR, Technical Cybernetics, (3)(1988) (in Russian).
14. H.A. Chan and P.J. Englert, Eds., Accelerated Stress Testing Handbook, IEEE Press, 2001.
15. G.K. Hobbs, Development of stress screens, Proceedings of the Annual Reliability and Maintainability Symposium, Philadelphia, PA, January 1987.
16. R.A. Evans, Reliability engineering, ancient and modern, IEEE Transactions on Reliability, 47(3), p. 209 (1998).
17. L.W. Condra, Reliability Improvement with Design of Experiments, Marcel Dekker, Inc., 2001.
18. W. Weibull, Statistical design of fatigue experiments, ASME Journal of Applied Mechanics, (March) (1952).
19. D. Kececioglu and J. Jack, The Arrhenius, Eyring, inverse power law and combination models in accelerated life testing, Reliability Engineering, 8 (1984).
20. D.C. Peck and O.D. Trapp, Accelerated Testing Handbook, Technology Associates, Portola Valley, CA, 1987.
21. W. Nelson, Accelerated Testing, John Wiley and Sons, New York, 1990.
22. D.J. Klinger, On the notion of activation energy in reliability: Arrhenius, Eyring and thermodynamics, Proc. of the Reliability and Maintainability Symposium, 1991.
23. L.F. Coffin, Jr., A study on the effect of cyclic thermal stresses on a ductile metal, ASME Journal of Applied Mechanics, 76(5) (1954).
24. S.S. Manson, Fatigue: a complex subject—some simple approximations, Experimental Mechanics, 5(7) (1965).
25. W. Engelmaier, Fatigue life of leadless chip carrier solder joints during power cycling, IEEE CPMT Transactions, CHMT-6 (3) (1985).

9

Micro-Deformation Analysis and Reliability Estimation of Micro-Components by Means of NanoDAC Technique

Bernd Michel[a] and Jürgen Keller[b]

[a] *Fraunhofer MicroMaterials Center, Berlin, Germany*
[b] *Fraunhofer Institute for Reliability and Micro Integration, IZM, Berlin, Germany*

9.1. INTRODUCTION

The manufacturing of microscopic and nanoscopic objects requires the quantification of their properties. While the measurement of geometrical and size data is more easily accessible by Scanning Force Microscopy (SFM) and related methods, kinematic and mechanical characterization is a general problem for micro- and nanoobjects and devices. Displacements and their derivatives are two basic properties to be measured for mechanical description. Until now only a few methods exist to make accessible quantified field data for these tiny regions.

For that purpose different kinds of SFM imaging have been used [1–5]. Among the published quantitative approaches two techniques exist—Moiré [6–8] and image correlation based methods [9–11].

In contrast to classical Moiré measurements correlation type measurements base on higher pixel resolution SFM scans. They allow to measure displacements and strains with moderate spatial resolution within the SFM scan area [11,12]. Digital Image Correlation (DIC) is the technique currently used by most of authors measuring object deformations from SFM images.

Considering DIC techniques SFM images are captured subsequently for different object states. Mechanical and/or thermal loading is carried out by special loading stages developed for SFM and SEM application. Locally applied cross correlation algorithms are utilized to compute displacement fields and the corresponding first order derivatives from SFM images [16].

9.2. BASICS OF DIGITAL IMAGE CORRELATION

Correlation analysis on gray scale images can be realized as *field measuring and characterization method* making use of digital image or pattern acquisition and subsequent digital image processing. Then a complete set of very local image pattern is tracked between two or more object states represented by different digitized images. One of the most outstanding advantages of field measuring methods in connection with digital image processing is the possibility to obtain full two-dimensional field information instead of only point wise data. The method of correlation analysis on gray scale image pattern offers a lot of interesting new possibilities for applications in many areas of materials science and production technology. With the description of materials deformation due to thermal and mechanical loading elastic properties such as Poisson's ratio and the coefficient of thermal expansion (CTE) are obtained by correlation techniques [17]. Other fields of application are the evaluation of system response of complete structures and components including material interfaces. Compared to other displacement and strain measurement techniques such as laser interferometric or Moiré methods

Digital Image Correlation has several advantages:

- In many cases only relatively simple low-cost hardware is required (optical measurements) or already existing microscopic tools like SEM and SFM can be utilized without any changes.
- Once implemented in a well designed software code, the correlation analysis of gray scale images is user friendly and easy to understand in the measuring and postprocessing process.
- For optical micrographs no special preparation of the objects under investigation is needed.
- According to its nature the method possesses an excellent downscaling capability. By using microscopic imaging principles, also very small objects can be investigated. Therefore, correlation analysis of gray scale images is predestined for qualitative and quantitative characterization of micromechanical and nanomaterial properties.

9.2.1. Cross Correlation Algorithms on Gray Scale Images

Digital image correlation methods on gray scale images were established by several research groups. Examples from different fields of applications can be found in various publications, e.g., in [9,16,18–23].

Modern SEM's allow to capture digital images and to apply correlation algorithms directly to them. This approach has been chosen by different research labs and is described in several publications [16,22,23]. The authors have developed and refined different tools and equipment in order to apply SEM images for deformation analysis on thermo-mechanically loaded electronics packages. The respective technique was established as *microDAC*, which means **micro D**eformation **A**nalysis by means of **C**orrelation algorithms [22].

The microDAC technique, by definition, is a method of digital image processing. Digitized micrographs of the analyzed objects in at least two or more different states (e.g., before and during mechanical or thermal loading) have to be obtained by means of an appropriate imaging technique. Generally, the utilized cross correlation algorithms can be applied to micrographs extracted from very different sources. Digitized photographs or video sequences but also images from e.g., optical microscopy, SEM, LSM or SPM are suitable

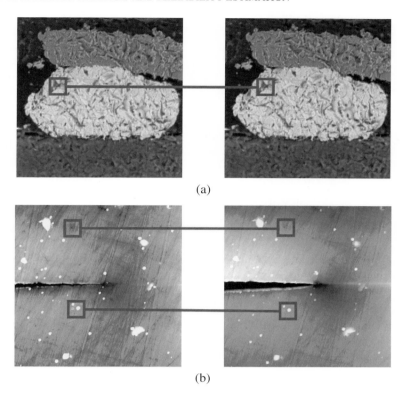

FIGURE 9.1. Appearance of local image structures (patterns) during specimen loading; (a) SEM images of flip chip gold bump; (left): at room temperature, (right): at 125°C; (b) SFM topography image of a crack in a thermoset polymer material for different crack opening displacements, scan size 15 × 15 µm.

for the application of digital image correlation. The basic idea of the underlying mathematical algorithms follows from the fact that images of different kinds commonly allow to record local and unique object patterns, within the more global object shape and structure. These pattern are maintained, if the objects are stressed by temperature or mechanically. Figure 9.1 shows two examples of images taken by SEM and SFM. Markers indicate typical local pattern of the images. In most cases, these patterns are of stable appearance, even if severe load is applied to the specimens. Just for strong plastic, viscoelastic or viscoplastic material deformation, local patterns can be recognized after loading, i.e., they can function as a local digital marker for the correlation algorithm.

The correlation approach is illustrated by Figure 9.2. Images of the object are obtained at a the reference load state 1 and at different second load state 2. Both images are compared with each other using a special cross correlation algorithm. In the image of load state 1 (reference) rectangular search structures (kernels) are defined around predefined grid nodes (Figure 9.2, left). These grid nodes represent the coordinates of the center of the kernels. The kernels themselves act as gray scale pattern from load state image 1 that have to be tracked, recognized and determined by their position in the load state image 2. In the calculation step the kernel window ($n \times n$ submatrix) is displaced inside the surrounding search window (search matrix) of the load state image 2 to find the best-match position (Figure 9.2, right).

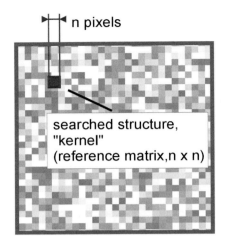 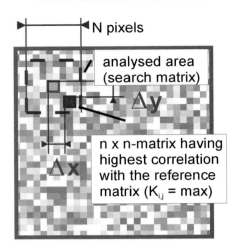

FIGURE 9.2. Displacement evaluation by cross correlation algorithm; (left) reference image at load state 1; (right) image at load state 2 used for comparison.

This position is determined by the maximum cross correlation coefficient, which can be obtained for all possible kernel displacements within the search matrix. The computed cross correlation coefficient K compares gray scale intensity pattern of load state images 1 and 2, which have the same size of the kernel. K is equal to:

$$K_{i',j'} = \frac{\sum_{i=i_0}^{i_0+n-1} \sum_{j=j_0}^{j_0+n-1} (I_1(i,j) - M_{I_1})(I_2(i+i', j+j') - M_{I_2})}{\sqrt{\sum_{i=i_0}^{i_0+n-1} \sum_{j=j_0}^{j_0+n-1} (I_1(i,j) - M_{I_1})^2 \sum_{i=i_0}^{i_0+n-1} \sum_{j=j_0}^{i_0+n-1} (I_2(i+i', j+j') - M_{I_2})^2}}. \quad (9.1)$$

$I_{1,2}$ and $M_{I_{1,2}}$ are the intensity gray values of the pixel (i, j) in the load state images 1 and 2 and the average gray value over the kernel size, respectively. i' and j' indicate the kernel displacement within the search matrix of load state image 2. Assuming quadrangle kernel and search matrix sizes $K_{i',j'}$ values have to be determined for all displacements given by $-(N-n)/2 \leq i', j' \leq (N-n)/2$.

The described search algorithm leads to a two-dimensional discrete field of correlation coefficients defined at integer pixel coordinates (i', j'). The discrete field maximum is interpreted as the location, where the reference matrix has to be shifted from the first to the second image to find the best matching pattern. Figure 9.3 shows an example of the correlation coefficients inside a predefined search window.

With this calculated location of the best matching submatrix an integer value of the displacement vector is determined.

9.2.2. Subpixel Analysis for Enhanced Resolution

As described in the previous section the calculated displacements by the cross correlation algorithm are evaluated for integer pixel coordinates. For the calculation of the displacement field with higher accuracy the displacement evaluation has to be improved.

In the reported calculation codes applied by different research groups, the accuracy of the cross correlation technique is improved in a second calculation step using a special

MICRODEFORMATION ANALYSIS AND RELIABILITY ESTIMATION

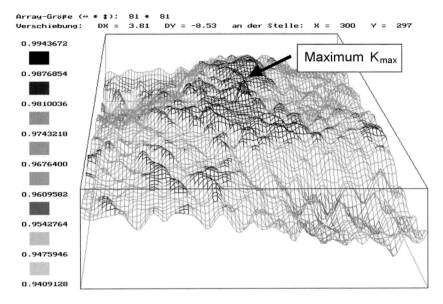

FIGURE 9.3. Discrete correlation function $K_{i',j'}$ defined at integer i', j' coordinates; the maximum of the coefficient of correlation is marked by an arrow.

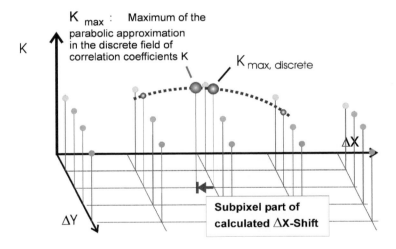

FIGURE 9.4. Principle of the parabolic subpixel algorithm.

subpixel algorithm. The presumably simplest and fastest procedure to find a value for the non-integer subpixel part of the displacement is realized in parabolic fitting. The algorithm searches for the maximum of a parabolic approximation of the discrete function of correlation coefficients in the close surrounding of the maximum coefficient $K_{max,discrete}$. The approximation process is illustrated in Figure 9.4.

The location of the maximum of the parabolic function defines the subpixel part of the displacement. This algorithm implemented quite often allows to get a subpixel accuracy of about 0.1 pixel. Even so it must be stated, that it can fail considerably and introduce large systematic errors under some circumstances. More advanced algorithms are more accurate, allow to reach subpixel accuracies up to 0.01...0.02 pixel for common 8 bit depth digitizing, but demand sophisticated analysis and depend on the kind image sources and of data to be treated.

9.2.3. Results of Digital Image Correlation

The result of the two-dimensional cross correlation and subpixel analysis in the surroundings of a measuring point primarily gives the two components of the displacement vector. Applied to a set of measuring points (e.g., to a rectangular grid of points with user defined pitches), this method allows to extract the complete in-plane displacement field. These results can be displayed in the simplest way as a numerical list which can be post-processed using standard scientific software codes. Commonly, graphical representations such as vector plots, superimposed virtual deformation grids or color scale coded displacement plots are implemented in commercially available or in in-house software packages. Figure 9.5 shows two typical examples of graphical presentations for the results at an SFM image.

Finally, taking numerically derivatives of the obtained displacement fields $u_x(x, y)$ and $u_y(x, y)$ the in-plane strain components ε_{ab} and the local rotation angle ρ_{xy} are determined:

$$\varepsilon_{xx} = \frac{\partial u_x}{\partial x}, \quad \varepsilon_{yy} = \frac{\partial u_y}{\partial y}, \quad \varepsilon_{xy} = \frac{1}{2}\left(\frac{\partial u_x}{\partial y} + \frac{\partial u_y}{\partial x}\right), \quad \rho_{xy} = \frac{1}{2}\left(\frac{\partial u_x}{\partial y} - \frac{\partial u_y}{\partial x}\right). \tag{9.2}$$

Derivation is included in some of the available correlation software codes or can be performed subsequently with the help of graphics software packages.

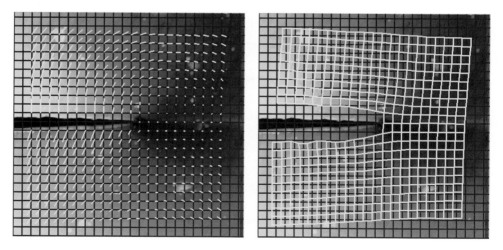

FIGURE 9.5. Digital image correlation results derived from SFM images of a crack tip, scan size [15 × 15 μm]; (left) image overlaid with user defined measurement grid and vector plot; (right) image overlaid with user defined measurement grid and deformed measurement grid, displacement vector and deformed grid presentation are enlarged with regard to the image magnification.

9.3. DISPLACEMENT AND STRAIN MEASUREMENTS ON SFM IMAGES

9.3.1. Digital Image Correlation under SPM Conditions

In comparison to DIC-based measurements treating optical or SEM micrographs, some more essential difficulties have to be overcome for SFM imaging. They correspond to the extreme magnification under SFM conditions. Because SFM image scans are taken over a time interval from one to several minutes smallest system drifts can cause significant artificial object deformations. Classifying different drift sources it can be distinguished between

- SFM scanner drifts, which are related to time dependent behavior of the piezo drives,
- relative movements between the scanner head and the sample fixture, mainly caused by temperature changes,
- drift of sample loading parameters (temperature, forces, load paths, etc.) within testing stages installed at the microscope, and
- incremental object deformations originating from viscous material behavior of test specimens, i.e., time dependent object deformations which take place even under constant loading parameters.

Substantial concerns regarding stability and reproducibility originate from drifts of the SFM scanner piezo and the thermo-mechanical loading parameters over time. As a consequence, the accurate selection of SFM equipment and loading stages is a crucial issue. Moreover, the development and implementation of methods of drift control and compensation may be a must for particular applications.

In the following drifts originating from SFM scanner drifts are considered. Figure 9.6 shows a typical result of a respective stability check carried out at a SFM equipment with activated feedback loop of the piezo scanner. For the stability measurement of Figure 9.6 a series of non-contact topography SFM scans have been picked up from an unloaded and stable mounted object. Topography data was extracted from one scan direction only, to suppress artifacts caused by scanner hysteresis. Displacement and strain values were computed for pairs of subsequent scans. The time period between the scans was negligible compared

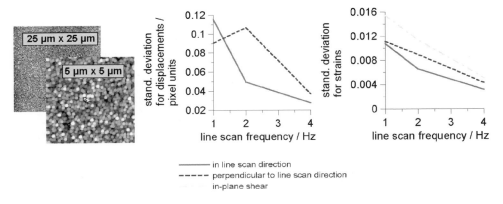

FIGURE 9.6. Estimation of measurement errors for nanoDAC by standard deviation of data determined over measurement points of a whole image, subsequent scans from an unloaded test specimen, non-contact scan mode; (left): scan from Si specimens (sample roughness approx. 10...20 nm); (right): displacement and strain standard deviation as a function of line scan frequency.

to the interval of scanning. Therefore, determined displacements and strains represent only the arbitrary measurement error of the DIC method superposed by the systematic error due to scanner drift. CMP (Chemical Mechanical Polishing) treated silicon surfaces which exhibit ideal pattern for correlation technique have been chosen for this analysis. All measurements were carried out after several hours of idle scanning in order to minimize piezo drifts.

By analyzing the data of Figure 9.6 obtained from an AutoProbe M5 device several conclusions can be made:

1. For suitable SFM choice and installation some of the commercially available equipment exhibits a scanner stability, which does not significantly reduce the possible measurement accuracy as limited by the correlation algorithms (see Section 9.2.3). For the SFM micrographs 0.1 pixel (and better) for local displacement values and 4×10^{-3} for local strain values are feasible. These results relate to topography scans obtained one immediately after the other.
2. Obviously, higher line scan frequencies, i.e., smaller scan time, improve accuracy. Values along the cantilever line scan direction are more accurate, what should be expected from the same stability considerations. The standard deviation for displacements keeps at levels as known from correlation analysis with SEM [10,24].
3. For strain values slightly higher systematic measurement errors are found than under SEM imaging. It is assumed that improvements of strain measurements are possible, because no optimizations of data refinement procedures (smoothing, grid building algorithms) have been included into the referred to analysis.

Drifts introduced by loading stages and the objects under investigation are a separate issue. Already slight drifts of loading forces and applied temperatures as well as material creep can result in large pseudo strains. The cause is a possibly accumulated displacement over the whole sample/stage size, which appears locally at the scan position as a large amount of "rigid body motion." In some cases this "rigid body drift" over the scan time leads to not negligible values of additional pseudo strains. For example, the accumulation of thermally induced displacements over an aluminum specimen of 1 cm length can give rise to already measurable pseudo strains for temperature drift rates as small as 1×10^{-3} K/min.

Figure 9.7 illustrates the impact of drift induced pseudo strains on real nanoDAC measurements. The 3D plot shows the displacement fields nearby a crack trip. The crack was opened by external forces into the direction perpendicular to the crack boundary (Mode I crack opening, see also Section 9.4.3). The comparison between the measured and the theoretical crack opening displacement fields reveals slight deviations.

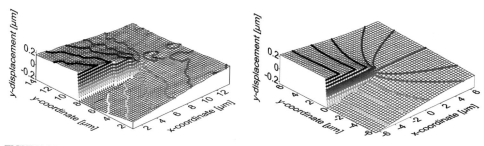

FIGURE 9.7. Crack opening displacement field near a crack tip (displacement component perpendicular to the crack boundary); (left): measured by SFM, (right): theoretical field, the incline of the overlaid displacement contourlines indicates a superposition of pseudo strains induced by drift.

In the measurement plots the displacement contourlines of equal y-displacement values are inclined on both sides of the crack boundary in different directions. This behavior is a result of a small, nearly constant pseudo strain in y-direction superposed to the real crack opening field. Nevertheless, the actual crack opening field dominates and is reproduced in the right way.

9.3.2. Technical Requirements for the Application of the Correlation Technique

There exist two main issues to be discussed if Digital Image Correlation (DIC) is supposed to be applied in SFM: suitable specimen loading must be realized within the equipment and high level scanning reproducibility must be provided.

The placement of thermal and/or mechanical loading stages in the SFM is mainly a question of compact design of loading stages, free access for SFM cantilevers to the specimen surface of interest and a scanning by cantilever-site actuation. In dependence on available equipment and loading stages it can be necessary to install additional spacers in between the cantilever head and the ground plate with x–y-stages for specimen adjustment.

I order to avoid scanning instabilities and different thermo-mechanical drifts (see Section 9.3.1) common tools of environmental isolation offered by equipment suppliers should be installed. This comprises active vibration compensation of equipment tables, acoustic enclosures against ambient sound and if possible also temperature stabilization. Advanced fast line scan equipment can be very helpful to meet this requirement. Because of the possible drift of object loading parameters it might be necessary to develop special tools for drift compensation. These measures can be taken actively by piezo driven displacement compensation or passively by numerical corrections of measured displacement and/or strain fields.

9.4. DEFORMATION ANALYSIS ON THERMALLY AND MECHANICALLY LOADED OBJECTS UNDER THE SFM

Most of the published work is aiming at the characterization of materials, either taking into consideration local material structures like grain size or property gradients (e.g., [7,25]) or focusing on the determination of material properties on microscopic or nanoscopic structures (e.g., [9,12,26]). The following two section present two examples, which give an impression about possible application fields of the DIC under SFM conditions.

9.4.1. Reliability Aspects of Sensors and Micro Electro-Mechanical Systems (MEMS)

Modern sensors and MEMS/NEMS devices consist of extremely fragile functional structures. Because of the desired device functionality, quite different materials in terms of material properties have to be combined with one another. Loading such structures thermally and/or mechanically means to implement severe material mismatch within submicron and nano-scale volumes. Therefore, functional or environmental loading causes local stresses and strains due to different material properties such as coefficient of thermal expansion (CTE), Young's modulus or time depended viscoelastic or creep properties. The smallest existing material imperfections or initial micro/nano-scale defects can grow under stress and strain and can finally lead to the failure of the device [10,27–30].

Thin layers used in sensor and MEMS technology undergo local stresses remote from elastic material behavior, where permanent device alterations are feared after each load cycle. Nowadays, responses of nanomaterials to applied external loads from temperature, vibrations, or chemical agents are not well understood. The same is true for actual failure mechanism and damage behavior.

The way to achieve this aim is the combination of displacement and strain measurements on the micro- and nano-scale with modeling techniques based on finite element analysis. Parameterized finite element models of MEMS are applied for faster prediction of life time and failure modes. The parameterization allows the variation of model geometries and materials in order to accelerate the MEMS design process [32].

9.4.2. Thermally Loaded Gas Sensor under SFM

Sensor applications with local temperature regulation such as the gas sensor shown in Figure 9.8 are usually thermally loaded with rapid and frequent change in temperature [29]. This thermal cycling and the temperature gradients over the structure imply thermal stresses and may cause failure of the component [33]. In the operation mode of the gas sensors thermal stresses are induced due to the activated micro-heater.

With in-situ SFM measurements on this micro system the capability of the nanoDAC approach is demonstrated measuring material deformation resulting from mismatch of material properties.

The gas sensor is designed to tolerate several hundreds of °C thermal loads. The thermal mismatch between the platinum electrodes (CTE = 9 ppm K^{-1}) and the SiO_2 substrate (CTE = 0.65 ppm K^{-1}) leads to high local stresses, if the entire device is heated up. Local displacements resulting from the thermal load have been measured by means of the nanoDAC technique.

9.4.2.1. In-plane Displacements
In-situ non-contact SFM scans on top of the gas sensor membrane have been carried out at room temperature and at 100°C. The area which was observed is illustrated in Figure 9.9 as location 2. At this area an overlap of the SiO_2 membrane by the platinum electrodes should result in a thermally induced stress/strain field. The temperature was achieved by applying a defined voltage to the microheater of the gas sensor.

The determined thermally induced displacement field shows that the platinum layer with its higher CTE value reveals an inherent expansion toward the edge of the layer. In supplementary tests with heating cycles with maximum temperatures in the range of 450°C severe delaminations of the platinum layer at the edges to the SiO_2 substrate layer were observed (Figure 9.11). Details of this testing cycle are described in more detail in [29].

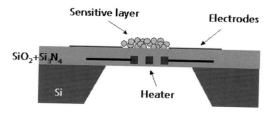

FIGURE 9.8. Layout of gas sensor.

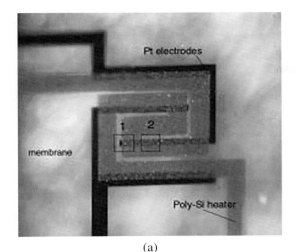

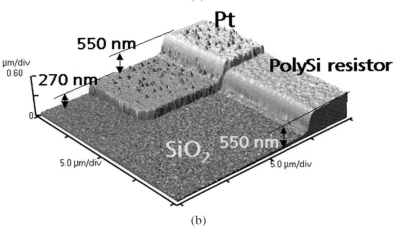

FIGURE 9.9. (a) Microscopic image of flow and gas sensor membrane, overall membrane thickness: approx. 2 μm, field of view: approx. 500 μm; (b) SFM topography scan of gas sensor depicting the Pt layer on top of the SiO_2 membrane and part of the Poly-Si heater embedded (detail 1 of Figure 9.9(a) source [29]).

9.4.2.2. Out-of-plane Displacements Besides the information on structural deformation in the $x–y$ plane the SFM measurement technique allows the determination of the out-of-plane displacement component. The height information of the SFM topography images before and after loading is analyzed for evaluation of movements or deformations in the z-direction.

Applying this technique to in-situ measurements of thermal deformations by SFM on the top of the sensor membrane have revealed a high value of remaining deformations even after a single heat cycle (25 to 100°C). Inelastic strains remain after cooling down to room temperature (Figure 9.12).

9.4.3. Crack Detection and Evaluation by SFM

Tiny defects or cracks in microelectronics components can lead to severe crack propagation and complete failure if electronic devices are stressed. Because of intrinsic stress

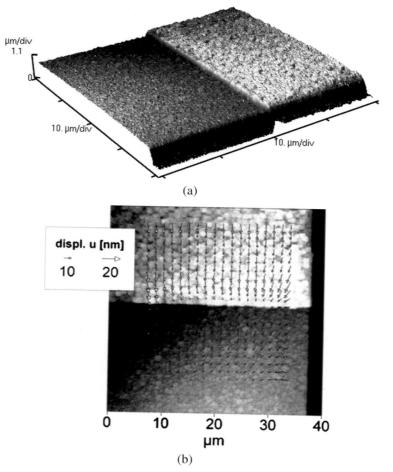

FIGURE 9.10. (a) SFM topography image of platinum and SiO$_2$ layers; (b) vector plot of displacements u measured by nanoDAC.

sources, like e.g., thermal mismatch, the changes in environmental conditions (temperature, pressure, mechanical vibrations) can initiate crack propagation and cause fatal damage. Experimental crack detection can be a crucial issue bearing in mind original crack sizes of about some micrometers. These cracks will open only some tens of nanometers or even less under sub-critical load. Their detection, however, is possible by DIC displacement measurements.

In the following, crack detection and evaluation will be shown at a cyanate ester resin polymer material. A typical application of this thermoset is the area of microelectronic systems, where it may be used as underfiller between chip and substrate or as matrix material for printed circuit boards. The unmodified resin has a high modulus of elasticity but poor resistance to fracture [3,34].

9.4.3.1. In-situ Measurement Technique For the crack detection experiments, a simple specimen configuration is selected to demonstrate the fundamental approach. With a compact tension (CT) crack test specimen as shown in Figure 9.13 Mode I (opening) loading of

MICRODEFORMATION ANALYSIS AND RELIABILITY ESTIMATION 245

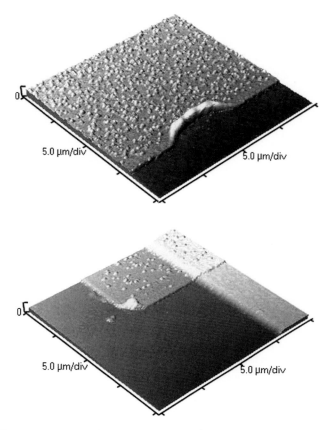

FIGURE 9.11. SFM topography scan of membrane layers after tempering at 450°C, Pt electrode destruction at edge and corners (compare to Figures 9.9 and 9.10).

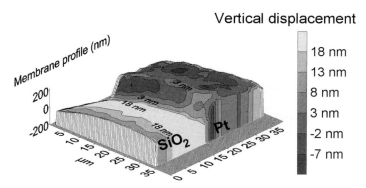

FIGURE 9.12. Residual sensor deformation after heat cycle (SFM based deformation measurement), 3D plot shows part of the membrane layer profile, the coloring (gray scale) indicates the remaining vertical deformation after a heat cycle.

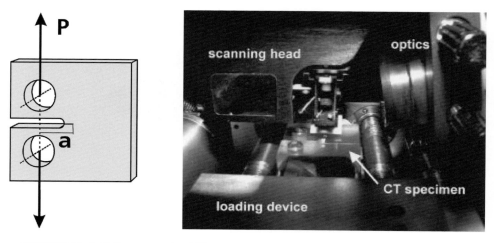

FIGURE 9.13. (left) Compact tension (CT) specimen; (right) In-situ loading of CT specimen under SPM.

the crack tip is enabled. Due to a high accuracy in the machining process of the specimen, in-plane and out-of-plane shear (Mode II and III) components are avoided to a considerable extent.

The CT specimen is loaded by a special tension/compression testing module, which can be utilized for in-situ SEM and SPM measurements. Figure 9.13 shows the CT specimen and parts of the loading device under the SPM.

SFM topography scans are taken at different locations of the crack face before and after loading. In the following presentation of measurement results the first location is approximately 50 µm away from the crack tip and the second is directly at the crack tip.

9.4.3.2. Crack Detection At the first location (crack face) of the CT specimen, the capability of the DIC methods for displacement measurements at nano-scale is demonstrated. SFM non-contact topography scans are taken at the crack face approximately 50 µm away from the crack tip. The CT specimen is loaded with a force far below the critical fracture load. The SFM images are taken before and after loading with a size of 33 × 33 µm and as 256 × 256 image arrays, i.e., the lateral resolution is approximately 130 nm/pixel.

Figure 9.14 shows the scans before and after loading with height profiles perpendicular to the crack.

A comparison of the two topography images and the height profiles of Figure 9.14 shows that the crack opening is not clearly recognizable due to the low load and the relatively coarse resolution of the scan. There are also scratches on the image which could be identified as cracks.

However, if the digital image correlation algorithm is applied to the images, the crack opening can easily be detected. Figure 9.15 shows the result obtained by nanoDAC analysis.

As illustrated in Figure 9.15 the crack opening due to loading of the CT specimen is about 200 nm. Obviously the crack cannot be identified from the SFM images themselves, because the crack opening is only in the order of 1 image pixel. This fact is illustrated by the height profiles from the topography plot (Figure 9.14), where scratches and the crack do not clearly differ from each other.

MICRODEFORMATION ANALYSIS AND RELIABILITY ESTIMATION

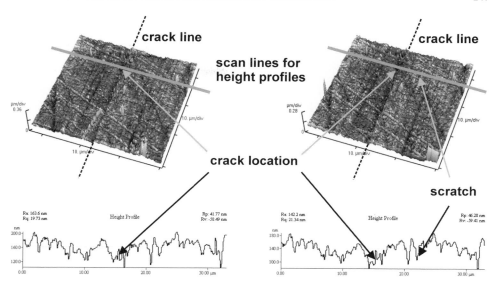

FIGURE 9.14. SFM topography images near a crack tip on cyanate ester resin specimen (33 μm × 33 μm image size); (left) SFM scan before crack opening; (right) SFM scan after crack opening.

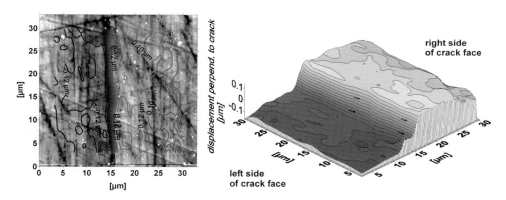

FIGURE 9.15. Displacement measurement from SFM images near a crack tip on a cyanate ester resin specimen; (left) SFM topography scan, with an overlaid contour plot showing the displacements in x-direction, u_x (component perpendicular to the crack boundaries); (right) Displacement field u_x as a 3D plot.

As a conclusion from these measurements, it can be stated that even for larger scan sizes (10...100 μm) displacements in the range of 10 nm can be measured. Cracks in the micron and submicron range can be detected and evaluated. Therefore, it is possible to apply the DIC technique for future reliability issues of MEMS and NEMS.

9.4.3.3. Crack Evaluation by COD Concept With successful detection of cracks at the micro and nano-scale the foundation is laid for a more detailed analysis of material faults and defects. The question if available fracture and damage criteria from macro or micro approaches can be transferred to a nanoscopic level is an important issue for reliability evaluation of MEMS and NEMS.

TABLE 9.1.
Crack opening displacement in LEFM for infinite bulk material and Mode I crack opening.

$$u_y^u = \frac{K_I}{2\mu}\sqrt{\frac{x}{2\pi}}(k+1), \quad u_y^l = -\frac{K_I}{2\mu}\sqrt{\frac{x}{2\pi}}(k+1)$$

for $x \leq 0$

$$u_y^u = u_y^l = 0 \quad \text{for } x > 0 \tag{9.4}$$

The classical stress intensity factor K is conventionally determined by means of macroscopic fracture tests at standardized fracture specimen such as the CT specimen. This specimen type which is also used for the crack detection test described in the Section 9.4.3.1 is used for the verification of the nanoDAC technique for crack evaluation [11].

A straightforward approach for crack evaluation in the SFM is the technique of crack opening displacement (COD) determination. For the combination of the COD concept and the K concept, the following assumptions have to be made:

- Linear Elastic Fracture Mechanics (LEFM) apply within the measurement area and for the applied load,
- the specimen consists of homogeneous material.

In order to determine the Mode I stress intensity factor K_I crack opening displacements u_y^u and u_y^l have been measured along both the upper and lower crack boundaries. If determined by LEFM they must equal the values of Table 9.1.

In Equation (9.4) μ is the shear modulus and k is a function of Poisson's ratio. For the surface of the specimen where plane stress predominates k is given by $k = (3 - \nu)/(1 + \nu)$ [35].

Taking the square of the difference of upper and lower displacements, we obtain a linear function of the x-coordinate or 0, in dependence, at which side of the crack tip we are:

$$\left(\frac{u_y^u - u_y^l}{2}\right)^2 = Cx, \quad x \leq 0$$
$$= 0, \quad x > 0. \tag{9.5}$$

The expression of Equation (9.5) does not change if specimen rotation due to load is included into the considerations. In this case, equal rotational terms on both sides of the crack boundary are subtracted from each other. For the equation above, the crack tip is set at location $x = 0$. The crack tip location on the real specimen can be found at the interception of a linear fit of the curve Cx with the x-coordinate axis. The slope C allows to estimate the stress intensity factor K_I, which is a measure of the crack tip load and is given by:

$$K_I = \frac{E}{1+\nu}\frac{1}{k+1}\sqrt{2\pi C}. \tag{9.6}$$

In Equation (9.6) E is the Young's modulus.

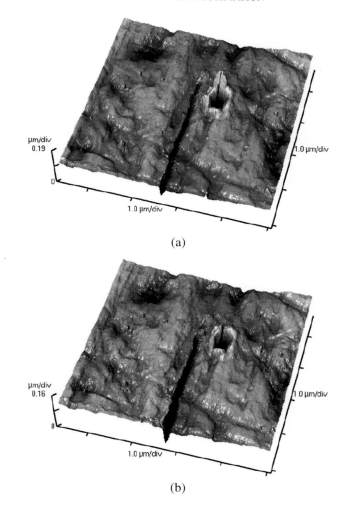

FIGURE 9.16. SFM image of crack tip (size: 4.6 × 4.6 μm) for the evaluation of stress intensity factors K_I by measuring crack opening displacement around the crack tip; (a) load state 1; (b) load state 2; (the indentation near the crack tip is a indentation caused by a cantilever approach).

Continuing, an example for this procedure will be presented. The crack tip of a cyanate ester resin CT specimen is mapped by the SFM equipment at two different load states. Figure 9.16 shows the corresponding images.

To demonstrate the down-sizing capabilities of the DIC approach the scan size of these images have been chosen smaller than that of the examples described in the previous section. For a scan size of 4.6 × 4.6 μm the crack opening is already recognizable in the SFM images. Figure 9.17 illustrates the displacement results in y-direction, u_y calculated by digital image correlation along with two lines marking the upper and lower crack face. The displacement results at these lines are u_y^l and u_y^u used for the determination of the slope C (Figure 9.17).

The determined value for K_I equals 0.033 MPa m$^{1/2}$ which is a value of about 1/20 of the critical fracture toughness K_{IC} for this type of cyanate ester thermoset.

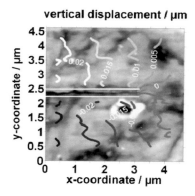
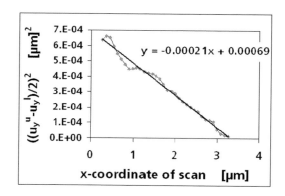

FIGURE 9.17. (left) SFM image of crack tip area (size: 4.6 × 4.6 μm) with overlaid displacement results in y-direction, u_y, lines for the upper and lower crack face are included (right) evaluation of slope C for the calculation of the stress intensity factor K_I.

9.5. CONCLUSION AND OUTLOOK

It has been shown that SFM images can be utilized to measure displacement and strain fields in very narrow regions of micro- and nanoobjects. In that way the versatile information stored in the micrographs can be converted into quantitative data. As a result characterization of materials behavior as a response to external macroscopic loading can be performed aiming at the consideration of the complex influence of the microscale and nanoscale structure. At present especially methods of Digital Image Correlation have been attracted to extract displacement and strain data from SFM images. On this occasion it must be mentioned that these tools have been developed within past very few years. One can expect, that capabilities of these tools will be extended within the next years. Stability and reproducibility issues restricting applications today may have less importance, as advanced SFM equipment and measurements approaches will come up. Also the determination of material properties by SFM based strain measurements is only at the very beginning. More advanced micro material failure analysis will be possible. Moreover, the better understanding of material behavior in the accessible tiny volumes should have an impact on the development of failure criteria.

REFERENCES

1. T. Kinoshita, Stress singularity near the crack-tip in silicon carbide: investigation by atomic force microscopy, Acta Materialia, 46(11), pp. 3963–3974 (1998).
2. K. Komai, K. Minoshima, and S. Inoue, Fracture and fatigue behavior of single crystal silicon microelements and nanoscopic AFM damage evaluation, Microsystem Technologies, 5(1), pp. 30–37 (1998).
3. C. Marieta, M. del Rio, I. Harismendy, and I. Mondragon, Effect of the cure temperature on the morphology of a cyanate ester resin modified with a thermoplastic: characterization by atomic force microscopy, European Polymer Journal, 36, pp. 1445–1454 (2000).
4. M.S. Bobji and B. Bushan, Atomic force microscopic study of the microcracking of magnetic thin films under tension, Scripta Materialia, 44, pp. 37–42 (2001).
5. C.J. Druffner and Sh. Sathish, Improving atomic force microscopy with the adaptation of ultrasonic force microscopy, Proc. of SPIE, 4703, pp. 105–113 (2002).
6. H. Xie, A. Asundi, C.G. Boay, L. Yungguang, J. Yu, Z. Zhaowei, and B.K.A. Ngoi, High resolution AFM scanning moiré method and its application to the micro-deformation in the BGA electronic package, Microelectronics Reliability, 42, pp. 1219–1227 (2002).

7. A. Asundi, X. Huimin, L. Chongxiang, C.G. Boay, and O.K. Eng, Micro-moiré methods—optical and scanning techniques, Proc. SPIE, 4416, pp. 54–57 (2001).
8. A. Asundi, H. Xie, J. Yu, and Zh. Zhaowei, Phase shifting AFM moiré method, Proc. of SPIE, 4448, pp. 102–110 (2001).
9. I. Chasiotis and W. Knauss, A new microtensile tester for the study of MEMS materials with the aid of atomic force microscopy, Experimental Mechanics, 42(1), pp. 51–57 (2002).
10. D. Vogel, J. Auersperg, and B. Michel, Characterization of electronic packaging materials and components by image correlation methods, Advanced Photonic Sensors and Applications II, Nov. 27–30, 2001, Singapore, Proc. of SPIE, 4596, pp. 237–247 (2001).
11. D. Vogel and B. Michel, Microcrack evaluation for electronics components by AFM nanoDAC deformation measurement, Proc. of IEEE-NANO 2001, Maui, Hawaii, Oct. 28–30, 2001, pp. 309–312.
12. D. Vogel, J. Keller, A. Gollhardt, and B. Michel, Displacement and strain field measurements for nanotechnology applications, Proc. of the 2002 2nd IEEE Conference on Nanotechnology, IEEE-NANO 2002, August 26–28, Washington D.C., 2002.
13. D. Post, B. Han, and P. Ilju, High Sensitivity Moiré, Springer-Verlag, Berlin, 1994.
14. R.S. Sirohi and F.S. Chau, Optical Methods of Measurement—Wholefield Techniques, Marcel Dekker Inc., N.Y., Basel, 1999.
15. B. Michel and R. Kuehnert, Mikro-Moiré-Methode und MikroDAC-Verfahren anwenden, Zeitschrift Materialprüfung, 38(6) (1996).
16. D. Vogel, J. Auersperg, A. Schubert, B. Michel, and H. Reichl, Deformation analysis on flip chip solder interconnects by microDAC, Proc. of Reliability of Solders and Solder Joints Symposium at 126th TMS Annual Meeting & Exhibition, Orlando, 1997, pp. 429–438.
17. D. Vogel, V. Grosser, A. Schubert, and B. Michel, MicroDAC strain measurement for electronics packaging structures, Optics and Lasers in Engineering, 36(2), pp. 195–211 (2001).
18. M.A. Sutton, W.J. Wolters, W.H. Peters, W.F. Ranson, and S.R. McNeill, Determination of displacements using an improved digital correlation method, Image and Vision Computing, 1(3), pp. 133–139 (1983).
19. Y.-J. Chao and M.A. Sutton, Accurate measurement of two- and three-dimensional surface deformations for fracture specimens by computer vision, in J.S. Epstein, Ed., Experimental Techniques in Fracture, VCH Publishers, N.Y., 1993, pp. 59–93.
20. M.A. Sutton, S.R. McNeil, J.D. Helm, and M.L. Boone, Measurement of crack tip opening displacement and full-field deformations during fracture of aerospace materials using 2D and 3D image correlation methods, IUTAM Symp. on Advanced Optical Methods and Applications in Solid Mechanics, 2000, pp. 571–580.
21. D.L. Davidson, Micromechanics measurement techniques for fracture, in J.S. Epstein, Ed., Experimental Techniques in Fracture, VCH Publishers, N.Y., 1993, pp. 41–57.
22. D. Vogel, A. Schubert, W. Faust, R. Dudek, and B. Michel, MicroDAC—a novel approach to measure in-situ deformation fields of microscopic scale, Proc. of ESREF'96, 1939-42, Enschede, 1996.
23. D. Vogel, J. Simon, A. Schubert, and B. Michel, High Resolution Deformation Measurement on CSP and Flip Chip, Technical Digest of the Fourth VLSI Packaging Workshop of Japan, 84–86, Kyoto, 1998.
24. D. Vogel, E. Kaulfersch, J. Simon, R. Kühnert, A. Schubert, and B. Michel, Measurement of thermally induced strains on flip chip and chip scale packages, Proc. of ITherm 2000, Las Vegas, USA, May 23–26, 2000, pp. 232–239.
25. E. Soppa, P. Doumalin, P. Binkele, T. Wiesendanger, B. Bornert, and S. Schmauder, Experimental and numerical characterization on in-plane deformation in two-phase materials, Computational Material Science, 21, pp. 261–275 (2001).
26. L. Cretegny and A. Saxena, AFM characterization of the evolution of surface deformation during fatigue in polycrystalline copper, Acta Mater., 49, pp. 3755–3765 (2001).
27. K. Pinardi, Z. Lai, D. Vogel, Y.L. Kang, J. Liu, Sh. Liu, R. Haug, and M. Willander, Effect of bump height on the strain variation during the thermal cycling test of ACA flip-chip joints, IEEE Trans. on Comp. and Pack. Techn., 23(3), pp. 447–451 (2000).
28. A. Schubert, R. Dudek, H. Walter, E. Jung, A. Gollhardt, and B. Michel, Lead-free flip-chip solder interconnects—materials mechanics and reliability issues, Proc. APACK: Int. Conf. on Advances in Packaging, Singapore, Dec. 5–7, 2001, pp. 274–287.
29. J. Puigcorbé, D. Vogel, B. Michel, A. Vilà, N. Sabaté, I. Gràcia, C. Cané, and J.R. Morante, High temperature degradation of Pt/Ti electrodes in micro-hotplate gas sensors, J. Micromech. Microeng., (13), pp. 119–124 (2003).
30. C.B. O'Neal, A.P. Malshe, W.F. Schmidt, M.H. Gordon, R.R. Reynolds, W.D. Brown, W.P. Eaton, and W.M. Miller, A study of the effects of packaging induced stress on the reliability of the sandia MEMS microengine, Proc. of IPACK2001: The Pacific Rim/ASME International Electronic Packaging Technical Conference and Exhibition, Hawaii, USA, July 8–13, 2001.

31. D. Vogel, Chen Jian, and I. de Wolf, Experimental validation of finite element modeling, in G.Q. Zhang, Ed., Benefiting from Thermal and Mechanical Simulations in Micro-Electronics, Kluwer Academic Publishers, Boston, 2000, pp. 113–133.
32. J. Auersperg, R. Döring, and B. Michel, Gains and challenges of parameterized finite element modeling of microelectronics packages, in B. Michel, Ed., Micromaterials and Nanomaterials, No. 1, Fraunhofer IZM, Berlin, 2002, pp. 26–29.
33. J. Puigcorbé, A. Vilà, J. Cerdà, A. Cirera, I. Gràcia, C. Cané, and J.R. Morante, Thermo-mechanical analysis of micro-drop coated gas sensors, Sensors and Actuators A, 97, pp. 379–385 (2002).
34. I. Hamerton, Chemistry and Technology of Cyanate Ester Resins, Blackie Academic and Professional, Glasgow, 1994.
35. T.L. Anderson, Fracture Mechanics, CRC Press LLC, Boca Raton, 1995.

10

Interconnect Reliability Considerations in Portable Consumer Electronic Products

Sridhar Canumalla[a] and Puligandla Viswanadham[b]

[a]*Mobile Devices Unit, Enterprise Solutions, Nokia, 6000 Connection Drive, IRVING, TX 75039, USA*
[b]*Nokia Research Center, Nokia, 6000 Connection Drive, IRVING, TX 75039, USA*

10.1. INTRODUCTION

Revolutionary changes have taken place in digital information processing in recent years—the world has gone wireless and life has gone mobile. Handheld computers, personal digital assistants, mobile phones, computer controlled domestic appliances and other portable consumer electronic hardware have become pervasive in daily life. These mobile consumer electronic products are considerably different from other consumer electronic products from a variety of perspectives. In addition to being function-rich, lightweight, and portable, they often serve as fashion accessories. Hence, in the design and construction of mobile electronic hardware, visual appeal needs to be considered in addition to durability. Indeed, industrial design and functionality guidelines often take precedence over design for durability.

Most hand held/portable consumer electronic products can be characterized as high volume, low cost devices. In a competitive business environment, manufacturing and product development costs, time-to-market, functionality, yields, customer satisfaction and product quality all have an impact on the business. Original equipment manufacturers face pressure to develop new, more advanced technology products in record time, while at the same time improving productivity, product field reliability and overall quality. Although product quality encompasses several measures, the relevant link between quality and reliability can be described as follows. Reliability is defined as the probability that a product will perform its intended function under encountered operating conditions for a specified period, whereas quality, in narrow terms of reliability alone, can be defined as the reliability at time zero. While there are other definitions to quality, there is general agreement that an unreliable product is *not* perceived as a high quality product [1].

TABLE 10.1.
Comparison of typical application conditions of desktop, mobile, and automotive hardware.

Hardware type	Product life/yrs	Power-on cycles/day	Power-on hours	Relative humidity/%	Environment temperature range/°C	Operational temperature range/°C	Voltage/ V
Desktop	5	1–17	13,000	10–80	10–30	20–60	12
Mobile terminal	5	20	43,800	10–100	−40–40	32–70	1.8–3.3
Automotive under-the-hood	15	5	8200	0–100	−40–125	−40–125	12

*Source JEDEC.

The highly personal use profile and mobility for these products implies that consumers will take these products with them wherever they go, and expect the same dependable performance irrespective of the exposure of the product to the elements, for example, rain, snow and accidental drop. Under such conditions, meeting the reliability expectations for portable products can be a challenge, especially since reliability expectations need to be met without compromising profitability. Therefore, one primary driver for product reliability is perceived quality and customer satisfaction. Another reason for ensuring reliability is that product field-failure rate, which plays a key role in controlling warranty and repair costs, tends to be higher for an unreliable product. In other words, all other factors remaining the same, a more reliable product will be more profitable. However, in reality, there is a level of optimum reliability beyond which additional reliability improvements have a decreasing rate of return. Therefore, it is prudent to develop products that meet specific business or customer requirements driven reliability target rather than aiming to have the most reliable product possible at the expense of profitability or time-to-market. A third reason to strive for product reliability is that reliability (and quality) could be employed as product differentiators in product marketing (advertising), which will only increase the business value of product reliability.

The operating environment for mobile electronic equipment also differs considerably from that for desktop or business computers, and is often more varied in terms of thermal excursions, and exposure to humidity and corrosive environments. The products are more prone to high humidity exposures in both non-condensing and condensing atmospheres. Additionally, portability makes the product more likely to experience mechanical loads such as drop, bend, twist, etc. Mechanical drops from such heights as a meter and half on hard surfaces are not uncommon. The number of power ON cycles, the operational voltages, and other conditions are also different. Table 10.1 shows a comparison of the typical operating environments for portable hand-held telecommunication devices with conventional desktop and automotive under-the-hood electronics.

Irrespective of the operating environment, programs for producing reliable products require quantitative methods for predicting and assessing various aspects of product reliability. This involves the collection of reliability data from the following [2]:

1. Laboratory life tests to assess product reliability.
2. Degradation tests of materials, devices, and components.
3. Design of experiments for reliability improvement.
4. Tests on early prototype units to learn about possible failure modes and mechanisms.

5. Monitoring of early-production units in the field.
6. Analysis of warranty data and samples from warranty population.
7. Systematic longer-term tracking of product in the field.

The need for shorter design cycle time is a driver for reducing the time and resources spent in reliability testing. Since non-accelerated tests can take an excessively long time to yield valuable data for reliability improvement, different kinds of accelerated tests have been developed to estimate relatively quickly the failure-time distribution or long-term performance of the product in the field, based on a careful study of the operating environment. However, in and of itself, accelerated life tests to *assess* reliability do not yield actionable data to *improve* reliability. Analysis of the failures to uncover failure mechanisms and the root causes of failure are crucial for formulating corrective actions that can improve reliability. Sometimes, a second round of reliability tests may be required to assess the reliability of the improved products.

The focus of this chapter is on the physics behind accelerated laboratory life tests to assess reliability in thermal, mechanical and electrochemical environments. Since failure analysis is ideally an integral part of any reliability assessment and improvement exercise, some representative failure mechanisms commonly observed in each of these reliability tests will also be discussed.

The study of interconnection reliability, until a few years ago, was driven primarily by the computer industry. Therefore, the vast majority of literature on electronic packaging reliability is comprised primarily of thermal cycling reliability, and to a lesser extent, corrosion and electromigration phenomena. In fact, there was a tacit assumption that reliability always implied thermal cycling reliability. As such, the titles of some reliability publications did not even indicate that the investigation pertained only to thermal cycling. Until recently, this did not cause any serious consternation among the packaging community in the days where much of the information processing hardware was confined to environments with controlled temperature and humidity.

The material discussed in this chapter is intended to introduce interconnection reliability issues in thermal, mechanical and electrochemical environments for portable, consumer electronic products to readers who are primarily familiar with similar issues in business, office and telecommunication applications. The scope of the chapter is limited to interconnection reliability and excludes important topics such as electromechanics or liquid crystal display issues, which are complex enough to justify a separate chapter.

10.2. RELIABILITY—THERMAL, MECHANICAL AND ELECTROCHEMICAL

10.2.1. Accelerated Life Testing

Some of the pitfalls of accelerated life tests (ALTs) need to be considered to avoid seriously incorrect inferences about the product reliability in the field based solely on laboratory tests [1]. The following aspects need to be recognized when interpreting the results of ALTs:

1. Multiple or unrecognized failure mechanisms—high levels of accelerating variables can induce failure mechanisms that would not normally be observed at operating conditions. For example, instead of just accelerating corrosion or electrochemical migration, higher temperatures may cause melting or material deformation or degradation. Higher humidity may cause swelling and delamination. In less extreme

cases, high levels of accelerating variables will change the relationship between life and the variable. If different failure mechanisms are operative at high levels of the accelerating variables, and this is recognized, failure times for that mechanism can be censored out. Sometimes, such censoring can result in inadequate data. If the presence of undesirable failure mechanisms is not recognized, it is possible that seriously incorrect inferences are drawn.
2. Failure to properly quantify uncertainty—it is important to recognize that all statistical estimates have some uncertainty associated with them. Using point estimates alone can be misleading in many cases. Uncertainty can result either from the experiment or from the model, and in general, statistical confidence intervals do not account for model uncertainty. Extrapolations, fundamentally, are fraught with errors, especially when based on inadequate sample sizes and point data. Performing a sensitivity analysis to assess model uncertainty or testing adequate number of samples is one solution.
3. Multiple time scales and degradation affected by more than one accelerating variable—in ALT, particularly when there is more than one failure mechanism, it should be recognized that all mechanisms may not be accelerated in the same manner. For example, when performing ALT of solder interconnections under accelerated conditions, creep and fatigue are accelerated differently depending on ramp rates and hold times at the different temperatures.
4. Masked failure mechanism—if there is more than a single failure mechanism, it is possible that one mechanism is accelerated more than others. In such cases, the masked failure mechanism will not show up in laboratory testing but can dominate field failures. It is not only prudent, but also cost effective, to verify that the failure mechanisms seen in the field are the same as the failure mechanisms observed in accelerated testing.
5. Faulty comparison—a popular use of ALT is in the comparison of alternative designs or materials from vendors, in addition to its use in predicting field reliability. The rationale behind is that if material from one vendor or one design performs better in laboratory tests, relative field reliability would follow a similar relationship. However, in cases where the reliability in the field is governed by a different failure mechanism than that observed in the laboratory test, ALT results can mask the actual field performance and serve as the basis for inaccurate prediction of field reliability.
6. Accelerating variables can cause deceleration—the most common examples involve failure mechanisms that require specific combinations of humidity, stress, and temperature. For example, when the usage rate is accelerated for a connector undergoing wear, the accelerated test can inhibit a secondary corrosion failure mechanism by continuously removing corrosion products and not giving enough time for the reaction to occur. Another example is failure due to tin whisker formation, which has a high propensity at a certain temperature and humidity for certain substrate and coating compositions and thicknesses. Optimum temperatures for tin whisker growth have been reported to be between 50 and 70°C by several researchers, for example [3]. Unfortunately, since the working temperature of most electronic equipment is relatively close to the optimum temperature for tin whisker growth, an injudicious selection of temperature acceleration can yield incorrect results.
7. Differences between prototype and production samples—It is important to test units manufactured under actual production conditions, using materials and parts that

will be employed in actual production samples. Sometimes, test methods capable of handling functional products may need to be developed. For example, ball-shear tests are widely used to assess the quality of the ball attachment process for area array packages such as ball grid array packages (BGAs) or chip scale packages (CSPs) [4]. However, the ball-shear test method cannot be applied to assess the interconnection quality or strength in a functional product after surface mount assembly because an individual ball is no longer accessible for test. To accommodate interconnection strength data requirements on functional products, tests such as the package-to-board interconnection strength test method (PBISS) can be used [5,6].

10.2.2. Thermal Environment

Historically, for office and business machines, accelerated thermal cycling tests are carried out in the 0 to 100°C range with 10–15 minute dwell times at ramp rates in the 10–15°C/minute range. A life requirement of 1000 cycles translates into a product life of about 7–10 years. These machines hardly experience other mechanical stresses in the operational environment. In contrast, hand held electronic hardware can experience extreme ambient temperature fluctuations in the range of $-30°C$ to $45°C$ depending on the geographic location. When the appliance is left in an automobile it can experience even more severe temperature conditions depending on the climate and diurnal variations. Thus, accelerated thermal cycling tests applicable to business machines will be not be severe enough to assess the performance of handheld electronic appliances. Another difference in regard to the portable hardware is the shorter product design life. The average product design life is in the range of 2 to 5 years instead of the 7 to 10 years in other consumer products such as desktop machines.

Owing to the aforementioned considerations, portable electronic PWB assemblies are generally subjected to accelerated thermal cycling of $-40°C$ to $125°C$ for 200 to 800 cycles in order to assess the product performance.

10.2.3. Mechanical Environment

One way to classify the mechanical environments for a portable electronic product is based on the rate of deformation: (a) low deformation—as experienced in bending and twisting, (b) medium to high deformation rate—as experienced in vibration or (c) high rate of deformation—as in case of drop or shock. Another way to characterize the environment is based on the life expectancy in number of fatigue cycles as being high cycle fatigue (vibration) or low cycle fatigue (drop, bending and twisting). In comparison to thermomechanical reliability, relatively little has been published in the public domain on reliability under mechanical loading. Broadly, mechanical loading can be divided into the following categories

(1) Drop or impact loading—typically high strain rate loading that can also cause bending and twisting of the product due to impact forces. The number of cycles to failure is generally low.
(2) Bending and twisting—typically low strain rate events such as encountered during key presses. The life expectancy is generally a few hundred cycles.
(3) Vibration loading—typically high strain rate loading with low amplitude. In general, vibration failures are of relatively less concern in portable electronic products.

In addition, reliability evaluations of portable electronic products can also involve either shear or pull testing performed at the interconnection or package level for purposes of determining the strength distribution. It is pertinent to include them in the discussion because shear and pull tests serve to define the strength of the interconnection between the package and PWB, which is closely related to reliability in drop, bend, twist or vibration loading.

10.2.3.1. Drop or Impact Environment When portable electronic products are subjected to mechanical drop or impact, it is important to recognize that failure can occur (a) at the solder or other interconnects, (b) connector or spring contacts, (c) inside the components such as LCD, housing, lens, etc. or (d) at the system level. Usually, these failures are due to the following causes:

(a) High inertial forces (g-forces) due to rapid change in velocity upon impact,
(b) Large strains in the solder interconnects between the PWB and package due to excessive dynamic buckling, flexure, or twisting of the PWB, and/or
(c) Shock waves that travel through the product assembly upon impact.

It is reasonable to assume that all three effects can co-exist during any single event and that the interactions among them can be relatively complex.

The drop tests carried out can be at the product level, or at the board assembly level. Product level drop tests involve tests on the entire product including the housing, while the board level drop tests are performed on just the PWB assembly with components mounted on it, as described below.

10.2.3.1.1. Product Level Drop or Impact Testing. Product level drop testing can be classified as constrained or free. In constrained drop testing, which is by far the most common, the product is clamped rigidly to a heavy table that is guided along vertical rails to have a single impact against a target surface.

Clatter is probably best understood in terms of drop impact of an elongated or flat object onto a surface. Invariably, one corner touches down first, the object begins to rotate, and clattering occurs as the various corners encounter the impact surface before the object finally comes to rest. In that sense clatter refers to the condition where the second or third impact of the object probably occurs before the deformation from the first impact has returned to zero. On an oscilloscope time readout, strain or acceleration data due to clatter will resemble an extended but single impact sequence. In contrast, multiple impacts refer to the condition when the object bounces up and lands at a different location and orientation. In the case of multiple impacts, it is probable that the deformation in the assembly has had a chance to return to zero after the first impact, and the second impact occurs a short time later. On a time scale, they appear as two distinct events rather than as a single event. A third type of secondary impact, chatter, refers to the condition when subsystems or components impact each other within the product, for example, a battery impacting the case or a component. On an oscilloscope display monitoring the deformation-time response, chatter will appear as two events superimposed on each other and not distinctly separated in time. One of the main effects of the resultant secondary impacts in real life situations is that, depending on the moment and the coefficient of restitution, the ends of the object can strike at much higher velocities than during the first impact. The increased amplitude of velocity shocks, the possibility of exciting resonant conditions and repetitive shocks are reasons why the damage in a "real life" drop can be significantly higher [7].

On the other hand, free fall testing replicates the abuse a portable product will experience in actual usage. The main disadvantage is that it is difficult to control the orientation of the product at impact and this affects the repeatability of the test results and ease of monitoring by instrumentation. There is little mention of experiments in literature where free fall drop testing has been automated. Goyal et al. [8] proposed that the object being tested be suspended onto the guided drop table in the precisely desired drop orientation, and that, just before impact, the object is released from its suspension. The intended result, theoretically, is that although the required orientation at first impact is maintained, the object is free to move unconstrained subsequent to the first impact. Another variation of the quest for greater repeatability in product level drop testing consists of using grippers to control the orientation of the product until just before impact [9].

Lim et al. [10] surveyed the response of several commercial portable products (Nokia 8250 and 8310, Sony Ericsson T68i, Compaq 3850, HP Palm m105 and m505) using strain gauges and accelerometers to monitor the response of the PWB during drop tests. Maximum strain values ranged from 500 to 2500 microstrain and varied considerably between the different drop orientations depending on the product. Although the horizontal drop orientations generally yielded the highest strains in the PWB and the highest accelerations, there was considerable variability, which indicates that the actual behavior is quite complex and eludes simple generalizations.

In a study of the role of the rigidity of the mobile housing in determining the impact tolerance [8], it was found that thin-walled clamshell case constructions, currently favored for its size and weight advantages, may not provide sufficient rigidity to impact induced loads. Housing modifications to increase the stiffness improved the drop reliability. In addition, it is believed that the drop tolerance of the mobile phone would improve if the battery pack were to remain firmly attached to the phone, minimizing velocity amplifications and possible chattering.

The Shock Response Spectrum (SRS) approach was applied in using compliant suspensions to reduce peak acceleration and increase drop impact performance [11]. Results from another study with a personal digital assistant (PDA) using accelerometers and strain gauges, located along both the longitudinal and transverse directions, suggests that although there is a reasonably good correlation between acceleration and strain, it is often very difficult to completely unravel the complex strain-time or acceleration-time data except in select orientations [12].

10.2.3.1.2. Board Level Drop Testing. Because of the complexities inherent in product level drop testing, alternative ways of estimating the product reliability from simpler tests have received much attention. One such technique is the board level drop test, where the PWB assembly is subjected to impact loads or high accelerations while measuring the acceleration, velocity and strain on the assembly. Such board level drop tests provide a common basis to evaluate the impact tolerance of electronic products if one assumes that the conditions during product level drop impact can be reproduced adequately by dropping a test PWB assembly. The advantages of board level testing are:

(a) the shock pulse amplitude can be fairly well controlled,
(b) the orientation of the PWB assembly is controlled closely, and
(c) the tests are relatively more repeatable.

The primary disadvantage is that the test is not a true reflection of reality because it does not include the effect of secondary effects such as clatter, chatter or multiple impacts, which can have a significant bearing on reliability.

Ong et al. [13] examined the relevance of a board level drop tester by comparing it with the data collected from an instrumented drop of a Nokia 3210 model phone. It was found that, for the product level drop test, depending on the orientation of drop, the impact force can vary by up to a factor of five. Their results indicate that an axial impact exerted the highest forces. Further, because of the possibility of multiple impacts, the damage induced in a single drop in a product level test may actually be much higher than the damage induced due to a single drop in a board level test. In addition, Ong et al. [13] report that in board level drop tests, flexure of the PWB can last much longer than in product level drop tests.

Despite these differences, board level drop tests are attractive for investigating package reliability and process quality issues. Mishiro et al. [14] observed a correlation between solder joint stresses and PWB strains in a study where numerical analysis and strain measurements were employed to assess CSP reliability for 3 different package constructions. Even if the PWB strain is the same, the package structure played a significant role in controlling the solder joint stresses and hence drop impact reliability. In particular, the package structure with a 0.15 mm thick elastomer between the die and polyimide substrate performed better than the package where the interposer consisted of a multilayer laminate, which in turn was better than package with only a polyimide substrate. Further, non-solder mask defined (NSMD) pad structure was shown to be significantly better than solder mask defined (SMD) pad structure for drop reliability. With regard to PWB build-up layer, aramid-epoxy PWBs with low adhesive strength performed poorly because of premature delamination in the build-up layer. Underfilling the package to board interspace was found to improve the reliability when the Young's modulus was sufficiently high. However, when the underfill modulus was low (5 MPa), however, drop test reliability was much worse. Similar results were also reported in another study, where underfilling the CSP improved the reliability significantly in drop loading and the degree of improvement depended on the underfill modulus [15]. However, if the underfill quality was not optimal, the presence of even a small void encompassing the corner solder joint can magnify the stresses in the solder joint, effectively negating any anticipated benefit of underfilling.

Recognizing the relative complexity of a product level drop test, the relatively simpler board level drop test has been used to quantify drop reliability in terms of the package structure, materials, and processing. For example, Hannan and Viswanadham [16] evaluated the drop reliability of CSPs with reworkable underfills for reliability enhancement. Kujala et al. [17] used a board level drop test to compare the relative performance of land grid array (LGA) package and CSPs under both thermal cycling and drop impact. The board level drop test was used as a means to study the reliability of a "corner-reinforced-only" CSPs for portable product applications [18], where the CSP was held down only at the corners with epoxy, without actually having any underfill surrounding the solder joint in the package to board interspace. The drop reliability of such corner reinforced CSPs was lower than in the case of complete, capillary underfill. However, the relatively modest 3–4× improvement in the performance may be sufficient for some portable product applications [19]. Board level drop tests have also been used to investigate the effect of PWB and component pad surface finish, and concomitant interfacial strength, on drop test performance, and this is discussed in a later section.

10.2.3.1.3. Simulation of Drop Test Behavior of PWB Assemblies and Products. Faced with the complexities of purely empirical product level drop testing, there have been several attempts to complement experimental studies with finite element simulation to better understand the drop phenomena. The key issues for a successful understanding of drop impact reliability are (a) sophisticated and consistent analysis tools, (b) test correlation for model validation and refinement, (c) specification to define reliability requirements, and (d) material property data, especially over a broad range of strain rates [20].

Simulation when combined with board level drop testing can enable accurate prediction of not only the failure location but also durability to within 10% [21]. It was found that drop orientation with the components oriented face down was a more stringent test condition than one with the components facing up. Results indicate that during the drop test, greater PWB bending induces larger stress to the solder joints. As anticipated, it was found that the outermost solder joints have larger stresses and that smaller PWBs enhance drop performance. More importantly, it was reported that the lead-free solder studied had better board level thermal cycling reliability but worse drop test reliability. It should be recognized that thermomechanical reliability alone does not assure product reliability under mechanical loads.

Relatively accurate correlation of model prediction and experimental data were reported using smeared property models [22]. Significant error may be introduced due to aliasing of the experimental and computational data and under-sampled experimental data acquisition may mask the recognition of peaks in strain or displacement. In addition, smeared property models may not capture structural degradation during successive drops produced due to progressive delamination between materials. A validated modeling technique can be used to accurately predict failures observed in portable electronic products, such as disengagement of snap-fit housings and CSP solder joint cracking [23].

While state-of-the-art simulation was shown by various people to accurately predict different aspects of the drop test, a combination of simulation and experiments can be expected to be the most effective approach for improving and predicting reliability under drop or impact loading.

10.2.3.1.4. Analytical Modeling of Drop Phenomena. In addition to numerical simulation, closed form analytical modeling has been employed to understand the physics behind drop related phenomena. Suhir [24] obtained formulae to calculate the maximum displacements, velocities and accelerations of surface mounted devices when a shock load is applied to a flexible PWB at its support contour. Consideration of the nonlinearity of the PWB vibrations was found to be important in the case of large shock-induced deflections. The dynamic response of a rectangular plate element assembly subjected to drop impact was simulated as a box within a box, with one gasket between the outer and inner boxes and another between the PWB and the inner box. Results suggest that lower g-forces can be ensured by having the lower natural frequency considerably different from the higher frequency [25]. For example, the inner cushioning gasket could be made substantially stiffer than the outer one. Probabilistic approaches could also be employed to ensure a low failure rate. The effect of the stiffness of a "spring" shock protector was also studied [26]. Because the possibility of a "rigid impact" needs to be avoided at all costs, if the maximum drop height is not known, the advantages afforded by a soft spring cannot be fully utilized. The effect of viscous damping on the maximum displacement and the acceleration of a 1-DOF linear system subjected to a shock load during drop impact was also studied [27]. Sometimes, the application of materials with high energy absorption can result in even higher acceleration levels, and this needs to be avoided by a careful consideration of the system's mass and spring constants.

Whether maximum acceleration is an adequate criterion of the dynamic strength of a structural element in an electronic product has been investigated using a simply supported beam and a cantilever with heavy end mass [28]. Surprisingly, it was found that even if the accelerations experienced are not severe, one can expect significantly high dynamic stresses. These results are supported by observations during product level drop tests, where the bending of the board plays a bigger role in controlling failure compared to purely inertial forces, especially for light components such as flip chips and CSP assemblies. Until recently however, acceleration has been measured preferentially because it is easier to measure. Regarding alternatives to drop testing, it was found that the applicability of shock tests to replace product level drop tests depends on whether the dominant frequency of the shock impulse (which is inversely proportional to duration) is sufficiently high in comparison to the fundamental frequency of the vulnerable structural element [29].

10.2.3.2. Bend or Twist Environment Most portable electronic products experience more severe PWB bending related stresses than thermal stresses. PWB bending failure in the creep regime can be caused by localized bending near a screw location or in the high cycle fatigue regime due to key press action. A third bending failure mode occurs when portable products are dropped [30]. Recognizing the importance of understanding the reliability under bending loads, several studies in recent years have been aimed at characterizing the deformation and failure of solder joints.

Darveaux and Syed [30] have used both 3-point and 4-point bending tests to examine the failure mechanisms under a range of conditions for different CSPs along with finite element simulation of the damage processes. For displacement controlled fatigue tests, life decreased with (a) reduction of span length, (b) increase in test board thickness, (c) increase in die size, and (d) increase in molding compound thickness. In load-controlled tests, which are more closely related to actual product reliability, opposite trends were observed. Simulation results indicate that the optimum component/PWB pad size ratio in bending is different than under thermal loading.

The failure modes observed can be summarized as (a) fracture in the solder or in the intermetallic layer at the component pad, (b) fracture in the solder or in the intermetallic layer at the PWB pad, (c) trace peeling and eventual laminate cracking of the PWB or the component, or (d) build-up layer fracture leading to trace cracks on the PWB. Since these are very similar to the failure modes frequently seen under drop or impact conditions, bend testing is generally perceived to be a relatively simpler alternative to more complicated drop tests. Improving the strength of the weakest failure link can offer improvements in performance. For example, in 3-pt bend and drop impact tests on CSPs, anchoring the pads with via-holes improved the performance over having no-via-in-pads [31].

A few studies have been reported on the effect of strain rate during the bending test. For example, Geng et al. [32] reported that the solder joint interconnection fails at approximately 50% lower board deflection when the test speed increases by two orders of magnitude (0.25 to 2.54 cm/s). It is relatively well known that although solder strength increases with increasing strain rate, strain to failure decreases. In that context, as long as failure occurs in the solder, solder joints can be expected to fail at lower strain rates in high displacement rate bending tests. However, the data does not show a very distinct trend at higher strain rates (25.4 cm/s), possibly due to experimental artifacts. In a different study, with increasing ram displacement rate in a 4-pt bending test, strain gages mounted on the PWB showed increasing strain at solder joint failure sites [33]. It was shown that Kirkendall voids at the intermetallic interphases between the Ni and the Ni-Sn-P layers degraded the interfacial strength enough to cause failure preferentially at these locations. It should be noted that Kirkendall-like voids were also reported at Cu-Sn interfaces in lead free solder joints on OSP pads by Chiu et al. [34], with severe drop performance degradation in strength upon thermal aging in the 100 to 150°C range. Some reliability studies also focused on testing methods for flexible or low stiffness PWBs. Rooney et al. [35] reported an offset bend test configuration that is useful for testing assemblies with thin PWBs (0.5 mm) having stiff components.

A planar 3-pt bending fatigue test method to assess the reliability of the CSP solder joints was recently proposed [36]. The applicability of this method was demonstrated for standard plastic ball grid array (PBGA) components mounted on a PWB. The same method to establish that via-in-pad structure by itself does not pose a reliability risk in bend fatigue [37]. This is in accordance with the results reported previously by Juso et al. (1998). The applied load could induce dielectric (build-up layer) cracking, which in turn can lead to trace and via failures. Although lead free solders (Pb-free) have been found to be more durable than tin-lead solders in bend tests [38], it should be remembered that different lead free solders can be expected to behave differently, and some can perform worse than Sn-Pb solders depending on surface finish, test conditions, sample history, and several other variables. Moire interferometry coupled with 4-pt bend testing can reveal the localized influence of solder ball interconnections on chip carrier and PWB deformation [39]. Large shear strains were found in solder balls across the entire array. It was found that maximum strains occur in the outermost row of the solder balls, which agrees with the observations from a study on underfilled CSPs with corner defects [15]. In another study, the effect of cyclic bending on CSP assembly reliability was investigated in addition to monotonic bend bending [40]. The average overstress limit for a CSP studied was determined to be 2550 N mm. It was concluded that the CSPs showed worse durability when the PWB assembly subjected to negative curvature (CSP mounted surface of the PWB is convex). This is understandable since negative curvatures would subject the corner joints of the CSP to tension and lead to premature failure.

Portable electronic products were also evaluated for reliability under twist loads in addition to bend loads. For example, Perera [41] reported on the effect of twist loads of 9% and 12% and observed that solder joint failures occurred mostly by fatigue processes.

10.2.3.3. Shear Tests Interconnection failure is a common mode in portable electronic products, and it is widely accepted that interconnection strength and solder joint quality can play a central role in determining product reliability. Thus, measurements of the interconnection strength are useful in understanding reliability of the product. The term interconnection strength in this context denotes the effective strength of the package-to-board-interconnection, and includes the strength of (a) the package-solder interface, (b) solder, (c) the solder-pad interface and (d) the build-up layer on the PWB. This interconnection strength plays a role in determining product reliability.

Conventionally, the ball shear strength is used to denote the strength of attachment of a solder ball of a BGA or CSP to the component *prior to board assembly* [4]. This measure of ball strength, although useful in measuring the quality of the ball attachment process, cannot be easily translated into a product level estimate of durability. This is because of the following reasons: (a) the bare component is no longer accessible for ball shear tests, and (b) the interconnection quality is determined not only by the solder/component bond but also by the solder strength, solder/printed wiring board interfacial strength and build-up layer quality.

Product level tests such as mechanical drop, twist and bend tests yield valuable information on the reliability in the field. However, the primary drawbacks of these tests are the complexity and the time required to analyze the results in terms of targeted improvement actions. Thus, there is a need for a product level interconnection strength test that can yield relatively rapid results and simultaneously provide targeted quality improvement actions.

One candidate method is a recently developed product level test, the package to board interconnection shear strength (PBISS) technique [5,42]. It was shown that the shear test is an effective tool to quantify the shear strength of CSPs and examine the effect of pad finish and build-up layer strength. Only low strain rate PBISS behavior was characterized because product level twist and bend tests are performed at a low strain rate.

However, the strain rates experienced by the solder joint during drop tests are significantly higher. Therefore, the shear strength behavior measured at slow deformation rates is not directly applicable as a proxy for drop reliability of portable electronic products. Solder behavior changes significantly with the rate of deformation, and the damage to the CSP interconnection can be expected to be significantly different also during high rate of strain [43,44,76]. In this context, it was demonstrated that high strain rate shear tests essentially mimic the failure mechanisms and relative performance observed in drop tests [6].

10.2.4. Electrochemical Environment

The failure mechanisms that are of importance in portable electronic products exposed to electrochemical environments can be described as:

1. Corrosion.
2. Electrochemical migration (ECM).
3. Conductive anodic filament (CAF).

The fundamental difference between the two is that corrosion involves the destructive attack of a metal by the environment as anodic oxidation without the necessity for electrical bias, whereas ECM involves the transport of metal ions from the anode to the cathode under the influence of an applied electric field. From a failure perspective, corrosion results in product failure primarily by causing electrical open or intermittent interconnections, while ECM results in failures primarily due to electrical shorts or intermittent connections. Some factors affecting these failure mechanisms are the environment (temperature, humidity, presence of corrosive elements), operating conditions (bias voltage, current density, temperature and conductor spacing), and materials (nature of metal or alloy, surface condition, ability to absorb humidity, coating composition and thicknesses).

10.2.4.1. Corrosion Corrosion, depending on the severity, results in the following failure pathways:

(a) Oxidative materials degradation resulting in loss of electrical continuity,
(b) Partial degradation of materials accompanied by the formation of conductive oxidation product, such as a salt, that could result in lower surface insulation resistance (SIR),
(c) Electrical shorts between adjacent conductive features, or
(d) Intermittent shorts or opens depending on the humidity levels and the ionic nature of the corrosion product.

Corrosion is often discussed in terms of half-cell reactions because all corrosion processes are essentially electrochemical reactions. The electrodes in question could be on the macro- or micro-scale. Macroscopic galvanic corrosion cells can occur when dissimilar metals are coupled electrically and exposed to a corrosive environment, while microscopic corrosion cells tend to occur on the scale of grains. In either case, oxidation occurs at the anode and reduction at the cathode. In other words, the metal dissolution occurs only at the anode. The medium or electrically conductive environment in which these chemical reactions proceed is usually referred to as the electrolyte even if the electrolyte may extend to a thickness of a few monolayers. Since all the cations produced by the anodic reaction are consumed by the cathodic reaction, both anode and cathode reactions proceed at the same rate for corrosion to occur in a continuous manner.

The propensity of a metal to undergo corrosion is described in terms of the standard electrode potentials, where the hydrogen electrode potential is arbitrarily assigned a value of zero. When two dissimilar metals are coupled, the less noble metal will corrode in relation to the more noble one. However, it is possible to promote corrosion of the more noble metal in a galvanic couple by electrical biasing, which makes the more noble metal the anode. Some forms of corrosion [45] that are relevant to portable electronic products are:

1. Uniform corrosion—this form of corrosion is evenly distributed over the surface, and the rate of corrosion is the same over the entire surface. A measure of the severity is the thickness or the average penetration.
2. Pitting and crevice corrosion—this localized form of corrosion appears as pits or crevices in the metal. The bulk of the material remains passive but suffers localized and rapid surface degradation. In particular, chloride ions are notorious for inducing pitting corrosion, and once a pit is formed, the environmental attack is locally autocatalytic.
3. Environmentally induced cracking—this form of corrosion occurs under the combined influence of a corrosive environment and static or cyclic stress. A static loading driven cracking is called stress corrosion cracking and a cyclic loading driven cracking is called corrosion fatigue. Residual stresses in electronic leads from lead bending operations were observed to cause stress corrosion cracking failures in the presence of moisture [46]. Stress corrosion cracking of package leads was also reported in the presence of solder flux residues [47].
4. Galvanic corrosion—this type of corrosion is driven by the electrode potential differences between two dissimilar metals coupled electrically. The result is an accelerated corrosive attack of the less noble material. Galvanic corrosion tends to be particularly severe if the anodic surface is small compared to that of the nobler cathode or cases where a nobler metal is coated onto a less noble one. For instance, when a porous Au plating over a Ni substrate is exposed to a corrosive environment, the gold coating acts as a large cathode relative to the small area of exposed Ni. This sets up a galvanic cell at the exposed substrate which experiences intense anodic dissolution. It has been observed that pore corrosion can be enhanced by a galvanic corrosion process when the substrate metal is less noble than the coating, and vice versa [48].

10.2.4.2. Electrochemical Migration The distinguishing feature of ECM from corrosion is the formation of dendrites that cause a short between adjacent conductors. There are some similarities to corrosion as well, and the oxidation of the metal at the anode is common to both processes. ECM, which is also known as migrated metal shorts [49,50], is probably best described as due to transport of ions between two conductors in close proximity, under applied electrical bias and along an electrically conductive medium. In general, three conditions are necessary and sufficient for ECM failures to occur, and they are (1) presence of sufficient moisture (sometimes as little as a few monolayers), (2) presence of an ionic species to provide a conductive medium, and (3) presence of an electrical bias to drive the ions from the anode to the cathode. In the presence of sufficient moisture, the process is accelerated by temperature, and several mechanisms of ECM have been in vogue.

The first step in the classical model of ECM consists of metal ion formation by anodic oxidation (similar to corrosion), which may be either direct electrochemical dissolution or a multi-step electrochemical process. At the anode, for example, where M represents a metal atom,

$$M \rightarrow M^{n+} + ne^-.$$

The second step is the transport of metal ions from the anode, through an electrolyte, toward a cathode. In the final step, at the cathode, the positively charged ions are reduced to a neutral metal atom. At the cathode

$$M^{n+} + ne^- \rightarrow M.$$

Successive cationic reductions facilitate the growth of dendrites toward the anode along energetically favorable crystallographic orientations. Therefore, the surface insulation resistance of the material progressively decreases as the migration advances toward the anode. Eventually, an electrical short results when the dendritic filament touches the anode. Silver [49], Cu, Pb, Sn [51,52,78], Mo and Zn [77] have all been observed to form dendrites by this process. The presence of flux containing ionic species is a known contributor to ECM and has been studied widely using surface insulation resistance measurements [53]. Following the migration ability of pure metalization systems, the propensity for ECM may be ranked as follows: Ag > Pb > Cu > Sn [54,55].

A second mechanism of ECM was proposed to explain the migrated metal short formation involving noble metals such as Au, Pd and Pt. Because of the relative chemical inertness of these metals, a halogen contaminant is needed to induce anodic dissolution [56,57]. In an acidic medium, a positively charged metal ion may form by the following route at the anode,

$$Au + 4Cl^- \rightarrow AuCl_4^- + 3e^-,$$

$$AuCl_4^- + H^+ \rightarrow H[AuCl_4^-] \rightarrow HCl + AuCl_3 \rightarrow H^+ + 4Cl^- + Au^{3+}.$$

These positively charged Au ions can migrate toward the cathode and form dendrites in a similar fashion as the classical model.

A third mechanism to explain the ECM of Ni starting at the anode involves the presence of a strongly alkaline electrolyte. The first step is the formation of a cation ($HNiO_2^-$) by anodic corrosion followed by a chemical process resulting in secondary ionic species [58].

$$Ni \rightarrow Ni^{2+} + 2e^-,$$

$$Ni^{2+} + 2OH^- \leftrightarrow Ni(OH)_2.$$

It is suggested that instead of migrating to the cathode, the Ni^{2+} ions thus formed undergo the following reaction to form an anionic complex

$$Ni^{2+} + 2H_2O \rightarrow HNiO_2^- + 3H^+.$$

This anion complex migrates through the electrolyte under the applied electrical field. Finally, the metal atoms are deposited at the anode in the form of metallic dendrites due to the electrochemical reaction of the cationic species with the H^+, Ni^{2+} or OH^- ions. Similar process could be operative for Co and Cu ECM as well in cases where anodic deposits of the metal are observed.

10.2.4.3. Conductive Anodic Filament (CAF) Formation CAF is the type of electromigration failure mechanism where the loss of insulation resistance between neighboring conductors is caused by the growth of a subsurface anodic filament along delaminated fiber/epoxy interfaces [59]. The first step in the formation of the CAF is the physical degradation of the fiber/epoxy bond. This is followed by an electrochemical reaction requiring both the presence of moisture and a potential gradient across the cathode and anode. The metal undergoes oxidation at the anode to yield a positively charged ion that migrates toward the cathode. As the metal species migrate toward the cathode, they precipitate at locations where the pH is thermodynamically conducive, and in time, the filament extends from the anode to the cathode causing a short. The CAF formation may occur along the surface of a PWB or between conductors in different layers separated by a dielectric or along the glass fibers in the weave [60].

10.2.5. Tin Whiskers

Single crystal whiskers, of several metals including Sn, Cd, Zn, Sb, In, Pb, Fe, Ag, Au, Ni and Pd have been reported (for example, [61–63]). While the mechanism for the growth of whiskers of different metals may possess similarities, the mechanism of Sn whisker growth has been studied extensively. However, due to recent emphasis on the implementation of Pb-free solders and the consideration of Sn as a component terminations and PWB finish, there has been an increased effort to study the reliability implications of Sn whiskers. Several reported field failures have been collected from medical, military, and space applications by Siplon et al. [62]. It is generally agreed that whisker growth occurs at the base of the whisker in response to imposed stresses or residual stresses below the surface. The formation of Cu_6Sn_5 or other intermetallic compounds at the interface between the tin and the substrate layer has been shown to result in a compressive stress in the Sn film [64,63]. Once the oxide layer covering the tin has ruptured, tin whiskers can be extruded as a means of releasing compressive stress. It has been demonstrated that the use of certain substrate-coating combinations, such as Ni over Cu, significantly reduces whisker growth [65]. It was also demonstrated that avoiding brighteners, annealing of any residual stresses, using thicker tin layers, and addition of Pb are beneficial in reducing the propensity for whisker growth. On the other hand, the use of brighteners, lack of annealing, tin layers thinner than 2 μm, copper-based substrates and addition of Zn were shown to promote tin whisker formation [3]. The study of Sn whisker related reliability issues in portable consumer electronic products is in its infancy insofar as published reports of whisker related failures. Owing to considerable variations in Sn plating formulations and test methodologies, estimation of product failure risk has not been easy. However, decreasing pitch and increasing circuit density coupled with the drive toward Sn-rich solder compositions can be expected to elevate the risk of failure due to Sn whisker related issues in the near future.

10.3. RELIABILITY COMPARISONS IN LITERATURE

Reliability testing and accompanying failure analysis that are needed to fully understand the magnitude and nature of reliability concerns can be expensive in terms of time and resources. As discussed earlier in this chapter, there is a constant business driven need to minimize or accelerate reliability tests. Therefore, it is only natural that every effort is made to utilize any available historical data to assess current reliability risks and minimize the reliability testing that needs to be performed. While the value of reliability comparisons is clear in terms of reducing the need for testing and saving time and money, comparison and utilization of reliability data from different sources is a difficult exercise at best, and one has to be cognizant of the multitude of factors that influence the final reliability projections. In this section, some of the relevant aspects in comparing reliability results from different sources are discussed.

10.3.1. Thermomechanical Reliability

Effects of thermal fatigue are generally evaluated through accelerated thermal cycling tests. Test units, in statistically significant numbers, are subjected to a predetermined thermal profile over a number of cycles until all or 50% of the samples fail, and failure distributions are determined. In evaluating technologies, comparisons of failure data from a variety of sources are attempted to verify, substantiate or discern significant variations in reliability and understand the mechanisms. There are several pit falls in this approach. The first one is the definition of failure. Some regard a percent change in the resistance of total risk net consisting of a number of solder joints. Others may consider resistance spikes of a given magnitude and lasting over a specified duration, and still others may consider only an open joint as constituting a failure. The number of joints in a risk net may be different from study to study as well as in the same study depending on the I/O s of the packages being studied. The actual value of the resistance change can be significantly different in each case, if only percent change in resistance is considered. In great many instances, the failure criterion is not even included. A comparison of the probability plots can lead to misleading conclusions if the failure criteria are not identical in all of them.

Test parameters are also crucial and need to be considered explicitly for meaningful reliability comparisons. For example, in a thermal cycling test, the important parameters are the ramp rates and dwell times at the temperature extremities. A ramp rate of 15°C/minute and a dwell time of 10 minutes at each extremity are generally considered appropriate in many instances. However, literature contains data with 6 cycles per hour all the way up to 2 cycles per hour. Differences in the dwell time at extremities can have significant influence on the thermal fatigue and creep behavior of interconnection alloys. The temperature that the package and board experience in a given profile can be different from settings of the temperature chamber. Many studies only indicate the temperature values involved and do not provide the actual temperature the product under test sees. It is only prudent to compare temperature profile of the chamber versus the actual temperature experienced by the product under test as a function of time.

Other important factors that influence the discrepancies between the two are: the number of layers, copper and epoxy content, thickness of the board, its heat capacity, nature and size of the components, presence and absence of heat sinks. For example, a high I/O large ceramic component may take a longer time to attain steady state in comparison to a thin small package such as a chip scale package. If the cycle profile is not set correctly, it can alter the dwell time on some packages. Thus, a package of high heat capacity is more likely to experience a shorter dwell than a smaller package. The net result is that the solder joints in the bigger package may not experience the anticipated creep relaxation, and hence the failure may be altered by an unpredictable amount. In addition, during the ramp-up portion of the cycle, temperature can overshoot the preset values and it takes some time for the temperature to reach the set value. If a number of boards are being tested in the chamber the location of the boards in the chamber, and their disposition can

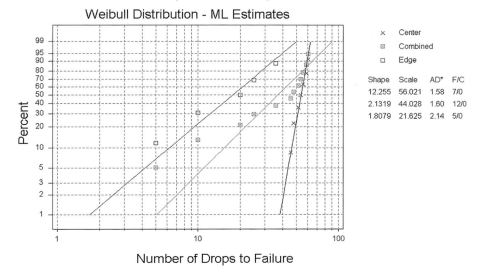

FIGURE 10.1. The presence of two different failure modes can be discerned in the data for the samples with no underfill in the drop test.

influence the temperature each board or package experiences. Boards stacked together and aligned perpendicular to the direction of air flow in the chamber will result in the boards immediately facing the air flow experiencing a different profile than other boards in the stack. In addition, the likelihood of blind spots in the chamber cannot be ruled out. Thus, a complete characterization of the thermal chamber to ensure that packages and the board attain the equilibrium temperature is very important.

Comparison of failure distributions can be complicated if the statistical distributions are not properly chosen and failure mechanisms are not well understood. The most popular solder joint failure distributions are the two-parameter Weibull distributions and occasionally three-parameter Weibull distributions. Even while using the two-parameter Weibull distributions, a single average line is often drawn through two apparently distinct distributions, as shown in Figure 10.1. This often leads to erroneous N_{50} values. In addition, a tacit assumption is made that there is only a single failure mechanism.

Sometimes, reliability results are reported without a failure analysis. Even when the failure mechanism is reported, the mechanism that is reported is based on the analysis done at the end of the test and not immediately following the detection of failure by electrical test. Thus, the understanding of the failure mechanism is corrupted or distorted by crack propagation, and micro structural changes occurred subsequent to the failure detection. When the distribution plots exhibit failures that indicate differing slopes, it is important to delineate them and conduct failure analysis to determine the exact failure mechanisms.

Thus, comparison of thermal cycling reliability tests has to be carried out with extreme care and caution taking into account all the factors that affect the inferences and conclusions. The current literature on reliability does not appear to readily lend itself to definitive correlations and comparisons.

10.3.2. Mechanical Reliability

Mechanical reliability comparisons for portable consumer electronics are more complicated and difficult than thermomechanical reliability comparisons because of dynamic and structural complexities. There are many more variables to be taken into account in the assessment of board-level mechanical reliability. These include package size, solder ball size, board structure and dimensions, drop height, orientation, impact duration, strike surface, etc. At the product-level, reliability comparisons are even more complicated due to additional dependence on the product form factor, weight distribution, impact orientation, occurrence of secondary impacts, and other test related variables. Therefore, the ability for comparison of mechanical drop test reliability is at its infancy.

Consistent test procedures with consistent acceleration and impact and failure criteria are critical in ensuring that results from one reliability test can be compared with results from another. For example, peak acceleration and the impact energy attained by the product depend on the frictional forces induced by the guide mechanism in the test equipment. Therefore, actual impact velocity can be different from the theoretically computed value.

The number of mounting screws and their location also has significant effect on the drop reliability. Boards mounted with only four screws can have lower impact life compared to those mounted with six screws under the same loading conditions due to greater bending. The type of screws and the torque applied to them can have a pronounced effect on the drop performance. The likelihood of screws loosening after subsequent drops cannot be ruled out. The dislodged screws can dramatically alter the board response during the drop. In addition, it needs to be verified that the failure locations and mechanisms are identical before attempting to compare reliability values. For example, failures that occur during drop can be due to interfacial brittle fracture at the package pad/solder interface, printed wiring board pad/solders interface, or the copper trace break at entry to the pad.

Location of the package on the PWB also plays an important role in determining reliability. Board bending and warpage can be very dependent on the board dimensions, and are usually greater along the longer dimension of the product. Typically, but not always, packages positioned at the center of the board are more susceptible to failure than the ones away from the center when the product is dropped on its face or back.

Package construction plays a significant role as well. Many portable electronics use low profile packages to accommodate the rather slim product form factors. These packages, such as ball grid array packages like Very-thin-profile Fine-pitch BGA (VFBGA), Thin-profile Fine-pitch BGA (TFBGA), and Quad Flat pack No-lead (QFN) packages, have low solder joint stand off, thinner die, and thinner molding compound. For example, VFBGAs have been shown to have slightly better performance than TFBGAs having the same I/Os [66]. In a different study, the 208 QFP package solder joints were observed to fail in a relatively small number of drops due to their mass and FLGA 300 (0.8 mm pitch) packages were relatively more durable [79].

Materials' aspects such as surface finish on the package and PWB pads can be expected to have a significant effect on the drop test reliability. Compatibility between PWB and component termination finishes, sometimes even inside the component module, can play a significant role in determining drop reliability. For example, incompatibility between Cu finish on resonators and ENIG finish on interposer PWB was found to severely degrade drop test performance [67]. In this case, the copper from solder/component interface migrated to the solder/interposer interface during the reflow and impeded the growth of Ni−Sn intermetallics, and instead, promoted the formation of a ternary Ni−Cu−Sn intermetallic phase. In the absence of a strong metallurgical bond between the Ni on the interposer PWB and the solder, premature failures occurred in drop testing.

Although, in general, Sn—Cu interfacial bond has been found to be superior to the Sn—Ni interfacial bond, recent evidence seems to suggest that Cu—Sn intermetallic bond can have risks as well. For example, the Kirkendall type of voiding found at the Cu/Cu_3Sn interface, especially after thermal aging, has been shown to impair board level drop performance [34]. Modification of the IMC bond strength by addition of trace amounts of some elements also needs to considered when comparing reliability results from different studies. For example, addition of 0.3%In and 0.04%Ni to Sn-Ag-Cu solder was shown to improve drop test reliability by as much as 20% even after 150°C thermal aging in comparison to the Sn-Ag-Cu solder [68].

10.4. INFLUENCE OF MATERIAL PROPERTIES ON RELIABILITY

10.4.1. Printed Wiring Board

The proliferation of portable electronic appliances in the form of mobile phones, personal digital assistants, pagers, etc., has brought about a "density revolution" in the printed wiring board technologies. Ever-smaller board features have necessitated new approaches to design, materials, fabrication, assembly, and testing. The consumer demand is for faster, cheaper, lighter, and more reliable electronic hardware. Conventional multilayer boards with 150 μm lines and 150 μm spaces with 325 μm drilled through hole vias cannot always accommodate the wiring densities for fine pitch high I/O area array devices such as ball grid array and chip scale packages. Therefore, weight reduction and high density requirements have resulted in the need for high density interconnect (HDI) boards. For portable electronic hardware with high density, thinner boards with finer lines and spaces with very small vias were needed. Thus evolved a completely new printed wiring board industry of HDI micro-via board technology featuring extremely thin laminates, and multilayer microvias. Several techniques such as Surface Laminar Circuitry (SLC), laser drilled micro-via techniques, Any Layer Inner Via Hole (ALIVH) technology have evolved. Buried, blind, and through-hole vias were needed to accommodate the product functionalities. These features are significantly different from the conventional printed wiring board technologies, and are approaching those used in the semiconductor industry. A semiconductor technology attitude is being cultivated by the printed wiring board industry to meet the new challenges. At the same time, the reliability requirements for portable electronic hardware are often more stringent than the conventional hardware. The only relaxation in the reliability requirement is one of product life; they are shorter than those required for desktop and business products. However, the mechanical and environmental requirements are more severe.

The complexity of the product varies considerably and may contain PWB assemblies that are either single sided or double sided. A double-sided assembly will be more rigid and display a different shock response. In some cases, depending on the product complexity, both buried and blind vias may be used simultaneously. The buried vias may be plated or filled with conductive paste and cured. The reliability of thin populated boards with blind and buried vias is inadequately understood under various mechanical loading conditions. Issues, such as mis-registration of the buried vias in the individual layers, can pose a reliability exposure.

Another important aspect of micro-via technology is the shape of the vias namely, square-well or bathtub, and the copper plating thickness and uniformity due to the variations in via shape. In addition, the registration of the micro-via on the capture pad is very important and crucial for product reliability. In case of poor registration the laser drilling may be partially off the pad and penetrate the adjacent laminate. This can result in voiding during reflow process due to the egress of the occluded moisture in the laminate, impacting the package to board interconnection integrity.

In a high density printed wiring board, different materials are used for the micro-via layer including non-reinforced epoxies, woven-fiber reinforced resins, chopped fiber reinforced resins, such as aramid-reinforced materials, and resin coated copper foils. The adhesion of the reinforcing material to the base resin can have a significant impact on reliability. Additionally, several Cu to laminate adhesion enhancement treatments, including mechanical abrasion have been in vogue. Each of these aspects can impact the reliability, especially under mechanical loading.

10.4.2. Package

In portable electronic products, package size and style can influence product concepts, and vice-versa. Packages have to fit the form factor of the product, which is usually very thin. Double sided surface mount assemblies with low standoff low profile packages are the order of the day. This limits the feasible options to chip scale packages, VSSOP, TSOP, lead less packages, LGAs, Quad Flat No-leaded package types, to name a few. With increasingly effective utilization of PWB real estate, an emerging trend is to explore the out-of-plane dimension to increase the packaging density within the constraints of the form factor and package height limitations. Device stacking and package stacking are becoming increasingly popular. An understanding of the failure modes and mechanisms of these packages on a variety of laminate materials, and their construction under thermal and mechanical loading is still in its infancy. Package size, materials and construction, die size and thickness, the order of the stacking, and the bonding methods used can all have significant impact on the failure nature and mechanisms. Failures can range from package damage such as popcorning, to silicon die damage, interconnection failures, delamination, laminate cracking, etc. Industry trends indicate that with thinner die, such as 50–70 μm thin die, packages with as many as six to seven die stacked together could be anticipated in the near future.

10.4.3. Surface Finish

Surface finish of printed wiring boards and the package termination play a significant role in the integrity and reliability of an interconnection. Hot-air-solder-leveling which has been the main PWB surface finish for well over half a century has outlived its usefulness since the advent of high I/O fine pitch surface mount and area array packaging technology. Several surface finishes have since come into use. Organic solderability preservatives (OSP) and electroless nickel-immersion gold (ENIG) have almost replaced solder leveling. ENIG has been used extensively owing to its long shelf life and for excellent solderability wherever co planarity requirements are stringent. However, as hardware integration and miniaturization continued, resulting in smaller feature sizes, problems related to defects in ENIG surfaced. The hypercorrosivity of immersion gold plating composition and attendant high phosphorous content can cause sporadic and unpredictable solderability problems (also referred to as black-pad). In addition, as portable electronic hardware is more subject to mechanical loading, intermetallic brittle fracture at the solder-pad interface is some times encountered. Also, as has been mentioned earlier, it is generally recognized that nickel tin intermetallics are more brittle than the copper-tin intermetallics. Often dual surface finishes are employed, with OSP to preserve solderability, and ENIG for electrical contact surfaces.

With ever increasing emphasis on the implementation of lead-free solders as the interconnection material, surface finishes of PWB, package leads, and terminations are being reexamined to arrive at acceptable alternatives. Immersion silver, immersion tin, palladium, nickel-palladium-gold etc., are being looked at. There does not appear to be a consensus on surface finishes. While each surface finish has its merits, the industry has to weigh the alternatives in terms of cost, performance and reliability for a given product group.

In the ensuing sections, several failure mechanisms pertaining to printed wiring boards, packages, and interconnections under a variety of loading conditions are described.

10.5. FAILURE MECHANISMS

As mentioned earlier, the failure mechanisms in handheld electronic products are different from those commonly encountered in desktop or mainframe business machine environments. Broadly, they may be categorized as those caused by (a) thermal loading, (b) mechanical loading (including mechanical drop, vibration, bending and twisting loads), and (c) electrochemical environments that induce corrosion and electromigration.

10.5.1. Thermal Environment

Failures induced due to thermal stresses in portable electronic hardware are in general similar to those in other electronic products. In portable electronic hardware, where use of HDI with multiple micro-via layers is prevalent, the shape of the micro-via, copper thickness, and the voids in the microvia influence the nature of the interconnect failure. In general, interconnect failures tend to occur on the package side of the solder joint, and are influenced by the coefficient of thermal expansion (CTE) of the package and sometimes aggravated by the solder mask defined pad geometry on the package side. In conventional Sn-Pb solders, the fracture generally occurs in the solder adjacent to the intermetallic layer, where the region is Pb rich in composition.

For Pb-free solder alloys, the interconnect failure mechanisms may display different kinds of deviations from the previously observed mechanisms for Pb-Sn alloys. Depending on the surface finish and the pad metallurgy, the interconnection can have multiple types of intermetallic phases dispersed in the bulk joint. In the case of tin-silver-copper system with OSP and Electroless Ni Immersion Au (ENIG) surface finish, Cu-Sn, Ag-Sn, Au-Sn intermetallics were found to be dispersed in the bulk of the joint or near the pads [69]. Solder joint failures due to thermal cycling are influenced by shear forces induced by to CTE mismatch between the component and PWB, with both fatigue and creep damage mechanisms operative at the same time. A damage accumulation map for Pb-free solders is discussed next.

In Pb-free solders, there are several significant differences in the microstructure compared to the Sn-Pb eutectic or near-eutectic solders, and these microstructural differences result in a very different damage evolution process. The primary microstructural differences are as follows:

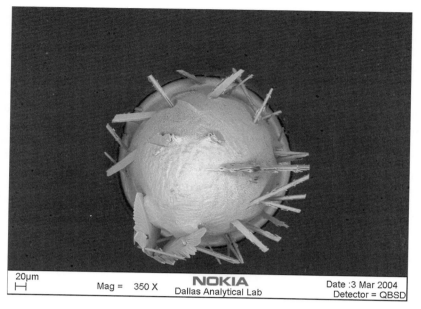

FIGURE 10.2. Partially etched solder ball of an unmounted revealing the IMC phase that can affect crack initiation and crack propagation thereby affecting thermomechanical reliability.

1. The intermetallic morphology is more complex and a multitude of small spheroidal Ag_3Sn IMCs are observed at the Sn dendrite boundaries. These could serve as initiation sites for voids and microcracks.
2. In addition to the increased presence of small particles in the interdendritic spacing, several large Cu_6Sn_5 IMCs and Ag_3Sn plates are distributed throughout the solder ball, which can effectively constrain the solder joint during shear deformation. A partially etched Sn3.5Ag0.7Cu solder ball on an unassembled CSP is shown in Figure 10.2 to illustrate how the IMCs in this solder system are distributed throughout the bulk of the solder joint to much larger extent than previously observed in Pb-Sn solders. A completely etched solder ball microstructure (in Figure 10.3) reveals the presence of Cu_6Sn_5 scallop shaped intermetallic phases adjacent to the Cu pad in addition to the IMCs distributed in the bulk of the solder.

These microstructural features can bring out damage mechanisms in Pb-free solders that were not a significant contributor to final failure in Pb-Sn solders under thermal fatigue/creep environments.

It should be noted that failure in thermal cycling in solders involves both fatigue and creep failure mechanisms. The relevant mechanisms of creep deformation are:

- Dislocation creep—involves the movement of dislocations which overcome barriers by thermally assisted mechanisms involving the diffusion of vacancies or interstitials ($10^{-4} < \sigma/G < 10^{-2}$).
- Diffusion creep—involves the flow of vacancies and interstitials under the influence of applied stress ($\sigma/G < 10^{-4}$).
- Grain boundary sliding—involves the sliding of grains past each other.
- Dislocation glide, which normally requires very high stresses, is probably not a major contributor to creep during thermal fatigue. Diffusion creep causes vacancies

FIGURE 10.3. Completely etched solder ball of an unmounted revealing the scallop shaped Cu_6Sn_5 IMC phases on the Cu pad along with rod shaped Cu_6Sn_5 IMCs. In addition, planar dendrites of Ag_3Sn can also be seen interspersed throughout the surface.

from grain boundaries experiencing tensile stresses to flow toward those that are experiencing compressive stresses.

In a solder joint with the microstructure and IMC morphology described in the earlier section, the driving force for failure is the imposed cyclic shear stress due to CTE mismatch and creep under this stress. Because of the low homologous temperature of solder, fatigue damage mechanisms are accompanied by creep damage mechanisms.

Consistent with previously reported damage mechanisms for Pb-Sn solder, inhomogeneous shear stress fields can result in recrystallization at pads, corners, and voids. Additional damage mechanisms not widely reported for Sn-Pb solder but observed for Pb-free solders by Dunford et al. [69] are described next.

Zones of recrystallized material were observed at locations with high strain gradients and strain incompatibilities, such as grain boundaries. These recrystallized zones grow with imposed cycling, and a multitude of smaller recrystallized grains form to relieve the strain. In parallel, creep driven damage mechanisms were observed to a degree not reported in previous studies. Another creep driven damage mechanism is the initiation of voids and cracks at locations of high strain incompatibility. For example, triple-point grain junctions and IMC-grain boundary junctions in the interior of solder balls and grain boundary (GB) junctions at the surface of the solder ball appeared to be the favored sites for crack initiation.

Further damage evolution is governed by the interaction of the localized damage (in the form of recrystallized zone) with the distributed damage (in the form of microcracks and voids). The severity of damage of all three types, namely (a) recrystallization zones, (b) microcracks and voids at recrystallized grain boundaries (RGB), and (c) cracks and voids at GB, grows with increased cycling.

Final failure, however, is dominated by the weakening of the material due to recrystallization and distributed microcracking in the damage zone. A macrocrack forms by the coalescence of the microcracks, primarily in the recrystallized zones. The propagation path of these macrocracks is very different from that observed for Pb-Sn solders. The IMC plates and rods sometimes serve to deflect the propagating macrocrack so that several macrocracks may exist in a solder joint without significantly impacting electrical continuity. These macrocracks coalesce with each other through the distributed damage, changing direction depending on local damage geometries and microcracks at the RGB or the cracks at the GB. Final failure occurs by propagation of the most dominant macrocrack traversing the solder ball, primarily near the pads on the board or the component. For example, in the solder joint of the CSP shown in Figure 10.4, one can see the tortuous path taken by the propagating macrocrack and the distribution of the microcracks near the fracture plane. Near the bottom of the solder joint, away from the component pad, an elongated void formed due to of creep related damage enlarging an initially small crack or void, is also seen. In the right half of the picture, the grain morphology with Ag_3Sn and Cu_6Sn_5 IMC particles interspersed in the interdendritic spaces is seen. A higher magnification picture of an elongated void caused by creep damage at grain boundaries in a different solder joint is shown in Figure 10.5. The damage evolution map for thermomechanical loading that brings together the different operative mechanisms just described is shown in Figure 10.6 [69].

10.5.2. Mechanical Environment

It is instructive to review the construction of a generic package mounted on a PWB before discussing the failure mechanisms. The PWB in portable electronic products serves not only as a carrier for the different electrical subsystems but also provides mechanical rigidity to the assembly. A typical PWB can have 4 to 12 electrical planes laminated between woven glass fiber reinforced epoxy layers that serve both a dielectric and mechanical support function. Electrical connection between these layers is often achieved through plated-through-hole vias, blind vias or buried vias. The outermost layer of the PWB, sometimes called the build-up layer, is the first interconnection layer between the solder joint and the PWB. Interconnection failures can be found at different levels as shown schematically in Figure 10.7, and can be classified as follows, based on the location of the crack:

- Die fracture within the package.
- Interposer level failure within the package.
- Solder joint fracture.
- Crack initiation inside the component and subsequent damage to the solder joint.
- Solder joint fracture.
- Interfacial failure—at the solder/PWB pad interface.
- PWB related failure—trace fracture.
- PWB related failure—micro-via fracture.

10.5.2.1. Die Fracture within the Package Sometimes, when the die is not supported optimally inside the package, almost all the flexure of the PWB can be transmitted to the relatively brittle semiconductor die inside the package. Cleavage fracture of the die can occur causing electrical failure. An example of this kind of failure is shown in the optical micrograph in Figure 10.8. The wire bonds on the die can also be seen along with the vertical crack in the die. The cracks at the inactive side of the die (bottom) are attributed to polishing damage during the grinding stage. Such artifacts have previously been observed in samples where excessive normal force was exerted on the sample, and should not be confused with cleavage type of cracking on the active side of the die.

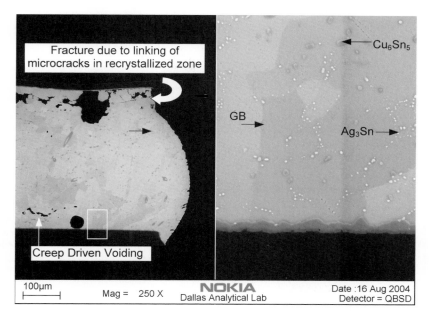

FIGURE 10.4. Interconnection fracture due to thermomechanical fatigue loading in Sn3.5Ag0.7Cu solder joint of CSP. The backscattered electron micrograph reveals the fatal crack near the component pad in addition to voiding and other damage near the PWB pad (Sample is a courtesy Michael Wellborn).

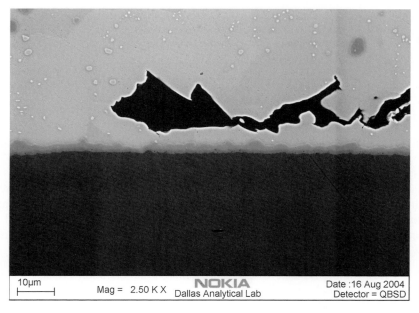

FIGURE 10.5. Creep driven damage resulting in elongated voids at grain boundaries.

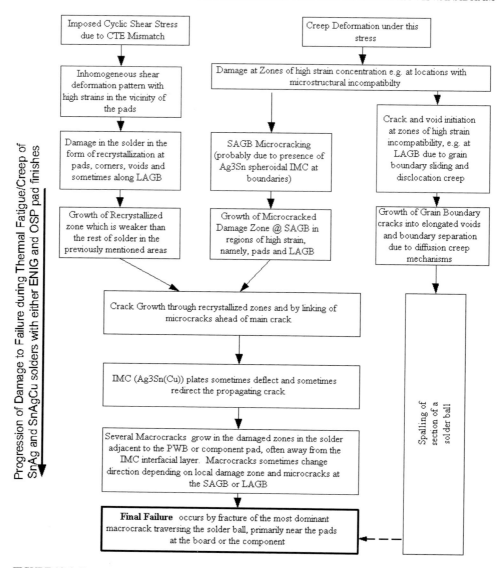

FIGURE 10.6. Damage mechanism evolution map for Pb3.5Sn0.7Cu solder under thermomechanical loading.

10.5.2.2. Interposer Level Package Failure The Cu circuitry inside the interposer can sometimes fail if the process conditions in the fabrication of the interposer are not optimal. The example shown in Figure 10.9 illustrates the particular case where sub-optimal adhesion between the via-barrel and the via-cap failed upon exposure to mechanical loading at the PWB level.

10.5.2.3. Crack Initiation Inside Component Leading to Solder Joint Damage Ceramic components, due to their weight and lower fracture toughness, are particularly susceptible to failure when the product is dropped. Local stress concentrations on the ceramic component, such as those created by machining can serve as crack initiation sites and cause

PORTABLE CONSUMER ELECTRONIC PRODUCTS

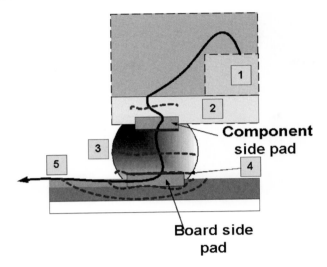

FIGURE 10.7. Simplified schematic of electrical interconnection from the Si die to multilayer PWB through different levels of packaging. Dashed line represents possible crack or open.

FIGURE 10.8. Die cracking due to mechanical loading.

premature failure as shown in Figure 10.10. The crack that originated at the machining groove caused an electrical open upon propagation. Apart from the machining on the ceramic component, a second factor contributing to the crack originating in the component is the relatively high strain rate of deformation during drop loading. Since solder deformation characteristics are highly strain rate dependent at room temperature, the solder joint is stiffer and stronger under higher deformation rates, thereby subjecting the ceramic compo-

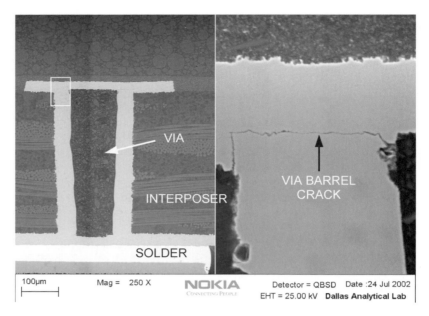

FIGURE 10.9. Via barrel cracking due to PWB level mechanical loading causing electrical failure.

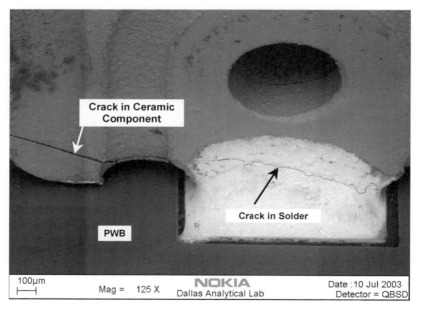

FIGURE 10.10. Crack in solder joint and ceramic component after mechanical shock (drop) reliability testing.

nent to proportionately higher stresses. For components operating at radio frequencies, a relatively minor partial crack, as shown in Figure 10.10, can sometimes cause parametric shift induced failures rather than a hard open.

PORTABLE CONSUMER ELECTRONIC PRODUCTS

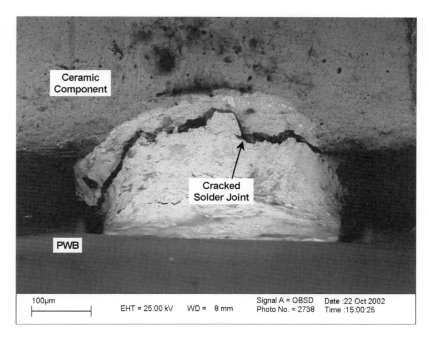

FIGURE 10.11. Crack in solder joint after twist testing.

10.5.2.4. Solder Joint Fracture due to PWB Level Twisting Bending and twisting are commonly encountered end-use environmental hazards for hand-held products. The deformation rates are much lower than those observed in mechanical drop. In such cases, the solder joint strength and stiffness are proportionately lower and promote fracture at the solder joint in contrast to locations within the ceramic component. An illustrative example is shown in Figure 10.11, where the solder joint is completely fractured without the damage extending into the ceramic. The lack of machining damage near the solder joint was probably a secondary factor in limiting damage to the solder joint without cracking the component.

10.5.2.5. Solder Joint Failure Related to Underfill Process It is a relatively common practice to provide additional reinforcement to a solder joint to improve its reliability under thermal and mechanical loads. For ball grid array (BGA) and chip scale packages (CSP) soldered onto PWBs, this reinforcement can be achieved by the use of a suitable underfill material in the package-to-board interspaces. This constrains the assembly against bending and thermal strains. One of the more commonly used procedures for underfilling a CSP soldered onto a board consists of dispensing liquid underfill along one or more edges of the CSP perimeter such that capillary action forces the underfill to fill the entire space between the CSP interposer and the PWB. Upon curing, the liquid underfill hardens and encapsulates the solder joints completely, thereby providing additional reliability by mitigating the deleterious effect of either thermal or mechanical strains.

The quality of the underfilling process is dependent on several variables such as temperature of the PWB or liquid underfill, cleanliness of the surfaces, speed of dispensing, etc. It has been shown that when the quality of the underfill is non-optimal and voids are present at the CSP corners, the benefit of the underfill is not realized even if the size of the void exposes only the corner solder joint [15]. An example of a partially underfilled CSP

FIGURE 10.12. (a) A partially underfilled CSP with a corner underfill void, and (b) A more severe underfill defect exposing a whole row of solder joints.

FIGURE 10.13. X-ray microscope image of a poorly underfilled CSP incorrectly indicating the lack of underfill defects. X-ray techniques can yield misleading results for certain kinds of defects.

is shown in Figure 10.12(a) and an optical micrograph of a more severe underfill defect is shown in Figure 10.12(b).

The true extent of an underfill defect cannot be ascertained by either visual or X-ray inspection. For example, Figure 10.13 shows a representative X-ray microscope picture of a CSP that does not reveal any underfill defect although visual inspection showed a substantial underfill defect at the perimeter. The scanning acoustic microscope, on the other hand, is very sensitive to voids and underfill defects. Difficulties encountered in acoustic inspection of CSP or BGA underfill include the signal to noise ratio due to material attenuation and uncertainty about the specific depth that the data includes. Both these problems are particularly severe for CSP and BGA underfill, unlike in flip chip underfill inspection. A judicious selection of transducer frequency, F# (ratio of focal length to diameter of the transducer), depth of focus and gating are essential for successful inspection. The acoustic image of the CSP in Figure 10.12(b) is shown in Figure 10.14, and the areas of incomplete underfill can be clearly identified in the top half of the acoustic image. A virtual cross-section along the dashed line is shown in the bottom half of the acoustic image, and the relative depths of the die, the interposer and the void can be seen. In addition, the bond wires extending from the die to the interposer are also visible. It is also useful to present the acoustic waveform along with the image to clarify the nature and location of defects.

An acoustic image of a different, improperly underfilled CSP is shown in Figure 10.15. The waveforms from three locations are presented alongside the acoustic image for ease of interpretation. The waveform from location 1 and 2 shows how the die and the interposer lie above the depth of the defect shown in location 3. The positive (upward) reflection from the top of the die and the Cu pads on the interposer are in contrast to the negative (downward) pulse from the underfill void. Thus, there is no ambiguity in concluding that the void lies below the interposer, where underfill would normally be expected in an underfilled sample. The lack of support for the solder ball can lead to failure of the interconnect that are now exposed to higher levels of loading. When exposed to ad-

PORTABLE CONSUMER ELECTRONIC PRODUCTS

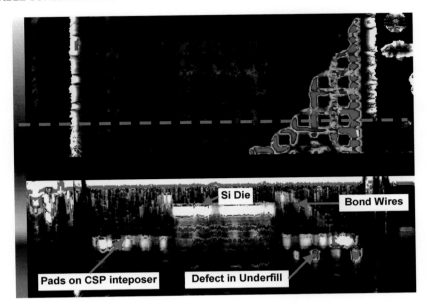

FIGURE 10.14. Acoustic image of the same CSP as in previous figure showing voiding in the underfill below the interposer of the CSP. A virtual cross-section (QBAM along the dashed line in the image) in the lower half of the image reveals that the underfill defect is below the interposer.

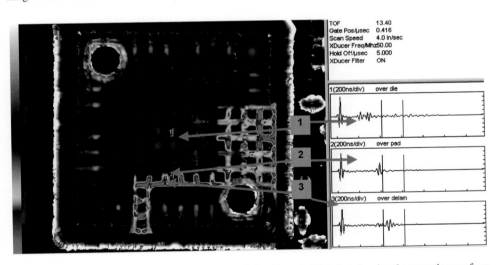

FIGURE 10.15. A more detailed acoustic image of a CSP with underfill defect showing the acoustic waveform traces over three locations: (1) the die, (2) the Cu pad on the interposer and (3) over the delamination.

verse environment such as mechanical loading, the solder joints or the build-up dielectric layer below the Cu pad on the PWB can develop cracks as shown in the scanning electron micrograph of the polished cross-section in Figure 10.16.

10.5.2.6. PWB Quality Related Fracture at Solder/PWB Pad Interface Electroless nickel/immersion gold (ENIG) plating of the Cu pads on the PWB gained considerable popularity as a pad surface finish in recent years. This is because it provides a cost ef-

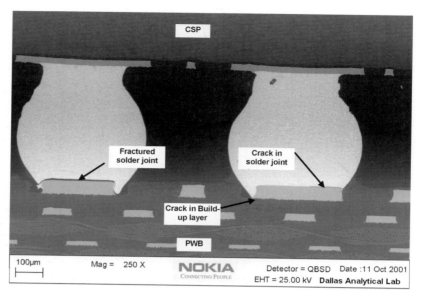

FIGURE 10.16. Scanning electron micrograph showing the fractured solder joint and concurrent damage at a neighboring solder joint.

fective means of ensuring coplanarity, which is a crucial requirement in high density, fine pitch assembly. The Ni layer was intended to provide a diffusion barrier between the gold and the Cu pad. The very thin gold layer (<0.5 μm) was intended to protect the Ni surface from oxidation and preserve its solderability until reflow. During reflow, when the solder melts and wets the gold surface, the gold layer dissolves instantly into the solder leaving a clean, solderable Ni surface for Ni-Sn metallurgical bond formation. Not so long ago, the interconnection pad sizes were relatively large because fine pitch packages were not widely used. When pad sizes were relatively large, quality variations in the ENIG plating did not immediately or always result in interconnection failures because the pad size-defect size ratio was substantial. Now, the pad size-defect ratio is smaller due to higher density of packaging. In addition to this, the increased use of less-aggressive organic solvent or water-soluble fluxes or no-clean fluxes can result in a pad surface that may not be as solderable. These trends in the industry have increased the risks due to ENIG surface finish related failures that are characterized by a brittle fracture at the solder/pad interface along with a dull, dark Ni surface exhibiting "mud crack" type of surface morphology [70,71].

One example of the brittle interfacial crack at the solder-PWB pad interface is shown in Figure 10.17. The fracture occurs below the intermetallic layer at the solder/ENIG pad, which indicates that the interfacial bond between the Ni-Sn intermetallic layer and the underlying Ni layer is weak. Indeed, the fracture surface on the PWB side pad is often devoid of any adhering solder as seen in the backscattered micrograph in Figure 10.18. The only evidence of solder adhering to the surface is seen in locations where a void in the solder offered easier crack propagation than fracture at the solder/Ni pad interface. The characteristic "mud cracking" types of features are also visible on the Ni fracture surface. A high magnification micrograph of the polished cross-section (Figure 10.19) reveals the high-P Ni layer and a transverse view of the hypercorrosion trenches in the Ni layer into which the solder has ingressed.

PORTABLE CONSUMER ELECTRONIC PRODUCTS

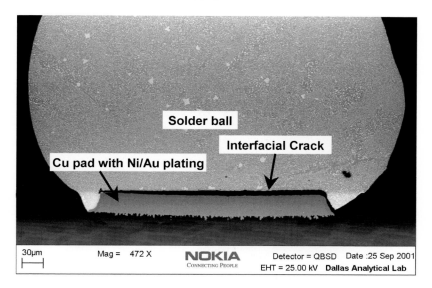

FIGURE 10.17. Interfacial fracture resembling brittle cleavage between solder ball and pad.

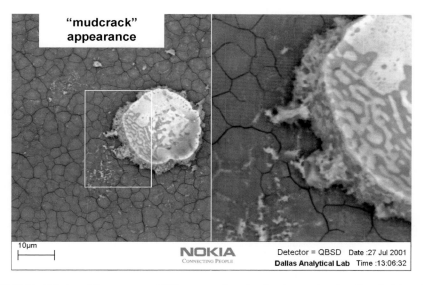

FIGURE 10.18. "Mud crack" appearance of Ni fracture surface showing the poor bond quality of solder to Ni/Cu.

10.5.2.7. PWB Build-up Layer Cracking Leading to Trace Fracture The PWB is usually a multi-layer laminate made up of layers of continuous, woven glass reinforced epoxy and Cu circuitry. The outermost layers sometimes referred to as the build-up or redistribution layers, serve both as the dielectric material and mechanical support for the Cu traces during PWB flexure or extension. Upon subjecting the assembled board to mechanical loading, such as encountered in mechanical drop, damage accumulates in the build-up layer in the form of cracking. Subsequent damage accumulation and electrical failure will depend on the redistribution method employed.

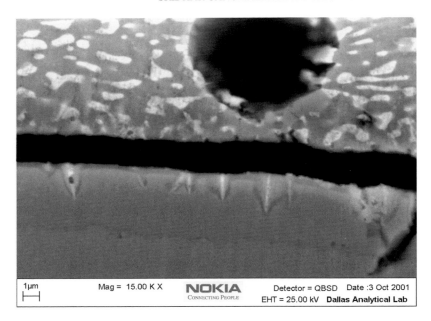

FIGURE 10.19. Hypercorrosion of Ni layer observed on a microsectioned sample with black pad defect.

The progression of damage for the case when redistribution is achieved through traces is described below. Initially, the damage in the build-up layer accumulates until the trace is no longer supported because of extensive cracking of the laminate under the Cu pad. The damage in the build-up layer is exemplified by the back-scattered electron micrograph in Figure 10.20(a). This weakening of the build-up layer forces the trace to shoulder an ever-increasing share of the mechanical loads imposed on the PWB, which eventually causes trace fracture by fatigue processes. An example of the fractured trace is shown in the top-view optical micrograph in Figure 10.20(b). The fracture process can be seen more clearly after a second microsectioning operation along the dotted line in Figure 10.20(b). The backscattered electron micrograph of the double-polished sample is shown in Figure 10.21(a) and a schematic explaining the crack under the pad causing the trace fracture is depicted in Figure 10.21(b).

The damage progression is similar in cases where a via-in-pad redistribution method is employed. The damage in the build-up layer, again, accumulates in the form of cracks. Once the support afforded by the build-up layer is diminished by the cracking, further mechanical flexure of the PWB subjects the via to increasingly higher stresses. Eventually, the via fractures due to fatigue, leading to an electrical open (Figure 10.22).

10.5.3. Electrochemical Environment

For portable and handheld electronic devices, two failure mechanisms related to electrochemical environments are of particular relevance—corrosion and electrochemical migration.

10.5.3.1. Corrosion Gold plating of connectors is a common practice designed to protect the underlying Cu and Ni layers from corrosive attack and promote good electrical contact. However, under the action of friction, the relatively thin and inert Au coating can be

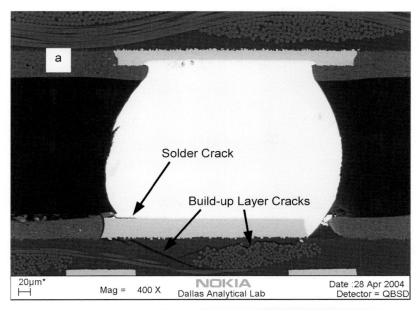
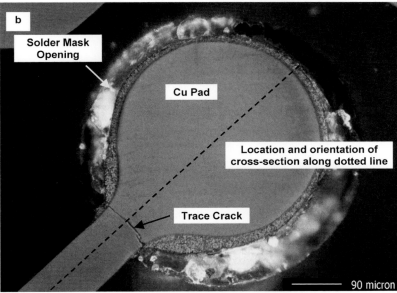

FIGURE 10.20. (a) Build-up layer cracking in a solder joint with trace and (b) Optical micrograph (top view) of a sample suspected to have a broken trace after the solder ball was removed by mechanical polishing. The dotted line represents the location and orientation of a second vertical microsectioning needed to show damage under the pad.

removed locally thereby exposing the Cu and Ni layers underneath. In such cases, fretting corrosion, pitting corrosion and localized galvanic corrosion can occur simultaneously, especially in the presence of an ionic species such as chlorides. This corrosion product, which is usually nonconductive, can cause electrical failure due to opens or intermittent. An ex-

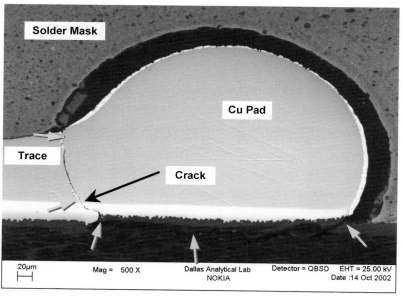
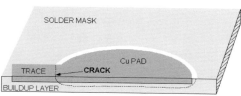

FIGURE 10.21. (a) Trace fracture accompanied by build-up layer cracking revealed in a double-cross sectioned sample (sample shown different than that depicted in Figure 10.18), (b) schematic showing the location of the crack in the build-up layer.

ample of gold plated connector corrosion is shown in Figure 10.23. The EDX elemental map for Au indicates that the coating is intact over the major portion of the area of interest. However, in the central portion of the image, the Au coating appears to have been removed completely, and the underlying Cu is exposed. This Cu surface, identified as a bright area in the Cu elemental map, also shows significant presence of O and Cl. The absence of any areas with high concentrations of Ni indicates that the mating surface of the connector has probably worn through the Ni layer in the area of contact.

10.5.3.2. Electrochemical Migration In several studies comparing the propensity for ECM of different metalization systems can be ranked as follows: Ag > Pb > Cu > Sn [54]. Although electrochemical migration phenomena have been observed with many metals, only Ag [49,51], and Cu to a limited extent [72], and perhaps Sn [73], have been found to exhibit this behavior in the presence of humid but non-condensing conditions. Indeed, Dumoulin et al. [73] concluded that silver migration presents the greatest risk because dendritic growth can occur whether Ag is outside the package or only partly exposed to humid air, on ceramic as well as on plastic substrates. Although Dumoulin et al. [73] suggested that Cu migration and Sn migration did not pose as big a risk based on their experimental data, in mobile electronic products which see a wide range of corrosive species during their lifetime, ECM of Cu and Sn can be as prevalent as Ag migration. In addition, residues

PORTABLE CONSUMER ELECTRONIC PRODUCTS 289

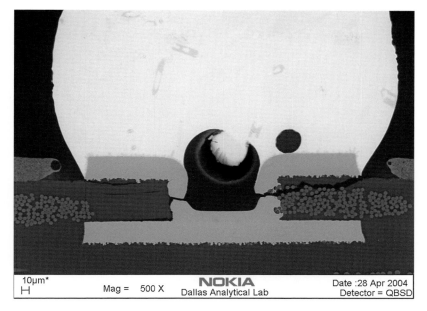

FIGURE 10.22. Build-up layer cracking in a solder joint with via-in-pad leading to via cracking upon further exposure to mechanical drop related stresses.

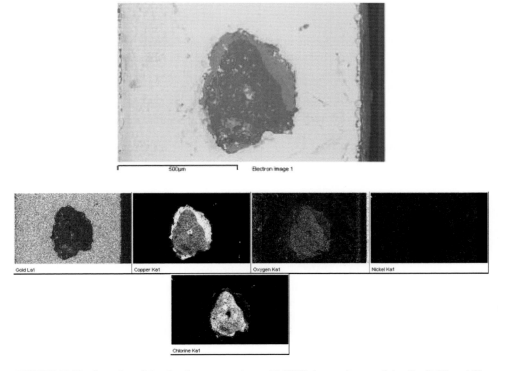

FIGURE 10.23. Corrosion of Au plated connector along with EDX elemental maps of Au, Cu, O, Ni, and Cl.

FIGURE 10.24. Cu electrochemical migration on a PWB between two through hole vias after damp heat exposure.

on the substrates that originate from the process, play an important role through water adsorption/conductivity behavior modification. One example each is provided next for ECM phenomena involving Cu, Sn and Ag.

Krumbein [74] noted that in practice, ECM can manifest itself as two separate, though not always distinct, effects that lead to impairment of the circuit's electrical integrity. Dendritic or filamentary bridging between the anode and cathode, which is one kind, has been discussed at length before. Colloidal staining is the second manifestation of ECM, which can also cause a short. Deposits of colloidal Ag, Cu or Sn have been observed to originate at the anode without necessarily remaining in contact with it. An example of this effect is also provided below.

Copper forms complex species such as $CuCl_4^{2-}$, $CuCl_2(H_2O)$, $Cu(H_2O)^{2+}$, etc. in the presence of halide containing species and moisture. An example of Cu electrochemical migration resulting in Cu dendrite formation is shown in Figure 10.24. If plated through hole vias or conductor pads are too close, Cu ECM can occur when the product is exposed to humid environments in the presence of an ionic contaminant.

Tin electrochemical migration mechanism is similar to that of Cu, but is much more prevalent because Sn constitutes a major portion of several commercial solder compositions such as 62SnPb2Ag, 10SnPb, Sn3.5Ag0.7Cu, etc. In addition, exposed Sn is more widespread on an assembled PWB as compared to Cu. The particular example shown in Figure 10.25 is from a test vehicle that failed upon exposure to damp heat testing. In this case, the potential difference between the terminals of a capacitor with Sn termination resulted in the migration of Sn from the anode toward the cathode. The right half of the picture shows a higher magnification view of the Sn dendrites at the cathode end of the termination. Elemental analysis mapping data of the surface of the capacitor is shown in Figure 10.26, where the Ba, Ti and O from the capacitor dielectric material can be seen clearly. In addition, the Sn map shows the presence of Sn between the terminations, where

PORTABLE CONSUMER ELECTRONIC PRODUCTS 291

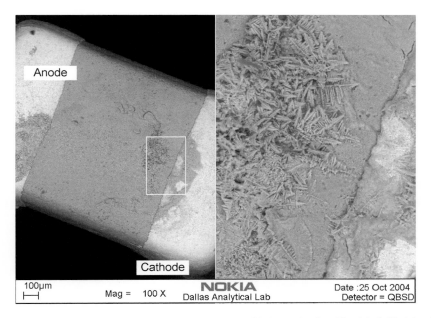

FIGURE 10.25. Tin electrochemical migration on a capacitor with tin termination. The right half of the picture shows a higher magnification view of the Sn dendrites at the cathode end of the termination.

there should be none. In several passives, Ni is used as a barrier layer between the silver adjacent to the dielectric and the tin termination. In this particular case, the Ni barrier layer at the anode is visible in areas where the Sn from the surface has been consumed by the ECM process. Another example of Sn ECM is shown in Figure 10.27, where colloidal form of ECM can be observed in addition to dendrite formation.

Silver ECM can occur on the PWB if there is exposed metal in the termination or pad finish, or it can occur on the surface of passive devices separate from the surface of the PWB. The occurrence of ECM on the surface of passive devices can potentially be a more serious reliability risk because of the current trend toward smaller size passives, which provides a ready site for ECM. A coating of Ag is commonly employed at the ends of the passive device to ensure that there is a good contact between the electrodes in a capacitor. However, since Ag is prone to ECM, it is advisable to isolate this Ag from the environment. Therefore, Ni is used as a barrier layer between the Ag base and the Sn outer layers. To be effective, this Ni layer should be continuous and free of cracks or gaps. In the event that the Ni layer is discontinuous, Ag can be exposed to the environment leading to dendrite formation as illustrated in Figure 10.28. Here, dendrites of Ag can be seen growing on the surface of the passive component after damp-heat reliability tests.

10.6. RELIABILITY TEST PRACTICES

Accelerated thermal cycling test practices are influenced not only by the design life and the operating environment but also by the nature of the PWB assembly. In a majority of cases, portable electronic hardware by its very nature has to be small, lightweight, and possess high I/O density. This implies the use of surface laminar circuitry or other HDI PWB technologies. Also, inherent is the use of small low profile packages.

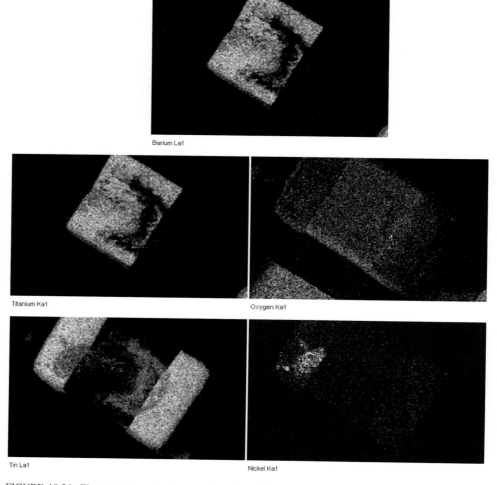

FIGURE 10.26. Elemental maps for the capacitor shown Figure 10.25 for (a) Ba, (b) Ti, (c) O, (d) Sn, and (e) Ni showing the presence of Sn ECM between the terminations and exposure of the Ni barrier layer under the consumed Sn surface at the anode.

It has been reported that the industry standard temperature cycle profile, where the upper and lower temperature dwells are invariant, leads to an underestimation of fatigue life [75]. It is well known that inelastic strain accumulation is generally proportional to fatigue life. It has been suggested that temperature fluctuations during upper dwell times can reduce elastic strain accumulation, and as such, using minicycles during dwell times will reduce the maximum inelastic strain. The magnitude of such inelastic strain reduction depends on the number of minicycles and their temperature ranges. Thus, selective superimposition of a judicious number of minicycles during the high temperature dwell may enable a more realistic fatigue life prediction.

In addition, portable electronic hardware involving radio communication features can have components such as power amplifiers and radio frequency devices that may run

FIGURE 10.27. Tin electrochemical migration involving both formation of dendrites and on a resistor with pure tin termination.

hotter during operation in addition to the thermal exposure imposed by the environment. The thermal effects in such cases can cause excessive growth of interfacial intermetallics, which may be deleterious to the interconnection integrity. Power cycling tests may be much more appropriate in such cases.

Thermal and mechanical stress exposures in portable electronic hardware are rather frequent and some times concurrent in contrast to desktop machines, and the effect on product performance can be significant. For example the interfacial intermetallic growth, which by itself may not affect the solder joint integrity due to the compliance of the alloy, can progressively degrade the mechanical reliability. Thus, separate thermal and mechanical reliability assessments may not reflect the true product performance, as the synergistic effect is not taken into account. The effect of thermal aging on the mechanical reliability can be significant and should be considered in all reliability assessments.

FIGURE 10.28. Silver electrochemical migration on a resistor with tin termination. Inadequate protection due to poor quality Ni barrier layers enabled the Ag to exhibit electrochemical migration.

10.7. SUMMARY

As portable consumer electronic hardware becomes more complex with multitudes of functions and increased data handling capacity, further miniaturization and higher levels of integration at all levels of packaging will be a natural trend. The reliability demands will be higher to ensure customer satisfaction and product acceptance. The implications for reliability, failure, and root cause analysis will be significant. More functions will be integrated into the device. The silicon device thickness will be in the range of 40–50 microns. Stacked devices, folded and stacked packages will be more prevalent with a combination of multiple levels of wire bonding and/or flip chip interconnection. Another emerging trend in packaging is the three dimensional integration at the wafer level. New materials that will have better mechanical properties and moisture resistance will be developed. More functions will be embedded into the printed wiring board and these may include active, passive and optical devices, attendant with new embedded interconnection schemes. The printed wiring board technology itself will witness revolutionary changes with thinner and improved materials capable of 10–25 μm vias, 10–20 μm lines and spaces, and structures involving several layers of stacked vias. Consequently, hitherto unknown failure mechanisms are likely to be encountered. As the feature sizes diminish the distinction between first and second level packaging becomes nebulous. Failure analysis, even at the printed wiring board PWB assemblies, will be a formidable challenge. With shorter product development cycles and faster to market business environment the need for more automated analytical tools with minimal operator intervention for rapid and repeatable root cause analysis will increase. Innovative reliability test practices will be needed to shorten the test durations to accommodate faster development schedules.

ACKNOWLEDGMENTS

The authors acknowledge the contributions of Sesil Mathew, Sambit K. Saha, Murali Hanabe, Steven Dunford, Nael Hannan and Laura Foss for technical discussions and permission to use their data, and appreciate the help provided by Leslie Landon, Tuula Stenberg, Elina Leivo, and Susanna Olli for technical discussions, and Michael Wellborn for providing a test sample. The management support of Timothy Fitzgerald was indispensable and appreciated.

REFERENCES

1. W.Q. Meeker and L.A. Escobar, Pitfalls of accelerated testing, IEEE Trans. on Reliability, 47(2), pp. 114–118 (1998).
2. W.Q. Meeker and M. Hamada, Statistical tools for the rapid development and evaluation of high-reliability products, IEEE Trans. on Reliability, 44(2), pp. 187–198 (1995).
3. R.J.K. Wassink, Soldering in Electronics, Electrochemical Publications Ltd., Great Britain, 1989.
4. R. Erich, R.J. Coyle, G.M. Wenger, and A. Primavera, Shear testing and failure mode analysis for evaluation of BGA ball attachment, IEEE/CPMT International Electronic Manufacturing Technology Symposium, 1999, pp. 16–22.
5. S. Canumalla, Test fixture and method, U.S. Patent #6,681,640, 2004.
6. M. Hanabe and S. Canumalla, Package to board interconnection shear strength (PBISS) behavior at high strain rates approaching mechanical drop, IEEE Electronic Components and Technology Conference, 2004, pp. 1263–1270.
7. S. Goyal and E.B. Buratynski, Methods for realistic drop testing, International Journal of Microcircuits and Electronic Packaging, 23(1) pp. 45–52 (2000).
8. S. Goyal, S. Upasani, and D.M. Patel, The role of case-rigidity in drop-tolerance of portable products, International Journal of Microcircuits and Electronic Packaging, 22(2), pp. 175–184 (1999).
9. C.T. Lim and Y.J. Low, Investigating the drop impact of portable electronic products, Proceedings of the IEEE Electronic Components and Technologies Conference, 2002, pp. 1270–1274.
10. C.T. Lim, C.W. Ang, L.B. Tan, S.K. Seah, W., and E.H. Wong, Drop impact survey of portable electronic products, Proceedings of the IEEE Electronic Components and Technologies Conference, 2003, pp. 113–120.
11. S. Goyal, E.B. Buratynski, and G.W. Elko, Role of shock response spectrum in electronic product suspension design, International Journal of Microcircuits and Electronic Packaging, 23(2), pp. 182–190 (2000).
12. S.K.W. Seah, C.T. Lim, E.H. Wong, V.B.C. Tan, and V.P.W. Shim, Mechanical response of PCBs in portable electronic products during drop impact, Proceedings of the IEEE Electronic Components and Technologies Conference, 2002, pp. 120–125.
13. Y.C. Ong, V.P.W. Shim, T.C. Chai, and C.T. Lim, Comparison of mechanical response of PCBs subjected to product-level and board-level drop impact tests, Proceedings of the IEEE Electronic Components and Technologies Conference, 2003, pp. 223–227.
14. K. Mishiro, S. Ishigawa, M. Abe, T. Kumai, Y. Higashiguchi, and K. Tsubone, Effect of the drop impact on BGA/CSP package reliability, Microelectronics Reliability, 42, pp. 77–82 (2002).
15. S. Canumalla, S. Shetty, and N. Hannan, Effect of corner-underfill voids on the chip scale package (CSP) performance under mechanical loading, 28th International Sympoium for Society for Testing and Failure Analysis, 3–7 November, Phoenix, AZ, 2002, pp. 361–370.
16. N. Hannan and P. Viswanadham, Critical aspects of reworkable underfills for portable consumer products, Proceedings of the IEEE Electronic Components and Technologies Conference, 2001, pp. 181–187.
17. A. Kujala, T. Reinikainen, and W. Ren, Transition to Pb-free manufacturing using land grid array packaging technology, Proceedings of the IEEE Electronic Components and Technologies Conference, 2002, pp. 359–364.
18. B.J. Toleno and J. Schneider, Processing and reliability of corner bonded CSPs, International Electronics Manufacturing Technology Symposium, 2003, pp. 299–304.
19. G. Tian, Y. Liu, P. Lall, R.W. Johnson, S. Abderrahman, M. Palmer, N. Islam, and J. Suhling, Drop reliability of corner bonded CSP in portable products, International Electronic Packaging Technical Conference, July 6–11, Hawaii, USA, 2003.

20. J. Wu, G. Song, C.-P. Yeh, and K. Wyatt, Drop impact simulation and test validation of telecommunication products, Intersociety Conference on Thermal Phenomena, 1998, pp. 330–336.
21. T.Y. Tee, H.S. Ng, C.T. Lim, E. Pek, and Z. Zhong, Impact life prediction modeling of TFBGA packages under board level drop test, Microelectronics Reliability, 43, pp. 1131–1142 (2004).
22. P. Lall, D. Panchagade, Y. Liu, W. Johnson, and J. Suhling, Models for reliability prediction of fine-pitch BGAs and CSPs in shock and drop impact, Proceedings of the IEEE Electronic Components and Technologies Conference, 2004, pp. 1296–1303.
23. L. Zhu, Modeling technique for reliability assessment of portable electronic product subjected to drop impact loads, Proceedings of the IEEE Electronic Components and Technologies Conference, 2003, pp. 100–104.
24. E. Suhir, Nonlinear dynamic response of a flexible printed circuit board to a shock load applied to its support contour, IEEE Electronic Components and Technology Conference, 1991, pp. 388–399.
25. E. Suhir and R. Burke, Dynamic response of a rectangular plate to a shock load, with application to portable electronic products, IEEE Trans. on Components, Packaging, and Manufacturing Technology, Part B: Advanced Packaging, 17(3), pp. 449–460 (1994).
26. E. Suhir, Shock protection with a nonlinear spring, IEEE Trans. on Components, Packaging, and Manufacturing Technology, Part A, 18(2), pp. 430–437 (1995).
27. E. Suhir, Dynamic response of a one-degree-of-freedom linear system to a shock load during drop tests: Effect of viscous damping, IEEE Trans. on Components, Packaging, and Manufacturing Technology, Part A, 19(3), pp. 435–440 (1996).
28. E. Suhir, Is the maximum acceleration an adequate criterion of the dynamic strength of a structural element in an electronic product? IEEE Trans. on Components, Packaging, and Manufacturing Technology, Part A, 20(4), pp. 513–517 (1997).
29. E. Suhir, Could shock tests adequately mimic drop test conditions? IEEE Electronic Components and Technology Conference, 2002, pp. 563–573.
30. R. Darveaux and A. Syed, Reliability of area array solder joints in bending, SMTA International Symposium, 2000.
31. H. Juso, Y. Yamaji, T. Kimura, K. Fujita, and M. Kada, Board level reliability of CSP, IEEE Electronic Components and Technologies Conference, 1998, pp. 525–531.
32. P. Geng, P. Chen, and Y. Ling, Effect of strain rate on solder joint failure under mechanical load, IEEE Electronic Components and Technologies Conference, 2002, pp. 97–978.
33. K. Harada, S. Baba, Q. Wu, H. Matsushima, T. Matsunaga, Y. Uegai, and M. Kimura, Analysis of solder joint fracture under mechanical bending test, IEEE Electronic Components and Technologies Conference, 2003, pp. 1731–1737.
34. T.C. Chiu, K. Zeng, R. Stierman, D. Edwards, and K. Ano, Effect of thermal aging on board level drop reliability for Pb-free BGA packages, Proceedings of the IEEE Electronic Components and Technologies Conference, 2004, pp. 1256–1262.
35. D. Rooney, Castello, N.T., M. Cibulsky, D. Abbott, and D.J. Xie, Materials characterization of the effect of mechanical bending on area array package interconnects, Microelectronics Reliability, 39, pp. 463–477 (2003) (in press).
36. L. Leicht and A. Skipor, Mechanical cycling fatigue of PBGA package interconnects, Proceedings of 30th International Symposium on Microelectronics, 14–16 Oct., 1998, pp. 802–807.
37. K. Jonnalagadda, Reliability of via-in-pad structures in mechanical cycling fatigue, Microelectronics Reliability, 42, pp. 253–258 (2002).
38. K. Jonnalagadda, M. Patel, and A. Skipor, Mechanical bend fatigue reliability of lead-free PBGA assemblies, The Eighth Intersociety Conference on Thermal and Thermomechanical Phenomena in Electronic Systems, (ITHERM 2002), 2002, pp. 915–918.
39. E.A. Stout, N.R. Sottos, and A.F. Skipor, Mechanical characterization of plastic ball grid array package flexure using moire interferometry, IEEE Trans. on Advanced Packaging, 23(4), pp. 637–645 (2000).
40. S. Shetty, A. Dasgupta, V. Halkola, V. Lehtinen, and T. Reinikainen, Bending fatigue of chip scale package interconnects, ASME International Mechanical Engineering Congress and Exposition, Orlando, Florida, Nov. 5–10, 2000.
41. U.D. Perrera, Evaluation of reliability of μBGA solder joints through twisting and bending, Microelectronics Reliability, 39, pp. 391–399 (1999).
42. S. Canumalla, H.-D. Yang, P. Viswanadham, and T. Reinikainen, Package to board interconnection shear strength (PBISS): Effect of surface finish, PWB build-up layer and chip scale package structure, IEEE Trans. on Components and Packaging Technologies, 27(1), pp. 182–190 (2004).
43. N.F. Enke, T.J. Kilinski, S.A. Schroeder, and J.R. Lesniak, Mechanical Behaviors of 60/40 Tin-lead solder lap joints, IEEE Trans. on Components, Hybrids, and Manufacturing Technology, 12(4), pp. 459–468 (1989).

44. T. Shohji, T. Yoshida, Takahashi, and S. Hioki, Tensile properties of Sn-Ag based lead-free solders and strain rate sensitivity, Materials Science and Engineering A, 366, pp. 50–55 (2004).
45. M. Tullmin and P.R. Roberge, Corrosion of metallic materials, IEEE Trans. on Reliability, 44(2), pp. 271–278 (1995).
46. J.D. Guttenplan, Corrosion in the electronics industry, ASM Metals Handbook, 9th edn, Vol. 13, ASM International, Metals Park, OH, USA, 1987.
47. A.J. Raffalovich, Corrosive effects of solder flux on printed circuit boards, IEEE Trans. on Components, Hybrids, and Manufacturing Technology, 7(4), pp. 155–162 (1971).
48. K. Yasuda, S. Umemura, and T. Aoki, Degradation mechanisms in tin- and gold-plated connector contacts, IEEE Trans. on Components, Hybrids, and Manufacturing Technology, 10(3), pp. 456–462 (1987).
49. G.T. Kohman, H.W. Hermance, and G.H. Downes, Silver migration in electrical insulation, Bell Systems Technolgy Journal, 34, p. 1115 (1955).
50. A. Shumka, and R.R. Piety, Migrated gold resistive shorts in microcircuits, Proceedings of the International Reliability Physics Symposium, 1975, pp. 93–98.
51. A. Dermarderosian, The electrochemical migration of metals, Proceedings of the International Society of Hybrid Microelectronics, 1978, p. 134.
52. G. Ripka and G. Harsanyi, Electrochemical migration in thick-film conductors and chip attachment resins, Electrocomponents Science and Technology, 11, p. 281 (1985).
53. L.J. Turbini, J.A. Jachim, G.A. Freeman, and J.F. Lane, Characterizing water soluble fluxes: Surface insulation resistance vs. electrochemical migration, IEEE/CHMT International Electronics Manufacturing Technology Symposium, 1992, pp. 80–84.
54. G. Harsanyi and G. Inzelt, Comparing migratory resistive short formation abilities of conductor systems applied in advanced interconnection systems, Microelectronics Reliability, 41, pp. 229–237 (2001).
55. T. Takemoto, R.M. Latanison, T.W. Eagar, and Matsunawa, A, Electrochemical migration tests of solder alloys in pure water, Corrosion Science, 39(9), pp. 1415–1430 (1997).
56. F.G. Grunhaner, T.W. Griswold, and P.J. Clendening, Migratory gold resistive shorts: Chemical aspects of failure mechanism, Proceedings of the International Reliability Physics Symposium, 1975, p. 99.
57. N.L. Sbar, Bias humidity performance of encapsulated and unencapsulated Ti-Pd-Au thin film conductors in an environment contaminated with Cl_2, IEEE Transactions on Parts, Hybrids, and Packaging, 12, p. 176 (1976).
58. G. Harsanyi, Electrochemical processes resulting in migrated short failures in microcircuits, IEEE Trans. on Components, Packaging, and Manufacturing Technology A, 18(3), pp. 602–610 (1995).
59. D.J. Lando, J.P. Mitchell, and T.L. Welsher, Conductive anodic filaments in reinforced polymeric dielectrics: Formation and prevention, International Reliability Physics Symposium, 1979, pp. 51–63.
60. P. Viswanadham and P. Singh, Failure Modes and Mechanisms in Electronics Packages, Chapman and Hall, New York, 1997.
61. M.E. McDowell, Tin whiskers: A case study, Aerospace Applications Conference, 1993, pp. 207–215.
62. J.P. Siplon, G.J. Ewell, E. Frasco, J.A. Brusse, and T. Gibson, Tin whiskers on discrete components: The problem, 28th International Sympoium for Society for Testing and Failure Analysis, 3–7 November, Phoenix, AZ, 2002, pp. 421–434.
63. K. Zeng and K.N. Tu, Six cases of reliability study of lead-free solder joints in electronic packaging technology, Materials Science and Engineering Reviews, 38, pp. 55–105 (2002).
64. B.-Z. Lee and D.N. Lee, Spontaneous growth mechanism of tin whiskers, Acta Materialia, 46(10), pp. 3701–3714 (1998).
65. R. Schetty, Minimization of tin whisker formation for lead free electronics finishing, IPC Works Conference, Miami, USA, 2000.
66. T.Y. Tee, H.S. Ng, D. Yap, X. Baraton, and Z. Zhong, Board level solder joint reliability modeling and testing of TFBGA packages for telecommunication applications, Microelectronics Reliability, 43, pp. 1117–1123 (2003).
67. S.K. Saha, S. Mathew, and S. Canumalla, Effect of Intermetallic Phases in Mechanical Drop Environment: 96.5Sn3.5Ag Solder on Cu and Ni/Au Pad Finishes, IEEE Electronic Components and Technologies Conference, 2004, pp. 1288–1295.
68. M. Amagai, Y. Toyoda, T. Ohinishi, and S. Akita, High drop test reliability of lead free solders, IEEE Electronic Components and Technologies Conference, 2004, pp. 1304–1309.
69. S. Dunford, S. Canumalla, and P. Viswanadham, Intermetallic morphology and damage evolution under thermomechanical fatigue of lead-free solder interconnections, IEEE Electronic Components and Technology Conference, 2004, pp. 726–736.

70. N. Biunno, A root cause failure mechanism for solder joint integrity of nickel/immersion gold surface finishes, IPC Printed Circuits Expo, Long Beach, CA, 1999, pp. 1–9.
71. E. Bradley and K. Banerji, Effect of PCB finish on the reliability and wettability of ball grid array packages, IEEE Transactions on Components, Packaging, and Manufacturing Technology, Part B, 19(2), pp. 320–330 (1996).
72. J.N. Lahti, R.H. Delaney, and J.N. Hines, The characteristic wearout process in epoxy-glass printed circuits in high density electronic packaging, Proceedings of the 17th Annual Reliability Physics Symposium, 1979, p. 39.
73. P. Dumoulin, J.-P. Seurin, and P. Marce, Metal migration outside the package during accelerated life tests, IEEE Trans. on Components, Packaging, and Manufacturing Technology, 5(4), pp. 479–486 (1982).
74. S.J. Krumbein, Metallic electromigration phenomena, IEEE Trans. on Components, Hybrids, and Manufacturing Technology, 11(1), pp. 5–15 (1988).
75. T. Dishong, C. Basaran, N. Cartwright, Y. Zhao, and H. Liu, Impact of temperature cycle profile on fatigue life of solder joints, IEEE Trans. on Components, Packaging, and Manufacturing Technology, Part B: Advanced Packaging, 25(3), pp. 433–438 (2002).
76. R. Darveaux, Constitutive relations for tin-based solder joints, IEEE Trans. on Components, Hybrids, and Manufacturing Technology, 15(6) (1992).
77. T. Kawanabe and K. Otsuka, Metal migration in electronic components, Proceedings of Electronic Components Conference, 1982, p. 99.
78. R.C. Benson, B.M. Romanesko, J.E. Weiner, B.N. Nall, and H.K. Charles, Metal electromigration induced by solder flux residue in hybrid microcircuits, IEEE Trans. on Components, Hybrids, and Manufacturing Technology, 11(4), pp. 363–370 (1988).
79. D. Xie, M. Arra, S. Yi, and D. Rooney, Solder joint behavior of area array packages in board level drop for hand held devices, IEEE Electronic Components and Technology Conference, 2003, pp. 130–135.

11

MEMS Packaging and Reliability

Y.C. Lee

Department of Mechanical Engineering, Campus Box UCB 427, University of Colorado, Boulder, CO80309-0427, USA

11.1. INTRODUCTION

Microelectromechanical systems (MEMS) technology enables us to create different sensing and actuating devices integrated with other microelectronic, optoelectronic, microwave, thermal, and mechanical devices for advanced microsystems. Semiconductor fabrication processes allow for cost effective production of these micro-sensing or actuation devices in the 1–100 µm size scale. Figure 11.1 illustrates a typical design and manufacturing process for a MEMS device. This illustration highlights some of the differences between MEMS and microelectronics fabrication and packaging. During the design, solid modeling is required since electro-thermal-mechanical coupling is essential to the functions of most of MEMS devices. The fabrication often involves deposition and etching of micron-thick layers with controlled mechanical and electrical properties [1,2]. In many devices, after the completion of the fabrication process, the sacrificial materials are removed by etching in order to release the device for mechanical movements. This release process is usually the first step in the MEMS packaging. The released device shown in the figure represents a configuration for pressure sensors or accelerometers or an element of an array for optical micro-mirrors and RF switches. After release, the devices can be tested on the wafer-level, followed by dicing. The released, diced device is assembled and sealed in a package. These testing, dicing, assembly and sealing steps are very challenging. Without proper protection, the micro-scale, movable features could be damaged easily during these steps [3]. As a result, it is always desirable to replace the process illustrated here by wafer-level packaging [4].

Hundreds of MEMS-based sensors and actuators and systems have been demonstrated and the number of their applications is growing. A few examples of their diverse applications are listed below [5]:

1. Pressure sensors: for sensing manifold air pressure and fuel pressure to decrease emission and fuel consumption; for measuring blood pressure.
2. Inertial sensors: accelerometers for measuring acceleration for launching air bags; gyros for measuring angular velocity to stabilize ride and to detect rollover.
3. Chemical micro sensors: for fast, disposable blood chemistry analysis; gas sensors.

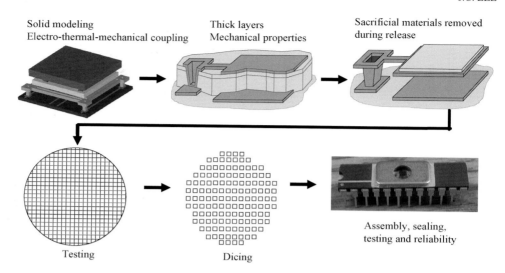

FIGURE 11.1. MEMS design, fabrication and packaging.

4. Optical MEMS: micromirrors for projection displays; optical switches for wavelength division multiplex switches; attenuators or micro-devices for active alignments for optical microsystems; micro-displays or paper-thin, direct-view displays.
5. Radio frequency (RF) MEMS: micro-resonators for integrated RF transceiver chips; RF switches for microwave systems.
6. Microfluidic MEMS: DNA hybridization arrays or similar lab-on-a-chips for biomedical and biochemical development, bio-analysis and diagnostic; printerheads for ink jet printing.
7. Power MEMS: on-chip power generation and energy storage for portable systems.
8. MEMS-based data storage: micro-positioning and tracking devices for magnetic, optical, thermal, or atomic force data tracks; micro-mirrors for optical beam steering.
9. Microsurgical instruments: for non-invasive techniques, intra-vascular devices, and laparoscopic procedures.

As new applications are developed, the MEMS market is experiencing a period of dramatic growth that is shown in Figure 11.2. Using Tire Pressure Monitoring System (TPMS) as an example, there is a need of 68 millions TPMS with a market value of $102M by year 2007. Similar high volume applications are for (a) mobile phones with microphones and acceleration sensors for human–machine interaction, gyroscopes for image stabilization and RF MEMS switches for transceivers; (b) hard disk drives with acceleration sensors for free fall detection; and (c) camcorders and cameras with gyroscope for image stabilization. In all these applications demanding low cost and small size, MEMS packaging is usually a major consideration.

MEMS packaging can be defined as all the integrations after the microfabrication of the device is complete. They include post-processing release, package/substrate fabrication, assembly, testing, and reliability assurance. Reliability is one of the performance measures that are strongly affected by the package as well as the device. Assurance of the reliability is considered as a packaging activity since packaging engineers rather than fabrication

MEMS PACKAGING AND RELIABILITY

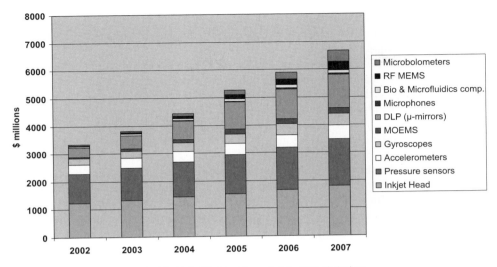

FIGURE 11.2. World wide market for MEMS devices.

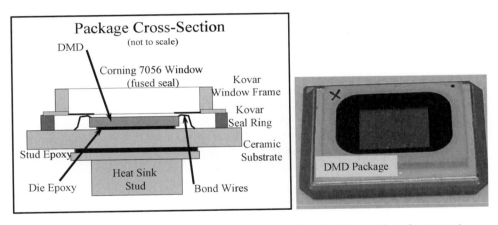

FIGURE 11.3. Package for digital mirror device (courtesy of John P. O'Connor, Texas Instruments).

engineers usually conduct environmental protection processes, burn-ins, and accelerated tests to ensure the production of a reliable MEMS device.

Figure 11.3 shows a package developed for Texas Instruments' digital mirror device (DMD). DMD has millions of micro-mirrors and is used for projection displays. This device has proven an important fact: mechanical devices can be switched over trillions of cycles while achieving the same reliability level as their electronic counterparts [6,7]. After release, a self-assembled monolayer (SAM) can be used to coat the device to avoid a moisture-induced striction problem. If needed, getters can be used to remove particles or moisture inside the package [8]. The DMD package is hermetically sealed with a Kovar ring. Particles can cause reliability failures, so the device has to be packaged in a Class-10 cleanroom. Outgassing of all the package materials should also be controlled. The large glass window is the critical optical interface between the DMD and other optical components for a projection system. Therefore, the window's alignment with the DMD is important [8]. Another concern is the hysteresis behavior of the mirror's aluminum material. The mirror may be difficult to move when it stays at one tilting angle for too long. This creep-related problem is temperature dependent; as a result, thermal management has to control the device temperature to avoid the hysteresis effect [7].

For inertial sensors, the above mentioned packaging approach is too expensive. It is replaced by another approach compatible with microelectronic packaging. The compatibility is achieved by the establishment of wafer-level capping [9–13]. As shown in Figure 11.4, silicon or glass caps are bonded onto a MEMS wafer for hermetic and/or vacuum seal. For a batching process, these caps are fabricated in another wafer as shown in Figure 11.5. The hermetic sealing is accomplished by wafer-to-wafer anodic bonding, soldering, or glass sealing. The capped MEMS devices are diced and packaged through injection plastic molding. As shown in Figure 11.6, this plastic packaging process is compatible with that used in microelectronic packaging. Therefore, packaging cost and size are reduced substantially. In addition to this example, there is an alternative with fully integrated device-to-cap fabrication. With a good design and custom-fabrication, a MEMS device and its encapsulation cover are fabricated in the same process run [14]. Another alternative is to replace the inorganic capping by liquid crystal polymer (LCP) [15,16]. Wafer-level capping and its improved version for wafer-level packaging are enabling technologies for cost and size reduction demanded by today's microsystems integrating MEMS and electronics.

MEMS packaging has been and continues to be a major challenge. The packaging cost is about 50% to 90% of the total cost of a MEMS product. Packaging should allow some moving parts to interact with other components through optical, electrical, thermal, mechanical, or chemical interfaces. As a result, many MEMS packaging problems are new

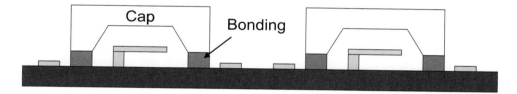

FIGURE 11.4. Wafer-level capping for hermetic sealing of MEMS.

FIGURE 11.5. A wafer with silicon caps.

MEMS PACKAGING AND RELIABILITY

to most of the electronic packaging engineers. In a National Science Foundation (NSF) workshop, several major MEMS packaging and reliability challenges have been identified [17]. Here are a few examples:

- Vacuum packaging may be needed when viscous damping is important.
- Die-attachment may create severe thermal stresses that affect the accuracy of pressure measurement.
- Thermal strains may affect the performance of membrane devices.
- Moisture can cause striction problems.

These new problems are usually dependent on specific MEMS functions. MEMS package provides functional interfaces between the MEMS device and the environment. These interfaces are directly related to the applications. Unfortunately (or fortunately), MEMS has a large number of diverse applications as listed above. As a result, a variety of functional interfaces are needed such as: optical, RF, thermal (radiation, conduction, or convection), fluids (liquids or gases), mechanical (body or surface loadings), and others (e.g., radiation, magnetic, etc.). Clearly indicated by this long list of interfaces, there will be no "standard" packages to meet the requirements of all the MEMS applications. The above mentioned wafer-level capping for inertia sensors is one of the best solutions to insert MEMS packaging into existing microelectronic packaging infrastructure.

In addition to functional interfaces, reliability is another major packaging consideration. Striction, fracture and fatigue, mechanical wear with respect to frequency and humidity, and shock and vibration effects are the major causes of MEMS failures. During the last 20 years, MEMS products have proven to be reliable [7,18,19]. The most reliable MEMS devices are hermetically packaged single-point contact or no-contact devices. Recently, novel MEMS devices with surface contacts have reached impressive reliability levels with billions or hundreds of billions of surface impacts. It is a significant improvement from the early studies on RF MEMS [20].

With impressive technology advancement, MEMS sensors and actuators are no longer niche applications. In the near future, every automobile will use 50 to 100 MEMS components and every cell phone will have at least 3 MEMS components. Every MEMS component has to be integrated with other microelectronic, optoelectronic or RF compo-

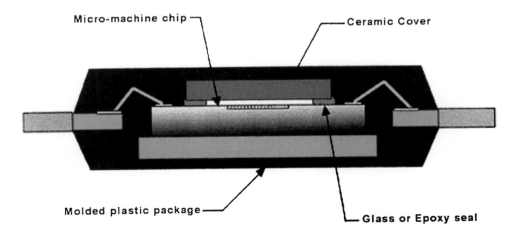

FIGURE 11.6. Plastic molding of MEMS devices capped.

nents. With such a large scale impact, we expect to see more advanced packaging technologies to be developed for the MEMS-based microsystems. In the following sections, we will describe (1) flip-chip assembly for hybrid integration, (2) soldered assembly for three-dimensional MEMS, (3) flexible circuit boards for MEMS, and (4) atomic layer deposition for reliable MEMS. They are different from the aforementioned MEMS packaging technologies being used to manufacture current products. The understanding of these new approaches will provide an insight into future MEMS packaging and reliability activities.

11.2. FLIP-CHIP ASSEMBLY FOR HYBRID INTEGRATION

MEMS devices have to be integrated with other electronic devices. Monolithic integration is always desirable; however, hybrid integration may be more practical due to its ability to integrate mixed-technology devices. For hybrid integration, flip-chip assembly could result in the smallest size while achieving superior performance. Such an assembly technology will be described in this section using PolyMUMPs-based MEMS as an example.

Figure 11.7 shows the cross-sections of PolyMUMPs (Polysilicon-Based Multi-User MEMS Processes), with its polysilicon and silicon oxide layers [1,21]. The oxide layers are sacrificial layers and are removed with HF after fabrication. An example of a typical design with over 50 different device layouts is shown in Figure 11.8. PolyMUMPs is only one of the MEMS foundry processes; there are quite a few other services using surface or bulk micro-machining or LIGA processes.

For hybrid integration, the MEMS devices manufactured in a foundry should be integrated with other devices on a new, common substrate. A flip-chip assembly process with silicon removal technology has been developed for such transfer and integration [22–24]. A 1D variable capacitor illustrated in Figures 11.9(a) and (b) is a good example. The MEMS fabricated in a silicon substrate was transferred to a ceramic RF substrate [25]. By using flexures having varying stiffness levels, the plates of the array would snap down indi-

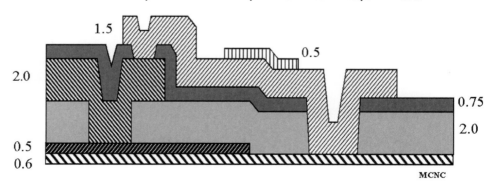

FIGURE 11.7. Cross-sections of PolyMUMPS (Multi-User MEMS Processes) foundry process.

vidually, and in sequence to change the capacitance. However, such an ideal operation was not achieved due to the following problems resulting from poor thermo-mechanical behaviors. As shown in Figure 11.10(a) for the capacitor with cantilever beams, bond height variations would result in non-repeatable capacitive performance. The warped beams would change their configurations away from two parallel plates defined by the bond and the dimple at the tip of each beam. With curved plates, the desirable digital snap-down sequence could not happen. The alternative fixed-fixed beams are shown in Figure 11.10(b), but this design still suffered large capacitance variations due to the warpage resulting from a thermal mismatch with different coefficients of thermal expansion (CTEs) between the MEMS device and the substrate. The digital increments in capacitance could be lost due to uncontrollable pull-in voltages associated with the varied bond heights and warpage.

In addition, this 1D variable capacitor flip-chip assembled required immersion of the chip in Hydrofluoric Acid after the assembly. Such immersion was slow and could damage some materials in the assembly [22–25]. Therefore, when the 1D variable capacitor was improved to a new two-dimensional (2D) device, we decided to improve the design and

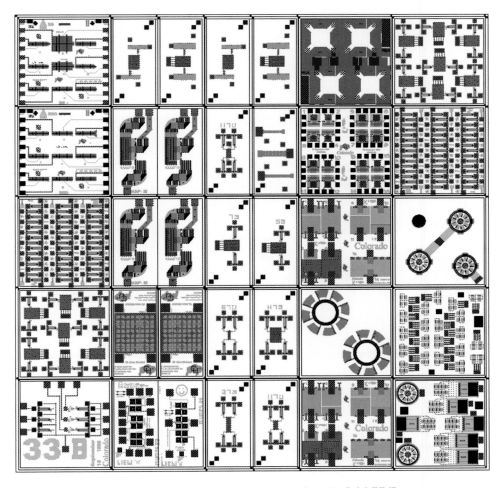

FIGURE 11.8. A typical 1 cm × 1 cm chip design using PolyMUMPs.

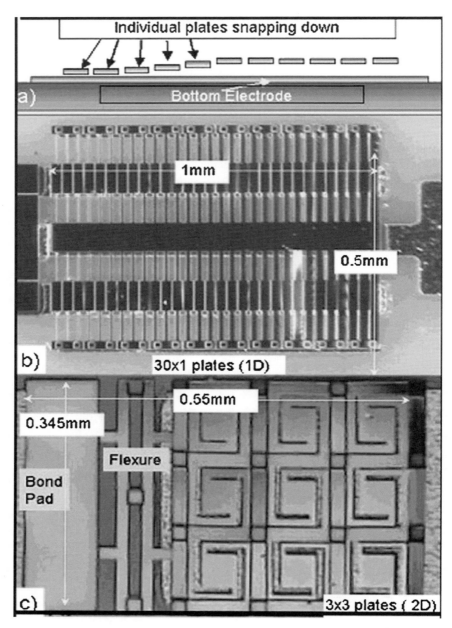

FIGURE 11.9. (a) Principle of digital pull-in in MEMS variable capacitors; (b) a 2-terminal, 1D variable capacitor (c) a 2D 3 × 3 MEMS variable capacitor.

the assembly process. The 2D device is shown in Figure 11.11. The device consisted of five components. Tethers connected the pre-assembly released MEMS to the silicon. The bonding pads joined the device to the new alumina substrate through solder bumps. Two compliant flexures accommodated the thermal mismatch between the silicon and alumina substrate during the flip-chip assembly. Arrays of 2 × 2, 3 × 3 or 4 × 4 capacitor plates

MEMS PACKAGING AND RELIABILITY

a) Existing MEMS variable capacitor developed at University of Colorado. illustration of ideal capacitance (top) and actual device (bottom) after fabrication and assembly Due to warpage in the top plate, C2 measurement was different from the ideal design.

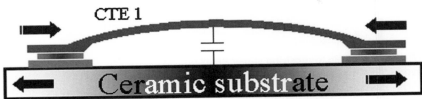

b) During flip-chip bonding where CTE difference resulting in thermal mismatch warping.

FIGURE 11.10. (a) Flip-chip assembly of MEMS with cantilever beams suffering from bond height variations and beam bending and (b) the assembly with fixed-fixed beams suffering from warpage due to a CTE mismatch.

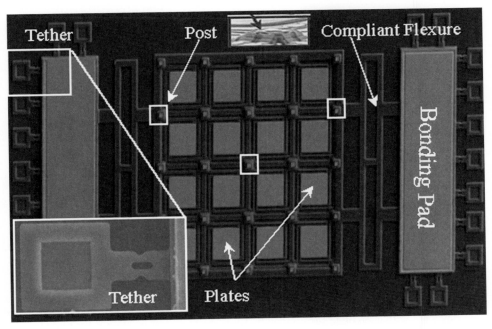

FIGURE 11.11. A variable capacitor featured with tethers, bonding pads, compliant flexures, 5 × 5 "posts" and 4 × 4 plates on the host silicon substrate.

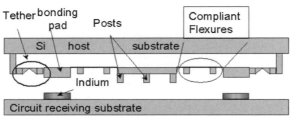

a) Align bonding pads to Indium bumps

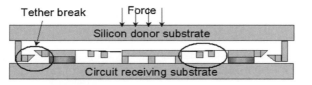

b) Bond MEMS chip to receiving substrate using thermo-compression bonding

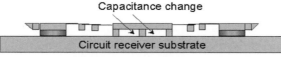

c) Remove MEMS host substrate. Capacitance does not depend on indium evaporation, it is controlled by the post.

(A)

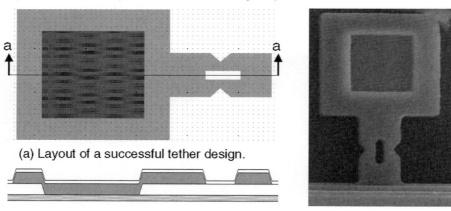

(a) Layout of a successful tether design.

(b) Cross-sectional view of the tether design in (a-a).

(c) SEM photograph of successful tethers.

(B)

FIGURE 11.12. Illustration of the flip-chip assembly process. The upper piece shown in a) in (A) is a pre-released MEMS chip with gold pads; the lower piece shown in a) in (A) is a patterned ceramic substrate with deposited indium bumps. Tethers are used to hold pre-assembly released MEMS. Posts are used for the precision gap control after the assembly.

were designed with each plate surrounded by four "posts" (legs) to support the plate and its flexures. The corresponding flip-chip assembly process with tethers and posts is described in Figure 11.12, where the receiving substrate could be any circuit or RF substrate.

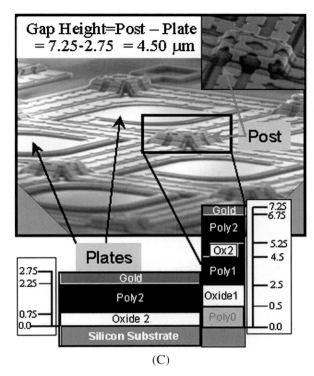

(C)

FIGURE 11.12. (Continued).

The use of tethers allowed us to transfer pre-assembly released MEMS devices onto a new substrate [26,27]. The tethers lightly connected a released MEMS device to its silicon donor substrate. They broke after delivering the device to the RF host substrate during or after the flip-chip assembly. A tether's design and photo are also shown in Figure 11.12. The use of posts enabled a precise gap control, which was critical to the operation of the capacitor plates. Before using posts, the gap was controlled by the solder joints. With evaporated indium, the gap height could vary up to ±25% [25]. To reduce such a variation, posts were created by stacking different layers during the design [15]. An example is shown in Figure 11.12. When the top plates were pulled down by the electrostatic force, each plate's pull-in voltage was controlled by the precise gap defined by the posts rather than the solder joints. In addition, posts also enabled us to design very compliant flexures to reduce thermal mismatch-induced warpage, which might degrade the electrostatic behavior of the MEMS by significantly increasing the pull-in voltage. With tethers and posts, the thermomechanical behavior of the 2D variable capacitors became controllable, and the desirable digital increments were demonstrated in the RF characterization [15].

11.3. SOLDERED ASSEMBLY FOR THREE-DIMENSIONAL MEMS

One of the most common methods for manufacturing MEMS devices is by using surface micro-machining, e.g., PolyMUMPs. Due to the nature of thin film deposition technology, a fundamental problem with surface micro-machining is its inability to produce highly

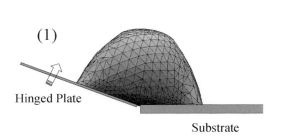
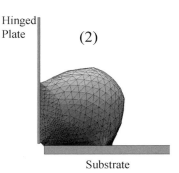

FIGURE 11.13. Illustration of solder self-assembly of a hinged MEMS plate and a solder assembled three-dimensional MEMS device with kickstands.

three-dimensional structures. A common solution is to fabricate flat, 2D hinged components that can be lifted or rotated into assembled structures [28]. Such structures are very common in many MEMS and microelectronics fields, namely micro-optics [29]. The draw back of hinged designs is that they need to be assembled after fabrication. The traditional way to perform this assembly is to do it manually or use additional MEMS mechanisms to assemble devices automatically [30,31]. Manual assembly usually consists of rotating the plates by hand using high precision micro-manipulators. This form of assembly is not

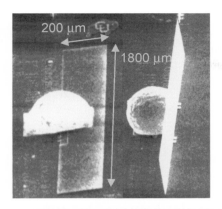
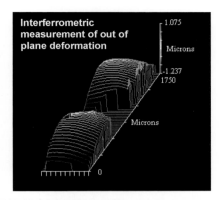
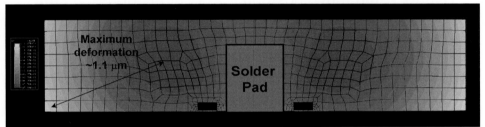

FIGURE 11.14. Validation example in which the model prediction (bottom) matches the interferometric measurement (top right).

practical for mass assembly and manufacturing though, and is rarely effective. Mechanism driven assembly is also insufficient because these MEMS mechanisms are often large and complex, and thus negate many advantages inherent in MEMS devices. An interesting solution is the use of the surface tension properties of molten solder or glass as the assembly mechanism [32–34].

The solder method involves using a standard hinged plate with a specific area metalized as solder wettable pads. Once the solder is in place, it is heated to its melting point, and the force produced by the natural tendency of liquids to minimize their surface energy pulls the free plate away from the silicon substrate (Figure 11.13). Solder is a predominant technology for electronics assembly and packaging. It is not only used for electrical connections, but also for sub-micron accuracy alignment in many packaging applications such as optoelectronic passive alignment [35]. Using solder, hundreds or thousands of precision alignments can be accomplished with a single batch reflow process, and the cost/alignment can be reduced by orders of magnitude. In addition, solder provides high quality mechanical, thermal, and electrical connections.

Figure 11.13 also shows an actual solder self-assembled plate that was 400 microns square and was assembled with an approximately 200 micron diameter solder sphere. But the plate was deformed due to process-induced stresses within the structure. Such thermo-mechanical deformations could be reduced by using finite element modeling and optimization algorithms.

The basic building block of a solder self-assembly structure consisted of a single solder sphere with a hinge and mechanical lock on either side (Figure 11.14). To optimize the structure, the parameters to be varied were: the contact position of the mechanical lock

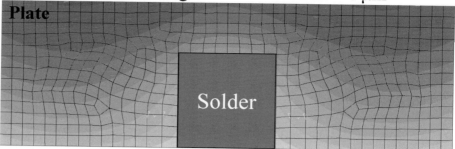
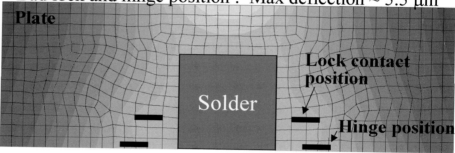
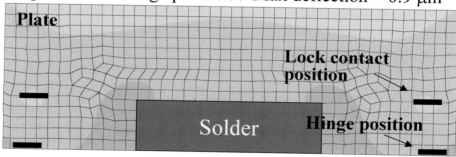

FIGURE 11.15. Comparison of results: (top) case without lock and hinges; (middle) case with lock and hinge poorly located; and (bottom) case with the optimum pad size and lock/hinge position.

and plate, the width and height of the solder pad, and the position of the hinge. The only constraint was that the solder pad should remain large enough to be practical for solder deposition. The finite element software ABAQUS was used to model the structure and extract relevant data, and the optimization algorithm NLPQL [36] was used to optimize the variables. The plate was modeled using composite shell elements, but the solder was simulated with standard three-dimensional solid elements. The interaction of the kickstand and hinges was modeled using contact surface approximations rather than by including the actual hinge and kickstand structure into the model. The accuracy of the model was gauged by comparing the predictions to experimental data. Figure 11.14 shows one such comparison in which a 200 by 800 μm solder self-assembly plate was modeled, fabricated, and measured interferometrically. All cases that the model predicted fell within the data variation.

After validation, the optimization program was able to generate a prediction that significantly reduced the deformation in the plate. The values to be minimized were the rms, average, and maximum deflections of the plate. Figure 11.15 shows three sample cases for one design optimization problem: (a) a case in which there was no lock or hinge contact, (b) a case in which the lock contact position and hinge were poorly placed, and (c) the optimum case. Interestingly, the case in which the lock and hinge placed poorly resulted in a more severe deformation than with no lock at all. The poor lock and hinge position resulted in a maximum deflection of ∼5.5 μm and a rms deflection of ∼3.4 μm, whereas, the prediction with no lock or hinge resulted in a max deflection of ∼4 μm and an rms deflection of ∼2.1 μm. Finally, the optimized structure showed a significant improvement with a maximum deflection of ∼0.9 μm and rms deflection of ∼0.6 μm.

The reason for the reduced deformation is likely due to the lock and hinge constraints working against the deformation resulting from solder shrinkage. The shrinkage tends to cup the plate around it like a shroud. By placing the hinge and lock near the edge of the plates, they restrict the plate and force it back toward the desired position. If the lock and hinge are placed too close to the solder, they only amplify the deformation. If there are too far out, the plate will bend significantly between them and the solder joint. In addition to the plate deformation, the deviation from the desirable angles can also be controlled [28]. Advanced thermo-mechanical analysis and optimization techniques are essential to design such complex MEMS structures.

11.4. FLEXIBLE CIRCUIT BOARDS FOR MEMS

Silicon processing is not the only means to fabricate micro-scale devices. In fact, we expect to see more and more micro-devices to be fabricated using polymer materials. Here is an example for a flexible circuit-based RF MEMS [37]. Photographs of different layers and assembled prototype of X/Ku band switch designs are shown in Figure 11.16. Coplanar waveguide (CPW) lines for mounting switches and on-wafer multi-line TRL calibration were patterned on the metalization layer of a Duroid substrate. Photosensitive benzocyclobutene (BCB) dielectric layer was spin-coated and patterned on CPW lines. Adhesive spacer film was milled to create slot-openings. The switch electrode metalization was patterned on Kapton-E polyimide film, which was machined using Excimer laser to create slot-openings. These layers were aligned using a fixture and assembled using a thermo-compression bonding cycle.

These switches are manufacturable using printed circuit board (PCB) facilities, and they can be integrated with PCB-based RF circuits and antennas. We expect them to have an impact on cost-demanding applications. However, with the large size, there are concerns about their RF losses and new reliability failure mechanisms different from those of thin-film based RF MEMS. The insertion loss could be less than 0.3 dB and the isolation could

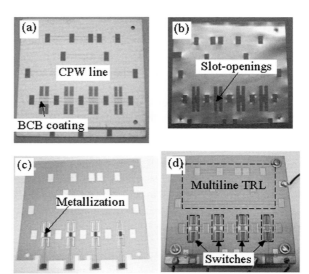

FIGURE 11.16. Photographs of (a) CPW line with BCB dielectric layer on Duroid substrate; (b) adhesive spacer film with milled slot-openings; (c) Kapton E polyimide film with switch top electrode metalization and laser machined slot-openings; and (d) assembled switch prototype.

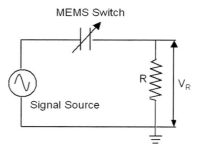

FIGURE 11.17. Circuit diagram of the reliability testing setup.

reach −50 dB at the designed frequencies [37]. Such performance is close to that achieved by thin-film based RF MEMS. The reliability has passed 75 millions of cyclic tests. This test is to be described in more details.

Figure 11.17 shows the circuit diagram for the reliability testing consists of the capacitive MEMS switch, i.e., the device under test (DUT), connected in series with a resistor. A function generator (Agilent 33250A) was used to generate the specified actuation waveform and the required amplitude was obtained by cascading the function generator with a power amplifier (Krohn-Hite 7600) [38,39]. The actuation waveform was applied to the switch and the voltage waveform across the resistor was the input to the data acquisition card (PCI-6035E, available from National Instruments) interfaced to a computer. The voltage waveform was recorded continuously using a LabVIEW program.

When the amplitude of the actuation waveform reached or exceeded the switch pull-down voltage, the switch capacitance changed from a low value (in up-position) to a high value (in down-position). On the other hand, the switch capacitance changed from a high

MEMS PACKAGING AND RELIABILITY

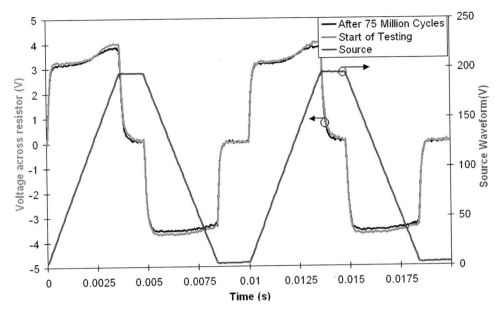

FIGURE 11.18. Reliability testing results showing the voltage waveform across the resistor after 75 millions of operations.

value (in down-position) to a low value (in up-position) when the amplitude of the waveform was less than the release voltage. This change in capacitance with time during the pull-down and release process caused a change in current passing through the test circuit. The voltage waveform across the resistor was proportional to the change in the current waveform and was observed to study the switch dynamic characteristics. Thus, this method could aid in studying the degradation and lifetime characteristics of the capacitive MEMS switches. In addition, switch failure due to mechanical failure, contact striction, fatigue, etc., could be investigated by analyzing the recorded waveform over time.

Switch reliability was studied by applying a triangular actuation waveform at a frequency and amplitude of 12 Hz and 200 V, respectively, for millions of continuous operations. The actuation waveform and the corresponding voltage waveform across the resistor were recorded simultaneously as shown in Figure 11.18. Switching speed of this switch was estimated to be in the millisecond range. This value was higher than those of other RF MEMS switches reported [20] and was due to a large switch up-position gap height of 50–70 μm compared to 2–4 μm in silicon based RF MEMS switches. A triangular wave at 12 Hz that had a rise time and fall time of 41.7 ms was used to ensure that there was enough time for the switch to respond to the actuation signal. The voltage waveform V_R was recorded every tenth of a million operations. The waveforms measured after 0.5 million and 0.7 million operations are also shown as dotted lines. These results coincide very well and indicate that the switch operated up to 1 million with no signs of degradation or failure. Such excellent responses survived after 75 millions of cyclic tests (see Figure 11.18).

Unfortunately, the switch failed right after 75 millions of cycles. After careful inspection, the failure was identified; it was caused by an electrical short resulting from the pinholes of the BCB dielectric layer. The polymer dielectric layer was not strong enough under 200 V across the 1-μm thickness. The RF MEMS device with reliability of 75 million cycles is good enough for switching bandwidths in a cell phone. But, it should be improved further for other applications. The most viable approach to enhance the BCB strength is to deposit an inorganic alumina dielectric layer through atomic layer deposition (ALD) technology. This inorganic, pin-hole free dielectric layer is expected to solve this problem for a more reliable switch. ALD is to be described in the next section on MEMS reliability.

11.5. ATOMIC LAYER DEPOSITION FOR RELIABLE MEMS

During the last 20 years, MEMS reliability has been improved significantly [7,40,41]. Successful, reliable products have proven an important fact: mechanical devices can be switched over trillions of cycles while achieving the same reliability level as the electronic devices [7]. The contact-associated failures are strongly affected by the contact modes and materials, and the effects can be changed significantly if there are minute environmental variations due to particles, charges, and moistures. Currently, self-assembled monolayer (SAM) surface coating is used widely to protect MEMS devices from failures [42]. This organic coating layer, however, provides limited protection and has its own reliability problems. With the advancement of nano-technologies, we now have the opportunity to design and fabricate nano-scale protective coatings to assure MEMS reliability. One of such technologies is atomic layer deposition (ALD). ALD can coat thin dielectric layer to protect MEMS from electrical shorts during operations [43,44]. ALD can coat nano-scaled multi-layers with conducting and dielectric materials for effective charge dissipation [45]. In addition, strong hydrophobic coating can be formed on the ALD coating to reduce moisture-induced adhesion even in a very high relative humidity environment. This technology will be introduced with an emphasis on the above-mentioned three reliability protection mechanisms.

Atomic Layer Deposition (ALD): ALD is a thin film growth technique allowing atomic-scale thickness control. ALD utilizes a binary reaction sequence of self-limiting chemical reactions between gas phase precursor molecules and a solid surface [45,46]. Films deposited by ALD are extremely smooth, pinhole-free and conformal to the underlying substrate surface. This conformity enabled successful coating to cover the entire MEMS device as shown in Figure 11.19. Furthermore, ALD is a low temperature process enabling deposition on thermally sensitive materials. For example, we can use photoresist to cover some patterned areas during deposition for selective instead of comprehensive coverage. ALD can be used to grow a variety of materials including oxides, nitrides, and metals.

Figure 11.20 illustrates the atomic layer deposition process. Reaction A deposits a monolayer of chemisorbed species on the surface. Because the resulting surface is inert to precursor A, further exposure generates no additional growth. Next, precursor B is introduced. This molecule reacts with the product surface from the A reaction in a self-passivating manner. Consequently, the B reaction terminates after the completion of one atomic layer. If reaction B regenerates the initial surface, then the two reactions can be repeated in an ABAB ... binary sequence to deposit a film of predetermined thickness.

One example of this process is the atomic layer deposition of Al_2O_3 consisting of the following binary reaction sequence in which the asterisks designate the surface species:

(A) $Al-OH^* + Al(CH_3)_3 \rightarrow Al-O-Al(CH_3)_2^* + CH_4$,

(B) $Al-CH_3^* + H_2O \rightarrow Al-OH^* + CH_4$.

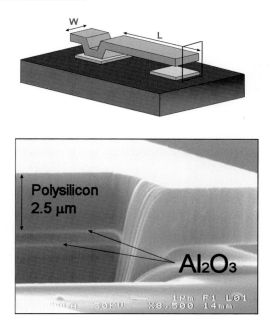

FIGURE 11.19. Illustration of beam and FIB cut section depicting deposited alumina layer.

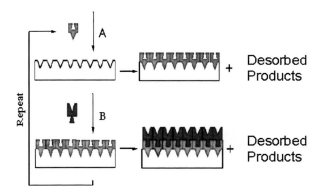

FIGURE 11.20. Description of ALD process.

In reaction (A), the $Al(CH_3)_3$ reacts with the surface hydroxyl groups to deposit a new monolayer of aluminum atoms terminated by methyl groups. In reaction (B), the methylated surface reacts with H_2O vapor, thereby replacing the methyl groups with hydroxyl groups. CH_4 is liberated in both the A and B reactions. The net result of one AB cycle is the deposition of one monolayer of Al_2O_3 onto the surface. The ALD Al_2O_3 film growth is extremely linear with the number of AB cycles performed and the growth rate is 1.29 Å/cycle. The deposition rate is about 0.12 nm (one AB cycle) in 6–10 seconds in a laboratory setup. In a manufacturing setup, the cycle time can be reduced by at least 10 times.

ALD can be used to coat many different nano-scaled, single-layer or multi-layer structures to protect MEMS from different reliability failures. In the following sections, we will illustrate such protection with three examples associated with electrical shorts and charge-induced and moisture-induced adhesion failures.

Dielectric Coating to Prevent Electrical Shorts: Particles are the top killers to most of MEMS devices with or without surface contacts. As a result, MEMS devices should be packaged in a Class-10 environment, and the outgassing inside the package should be controlled by selecting right materials or using getters [17]. One of the particles-induced failures is shorting between conducting parts. A conformal layer of dielectric material coated can prevent this electrical short problem. As shown in Figure 11.19, a conformal layer of alumina (Al_2O_3), an excellent dielectric, was deposited onto released MEMS devices [43,44]. The ALD films cover all sides of a released MEMS device including bottom surfaces, such as underneath cantilever beams. This process was carried out at temperatures down to 150°C—significantly cooler than typical CVD temperatures. This allows for the coating of composite devices made from materials such as poly-silicon and gold without the risk of damaging the individual layers in the MEMS device. In addition, polymer based MEMS devices could also be coated at temperatures as low as 70°C. The deposition technique is compatible with integrated circuit devices as well as thermally sensitive packaged systems.

In addition to cantilever shown above, such a dielectric coating can be used for comb drive actuators and other sensors and actuators [47]. The conformable, selectable, nano-scaled coating can protect these devices from shorts caused by unexpected contacts or by particles.

Charge Dissipation for Reliable MEMS: Charge accumulation is the leading failure mode for RF MEMS switches with dielectric contacts. Figure 11.21 illustrates this charging effect: switch lifetime was about 10,000,000 cycles with 50 Volts applied, however, it reduced substantially to only 10,000 cycles when the voltage increased to 65 Volts [20]. The charge accumulation in the dielectric layer after cyclic loading with high electric fields is proportional to the voltage applied with an exponential function. The charge detrapping is governed by different mechanisms: Schottky emission, Frenkel-Pool emission, tunnel or field emission, space-charge emission, ohmic conduction, and ionic conduction. These mechanisms are very complicated and process- and material-dependent. One simple solution is to increase the effective electrical conductivity of the dielectric layer by doping. ALD can deposit a multilayer composite with hundreds of Al_2O_3 and ZnO layers. Figure 11.22 presents the resistivity values of Al_2O_3/ZnO multilayers with different Zn contents. The resistivity can be changed from 10^{16} to 10^{-3} ohm-cm by choosing a specific content [45]. This accurate control of the resistivity was proven critical to assure reliable RF MEMS switches [48].

Hydrophobic Coating for Reliable MEMS: MEMS reliability is seriously impaired by interfacial interactions. Humidity plays a key role in determining the character of interfacial adhesion. At high relative humidity, water capillary condensation in high aspect ratio micron-sized MEMS structures can cause striction and MEMS failure. To minimize water capillary condensation, the MEMS device can be coated with a hydrophobic film. These hydrophobic films are generally deposited in solution using chlorosilane attachment of alkylsilanes or perfluoroalkylsilanes onto surface hydroxyl groups [42]. Under optimum conditions, the attached alkylsilanes can form a self-assembled monolayer (SAM). However, SAM coating has its own processing challenges and reliability problems [49]. In order to achieve reliable SAM coating, ALD was used to deposit an alumina seed layer before ALD-SAM coating [50,51]. The seed layer could optimize the hydrophobic precursor attachment by: (1) covering the MEMS surface uniformly with a continu-

MEMS PACKAGING AND RELIABILITY

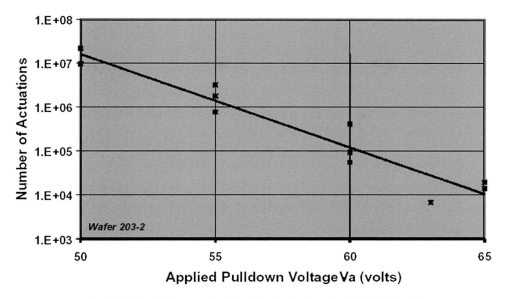

FIGURE 11.21. Lifetime as a function of applied voltage for a RF MEMS switch.

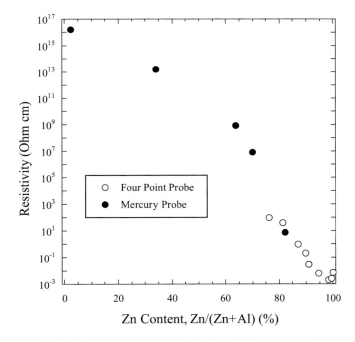

FIGURE 11.22. Electrical resistivity of ALD-coated ZnO/Al$_2$O$_3$.

ous adhesion layer; (2) providing a high surface coverage of hydroxyl groups for maximum precursor attachment; and (3) smoothing and removing nanometer-sized capillaries that may otherwise lead to moisture-induced striction problems. Additionally, polymerization was avoided by using alternative precursors, such as alkylaminosilanes, instead of the traditional chlorosilanes. These alternative precursors reacted more completely and effectively with the surface hydroxyl groups without initial reaction with H_2O. Furthermore, non-chlorinated alkylaminosilanes would not produce HCl, a byproduct that might corrode metal surfaces. Tridecafluoro-1,1,2,2-tetrahydrooctylmethylbis(dimethylamino)silane (FOMB(DMA)S, $C_8F_{13}H_4(CH_3)Si(N(CH_3)_2)_2$) was chosen as the hydrophobic precursor. The hydrophobic film generated by ALD was proven to be more reliable [50]. Its effectiveness creates an opportunity to control the moisture-induced striction reliability problems and pave the way for non-hermetic MEMS packaging.

11.6. CONCLUSIONS

This chapter reviewed MEMS packaging and reliability. Wafer-level capping was introduced as a good example to develop a MEMS packaging technology that is compatible with microelectronic packaging. Such compatibility is essential to reduce packaging cost and size to meet the demands of large scale applications of MEMS. In addition, advanced studies were described with an emphasis on the flip-chip assembly to integrate MEMS with other components, solder assembled three-dimensional MEMS, flexible circuit-based MEMS, and atomic layer deposition for reliable MEMS. These studies are different from current practices focusing on wafer-level capping and packaging. They are good examples, however, to illustrate new packaging technologies being developed in the laboratories.

With advancement of MEMS technologies, hundreds of novel microsystems are demonstrated every year. The impressive insertion of various MEMS sensors and actuators in automobiles, cell phones, and biomedical applications represents only the beginning of a new era. MEMS can be further improved by integrating its micro-scale components with nano-scale devices, i.e., nano-electromechanical systems (NEMS). Integrated MEMS/NEMS will rival, and perhaps even surpass, the societal impact of integrated circuits (ICs). There are many opportunities for packaging engineers to make significant contributions and lead the advancement.

ACKNOWLEDGMENTS

Most of the studies reviewed in this paper were conducted at the University of Colorado—Boulder. The author would like to thank his colleagues: Professors Victor M. Bright, Steve M. George, Martin L. Dunn, and K.C. Gupta and former and current research associates/graduate students: F.F. Faheem, K.F. Harsh, R. Ramadoss, Simone Lee, N.D. Hoivik, J.W. Elam, C.F. Herrmann, F.W. DelRio, and A. Laws. The author is partially supported by a project sponsored by the DARPA (Chip-scale atomic clock program) and managed by the Department of Interior (NBCH1020008).

REFERENCES

1. D. Koester, A. Cowen, R. Mahadevan, M. Stonefield, and B. Hardy, PolyMUMPs Design Handbook, Revision 9, MEMSCAP, San Jose, CA, USA.
2. E.J. Garcia and J.J. Sniegowski, Surface micromachined microengine, Sensors and Actuators A, 48, pp. 203–214 (1995).
3. C.M. Roberts, Jr., L.H. Long, and P.A. Ruggerio, Method for separating circuit dies from a wafer, U.S. Patent #5,362,681, Nov. 1994.

4. T.W. Kenny, R.N. Candler, H.J. Li, W.T. Park, J. Cho, H. Li, A. Partridge, G. Yama, and M. Lutz, An integrated wafer-scale packaging process for MEMS, Proc. ASME International Mechanical Engineering Congress & Exposition, New Orleans, Louisiana, November 17–22, 2002.
5. K. Petersen, Bringing MEMS to Market, Proc. Solid-State Sensor and Actuator Workshop, Hilton Head Island, South Carolina, June 4–8, 2000, pp. 60–64.
6. L.J. Hornbeck, From cathode rays to digital micromirrors: a history of electronic projection display technology, TI Technical Journal, (July–September), pp. 7–46 (1998).
7. Sontheimer and M. Douglass, Identifying and eliminating digital light processing TM failure modes through accelerated stress testing, TI Technical Journal, (July–September), pp. 128–136 (1998).
8. J.P. O'Connor, Packaging design considerations and guidelines for the digital micromirror device, Proc. of IPACK'01, The Pacific Rim/ASME International Electronic Packaging Technical Conference and Exhibition, Kauai, Hawaii, July 8–13, 2001.
9. G.A. Riley, Wafer-level hermetic cavity packaging, Advanced Packaging, 13(5), pp. 21–24 (2004).
10. D. Sparks, et al., Reliable vacuum packaging using NanoGettersTM and glass frit bonding, reliability, testing and characterization of MEMS/MOEMS III, Proc. SPIE, 5343 (2004).
11. Y.T. Cheng, L. Lin, and K. Najafi, Localized silicon fusion and eutectic bonding for MEMS fabrication and packaging, J. Microelectromech. Syst., 9, pp. 3–8 (2000).
12. L.E. Felton, M. Duffy, N. Habluzel, P.W. Farrell, and W.A. Webster, Low-cost packaging of inertial MEMS devices, International Symposium on Microelectronics, Proc. SPIE, 5288, pp. 402–406 (2003).
13. L.E. Felton, N. Hablutzel, W.A. Webster, and K.P. Harney, Chip scale packaging of a MEMS accelerometer, Proc. 54th Electronic Components and Technology Conference, 2004, pp. 869–873.
14. R.N. Candler, W.-T. Park, H. Li, G. Yama, A. Partridge, M. Lutz, and T.W. Kenny, Single wafer encapsulation of MEMS devices, IEEE Transactions on Advanced Packaging, 26(3), pp. 227–232 (2000).
15. F.F. Faheem and Y.C. Lee, Tether- and post-enabled flip-chip asssembly for manufacturable RF-MEMS, Sensors and Actuators, A-114(2-3), pp. 486–495 (2004).
16. F.F. Faheem, K.C. Gupta, and Y.C. Lee, Flip-chip assembly and liquid crystal polymer encapsulation for variable MEMS capacitors, IEEE Transactions on Microwave Theory and Techniques, pp. 2562–2567 (2003).
17. A. Tseng, W.C. Tang, Y.C. Lee, and J. Allen, NSF 2000 workshop on manufacturing of micro-electromechanical systems, Journal of Materials Processing & Manufacturing Science, 8(4), pp. 292–360 (2001).
18. R. Maboudian, W.R. Ashurst, and C. Carraro, Tribological challenges in micromechanical systems, Tribology Letters, 12, pp. 95–100 (2002).
19. D.M. Tanner, Reliability of surface micromachined MicroElectroMechanical actuators, 22nd Int. Conf. Microelectronics, Nis, Yugoslavia, 2000, pp. 97–104.
20. C. Goldsmith, et al., Lifetime characterization of capacitive RF MEMS switches, 2001 IEEE MTT-S International Microwave Symposium Digest, May 2001, pp. 227–230.
21. Y.C. Lee, B. McCarthy, J. Diao, Z. Zhang, and K.F. Harsh, Computer-aided design for microelectromechanical systems (MEMS), International J. of Nano Technology, 18(4/5/6) (2003).
22. Z. Feng, H. Zhang, W. Zhang, B. Su, K.C. Gupta, V.M. Bright, and Y.C. Lee, MEMS-based variable capacitor for millimeter-wave applications, Sens. Actuator A Phys., 91, pp. 256–265 (2001).
23. M.A. Michalicek and V.M. Bright, Flip-chip fabrication of advanced micromirror arrays, The 14th IEEE Int. Conf. Microelectromech. Syst., Jan. 21–25, 2001, pp. 313–316.
24. N. Hoivik, Y.C. Lee, and V.M. Bright, Flip-chip variable high-Q MEMS capacitor for RF spplications, ASME InterPACK'01, Kauai, Hawaii, USA, July 8–13, 2001.
25. N. Hoivik, M.A. Michalicek, Y.C. Lee, K.C. Gupta, and V.M. Bright, Digitally controllable variable high-Q MEMS capacitor for RF applications, IEEE Int. MTT-S, Phoenix, AZ, USA, May 20–25, 2001, pp. 2115–2118.
26. D.A. Singh, M.B. Horsley, A. Cohn, P. Pisano, and R.T. Howe, Batch transfer of microstructures using flip-chip solder bonding, J. Microelectromech. Syst., 8(3), pp. 27–33 (1999).
27. J.Y. Chen, L.S. Huang, C.H. Chu, and C. Peizen, A new transferred ultra-thin silicon micropackaging, J. Micromech. Microeng., 12, pp. 406–409 (2002).
28. K.F. Harsh, V.M. Bright, and Y.-C. Lee, Design optimization of surface micro-machined self-assembled MEMS structures, The ASME International, Intersociety Electronic & Photonic Packaging Conference & Exhibition (InterPACK'01), Kauai, Hawaii, July 8–13, 2001.
29. N.C. Tien, M. Kiang, M.J. Daneman, O. Solgaard, K.Y. Lau, and R.S. Muller, Actuation of polysilicon surface-micro-machined mirrors, Proc. SPIE, 2687, p. 27 (1996).
30. T. Akiyama, D. Collard, and H. Fujita, Scratch drive actuator with mechanical links for self-assembly of three-dimensional MEMS, Journal of Micro-Electromechanical Systems, 6(1), pp. 10–17 (1997).

31. L. Fan, M.C. Wu, and K.D. Choquette, Self assembled micro-actuated XYZ stages for optical scanning and alignment, Transducers 97: 1997 International Conference on Solid-State Sensors and Actuators, Chicago, June 16–19, 1997.
32. K.F. Harsh, V.M. Bright, and Y.C. Lee, Solder self-assembly for three-dimensional micro-electromechanical systems, Sensors and Actuators A, 77, pp. 237–244 (1999).
33. P.W. Green, R.R.A. Syms, and E.M. Yeatman, Demonstration of three-dimensional microstructure self-assembly, Journal of Micro-electromechanical Systems, 4(4), pp. 170–176 (1995).
34. R.R.A. Syms, Rotational self-assembly of complex microstructures by surface tension of glass, Sensors and Actuators A, 65, pp. 238–243 (1998).
35. Q. Tan and Y.C. Lee, Soldering for optoelectronics packaging, IEEE Electronic Components and Technology Conference, Orlando, FL, May 28–30, 1996, p. 26.
36. K. Schittkowski, NLPQL: A Fortran subroutine for solving constrained nonlinear programming problems, Annals of Operations Research, 5, pp. 485–500 (1985/86).
37. R. Ramadoss, S. Lee, V.M. Bright, Y.C. Lee, and K.C. Gupta, Polyimide film based RF MEMS capacitive switches, 2002 IEEE/MTT-S International Microwave Symposium—MTT 2002, 2–7 June 2002, Seattle, WA, IEEE MTT-S CDROM, 2002, pp. 1233–1236.
38. S. Lee, R. Ramadoss, K.C. Gupta, Y.C. Lee, and V.M. Bright, Reliability testing of flexible circuit-based RF MEMS capacitive switches, Microelectronics Reliability, 44, pp. 245–250 (2004).
39. Advanced Design System 2001, Agilent Technologies, CA, USA.
40. MEMS Industry 2004 Report Focus on Reliability, MEMS Industry Group, Pittsburgh, PA, USA.
41. D.M. Tanner, Reliability of surface micromachined MicroElectroMechanical actuators, 22nd Int. Conf. Microelectronics, Nis, Yugoslavia, 2000, pp. 97–104.
42. R. Maboudian, W.R. Ashurst, and C. Carraro, Tribological challenges in micromechanical systems, Tribology Letters, 12, pp. 95–100 (2002).
43. N.D. Hoivik, J.W. Elam, R.J. Linderman, V.M. Bright, S.M. George, and Y.C. Lee, Atomic layer deposited protective coatings for micro-electromechanical systems, Sensors and Actuators, A-103, pp. 100–108 (2003).
44. N. Hoivik, J. Elam, S. George, K.C. Gupta, V.M. Bright, and Y.C. Lee, Atomic layer deposition (ALD) technology for reliable RF MEMS, 2002 IEEE/MTT-S International Microwave Symposium—MTT 2002, 2–7 June 2002, Seattle, WA, IEEE MTT-S CDROM, 2002, pp. 1229–1232.
45. J.W. Elam and S.M. George, Growth of ZnO/Al_2O_3 alloy films using atomic layer deposition techniques, Chem. Mater., 15, p. 1020 (2003).
46. S.M. George, A.W. Ott, and J.W. Klaus, Surface chemistry for atomic layer growth, Journal of Physical Chemistry, (100), pp. 13121–13131 (1996).
47. J.J. Yao, RF MEMS from a device perspective, J. Micromech. Microeng., 10(4), pp. R9–R38 (2000).
48. F.W. DelRio, C.F. Herrmann, N. Hoivik, S.M. George, V.M. Bright, J.L. Ebel, R.E. Strawser, R. Cortez, and K.D. Leedy, Atomic layer deposition of Al2O3/ZnO nano-scale films for gold RF MEMS, IEEE MTT-S International, Volume 3, 6–11 June 2004, pp. 1923–1926.
49. M.P. de Boer, J.A. Knapp, T.A. Michalske, U. Srinivasan, and R. Maboudian, Adhesion hysteresis of silane coated microcantilevers, Acta Mater., 48, pp. 4531–4541 (2000).
50. C.F. Herrmann, F.W. DelRio, V.M. Bright, and S.M. George, Hydrophobic coatings using atomic layer deposition and non-chlorinated precursors, 17th IEEE International Conference on MEMS, 2004, pp. 653–656.
51. U. Srinivasan, M.R. Houston, and R.T. Howe, Alkyltrichlorosilane-based self-assembled monolayer films for stiction reduction in silicon micromachines, J. MEMS, 7, p. 252 (1998).

12

Advances in Optoelectronic Methodology for MOEMS Testing

Ryszard J. Pryputniewicz

NEST—NanoEngineering, Science, and Technology, CHSLT—Center for Holographic Studies and Laser Micro-MechaTronics, Mechanical Engineering Department, Worcester Polytechnic Institute, Worcester, MA 01609, USA

Abstract   Continued demands for delivery of high performance micro-optoelectromechanical systems (MOEMS) place unprecedented requirements on methods used in their development and operation. Metrology is a major and inseparable part of these methods. Optoelectronic methodology is an essential field of metrology that facilitates development of MOEMS because of its inherent advantages over other methods currently available. Due to its scalability, optoelectronic methodology is particularly suitable for testing of MOEMS where measurements must be made with ever increasing accuracy and precision. This was particularly evident during the last few years, characterized by miniaturization of devices, when requirements for measurements have rapidly increased as the emerging technologies introduced new products, especially, optical MEMS. In this chapter, a novel optoelectronic methodology for testing of MOEMS is described and its application is illustrated with representative examples. These examples demonstrate capability to measure submicron deformations of various components of the micromirror device, under actual operating conditions, and show viability of the optoelectronic methodology for testing of MOEMS.

12.1. INTRODUCTION

Advances in technology are frequently based on miniaturization of electronics while simultaneously increasing their capabilities and reducing cost. These advances have led to development of microelectromechanical systems (MEMS). Now, MEMS defines both the technologies to make these systems and the systems themselves.

One of the systems that were made possible by the MEMS technology is a micromirror system for optical applications [1]. This microsystem is a part of a group of micro-optoelectromechanical systems (MOEMS). MOEMS, fabricated using silicon and polysilicon micromachining processes, have widespread applications [2–5], including, but not limited to, optical beam steering, scanners, adaptive optical arrays, flat panel displays, optical interconnects, etc. Specific advantages that MOEMS have over their larger, conventional, counterparts are lower mass, faster response speeds, lower operating power, compact design, and the potential for large arrays of micro-optical elements [6,7].

Development of MEMS, including MOEMS, and structures they interact with, requires sophisticated design, analysis, fabrication, testing, and characterization tools. These tools can be categorized as analytical, computational, and experimental. Solutions using the tools from any one category alone do not usually provide necessary information on MEMS/MOEMS and extensive merging, or hybridization, of the tools from different categories is used [8–10]. One of the approaches employed in the development of MEMS/MOEMS, as well as other complex structures of current interest, is based on a combined use of analytical, computational, and experimental solutions (ACES) methodology [11–13]. In fact, ACES methodology provides solutions where they would not otherwise be possible, or at best be difficult to obtain, while using either only analytical, or only computational, or only experimental tools alone.

In general, analytical tools are based on exact, closed form solutions. These solutions, however, are usually applicable to simple geometries for which, boundary, initial, and loading (BIL) conditions can be readily specified. Analytical solutions are indispensable to gain insight of overall representation of the ranges of the anticipated results. They also facilitate determination of the "goodness" of the results based on the uncertainty analyses. Computational tools, i.e., finite element methods (FEMs), boundary element methods (BEMs), and finite difference methods (FDMs), provide approximate solutions as they discretize the domain of interest and the governing partial differential equations (PDEs). The characteristics of discretization, in conjunction with the corresponding BIL conditions, influence degree of approximation and careful convergence studies [14] should be performed to establish correct computational solutions and modeling. It should be noted that both analytical and computational solutions depend on material properties. If material properties are well known, then solutions will give correct results, providing convergence was achieved subject to properly specified BIL conditions; if material properties are not well known, in spite of having a good knowledge of other modeling parameters, erroneous results will be obtained [15]. Experimental tools, however, in contrast to analytical and computational tools, evaluate actual objects, subjected to actual operating conditions, and provide ultimate results characterizing the objects being investigated. The experimental tools, used in this chapter, employ recent advances in optoelectronic laser interferometric microscope (OELIM) methodology [16,17].

In this chapter, hinged micromirror devices, actuated by electrostatically driven microengines, are considered, as detailed in Section 12.2.

12.2. MOEMS SAMPLES

The MOEMS considered in this chapter are micromirror devices, actuated by electrostatically driven microengines, Figure 12.1. These microsystems were fabricated at Sandia National Laboratories (SNL) using Sandia's Ultraplanar MEMS Multilevel Technology [18] (SUMMiT™ V). The entire micromirror system is made of polysilicon by surface micromachining. The process does not rely on assembly of the microsystem out of separately fabricated pieces, but produces the finished device by batch fabrication [19]. That is, at the end of the fabrication process, the microsystem is ready for use, e.g., in an optical interconnection application, Figure 12.2.

The SUMMiT™ process is based on a set of specific design tools [20,21]. These tools have been developed and validated for use with the multilayer surface micromachining,

ADVANCES IN OPTOELECTRONIC METHODOLOGY FOR MOEMS TESTING

FIGURE 12.1. Sandia micromirror device actuated by electrostatically driven microengine.

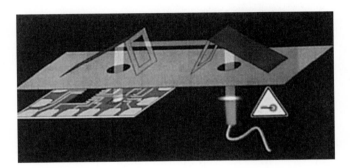

FIGURE 12.2. Interconnection concept based on the Sandia micromirror device.

Figure 12.3. Their use is facilitated by availability of standard components library. For example, to design a hinged micromirror, *Optical Components* are pulled down from the *Components Library*, Figure 12.4. Then, specific components, e.g., anchor (i.e., ground) hinge, is selected, Figure 12.5. The library component shown in Figure 12.5 contains all design details, Figure 12.6, necessary to fabricate a functional hinge using the SUMMiT™ process. Integrating other optical components, available in the library, a hinged micromirror of desired/specific dimensions can be designed, Figure 12.7, and fabricated, Figure 12.8. The micromirror integrated with other parts forms the microsystem, Figure 12.1.

The microsystem is fabricated using surface micromachining of multiple (structural) polysilicon films with intervening (sacrificial) oxide films, Figure 12.3. All structural films are made using low-pressure chemical vapor deposition (LPCVD) of structural polycrystalline silicon. The sacrificial silicon-dioxide films are also deposited by LPCVD. Fabrication of the micromirror system, including the electrostatic comb drives, gears, and the interconnecting linkages, requires four polysilicon structural-layers; in the SUMMiT™ V process up to 5 structural polysilicon layers are available if needed. In this process, the first polysilicon layer (POLY0) provides a voltage reference plane and electrical interconnections, while the remaining three (or four, as the particular case may be) polysilicon layers

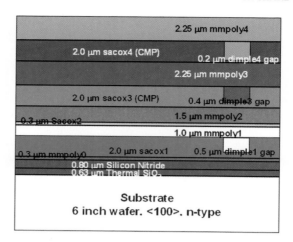

FIGURE 12.3. Film stack of the SUMMiT™ process.

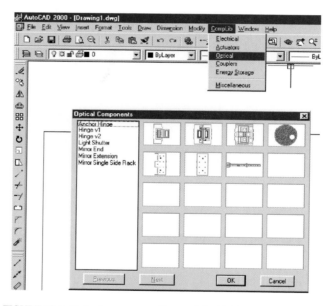

FIGURE 12.4. Optical components library of the SUMMiT™ design tools.

(i.e., POLY1, POLY2, POLY3, and POLY4) are used to form the mechanical/structural components.

In the micromirror device, the microengine converts electrical energy to kinetic energy. This microengine is controlled by two mutually orthogonal linear comb drive actuators. These comb drives consist of two sets of fingers, one stationary and the other movable. At rest, the fingers are in as fabricated position. When a voltage is induced, the movable set of fingers is attracted toward the stationary set thus producing a motion with respect to the base (or reference) while deforming the folded elastic suspension springs supporting the comb drives. When the voltage is suppressed, the elastic forces produced by the elastic springs, which are integral parts of the comb drives, restore any deflections, or movements,

ADVANCES IN OPTOELECTRONIC METHODOLOGY FOR MOEMS TESTING

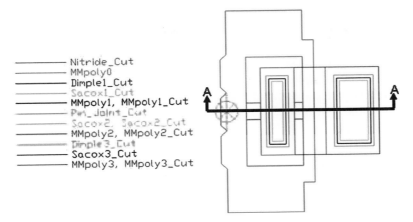

FIGURE 12.5. Anchor hinge as selected from the standard components library. Detailed cross section along A–A is shown in Figure 12.6.

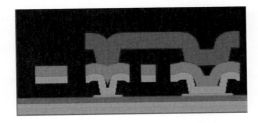

FIGURE 12.6. 2D visualization of details of cross section along the line A–A of Figure 12.5.

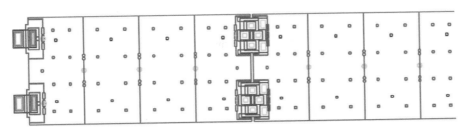

FIGURE 12.7. Hinged micromirror designed from standard components in the library of the SUMMiT™ design tools.

FIGURE 12.8. Hinged micromirror fabricated using the SUMMiT™ process based on the design shown in Figure 12.7.

of the actuator. The comb drives are connected to linkage arms. The linkage arms, in turn, are connected to the drive gear via pins. The drive gear is about 50 μm in diameter with perfectly formed teeth.

To understand dynamics of the micromirror device, deformations of its various components should be measured as a function of operational speeds. These deformations should then be correlated with parameters defining kinematics and kinetics of various components of the micromirror device, which are based on analysis discussed in Section 12.3.

12.3. ANALYSIS

An initial goal of the analysis is to determine accelerations of all moving parts of the microengine [22,23]. Then, using Newton's Second Law, calculate dynamic forces acting on the microengine. Once the dynamic forces are known, we can determine whether the microengine will perform as anticipated under expected operating conditions, or not?

Dynamic forces are based on accelerations, both linear and angular. In order to calculate accelerations we must first determine positions of all components in the microsystem for each increment of the input motion. Once equations defining positions are known, we differentiate them with respect to time to calculate velocities, and then differentiate again to obtain accelerations.

One way to develop equations defining positions of components is to write vector-loop-equations (VLEs) with reference to the kinematic diagram of the microengine [22]. A VLE starts at a specific point on the microengine and follows a loop, via other characteristic points, to end up at the point where it started. That is, the magnitude between the start and the stop points of a given VLE is zero, or a null vector. Thus, because of the nature of the microengine, two VLEs completely define its kinematics, based on which the corresponding equations describing displacements, velocities, and accelerations of characteristic points of the microsystem can be determined as functions of time [22]. That is, the equations defining linear and angular positions are

$$\theta_6 = \cos^{-1}\left[\left(\frac{R_2}{R_6}\right)\cos\theta_2\right], \tag{12.1}$$

$$R_5 = R_2 \sin\theta_2 - R_6 \sin\theta_6, \tag{12.2}$$

$$\theta_4 = \sin^{-1}\left(\frac{R_2 \sin\theta_2 - R_3 \sin\theta_6 - R_8}{R_4}\right), \tag{12.3}$$

and

$$R_7 = R_2 \cos\theta_2 - R_3 \cos\theta_6 - R_4 \cos\theta_4, \tag{12.4}$$

where R and θ denote linear and angular positions, respectively, of linkages which are identified by subscripts. In Equation (12.4),

$$\theta_2 = \theta_{20} + \omega_2 t \tag{12.5}$$

with the subscript 0 denoting initial, i.e., at $t = 0$, angular position, and the angular speed ω_2 is defined as

$$\omega_2 = \frac{2\pi}{60} N \qquad (12.6)$$

in which N is the *rotational speed* in *revolutions-per-minute* (rpm). It should be noted that one of the programs at SNL was dedicated to development of a microengine capable of sustained operation at 1,000,000 (i.e., one-million) rpm!

Using Equations (12.1) to (12.4), linear and angular velocities, V and ω, respectively, can be determined to be

$$\omega_6 = \left(\frac{R_2}{R_6}\right) \omega_2 \frac{\sin\theta_2}{\sin\theta_6}, \qquad (12.7)$$

$$V_5 = R_2 \omega_2 \cos\theta_2 - R_6 \omega_6 \cos\theta_6, \qquad (12.8)$$

$$\omega_4 = \frac{R_2}{R_4} \omega_2 \frac{\cos\theta_2}{\cos\theta_4} - \frac{R_3}{R_4} \omega_6 \frac{\cos\theta_6}{\cos\theta_4}, \qquad (12.9)$$

and

$$V_7 = -R_2 \omega_2 \sin\theta_2 + R_3 \omega_6 \sin\theta_6 + R_4 \omega_4 \sin\theta_4. \qquad (12.10)$$

Based on Equations (12.7) to (12.10), equations for linear and angular accelerations, a and α, respectively, become

$$\alpha_6 = \frac{R_2}{R_6} \omega_2^2 \frac{\cos\theta_2}{\sin\theta_6} - \omega_6^2 \frac{\cos\theta_6}{\sin\theta_6}, \qquad (12.11)$$

$$\alpha_4 = \frac{1}{R_4 \cos\theta_4} \left(-R_2 \omega_2^2 \sin\theta_2 + R_3 \omega_6^2 \sin\theta_6 - R_3 \alpha_6 \cos\theta_6 + R_4 \omega_4^2 \sin\theta_4\right), \qquad (12.12)$$

$$a_5 = -R_2 \omega_2^2 \sin\theta_2 + R_6 \omega_6^2 \sin\theta_6 - R_6 \alpha_6 \cos\theta_6, \qquad (12.13)$$

and

$$a_7 = -R_2 \omega_2^2 \cos\theta_2 - R_3 \omega_6^2 \cos\theta_6 + R_3 \alpha_6 \sin\theta_6 + R_4 \omega_4^2 \cos\theta_4 + R_4 \alpha_4 \sin\theta_4. \qquad (12.14)$$

Kinetic analysis is based on applying Newton's Second Law of motion to the components of an operating microengine. In the foregoing analysis of the microengine, we have linkages 4 and 6 as well as horizontal and vertical comb drives that are moving. Therefore, the x and y components of the forces acting on the pin of the drive gear, due to the linkages and comb drives, can be written as

$$F_x = FD4_x + F4_x + F6_x, \qquad (12.15)$$

and

$$F_y = FE6_y + F4_y + F6_y, \tag{12.16}$$

where the components of the forces $F4$, $F6$, $FD4$, and $FD6$ are determined based on the dimensions, materials, operating conditions, and relative motions of specific components with respect to the other components of the micromirror device.

Using Equations (12.15) and (12.16), magnitude of force acting on the pin of the drive gear can be computed as

$$F = \sqrt{F_x^2 + F_y^2}, \tag{12.17}$$

which is a function of time. Sample calculations of the forces acting on the pin during operation of the micromirror device follow in subsequent sections.

In addition to analytical modeling of kinematics and kinetics of the microsystem, computational modeling of its dynamics was also performed using multiphysics approach [24,25].

12.4. OPTOELECTRONIC METHODOLOGY

Optoelectronic methodology, as presented in this chapter, is based on the principles of optoelectronic holography (OEH) [10,16,17,26]. Basic configuration of the OEH system is shown in Figure 12.9. In this configuration, laser light is launched into a single mode optical fiber by means of a microscope objective (*MO*). Then, a single mode fiber is coupled into two fibers by means of a fiber optic directional coupler (*DC*). One of the optical fibers comprising the *DC* is used to illuminate the object along the direction $\mathbf{K}_1$, while the output

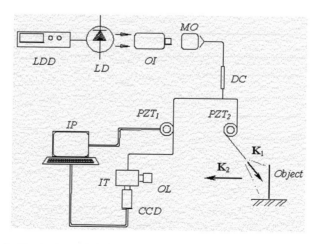

FIGURE 12.9. Single-illumination and single-observation geometry of a fiber optic based OEH system: *LDD* is the laser diode driver, *LD* is the laser diode, *OI* is the optical isolator, *MO* is the microscope objective, *DC* is the fiber optic directional coupler, PZT_1 and PZT_2 are the piezoelectric fiber optic modulators, *IP* is the image-processing computer, *IT* is the interferometer, *OL* is the objective lens, *CCD* is the camera, while $\mathbf{K}_1$ and $\mathbf{K}_2$ are the directions of illumination and observation, respectively.

from the other fiber provides reference against which signals from the object are recorded. Both, the object and reference beams are combined by the interferometer (IT) and recorded by the system camera (CCD).

Images recorded by the CCD are processed by the image-processing computer (IP) to determine the fringe-locus function, Ω, constant values of which define fringe loci on the surface of object under investigation. The values of Ω relate to the system geometry and the unknown vector $\mathbf{L}$, defining deformations, via the relationship [27]

$$\Omega = (\mathbf{K}_2 - \mathbf{K}_1) \cdot \mathbf{L} = \mathbf{K} \cdot \mathbf{L}, \tag{12.18}$$

where $\mathbf{K}$ is the sensitivity vector defined in terms of vectors $\mathbf{K}_1$ and $\mathbf{K}_2$ identifying directions of illumination and observation, respectively, in the OEH system, Figure 12.9.

Quantitative determination of structural deformations due to the applied loads can be obtained, by solving a system of equations similar to Equation (12.18), to yield [27]

$$\mathbf{L} = [\tilde{\mathbf{K}}^T \tilde{\mathbf{K}}]^{-1} (\tilde{\mathbf{K}}^T \Omega), \tag{12.19}$$

where $\tilde{\mathbf{K}}^T$ represents a transpose of the matrix of the sensitivity vectors $\mathbf{K}$.

Equation (12.19) indicates that deformations determined from interferograms are functions of $\mathbf{K}$ and Ω, which have spatial, i.e., (x, y, z), distributions over the field of interest on the object being investigated. Equation (12.19) can be represented by a phenomenological equation [28]

$$\mathbf{L} = \mathbf{L}(\mathbf{K}, \Omega), \tag{12.20}$$

based on which the RSS-type (where RSS represents *the Square Root of the Sum of the Squares*) uncertainty in $\mathbf{L}$, i.e., $\delta \mathbf{L}$, which can be determined to be [28]

$$\delta \mathbf{L} = \left[\left(\frac{\partial \mathbf{L}}{\partial \mathbf{K}} \delta \mathbf{K} \right)^2 + \left(\frac{\partial \mathbf{L}}{\partial \Omega} \delta \Omega \right)^2 \right]^{1/2}, \tag{12.21}$$

where $\partial \mathbf{L}/\partial \mathbf{K}$ and $\partial \mathbf{L}/\partial \Omega$ represent partial derivatives of $\mathbf{L}$ with respect to $\mathbf{K}$ and Ω, respectively, while $\delta \mathbf{K}$ and $\delta \Omega$ represent the corresponding uncertainties in $\mathbf{K}$ and Ω, respectively. It should be remembered that $\mathbf{K}$, $\mathbf{L}$, and Ω are all functions of spatial coordinates (x, y, z), i.e., $\mathbf{K} = \mathbf{K}(x, y, z)$, $\mathbf{L} = \mathbf{L}(x, y, z)$, and $\Omega = \Omega(x, y, z)$, respectively, when performing partial differentiations. After evaluating, Equation (12.21) indicates that $\delta \mathbf{L}$ is proportional to the product of the local value of $\mathbf{L}$ with the RSS value of the ratios of the uncertainties in $\mathbf{K}$ and Ω to their corresponding local values, i.e.,

$$\delta \mathbf{L} \propto \mathbf{L} \left[\left(\frac{\delta \mathbf{K}}{\mathbf{K}} \right)^2 + \left(\frac{\delta \Omega}{\Omega} \right)^2 \right]^{1/2}. \tag{12.22}$$

For typical geometries of the OEH systems used in recording of interferograms, the values of $\delta \mathbf{K}/\mathbf{K}$ are less than 0.01. However, for small deformations, of the magnitudes encountered while studying MEMS/MOEMS, the typical values of $\delta \Omega/\Omega$ are about one order of magnitude greater than the values for $\delta \mathbf{K}/\mathbf{K}$. Therefore, the accuracy with which the fringe orders are determined influences the accuracy in the overall determination of deformations [29,30]. To minimize this influence, a number of algorithms for determination of Ω have been developed. Some of these algorithms require multiple recordings of each of the two states, in the case of double-exposure method, of the object being investigated with introduction of a discrete phase step between the recordings [10,31,32].

For example, the intensity patterns of the first and the second exposures, i.e., $I_n(x, y)$ and $I'_n(x, y)$, respectively, in the double-exposure sequence can be represented by the following equations:

$$I_n(x, y) = I_o(x, y) + I_r(x, y) + 2\{[I_o(x, y)][I_r(x, y)]\}^{1/2} \cos\{[\varphi_o(x, y) - \varphi_r(x, y)] + \theta_n\}, \tag{12.23}$$

and

$$I'_n(x, y) = I_o(x, y) + I_r(x, y) + 2\{[I_o(x, y)][I_r(x, y)]\}^{1/2} \cos\{[\varphi_o(x, y) - \varphi_r(x, y)] \\ + \theta_n + \Omega(x, y)\}, \tag{12.24}$$

where I_o and I_r denote the object and reference beam irradiances, respectively, with (x, y) denoting spatial coordinates, φ_o denotes random phase of the light reflected from the object, φ_r denotes the phase of the reference beam, θ_n denotes the applied n-th phase step, and Ω is the fringe-locus function relating to the displacements/deformations the object incurred between the first and the second exposures; Ω is what we need to determine. When Ω is known, it is used in Equation (12.19) to find [27] **L**.

In the case of 5-phase-steps algorithm with $\theta_n = 0, \pi/2, \pi, 3\pi/2$, and 2π, the distribution of the values of Ω can be determined using [32]

$$\Omega(x, y) = \tan^{-1}\left\{\frac{2[I_2(x, y) - I_4(x, y)]}{2I_3(x, y) - I_1(x, y) - I_5(x, y)}\right\}. \tag{12.25}$$

Results produced by Equation (12.25) depend on the capabilities of the illumination, the imaging, and the processing subsystems of the OEH system. Developments in laser, fiber optics, CCD camera, and computer technologies have led to advances in the OEH metrology; in the past, these advances almost paralleled the advances in the image recording media [30]. A fiber optics based OEH system, incorporating these developments, is depicted in Figure 12.10. In addition to being able to measure static and dynamic deformations of objects subjected to a variety of BIL conditions, the system shown in Figure 12.10 is also able to measure absolute shape of the objects using multiple-wavelength optical contouring [10]. This dual-use is possible because of rapid tuning of the laser and real-time monitoring of its output characteristics by the wavelength meter (WM) and the power meter (PM)—both-integrated into the OEH system. In the configuration shown in Figure 12.10, the image-processing computer (IP) controls all functions of the OEH system.

In response to the needs of the emerging MEMS technology, an optoelectronic laser interferometric microscope (OELIM) system for studies of objects with micron size features was developed [16,33]. In the OELIM system, Figure 12.11, the light beam produced by a laser is directed into an acousto-optic modulator (AOM) and then into a single mode optical fiber. The output of the fiber is collimated by the collimating illumination lens subsystem (C). The resulting light field is then divided into reference and object beams by the beam splitter (BS). The reference beam is directed toward a PZT actuated mirror (M) and back to the beam splitter. The object beam is directed toward the MEMS under study and

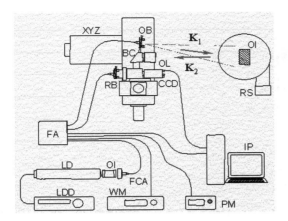

FIGURE 12.10. Fiber-optic based OEH setup arranged to perform high-resolution surface shape and deformation measurements: LDD is the laser diode driver, WM is the wavelength meter, PM is the optical power meter, LD is the laser diode, OI is the optical isolator, FCA is the fiber coupler assembly, IP is the image-processing computer, FA is the single-mode fiber optic directional coupler assembly, RB is the FC-connectorized reference beam fiber, CCD is the digital CCD camera, RS is the rotational stage, BC is the beam combiner, OL is the objective lens, XYZ is the X-Y-Z translational stage, OB is the FC-connectorized object or illumination beam fiber, OI is the object under investigation, while K_1 and K_2 are the vectors defining illumination and observation directions, respectively.

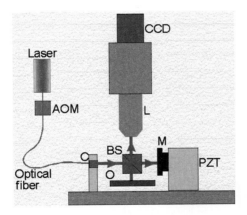

FIGURE 12.11. Optical configuration of the OELIM setup: AOM is the acousto-optic modulator, C is the collimating illumination lens subsystem, BS is the beam splitter, L is the long working distance microscope objective, M and PZT comprise the phase stepping mirror, and O is the MEMS object.

is reflected back to the beam splitter. The two beams recombine at the BS and are imaged by the long working distance microscope objective (L) onto a sensing element of the CCD camera, which records the resulting interference patterns. These patterns are, finally, transferred to the system computer for subsequent quantitative processing and display of the results.

Using the systems shown in Figures 12.10 and 12.11, issues relating to the sensitivity, accuracy, and precision, associated with application of the algorithm defined by Equation (12.25), were studied while evaluating effects that the use of high-spatial and high-digital resolution cameras would have on the results produced [34]. In addition, development of optimum methods for driving lasers is conducted. This development is closely coupled with the development of fiber optic couplers and corresponding subsystems for efficient beam delivery.

12.5. REPRESENTATIVE APPLICATIONS

Using the analytical model presented in Section 12.3, forces acting on the microengine during its operation were calculated. Representative results of these calculations are shown in Figures 12.12 to 12.14 for the pin connecting the electrostatic comb drive linkages to the drive gear. More specifically, Figures 12.12 and 12.13 show polar plot representations of the x and y components of the forces acting on the drive gear pin and indicate that the magnitude of these forces increases from 4 nN, when the microengine operates at 6000 rpm, to 27 µN, when the microengine runs at 500,000 rpm. Figure 12.14 shows the magnitude of the drive gear pin force as a function of rotational speed of the microengine. Clearly, this force increases nonlinearly at an increasing rate as the rotational speed of the microengine increases.

The forces generated during operation of the microengine, load the drive gear and make it wobble as it rotates around its shaft. A unique capability to measure this wobble is provided by the OELIM methodology. Typical results obtained for two different positions in the rotation cycle of the drive gear are shown in Figure 12.15, where fringe patterns vividly display changes in magnitude and direction of the displacements/deformations of the microgears. Displacements of the drive gear, corresponding to the fringe patterns shown

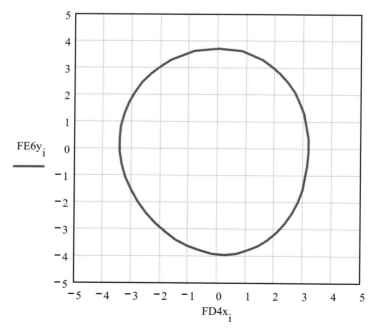

FIGURE 12.12. Polar representation of the x and y components of the force acting on the pin connecting the electrostatic comb linkages to the drive gear, for the microengine operating at 6000 rpm. Force is shown in nN.

ADVANCES IN OPTOELECTRONIC METHODOLOGY FOR MOEMS TESTING

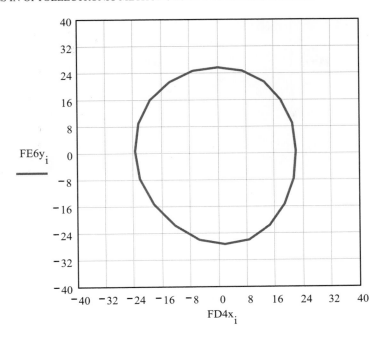

FIGURE 12.13. Polar representation of the x and y components of the force acting on the pin connecting the electrostatic comb linkages to the drive gear, for the microengine operating at 500,000 rpm. Force is shown in μN.

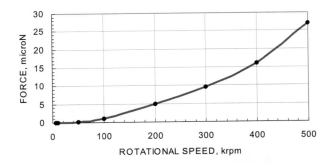

FIGURE 12.14. Force acting on the pin of the drive gear versus rotational speed of the microengine.

in Figure 12.15, are displayed in Figure 12.16 and are seen to vary in magnitude from 0.8 μm to 1.6 μm (it should be realized that thickness of the microgears in the direction of measured deformations is about 2 μm. These variations are due to kinematics and kinetics caused by operational impulsive loading forces generated by the input signals during a typical rotation cycle. In addition, the experimental results show that the wobble depends on the angular position in the rotation cycle, which can be related to the forces exerted on the drive gear by the pin during the cycle, Figures 12.12 and 12.13.

Representative deformations of the microgears, when the microengine operates at 360,000 rpm, are shown in Figure 12.17 and indicate maximum deformations of 1.8 μm.

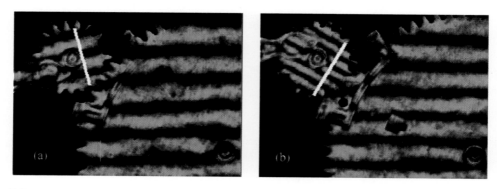

FIGURE 12.15. Representative OELIM fringe patterns recorded during a study of dynamic characteristics of microengines, at two different positions in a rotation cycle. White lines indicate locations (a) and (b) where measurements of displacements shown in Figure 12.16 were made.

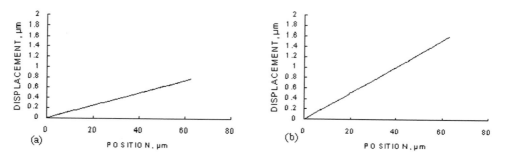

FIGURE 12.16. Displacements of the drive gear of the microengine along the white lines: (a) of Figure 12.15(a), and (b) of Figure 12.15(b).

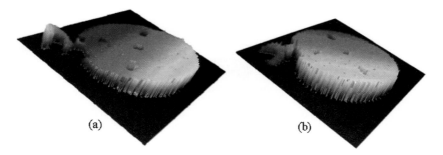

FIGURE 12.17. Deformations of the microgears of the microengine operating at 360,000 rpm: (a) 0° position, (b) 90° position.

Operational functionality of the micromirror system depends also on the quality of motions of section AB of the hinged micromirror, Figure 12.18. OELIM was used to measure these motions by recording fringe patterns, Figure 12.19, which were, in turn, interpreted to determine displacements/deformations of the section AB of the micromirror, Figure 12.20.

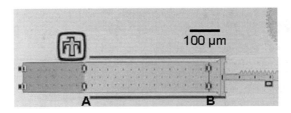

FIGURE 12.18. Measurements were made on the AB section, 100 μm wide and 400 μm long, of the hinged micromirror.

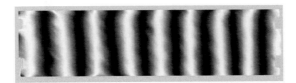

FIGURE 12.19. Representative OELIM fringe pattern of the AB section of the hinged micromirror, shown in Figure 12.18.

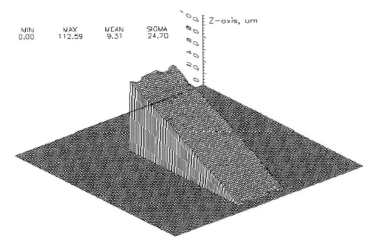

FIGURE 12.20. Wireframe representation of absolute shape and deformations of the section AB of the hinged micromirror, corresponding to the upright position displayed in Figure 12.1. Measurements show that the micromirror displacements range from 0 μm at hinge B to 113 μm at hinge A at which there is also a tilt of 18 mrad.

Figure 12.20 shows that, for the operating conditions for which the fringe pattern of Figure 12.19 was recorded, the out of plane displacement of the section AB of the micromirror was about 113 μm.

Using Figure 12.20, detailed information about deformations of the micromirror can be obtained, Figure 12.21. In this figure, traces (made parallel to the long edges of section AB of the micromirror) are shown. Noticeable differences between the two traces were observed and led to calculation of a tilt amounting to about 18 mrad at the hinge A.

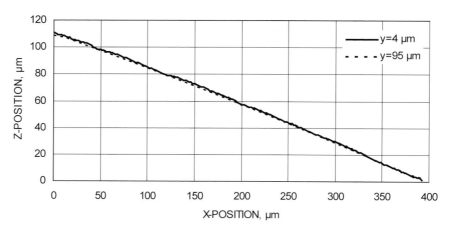

FIGURE 12.21. Vertical displacements (Z-POSITION) of section AB as a function of position along the length (X-POSITION) of the micromirror as determined from traces parallel to the long edges of the wireframe section shown in Figure 12.20.

Following procedures used to obtain representative results shown in this section, deformations and motions of other MEMS/MOEMS can also be determined. Results of these future studies will be reported on in subsequent publications.

12.6. CONCLUSIONS AND RECOMMENDATIONS

Novel optoelectronic methodology for testing of MOEMS was presented. This methodology is based on the optoelectronic laser interferometric microscopy (OELIM) and provides remote submicron measurements in near real-time in full-field-of-view and under actual operating conditions.

Representative results indicate that various components of the micromirror device deform up to 1.8 μm, depending on the position in the rotation cycle and the corresponding force system acting on the component. Although on the micrometer-scale, these deformations are rather large when compared with a nominal thickness of 2 μm of the gears and other moving components of the microengines considered herein.

To complement measurements, a vector based analytical model was developed to determine forces acting on various components of the micromirror device. Using this model, it was shown that the forces acting on the pin connecting the comb drive linkages to the drive gear range from 4 nN, when the microengine is running at 6000 rpm, to 27 μN, when the engine is operating at 500,000 rpm. Furthermore, polar representations of the Cartesian components of the forces acting on the pin while the engine is operating at a constant speed, show variation in magnitude of these components. This, together with fabrication tolerances, gives rise to wobble of the gears as they rotate around their hubs, which is not desirable for sustained operations at any speed.

At even higher speeds, up to 1,000,000 rpm as dictated by requirements in specific programs, forces will substantially increase and will lead to even larger motions and instabilities than measured thus far. Therefore, work of the type presented in this chapter is very timely and will contribute to further advances of the MOEMS.

The results presented in this chapter indicate that vector mechanics approach combined with noninvasive OELIM methodology is a viable hybrid-tool (consisting of the experimental and analytical/computational methods) for characterization of dynamic effects in the micromirror devices as well as in other MOEMS. By understanding the details of MOEMS performance in three-dimensions, we can make specific suggestions for improvements in their design and fabrication based on the SUMMiT™ technology [35,36].

Need for remote and noninvasive measurements in full-field-of-view providing data in three-dimensions and in real-time that optoelectronic methodology is capable of will be ever increasing as the emerging technologies (ET) evolve into mature technologies. This need will continue to be over multiscales ranging from milliscale to nanoscale and even down to picoscale as the "building blocks" out of which large structures will be made in the future, as the ET evolve with advances in Nanotechnology, will be shrinking in size. To be ready to satisfy the testing and characterization demands that ET will generate, development of metrology, specifically optoelectronic methodologies, should be continued.

ACKNOWLEDGMENTS

The micromirror devices used in this study were fabricated at and provided by Sandia National Laboratories. Sandia is a multiprogram laboratory operated by Sandia Corporation, a Lockheed Company, for the United States Department of Energy under Contract DE-AC04-94AL85000.

REFERENCES

1. D.L. Hetherington and J.J. Sniegowski, Improved polysilicon surface-micromachined micromirror device using chemical-mechanical polishing, Proc. Internat. SPIE Symp. on Optical Sci., Eng., and Instrumentat., San Diego, CA, 1998.
2. S.C. Gustafson, G.R. Little, D.M. Burns, V.M. Bright, and E.W. Watson, Microactuated mirrors for beam steering, Proc. SPIE, 3008, pp. 91–99 (1997).
3. M. Ikeda, H. Goto, H. Totani, M. Sakata, and T. Yada, Two-dimensional miniature optical scanning sensor with silicon micromachined scanning mirror, Proc. SPIE, 3008, pp. 111–122 (1997).
4. R.L. Clark, J.R. Karpinski, J.A. Hammer, R. Anderson, R. Lindsey, D. Brown, and P. Merritt, Micro-optoelectromechanical (MOEM) adaptive optic system, Proc. SPIE, 3008, pp. 12–24 (1997).
5. S. Kurth, R. Hahn, C. Kaufman, K. Keher, J. Mehnerm, U. Wollman, W. Dotzel, and T. Gessner, Silicon mirrors and micomirror arrays for spatial laser beam modulation, Sensors and Actuators, A66, pp. 76–82 (1998).
6. T. Gessner, W. Dotzel, D. Billlep, R. Hahn, C. Kaufmann, K. Kehr, C. Steiniger, and U. Wollman, Silicon mirror arrays fabricated using bulk- and surface-micromachining, Proc. SPIE, 3008, pp. 296–305 (1997).
7. J.B. Sampsell, Digital micromirror device and its application to projection displays, J. Vac. Sci. Technol., B12, pp. 3242–3246 (1994).
8. R.J. Pryputniewicz, A hybrid approach to deformation analysis, Proc. SPIE, 2342, pp. 282–296 (1994).
9. C. Furlong and R.J. Pryputniewicz, Hybrid computational and experimental approach for the study and optimization of mechanical components, Opt. Eng., 37, pp. 1448–1455 (1998).
10. C. Furlong, Hybrid, experimental and computational, approach for the efficient study and optimization of mechanical and electro-mechanical components, Ph.D. Dissertation, Worcester Polytechnic Institute, Worcester, MA, 1999.
11. D.R. Pryputniewicz, ACES approach to the development of microcomponents, MS Thesis, Worcester Polytechnic Institute, Worcester, MA, 1997.
12. R.J. Pryputniewicz, P. Galambos, G.C. Brown, C. Furlong, and E.J. Pryputniewicz, ACES characterization of surface micromachined microfluidic devices, Internat. J. of Microelectronics and Electronic Packaging (IJMEP), 24, pp. 30–36 (2001).

13. D.R. Pryputniewicz, C. Furlong, and R.J. Pryputniewicz, ACES approach to the study of material properties of MEMS, Proc. Internat. Symp. on MEMS: Mechanics and Measurements, Portland, OR, 2001, pp. 80–83.
14. P.J. Saggal, V. Steward, C. Furlong, and R.J. Pryputniewicz, Analytical and experimental study of dynamics of a MEMS accelerometer, MRS Proc. Nano- and Micro-Electromechanical Systems (NEMS and MEMS) and Molecular Machines, Boston, MA, 2002.
15. C. Furlong and R.J. Pryputniewicz, Computational and experimental approach to thermal management in microelectronics and packaging, J. Microelectronics Internat., 18, pp. 35–39 (2001).
16. G.C. Brown, Laser interferometric methodologies for characterizing static and dynamic behavior of MEMS, Ph.D. Dissertation, Worcester Polytechnic Institute, Worcester, MA, 1999.
17. R.J. Pryputniewicz, M.P. de Boer, and G.C. Brown, Advances in optical methodology for studies of dynamic characteristics of MEMS microengines rotating at high speeds, Proc. IX Internat. Congress on Exp. Mech., SEM, Bethel, CT, 2000, pp. 1009–1012.
18. M. Rogers and J.J. Sniegowski, 5-level polysilicon surface micromachining technology: application to complex mechanical systems, Tech. Digest of the Solid State Sensor and Actuator Workshop, Hilton Head Island, SC, 1998.
19. E.J. Garcia and J.J. Sniegowski, Surface micromachined microengine, Sensors and Actuators, A48, pp. 203–214 (1995).
20. M.S. Rogers, S.L. Miller, J.J. Sniegowski, and G.F. LaVigne, Designing and operating electrostatically driven microengines, Proc. 44th Internat. Instrumentation Symp., Reno, NV, 1998, pp. 56–65.
21. V.R. Yarberry, Meeting the MEMS "design-to-analysis" challenge: the SUMMiT™V design tool environment, Paper No. IMECE2002–39205, Am. Soc. Mech. Eng., New York, NY, 2002.
22. E.J. Pryputniewicz, ACES approach to the study of electrostatically driven MEMS microengines, MS Thesis, Worcester Polytechnic Institute, Worcester, MA, 2000.
23. E.J. Pryputniewicz, S.L. Miller, M.P. de Boer, G.C. Brown, R.R. Biederman, and R.J. Pryputniewicz, Experimental and analytical characterization of dynamic effects in electrostatic microengines, Proc. Internat. Symp. on Microscale Systems, Orlando, FL, 2000, pp. 80–83.
24. R.J. Pryputniewicz, Integrated approach to teaching of design, analysis, and characterization in micromechatronics, Paper No. IMECE2000/DE-13, Am. Soc. Mech. Eng., New York, NY, 2000.
25. A.J. Przekwas, M. Turowski, M. Furmanczyk, A. Hieke, and R. J. Pryputniewicz, Multiphysics design and simulation environment for microelectromechanical systems, Proc. Internat. Symp. on MEMS: Mechanics and Measurements, Portland, OR, 2001, pp. 84–89.
26. C. Furlong and R.J. Pryputniewicz, Characterization of shape and deformation of MEMS by quantitative optoelectronic metrology techniques, Proc. SPIE, 4778, pp. 1–10 (2002).
27. R.J. Pryputniewicz, Quantitative determination of displacements and strains from holograms, in Holographic Interferometry, Vol. 68 of Springer Series in Sciences, Springer-Verlag, Berlin, 1995, Ch. 3, pp. 33–72.
28. R.J. Pryputniewicz, Engineering Experimentation, Worcester Polytechnic Institute, Worcester, MA, 1993.
29. R.J. Pryputniewicz, High precision hologrammetry, Internat. Arch. Photogramm., 24, pp. 377–386 (1981).
30. R.J. Pryputniewicz, Hologram interferometry from silver halide to silicon and ... beyond, Proc. SPIE, 2545, pp. 405–427 (1995).
31. C. Furlong and R.J. Pryputniewicz, Absolute shape measurements using high-resolution optoelectronic holography methods, Opt. Eng., 39, pp. 216–223 (2000).
32. R.J. Pryputniewicz, P. Hefti, A.R. Klempner, R.T. Marinis, and C. Furlong, Hybrid methodology for the development of MEMS, J. Strain Analysis for Engineering Design, 41, pp. 708–718 (2006).
33. C. Brown and R.J. Pryputniewicz, Holographic microscope for measuring displacements of vibrating microbeams using time-average electro-optic holography, Opt. Eng., 37, pp. 1398–1405 (1998).
34. C. Furlong, J.S. Yokum, and R.J. Pryputniewicz, Sensitivity, accuracy, and precision issues in optoelectronic holography based on fiber optics and high-spatial and high-digital resolution cameras, Proc. SPIE, 4778, pp. 216–223 (2002).
35. R.J. Pryputniewicz, MEMS design education by case studies, Paper No. IMECE2001/DE-23292, Am. Soc. Mech. Eng., New York, NY, 2001.
36. R.J. Pryputniewicz, E. Shepherd, J.J. Allen, and C. Furlong, University—National Laboratory alliance for MEMS education, Proc. 4th Internat. Symp. on MEMS and Nanotechnology (4th-ISMAN), Charlotte, NC, 2003, pp. 364–371.

13

Durability of Optical Nanostructures: Laser Diode Structures and Packages, A Case Study

Ajay P. Malshe[a] and Jay Narayan[b]

[a] Department of Mechanical Engineering, MEEG 204, University of Arkansas, Fayetteville, AR 72701, USA
[b] Department of Materials Science and Engineering, North Carolina State University, Raleigh, NC 21695-7907, USA

Abstract   Durability is a synergistic reliable response of subsystems in integrated (packaged) systems, which in this case under discussion are nanostructured integrated optical systems. Understanding science and engineering aspects of these optical nanostructures integrated systems through design, fabrication, packaging and reliability testing are of paramount importance to obtain durable optical nanostructured packaged systems. To communicate specific aspects, this chapter addresses durability of optical quantum structures through carefully selected case studies in two parts. The part one includes novel design and deposition of quantum structures and the part two includes discussion of reliability of packaged quantum layered laser diode structures.

In the first case study, it is demonstrated that, $In_xGa_{(1-x)}N$ based multiquantum well (MQW) light emitting diodes and lasers (LEDs and LDs) have been fabricated and it is shown that high optical efficiency in these devices is related to thickness variation (TV) of $In_xGa_{(1-x)}N$ active layers. The thickness variation of active layers is found to be as important as In composition fluctuation in quantum confinement of excitons (carriers) in these devices. In this work, MQW $In_xGa_{(1-x)}N$ layers are produced with a periodic thickness variation, which results in periodic fluctuation of bandgap for the quantum confinement of excitons. Detailed STEM-Z contrast analysis, where image contrast is proportional to Z^2 (Z = atomic number), was carried out to investigate the spatial distribution of In. It is discovered that there is periodic variation in thickness of $In_xGa_{(1-x)}N$ layers with two periods, one short range (SR-TV, 30 to 40 Å) and other long-range thickness variation (LR-TV, 500 to 1000 Å). It is envisaged that LR-TV is the key to quantum confinement of the carriers and enhancing the optical efficiency and at the same time offering excellent reliability. The SR-TV is caused by In composition fluctuation. It was also found that the variation in In concentration is considerably less in the LED and LD structures which exhibit high optical efficiency. A comparative microstructural study between high and low optical efficiency MQW structures is presented to show that thickness variation (SR-TV) of $In_xGa_{(1-x)}N$ active layers is the key to their enhancement in optical efficiency.

Once quantum structures are engineered and devices are fabricated packaging and its reliability become important. Hence, in the second case study presented, authors discuss the laser diode package reliability for as applications continue to demand increasingly higher optical output power and longer lifetime, thermo-mechanical stresses on dissimilar materials interfaced for packaging pose an ever-growing challenge for the realization of a durable system. Particularly important for an epitaxy-down configuration is the die-attachment interface, which is desired to be defect-free and stress-managed for reliable optical alignment. A knowledge of the changes in the physical defect density and magnitude of the thermo-mechanical stress present in the active region as a function of the fabrication process and aging is crucial to an understanding of the influence of the process parameters and operating conditions on device performance and reliability. In this case study, we discuss investigation of high power laser diode array packages aged under various conditions. Microscopic defect analyses of the die attachment interface and device stress were carried out using primarily metallography, scanning electron microscopy (SEM), scanning acoustic microscopy (SAM), micro-hardness, and micro-Raman spectroscopy. It was noted that the intermetallic compounds and microscopic physical defects at the die attach interface are detrimental to transient heat transfer, and thus, overall package reliability. Using micro-Raman spectroscopy, we found that tensile stress near the bar-package interface increases with aging for the first few hundred hours and then decreases with further aging.

In conclusion, this chapter discusses synergistic engineering of nano structures along with micro interfaces in a macroscopic packaging which is essential for realizing a durable nanostructured integrated optical system.

13.1. HIGH EFFICIENCY QUANTUM CONFINED (NANOSTRUCTURED) III-NITRIDE BASED LIGHT EMITTING DIODES AND LASERS

13.1.1. Introduction[*]

The III-nitrides and their alloys have assumed a special importance due to their tremendous potential for fabricating the light emitting diodes and lasers (LEDs and LDs) operating in the red to ultraviolet (UV) energy range. The active layer in these devices with a composition of $In_xGa_{(1-x)}N$ has been the key to obtaining a high optical efficiency. The alloying with In is considered to be important, however, its role has not been clarified. Some studies have suggested In composition fluctuation leading to a phase separation to be responsible for high optical efficiency. Since the In content controls the bandgap in $In_xGa_{(1-x)}N$ alloys, it is envisaged that the composition fluctuation leads to quantum confined (QC) regions whose size is smaller than the dislocation separation (DS). These QC regions trap the bound excitons which recombine to produce photons, and thus recombination of excitons is not affected by the presence defects such as dislocations.

The evidence for indium composition fluctuation in $In_xGa_{(1-x)}N$ layers in MQW structures has been largely circumstantial. There is some evidence for phase separation (indium rich and indium poor phases) in $In_xGa_{(1-x)}N$ ($x > 0.3$) only in relatively thick layers (300–400 nm), which were grown by ECR-MBE. However, no phase separation is observed in $GaN/In_xGa_{(1-x)}N/GaN$ double heterostructures with $x > 0.3$, grown under similar conditions. Similarly, in MOCVD grown samples, phase separation has been reported only in thick $In_xGa_{(1-x)}N$ layers for $x > 0.28$. In the case of multiple-quantum-well structures (sapphire/4000 nm GaN:Si/10 period 2 nm InGaN/4 nm GaN/200 nm GaN:Mg),

---

[*]J. Narayan et al., Appl. Phys. Lett. 81, 841 (2002); U.S. Patent #US 6,881,983 B2, April 19, 2005, and references there in.

phase separation was observed only after prolonged (40 h) post annealing above 950°C. Some authors have claimed the formation of indium-rich and indium-poor regions (2–5 nm size) in MOCVD grown MQW structures using diffraction contrast transmission electron microscopy (TEM) techniques. Since the image contrast in these techniques is sensitive to diffraction of atomic planes, these observations do not provide reliable information on composition fluctuation.

The LEDs and LDs based upon $In_xGa_{(1-x)}N$ multiquantum well (MQW) structures, which exhibit high optical efficiency, are found to show subband emissions. These emissions have been explained on the basis of In composition fluctuation in $In_xGa_{(1-x)}N$ layers. The energy separation of each subband emission in these samples is typically about 2 meV which is much smaller than expected for transition between $n = 1$ and $n = 2$ energy levels in 2–5 nm quantum dots. Thus, there is a urgent need to clarify the role of In composition fluctuation or any other effects leading to the formation of quantum confined regions. In this study, we have used high-resolution TEM and STEM-Z contrast techniques to investigate the In composition fluctuation and thickness variation and correlate them with optical efficiencies.

The $In_xGa_{(1-x)}N$ multiquantum well structures (10 period $In_xGa_{(1-x)}N/GaN//$ 2/10 nm) were grown by the MOCVD technique at a temperature of 800°C and the GaN capping layer at 950°C for less than 30 min. The details of growth of these structures are reported elsewhere. These wafers were used to prepare cross-section specimens by a standard ion milling procedure. A special care was necessary to used in terms of low-temperature, low voltage and shallow angle thinning to minimize the surface damage for STEM-Z studies. For STEM-Z (scanning transmission electron microscopy-atomic number contrast studies, we used atomic resolution JEOL 2010 field emission electron microscope with GIF (Gatan Image Filter attachment. In the STEM-Z mode, a small electron probe (1.6 nm) is scanned across the thin cross-section specimen and the Z-contrast image results from mapping the intensity of electrons reaching the annular detector. The detector performs the function of Lord Rayleigh's condenser lens. It enforces high scattering angles, so that Rutherford scattering dominates and atoms contribute to the image with a brightness determined by their mean square atomic number (Z) and with a resolution of the probe size (1.6 nm). Since the atomic number of In (49) is much higher than that of Ga (31), the image contrast is dictated by In concentration.

Thus thickness variations can result in the formation of QC (quantum confined) regions. The InGaN/GaN quantum-well structure of high-efficiency LEDs is shown in Figure 13.1. The details of quantum-well structure and associated thickness variation to produce quantum-confined (QC) regions are shown in Figures 13.2 and 13.3 as a cross-section STEM-Z contrast image of $In_{0.2}Ga_{0.8}N/GaN$ layers are shown in Figures 13.2 and 13.3. In the STEM-Z images, the contrast is proportional to Z^2 (Z = atomic number). Our contrast analysis reveals that the variation in In concentration is not as significant, and that the enhanced efficiency results from the thickness variation. The $In_{0.2}Ga_{0.8}N$ layers with similar composition but with uniform thickness resulted in considerably less optical efficiency.

The contrast due to indium is enhanced by two and half times compared to the gallium concentration. The SR-TV (short range thickness variation) period in Figures 13.2 and 13.3 is estimated to be 30 to 40 Å. Figures 13.2 and 13.3 show a STEM-Z (transmission electron microscopy) micrographs from specimens, which consistently showed higher optical efficiencies. These specimens showed a short range (30 to 40 Å period) and a long range (500 to 1000 Å period) thickness variation. In contrast to the high optical efficiency from specimens (shown in Figures 13.1– 13.3), the specimens with relative low optical efficiencies are shown in Figure 13.4. In these specimens, where optical efficiencies are lower

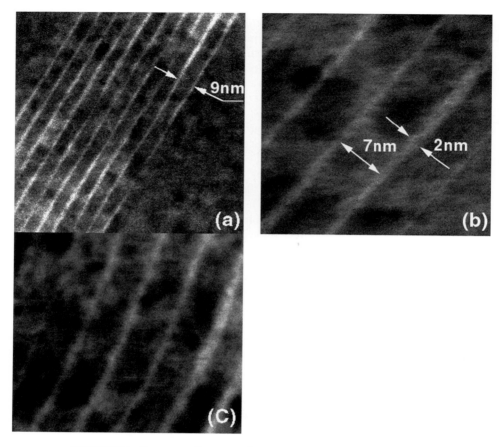

FIGURE 13.1. Quantum well structure of novel high-efficiency InGaN/GaN LEDs.

as much as a factor of two and three, superlattice thickness as well as indium concentration are quite uniform. Thus our experimental results on comparative study of high- and low-efficiency LEDs and LDs (as shown in Figure 13.5) clearly demonstrate that thickness variation coupled with indium concentration variation is the key to enhancing the optical efficiencies in LEDs and LDs.

In these studies, it is envisaged that the In composition fluctuation results in quantum-dot like structures from which subband emission occurs. The quantum well trap excitons whose radiative recombination is responsible for efficient spontaneous emission in MWQ LEDS and LDs. High optical efficiency results despite high dislocation density ($\sim 10^{10}$ cm^{-2}) because the localization of excitons is within a region less than the dislocation separation (DS) in these structures. The DS is given by $\rho^{-1/2}$, where ρ is the density of dislocations (number/cm^2). Thus, the loss of excitons due to nonradiative recombination at the dislocations is avoided resulting in high optical efficiency of LEDs and LDs. In our investigation, we have produced periodic thickness variations (short-range, ST-TV; and log-range, LR-TV) which result in the formation of QC regions for the excitons which recombine without being affected by the presence of dislocations. Our detailed STEM-Z contrast analysis shows that thickness of $In_xGa_{(1-x)}N$ layers are equally important as In composition fluctuation in producing quantum confined regions for excitons leading to en-

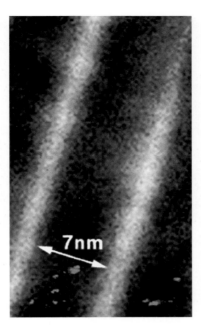

FIGURE 13.2. Thickness variation in InGaN/GaN quantum wells to confine the carriers.

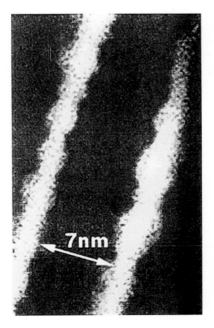

FIGURE 13.3. Another example of high resolution (STEM-Z) micrograph showing details of quantum confinement of carriers in InGaN nanopockets.

hanced optical efficiency of LEDs and LDs. In the previous studies, the fluctuation in In concentration $In_x Ga_{(1-x)} N$ layers has been investigated by cross-section TEM (using con-

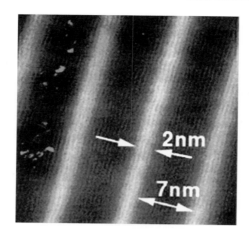

FIGURE 13.4. Uniform structure of InGaN/GaN quantum wells.

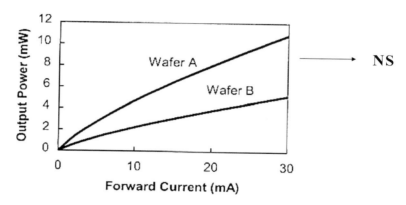

FIGURE 13.5. Comparison of LED efficiencies: (A) nonuniform quantum wells; (B) uniform quantum wells.

ventional phase contrast transmission electron microscopy), photoluminescence (PL) and Raman spectroscopy techniques. From these studies, the size of these regions was estimated to be 2–5 nm. The time-resolved photoluminescence spectroscopy and electroreflectance studies have revealed that a small In addition the GaN active layers plays a key role in suppressing the nonradiative recombination processes. The Raman Stokes shift between the exciting and emission in the range of 100 to 250 meV at RT was attributed to energy depth of localized states of the carriers (excitons) in the $In_x Ga_{(1-x)}N$ layers.

The change in bandgap of $In_x Ga_{(1-x)}N$ alloys can occur as function of the composition "x" and the thickness "L_z" of the superlattice. For a typical active layer composition ($x = 0.2$) the change in bandgap is estimated to be as follows: $x = 0.2$, bandgap $= 3.0$ eV; $x = 0.1$, bandgap $= 3.2$ eV; $x = 0.3$, bandgap $= 2.8$ eV. This amounts to a 50% change (from $x = 0.2$) in active layer composition. Experimentally observed composition fluctuations are less than 10% which should lead to a less than 0.07 eV change in the bandgap. (For $x = 0.35$, bandgap $= 2.70$ eV; $x = 0.40$, bandgap $= 2.60$ eV; $x = 0.45$, bandgap $= 2.55$ eV; $x = 0.50$, bandgap $= 2.47$ eV.) Since the bandgap is dictated by the thickness (L_z) via: $E_n = h^2 n^2 /(8m^* L_z^2)$, where E_n is the allowed energy level (n), h is Planck's constant, and m^* is effective mass. Thickness variations lead to changes in ΔE

proportional to L_z^{-2}. Thus, 10 to 20% in thickness variation can result in 20 to 40% change in ΔE (using $m_e^* = 0.11 m_0$ for InN, $m_e^* = 0.20 m_0$ for GaN, Refs. [8,9]). Experimentally observed thickness variations (long-range, LR-TV) are in the range of 10 to 50%, and short-range variations SR-TV are less than 10%.

It is proposed that the thickness variation is caused by two-dimensional strain in the $In_x Ga_{(1-x)}N$ layer below its critical thickness. Since strain energy increases with thickness, the uniform thickness breaks into a periodic variation by which the free energy of the system can be lowered. Since the strain also increases with In concentration, some fluctuation in In concentration is also expected. However, this phenomenon of thickness variation has been well documented for pure germanium thin film growth on (100) silicon below its critical thickness where no composition fluctuation is involved. We have modeled the thickness variation and derived the following relation for TV period (λ)

$$\lambda = \pi \gamma (1-\nu)/[2(1+\nu)^2 \mu \varepsilon^2], \tag{13.1}$$

where γ is the surface energy, ν is the Poisson's ratio, μ is the shear modulus of the film, and ε is the strain normal to the film surface. To avoid nonradiative recombination at the dislocations (density ρ), we derive the optimum structure to be

$$\rho^{-1/2} > \pi \gamma (1-\nu)/[2(1+\nu)^2 \mu \varepsilon^2] \text{ or } \rho < \{\pi \gamma (1-\nu)/[2(1+\nu)^2 \mu \varepsilon^2]\}^{-2}. \tag{13.2}$$

It is estimated that a typical value of λ using the following parameters for our growth conditions. For $In_{0.4}Ga_{0.6}N$, shear modulus is estimated to be 82 GPa, Poisson's ratio to be 0.3, surface energy 4000 ergs/cm^2, strain 2%, this results in λ of 793 Å or 800 Å, which is in good agreement with observed LR-TV.

The high efficiency MQW structured LEDs exhibit characteristic subband emission separated by 1–2 meV. In addition, cathodoluminescence measurements show Stokes like shift between the exciting and emission in the range of 100 to 250 meV. A spherical potential treatment suggested by Brus was used to make an estimate of the QC region corresponding to confinement energy (ΔE) expressed as $\pi^2 \hbar^2/(2m^* R^2)$, where m^* is the reduced effective mass of electron-hole pairs, $\hbar$ is the Planck's constant, and R is the dot radius. Using this model we estimated the transition energy ΔE to be 2 meV corresponding to the 50 nm quantum dot radius. This is consistent with experimentally observed LR-TV period of about 80 nm. Similarly, Stokes like shift in the CL measurements of 200 meV is expected to arise from the quantum confined region of 5 nm, which is closer to SR-TV of 3–5 nm observed in our STEM-Z contrast experiments. The indium composition fluctuation can also result from surface diffusion flux of vacancies and lead to short range thickness variation of the order of 2–5 nm.

In summary, $In_x Ga_{(1-x)}N$ based multiquantum well (MQW) light emitting diodes and lasers (LEDs and LDs) are fabricated and it is shown that high optical efficiency in these devices is related to thickness variation (TV) of $In_x Ga_{(1-x)}N$ active layers. It is discovered that there is a periodic variation in thickness of $In_x Ga_{(1-x)}N$ layers with two periods, one short range (SR-TV, 30 to 40 Å) and other long-range thickness variation (LR-TV, 500 to 1000 Å). It is envisaged that LR-TV, which may be related 2 meV subband emission is the key to quantum confinement of the carriers and enhancing the optical efficiency. It is envisaged that the SR-TV is caused by In composition fluctuation. It was also found that the variation in In concentration is considerably less in the LED and LD structures which exhibit high optical efficiency. The reliability in forming these nanostructures is the key to obtain sustained high-efficiency of LEDs and LDs.

13.2. INVESTIGATION OF RELIABILITY ISSUES IN HIGH POWER LASER DIODE BAR PACKAGES

13.2.1. Introduction

Increasing optical efficiency, new package designs, better optical coupling methods, a growing number of applications, and steadily declining prices have accelerated the transition of high power laser diodes from research and development into mainstream applications. As applications continue to demand increasingly higher optical output power and longer lifetime, thermo-mechanical stresses on dissimilar materials interfaced for packaging pose an ever-growing challenge to the realization of a durable system. Thus, it has become increasingly important to analyze the root causes of specific degradation modes at optimum laser operating conditions so as to manufacture reliable systems.

An edge emitting high power laser diode packaged system studied in this section is a combination of quantum-well laser diode arrays and multilayered integrated metalization schemes that combine dissimilar materials for die attachment on a copper heat sink, all functioning under high transient temperature conditions. Typically, failure of a packaged system is due to interrelated electro-thermo-mechanical-material reasons. For example, temperature cycling of a packaged laser bar attached using a soft solder results in creep and stress relaxation at the die attach interface causing mechanical deformation of the laser diode array. Such deformation causes variations in the optical emission across the bar. During product development and manufacturing, understanding the influence of packaging parameters and operating conditions on optical device performance, thermo-mechanical stresses on the device at the die attach interface, and physical defect density and microstructural changes in the die attachment material under continuously evolving/degrading conditions is crucial.

Over the years, many analytical and experimental studies have been performed to assess die-attachment joint integrity from a physics-of-failure perspective [1–12]. However, detailed analyses, correlating device performance and failure modes, packaging and operating conditions, and thermo-mechanical stresses and material behavior are necessary in order to realize reliable, application-specific, durable systems. Many inconsistencies result from a lack of knowledge of the unique properties of each solder, such as age and cycle softening, hardening due to intermetallic grain-growth, dynamic recrystallization, strain-rate hardening, superplasticity, etc. Many analytical inconsistencies are traced to differing interpretations of the effects of the temperature, the current, the cycle frequency, and the period [13].

Various kinds of problems are associated with thermo-mechanical stress, which appears in the active region of a device during its growth, packaging, and also, during its operation. Stress may directly trigger the nucleation and propagation of dislocations and the formation of voids and cracks. The absorption of photons by dislocations and the migration of carriers toward such dislocations in the active region of a diode result in additional thermo-mechanical stresses, and hence, can be a major reliability problem. Furthermore, even when stress is not severe enough to destroy the functionality of a diode array, its presence can influence routine performance by modifying the semiconductor band structure, which, in turn, affects the output wavelength, threshold current, and quantum efficiency. The problem becomes more acute with shrinking device size, increasing complexity of the integrated system, and increasing output power [14–23].

Consequently, the objective of this research work was to explore and understand various failure mechanisms affecting packaged high power laser diode bars under different testing/ operating conditions.

13.2.2. Preparation of Packaged Samples for Reliability Testing

The laser diode array packages investigated are continuous wave (CW) laser diode bars. A schematic diagram of a packaged bar is shown in Figures 13.6(a) and (b). The GaAs multiple quantum-well bars (19 emitters; 1 mm × 1 cm × 100 µm) were mounted on a polished copper heat sink in an *epitaxy side down* configuration using soft indium solder and Ti-Pt-Au interface metallurgy. A copper heat sink provided anode contact to the device, while a metal foil mounted on top of the bar provided cathode contact. Ridges of 150 µm widths were formed in the p-side of the bar. During die attachment reflow, a load of about 50 grams was applied in the direction normal to the plane of the bar. The reflow process was carried out in a vacuum furnace. The bars were then aged for 0, 96, 744, 1000 and 7000 hours at 40 A operating current for an output power of 30 W. Some bars succumbed to infant mortality, while most of the bars survived and operated satisfactorily for over 7000 hours.

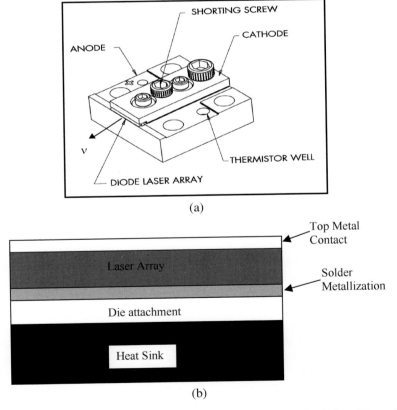

FIGURE 13.6. (a) Schematic view of the package. (b) Schematic cross-sectional view of the package.

In order to identify and understand the defects and related failure modes, samples were characterized using the following complementary analytical techniques: optical microscopy and scanning electron microscopy (SEM, Hitachi) to observe interface microstructure and micro-defects, scanning acoustic microscopy (SAM, Sonoscan) to explore void distribution at the device bar and die-attachment interface, energy dispersive spectroscopy (EDS, Kevex) for chemical analysis of the interface, and micro-hardness (Buehler) to test the mechanical response as a function of thermal degradation. The purpose of this analysis was to determine the relationship between material microstructure, micro-hardness, and chemical changes in the die-attachment as a function of aging of the package.

Thermo-mechanical stresses present in the active region of the device were measured using micro-Raman spectroscopy (Renishaw) by observing the shift of the characteristic GaAs peak (the stress measurement error was $\pm 5\%$) [24–29]. Micro-Raman measurements were performed using a 488 nm Ar^+ laser with a spatial resolution of about 1 µm. Observations were made at various locations along and across the active emitter regions of the laser bars. Precisely scanning of the quantum-well along the 1 cm width with a spatial resolution close to 1 µm was practically difficult. Thus, the package-aging induced stress profile was measured across the width of the active emitter region of the bar. Observations were made on the front [110] facet of the device. In this configuration, scattering from the transverse optical (TO) phonons was allowed in accordance with the symmetry selection rules [15]. The shift in the TO peaks was studied as a function of aging. Positive and negative peak shifts correspond to compressive and tensile stress in the GaAs material, respectively. Since the peak shifts are small, peak fitting was employed systematically. The Gaussian peak fit was used for the laser peak and mixed Gaussian and Lorenztian peak fits were used for the Raman peak. The change in the full width at half maximum (FWHM) of the Raman peak was measured and plotted as a function of spatial position [15]. Effects of laser heating of the sample, in our case, were minimal because of the confocal Renishaw Raman spectroscopy system, which facilitates measurements with low laser power at high speed, thereby reducing the laser heating effects. It is reported that lattice heating caused by a cw excitation laser does not exceed 10 K [18].

13.2.3. Finding and Model of Reliability Results

13.2.3.1. Physical Defect and Morphological Observations The following is a detailed discussion of the various micro-defect structures observed in the aged laser bars.

Non-uniform physical contact between the top cathode metal foil contact and the laser diode bar, the bar and the multilayer metalization, the metalization and the die-attachment, and the die-attachment and the copper heat sink were observed at various locations across the width of a bar. This non-uniform contact varied in morphology from delamination (Figure 13.7) to crack propagation in metalization (Figure 13.8) to cavitation at the ridges to crack propagation near the ridge structures. Figure 13.9 provides evidence that non-uniform physical contact, resulting from non-uniform microloading on the GaAs bar along the metalization and die-attach interface during the die-attachment process, particularly at the ridges, can cause cavitations and stress gradients. This results in cracking of the bar.

Coefficient of thermal expansion (CTE) mismatches between the metalization, die attachment, and copper heat sink are responsible for delaminations and cracks that run along the interface. Such non-uniform physical contact between a high power laser diode

DURABILITY OF OPTICAL NANOSTRUCTURES: LASER DIODE STRUCTURES AND PACKAGES 351

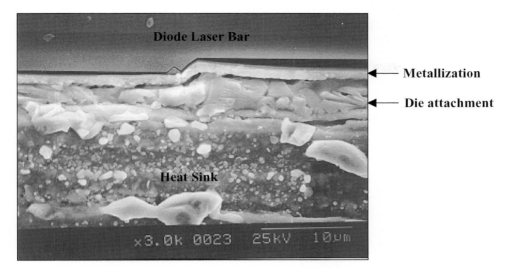

FIGURE 13.7. Delamination of metalization.

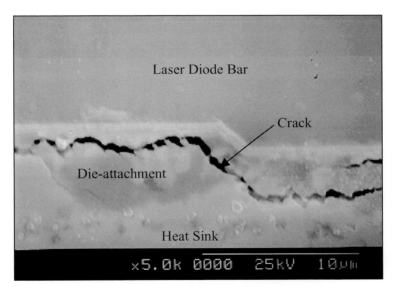

FIGURE 13.8. Crack running along the die-attachment.

bar and a copper heat sink causes excessive heating at the device junction, resulting in occasional burning at the bar emitter surface.

Figure 13.10 shows the output of the laser emission analyzer for a mounted bar, aged for 1000 hrs. Physical bending/displacement of the bar can be clearly seen. This is a representative case of various observations on different samples for which random bending, along with non-uniform optical emission from the bars across their widths, was observed. This problem, commonly known as the "smile problem," was observed to initiate during the die attachment process.

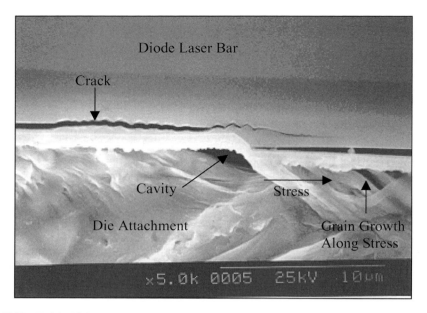

FIGURE 13.9. Higher concentration of physical defects and stress induced grain growth at ridges.

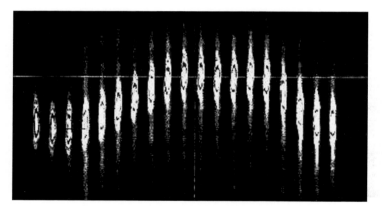

FIGURE 13.10. Beamview analyzer picture of the "Smile Problem" (cross-sectional view).

Interestingly, dynamic recrystallization was observed to present in the die-attachment at the interface, particularly in the 7000 hours aged samples. Since the heat generated is concentrated in narrow regions, significant thermal gradients contribute to the nucleation and grain boundary motion required for dynamic crystallization. Figure 13.9 shows grain growth in the direction of the shear stress. The crack seen in the bar is due to excessive stress.

Figures 13.11(a) and 13.11(b) show typical distributions of voids in a good and a defective bar, respectively. It was observed that voids present in the die-attachment region are distributed randomly, but typically, void density is relatively higher at the center of the bars (along length). The presence of voids directly under the facet region was observed to be responsible for the emitter burn spots, and occasionally, for vertical cracking of the bar.

DURABILITY OF OPTICAL NANOSTRUCTURES: LASER DIODE STRUCTURES AND PACKAGES

FIGURE 13.11a. SAM picture of randomly distributed (light regions) lesser voids in solder region in a good bar.

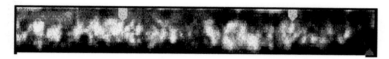

FIGURE 13.11b. SAM picture of randomly distributed (light regions) larger voids in solder region in a defected bar.

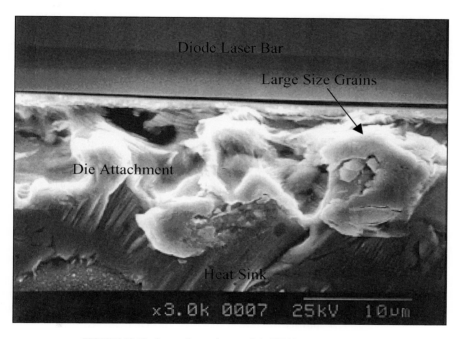

FIGURE 13.12. Large size grain growth in 7000 hours aged sample.

After an important synchronized set of observations, we conclude that both the increase in void density and the occurrence of delamination at the bar and die attachment interface increase with aging. Presence of voids is detrimental since they are the major barriers for efficient heat transfer and can cause accelerated failure of the packaged device. However, after the first few hundred hours of aging, few regions in the die-attachment appeared stable. We believe that this may be due to the plastic flow of die-attachment as a result of local heating.

Figure 13.12 shows the large grain growth for a sample aged for 7000 hours. The grain size at the interface of the copper (Cu) and indium (In) increased with aging time. We suggest that excessive heating due to insufficient physical contact between the device and heat sink gives rise to grain growth at the interface as a function of aging time.

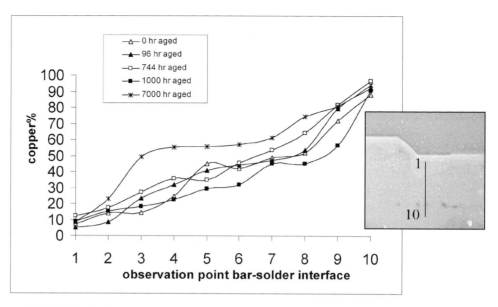

FIGURE 13.13. Copper concentration is highest near die-attach interface for 7000 hours aged sample.

As expected, physical defects were observed more frequently at the ends of the laser diode bars (along width), owing to higher stresses near the ends of the bars. Further, out of total 40 samples, in more than 90% of the samples it was observed that the die-attachment layer and the heat sink front surface were misaligned up to few microns along the length. This misalignment varied from sample-to-sample, and along the width of a bar.

13.2.3.2. Chemical and hardness observations   Figure 13.13 provides energy dispersive X-ray analysis (EDS) data for aged samples. The measurements were performed on the die-attachment region. Observation points 1–10 on the x-axis represent sampling points where point 1 is nearest to the device-metalization interface and point 10 is close to the copper heat sink, within approximately 0.5 μm. The measurements were performed on various emitter regions across the bar to collect better statistics. We observed copper diffusion from the copper heat sink into the indium solder. We further observed that the diffused copper concentration and profile varies along the width of the bar (data not shown). Based on the Cu-In phase diagram and the previously discussed grain growth, we conclude that, at the interface, there is intermetallic formation, which is known to be hard and brittle and can contribute to crack development under stress conditions. We also observed that copper diffusion occurs at higher rates near the metalization-die attach interface during the initial hours of aging, and that the change is relatively small, though gradual, over the remaining aging period. A similar study was performed for gold diffusion into the die-attachment region. Gold diffusion into die-attachment was also observed to increase with aging (data not shown). However, due to the small thickness of the die-attach layer, X-ray diffraction (XRD) could not be employed to identify the various intermetallic phases in the die-attachment layer.

To confirm the increase in hardness caused by the intermetallic formation, we performed microhardness analysis on the interface metallurgy. Table 13.1 gives Knoop microhardness data, measured at 50 g load, as a function of aging. The hardness of the die-

DURABILITY OF OPTICAL NANOSTRUCTURES: LASER DIODE STRUCTURES AND PACKAGES   355

TABLE 13.1.
Knoop micro-hardness as a function of aging.

Sample description	Average Knoop micro-hardness (at 50 g)
0 hr aged	157
96 hr aged	160
744 hr aged	196
1000 hr aged	203
7000 hr aged	214

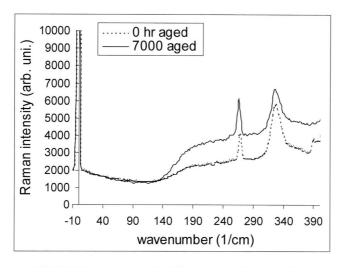

FIGURE 13.14. Typical profile of Raman peak shift of laser diode.

attachment region increased with aging time, which agrees with the observation of intermetallic phase formation as discussed previously.

13.2.3.3. Micro-Raman Observations   Figure 13.14 shows a typical Raman spectrum in the active region of a high power laser diode studied in the present work. Micro-Raman measurements were performed on the optically sensitive GaAs device, across the width of the bar, at about 1 μm spacing. Sampling points 1 to 6 on the x-axis of the plot in Figures 13.15 and 13.16 represent measurement locations across the width of the front face of the active region of the GaAs bar. Point 1 is the point nearest the device-heat sink interface and point 6 is the farthest away. For analysis, we used the GaAs TO Raman peak at 269.3 cm^{-1} [16].

Figure 13.15 is a typical graph of micro-Raman shift and stress profile as a function of aging time. The graph clearly shows a shift in the peak, particularly toward lower wavenumbers for regions of the device near the die-heat sink interface, and high wavenumbers for the region away from the interface. The negative and positive shifts are a result of tensile and compressive stresses, respectively, in the active region of the laser diode array. Using the known pressure dependence of the TO-mode shift (0.5 cm^{-1}/100 MPa), these experimental mode shift profiles can be converted to stress profiles, and this is shown in Figure 13.15 [15,18]. A maximum tensile stress of ∼800 MPa and a compressive stress

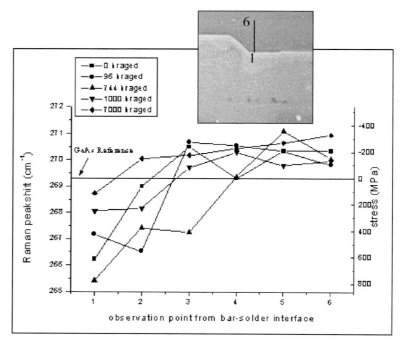

FIGURE 13.15. A typical profile of Raman peak shift and stress as a function of distance from bar–solder interface.

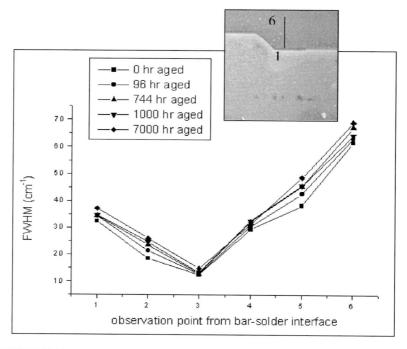

FIGURE 13.16. A typical graph of FWHM as a function of distance from bar–solder interface.

of ~420 MPa are observed. Two stress zones, one under compressive and the other under tensile stress, can be seen. The zone near the GaAs bar-heat sink interface is under tensile stress, while the zone near the quantum-well active layer is under compressive stress. As a function of aging there is measurable change in the tensile stress value, unlike little or no change in the compressive stress value near active layer. This significant change in the tensile stress value has clearly affected reliability of the packaged bar. Further the randomness the variation, we believe is caused by the multiple defects that can arise during aging and needs more detail investigation. Also, we believe that, under the influence of thermal cycles during laser diode operation, the hardness of the die-attachment increases, which causes the tensile stress to decrease [30].

The FWHM of the GaAs peak is an indicator of the amount of disorder present in the lattice. The disorder can be due to rearrangement of atoms under the influence of high temperatures and increasing threshold current. Hence, the magnitude of the FWHM is related to lattice degradation caused by thermo-mechanical stress. From Figure 13.16, we can see that the FWHM is high (~ 35 cm^{-1}) at the bar-die-attachment interface, decreases as we move further into the bar, and is smallest (~ 12 cm^{-1}) near the active layer, which is the quantum-well structure. We can infer that the quantum-well structure is more robust. This observation is also in contrast to the usual notion that in the quantum-well structure, where maximum intermixing is present, disorder, and hence, FWHM should be maximum [27,31,32]. We believe that the lower FWHM at the quantum well region is an artifact of confinement. It is observed that, although the stress decreases as we go from bar-die-attachment interface into the substrate, the FWHM first decreases and then increases. This observation is surprising since it is contrary to the commonly held belief that aging related defects in the crystal are predominantly observed near the bar-package interface and decrease into the substrate. Although the change is small, it is clear from Figure 13.16 that the FWHM, and hence, the amount of disorder in the lattice increases with aging.

It is important to note that both the stress-induced peak shift and the FWHM increase as we approach the ridge edges from both sides. The camel hump-like stress profile, which is characteristic of ridge lasers, was not observed [15,18]. This is due to a very high width to height ratio of the emitter. It is also worth noting that the shape and strength of the stress profile curve and the FWHM change from one emitter to another in an array.

13.3. CONCLUSIONS

In summary, it is demonstrated that, in high power laser diode systems, various materials, optical, mechanical, and thermal parameters are inter-related and form the basis of a synergistic analysis of the failure modes. Complementary analytical techniques have been applied to evaluate various unique nanostructures, physical defects at the die attach-device interface and aging-induced grain growth and change in the chemistry of the interface metallurgy. It is demonstrated that both physical defects and microstructural changes in the die-attach, during device operation, affect the reliability of packaged high power laser bars. It is observed significant copper diffusion into the die attach layer, which caused the formation of hard and brittle intermetallics. Further, it is concluded that stress relief during aging and excess heating caused by inadequate physical contact between the bar and the heat sink result in uneven bending of the bar causing non-uniform optical alignment. It was demonstrated that micro-Raman spectroscopy is a very useful tool for studying packaging-induced local mechanical stress. Although there is certain failure related randomness in the interface and active regions, a definite failure trend exists as the package undergoes aging.

Physical defects, optical response, and chemical composition analysis, along with micro-hardness measurements, provide valuable insight into packaging and operation-induced defects affecting the reliability of high power laser diodes. It is believed that the results of the work discussed here should ultimately provide assistance in predicting packaging failure modes that will lead to better processing and packaging schemes in the future. The ability to map the distribution of stress within the device and then correlate the resulting data with points of failure should lead to more reliable design and manufacturing.

A more fundamental study of the evolution of die-attachment metallurgy, in which intermetallic formation, together with softening effects and the finite size under the thermal fatigue conditions are considered, is essential. Such a study should provide information that lead to a much better understanding of the random and unpredictable behavior of some of the failure mechanisms observed in this work. In addition, *in situ* micro-Raman analysis, along with photoluminescence measurements, and numerical modeling during the aging process are essential to a better understanding of the gradual changes in the stresses as a function of aging time.

ACKNOWLEDGMENTS

One of the authors acknowledge Ajit Dhamdhare (previously at the University of Arkansas) for his MS contributions, and Dr. John Nightingale, Mr. Robert Miller, and Mr. John Morales of Coherent, Inc. for valuable technical discussions. We he wishes to acknowledge Dr. Richard Bormett and Ms. Diane Allen of Renishaw Inc. and Dr. John Shultz of Arkansas Analytical Laboratory for micro-Raman and other analytical support work.

REFERENCES

1. C.E. Ho, W.T. Chen, and C.R. Kao, Interactions between solder and metallization during long-term aging of advanced microelectronic packages, Journal of Electronic Materials, 23(7), pp. 379–385 (2001).
2. P.L. Tu, Y.C. Chan, and J.K.L. Lai, Effect of intermetallic compounds on the thermal fatigue of surface mount solder joints, IEEE Transactions on Components Packaging and Manufacturing Technology, Part B, 20(1), pp. 87–92 (1997).
3. S.A. Merritt, P.J.S. Heim, S.H. Cho, and M. Dagenais, Contrlled solder interdiffusion for high power semiconductor laser diode die bonding, IEEE Transactions on Components Packaging and Manufacturing Technology, Part B, 20(2), pp. 141–145 (1997).
4. J.M. Parsey, S. Valocchi, W. Cronin, J. Mohr, B.L. Scrivner, and K. Kyler, A metallurgical assessment of SnPbAg solder for GaAs power devices, Journal of Materials, 31(3), pp. 28–31 (1999).
5. W.L. Phillipson, and D.J. Diaz, Microstructural examination and failure analysis of microelectronic components, Materials Developments in Microelectronics Packaging Conference Proceedings, 1991, pp. 289–297.
6. A. Zubelewicz, R. Berriche, L.M. Keer, and M.E. Fine, Lifetime prediction of solder materials, Journal of Electronic Packaging, 111, pp. 179–182 (1989).
7. N. Strifas, and A. Christou, Die attach adhesion and void formation at the GaAs substrate interface, Mat. Res. Soc. Symp. Proc., 356, pp. 869–874 (1995).
8. N. Zhu, Thermal impact of solder voids in the electronic packaging of power devices, IEEE 15th SEMI-THERM Symposium, 1999, pp. 22–29.
9. W. Nakwaski, An additional temperature increase within GaAs/(AlGa)As diode lasers caused by the deterioration of an indium solder, Electron Technology, 23(1/4), pp. 33–38 (1990).
10. V.D. Bo, F. Bartels, A. Scandurra, Ch. Luchinger, and S. Radeck, More insights into the soft die attach, 11th European Microelectronics Conference, 1997, pp. 308–319.
11. Q. Tan and Y.C. Lee, Soldering technology for optoelectronic packaging, IEEE Electronic Components and Technology Conference, 1996, pp. 26–36.

12. A.Y. Kuo and K.L. Chen, Effect of thickness on thermal stresses in a thin solder or adhesive layer, Conference Proceedings of ASME Winter Annual Meeting, 1991, pp. 1–6.
13. L. Wen, G.R. Mon, and R.G. Ross, Design and reliability of solders and solder interconnections, Symposium Proceedings of TMS Annual Meeting, 1997, pp. 219–226.
14. M.E. Polyakov, Mechanical stresses in AlGaAs/GaAs quantum-well heterolasers, Sov. J. Quantum Electronics, 19(1), pp. 26–29 (1989).
15. P.W. Epperlein, Temperature, stress, disorder and crystallization effects in laser diodes: measurements and impacts, Proc. SPIE, 3001, pp. 13–28 (1997).
16. P. Puech, G. Landa, R. Carles, and C.J. Fontaine, Strain effects on optical phonons in <111> GaAs layers analyzed by Raman scattering, J. Appl. Phys., 82(9), pp. 4493–4499 (1997).
17. E. Anastassakis, Stress measurements using Raman scattering, Analytical Techniques for Semiconductor Materials and Process Characterization: Proceedings of the Satellite Symposium to ESSDERC 1989, Berlin, 1989, pp. 298–326.
18. P.W. Epperlein, G. Hunziker, K. Datwyler, U. Deutsch, H.P. Dietrich, and D.J. Webb, Mechanical stress in AlGaAs ridge lasers: its measurement and effect on the optical near field, Proceedings of the 21st International symposium on Compound Semiconductors, Vol. 21, 1994, pp. 483–488.
19. R. Puchert, J.W. Tomm, A. Jaeger, A. Barwolff, J. Luft, and W. Spath, Emitter failure and thermal facet load in high power laser diode arrays, Appl. Phys. A, 66, pp. 483–486 (1998).
20. I. De Wolf, J. Chen, M. Rasras, W.M. van Spengen, and V. Simons, High-resolution stress and temperature measurements in semiconductor devices using micro-Raman spectroscopy, Proc. SPIE, 3897, pp. 239–252 (1999).
21. I. DeWolf and H.E. Maes, Mechanical stress measurements using micro-Raman spectroscopy, Microsystems Technologies, 5, pp. 13–17 (1998).
22. A. Barwolff, J.W. Tomm, R. Muller, S. Weiss, M. Hutter, H. Oppermann, and H. Reichl, Spectroscopic measurement of mounting-induced strain in optoelectronic devices, IEEE Transactions on Advanced Packaging, 23(2), pp. 170–175 (2000).
23. J.W. Tomm, R. Muller, A. Barwolff, T. Elsaesser, A. Gerhardt, J. Donecker, D. Lorenzen, J. Daiminger, S. Weiss, M. Hutter, E. Kaulfersch, and H. Reichl, Spectroscopic measurement of packaging-induced strains in quantum well laser diodes, J. Appl. Phys., 86(3), pp. 1196–1201 (1999).
24. D. Wood, G. Cooper, D.J. Gardiner, and M. Bowden, Raman spectroscopy as a mapping tool for localized strain in microelectronics structures, Journal of Materials Science Letters, 16(14), pp. 1222–1223 (1997).
25. P.W. Epperlein, Raman spectroscopy of semiconductor lasers, Proceedings of Conference on Lasers and Electro-Optics, Vol. 9, 1996, pp. 108–109.
26. B. Dietrich and K.F. Dombrowski, Experimental challenges of stress measurements with resonant micro-Raman spectroscopy, Journal of Raman Spectroscopy, 30, pp. 893–897 (1999).
27. P.S. Pizani, F. Lanciotti, Jr., R.G. Jasinevicius, J.G. Duduch, and A.J.V. Porto, Raman characterization of structural disorder and residual strains in micromachined GaAs, Journal of Applied Physics, 87(3), pp. 1280–1283 (2000).
28. K. Iizuka, T. Yoshida, I. Matsuda, H. Hirose, and T. Suzuki, Micro-Raman study of the residual stress in molecular-beam-epitaxy-grown $Al_xGa_{1-x}As$/GaAs multilayer structures, Materials Science and Engineering, B5, pp. 261–264 (1990).
29. J.P. Landesman, A. Flore, J. Nagle, V. Berger, E. Rosencher, and P. Puech, Local stress measurements in laterally oxidized GaAs/Al_xGa_{1-x}As heterostructure by micro-Raman spectroscopy, Appl. Phys. Lett., 71(17), pp. 2520–2522 (1997).
30. R.R. Varma, Bonding induced stress in semiconductor laser, Proceedings of IEEE Electronic Components and Technology Conference, 1993, pp. 482–484.
31. A.S. Helmy, A.C. Bryce, C.N. Ironside, J.S. Aitchison, and J.H. Marsh, Raman spectroscopy for characterizing compositional intermixing in GaAs/AlGaAs heterostructures, Appl. Phys. Lett., 74(26), pp. 3978–3980 (1999).
32. G. Attolini, L. Francesio, P. Franzosi, C. Pelosi, S. Gennari, and P.P. Lottici, Raman scattering study of residual strain in GaAs/InP heterostructures, J. Appl. Phys., 86(11), pp. 6425–6430 (1994).

14

Review of the Technology and Reliability Issues Arising as Optical Interconnects Migrate onto the Circuit Board

P. Misselbrook[a,c], D. Gwyer[b], C. Bailey[b], P.P. Conway[c] and K. Williams[c]

[a]*Celestica, Kidsgrove, Stoke-on-Trent, UK*
[b]*Centre for Numerical Modelling and Process Analysis, University of Greenwich, London, UK*
[c]*Interconnection Group, Loughborough University, Loughborough, UK*

Abstract   Light has the greatest information carrying potential of all the perceivable interconnect mediums; consequently, optical fiber interconnects rapidly replaced copper in telecommunications networks, providing bandwidth capacity far in excess of its predecessors. As a result the modern telecommunications infrastructure has evolved into a global mesh of optical networks with VCSEL's (Vertical Cavity Surface Emitting Lasers) dominating the short-link markets, predominately due to their low-cost. This cost benefit of VCSELs has allowed optical interconnects to again replace bandwidth limited copper as bottlenecks appear on VSR (Very Short Reach) interconnects between co-located equipment inside the CO (Central-Office).

Spurred by the successful deployment in the VSR domain and in response to both intra-board backplane applications and inter-board requirements to extend the bandwidth between IC's (Integrated Circuits), current research is migrating optical links toward board level USR (Ultra Short Reach) interconnects. Whilst reconfigurable Free Space Optical Interconnect (FSOI) are an option, they are complicated by precise line-of-sight alignment conditions hence benefits exist in developing guided wave technologies, which have been classified into three generations. First and second generation technologies are based upon optical fibers and are both capable of providing a suitable platform for intra-board applications. However, to allow component assembly, an integral requirement for inter-board applications, 3rd generation Opto-Electrical Circuit Boards (OECB's) containing embedded waveguides are desirable.

Currently, the greatest challenge preventing the deployment of OECB's is achieving the out-of-plane coupling to SMT devices. With the most suitable low-cost platform being to integrate the optics into the OECB manufacturing process, several research avenues are being explored although none to date have demonstrated sufficient coupling performance. Once in place, the OECB assemblies will generate new reliability issues such as assembly configurations, manufacturing tolerances, and hermetic requirements that will also require development before total off-chip photonic interconnection can truly be achieved.

14.1. BACKGROUND TO OPTICAL INTERCONNECTS

Whilst there are many important characteristics of data carrying interconnects such as security, speed, and reliability, independent of link distance: the fundamental link property is unquestionably information-carrying capacity. As demonstrated by the Shannon-Hartley theorem, capacity is proportional to the channel bandwidth, which in turn is proportional to the frequency of the carrier [4]. This formula, drawn from information theory, is true regardless of specific technology and highlights the issue that the bandwidth capacity of transmission mediums is ultimately limited, not by technological advances, but by frequency, a physical property of the medium itself. Defined by this, copper, the foremost transmission medium, has the lowest potential bandwidth, increasing with twisted pairs, RF, and microwaves (satellite channels), through to light that has the highest frequency and therefore the greatest bandwidth potential [4].

The vast bandwidth potential offered by optical links over more traditional electrical links first became a commercial reality during the 1970s as technological advances, such as the development of edge-emitting diodes and single mode fibers, enabled optical links to supersede copper in long distance telecommunication links [5–7]. Through the 1980s and 90s, as demand on link capacity increased across the network, copper became increasingly redundant over reducing distances as it struggled to provide for the bandwidth explosion generated by three driving factors: the increasing base of global end-users, popularity of technologies such as the Internet, and the emergence of data intensive multimedia services such as video conferencing [5]. The modern telecommunications infrastructure is a global mesh of optical networks offering a plethora of multimedia services. Therefore to ensure global compatibility, various transmission standards have been adopted such as SONET, ATM, and Ethernet. These standards govern specifications from data protocol to loss budgets for each of the individual interconnecting networks, which are typically organized by function and link distance into three market-segments: long-haul, Metropolitan Access Networks (MAN), and local access networks. This hierarchy allows for the aggregation of the lower bandwidth access traffic, generated by the user, through the regional MAN to the corresponding MAN CO, also called the Point of Presence (POP) [8,9]. Each CO contains switches and routers that interconnect with other POP's through the long-haul network to provide complete inter-networking of all end users.

Although the link distances reduce, each market segment cannot simply be a scaled down version of the larger due to the varying requirements based on the traffic each network handles. During their evolution, each market segment has therefore developed specific requirements on the cost and performance of the transmission equipment utilized.

14.2. TRANSMISSION EQUIPMENT FOR OPTICAL INTERCONNECTS

Long-haul networks span both regional and extended geographic distances, connecting MANs to extend global connectivity between regional domains [10,11]. Due to the long distances involved and the deployment of Dense Wavelength Division Multiplexing (DWDM) systems, to provide for the huge bandwidth required at each link, high performance single-mode transmitters are required (Figure 14.1). For this reason, edge emitting Distributed Feedback (DFB) laser diodes transmitting around the 1550 nm wavelength have established themselves as the technology of choice for long-haul applications, giving unrivaled performance in areas such as: single mode stability, power output, line-width, and

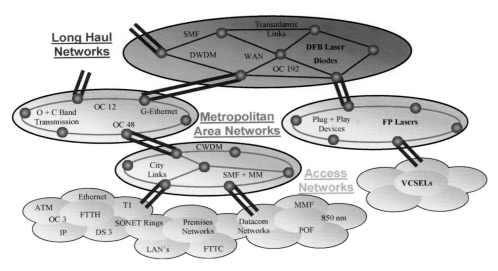

FIGURE 14.1. Optical network hierarchy: The three market segments, long-haul, MAN, and access networks interconnect at CO switches to route traffic between global end users.

wavelength selection [11]. Achieving such performance comes at a trade-off with price, with components being dominated by performance considerations rather than cost.

Metro networks operate over much reduced link distances and provide the connection between Local Area Networks (LAN) or access networks to each other and to the backbone for global communications between end users. Although DFB lasers meet many of the performance requirements of the metro space, they are uneconomical due to the cost of packaging the components into industry standard 14-pin butterfly devices. Consequently cheaper Fabry-Perot (FP) lasers transmitting at the shorter 1310 nm wavelength are more commonly used as link distances decrease [12].

As access networks have predominately the shortest link-lengths of the three telecommunication network segments, between several hundred to several thousand meters, interconnects deployed here are much more sensitive to cost [13]. Thus, a high premium for additional performance would not persuade operators to switch to optical channels from simpler, and cheaper copper-based interconnects with proven reliability. From this standpoint the most important breakthrough in allowing optical interconnects into the domain of shorter link lengths was the development of the VCSEL in the late '90s. Operating in the 850 nm window, short-wavelength multimode VCSEL's provide a very cost effective solution with ample performance density to operate over the short link lengths involved in access networks. The fundamental difference between edge emitting diodes and surface emitting VCSEL's is just that; VCSEL's emit the light beam from the surface of the wafer where as previously the light was generated in the plane of the wafer, only becoming accessible after wafer dicing.

VCSEL construction has seen significant development [14–16] since their inception from the first design with the active area sandwiched between two Distributed Bragg Reflector (DBR) mirror stacks, fabricated predominately on gallium arsenide (GaAs) wafers. The light beam is created when electrical current is applied across the active layer, via intra-cavity contacts, generating photons that are then reflected by the DBR mirrors before being emitted from a circular aperture, about 14 µm in diameter [17], on the top surface

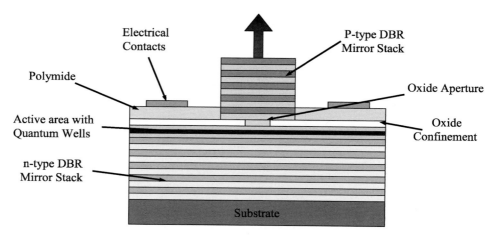

FIGURE 14.2. VCSEL cross-section: Layers of material are thinly deposited onto GaAs wafers to form DBR mirrors sandwiched either side of an active area. The quantum wells convert electrical energy into photons, emitting a light beam perpendicular to the surface of the substrate.

of the wafer. Before the wafer is diced a polyimide coating is applied between the VCSEL structures to protect the sides from oxidization [4]. A single VCSEL die is typically 250 μm square although wafers can be diced into any pattern or array. The electrical contacts can be created to enable the die to be wire bonded or flip-chip bonded with the aperture on the top or bottom side (Figure 14.2). The unique manufacturing process of VCSEL devices has enabled them to dominate the optical access market in recent years, their many inherent advantages include: low-cost due to their ability for high volume manufacture and in-production testing at the wafer level, ultra-high modulation rates, low power consumption with threshold currents less than a milliamp, and low coupling tolerances due to the circular beam output.

Although DFB, FP, and VCSEL transmission lasers dominate their respective markets, the downturn in the telecommunications industry, just after the turn of the century, has resulted in a necessary rationalization in the performance and packaging of components. This has opened the door for new technologies and the spread of the developed sources into new, previously inhibited sectors. The most important issue that this rationalization addressed was an effort to reduce costs across all parts of the manufacturing process. It is widely accepted that of the final laser module's cost, packaging constitutes between 85%, for laser diodes in butterfly packages [11], and 33%, for lower specification VCSEL's [18], hence laser module packaging has been most affected by cost reduction initiatives. This has primarily seen an underlying trend across the market segments toward vertical integration into standardized transceiver packages, which incorporates the transmitter, receiver, and electronics into a standard Small Form Factor Pluggable (SFP) module [19].

This drive to reduce the cost of optical components has also had additional benefits, allowing VCSEL's to become competitively priced offering an alternative to copper in VSR interconnects, between 10–300 m, as copper yet again creates bottlenecks in the COs and POPs of telecommunication networks. As explained further below, the successful emergence of VCSEL's in the VSR arena has also renewed the long anticipated wait for optical solutions to USR interconnects, distances less than 10 m and predominantly based on circuit boards.

14.3. VERY SHORT REACH OPTICAL INTERCONNECTS

The expanding demand on networking capacity has meant service providers need to connect core routers and optical transport equipment with multiple high-speed links to prevent bottlenecks developing at the POP. Traditionally, these interconnects have been deployed across copper based-interconnects, but the move toward data rates of 10 Gbps and beyond is pushing these links to their limitations, so increasingly optical interconnects are being considered. Since the majority of POP equipment tends to be physically located within the same building, a significant proportion of these links are less than 300 m, where it is uneconomical to deploy optical interconnects operating over standards optimized for longer distances [3]. Subsequently, a set of VSR optical interconnection standards has been developed by the Optical Networking Forum (OIF) aimed at low-cost interconnects between co-located equipment (<300 m) [20].

VSR leverages technology developed for Gigabit-Ethernet, to transport OC-192 (10 Gbps) traffic. The first implementation agreement developed for the VSR arena was OIF-VSR4-01, which employs parallel optics with the signal distributed over twelve 1.25 Gbps VCSEL channels operating over multimode fiber (MMF). As an alternative to the 12 fiber parallel optics standard, OIF-VSR4-03 a four-fiber standard was implemented with the VCSEL's operating at 2.5 Gbps. For VSR applications where single mode fiber (SMF) has already been laid, a significant cost advantage is made by utilizing this, so a standard was also developed for 1310 nm VCSEL's and FP lasers over SMF, which has now been developed as OIF-VSR4-05 (Figure 14.3). Finally, in appreciation of the advances being made in VCSEL technology, the set of OC 192 VSR implementation agreements was completed by OIF-VSR4-04, based on a single 850 nm MM fiber channel, for distances up to 85 m.

FIGURE 14.3. VSR optical interconnects: Discrete optical fibers connect co-located equipment over distances up to 300 m, using OIF implementation agreements. Figure courtesy of BPA [2].

This comprehensive set of VSR standards has enabled low-cost optical interconnects, based on VCSEL technology, to replace copper as the technology of choice for high-speed connections inside the CO and POP. However, this migration from long haul to VSR, despite repeated predictions, has taken over two decades. The main reasons for this delay have been in over-coming the technological challenge of manufacturing optical interconnects with strong manufacturability, greater performance, and at a lower cost than electrical interconnects. Now that this has been achieved down to a few hundred meters, the push is on to produce board-level USR optical interconnects for distances less than 10 m. The drive toward this goal is currently coming from two main directions, although the end impact of low-cost on-board optical interconnects is expected to have a much broader impact across the electronics market. The first factor stems from the successful implementation of VSR optical links between telecommunications equipment, which has now lead to a bottleneck forming on the backplanes. Low-cost inter-board optical links to transport OC 192 traffic between daughter-cards in a rack configuration are required.

The second driving force toward board-level USR optical interconnects is concerned with intra-board applications, specifically the increase in bandwidth capacity between IC's, which is being identified as an issue as chip manufacturers commit to the extension of Moore's law well into the future [21]. Based on this trend, the doubling of transistors on IC's every eighteen months, it is predicted that within five years CMOS-based transistors will be fast enough for transceivers to operate at clock speeds of around 14 GHz with data-transfer rates in the region of 20 Gbs [22]. Since copper interconnects on FR4 based Printed Circuit Boards (PCB's) become bandwidth limited beyond 10 GHz, primarily due to frequency dependent losses such as the skin effect in conductors and dielectric losses in the substrate, manufacturers have sought to address the bandwidth limitation of copper traces. However, the potential copper-trace based solutions such as new board substrates and sophisticated encoding techniques tend to be expensive making optical solutions increasingly attractive.

History shows that the barrier to implementing optical links over established copper-based interconnects is overcoming the technical challenges to achieve higher performance for lower cost. These technology issues are similar for both the driving forces behind USR optical interconnects and can currently be divided into two research areas: free space optics, and guided waves.

14.4. FREE SPACE USR OPTICAL INTERCONNECTS

The most basic, although by no means the simplest method of applying optical interconnects at the board level is to use lenses and collimators to expand a VCSEL beam suitably so that it can be sent through the air to a corresponding configuration, for detection by a Photo Detector (PD) [23–26]. So called FSOI's can boost the bandwidth between chips due to the combination of both high data rates and the fact that FSOI's can be densely packed on the circuit board. The addition of diffractive optics to the configuration allows the signal to be routed to different detectors, so FSOI's can benefit from being reconfigurable [27–29].

Complications to this basic principal develop due to the requirement to maintain a precise line-of-sight rule, in that if the VCSEL's and PD's become misaligned or blocked for any reason the signal is lost. Although research is ongoing into ways of tracking and actively maintaining the links as the transmitter and detector arrays move relative to one

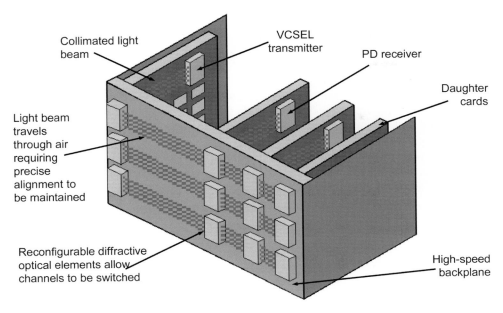

FIGURE 14.4. FSOI based backplanes: VCSEL's and PD's on separate daughter cards communicate across an optically enabled high-speed backplane via free space channels.

another [30,31], the occurrence of catastrophic signal loses due to environmental issues can not be eliminated without hermetically sealing the entire backplane, which would certainly compromise the cost advantage. Greater control of the signal between the transmitter and receiver is beneficial and can be achieved by guiding it between geometric boundaries such as a fiber or waveguide, as is the case in telecommunications systems. Consequently, there is an increased interest in developing PCB based guided wave USR optical interconnect solutions that permit greater freedom in the routing between devices, without strict line-of-sight rules, and using fewer elements than FSOI systems (Figure 14.4).

14.5. GUIDED WAVE USR INTERCONNECTS

Optical fiber and planar waveguides are two examples of guided wave interconnects. Although the end product and manufacturing processes are markedly different, with fibers having a circular cross section and waveguides square, the physics behind light transmission is similar. The basic principal consists of the light signal propagating down a core that is surrounded by a cladding material with a slightly lower refractive index. For multimode systems the light is retained in the core material due to Total Internal Reflection (TIR) at the core-cladding interface.

To date, the most successful uptake of board-level guided wave USR interconnects has been on high-speed backplanes inside CO telecommunication equipment. As identified by BPA [32], there are three methods for the implementation of these interconnects over backplanes. The 1st generation consists of discrete optical fiber interconnects. Each fiber is individually routed at the back of the switch for point-to-point optical links, a direct descendant of the VSR links that are currently being deployed to interconnect backplanes. However, although separate fiber interconnects have been widely implemented across VSR

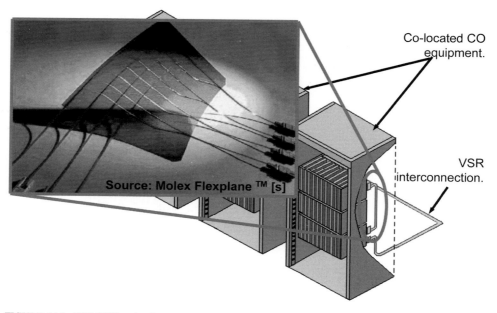

FIGURE 14.5. FOB USR technology: Inter board fiber management problems are reduced by combining the fiber into a flexible harness or PCB. Inset, an example of Molex's Flexplane™.

links between systems and backplanes, attempting to extend 1st generation technology to address links across the backplane have been thwarted by fiber management problems generated by the resulting increase in the density of fiber terminations. This greater density creates a "rat's nest" of fibers at the backplane and also requires significant resources to manually route the fibers, outside of the manufacturing environment.

To enable greater integration of the USR optical interconnects into the board and to counter some of the problems associated with discrete fiber interconnects, various institutions and companies [1,33,34] are developing so called 2nd generation technologies which combine the fiber into a rigid board, termed fiber-on-board (FOB). The essence of FOB technology is to embed standard optical fiber into a PCB harness (Figure 14.5). Whilst fiber handling and its associated costs are clearly still involved in the manufacturing process, the method is capable of yielding low-loss connections with relatively few technology hurdles. Fiber management still creates problems though, particularly in maintaining the minimum bend radius of the fiber, typically a few centimeters, which if exceeded would result in unacceptable attenuation due to the conditions for TIR being negated. Exceeding the minimum bend radius will mostly occur when routing the fibers around the board and prevents more than two fibers from crossing on the same plane and also fibers from being bent out-of-plane to the surface of the board, which is required if optical Surface Mount Technology (SMT) components are to be assembled. Coupling light into the embedded fibers is further complicated due to issues concerned with mounting and joining the output of transmitter and receivers to non-connectorized fiber terminations. Predominately non-connectorized fibers are spliced together using a fusion process, however this process requires a reasonable amount of accessible fiber preventing splices in close proximity to the board edges. This also limits the number of rework options available, since if a fiber exiting the board is

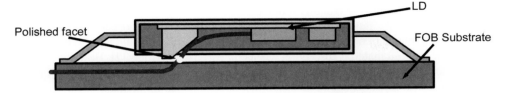

FIGURE 14.6. FOB component coupling issues: Maintaining the minimum bend radius inhibits the in laid fiber from being bent upward. The assembly also requires precise cleave and polish location of fiber termination.

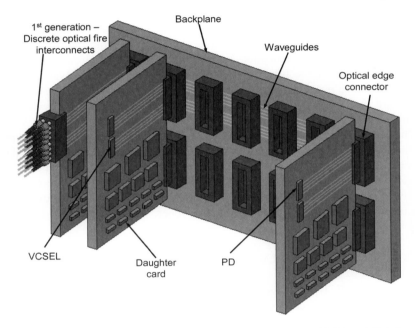

FIGURE 14.7. OECB based optical backplane architecture: SMT assembled VCSEL's emit light into embedded waveguides, which route onto the backplane to be distributed onto other daughter cards or in to fiber for aggregation over larger areas.

broken within a close proximity to the PCB harness the entire board must be re-fabricated (Figure 14.6).

Due to the associated costs and problems with fiber management and component coupling, a significant cost advantage can be realized by eliminating fibers from the heart of the manufacturing process. This is the main advantage in 3rd generation technologies, which look to replace fibers with planar optical waveguides. Waveguides allow the optical links to be fully integrated into the PCB, with easier coupling routes for SMT devices, which allows the solution to address inter-board as well as intra-board interconnects. It is anticipated that most 3rd generation technologies will sandwich an optical layer, containing the waveguide, into a standard FR4-stack. The board will remain FR4 based since not all interconnects are required to be high speed so some copper tracks will be retained thus creating a hybrid OECB with both electrical and optical interconnects on one board. OECB fabrication builds on a core FR4 board produced using current fabrication methods with the copper tracks being routed through layers in the board by vias and blind vias (Figure 14.7).

There are currently two main candidate materials being investigated to form the optical waveguide layers: glass, and polymer. BPA's backplane report suggested that to achieve the necessary attenuation (less than 5 dB/cm), glass waveguides would have to be adopted due to their reduced material absorption when compared to polymers [2]. However, against this prediction, glass manufacturing issues have prevented the production of high quality waveguides with low attenuation levels, whilst developments in the manufacture of polymer based layers have produced attenuation figures under 0.03 dB/cm [35] and are therefore becoming the technology of choice. Polymer waveguide processing techniques also have the advantage of being compatible with current PCB fabrication methods, allowing low-cost manufacture, an important specification for OECB's.

With the addition of optical edge connectors, OECB based interconnects enable high-speed point-to-point optical links to be established between transmitting and receiving equipment satisfying the specification for both inter-board optical backplanes and high bandwidth inter-board connections between ICs. Due to their inherent advantages, the technology set of choice is anticipated to comprise of readily available VCSEL and PD pairs, with the light signals being routed between the pairs via embedded waveguide structures, which are becoming commercially available through companies such as [35,36].

The major OECB technology challenge remaining, currently surrounds coupling between the waveguides and optical SMT components. With the principal board fabrication issue being the 90° out of plane coupling, that would enable light incident normal to the board's surface to be coupled and transmitted along the waveguide. Subsequent research into assembly configurations, manufacturing tolerances, and hermetic requirements, will also be required before OECB systems can be commercially exploited.

14.6. COMPONENT ASSEMBLY OF OECB'S

The major OECB fabrication issue currently under investigation is the assembly, and 90° out-of-plane interconnection between the SMT components and the embedded waveguides. Research is also proceeding into the possibility of opening up pockets in the optical layers, sinking edge emitting components to emit directly into the waveguide facet therefore negating the requirement for the 90° change in direction. However, methods that couple signals to the waveguides without altering the direction of the light are limited in their application to other components, including low-cost VCSEL's, and therefore do not provide the overriding advantage of deploying 3rd generation interconnects, and the low-cost assembly of components. It is therefore essential to develop a low-cost method to change the direction of the light from being parallel with the board to projecting 90° out of plane to the top surface, thereby enabling compatibility with SMT devices and fiber ferrules. Literature suggests that there are two main distinguishable research trends for coupling out of plane to OECB's: assembled techniques, and integrated mirrors.

Assembled techniques form the 90° optical interconnection by first creating a trench in the OECB to expose the vertical facet of the waveguides. An optical sub-assembly based on a Ball Grid Array (BGA) interposer is then placed onto the OECB with a carrier inserted into the trench. Light is then coupled to and from the waveguides either by direct alignment of a VCSEL or by using a mirror to direct the light upward to the surface using either a free space or guided wave approach. Whilst assembled techniques provide a simple process route with relatively few technological issues to resolve, the optical coupling relies on manufacturing tolerances that are currently unable to provide sufficient repeatability. This

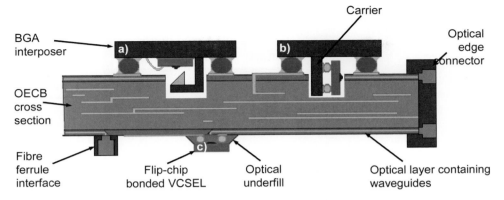

FIGURE 14.8. OECB component coupling techniques: Typical configurations for coupling components to OECB's showing, on the top, two examples of assembled mirrors with (a) the VCSEL mounted on an interposer allowing easier progression for other components such as fiber ferrules and (b) the VCSEL aligned to the waveguide supported by a carrier providing electrical contact. The bottom side of the OECB diagram shows (c) an example of an integrated mirror with routes for connecting a wide variety of components.

is due to the reliance on tensions in the solder joints to self align the components to the waveguides, assuming that the bond pads are correctly positioned in the outset. The depth to which the carrier is inserted is also a critical alignment factor that cannot be guaranteed since the reference plane is FR4.

Integrated mirror techniques combine the 90° deflecting mirrors on the ends of the waveguides, allowing active devices to be flip-chip bonded directly on top of the OECB. This technology not only removes the need for additional assemblies and their associated part costs and manufacturing processes, but also enables the OECB to drive toward low-cost volume manufacture through SMT assembly. It is this move toward traditional EMS (Electronic Manufacturing Service) competencies, coupled with the perceived benefits in the re-work and test areas, which makes integrated mirrors the solution that is the most likely to break the significant USR market barriers of cost and performance (Figure 14.8).

To date, several organizations have reported the development of integrated mirror solutions that allow the direct coupling of components into the waveguides, with examples being published work from Intel and Optical Cross-Connect (OXC) [36]. These two similar demonstrators are manufactured based on a standard core board containing electrical contacts for the components to which a section containing the waveguides is laminated. The waveguide section is pre-fabricated with 45° metalized Input/Output (IO) mirrors on the ends. The section is then aligned on the core board using fiducial marks to ensure the precise position of the mirrors to the VCSEL pads and bonded using an Ultra Violet (UV) curable epoxy. To enable SMT assembled devices to stand clear of the waveguide section it is necessary to sink the layer into a trench formed in the original board so as to sit the top surface of the waveguide insert within tolerance of the top surface of the core substrate. Using this approach Intel have reported that when coupling VCSEL's and PD's to the waveguides a misalignment in excess of ±10 μm still yielded 80% efficiency from the maximum coupling. This tolerance range allows passive alignment, crucial in achieving the low cost manufacturing required for the systems.

With no set standards defining USR optical interconnect performance the important characteristic of the link is to maintain the signal integrity over the distances involved,

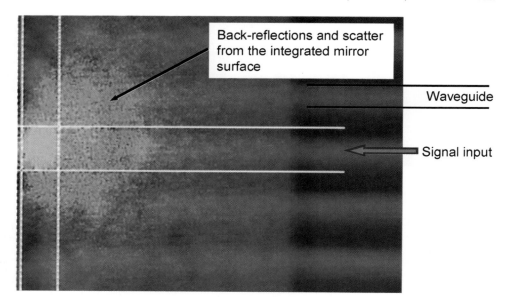

FIGURE 14.9. Plan view of an OECB demonstrating capability of a novel integrated mirror manufacturing process: For display purposes a visible laser is used to illuminate a waveguide with an integrated mirror formed on one end. Backscatter and divergence need to be reduced to improve attenuation.

thereby ensuring all the bits sent are correctly received. Although bit errors are generated by many factors, large link attenuation is a key contributor. Intel's demonstrator reported 1 Gbts error free transmission over link distances of approximately 20 cm with a loss budget of between 7 and 12 dB. Of this loss budget the IO coupling between waveguide and component constituted signal attenuation of between 1.5 and 3 dB. Despite demonstrating significant performance capability, the separate alignment and bonding steps required to attach the waveguide layer to the board contributes to additional tooling and process costs since the assembly methods are not aligned to that of current PCB fabrication. It is therefore necessary to further integrate the manufacture of the waveguides and IO mirrors into the substrate (Figure 14.9).

To that end, research by the authors is investigating a novel manufacturing approach utilizing current PCB fabrication techniques to manufacture OECB's with integrated mirrors. The substrates require no post processing allowing VCSEL's and PD's to be directly bonded to the OECB substrate. Although the integrated mirror OECB samples have not yet reliably yielded insertion losses of less than 3 dB, this is caused by the prototype nature of the manufacturing method and research is ongoing. However, once reliably achieved, significant research and development of the OECB is required to ensure sufficient reliability and performance measures are met. The remaining uncertainties include assembly issues, manufacturing tolerances, and hermetic requirements.

Although it would be advantageous to keep the assembly processes similar to those currently used for high volume SMT production, some differences are necessary. For example, it will probably be necessary to assemble the components using fluxless processes due to the risk of the flux coating the lenses and optical elements; this is common practice in optical component assembly. It is also envisaged that to remove the risk of particles permeating under the SMT devices, an optical underfill will be required. The underfill's primary function will be to add hermetic protection to the signal by removing any free-space transmission rather than to aid mechanical reliability of the package. Removing the air-waveguide interface by index matching the underfill to the waveguide may also decrease optical losses by removing Fresnel reflections.

14.7. COMPUTATIONAL MODELING OF OPTICAL INTERCONNECTS

The issues met during the manufacturing and operation of optical interconnects can be aided by the use of the finite element modeling. This valuable tool can be used for the prediction of thermal, thermo-mechanical, and optical behavior during the manufacturing processes. Thermal simulations can show how much heat diffuses through a VCSEL package, and the rate at which it does so. Thermo-mechanical models predict the build up of stress and strain in the package due to CTE (Coefficient of Thermal Expansion) mismatches in the various materials present, which gives an insight into how close the CTE values of local materials must be to avoid undesirable stress values. Finally, optical simulations can help calculate the potential optical losses in the package by modeling a range of scenarios to see their effects on the transmitted signal.

Modeling work has been undertaken to aid the analysis of the OECB assemblies described above. A thermo-mechanical model was constructed to investigate misalignment and stress caused by machine component placement tolerance and CTE mismatches in the OECB, VCSEL's, and bonding materials [38]. Secondly, an optical model was used to predict the coupling efficiency of the optical interconnect as a signal is transmitted through it. The optical signal will be subject to attenuation losses when the VCSEL to waveguide coupling deteriorates. Computational modeling can help in identifying optimum process parameters that should be used to assemble the OECB's, giving an appropriate compromise between mechanical reliability and optical efficiency [37].

The thermo-mechanical model is constructed from the proposed OECB assembly, and dimensions used are accurate to manufacturing specifications. The model is comprised of a 1×4 VCSEL array, which is flip-chip mounted onto an OECB, bonded by Au stud bumps. The GaAs die of the VCSEL array, copper contact pads, and the Au stud bumps are all surrounded by an underfill material (Figure 14.10).

The thermo-mechanical model simulates the four VCSEL's in the array heating up due to normal operation. Localized heating in the area surrounding each VCSEL device is seen at around the expected temperature of 85°C. With the correct heating profile, the resultant stresses can be seen due to these thermal effects and the CTE mismatch between the various materials present (Figure 14.11). For all the simulations the greatest stress and deformation occurred in the outermost VCSEL's in the 1×4 array, and so attention was focused on these when measuring the in-plane misalignment (between VCSEL aperture and waveguide entrance).

It should be noted that all but one material used in the thermo-mechanical model is isotropic. FR4 material is orthotropic, having material properties in the plane different to that out of the plane. In the plane the CTE value is 21 ppm compared to that of 60 ppm out of the plane (Figure 14.12).

Two underfill types are used in the simulations to compare the impact each one has on the model. From Figure 14.11, "Underfill 1" is standard underfill and "Underfill 2" is an optical underfill which does not contain filler particles as would be found with a traditional underfill. Such an optical underfill is similar to the so-called no-flow underfill. The CTE

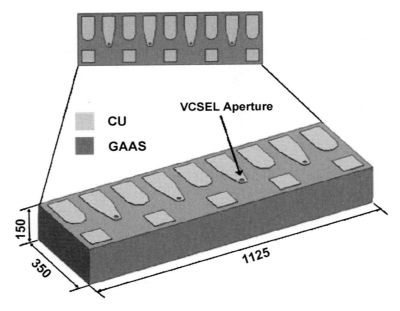

FIGURE 14.10. VCSEL array geometry: 1 × 4 VCSEL array geometry with copper pad detail.

	Thermal Cond. (W/mK)	Specific Heat (J/kgK)	Density (kg/M^3)	CTE (K^{-1})*10^{-6}	Young's Mod (N m^{-2}) GPa	P. Ratio
FR4	0.3	100	1900	21	21	0.3
Copper	400	384	8960	16.5	130	0.34
Gold	8	129	19280	14.2	78	0.44
GaAs	45	350	5320	5.8	85	0.31
Underfill 1	250	900	8000	30	6	0.35
Underfill 2	250	900	8000	75	2	0.35
Epoxy	29	150	2100	60	0.179	0.32
Waveguide	29	150	2100	60	0.179	0.32

FIGURE 14.11. Material properties: Thermo-mechanical material properties used for the VCSEL array computational model.

value for the optical underfill is much higher and is less stiff (due to a lower Young's modulus). As expected, much more movement and deformation was seen with the no-flow underfill material during simulations, due to the variation in material properties between the two underfill types. For this reason the optical underfill material was used for all further simulations (where underfill is required), as this shows the worst case scenario for using an underfill from the two choices available.

The thermo-mechanical model has boundary conditions in place to allow for natural deformation and stress build up to occur. For a single VCSEL array, by fixing two whole side faces and the entire bottom face it is assumed that the array is one out of a block of four. The fixed side faces define the simulation symmetry planes. This has the inherent advantage that only one VCSEL array is actually needed to modeled, significantly reducing pre-processing model development and simulation run time. The computational mesh is comprised of 46,659 individual elements and is unstructured (Figure 14.13).

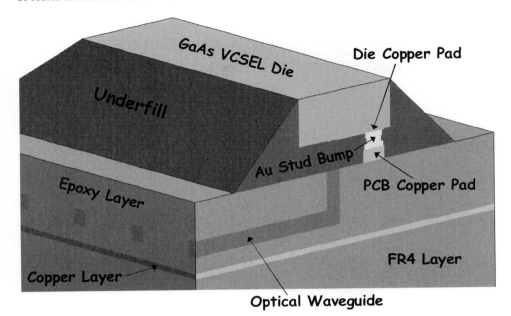

FIGURE 14.12. Material layout: the various materials and their locations on the model geometry.

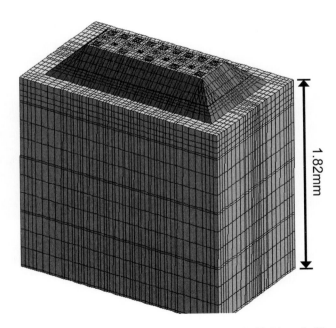

FIGURE 14.13. Computational model geometry: the VCSEL array embedded in underfill and mounted on an OECB. The mesh is shown.

During initial testing the thermo-mechanical model showed the greatest general deformation in a single direction, along the x-axis (Figure 14.14). Deformation in the other two axes directions movement was negligible. The misalignment, dx, is the measure of the

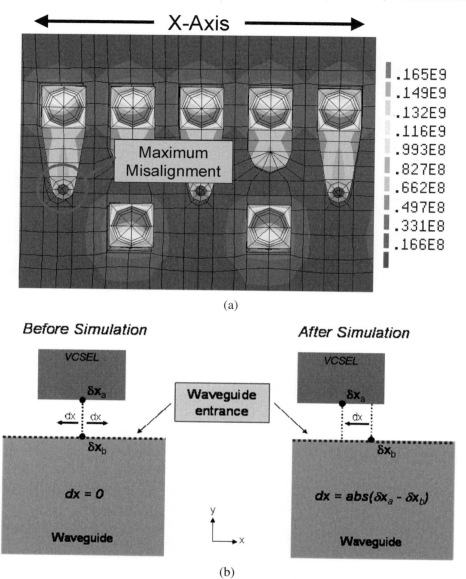

FIGURE 14.14. Misalignment results: (a) Maximum misalignment, in the x-direction, shown to be on the outer VCSEL of the array. GaAs die and underfill materials removed for clarity. (b) Method of calculating misalignment during simulations in the x-direction.

absolute difference between the center of the VCSEL aperture and center of the corresponding waveguide entrance. After simulation dx gives the amount of deformation occurring during the simulation.

Thermo-mechanical simulations induced stress in the VCSEL array by increasing the localized temperature surrounding each VCSEL and allowing the differences in the CTE values of the various materials to act. "Delta t" (Δt) is the temperature increase above the ambient starting temperature, which is 25°C at the beginning of each simulation. The

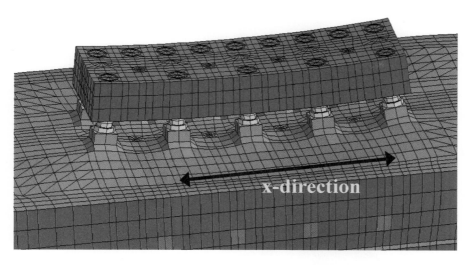

FIGURE 14.15. Mesh deformation: maximum misalignment is in the x-direction, and for the outer most VCSEL's in the array. GaAs die and underfill materials removed for clarity.

value of Δt, which is 60°C for the thermo-mechanical model, together with the CTE value for each material will give the maximum amount of displacement possible. For example, taking the CTE value for epoxy material to be 60 ppm (parts per million), the maximum linear displacement possible for $\Delta t = 60°C$ (in any axial direction) is given as follows:

$$\Delta L = \alpha \cdot L_0 \cdot \Delta t,$$
$$\Delta L = (60e - 6) \cdot (1e - 3) \cdot (60),$$
$$\Delta L = 3.6 \, \mu m.$$

Taking the original length (L_0) to be 1 mm, which is the approximate length of the VCSEL array, we have the maximum possible linear expansion for epoxy material to be just 3.6 µm. With other materials having much smaller CTE values, the actual misalignment expected would be significantly less than for the calculated epoxy material alone. In fact, simulations results show sub micron misalignment for dx (Figure 14.15). This is consistent with expected results.

To fully understand the impact of optical underfill on the VCSEL array coupling to waveguide, simulations were run for two different underfill scenarios. One simulation had the encapsulant all the way around the sides of the GaAs die, and throughout the gaps between the stud bumps (see Figure 14.12). The other simulation scenario had no underfill present on the model, leaving the die and stud bumps effectively exposed [Figure 14.16(a)]. It was found that the higher CTE value for the optical underfill deforms the VCSEL array package significantly more than when the underfill is removed. This results in higher stress levels around the stud bump joints [Figures 14.16(a) (b)].

Also, further simulations were performed with optical and standard underfill with both traditional tin/lead solder (SnPb) joint material and Au stud bumps. For simulations where the Au stud bumps were used, an epoxy layer was present; and for the solder (SnPb) material, FR4 is used in place of the epoxy layer. From the results, simulations showed that standard underfill (when used with SnPb solder) acts as a support for the package and helps to reduce deformation and stress levels when compared to no underfill. This is an expected

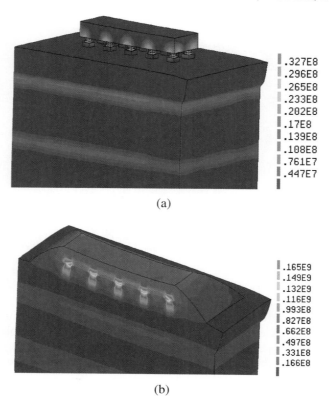

FIGURE 14.16. Stress contour plots for the two underfill scenarios: (a) no underfill model; (b) complete underfill model (optical underfill), cutaway view to see the stress contours in the stud bumps.

Underfill Type	Au Bump Stress (MPa)	SnPb Stress (MPa)
Traditional	57.18	27.44
Optical	170.26	96.10
None	27.89	35.11

FIGURE 14.17. Stress results: comparison of the stress values for the Au stud bumps and solder stud bumps.

trend for traditional die, underfill, solder and FR4 materials. But for the simulations when using the optical underfill (with Au stud bumps) it was found that the presence of the optical underfill and epoxy layer acts as a stress enhancer, when compared to no underfill (Figure 14.17).

The optical model simulates the light coupling from each VCSEL aperture into the corresponding waveguide entrance. The waveguide is constructed of two straight sections

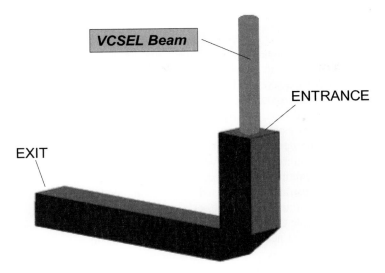

FIGURE 14.18. Optical waveguide: schematic diagram of the optical model setup.

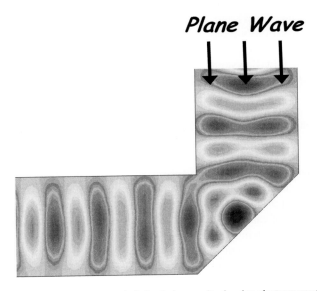

FIGURE 14.19. Optical model simulation: optical simulation results showing electromagnetic plane wave propagation through the waveguide.

connected at right angles via an integrated mirror (Figure 14.18). The governing equation solved in this analysis is the Helmholtz equation, which is a subset of Maxwell's equations, and assumes the light source is monochromatic and is propagated as a plane wave [39].

The aim of the optical simulations is to predict the coupling efficiency of the VCSEL beam. This is characterized by the attenuation value which is calculated by comparing the level of the optical signal at the VCSEL aperture to that at the waveguide exit. Any attenuation observed will be as a result of the misalignment between VCSEL and waveguide, geometry of the waveguide, and the polymer material properties used (Figure 14.19).

14.8. CONCLUSIONS

For decades optical interconnects have been replacing copper throughout the telecommunications network in order to provision for the continually increasing demand for greater bandwidth capacity. As this trend becomes to be pertinent even at the shortest link distances, the successful deployment of optical links becomes dependent on the technological advances necessary to create optical links with strong manufacturability and greater performance over established electrical interconnects whilst, most importantly, at a lower cost. Just as the development of low-cost VCSEL's allowed optical links to replace copper over VSR interconnects between co-located CO equipment, the development of hybrid OECB's, integrating both electrical interconnects and 3rd generation embedded waveguides into one substrate, are expected to unlock the path toward high speed optical backplanes and optically interconnected ICs.

Prior efforts to create OECB's have centered on laminating optical layers to FR4 and although they have yielded acceptable out-of-plane coupling losses of around 3 dB, the additional lamination process is not capable of achieving the required cost target. However, the authors have investigated a novel manufacturing method for creating low-cost OECB's by aligning the fabrication processes much closer to that of those already used inside PCB plants. With the research results to date suggesting that the resulting integrated mirrors are capable of coupling fibers and SMT components with attenuation losses of less than 3 dB, the future for OECB's looks promising. However, consistently achieving and improving this low-cost out-of-plane interconnection attenuation is only part way to seeing commercially viable OECB systems. Significant development is still required, specifically centered on the passive assembly of optical components to the OECB's and their long term reliability.

To aid the development of the OECB assemblies, computational modeling is being used. The computational models for the thermo-mechanical and optical simulations are both in place and producing positive results with close synergy to the experimental work. Initial results from the modeling work have shown that the temperature of the VCSEL will reach 85°C producing some misalignment of the VCSEL due to CTE mismatches. The optical model is now nearing completion and the aim here will be to predict attenuation for a particular set of design parameters. Once completed, a virtual DOE will be run to help understand the impact of each design parameter in an effort to reach the optimal conditions that minimize the signal attenuation.

Assuming the development of low-cost OECB substrates continues to progress and that there are no major component assembly and reliability issues to overcome. OECB enabled, high-speed passive optical backplanes should be a reality on CO telecommunications equipment within the near future. Future development in the assembly of MOEMSs to the OECB's will enable the backplanes to become reconfigurable, paving the way toward the low-cost all optical switch. OECB assemblies will also enable on-board optical clocks to time future generations of IC chips. Bringing ever closer the holy grail of photonic computing.

ACKNOWLEDGMENTS

The authors would like to acknowledge the financial support provided by both the EPSRC and Celestica in the form of two Industrial Case Awards. In addition they would also like to acknowledge the assistance provided during the development of the OECB integrated mirrors by both Terahertz, in supplying the waveguides, and Ablestik, for the supply of the optical adhesives.

REFERENCES

1. MOLEX Website, www.ttieurope.com/microsites/molex/products_items/backplane/flexplane/overview.cfm (accessed 12th Dec. 2004).
2. BPA, Optical backplanes—A global market and technology review 2000–2005, BPA Consulting: Dorking, Surrey, UK, Report No. 762, 2001.
3. P. Van-Daele, Optics in PCB's: a Dream Coming True in Short Distance Optical Interconnects—From Backplane to Intra-chip Communication, CHACH, Chalmers University, Sweden, March 7th 2003.
4. D.K. Mynbaev and L.L. Scheiner, Fibre-Optic Communications Technology, Prentice-Hall, Inc., Upper Saddle River, New Jersey, USA, 2001.
5. G.P. Ryan, Ed., Dense Wavelength Division Multiplexing, ATG's Communications & Networking Technology Guide Series, The Applied Technologies Group: One Apple Hill, Suite 216, Natick, MA, USA, 1997.
6. D.A.B. Miller, Rationale and challenges for optical interconnects to electronic chips, Proc. IEEE, 88(6), pp. 728–749 (2000).
7. D.A.B. Miller, Optical interconnects to silicon, IEEE Journal: Selected Topics in Quantum Electronics, 6(6), pp. 1312–1317 (2000).
8. Cisco Systems Inc., Cisco Very Short Reach OC-192/STM-64 Interface: Optimizing for Network Intra-POP Interconnects, White Paper, 170 West Tasman Drive, San Jose, CA, USA, Cisco Systems, Inc., 2001.
9. M. Fuller, VSR optics acheives working concept and rough consensus, Lightwave, (February) (2001).
10. M.K. Dhodhi, S. Tariq, and K.A. Saleh, Bottlenecks in next generation DWDM-based optical networks, Computer Communications, 24, pp. 1726–1733 (2001).
11. Photonami, The Surface Emitting DFB M3 Laser, Photonami, 50 Mural Street, Richmond Hill, Onatario, Canada, June, 2003.
12. P.M. Henderson, Introduction to Optical Networks, January 29th, 2001, Mindspeed Technologies, Inc., Newport Beach, CA, USA, p. 36.
13. D. Welch, VCSELs: driving the cost out of high speed fibre optic data links, Compound Semiconductor Magazine, (July) (2000).
14. F. Mederer, et al., High performance selectively oxidized VCSELs and arrays for parrallel high-speed optical interconnects, IEEE Transactions on Advanced Packaging, 24(4), p. 442–449 (2001).
15. M. Peach, ULM refines VCSEL manufacturing technique, Lightwave Europe, June 2003.
16. Y.-C. Ju, Vertical-Cavity Surface Emitting Lasers (VCSEL), ETRI, KyungPook National University: Department of Physics Education, Teachers College, 2004.
17. Ulm Photonics, Product Datasheet: 850 nm multimode 2.5/3.125/5 Gbs N × M VCSEL array for direct FlipChip. Accessed (14th Dec. 2004), www.ulm-photomics.com.
18. E. Mohammed, et al., Optical interconnect system integration for ultra-short-reach applications, Intel Technology Journal: Optical Technologies and Applications, 8(2), pp. 115–128 (2004).
19. J. Theodoras, L. Paraschis, and A.C. Houle, Emerging Optical Technologies for Multi-Service Metro Networks, Advanced Technology Group, Optical Networking, Cisco Systems, 170 West Tasman Drive, San Jose, CA, USA.
20. Optical Internetworking Forum, www.oiforum.com. Accessed (8th Dec. 2004).
21. M.L. Hammond, Moore's law: the first 70 years, Semiconductor International, (April) (2004).
22. I. Young, Intel introduces chip-to-chip optical I/O prototype, Technology @ Intel Magazine, (April), p. 7 (2004).
23. N. Savage, Linking with light, IEEE Spectrum, (Aug.), pp. 32–36 (2002).
24. B. Layet and J.F. Snowdon, Comparison of two approaches for implementing free-space optical interconnection networks, Optics Communications, 189, pp. 39–46 (2001).
25. M. Chateaunenf, et al., 512-channel vertical-cavity surface-emitting laser based free-space optical link, Applied Optics, 41, pp. 5552–5561 (2002).
26. S. Mukherjee, The quest for an affordable replacement for inter-processor switched free-space interconnect, Short Distance Optical Interconnects—From Backplanes to Intrachip Communication, CHACH, Chalmers University, Sweden, March 7th 2003.

27. G.C. Boisset, B. Robertson, and H.S. Hinton, Design and construction of an active alignment demonstrator for a free-space optical interconnect, IEEE Photonics Technology Letters, 7(6), pp. 676–678 (1995).
28. K. Hirabayashi, et al., Optical beam direction compensating system for board-to-board free space optical interconnection in high-capacity ATM switch, Lightwave Technology, 14(5), pp. 874–882 (1997).
29. D.P. Resler, et al., High-efficiency liquid-crystal optical phase-array beam steering, Optics Letters, 21(9), pp. 689–691 (1996).
30. T.-Y. Yang, J. Gourlay, and A.C. Walker, Adaptive alignment packaging for 2-D arrays of free-space optical interconnected optoelectronic systems, IEEE Transactions on Advanced Packaging, 25(1), pp. 54–64 (2002).
31. T.D. Wilkinson and W.A. Crossland. Reconfigurable optical interconnects, MicroTech 2004, Moller Centre, Cambridge, IMAPS, UK, March 3rd–4th 2004.
32. Dickens, Optical backplanes—A global market and technology review 2000–2005, BPA Consulting, Dorking, Surrey, UK, 2001.
33. NTT-AT Website, www.ntt-at.com/products_e/opticalfiber/ (accessed 12th December 2004).
34. S. Agelis, et al., Modular interconnection system for optical PCB and backplane communication, Proc. International Parrallel and Distributed Processing Symposium, Fort Lauderdale, FL, USA, April 19th 2002.
35. Terahertz Photonics, Truemode Backplane: Product Information, Rosebank Park, Livingston, Scotland, UK, 2003.
36. OXL Website, www.opticalcrosslinks.com (accessed 14th December 2004).
37. J.J. Morikuni, P.V. Mena, A.V. Harton, and K.W. Wyatt, The mixed-technology modeling and simulation of opto-electronic microsystems, Journal of Modeling and Simulation of Microsystems, 1(1), pp. 9–18 (1999).
38. PHYSICA, Multi-physics software Limited, University of Greenwich, London. http://www.gre.ac.uk/~physica.
39. J. Piprek, Semiconductor Optoelectronic Devices—Introduction to Physics and Simulation, Academic Press, California, 2003.

15

Adhesives for Micro- and Opto-Electronics Application: Chemistry, Reliability and Mechanics

D.W. Dahringer

153 Hawthorne Avenue, Glen Ridge, New Jersey 07028, USA

15.1. INTRODUCTION

The performance and reliability of micro-and opto-electronic devices has been crucial to the recent advances in technology. The successful incorporation of adhesives into the design and manufacturing operations has been a major contributor to those advances. In many innovative optical and microelectronic applications, adhesives may be the only practical, cost-effective choice for producing such devices and components. Despite the ubiquiteness of adhesives in our daily lives, only sometimes obvious, there is a persistent concern and distrust for their use. Much of this can be attributed to causal personal experiences with so-called "miracle glues" and the associated outlandish performance claims. In reality, the overall aversion to adhesives can be attributed more to the misuse of them than to the actual material itself. This chapter will attempt to underscore the importance of proper joint design and assembly processing as well as the adhesive characteristics necessary for the ultimate success of a bonded joint.

15.1.1. Use of Adhesives in Micro and Opto-Electronic Assemblies

There are a number of reasons why adhesives are selected for assembly operations, even when there may be alternative assembly techniques available. Some of these reasons include performance, reliability, cost, ease of manufacturing and form factor.

15.1.1.1. Performance The high performance of adhesives in properly designed and manufactured devices has allowed designers to take maximum advantage of the precise part alignment and thermal and environmental stability achievable. Submicron assembly tolerances in many optical devices are commonplace, often achieving very high coupling efficiencies that are stable over broad temperature ranges. Adhesives are used to bond glass, ceramics and metals in configurations and dimensions frequently incompatible with other fastening techniques. For micro-electronic components, adhesives find use in enhancing the performance and reliability of its discreet chip packaging, BGA's, ferrite magnetics, etc.

15.1.1.2. Reliability Once established, a proper design and manufacturing process using adhesives as the attachment medium, can be both reproducible and reliable. Awareness of potential failure mechanisms and the conditions to which an assembled joint will be exposed, can aid a physical designer in developing the most reliable design and process. Examples can readily be seen in poor joint designs that fail just on cooling to room temperature after curing the adhesive; and other bonded structures that have survived without degradation in performance or integrity over many years of extreme environmental exposure.

15.1.1.3. Cost Adhesives are normally a small contribution to the overall cost of an assembly, despite what may seem like high prices for relatively small quantities of material. Part of this cost comes from the specialty nature of both optical and electro-mechanical bonding materials and partly from the small volume usage over which development costs must be recovered by the manufacturer.

As device dimensions get smaller and bondlines thinner, the quantity of adhesives per device and its cost, almost become irrelevant. Certainly, precious metal-filled (conductive) adhesives used on ground planes could be an exception and certainly the costs associated with the bonding and curing process are part of the total adhesive related costs, nevertheless, they are usually significantly below comparable costs for alternative assembly methods.

15.1.1.4. Ease of Manufacturing The ease of manufacturing a product with adhesives can only be appreciated when the assembly process is compared to a viable alternative. From an absolute perspective, the essence of a bonded joint is to prepare the substrates, fixture, apply the adhesive and cure the adhesive. Each of these steps can be simple or complicated depending on process and performance needs, but are usually less difficult than competing processes.

15.1.1.5. Form Factor Because adhesives will allow very thin bond lines and distribute stresses over relatively large areas, designers usually have wider options in deciding the final shape and size of a device compared to other assembly techniques. This is particularly true in the opto-electronic field where small size has become the industry standard.

15.1.2. Specific Applications

15.1.2.1. Micro-Electronics Most applications for adhesives in micro-electronics fall into the areas of die attach, BGA underfill, component attach (as in surface mount), and component assembly. For die attach, much but not all of the work is with conductive adhesives especially in the manufacture of discreet devices. Underfill applications are designed to improve the reliability of the solder interconnections which would otherwise be stressed by the thermal coefficient of expansion (TCE) difference between the BGA component and the printed circuit board substrate. Component assembly would include devices such as ferrite transformers and inductors, displays, switches and contact pads.

15.1.2.2. Opto-Electronics The opto-electronic field has many specific applications for adhesives, including the ubiquitous fiber to ferrule, V-groove arrays, splitters, gratings, optical bench component assemblies, integrated assemblies and protective packages.

15.2. ADHESIVE CHARACTERISTICS

15.2.1. General Properties of Adhesives

15.2.1.1. Data Sheets Most adhesive suppliers provide data sheets describing their products in an attempt to simplify the customer's selection process by indicating some measure of both application and performance related properties. Unfortunately, from a practical point of view, much of the data is of limited value. Material properties, such as filler content, color, viscosity of one or more parts, storage conditions and cure conditions can be important for the selection process, but in-depth comparisons between suppliers may be much more difficult to realize. Part of this is due to lack of standardization of test methods between suppliers; and part is due to insufficient description of the conditions of the material and the test process; and part to the lack of appreciation by the consumer of the effect different test conditions can have on the reported data. For example, a supplier may indicate that an adhesive is suitable for use to, let's say 250°C based on a TGA weight loss of 5 or 10% at that temperature. In reality, the test sample (whose shape is typically that of a pellet) is cured in bulk, and could behave totally different than a thin adhesive bondline between substrates. For example, volatiles, either in the uncured adhesive or the substrate, or even generated as part of the curing process, might be trapped in the bondline but not necessarily in the bulk specimens. Furthermore, a loss of 5–10% weight (as a gaseous degradation product) could have severe (mechanical) implications in a bondline, especially if the adhesive is likely above its glass transition temperature.

Another example of data sheet confusion is the glass transition, supposedly an indicator of the point where a material switches between a glassy solid and a rubber. In addition to the fact that it is usually a range instead of a "point," one can measure different T_g's on the same sample, depending on the test method, DSC, TMA or DMA. Even with a single test method, variations in the test parameters (e.g., sample size, scan rate) can result in different values. Interpretation of the collected data can also add variability, e.g., some DSC methods pick the inflection point midway between the change in measured heat capacity, while others may select the first deviation. With TMA analysis, either the initial point of change in TCE, or the intersection of the two slopes is considered the T_g.

Additionally, different sample preparation schemes can cause major variations in T_g even if all of the above considerations have been standardized. For example, if a test sample is cured dynamically in a DSC at some rate to some maximum temperature and then rescanned for T_g, the result could be significantly different depending on the cure scan rate, maximum temperature and any thermal dwells. Isothermal cures can add additional possibilities when both time and temperature factor into the results.

To further complicate this matter, the actual thermal profile of a specific adhesive in an actual joint is often quite different than a design process engineer's expectation. In oven cures, the bondline (adhesive) temperature can be very different than the oven temperature depending on thermal mass of the substrates and recovery time of the oven. This can be a major factor in determining the zero stress condition/temperature for a joint, especially for rapid curing adhesives.

Other typical data sheet properties, involving less controversy, are color, density, specific gravity, mix ratio (for multi-component system), storage requirements and cured electrical properties. Viscosity is a data sheet parameter having a major impact on process adaptability for some applications, especially those requiring capillary flow, such as chip underfill; or high thixotropy such as surface mount adhesives. For non-thixotropic (shear insensitive) adhesive systems, the viscosity reported is usually meaningful (at least at room or reported temperature), however, for the systems with filler and the possibility of shear sensitivity, the actual reported viscosity may be significantly different from the real viscosity at the use condition. Thixotropic index (a ratio of viscosities at two different shear rates) can sometimes provide useful information (although mostly intuitive for the practitioner). Typically missing from the data sheet viscosity equation is the affect of temperature (either recommended application or cure) on this property.

Data sheet representation of moisture absorption is often misleading in that influential factors in specimen preparation, such as cure condition, size, shape, porosity and exposure conditions (such as time and temperature) are often missing. Generally, an individual supplier will be consistent with data sheet report of moisture pick-up values for different materials in their line, but comparing one supplier to another can be difficult at best. A further possible distortion of moisture data can come from materials that may have a water-leachable component. This can best be determined by a redried weight loss method, which is rarely reported and quite possibly be the explanation for some claims of low moisture absorption adhesives.

15.2.1.2. Physical Properties The physical properties of adhesives can and should vary with the specific process needs and the end use application, thus largely preempting the possibility of a "universal adhesive." Typical bulk material mechanical properties, like tensile strength, elongation and modulus, unfortunately provide little in the way of practical help in choosing an adhesive, although a low modulus product is sometimes selected in an attempt to reduce joint stresses where the substrates differ in TCE.

Tensile strength of an adhesive is unusually difficult to measure, mainly because of sample preparation problems. Sometimes a butt tensile test can be used, but rarely would one expect true, cohesive failure within a bondline. On the other hand, a non-axial loaded butt tensile failure is probably a better simulation of a cleavage test and can relate to some of the common component attachment applications in optical assemblies. Many companies have developed QC tests based on cleavage stress using a simple glass rod to glass slide test joint. Fixed length rods pushed at the top, parallel to the bondline, can be a convenient test specimen for not only measuring initial properties of an adhesive but also the effect of environmental conditions such as high temperature and/or high humidity aging or thermal cycling.

The property of elongation is somewhat useful in guessing the performance of a large fillet, calculating the edge stresses of a thin bondline or figuring the likelihood of a mismatched TCE joint failing. Elongation does come into play in joints with significantly thick bondlines and situations where a high elongation (and low modulus) adhesive is used for a potting type application. Factors often overlooked in these applications are the necking affect perpendicular to stress induced elongation and the usually high effective volume TCE of low modulus materials. For most well designed, thin bondline joints, elongation is of little practical importance, primarily due to the physical constraints imposed on the z-axis by the substrates' influence on the x and y-axis and, of course, the small "h" of the adhesive.

Specific gravity or sometimes density is useful in calculating cost per unit volume and volume ratios of multi-component adhesives when only weight ratios are provided. In general, unfilled adhesives average close to 1.0 g/cm^3 but can reach much higher (or even lower) densities with the addition of fillers and/or air. Air can be unintentionally added during hand or machine mixing without appropriate degassing. A simple method of confirming the existence of entrapped air (even with heavily filled adhesives) is to squeeze a drop between two glass microscope slides and inspect under magnification for typical spherical (usually appearing dark) voids. Shims or large particles (spanning the bondline) can cause shrinkage voids on cure which differ in appearance from entrapped air. These voids often have a "crowfoot" appearance, a halo effect around a discreet, large particle or sometimes cracks between particles or columns as can occur in underfill encapsulants.

Refractive index typically averages 1.56 for epoxies and 1.42 for acrylics and perhaps even lower for silicones. What may need clarification in refractive index applications is that numbers reported are often obtained on an uncured specimen (for easy cleanup of the instrument), at a wavelength and temperature different from the final use conditions. Each of these factors, although quite small, can have an impact on some in the light path applications.

The change in volume of an adhesive as it cures is called cure shrinkage and results from the fact the curing agent and resins in a typical adhesive have a lower density (take up more volume) than the final cured material. A typical data sheet value can be determined by using density ratio (before and after cure) however, this may not reflect accurately on an application requirement or on other shrinkage test methods which may report lineal instead of volume shrinkage.

In a real life application, where an epoxy is placed between two substrates and heated to some curing temperature, a very complex scenario is played out. First, the liquid adhesive expands as the temperature rises and excess material exudes from the joint (or the substrates move slightly apart if they are not constrained). Second, as the cure mechanism begins, volume shrinkage causes the still liquid exudate to backfill up to the point of gelation. Gelation occurs roughly in the 30–40% of full cure range. At that point, no more backfill can occur and further shrinkage must develop stress in the z-axis of a constrained thickness bondline. What is even more significant is that after gelation, the x and y-axis of the adhesive in the bondline are constrained by the substrates and therefore all the remaining volume shrinkage is forced into the z-axis. The final step of the process is cooling the assembly from cure temperature to the use temperature which further adds to the z-axis shrinkage and the stress on a constrained assembly. In the case of an unconstrained assembly of two planer substrates of the same TCE (visualize two glass microscope slides), the bondline thickness will vary to compensate for each of the steps and will provide an essentially stress-free joint. Unfortunately, many practical bonding applications do not consist of both planer and same TCE substrates.

15.2.1.3. Thermal Properties Thermal properties of adhesives can be examined from both a chemical and a mechanical viewpoint as well as the short term and long term effects. The chemical effects tend to be relatively straightforward whereas the thermomechanical effects can be quite complex. The appreciation and understanding of these thermomechanical effects can dramatically improve a practitioner's success in designing high performance and reliable adhesive joints.

In general, most thermosetting epoxy adhesives undergo few chemical changes below the temperature at which they were cured. Exceptions to this include adhesives that have been incompletely cured at the "cure" temperature, adhesives cured at very high temperatures (sometimes called snap cure), adhesives that have been mis-formulated (i.e., too much or too little curing agent) and adhesives that contain unreacted or unreactive components. Subsequent heating of an incompletely cured adhesive can cause additional cross-linking which will generally raise T_g, cause additional shrinkage and improve chemical and moisture resistance. In some adhesives, additional heating causes slight chemical rearrangement where strained bonds are broken and new ones form. At higher temperatures several different forms of degradation can take place. Acrylics can undergo a depolymerization leading to volatilization and weight loss. Epoxies can develop color followed by charring and weight loss or gain possibly due to bond breakage or oxidation. Small amounts of even the most aggressive degradation associated with very brief high temperature exposure can, in some cases, be tolerated from a chemical perspective, if the mechanical aspect of the joint does not fail or degrade significantly.

15.2.1.4. Thermomechanical Properties One of the most interesting aspects of adhesive engineering and probably the most significant in terms of joint design for reliability in non-ideal assemblies (as opposed to the "ideal" joint having an unconstrained planer bondline between substrates of identical TCE) involve the interaction of thermomechanical properties of an adhesive and its substrates during thermal excursions. The TCE of an adhesive is typically defined as the change in length of a material, per degrees C change in temperature, and usually expressed as parts per million per degree C, or ppm.

TCE's are typically measured in a dilatometer where a specimen height (length) is recorded as a function of or programmed temperature ramp and the slopes of the curve are used to calculate $\alpha 1$ and $\alpha 2$. At the T_g of thermoset epoxies, a change in expansion coefficient takes place, with a new slope $\alpha 2$ and value approximately three times that below the T_g. The intersection of the two slopes provides the typical TMA value for T_g and from a practical point of view, the more useful data point for determining joint thermomechanical properties compared to the value obtained in a DSC. Typical TCE's of unfilled epoxies normally run around 60–70 ppm/°C below T_g and 180–210 ppm/°C above T_g.

The incorporation of fillers into adhesive formulations can be used to modify the TCE's of the adhesives. In general, the amount of TCE reduction tracks the volume fraction and individual TCE of the components. When lowering the adhesive's TCE is a prime concern, silica is generally the filler of choice because of its very low TCE. There is a practical limit to how low in TCE one can go and still maintain a fully "wet out" filler and void-free space between the filler particles. The use of spherical silica particles with precise (multi-modal size distribution) has enabled some adhesive formulators to manufacture adhesives with TCE's of 12–15 ppm/°C. This range can be very helpful in some applications where joint design is less than ideal. There are, however, trade-offs in the use of such materials, mainly in the handling where these heavily filled adhesives are likely to be stiff pastes, making adhesive dispensing and placement difficult. For a balance of handling and low TCE, values in the range of 30 ppm/°C seem to be popular, especially in the BGA underfill market.

The use of low TCE materials to control unconstrained z-axis movement in optical devices may actually be counter-productive. This is because the paste-like nature results in a thick bondline versus the extremely thin bondline achievable with some unfilled materials. For example, one might be able to squeeze a 15 ppm adhesive to 100 microns, but an unfilled version of the same adhesive could be capable of forming a joint with a thickness of a micron or less. In this example, the low TCE adhesive could cause a joint movement (z-axis) approximately 20 times that of the unfilled adhesive. It is also important to realize that a TCE number obtained from a TMA will probably not calculate the same as the z-axis expansion of the adhesive in a joint. The reason for this is that the substrates will constrain the adhesive in the x and y axes, forcing the "unused" portion of the volumetric expansion into the z-axis. In the case of two quartz substrates with TCE's close to zero, virtually all of the volumetric expansion of the adhesive will go into the z-axis. It should also be noted/repeated that this TCE "accommodation" does not cause temperature cycle stress problems for unconstrained planer joints with matching TCE substrates.

Thermal conductivity is another characteristic of an adhesive that can be important to assemblies requiring heat sinking or heat dissipation. Unfilled epoxies typically have very low thermal conductivity, roughly 1 to 5% of copper. Thermal conductivity can be enhanced by incorporating select fillers, such as aluminum oxide or boron nitride if electrical insulation is important; or metal powders if electrical conductivity is desired or not a concern. Even with the highest conductivity fillers, it is rare to achieve bulk thermal conductivity as high as 15% of copper. One can again make an analogy similar to that of TCE above where higher thermal flux may be achievable with a thin unfilled bondline versus a much thicker, higher TC material.

Thermal conductivity, like electrical conductivity, does not appear to follow volume fraction averages, but seems to require some minimum concentration, where the filler particles are sufficiently numerous so that each particle has a statistical probability of touching at least two other particles. This, like any filler addition, can raise handling and application issues.

15.2.1.5. Rheology From a practical point of view, the rheological properties of an adhesive define the way it can or needs to be used, viscosity and thixotropy being the most common and generally useful properties. A low viscosity will permit easy mixing if necessary, syringe dispensing, flow into confined spaces and fast wicking, however, it can also be the cause of unwanted spreading, dripping, sagging and drooling. High viscosity may reverse the advantages mentioned above but not necessarily prevent spread, drooling, etc., especially if a high viscosity material is applied at room temperature and subsequently heated for cure. As indicated earlier, very few data sheets mention the affect of temperature on viscosity. Perhaps the closest practical indicator is for underfill type materials where a flow temperature is suggested along with some measure of time to fill a simulated BGA cavity.

Viscosity can be controlled (up or down) in a number of ways; molecular weight of the resins and curing agents (which can include adducting), both reactive and non-reactive additives, inert fillers (usually only up) and of course temperature. Another technique used to raise viscosity by some end users is to allow a room temperature reactive mixed adhesive to advance (pre-react). Barring historical reasons, the interdependency of time and temperature can cause this method to be technologically challenging for both product consistency and process operating window.

Unlike viscosity, where one can easily develop a sense of a material from a single number, thixotropy is a more difficult concept to grasp. Thixotropy refers to a material's apparent viscosity that becomes lower under the influence of shear. Data sheet references to thixotropic index (TI) are typically based on the ratio of viscosities at two different viscometer spindle speeds. Little standardization exists in setting the parameters for TI, either between suppliers or within adhesive categories. These numbers rarely conjure up a usable mental picture of a material unless at least one of the actual viscosity and test speed conditions are indicated. The TI, however, can be a very useful QC tool for suppliers as it can be significantly influenced by formulation precision, the manufacturing process and air incorporation, as well as storage conditions. For the novice, one of the best ways to describe thixotropy is by a familiar example such as mayonnaise, where it spreads easily and doesn't run or drip. Another example could be the latest in decorator paints where the material will brush or roll easily yet not drip between the bucket and wall or ceiling.

Thixotropy can be a major contribution to the successful processing of electronic and optical assemblies. And perhaps one of the most obvious illustrations of thixotropy in electronics is the typical surface mount adhesive. This is dispensed onto a board, often between electrical contact pads to hold a surface mount component in place during a solder wave step (often in defiance of gravity). The material is formulated to dispense through very fine needles and to retain its shape until component placement without slumping or contaminating nearby solder pads. The adhesive must also not trail (a like tail) "strings" as the needle is moved from place to place on the board. Other versions of the functional type of adhesive may have slightly different levels of thixotropy to allow pin transfer or stencil application techniques.

15.2.2. Adhesive Chemistry

15.2.2.1. General Many, perhaps many hundreds, of attempts have been made to define adhesives, for the obvious reason that the perfect definition has yet to be recorded. From a practical standpoint, let us consider an adhesive to be "a material that can be applied between two substrates and which will then change to a solid material having desirable (functional) properties." Interestingly, this "definition" would include solders and even water for sub-freezing applications but would exclude a significant portion of the pressure-sensitive adhesive field.

15.2.2.2. Classifications Adhesive can be classified by; one of their reactive groups (e.g., epoxy, acrylic, cyanoacrylate); by reaction product (e.g., urethane, silicone, polyamide); by end use (e.g., construction, paper, medical); by application technique (e.g., spray, screen, hot melt); and by cure characteristics (e.g., thermosetting, UV, anaerobic) to name a few. These casual classifications can lead to confusion outside the specific area of use and especially when used incorrectly as in the case of "epoxy" being used to describe many non-epoxy adhesives or "UV epoxy" frequently used to describe a UV curable acrylic.

In the micro and opto-electronic fields, the most prominent categories of interest are the thermosetting epoxy and UV curable acrylic and epoxy formulations. In addition, some amount of silicone, anaerobic, thermoplastic and cyanoacrylate can be found along with pressure-sensitive types.

15.2.2.3. Cure Specifics A number of scholarly texts exist on the detailed chemistry of adhesives (1–4) and will not be part of this chapter. However, we will try to cover the more practical aspects of how one gets from the pure chemistry to the usable adhesive product. The most common forms of adhesives are generally based on a resin and a curing agent whose primary purpose is to change the resin from a dispensable state (usually fluid) to a solid, capable of holding the two substrates together.

Epoxy resins can vary between low viscosity fluids to meltable solids and can contain two or more reactive epoxy groups per molecule. The most popular resin is the diglycidyl ether of bisphenol A with two reactive groups, and a viscosity similar to heavy motor oil or honey. Other resins include; cycloaliphatic, aliphatic, novolacs and other multifunctional epoxies. In the curing or cross-linking process, a curing agent activates the reactive groups, causing them to combine with the curing agent or themselves, leading to cross-linked networks or higher molecular weight resins.

Much of the sophistication in epoxy formulation lies in the selection of the curing agent, for that mostly determines shelf life, pot life, cure condition (time and temperature), packaging options and to some degree, dispensing options. Resin selection can also influence the same properties but usually to a lesser degree. The most common curing agents for the typical epoxy resin adhesives are based on amines and their relatives including aliphatic, aromatic, cycloaliphatic, tertiary, adducts, amides, imidazoles, etc. On occasion, more than one curing agent can be used to achieve desired handling or performance characteristics.

Another class of epoxy curing agents include the acid anhydrides, but these are used more for potting and casting applications than adhesives, mainly because of longer cure times. The (acid) anhydride curing systems usually require an accelerator such as a tertiary amine. Lewis acid complexes and mercaptans are additional categories of curing agents for epoxy resins. Several of these curing agents can be considered latent by virtue of no (or almost no) reactivity at room temperature, but activate to cure the resin at some elevated temperature. The ways that this may be achieved include; chemical blockage with a sharp disassociative temperature; mechanical blockage as with a meltable wax coating; a solid curing agent with a sharp melting point and a very slow reactivity at room temperature. Mechanical blocking systems intuitively result in a coarser dispersion than can be achieved with liquid curing agents or very finely powdered curing agents dispersed under high shear. The mix ratio of resin to curing agent is usually based on chemical stoichiometry, but can be modified to optimize certain properties and performance.

Another class of curing agents for epoxy resins (mostly cycloaliphatic) are the UV cationic and more recently the thermal cationic agents. These curing agents can be relatively stable (protected from UV) at room temperature, yet solidify rapidly when exposed to intense sources of UV, or elevated temperature for the thermal cationics. Certain chemical compounds (e.g., amines and materials containing sulfur) can inhibit the cure of adhesives based on this cure chemistry. The popularity of these UV systems in the opto-electronics assembly area can be attributed to the fast cure, which sometimes can then be supplemented with additional thermal cure to improve properties such as T_g and chemical resistance.

UV acrylics have been used as optical adhesives for a relatively long time, but have been significantly replaced with thermal or UV epoxies due to better performance. The acrylate monomers (resin) are most often cured by a free radical mechanism which, depending on the functionality of the resin, can yield a linear (thermoplastic) polymer or cross-linked (thermoset) polymer. The source of free radicals could be thermal (e.g., peroxide); UV activators or electron beam. Disadvantages of the early acrylates included high cure shrinkage, outgassing, limited environment resistance, poor high temperature performance and possible air inhibition. Newer generations of acrylate formulations have enhanced the ease of use and the performance, but generally not to the level of thermal epoxies.

Silicones are usually rubbery materials mostly used as sealants and shock and vibration mounting materials, although occasionally used as an adhesive. They are available in a two part, room temperature and elevated curing systems as well as one part moisture, thermal and UV curable systems. The cured silicone rubber typically has very high TCE and low modulus over an extended temperature range, and is resistant to water and many chemicals, especially acids and bases. Exposure to many organic chemicals (e.g., solvents) will most likely cause swelling, sometimes severe. As a sealant, silicones provide excellent water barriers but allow moisture to pass quite readily.

Anaerobic adhesive/sealants are close relatives of the acrylate adhesives that are formulated to cure in the absence of oxygen (e.g., when placed between two substrates) and quite rapidly with an initiator (usually on an at least one of the surfaces). Originally employed as thread locking materials, extensive development efforts have led to other types of adhesive applications. Most notable use in the optics industry is in the field connectorization of optical fiber (inside a ferrule), where a no-mix adhesives and rapid room temperature cures are of interest; oxygen inhibition at the air interface can negatively impact the polishing process.

Thermoplastics or hot melt adhesives, that are applied hot and form a bond when they cool and solidify, are sometimes used for non-critical or temporary "hold down during assembly" applications but they usually suffer from creep/flow under mildly elevated temperatures. The traditional hot melt adhesive is based on a linear polymer chain with additives for stability and tack. Advanced hot melt systems can incorporated some cure capability that proceeds, usually slowly, after application. The most common mechanism seems to be moisture induced cross-linking.

Cyanoacrylate adhesives, the so-called "super glues," are one part, with or without a primer/activator) that cures very rapidly when pressed between catalytically active surfaces. Issues of performance, such as heat and moisture resistance and poor peel and impact resistance, have limited the usefulness of this class of adhesive in the optical and electronic industries, except for temporary fixturing or holding.

15.2.2.4. Additives   Adhesives are rarely just combinations of pure chemicals (e.g., resins and curing agents) as suggested above. Adhesive suppliers devote much of their development energy to modifying the "two chemicals" to enhance either the ease of use or the performance of the cured material and assembly. Some of the additives used include; fillers, flexibilizers, tougheners and adhesion promoters.

15.2.2.4.1. Fillers.   Fillers are used for a variety of purposes in adhesives, including density control, viscosity, thixotropy, TCE, mix ratio modification, cost, opacity, color, compressive strength, heat distortion temperature, electrical conductivity and thermal conductivity. Most of the fillers used in the industry are mineral or metal but organic fillers such as ground rubber or wood flour have been used. Silica in numerous variants is probably the most common filler material in high technology epoxy systems. Calcium carbonate, talc, aluminum powder, silver flake, boron nitride and chopped glass fiber are also used for specific purposes. The usual form is a fine particle with a fairly narrow size distribution (achieved by screening, air flotation or other methods). Particle size and surface area will affect the amount of filler that can be added to a specific resin or curing agent without exceeding a rheology requirement. Incorporation of a filler into a liquid adhesive component becomes more difficult as the average particle size of the filler decreases and the viscosity of the liquid increases. Most adhesive manufacturers will incorporate fillers under vacuum and high shear conditions, primarily to eliminate voids (air) and particle agglomerates.

15.2.2.4.2. Flexibilizers.   Flexibilizers can be added to resins or curing agents if the adhesive is deemed to be too brittle. Choices for flexibilizing agents include higher molecular weight epoxy resins, non-aromatic backbone epoxy resins (both mono and difunctional), compatible but non-reactive plasticizers and high molecular weight curing agents. Most of the approaches to increasing flexibility will lower T_g of the cured adhesive compared to the non-flex version. In addition, most of the flexibilizing additives will change the pre-cure viscosity of the adhesive, some significantly lower, others higher. The addition of most flexibilizing agents to traditional epoxy systems generally causes the cure rate to slow.

15.2.2.4.3. Tougheners.   In some respects, toughening agents appear similar to flexibilizers but in reality, their function is to act as crack stoppers. This is usually accomplished by the formation of discreet particles of a "rubber" within the adhesive matrix. Two approaches have been used; in situ formation of rubber particles separated from a compatible mixture as the resin begins its cure process, and direct incorporation of fine rubber particles into the adhesive matrix as a filler.

15.2.2.4.4. Adhesion Promoters.   Adhesion promoters can be added to an adhesive to enhance joint strength, durability and environmental resistance. They can also act to enhance or modify the surface of fillers to change the amount of reinforcement realized or even to modify the interaction of the filler surface with the environment (such as moisture). The most common adhesion promoters are based in silanes but other chemistries, such as titanates and zirconates have also been explored.

In theory, the silane is capable of reacting with the surface of a substrate and also provide a "tail" capable of reacting with the adhesive matrix. The result is a substrate interface that theoretically should have more resistance to attack by aggressive environments, such as high temperature and humidity. The best utilization of an adhesion promoter is to apply it directly to the substrate before applying the adhesive, however, it can also be incorporated into the adhesive so that upon adhesive application, the silanes have a chance to get to and react with the substrate surface. Care must be exercised in selecting the silane reactive "tail" to avoid premature reaction with the adhesive and causing shortened shelf life.

15.3. DESIGN OBJECTIVE

15.3.1. Adhesive Joint Design

15.3.1.1. Stress Sources   Whenever possible, it should be the goal of a physical designer to minimize inadvertent stresses on a bonded joint. In the real world, more often than not, the design of a joint is fixed by system requirements and the adhesive must try to make the accommodation. This can, in fact, create situations in which no adhesive could survive and lead to extensive (and expensive) screening of materials (and perhaps processes) from many suppliers; and in the end contribute to the questionable reputation of adhesives.

Stress on a joint can emanate from two distinct sources; external mechanical and "internal," resulting from the interaction of substrate properties, adhesive properties and the changes in the adhesive during cure. Externally induced stresses are usually used to measure a joint's performance, which is in essence the "true" joint resistance to stress minus the preexisting internal stress.

Once a joint is assembled and the adhesive cured, stress in the joint can develop as a result of:

1. Adhesive post gel cure shrinkage.
2. Non-planar adhesive bondline.
3. Substrate TCE differences.
4. Static thermal gradients.
5. Dynamic thermal gradients.
6. Moisture or chemical induced adhesive swelling.
7. Chemical changes at the substrate interface.
8. Presence of bondline shims.

Some of the causes of stress can be reversible (usually on temperature change), others irreversible and even a combination of the two as when a partial delamination or disbond occurs.

15.3.1.2. Substrate Properties   The properties of each substrate in an adhesive joint are in many respects as or more important than those of the adhesive, certainly the substrates are the means through which external stress is applied to the joint, and the means which cause or allow internal stresses to be created.

The internal stress can be thought of as composed of two major components; those created by the combination of adhesive cure shrinkage and dimensional constraints of the substrate design; and those created after joint formation (adhesive cure) by the relative dimensional changes of the substrates (and adhesive) as the temperature of the joint deviates from the zero stress condition*.

The TCE of each substrate will determine its actual size (adjusted for stress) at any given temperature, so that substrates with different TCE's will change size relative to each other as the temperature changes. This differential will increase proportionally to the change in temperature (in either direction) from the zero stress condition/temperature. Depending on the rigidity of the substrates (section modulus), the bonded assembly may simply warp (as in a bimetallic element) or place shear stress on the adhesive if the substrates are incapable of bending.

In the so-called ideal joint, where the bondline is planer and the TCE's of the substrate are identical, there is essentially no internal stress on the assembly. In this type of a joint (where the substrates are *unconstrained* by fixtures or shims, etc.) the adhesive post gel shrinkage will simply move the substrates closer together and not generate stresses. Since the post gel adhesive is effectively attached to the substrates in the x and y planes of the adhesive, the total volume shrinkage (less the substrate x and y change) must occur in the z-axis or normal to the adhesive plane.

One can demonstrate both the effects of constrained and essentially unconstrained adhesive shrinkage in a simple experiment. By assembling two glass microscope slides on a hot plate with two 5 mil shims near the ends (broken cover slips work well) and wicking an adhesive (e.g., Epo-tec 353 ND or Zymet F-711) into the pre-heated (~100–125°C) assembly gap. Just before the adhesive begins to gel, voids will form around the shims during the brief period when the material is still fluid by too viscous to backfill, followed by gelation and cure. After cure, one can actually measure the difference in bondline thickness at the center of the slide compared to the ends over the shims. The elasticity of the microscope slides allows the bending to occur near the shims leaving the central portion of the assembly relatively stress free.

In the non-ideal joint configuration, where the bondline is planer, but the two substrates differ in TCE and have a high section modulus, there will be shear stress at the adhesive layer/substrate interface as the assembly temperature deviates from the zero stress temperature. Depending on the thickness and modulus of the adhesive layer, some of that stress may be dissipated, however, since most bondline (especially for optical assemblies) are usually quite thin, the adhesives' ability to absorb the differential expansion is quite limited. This situation can lead to progressive delaminating of the adhesive from one of the substrates, and is most evident upon repeated low temperature exposure such as during temperature cycling. High temperature exposure is less likely to cause delamination in this type of joint design because the higher temperatures are usually closer to the zero stress temperature (therefore minimizing ΔT) and also closer to T_g which significantly lowers the modulus of the adhesive.

More complex joint designs, for example, those with non-planer bondlines and those with the equivalent of shims can become more vulnerable to failure during temperature cycling, elevated temperature exposure and even moisture uptake. Two common examples of poor design and materials choices in the fiber optic arena include the V-groove fiber array with a flat glass plate cover, and a typical fiber ferrule with a funnel end mainly

---

*The zero stress condition is normally thought of as the temperature at which adhesive cure takes place, however, this is not exactly correct. Firstly, the major reference point for zero stress condition is the temperature at which gelation occurs and then modified upward to compensate for post gel shrinkage in the adhesive. In cases where the substrates differ in TCE or the bondline is non-planer, it may be impossible to attain a true zero stress condition.

used for fiber guidance and after assembly strain relief. In the v-groove array assembly, the etched silicon array is used to provide highly accurate fiber positioning in the x, y and θ orientations. The flat plate makes the third point contact on each fiber (the first two from the sides of each v-groove); and is sometimes used to allow UV light to cure a UV adhesive and to inspect for voids which could compromise end polishing of the fibers. The relatively large volumes of adhesive at the apex of the v-groove and between the fibers, cover plate and wafer enhance the possibility of voids forming as a result of cure shrinkage; as well as disbonds between the mechanically constrained (by the fibers) wafer and plate. The reverse will occur at temperatures above the zero stress temperature where the layer of adhesive will expand and actually lift the plate, causing disbonds around the fibers. This last mechanism can be exacerbated by moisture uptake (whose effect can simulate much higher temperatures in the adhesive).

The second example is the widely used fiber ferrule, where a short length of capillary tube (made from glass, ceramic or metal) is used to mechanically support/protect the end of a fiber and to allow end polishing, handling and sometimes passive alignment. In the most common ferrules, the tube has a center bore with micron or sub micron level clearance for a fiber and a "funnel" like end which is used for a variety of purposes including ease of adhesive injection, fiber entry guidance, fiber buffer termination and fiber strain relief. Depending on the adhesive and the application technique, the annulus (between fiber and ferrule) can be from 50–95% full, which generally prevents movement of the fiber within the ferrule. The incomplete fill of the capillary can be attributed to poor application technique, air entrapment in the adhesive and shrinkage of the adhesive upon cure and cool down. Problems can arise if the funnel area is filled with a rigid adhesive and/or one that has a high moisture uptake. Firstly, if the assembly exceeds the zero stress temperature, the expansion of the cone of adhesive can put a strong tensile force on the fiber and actually cause a fracture in the capillary very near the apex of the cone. This, of course, assumes that the cone adhesive is strongly bonded to the fiber and can be explained by the high TCE of the typical unfilled adhesive in the cone and the resultant movement of cone bulk away from the apex as it expands. Again, moisture uptake can provide the same effect or exacerbate the temperature effect. Another problem related to this assembly is the that the rigid "strain relief" is in fact a stress concentration that can cause fiber fracture under mild side loading of the buffered fiber relative to the ferrule.

15.3.1.3. Non-Planar Bondlines In adhesive joint designs where the adhesive bondline is not planar, both the TCE and cure shrinkage of the adhesive becomes significant reliability factors. Depending on the actual geometry of the joint design, a non-planar bondline will usually cause a partial disbond to occur during excursions from the zero stress temperature. Exceptions can arise from compliant substrates and some cases of bondline fold symmetry (visualize two pieces of angle iron bonded together).

In a special case of two flat substrates bonded with a wedge of adhesive (substrate tilt), the joint would likely resist temperature cycling but will cause temperature dependent relative movement of the substrates and account for poor optical assembly performance. To illustrate a different effect of non-planer joints, imagine a block of, let's say aluminum, bonded into the apex of an uneven angle of the same material. The longer leg of the angle will constrain movement along that leg, preventing the short leg from following the normal shrinkage of the adhesive between it and the block. The result is that the shorter bondline will experience severe normal stresses, not only just as a result of cure, but also the added TCE shrinkage as the assembly cools from the zero stress temperature.

15.3.1.4. Thermal Gradients The designer of an adhesive joint needs to be aware of the potential for thermal gradients to raise stress levels in bonded assemblies. Even ideal adhesive joints assembled at a uniform temperature can experience stresses if exposed to thermal gradients where one substrate is at a different temperature than the other substrate. In reality, this situation can mimic that of a substrate TCE mismatch. An even more potentially damaging scenario can be encountered under thermal shock conditions where a severe dynamic gradient can sometimes damage both bondline and substrate.

15.3.1.5. Moisture or Chemical Absorption Most adhesives (and other organic materials) will absorb moisture (and other chemicals) when environmentally exposed. Conditions of exposure, such as time, temperature, pressure and concentration can influence the total amount of absorption. In addition to those factors, geometry and diffusion rate will affect equilibration time. The absorption of moisture or solvents usually results in an increase in volume or swelling and can also affect other properties such as modulus and T_g through a plasticization mechanism. Most of these changes are reversible, although sometimes T_g can be permanently lowered. In the case where exposure to moisture actually leaches material from the adhesive (as demonstrated by lower re-dry weight than the original pre exposure weight), reversibility is less likely and T_g might even increase.

Stress on a joint can change due to moisture absorption in a fashion similar to raising the temperature. Another way of looking at this is to expect that the zero stress temperature of a moisture saturated bondline would be considerably lower than that of a dry sample. If one then adds elevated temperature to the assembly, stress conditions can increase dramatically over a dry sample. Some studies in the past have shown that diffusion into a bondline (even an ideal joint) can cause a swelling gradient apparently related to the diffusion characteristics of the adhesive. This "wave front" separating the dry from the damp (swollen) adhesive can add to joint stresses.

15.3.1.6. Substrate Surface Chemical Changes Many substrates are bonded with the surface in a relatively active state. This provides the opportunity for post bond reaction with chemical species present in the environment. Probably the most common example would be moisture-induced oxidation (corrosion) of the substrate at the adhesive/substrate interface, similar to rusting of steel under a layer of paint. Most pre-bond surface preparations (at least of metals) try to stabilize the surface and then try to prevent corrosion, even if it means replacing a native oxide layer with a more tenacious and less reactive oxide or other chemical species.

When the chemistry of a substrate surface at an adhesive interface changes, the original bond strength will usually be compromised. One can easily envision the case of oxide growth, but there is also the situation where the adhesive may be capable of reducing substrate oxides to which the adhesive was originally bonded. Either case can cause weakness and/or failed joints.

15.3.1.7. Performance Qualifications During the early development stage of a new optical device or component and occasionally a microelectronic component, it is often necessary to build prototype or even proof of concept designs to gage performance and ultimately to prepare performance specifications for the commercial product. This process is generally interactive in both design and assembly as its environmental window (e.g., temperature, humidity, shock, etc.) migrates from a room temperature operating prototype to a viable commercial range. Typically, the change in performance of a device measured over a temperature range can be useful in analyzing the design factors responsible and thus the remedy. Subtleties in the repeatability, or lack thereof, of the measured performance on temperature cycling can further aid the design iteration process.

Another part of the design process involves the commercial viability of the assembly steps and of course the materials selection process. Ideally, timely expertise in all of these areas would be available for any new design project, but unfortunately, this is usually not the case, often resulting in significant project and development cost overruns.

15.3.2. Manufacturing Issues

15.3.2.1. General The goals of manufacturing engineers are on occasion somewhat diverse from those of the design, quality and sales portions of the organization. Typically, a cost driven position in manufacturing can lead to compromises in a products' performance and reliability. It is very important that each of those organizational areas work together to achieve a product that will meet or surpass a customers needs. This may mean the use of a more costly or longer manufacturing step to achieve better performance and reliability. There are, however, cases where a lower manufacturing costs process can still provide a product that will meet the reliability needs of a customer and makes economical sense. An example might be the use of a non hermetic package for a central office type application. Many times the use of extra time or care in an adhesive bonding process can lead to yield improvements and perhaps even lower manufacturing costs.

The earlier on in a new project that physical designers and manufacturing engineers can work together, usually the more successful the result, especially if each group gains a broader perspective of the others challenges. Early and frequent design reviews can be a productive approach, perhaps using a facilitator with an interdisciplinary background.

The following sections will address manufacturing issues specific to the practice of using adhesives in the assembly of optical and electronic devices.

15.3.2.2. Fixturing The design of bonding fixtures can be a challenge, especially in the optical assembly field where submicron positioning of components is often required. Additionally, an adhesive must be applied to the joint and then the assembly must undergo a cure step before unfixturing. Consideration must also be paid to the affect of adhesive shrinkage on the joint stress as well as thermal movement of the fixturing during the cure step.

Fixturing tools can be quite expensive and have a major impact in a volume manufacturing environment where cycle time requirements and cost influence the number of fixtures needed.

Fixturing in micro-electronic systems is usually not as elaborate as with optical components and sometimes even rely on the self-fixturing afforded by the rheology of adhesives; as in the placement of surface mount components prior to cure. In the case of chips/BGA underfill, the component is already fixtured by the solder interconnections.

In the design of fixturing for optical components assembly, some of the following may be considered important:

1. Gripping interface (part holders).
2. Degrees of movement.
3. Robustness.
4. Precision of movement.
5. Repeatability.
6. Backlash.

7. Thermal stability.
8. Ease of loading.
9. Ease of alignment.
10. Mechanical stability.

15.3.2.3. Curing there are several approaches to curing adhesives in a manufacturing environment. Some of the factors influencing the cure selection will include those associated with the adhesive, the fixturing, the components and the production schedule.

Thermally cured adhesives, be they two component room temperature curable or one component (premixed), will generally cure faster at higher temperatures. If, for some reason, a low temperature cure is necessary, relatively long cure times will be encountered (e.g., 20 minutes to 24 hrs), obviously depending on the formulation. For some applications where fixturing is quite simple (as for instance a small clip), many units can be assembled and allowed to cure overnight. The shorter low temperature cure time comes with the typical trade off of short pot life, frequent mixing and usually significant material waste. By using a higher cure temperature, adhesives with longer pot lives can be selected. Again, with simple fixturing, this approach allows both batch oven and continuous or tunnel oven curing. Cure times of relatively long pot life premixed adhesive compositions can be as short as several seconds to an hour or so, again, depending on the formulation and temperature. Some of the means used to heat assemblies include focused IR, forced hot air, resistance wire "ovens," inductive heating, magnetic (eddy current) heating, embedded heating elements, as well as the simple hot plate. Depending on the heating method, attempts to achieve very short cure times may cause thermomechanical problems in the assembly due to the possibility of severe thermal gradients in both the substrates and the adhesive during the pre and post gel stages of the adhesive.

Adhesives designed for UV/visible activation generally appeal to the process engineers because of no mixing, long pot life and rapid cure and usually no heating.

Some formulations that have dual cure capability allow gel of the adhesive with UV exposure and subsequent thermal post cure after removal from fixturing so that the adhesive may develop better properties. Implicit in the use of this type of adhesive is the need for at least one substrate to be transparent to the radiation that activates the adhesive. An issue of concern for these adhesives is the possibility of cure inhibition due to the presence of certain chemical species such as amines and sulfur compounds. One must also be aware of the possibility of thick bondlines causing significant part movement due to shrinkage and the possibility of variable properties from run to run or even within a single bondline if the UV source varies in intensity, duration and even temperature.

Other curing mechanisms such as surface activation/catalysis, moisture reaction and anaerobic are less common in the optical bonding field and will be bypassed here.

15.3.2.4. Bondline Thickness Thick bondlines (e.g., over 3 mils) are usually less desirable than thin ones. Part of the reason is the control of that thickness becomes difficult along with possibility of tilted substrates and voids in the adhesive. With a thick bondline in an optical assembly, one can expect shrinkage movement in the z-axis (normal to the adhesive bondline). After cure, temperature and humidity will affect z-axis as well as pitch and roll if there is a wedge of adhesive. Obviously, the thinner the bondline, the smaller these effects.

15.3.2.5. Repeatability In any manufacturing operation, once a process/procedure has been optimized, it is usually desirable to be able to repeat that on a day to day basis. Some of the adhesive related variables that could lead to poor repeatability include; adhesive consistency (formulations), mixing accuracy; property variation within the adhesives' pot life. Other factors that can influence manufacturing repeatability include; size of gap (positional uniformity), cure environment consistency, substrate surface contamination, excessively moist or dry humidity, static electricity, voids and trapped air.

15.3.2.6. Voids There are five sources of voids in an adhesive bondline; air mixed into the adhesive, air trapped in the bondline by the assembly process, boiling of an adhesive component, voids created by pre gel shrinkage of a constrained bondline and delamination caused by post gel shrinkage of a constrained bondline. The first three sources of voids are process related and usually easily corrected, while the latter two are related to the physical design of the joint, and can usually only be fixed by modifying that design. The exception being a change in the filler content of the adhesive or gel temperature to reduce overall shrinkage. Air voids are generally less damaging than shrinkage voids, unless they represent a significant portion of the bondline and compromise the joint strength. One must also be concerned about he possibility of water transpiration for seriously porous adhesive bondlines.

15.3.2.7. Surface Preparation Generally considered to be the cornerstone of good bonding practice, surface preparation is usually costly and time consuming from a manufacturing point of view. And it is frequently a subject of debate between proponents of quality, reliability and manufacturing costs, especially when the development and qualification process time constraints limit the collection of proof test data. Unlike the aerospace and automotive industries, where resources are more apt to be available and time to market is not as critical; the micro and opto-electronic fields are often plagued with judgment calls and trade-offs. That said, the obvious approaches to surface preparation are (in order of ascended cost):

1. None to minimal.
2. Just enough to pass requirements.
3. Best possible.

Some of the surface preparation techniques used include; air blow, sanding/sandblast, solvent wipe/rinse, detergent wash, chemical etch, firing, plasma etch, silane treatment, surface oxidation, conversion coatings, plating, priming, etc. The choice or appropriateness of techniques or combination of techniques will also depend on the nature of the substrate.

15.3.2.8. Adhesive Mixing Most adhesives are mixtures of at least two chemicals, generically called resin and curing agent (plus additives). Depending on the stability or pot life of the mixed adhesive, a manufacturer may supply the adhesive as two separate components (each of which could contain at least one active and none to many inactive materials). The separate component systems are usually (but not exclusively) those with short pot lives and that can cure under fairly mild conditions. The main disadvantages in using two-part adhesives are that proportioning (weighing), mixing and air entrapment can lead to suboptimum results. For example, many room temperature adhesives have mixing ratios in the 10–1 range, where weighing accuracy of small batches can be challenging, especially in a manufacturing environment. Add to that the problem of trying to get a homogeneous mixture with tools and containers all wetted by the components, as well as, the problem of air mixed into the adhesive.

The use of pre-weighed plastic packages can eliminate some of the proportioning inaccuracies and offer a convenient mix container. Issues of air bubbles and homogeneity can still be a concern depending on component viscosity, package wall thickness/stiffness and the operators dedication. In volume operations, the use of a mix-metering machine, that can proportion and mix two component systems from reservoirs then dispense precise amount of adhesive for each part or group of parts, can be cost effective. If mechanical mixers are used, maintenance of the mixer can be a problem, otherwise, static mixers can be used and replaced as needed. Although the initial cost of these machines can be high, the wide variety of material properties and mix ratios capable of being handled provides a great deal of application flexibility for large production runs.

Although usually more costly than two (reactive) component systems, pre-mixed and pre-mixed frozen versions may actually be very cost effective. Many manufacturers will mix their adhesives under vacuum to prevent air incorporation, prepare large batches (thus increasing proportion accuracy) and load multiple syringes or other containers, provide a batch quality control for all the containers and quick-freeze them to preserve pot life. Some of the more latent curing agents can be used to make adhesive formulations that are stable at room temperature for long periods of time, and therefore not require low temperature storage. Variations of the latter include the pre-form, sometimes available as a shaped pressed powder that melts and flows (controlled) at cure temperature; and the thin bonding film with or without reinforcement.

15.3.2.9. Application Techniques The placement of the correct amount of adhesive in the correct place and in the correct (optimum) fashion can involve an interesting number of options. Some of the techniques for adhesive placement include; dab, spatula, trowel, syringe, screen, stencil, preform, pin transfer, platen transfer, ink jet. The first four are generally used in low volume applications while the rest fit better into multiple patterned deposits as in a surface mount adhesive. The selection of technique may be limited by the properties of the adhesive and vice versa. For example, a low viscosity adhesive, which might work well in a syringe application, might also work with a fine screen but probably not with a stencil or an ink jet.

When a manual drop/dab/blob of adhesive is placed on a substrate and the second substrate pressed into place, several, less than ideal, bondline characteristics may result. First, the odds strongly favor either an inadequate amount or an over-abundance of adhesive leaving peripheral voids or messy fillets. Second, depending on viscosity and the aspect ratio (bond area over bondline thickness), pressure on the substrate may be insufficient to achieve the desired thickness. Third, the substrate pressure may be off normal, yielding a wedge of adhesive. Fourth, voids due to trapped air or mixed-in air can expand on elevated temperature cure and result in partial bondlines. And fifth, absorbed moisture and other gas on the substrates can influence the integrity of the adhesive substrate interface.

In the optical component field, a particularly attractive application technique exists that eliminates most of the above problems, and has at lest three other significant advantages. It is the "hot wicking" technique. It works extremely well in critical alignment light path joints with micron or sub-micron bondlines and with adhesives that cure rapidly at a temperature where the viscosity drops precipitously before gel. The process is quite simple; requiring pre-heating of the substrate (and presumably the immediate fixturing) so that uniform and stable elevated temperatures exist at the bondline (usually 100–120°C); the substrates are brought into close contact (co-planar); alignment achieved and then an edge of the bondline is touched with a drop of adhesive. The adhesive will wick across and fill the entire joint without voids, flushing out volatile contaminants from the surface and then cure rapidly (assuming the correct choice of adhesive) usually within 1–5 minutes and without relative substrate movement except in the direction normal to the plane of the adhesive to compensate for post-gel volume shrinkage. The extra advantages of the hot wicking process are rapid gradient-free cure, the bondline is extremely thin and coplanar, the alignment process is set at the cure temperature and post cure is usually not required.

15.4. FAILURE MECHANISM

15.4.1. General

Much of the discussion here will be a consolidation of information detailed in other parts of the chapter. Basically an adhesive joint fails or degrades because:

1. The adhesive changes.
2. The interface changes.
3. Interfacial stresses exceed the adhesives' capability.
4. Externally applied stresses exceed the adhesives' capability.
5. Combination of the above.

Some of these mechanisms are strictly adhesive related, others are a function of the environment and still others are related to an inappropriate joint design. An experienced practitioner can often predict a failure mode just from viewing an assembly design, observing performance data under temperature and environmental exposure or dissecting failed parts.

15.4.2. Adhesive Changes

Changes in an adhesive can be four basic types; thermal degradation due to temperatures that cause irreversible changes in the chemistry of the adhesive, changes in the volume/density of the adhesive as a result of crosslinking (also irreversible); changes due to TCE after cure, which are generally reversible; and changes caused by exposure to chemical environments (e.g., water or solvents) which can be both reversible or non-reversible if, for example, a solvent extracts some material from the adhesive. All of these changes can affect interfacial stresses as a function of the joint design.

15.4.3. Interfacial Changes

Changes at an interface can be caused by chemical reaction with moisture, oxygen, reactive materials in the adhesive or even with the substrate. The most common effect is one of surface oxidation similar to rusting, however, the opposite effect of reduction has also been observed. Physical change in the structure of plated surfaces is another possibility for joint failure. All of these changes can lead to diminished joint strength and ultimate joint failure simply because the adhesive is essentially being "pushed off" the substrate to which it was originally attached at the time the bond was made.

15.4.4. Interfacial Stress

Reasons noted in the preceding paragraphs can cause interfacial stress, but in addition to those noted factors, there is the inherent joint design contribution that can exacerbate both adhesive and interface stress. Examples of joint design stresses include shims two step bonding, adhesive wedges and cross joint thermal gradients.

15.4.5. External Stress

Adding external applied stress to a joint simply enhances the effect of all the other sources of stress present. This is why, for example, that the joint strength of a non ideal joint (different TCE substrates and/or non planer bondline) will decrease as the test temperature gets farther away from the zero stress temperature. Shock and vibration testing is essentially another form of external stress that can add to the "internal" stresses.

REFERENCES

1. S.R. Hartshorn, Structural Adhesives, Chemistry and Technology, Plenum Press, New York, 1986.
2. L.-H. Lee, Adhesive Chemistry Developments and Trends, Plenum Press, New York, 1984.
3. C.A. May and Y. Tanaka, Epoxy Resins Chemistry and Technology, Marcel Dekker, New York, 1973.
4. A. Pizzi and K.L. Mittal, Handbook of Adhesives Technology, Marcel Dekker, New York, 1996.

16

Multi-Stage Peel Tests and Evaluation of Interfacial Adhesion Strength for Micro- and Opto-Electronic Materials

Masaki Omiya, Kikuo Kishimoto, and Wei Yang

Tokyo Institute of Technology, Graduate School of Science and Engineering, Department of Mechanical and Control Engineering 2-12-1, O-okayama, Meguro-ku, Tokyo 152-8552, Japan

16.1. INTRODUCTION

Rapidly development of the electronic products requires small, high density and functional devices. Many layers deposited on silicon substrate can accomplish several kinds of functions in one package. Moreover, in these days, 3-dimensional packaging and assembling technologies have been developing all over the world (e.g., [1]). Multi-layer technology is a key for developing electronic products in future.

The ensuring of reliability is the one of the critical issues in micro- and opto-electronic devices. Those electronic devices contain several kinds of metal or polymer thin films. Due to intrinsic/thermal residual stresses in films or substrates, or elastic/lattice mismatch between film and substrate, the delaminations between layers sometimes occur. The debonding between layers brings about the failure of devices and it might be the source of a tragic accident, since, nowadays, the electronic devices are closely related to human life. One need to design those devices to work well through its entire life time and the information of interfacial strength is essential to designing the reliable devices. Therefore, it is important to evaluate the interfacial strength precisely and the development of reliable testing methods for evaluating the interfacial strength is needed.

Meanwhile, attention has been directed to an electrically conductive ceramic film, which is deposited on a polymeric substrate [2–4]. The applications of those conductive films are used for the display of mobile computers, cellular phones or the flexible paper type display. Those conductive films have the advantage of low power consumption or flexibility to deformation. The popular components of those films especially for the display use are ITO (indium tin oxide) and PET (poly(ethylene terephthalate)). It is well known that the mechanical properties of polymers, such as tensile strength or rupture strain, are degraded by the irradiation of ultraviolet (UV) rays [5]. When the polymer-based conductive films are used in the open air, it is necessary to consider the degradation of the mechanical properties in designing of such products. The bonding mechanism between ceramic and polymer is mainly intermolecular force. When UV rays irradiate polymeric materials, the principal chains are broken down and oxidized. Those reactions may affect the bonding strength between ceramics and polymers.

In view of interface mechanics, interfacial strength depends on the phase angle of loading. To assess the interface strength, one need to conduct the interfacial fracture tests under a wide range of phase angle. A lot of researches related to the interfacial fracture tests have been published and a detailed review of the appropriate mechanics has been given by Hutchinson and Suo [6], Evans and Hutchinson [7], Evans et al. [8].

Especially for the adhesion strength of thin films on substrates, the conventional methods can not be applied. Thin film structures are widely used in nano-machines or electronic devices. Therefore, the development of the testing method to evaluate the adhesion strength of thin film and multilayer thin films is needed. Russel et al. [9] have used tape testing and scratch testing methods to measure the interfacial strength of Cu/SiO_2. The tape test is qualitative and often not a reliable test, useful only for testing weakly adhering film. The scratch test is semi-qualitative, in that the normal load at which a predefined failure event or morphology occurs is defined as a measure of adhesion. While these semi-qualitative tests are simple and informative, they are incapable of incorporating all the relevant parameters.

One such quantitative test that retains the same ease of preparation and test conduction as the scratch test is the indentation-induced delamination test. Marshall and Evans and Marshall et al. [10,11] have presented the fracture mechanics analysis of indentation-induced delamination of thin films. Evans and Hutchinson [12] have analyzed the mechanics of the delamination and spalling of compressed films or coatings by using a combination of fracture mechanics and post-buckling theory. Rosenfield et al. [13] have also conducted the indentation-induced delamination tests. They have compared the indentation tests with double-cantilever-beam technique, four-point flexure-beam technique and finite element analyses. Bahr et al. [14] have conducted nano-scale indentation-induced delamination tests. They have used an acoustic emission in conjunction with nanoindentation tests to monitor a delamination or cracking event. Indentation methods generally rely on the formation of a dilated plastic zone in the film to cause the film to blister [15]. Values of G_c, which is fracture toughness of the interface, are related to the indentation volume (or plastic zone size) and extent of debonding. Therefore, this test method suffers from a limitation similar to that the tape test in that it is limited to a very weakly adhered film. In well-adhered films, indentation fails to produce a delamination unless in ordinarily, high loads or depths are used, in which case the substrate deformation renders deconvolution of the adhesion energy from test parameters impossible. However, a modification of this test method has been developed by Kriese et al. [16,17], Gerberich et al. [18] in which the use of a thin hard coating film on the original film to constrain the plastic deformation and brings about the delamination between the original film and substrate.

Other quantitative method for interfacial strength is the pressurized blister test (Jensen [19], Jensen and Thouless [20]). This testing method has been successfully developed and analyzed for thin polymer films, but is often compromised by the inherently compliant loading system, chemical interactions between the debond and the pressurized environment (stress-corrosion cracking), and the etching or machining procedures are needed to produce the cavity.

Some novel testing methods have also proposed. Bagchi et al. [21], Bagchi and Evans [22], Zhuk et al. [23] have developed "superlayer tests," to measure the debonding energy for thin metalization lines on dielectrics. They fabricated Cu films on various dielectric substrates and evaporated superlayers on Cu films to constrain the plastic deformation of Cu layer. Kitamura et al. [24–26] have used a sandwich specimen, where the deformation of thin films is tightly constraint by substrates for preventing the plastic deformation and fracture of thin films and measured the interfacial strength of nanoscale thickness film. Nakasa et al. [27], Zhang et al. [28] have developed the edge-indentation method to measure the delamination strength of thermally sprayed coatings.

The most well known and straightforward method to delaminate the film from the substrate is the peel test. Peel test is a simple mechanical test to measure the adhesion strength, especially for the case of a thin film deposited on a substrate. Many experimental efforts and analyses have been devoted and a comprehensive survey on the earlier development of the subject can be found elsewhere, see Kim and Aravas [29]. For purely elastic case, all earlier works [30–40] identified the following relation among the adhesion strength Γ, the peeling force P and the peeling angle ϕ formed between the interface and the peeling force:

$$\Gamma = P(1 - \cos\phi), \tag{16.1}$$

where P is the forces per unit width of film. When evaluated according to (16.1), the symbol Γ contained a contribution from the residual stress. Moreover, the possible dependence of the adhesion strength on the peeling angle ϕ was not clear at that time. Extensive works [41–47] were devoted to the plastic deformation of the peel as it detached from the substrate, and bend through the moment-curvature hysteresis loop (including plastic loading, elastic unloading, plastic reverse loading, and elastic reverse unloading). A cohesive strip along the leading portion of the interface gives a new twist to the problem. The work by Wei and Hutchinson [48,49] emphasized the influence of cohesive strength improving the nearby plastic dissipation in film and substrate. Their cohesive law [50], however, is normalized in such a way that only an isotropic response with respect to decohesion direction can be accommodated. Other interface cohesive laws (e.g., [51–53]) elaborated several delicate issues of interface debonding. The anisotropic cohesive law by Ma and Kishimoto [51] has the potential to predict a concave adhesive strength versus phase angle curve.

For the interfacial strength of materials used in micro- and opto-electronic devices, Park et al. [54–56] measured the interfacial fracture energy of Cu/Cr/polyimide system by 90° peel test. However, they conducted only one peel angle and it is not enough to discuss the interfacial strength, which is depend on the phase angle. A method of multi-stage peel test (MPT) is proposed in this paper to tackle the issue of measuring adhesion strength as a curve of phase angles. A testing fixture is presented that allows the application of different lateral loads. Balance between the lateral loading and that projected by the peeling load gives rise to a specific peeling angle. Different peeling angles may result for a single film-substrate specimen if one deliberately varies the lateral loads. A steady state peeling load can be achieved after certain amount of decohesion under a prescribed lateral load. MPT involves the measurements on steady state peeling loads for an incremental sequence of lateral loads. These steady state peeling loads are used to correlate the adhesion strength versus phase angle curve.

The plan of this paper is outlined as follows. Testing scheme for MPT will be described in the next section. In Section 16.3, peel tests for copper/chromium/polyimide/silicon substrate will be presented. Those structures are widely used in chip scale packages

(CSP). We will discuss the effect of copper thickness on the interfacial strength and the multi-layer delamination will be considered. The interfacial strength between the conductive thin ceramic film and polymer substrate will be presented in Section 16.4. Those polymer-based films are damaged by ultraviolet rays. We will discuss the degradation of mechanical properties of polymer film and interfacial strength between ceramics and polymer substrates. We also carried out the *in situ* observation of surface cracks on the ceramic layer during the tensile test. The interfacial strength affects the crack formation on the ceramic layer. The concluding remarks with the limitation of multi-stage peel test and future studies will be shown in Section 16.5.

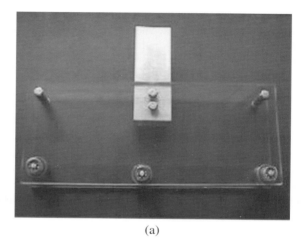

(a)

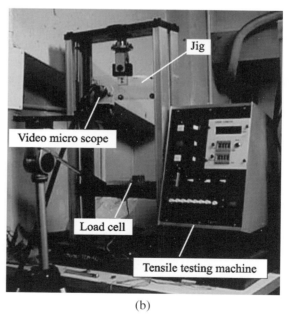

(b)

FIGURE 16.1. The apparatus of multi-stage peel test. (a) Special jig for multi-stage peel test. (b) Setup of the testing machine.

16.2. MULTI-STAGE PEEL TEST (MPT)

Peel tests have been developed for the evaluation of interfacial strength in practical usage, for example, adhesive strength of thin films, coating films. Many papers related to the peel test have also been published (e.g., [29,46]). The most attractive feature of multi-stage peel test is that it is possible to evaluate the interfacial strength of thin films under various phase angles for only one specimen.

16.2.1. Testing Setup

A specially designed apparatus shown in Figure 16.1(a) facilitates the MPT. The special jig is attached to the upper cross head [Figure 16.1(b)] that moves upward at a controlled peeling rate. The movement is recorded by an extensometer. The peeled film is calmed to the lower cross-head that is immobile during a test. A load cell is installed adjacent to the peeling end that records the history of peeling force P. The schematic representation of the MPT is drawn in Figure 16.2. The specimen is put on two roller bearings and by pulling the film, the film delaminates from the substrate. A film/substrate assembly can move horizontally on the bearings with suppressed friction. The friction between rollers and specimen is controlled within less than 0.1 N. During the MPT, the peel front keeps staying near the central bearing and it makes it possible to measure the peel angle continuously. The peel angle was measured and recorded by using the video-microscope and digital video recorder.

As a departure from the conventional peel test, the film/substrate assembly is stretched horizontally by a dead weight P_h through a pulley system that is controlled during the test (in Figure 16.3). Under quasi-static peeling, balance of forces in the horizontal direction predicts the following peel angle:

$$\phi = \cos^{-1}\left(\frac{P_h}{P}\right). \tag{16.2}$$

Vertical component of the peel force is countered by roller supports. To conduct peeling, the force P should be larger that the horizontal stretching force, P_h. The peel angle

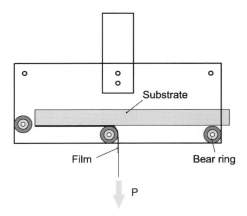

FIGURE 16.2. The schematic representation of peel test.

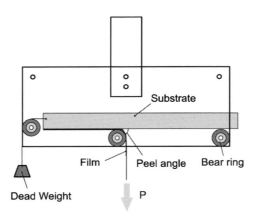

FIGURE 16.3. The dead weight attached on the specimen for changing the peel angle.

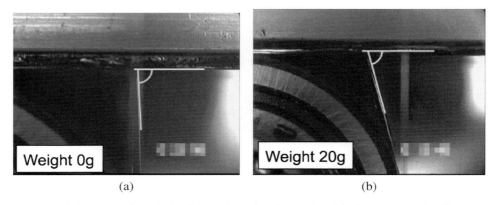

FIGURE 16.4. The effect of the dead weight on the peel angle. (a) No weight. The peel angle is 89°. (b) The dead weight is 20 g. The peel angle is 75°.

decreases monotonically with the increase of P_h. By varying P_h, a wide range of peel angles (between 30 degrees under a large horizontal force to almost 90 degrees for negligible horizontal force) can be achieved. The peel angle can be measured independently and *in situ* by a microscope horizontally mounted on the side facing the test machine, see Figure 16.1(b). A typical image is shown in Figure 16.4. When the dead weight is attached to the specimen, the peel angle changed due to the horizontal force. The relationship between the horizontal force and the peel force decide the peel angle. This effect causes the phase angle shift at the peel front. Therefore, the mixed mode delamination tests under the wide range can be possible by the MPT.

16.2.2. Multi-Stage Peel Test

Figure 16.5 describes a typical curve for the evolution of peel force when the peeling rate maintains at 5 mm/min. A steady state emerges after about 5 mm of peeling length. Samples peeled at other rates (from 1 mm to 5 mm/min) deliver similar results of steady state peeling convened at a peeling length about 5 mm. If the dead load (that controls

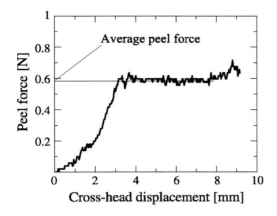

FIGURE 16.5. Typical example of peel force evolution during the MPT.

the horizontal stretching force) is applied incrementally, and a minimum peeling length of 5 mm is imposed on each peeling increment, one can have a multi-stage peeling test in one pass that records the steady state peeling forces at different peeling angles. The apparatus has a lateral span of 220 mm, and the range available for peeling is about 80 mm. Accordingly, over 10 peeling stages can be accommodated in one testing block. Those steady state peeling forces will correlate to the phase angle dependence of adhesion strength in the next section.

16.2.3. Energy Variation in Steady State Peeling

After the attainment of a steady state, the peeling configuration stabilizes. The load point displacement can be chosen as the time variable. Possible rate dependences of the film and the adherent become implicit. Analysis for steady state peeling is further simplified by considering the energetic aspect of the system, as schematically shown in Figure 16.6. During each peeling increment of Δl in a steady state, neither the peeling configuration nor the energy storage and dissipation change for most portion of the system for an observer fixed spatially, say, to the central roller. The only difference in energy exchange consists of a segment of length Δl far behind the peeling edge converting to a segment of the same length far ahead of the peeling edge. The corresponding change in the substrate is negligible since the substrate parts of both segments are essentially stress-free.

Stored in the film far behind the peeling edge is the elastic energy caused by the residual stresses generated during film deposition, the density W_{res} of this energy can be computed as:

$$W_{res} = \frac{\alpha(\nu)}{E} \int_0^h \sigma_{res}^2(y)dy, \tag{16.3}$$

where E and ν denote the Young's modulus and the Poisson's ratio of the film, and σ_{res} denotes the horizontal residual stress stored in the film during deposition. If σ_{res} is uniform across the film thickness h, then

$$W_{res} = \frac{\alpha(\nu)}{E} \sigma_{res}^2 h. \tag{16.4}$$

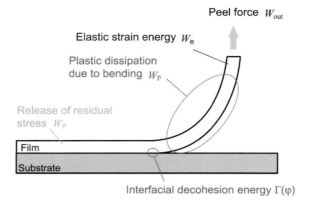

FIGURE 16.6. Energy balance under steady state peeling.

The coefficient $\alpha(\nu)$ depends on the stress state in the film and the Poisson's ratio. It equals to $(1 - \nu^2)/2$ for a plane strain film, i.e., see Yu et al. [57] and equals to $1 - \nu$ for an equal biaxial stress state caused by thermal mismatch, see Yang and Freund [58].

Near the film end far away from the debonding edge, the energy density comes from the following sources: (1) the decohesion energy, $\Gamma(\varphi)$; (2) the plastic dissipation, W_p, of a certain film moment-curvature hysteresis terminated at the film state near the pulling end; (3) the strain energy due to the residual stress in the film; and (4) the elastic strain energy W_e of the film near the pulling end. For a plane strain situation, the last contribution includes a stretching part of $[(1 - \nu^2)/(2Eh)]P^2$ and a coiling part of

$$\frac{Eh^3}{24(1 - \nu^2)K_{coil}^2},$$

where K_{coil} is the coiling curvature at the end of the film when it is released from the peeling grip. Contribution from the elastic strain energy W_e is usually small when compared with the others.

Beside the variation of the energies stored or dissipated in the system, the work of the peel force P and that of the dead weight P_h contribute to the potential energy of the system. The rate of those works is denoted as W_{out}, and is given by:

$$W_{out} = P - P_h, \tag{16.5}$$

where, θ is the peel angle measured in the MPT. Apart from the steady state assumption, Equation (16.5) is derived under the ignorance on the work done by all frictional forces. The absence of friction also leads to the equivalence between the raising rate of the dead load and the pulling rate of the lower cross-head when a steady state prevails. Conservation of the potential energy gives:

$$W_{out} = \Gamma(\varphi) + W_p + W_e + W_{res}, \tag{16.6}$$

that is,

$$P - P_h = \Gamma(\varphi) + W_p + \frac{Eh^3}{24(1 - \nu^2)K_{coil}^2} + \frac{1 - \nu^2}{2E}\left[\frac{P^2}{h} - \int_0^h \sigma_{res}^2(y)dy\right], \tag{16.7}$$

when subjected to a plane strain condition. In Equation (16.7), the material properties such as the Young's modulus E, the Poisson's ratio ν, and those needed in determining W_P can be measured independently. The film thickness h and the dead weight P_h are known *a priori*. The steady state peeling force P and the residual coil curvature K_{coil} can be measured during the test. Appropriate analyses [29,47,48] can be invoked to evaluate W_p. However, one still has the difficulty to distinct the adhesion $\Gamma(\varphi)$ from the contribution of the unknown deposition stress σ_{res} and this value should be measured by another experiments.

The difficulty to obtain the interfacial strength from the above equation is how to estimate the plastic dissipation in detached film. Kim and co-workers [29,46] have proposed a generalized elastic-plastic slender beam theory for the analysis of the detached part of the film in a peel test. They have taken account of elastic unloading and reverse plastic bending of the film and given a closed form solution for the maximum curvature (root curvature) and hence the plastic dissipation, attained by an elastic-perfectly plastic film. Kinloch et al. [47] have studied the peeling of bilinear work hardening materials and found a good agreement with experiment. Since copper thin films are well approximated by bilinear work hardening constitutive equation, we followed the Kinloch et al. [47] approach to estimate the plastic dissipation.

The stress–strain curve at a point on the film cross-section and the moment-curvature diagram are shown in Figures 16.7 and 16.8, respectively. When the loading and unloading of the peeling film both involve plastic deformation, the plastic dissipation energy is correspond to the total energy loss in the loading and unloading cycle, the area [OABC] in Figure 16.8. The plastic dissipation W_p is,

$$W_p = \frac{\Delta OABC}{b}. \tag{16.8}$$

Thus,

$$W_p = G^e_{max} f_1(k_0), \tag{16.9}$$

$$W_{out} = G^e_{max} \frac{1 - \cos\theta}{1 - \cos(\theta - \theta_0)} f_2(k_0), \tag{16.10}$$

where,

$$f_1(k_0) = \frac{4}{3}\alpha(1-\alpha)^2 k_0^2 + 2(1-\alpha)^2(1-2\alpha)k_0$$
$$+ \frac{2(1-\alpha)}{3(1-2\alpha)k_0}\left[1 + 4(1-\alpha)^3\right] - (1-\alpha)\left[1 + 4(1-\alpha)^2\right], \tag{16.11}$$

$$f_2(k_0) = \frac{1}{3}\alpha\left[1 + 4(1-\alpha)^2\right] + 2(1-\alpha)^2(1-2\alpha)k_0$$
$$+ \frac{8}{3}\frac{(1-\alpha)^4}{(1-2\alpha)k_0} - 4(1-\alpha)^3. \tag{16.12}$$

G^e_{max} is the maximum stored elastic energy as defined by,

$$G^e_{max} = \frac{1}{2}E\varepsilon_y^2 h, \tag{16.13}$$

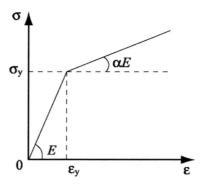

FIGURE 16.7. The stress–strain curve for bilinear, work-hardening material.

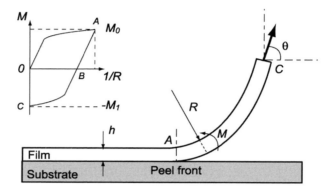

FIGURE 16.8. Deformation of the peeling film and the moment-curvature diagram undergone by peeling film.

and α is the strain hardening coefficient and ε_y is the yield strain. The term k_0 is given by,

$$k_0 = \frac{R_1}{R_0}, \qquad (16.14)$$

where R_0 is the actual radius of curvature at the peel front and R_1 is the radius of curvature at the onset of plastic yielding and is given by,

$$R_1 = \frac{h}{2\varepsilon_y}. \qquad (16.15)$$

It is noted that the actual radius of curvature at the peel front is difficult to determine, but it is an important parameter. The attached part of the film has been modeled as an elastic beam on an elastic foundation of thickness $h/2$. Applying a beam theory, the relationship of the root angle and k_0 has been obtained as,

$$\theta_0 = \frac{1}{3}(4\varepsilon_y)k_0. \qquad (16.16)$$

From Equations (16.10) and (16.16), the root angle and k_0 can be obtained. Then, the plastic dissipation can be calculated from Equation (16.8). The detail of derivation of those equations can be seen in Kinloch et al. [47].

16.3. INTERFACIAL ADHESION STRENGTH OF COPPER THIN FILM

16.3.1. Preparation of Specimen

The specimen used in this paper is composed of copper, chromium, polyimide layers deposited on silicon substrate which is fabricated by the standard method of CSP packages. The cross-sectional view of the specimen is shown in Figure 16.9. The thickness of silicon wafer is about 1 mm, polyimide layer which is coated on Si substrate by spin-coating method is 11 μm and chromium layer which is sputtered on polyimide layer is 0.2 μm. The copper layers those are plated on chromium layer are changed as 5, 10 and 20 μm to investigate the effect of copper film thickness on the interfacial strength. The copper spattering layers at the edge of specimen are installed as the scarified layer for the crack initiation. On the multi-stage peel test, the specimen is attached on an aluminum bar (which cross-section is 10 mm × 10 mm) with an epoxy adhesive to prevent the bending of silicon wafer during the peel test.

During the manufacturing process, the residual stresses would be induced in the specimen. Therefore, before peel tests, the residual stresses in copper films were measured by X-ray analysis. The measured residual stresses are shown in Figure 16.10. From this figure, the residual stresses in copper films are less than 3 MPa. The stress relaxation would be occurred in polyimide layer and reduce the residual stresses in copper film. When the residual stresses are assumed to be constant over the film thickness, the dissipation energy needed to release the residual stress can be calculated from Equation (16.4) as shown in Figure 16.10. The dissipation energy is so small that one can neglect the effect of residual stresses in this case.

The stress–strain curve of copper film is necessary to estimate the plastic dissipation as described in previous section. The tensile tests of copper film which thickness is 15 μm were carried out and obtained Young's modulus and 0.2% yield stress. The properties of each material are shown in Table 16.1. Poisson's ratio of copper film and other properties are referred from those of bulk materials.

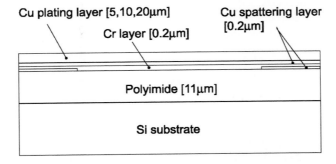

FIGURE 16.9. The cross-sectional view of the specimen.

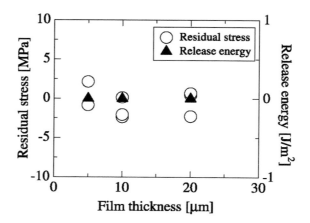

FIGURE 16.10. The residual stresses and the dissipation energy needed to release the residual stresses.

TABLE 16.1.
Material properties of the specimen.

Material	Young's modulus (GPa)	Poisson's ratio	Yield stress $\sigma_{0.2}$ (MPa)
Copper	30	0.3	440
Chromium	115	0.3	—
Polyimide	3	0.45	—

16.3.2. Measurement of Adhesion Strength by the MPT

The characteristic load–displacement curve obtained by the multi-stage peel test is shown in Figure 16.11. After the onset of the peeling, the peel force becomes constant and the delamination is in steady-state condition. Even in steady-state condition, the peel force slightly scatters since the adhesion strength is not locally constant. Hence, we averaged the peel force on some time period for the calculation of the decohesion energy. Figure 16.12 shows the top view of the specimen after the peel test. From the observation of the peeled specimen, the delamination mostly occurred at copper/chromium interface. Therefore, we discuss the interfacial strength of the copper/chromium interface in this paper.

From the averaged peel force and the peel angle, the work done by peel force can be calculated by Equation (16.5). Figure 16.13 shows the work done by peel force, i.e., the energy from outside into the peel front, in no dead weight case. As the film is thinner, the work needed to delaminate becomes larger. From the energy balance of Equation (16.6) and considering the dissipations due to the plastic deformation and the residual stresses, the decohesion energy can be calculated as shown in Figure 16.13. Comparing to the results of the work done by peel force, the effect of film thickness becomes small. When the film thickness is over 10 μm, the decohesion energy becomes constant and is approximately 20 J/m^2. This value is considered to be the interfacial strength between copper and chromium layers. By changing the dead weight attached on the specimen, the dependence of the decohesion energy on the peel angle was obtained in 10, 20 μm cases as shown in Figure 16.14. The peel angles varied from about 45 degrees to 90 degrees and the decohesion energy increase with the peel angles.

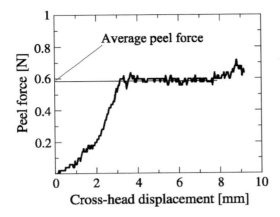

FIGURE 16.11. The characteristic peel force and displacement curve.

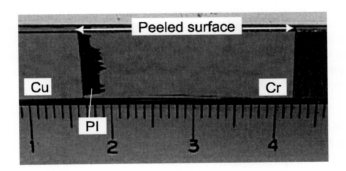

FIGURE 16.12. The top view of the specimen after the peel test. The thickness of Cu layer is 20 μm.

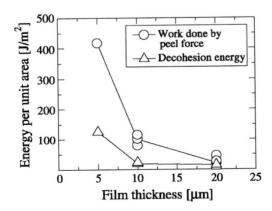

FIGURE 16.13. The work done by peel force and the decohesion energy of Cu/Cr layer.

16.3.3. Discussions

In previous section, we obtained the decohesion energy of copper and chromium interface by the multi-stage peel test. However, it should be considered whether this obtained

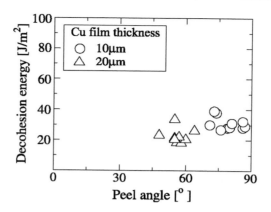

FIGURE 16.14. The peel angle effects on the decohesion energy.

value is reasonable or not. To confirm it, we have carried out finite element analysis and compare experimental results with numerical results. During the steady-state condition, the peel problem can be divided to the detached part and attached part as shown in Figure 16.15. Our interest is the energy flow around the peel front. Then, we can only consider the attached part with appropriate boundary conditions. From the equilibrium condition of force and moment between detached and attached part, the boundary conditions can be obtained as:

$$\left.\begin{array}{l} N = F\cos\theta - F_h \\ Q = F\sin\theta \\ M = F\sin\theta \cdot d \end{array}\right\}, \qquad (16.17)$$

where, F is the averaged peel force, F_h is the dead weight, θ is the peel angle and d is the reference length from the peel front at which the peel angle is measured. Then, the numerical model of the multi-stage peel test is reduced to the equivalent interface crack problem as shown in Figure 16.16. In this case, silicon substrate is assumed to be rigid for simplicity.

From experiments, the averaged peel forces and peel angles during the steady-state condition were measured. Those values are used as the boundary conditions for numerical simulations. The interfacial fracture energy, which is correspond to the decohesion energy, was evaluated by J-integral in numerical simulations. Figure 16.17 shows the comparison between the results of the MPT method and J-integral calculation. In elastic analyses, the results of the J-integral calculation agree well with the MPT results, but the obtained results depend on film thickness. On the other hand, in elastic-plastic analyses, the J-integral values become constant, about 20 J/m^2, and are independent of film thickness. J-integral value for elastic-plastic material still has the meaning of energy release rate. Hence, this value is considered to be the interfacial strength of Cu/Cr interface. The difference between the MPT results and J-integral values stems from the formation of plastic zone around the peel front. Not only bending of the film but also stress concentration at the peel front induced the plastic deformation at the peel front. The plastic zone ahead of peel front is shown in Figure 16.18. In 20 μm case, the plastic zone is formed at the vicinity of the peel front and can be neglected like the small-scale yielding condition. On the contrary, in 5 μm case, the copper film is largely bended and the plastic zone is formed around the peel front

MULTI-STAGE PEEL TESTS AND EVALUATION OF INTERFACIAL ADHESION STRENGTH 417

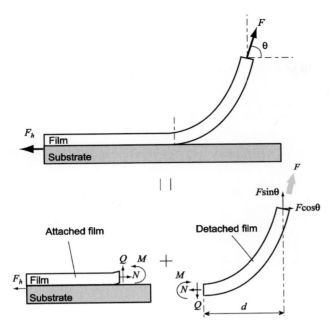

FIGURE 16.15. The reduction of solved problem into the equivalent interface crack problem.

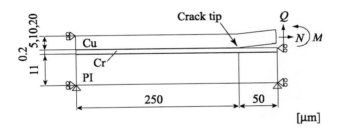

FIGURE 16.16. Numerical model for multi-stage peel test.

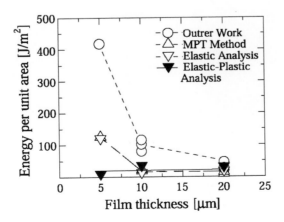

FIGURE 16.17. Comparison with the results of MPT and numerical simulation.

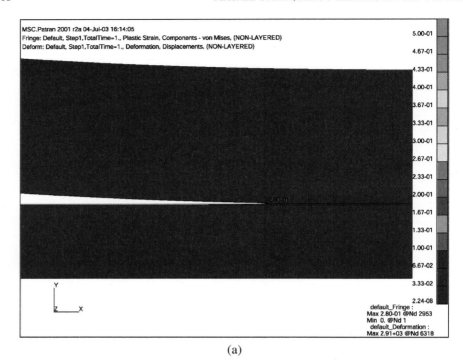

FIGURE 16.18. The plastic strain distribution around the peel front. (a) Copper film thickness is 20 μm. (b) Copper film thickness is 5 μm.

widely. This case is corresponding to the large-scale yielding condition. The energy evaluation in the MPT method does not cover the large-scale yielding condition. Therefore, the MPT evaluation includes the energy dissipation due to the plastic deformation.

The J-integral evaluation in 5 μm case is slightly smaller than the other cases. This means that the decohesion energy between copper and chromium is smaller than the other cases. To consider this difference, the surface observation of peeled specimen was carried out by Atomic Force Microscope (AFM). Figure 16.19 shows the results of surface observation after the multi-stage peel test. The chromium surface is very flat and the roughness is less than a few nano-meter in 20 μm case. On the other hand, in 5 μm case, small hills those heights were about 40 nm were observed. These results imply that the delamination of Cr/PI and Cu/Cr occurred simultaneously during the peel test. Figure 16.20 shows the mean stress distribution on Cr/PI interface and high mean stress concentration occurred ahead of the peel front. This mean stress may expand micro or nano voids on Cr/PI interface and induce the delamination on Cr/PI interface. The decohesion energy measured by the MPT method for 5 μm case includes the decohesion energy both Cu/Cr and Cr/PI interfaces. The delamination occurred between Cr and PI layers may influence the stress distribution and constrain the plastic deformation ahead of the peel front. The finite element analyses did not consider the multiple delamination and it overestimated the plastic dissipation than the realistic case. Hence, the J-integral evaluation in 5 μm case is slightly smaller than the other cases.

The MPT evaluation gives us only the energy flow into the peel front and does not eliminate the energy dissipation around the peel front, such as plastic dissipation. Therefore, the MPT method can be applied to the peeling under small-scale yielding condition. Nowadays, the film thickness becomes thinner and thinner going down to nano or sub-nano thickness in electronic devices. In those situations, the multiple delamination would be one of critical issues. The development of precious evaluation methods for multi-layer systems are needed and left in future works.

16.4. UV-IRRADIATION EFFECT ON CERAMIC/POLYMER INTERFACIAL STRENGTH

It is well known that the mechanical properties of polymers, such as tensile strength or rupture strain, are degraded by the irradiation of ultraviolet (UV) rays. When the polymer-based conductive films are used in the open air, it is necessary to consider the degradation of the mechanical properties in design of products. The bonding mechanism between ceramic and polymer is mainly intermolecular force. When UV rays irradiate polymeric materials, the principal chains are cut and oxidized. Those reactions may affect the bonding strength between ceramics and polymers. That is the motivation of this research and we investigated the effects of UV irradiation on the interfacial adhesion strength between ITO (indium tin oxide) coating layer and PET (poly(ethylene terephthalate)) substrate by multi-stage peel test.

16.4.1. Preparation of PET/ITO Specimen

Two types of specimens were prepared. One type is composed of ITO and PET and the other is composed of ITO, TiO_2 and PET. All specimens were fabricated by spattering ITO or TiO_2 on PET substrate. The thickness of each layer is summarized in Table 16.2.

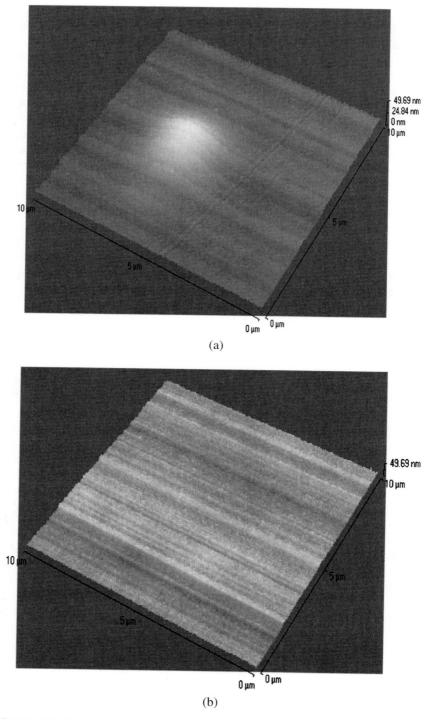

FIGURE 16.19. The Cr surface configuration after the peel test. (a) The thickness of Cu layer is 5 μm. (b) The thickness of Cu layer is 20 μm.

To investigate the effect of UV irradiation, half of specimens were kept in a fade meter for 120 hours. The UV ray is exposed on ITO face and the intensity of UV ray is about 30 W/m². The cross-sectional view of the specimen is shown in Figure 16.21. The specimen was attached on aluminum bar with an epoxy adhesive, since the specimen was too thin to conduct the peel test directly. The stress–strain curves of PET film needed to estimate the plastic dissipation are shown in Figure 16.22. The stress–strain relation depends on the direction of loading. Young's modulus and yield stresses are summarized in Table 16.3. After UV irradiation, the rupture strain gradually decreased. The relation between rapture strain and irradiation period is represented in Figure 16.23. The rupture strain becomes smaller and smaller, as the irradiation time become long. After 120 hour irradiation, PET film is ruptured almost within elastic region.

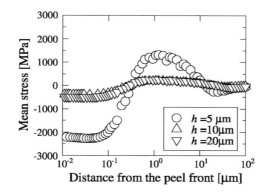

FIGURE 16.20. The mean stress distribution on chromium and polyimide interface.

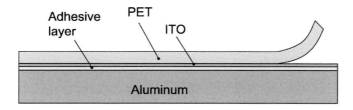

FIGURE 16.21. The cross-sectional view of test specimen.

TABLE 16.2.
The condition of test specimen.

Name	PI	PIUV	PIT	PITUV
Substrate	PET 100 μm	PET 100 μm	PET 100 μm	PET 100 μm
Coating layer	ITO 108 nm	ITO 108 nm	ITO/TiO$_2$ 100/20 nm	ITO/TiO$_2$ 100/20 nm
UV	—	120 hours	—	120 hours

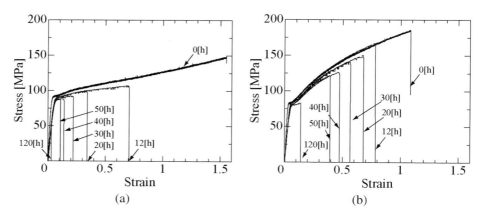

FIGURE 16.22. The stress–strain relation of PET film. (a) Parallel to the rolling direction. (b) Vertical to the rolling direction.

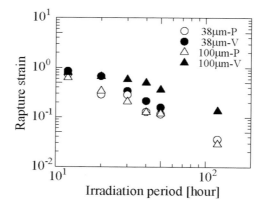

FIGURE 16.23. The irradiation effects on rupture strain of PET film.

TABLE 16.3.
Mechanical properties of PET film.

Name	LT	TL	Name	LT	TL
Young's modulus	3.8 GPa	4.1 GPa	Strain hardening coefficient, α	0.105	0.194
Poisson's ratio	0.41	0.49	Yield strain, ε_y	0.028	0.02

16.4.2. Measurement of Interfacial Strength by MPT

The load history during the peel test for PET/ITO specimen is shown in Figure 16.24. and the peel angle at the peel front during the peel test is shown in Figure 16.25. When the dead weight is about 0.83 N, the peel angle is about 55°. Without the dead weight, the peel angle is about 71°. From the load history (Figure 16.24) and the peel angle at that time, the work done by peel force can be calculated from Equation (16.5). Also, the plastic dissipation in PET film can be estimated by Equation (16.8). When neglecting the residual stress in PET film because of the relaxation of polymeric material, the decohesion energy, then, can be evaluated from Equation (16.7). The decohesion energy by MPT method is shown in Figure 16.26 and 16.27. Figure 16.26 represents the decohesion energy of PET/ITO interface. For virgin films, the interfacial strength is about 20 J/m^2 and slightly depends on the peel angle. On the other hand, after UV irradiation, the interfacial strength is drastically decreased about 1 to 10 J/m^2 and clearly depends on the peel angle. Figure 16.27 represents the decohesion energy for PET/TiO$_2$/ITO specimen. The delamination occurred between PET and TiO$_2$ interface and the decohesion energy correspond to the interfacial strength between PET and TiO$_2$. In this case, for virgin films, the decohesion energy is smaller than that of PET/ITO interface. However, the degradation of interfacial strength due to UV irradiation is smaller than that of PET/ITO interface. It is considered that TiO$_2$ layer works as a filter and it makes the intensity of UV ray transmitted to PET layer smaller. Therefore, inserting TiO$_2$ layer between ITO and PET is useful to prevent the degradation of interfacial strength.

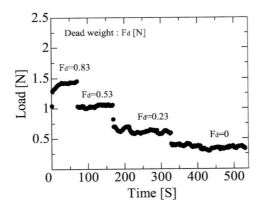

FIGURE 16.24. Load history of multi-stage peel test (PET/ITO$_{LT}$).

FIGURE 16.25. Measurement of peel angle: (a) dead weight is 0.83 N, (b) no dead weight.

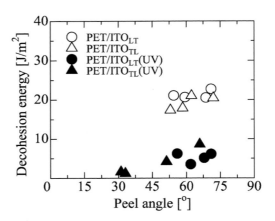

FIGURE 16.26. Decohesion energy of PET/ITO interface and effect of UV irradiation.

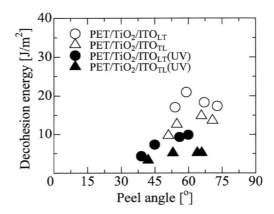

FIGURE 16.27. Decohesion energy of PET/TiO$_2$ interface and effect of UV irradiation.

16.4.3. Surface Crack Formation on ITO Layer under Tensile Loading

Cracks easily formed on ITO layer under tensile loading since ITO is a brittle ceramic. Once the surface crack is formed, the functionality of the complex film will be lost. Our interest is how the interfacial strength affects the crack formation on ITO layer. Therefore, in this sub-section, we carried out the tensile tests for PET/ITO complex film and the *in situ* observation of the ITO surface.

The small tensile testing machine was used under the confocal laser scanning microscope. The rectangular specimen was prepared which width is 1 mm and length is 50 mm. The stress–strain relation of the complex film is shown in Figure 16.28. It is noted that the strain is measured from the cross head displacement divided by the initial length between cramps. The stress–strain relation is almost same with that of PET film and the mechanical properties of these complex films are decided by that of PET film. However, the rupture strain is larger than that of PET film. This is because ITO layer make the intensity of UV ray reached to PET film weaker.

The *in situ* observation of surface crack formation is shown in Figure 16.29. When the strain reached 2.8%, "vertical cracks" can be observed on ITO layer. The distance between

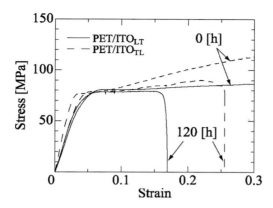

FIGURE 16.28. The stress–strain relation of PET/ITO specimen.

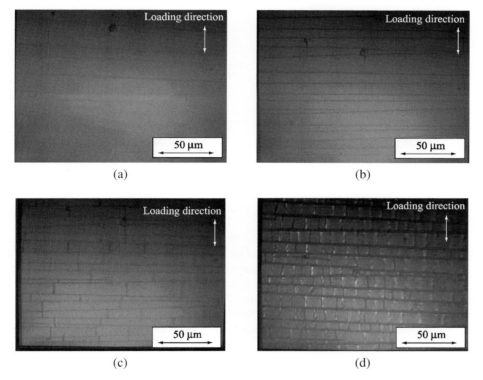

FIGURE 16.29. *In situ* observation of PET/ITO$_{LT}$ surface during tensile test. (a) $\varepsilon = 2.8\%$, (b) $\varepsilon = 6.4\%$, (c) $\varepsilon = 12\%$, (d) $\varepsilon = 30\%$.

cracks is almost constant. The density of cracks increased with applied strain as shown in Figure 16.29(b). The driving force of crack formation is shear force between ITO and PET layers [59,60]. When the interfacial strength is large, the interfacial shear rigidity is stiff and the same strain within PET film will be induced in ITO layer. That brings about high tensile stress in ITO layer and cracks are formed even in small strain range. After the strain reached about 12%, "parallel cracks" to the loading direction were observed. Due

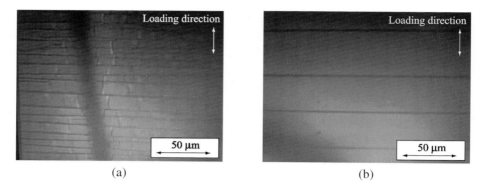

FIGURE 16.30. The effect of UV irradiation on the fracture pattern of PET/ITO$_{TL}$ surface: $\varepsilon = 10\%$. (a) Virgin specimen. (b) UV-irradiated specimen (120 hours).

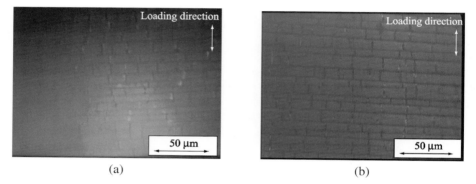

FIGURE 16.31. The effect of UV irradiation on the fracture pattern of PET/TiO$_2$/ITO$_{TL}$ surface: $\varepsilon = 10\%$. (a) Virgin specimen. (b) UV-irradiated specimen (120 hours).

to the Poisson's effect, the compressed stress vertical to the loading direction induced and buckling cracks occurred in ITO layer. After the initiation of buckling crack, the density of "vertical cracks" is constant and only the density of "parallel cracks" increased.

The comparison of UV irradiation effects on crack formation is shown in Figure 16.30 and 16.31. Figure 16.30 represents of PET/ITO$_{TL}$ specimen at the tensile strain is about 10%. After UV irradiation for 120 hours, only "vertical cracks" can be observed. This is because the interfacial strength between PET and ITO decreased by UV irradiation as shown in Figure 16.26 and then, the interfacial shear rigidity also decreased. Even when PET layer deformed largely, the strain induced in ITO layer can not be enough to form "parallel cracks." On the other hand, the crack formation of PET/TiO$_2$/ITO specimen is almost same as that of virgin specimen as shown in Figure 16.31. In this case, the degradation of the interfacial strength is smaller than that of PET/ITO case. From these results, the interfacial strength closely related the crack formation behavior on ITO layer.

16.5. CONCLUDING REMARKS

In this chapter, the multi-stage peel test method is introduced and is applied to the measurement of the adhesion strength for copper thin film and conductive ceramics film.

During the steady-state peeling, the interfacial strength can be evaluated by the energy balance of peeling system. When the film thickness is thick enough to neglect the plastic zone size at the peeling tip, i.e., under small-scale yielding condition, MPT method can be valid to evaluate the interfacial strength precisely. However, when the film thickness is thinner and the plastic zone size can not be neglected, i.e., under large-scale yielding condition, the MPT method can not be applied directly and needs the help of some numerical calculations. More, in multi-layer system, the delamination on other layers would be occurred during peel tests. In those cases, the delamination of other layers will affect the stress condition at the peeling tip and the conventional evaluation techniques can not be applied. Those multiple delamination would be important issues in nano- and subnano-thickness film structures and the solution of those problems has been left in future works.

We also applied MPT method to the measuring of the interfacial strength between conductive ceramics thin film (ITO) and polymer substrate (PET). The stress–strain relations of PET film are significantly degraded by UV irradiation. The interfacial strength between ITO and PET was also degraded. When TiO_2 layer inserted between ITO and PET layers, the degradation of interfacial strength became smaller than that of PET/ITO interface. TiO_2 layer works as a filter and protect the interfacial bonding between ITO and PET from UV attack. The interfacial strength is closely related to the crack formation of ITO layer under tensile loadings. However, it has not been clear how much molecules on the interface or under the interface are damaged by UV irradiation and how to connect the number of damaged molecules and the degradation of interfacial strength quantitatively. The bottom-up approaches, such as molecular dynamics, are necessary to solve these problems. Moreover, when supplying the power voltage on ITO layer, the interfacial strength and crack formation would be different from obtained results. Therefore, it is necessary to conduct those experiments under more realistic situations.

ACKNOWLEDGMENT

The authors would like to acknowledge Dr. Amagai of Texas Instruments Japan and Dr. Yanaka of Toppan Printing Co. Ltd., for their preparation of test specimen.

REFERENCES

1. S.F. Al-sarawi, D. Abbott, and P.D. Franzon, A review of 3-D packaging technology, IEEE Transactions on Components, Packaging and Manufacturing Technology, Part B, 21, pp. 2–14 (1998).
2. E. Harlev, T. Gulakhmedova, I. Rubinovich, and G. Aizenshtein, New method for the preparation of conductive polyaniline solutions: Application to liquid crystal devices, Advanced Materials, 8, pp. 994–997 (1996).
3. G. Gu, P.E. Buroows, S. Venkatesh, S.R. Forrest, and M.E. Thompson, Vacuum-deposited, nonpolymeric flexible organic light-emitting devices, Optics Letters, 22, pp. 172–174 (1997).
4. Z. Chen, B. Cotterell, W. Wang, E. Guenther, and S.J. Chua, A mechanical assessment of flexible optoelectronic devices, Thin Solid Films, 394, pp. 201–205 (2001).
5. M. Notomi, T. Gotoh, K. Kishimoto, and T. Shibuya, Evaluation of ultra-violet degradation on PP and PC by layering-films exposure test, Transactions of the Japan Society of Mechanical Engineers, Part A, 63, pp. 437–444 (1997).
6. J.W. Hutchinson and Z. Suo, Mixed mode cracking in layered materials, Advances in Applied Mechanics, 29, pp. 63–191 (1992).
7. A.G. Evans and J.W. Hutchinson, The thermomechanical integrity of thin films and multilayers, Acta Metallurgica Materialia, 43, pp. 2507–2530 (1995).

8. A.G. Evans, J.W. Hutchinson, and Y. Wei, Interface adhesion: effects of plasticity and segragation, Acta Metallurgica Materialia, 47, pp. 4093–4113 (1999).
9. S.W. Russel, S.A. Rafalski, R.L. Spreitzer, J. Li, M. Moinpour, F. Moghadam, and T.L. Alford, Enhanced adhesion of copper to dielectrics via titanium and chromium additions and sacrificial reactions, Thin Solid Films, 262, pp. 154–167 (1995).
10. D.B. Marshall and A.G. Evans, Measurement of adherence of residually stressed thin films by indentation. I. Mechanics of interfaced delamination, Journal of Applied Physics, 56, pp. 2632–2638 (1984).
11. C. Rossington, A.G. Evans, D.B. Marshall, and B.T. Khuri-Yakub, Measurement of adherence of residually stressed thin films by indentaion. II. Experiments with ZnO/Si, Journal of Applied Physics, 56, pp. 2639–2644 (1984).
12. A.G. Evans and J.W. Hutchinson, On the mechanics of delamination and spalling in compressed films, International Journal of Solids and Structures, 20, pp. 455–466 (1984).
13. L.G. Rosenfeld, J.E. Ritter, T.J. Lardner, and M.R. Lin, Use of the microindentation technique for determining interfacial fracture energy, Journal of Applied Physics, 67, pp. 3291–3298 (1990).
14. D.F. Bahr, J.W. Hoehn, N.R. Moody, and W.W. Gerberich, Adhesion and acoustic emission analysis of failures in nitride films with a metal interlayer, Acta Materialia, 45, pp. 5163–5175 (1997).
15. M.D. Drory and J.W. Hutchinson, Measurement of the adhesion of a brittle film on a ductile substrate by indentation, Proceedings of the Royal Society of London Series A—Mathematical Physical and Engineering Sciences, 452, pp. 2319–2341 (1996).
16. M.D. Kriese, N.R. Moody, and W.W. Gerberich, Effects of annealing and interlayers on the adhesion energy of copper thin films to SiO2/Si substrate, Acta materialia, 46, pp. 6623–6630 (1998).
17. M.D. Kriese, D.A. Boismier, N.R. Moody, and W.W. Gerberich, Nonomechanical fracture-testing of thin films, Engineering Fracture Mechanics, 61, pp. 1–20 (1998).
18. W.W. Gerberich, D.E. Kramer, N.I. Tymiak, A.A. Volinsky, D.F. Bahr, and M.D. Kriese, Nanoindentation-induced defect-interface interactions: phenomena. Methods and limitations, Acta Materialia, 47, pp. 4115–4123 (1999).
19. H.M. Jensen, The blister test for interface toughness measurement, Engineering Fracture Mechanics, 40, pp. 475–486 (1991).
20. H.M. Jensen and M.D. Thouless, Effects of residual stresses in the blister test, International Journal of Solids and Structures, 30, pp. 779–795 (1993).
21. A. Bagchi, G.E. Lucas, Z. Suo, and A.G. Evans, A new procedure for measuring the decohesion energy for thin ductile films on substrates, Journal of Material Research, 9, pp. 1734–1741 (1994).
22. A. Bagchi and A.G. Evans, Measurements of the debond energy for thin metallization lines on dielectrics, Thin Solid Films, 286, pp. 203–212 (1996).
23. A.V. Zhuk, A.G. Evans, J.W. Hutchinson, and G.M. Whitesides, The adhesion energy between polymer thin films and self-assembled monolayers, Journal of Material Research, 13, pp. 3555–3565 (1998).
24. T. Kitamura, T. Shibutani, and T. Ueno, Development of evaluation method for interface strength between thin films and its application on delamination of Cu/TaN in an advanced LSI, Transactions of the Japan Society of Mechanical Engineers, 66, pp. 1568–1573 (2000).
25. T. Kitamura, H. Hirakata, and T. Itsuji, Delamination strength of Cu thin film characterized by nanoscale stress field near interface edge, Transactions of the Japan Society of Mechanical Engineers, Part A, 68, pp. 119–125 (2002).
26. T. Kitamura, H. Hirakata, and Y. Yamamoto, Interface strength of tungsten micro-component on silicon substrate by means of AFM, Transactions of the Japan Society of Mechanical Engineers, Part A, 69, pp. 1216–1221 (2003).
27. K. Nakasa, M. Kato, D. Zhang, and K. Tasaka, Evaluation of delamination strength of thermally sprayed coating by edge-indentation method, Journal of the Society of Material Science, Japan, 47, pp. 413–419 (1998).
28. D. Zhang, M. Kato, and K. Nakasa, Fracture mechanics analysis of edge-indentation method for evaluation of delamination strength of coating, Journal of the Society of Material Science, Japan, 49, pp. 572–578 (2000).
29. K.S. Kim and N. Aravas, Elasto-plastic analysis of the peel test, International Journal of Solids and Structures, 24, pp. 417–435 (1988).
30. G.J. Spies, The peeling test on redux-bonded joints, Journal of Aircraft Engineering, 25, pp. 64–70 (1953).
31. J.J. Bickerman, Theory of peeling through a hookean solid, Journal of Applied Physics, 28, pp. 1484–1485 (1957).
32. D.H. Kaeble, Theory and analysis of peel adhesions: mechanism and mechanics, Transaction of Society of Rheology, 3, pp. 161–180 (1959).

33. D.H. Kaeble, Theory and analysis of peel adhesions: bond stresses and distributions, Transaction of Society of Rheology, 4, pp. 45–73 (1960).
34. C. Jouwersma, On the theory of peeling, Journal of Polymer Sciences, 45, pp. 253–255 (1960).
35. S. Yurenka, Peel testing of adhesive bonded metal, Journal of Applied Polymer Science, 6, pp. 136–144 (1962).
36. J.L. Gardon, Peel adhesion. II. A theoretical analysis, Journal of Applied Polymer Sciences, 7, pp. 643–664 (1963).
37. E.B. Saubestre, L.J. Durney, J. Haidu, and E. Bastenbeck, The adhesion of electrodeposits to plastics, Plating, 52, pp. 982–1000 (1965).
38. K. Kendall, The shapes of peeling solid films, Journal of Adhesion, 5, pp. 105–117 (1973).
39. A.N. Gent and G.R. Hamed, Peel mechanics, Journal of Adhesion, 7, pp. 91–95 (1975).
40. D.W. Nicholson, Peel mechanics with large bending, International Journal of Fracture, 13, pp. 279–287 (1977).
41. M.D. Chang, K.L. Devries, and M.L. Williams, The effects of plasticity in adhesive fracture, Journal of Adhesion, 4, pp. 221–231 (1972).
42. A.N. Gent and G.R. Hamed, Peel mechanics for an elastic-plastic adherent, Journal of Applied Polymer Sciences, 21, pp. 2817–2831 (1977).
43. A.D. Crocombe and R.D. Adams, Peel analysis using the finite element method, Journal of Adhesion, 12, pp. 127–139 (1981).
44. A.D. Crocombe and R.D. Adams, An elasto-plastic investigation of the peel test, Journal of Adhesion, 13, pp. 241–267 (1982).
45. A.G. Atkins and Y.W. Mai, Residual strain energy in elastoplastic adhesive and cohesive fracture, International Journal of Fracture, 30, pp. 203–221 (1986).
46. K.-S. Kim and J. Kim, Elasto-plastic analysis of the peel test for thin film adhesion, Journal of Engineering Materials and Technology, 110, pp. 266–273 (1988).
47. A.J. Kinloch, C.C. Lau, and J.G. Williams, The peeling of flexible laminates, International Journal of Fracture, 66, pp. 45–70 (1994).
48. Y. Wei and J.W. Hutchinson, Interface strength, work of adhesion and plasticity in the peel test, International Journal of Fracture, 93, pp. 315–333 (1998).
49. Y. Wei and J.W. Hutchinson, Peel test and interfacial toughness, in W. Gerberich and W. Yang, Eds., Interfacial and Nanoscale Failure, Volume 8, Comprehensive Structural Integrity, I. Milne, R.O. Ritchie, and B. Karihaloo, Editors-in-Chief, Elsevier, Amsterdam, 2003, pp. 181–217.
50. V. Tvergaard and J.W. Hutchins, The influence of plasticity on mixed mode interface toughness, Journal of the Mechanics and Physics of Solids, 41, pp. 1119–1135 (1993).
51. F. Ma and K. Kishimoto, A continuum interface debonding model and application to matrix cracking of composite, JSME International Journal, Series A, 39, pp. 496–507 (1996).
52. C. Zhou, W. Yang, and D. Fang, Damage of short-fiber-reinforced metal matrix composite considering cooling and thermal cycling, Journal of Engineering Materials and Technology, 122, pp. 203–209 (2000).
53. M. Omiya, K. Kishimoto, and W.M. Yang, Interface debonding model and it's application to the mixed mode interface fracture toughness, International Journal of Damage Mechanics, 11, pp. 263–286 (2002).
54. I.S. Park and J. Yu, An X-ray study on the mechanical effects of the peel test in a Cu/Cr/polyimide system, Acta Materialia, 46, pp. 2947–2953 (1988).
55. Y.B. Park, I.S. Park, and J. Yu, Interfacial fracture energy measurements in the Cu/Cr/polyimide system, Materials Science and Engineering A: Structural Materials: Properties, Microstructure and Processing, 266, pp. 261–266 (1999).
56. Y.B. Park, I.S. Park, and J. Yu, Phase angle in the Cu/polyimide/alumina system, Materials Science and Engineering A: Structural Materials: Properties, Microstructure and Processing, 266, pp. 109–114 (1999).
57. H.-H. Yu, M.Y. He, and J.W. Hutchinson, Edge effects in thin films, Acta Materialia, 49, pp. 93–107 (2001).
58. W. Yang and L.B. Freund, Shear stress concentration near the edge of a thin film deposited in a substrate, Brown Technical Report, November 1984.
59. M. Yanaka, Y. Kato, Y. Tsukahara, and N. Takeda, Effects of temperature on the multiple cracking progress of sub-micron thick glass films deposited on a polymer substrate, Thin Solid Films, 355-356, pp. 337–342 (1999).
60. B.F. Chen, J. Hwang, G.P. Yu, and J.H. Huang, In situ observation of the cracking behavior of TiN coating on 304 stainless steel subjected to tensile strain, Thin Solid Films, 352, pp. 173–178 (1999).

17

The Effect of Moisture on the Adhesion and Fracture of Interfaces in Microelectronic Packaging

Timothy P. Ferguson[a] and Jianmin Qu[b]

[a] Southern Research Institute, 757 Tom Martin Drive, Birmingham, AL 35211, USA
[b] G.W. Woodruff School of Mechanical Engineering, Georgia Institute of Technology, Atlanta, GA 30332-0405, USA

Abstract

A significant problem in the microelectronic packaging industry is the presence of moisture-induced failure mechanisms. Moisture is a multi-dimensional concern in packaging, having an adverse effect on package reliability by introducing corrosion, development of hygro-stresses, and degradation of polymers present in the package. Moisture can also accelerate delamination by deteriorating the polymer interfaces within the package. As the interfacial adhesion between the chip, underfill, and substrate decreases, the likelihood of delamination at each encapsulant interface increases. Once the package delaminates, the solder joints in the delaminated area are exposed to high stress concentrations, resulting in a reduction of overall package life.

Moisture can affect interfacial adhesion through two primary mechanisms. The first mechanism is the direct presence of moisture at the interface altering the interfacial integrity of the adhesive joint. The second mechanism is the absorbed moisture in either the adhesive and/or substrate altering the mechanical properties of those materials, which changes the response of the adhesive structure in the presence of an externally applied load. Inevitably, the effect of moisture on the adhesion and fracture of interfaces entails a multi-disciplinary study, and several aspects should be considered. From a global perspective, the primary aspects include moisture transport behavior, changes in bulk material properties from moisture absorption, effect of moisture on interfacial adhesion, and recovery from moisture upon fully drying, although several subsections within each major group occur due to the complexity of the problem.

In this chapter, a systematic and multi-disciplinary study is presented to address the fundamental science of moisture-induced degradation of interfacial adhesion. First, the moisture transport behavior within underfill adhesives is experimentally characterized. The results are incorporated into a finite element model to depict the moisture ingress and interfacial moisture concentration after moisture preconditioning. Second, the effect of moisture on the variation of the adhesive elastic modulus is demonstrated and the physical mechanisms for the change identified. Third, the aggregate effect of moisture on the interfacial fracture toughness is determined. This includes the primary effect of moisture being physically present at the interface and the secondary effect of moisture changing the elastic modulus of the adhesive when absorbed. Both reversible and irreversible components of the interfacial moisture degradation are evaluated. Using adsorption theory in conjunction with fracture mechanics, an analytical model is developed that predicts the loss in interfacial fracture toughness as a function of moisture content. The model incorporates key parameters relevant to the problem of moisture in epoxy joints identified from the experimental portion of this research, including the interfacial hydrophobicity, active nanopore density, saturation concentration, and density of water.

17.1. INTRODUCTION

It is inevitable that an electronic device will be exposed to varying degrees of moisture. Since many microelectronic packages utilize epoxy based materials such as underfill and molding compounds, they are highly susceptible to moisture absorption. The moisture uptake can lead to undesirable changes in mechanical performance, interfacial adhesion, and reliability [46,50].

Long term reliability and life prediction of microelectronic assemblies requires a rooted understanding in the interfacial failure mechanisms and associated debonding behavior of adhesive joints within these assemblies. With the advent of flip-chip technology, the need for improved understanding of delamination in these assemblies has taken on added importance. One of the keys to the success of flip-chip technology lies in development of underfill, which is an epoxy-based encapsulant that mechanically couples the chip to the board. Underfill drastically enhances the reliability of microelectronic assemblies when compared to unencapsulated devices [51], provided the structural integrity of the adhesive bond is maintained. Although delamination of the underfill in the microelectronic assembly tends to cause near immediate failure as soon as it reaches a solder joint, until recently the factors that affect the strength and durability of these interfaces have not been investigated and are the focal points of current studies in reliability research in microelectronic packaging. One of the most detrimental of these factors is moisture, which can significantly compromise the interfacial adhesion and accelerate the onset of delamination.

Another major area of concern in microelectronic packaging occurs at the interface between the copper alloy lead frame and the epoxy mold compound. Due to its relatively low cost in conjunction with its high electrical and thermal conductivity, copper alloys are widely used as a lead frame material. However, the epoxy/copper interface has poor interfacial adhesion strength and relatively high residual stress, which predisposes it to delamination. The copper surface is also highly susceptible to oxidation, which is an additional consideration when evaluating the interfacial adhesion of interfaces involving copper [9,28]. The delamination between the copper lead frame and the mold compound adversely affects the durability of these packages and is a common failure mode during the qualification process. In addition to further compromising the interfacial integrity of the interface, the delamination can also affect long term package reliability by yielding enhanced transport of moisture along the interface resulting in corrosion. The corrosion process will be accelerated if the absorbed moisture is a carrier of ionic impurities from the surrounding external environment [62]. Consequently, the epoxy/copper interface is another significant area of concern in microelectronic packaging reliability. Several studies continue to investigate this topic to better understand the durability and failure mechanisms, including the loss of adhesion in the presence of moisture.

17.2. MOISTURE TRANSPORT BEHAVIOR

17.2.1. Background

Central to understanding the effect of moisture on interfacial adhesion is to first identify the rate at which moisture is delivered to the interface. The three primary parameters that have the greatest effect on diffusion rates are the size of the diffusing particles, temperature, and viscosity of the environment. Lighter particles have a higher velocity for the same kinetic energy as heavier particles, thus lighter particles diffuse faster than heavier particles. Similarly, an increase in temperature will produce a higher kinetic energy yielding an increase in velocity, thus particles will diffuse more rapidly at elevated temperatures. Last, diffusion is more rapid in a gas than in a solid as a result of less atomic interactions, which retard the diffusion process.

Since the vast majority of underfills are epoxy based, they are highly susceptible to moisture absorption. A standard epoxy formulation can absorb between 1 and 7 wt% moisture [48]. Additional considerations that apply specifically to moisture absorption in epoxies include the epoxy surface topology and resin polarity. Soles et al. [47] have found that water initially enters the epoxy network through the nanopores that are inherent in the epoxy surface topology. They have determined the average size of a nanopore diameter to vary from 5.0 to 6.1 Å and account for 3–7% of the total volume of the epoxy material. Since the approximate diameter of a kinetic water molecule is just 3.0 Å, moisture can easily traverse into the epoxy via the nanopores.

Although surface topology can influence moisture penetration into an epoxy, of primary importance is the resin polarity, with the high polarity of the water molecule being susceptible to specific epoxy–water interactions. Less polar resins such as non-amine resins have more enhanced moisture diffusion coefficients than amine-containing resins. Soles and Yee [48] have shown that polar sites, such as amine functional groups, provide low energy wells for the water molecules to attach. Consequently, polar hydroxyls and amines can regulate transport through the nanopores by either blocking or allowing moisture to traverse the epoxy resin depending on the orientation of the resin with respect to nanopore position. Conversely, the absence of hydroxyls and amines in a non-amine resin leads to an enhanced moisture diffusion coefficient. In addition, non-amine resins absorb very little water relative to more polar resins, such as amine resins. Soles and Yee [48] have shown that by increasing the cross-link density, the intrinsic hole volume fraction is increased, which yields an increase in the equilibrium moisture content. Steric hindrances located at cross-link junctions open the epoxy matrix to facilitate interactions of water with polar groups, thus increasing the moisture uptake. Depending on the various chemical conformations of the epoxy resin in association with the inherent nanopores present in the epoxy structure, water molecules will behave differently in various epoxy resins.

17.2.2. Diffusion Theory

Since the transfer of heat by conduction is also attributed to random molecular motions, it is clear that diffusion is analogous to heat conduction. Fick adopted Fourier's mathematical expression for heat conduction to quantify diffusion. Fick's first law states that the rate of transfer of diffusing particles per plane of unit area is proportional to the concentration gradient measured normal to the plane:

$$F_x = -D\frac{\partial C}{\partial x}, \tag{17.1}$$

where F_x is the diffusion flux in the x direction, D is the diffusion coefficient, and $\partial C/\partial x$ is the concentration gradient. The negative sign in the above expression accounts for the fact that diffusion occurs in the opposite direction of increasing concentration. In addition, the expression is only valid for an isotropic medium.

Fick's second law of diffusion describes the nonsteady state diffusion of a substance and can be derived using Equation (17.1). Utilizing Equation (17.1) and a differential volume element in Cartesian coordinates, Crank [11] has shown that the following expression can be obtained assuming a constant diffusion coefficient:

$$\frac{\partial C}{\partial t} = D\left(\frac{\partial^2 C}{\partial x^2} + \frac{\partial^2 C}{\partial y^2} + \frac{\partial^2 C}{\partial z^2}\right), \tag{17.2}$$

where C is the concentration of the diffusing substance and D is the diffusion coefficient. For one-dimensional diffusion along the x-axis, the previous relation reduces to the following form:

$$\frac{\partial C}{\partial t} = D\left(\frac{\partial^2 C}{\partial x^2}\right). \tag{17.3}$$

The solution of Equation (17.3) for the concentration of a diffusing substance in an isotropic plane sheet of finite thickness as a function of time and space is given by [11]:

$$\frac{C(x,t)}{C_1} = 1 - \frac{4}{\pi}\sum_{n=0}^{\infty}\frac{(-1)^n}{2n+1}\exp\left[\frac{-D(2n+1)^2\pi^2 t}{4\ell^2}\right]\cos\frac{(2n+1)\pi x}{2\ell}, \tag{17.4}$$

where D is the diffusion coefficient, ℓ is the half-thickness of the sheet ($-\ell < x < \ell$), C is concentration of the diffusing substance absorbed by the sample at position x and time t, and C_1 is the saturation concentration of the absorbed substance. The application of Equation (17.4) assumes that immediately after the sheet is placed in the vapor both surfaces obtain a concentration that is equivalent to the equilibrium uptake, remaining constant. In addition the equation assumes that the diffusion coefficient remains constant throughout the diffusion process and that the initial concentration of the diffusing substance in the specimen is zero. An analogous expression given on a mass basis has been shown by Crank [11] to be the following:

$$\frac{M_t}{M_\infty} = 1 - \frac{8}{\pi^2}\sum_{m=0}^{\infty}\frac{1}{(2m+1)^2}\exp\left[\frac{-D(2m+1)^2\pi^2 t}{h^2}\right], \tag{17.5}$$

where D is the diffusion coefficient, h is the total sheet thickness, M_t is total mass of the diffusing substance absorbed by the sample at time t, and M_∞ is the equilibrium mass of the absorbed substance. In the initial stages of absorption where $M_t/M_\infty < 1/2$ and assuming a constant diffusion coefficient, D, Equation (17.5) can be shown to be approximated by the following:

$$\frac{M_t}{M_\infty} = \frac{4}{h}\sqrt{\frac{Dt}{\pi}}. \quad (17.6)$$

If absorption data is plotted with M_t/M_∞ as a function of $(t/h^2)^{1/2}$ and exhibits linear behavior for $M_t/M_\infty < 1/2$, the diffusion coefficient can be determined by rearranging Equation (17.6) to the following form:

$$D = \frac{\pi}{16}\left(\frac{M_t/M_\infty}{\sqrt{t}/h}\right)^2. \quad (17.7)$$

The diffusivity, D, can now be experimentally determined using absorption data with Equation (17.7). Again, Equations (17.4), (17.5), (17.6), and (17.7) all assume that the one-dimensional absorption occurs on both sides of the plane sheet with a concentration-independent, constant diffusivity. If absorption results in a diffusion coefficient that is variable rather than constant, explicit analytical solutions are no longer available.

17.2.3. Underfill Moisture Absorption Characteristics

Being epoxy-based, underfill resins are highly susceptible to moisture ingress. It is important to ascertain the fundamental moisture absorption behavior of each underfill when evaluating reliability performance in moist environments. In the case of adhesion testing, it is essential to accurately quantify the change in adhesion for a particular interfacial moisture concentration. This provides insight into the constitutive behavior of adhesion in the presence of moisture.

Two no-flow underfills were evaluated to determine their absorption behavior to select an ideal candidate for a fundamental study in the effect of moisture on interfacial adhesion. Underfill resin A (UR-A) was developed at the Georgia Institute of Technology. Underfill resin B (UR-B) was supplied by a commercial manufacturer. It should be noted that since both underfills were formulated for no-flow assembly, neither contained any filler content. Test samples were made approximately 60 mm in diameter and 2 mm thick, hence promoting predominantly one-dimensional diffusion through the thickness of the sample. The samples were baked at 115°C for at least 12 hours to remove moisture prior to being placed into a humidity chamber for moisture preconditioning. The atmosphere within the humidity chamber was maintained at a constant temperature ($85 \pm 1°C$), humidity ($85 \pm 1\%$RH), and pressure (P_{atm}). Moisture uptake profiles for each underfill are shown in Figures 17.1 and 17.2.

It is evident from Figures 17.1 and 17.2 that UR-A had not reached saturation after 168 hours of exposure, whereas UR-B had approached a saturated state within the same timeframe. In fact, samples constructed from UR-A did not reach saturation even after 725 hours of exposure. This absorption behavior is not uncommon, with Vanlandingham et al. [54] noting that some of the epoxies evaluated in their study had not reached saturation even after 3000–4000 hours of exposure at 50°C/85%RH. Similarly, Ardebili et al. [1]

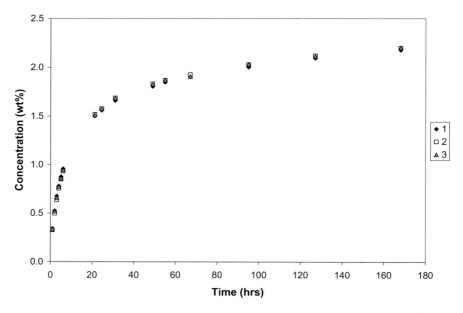

FIGURE 17.1. Moisture uptake profile for UR-A at 85°C/85%RH. (Ferguson and Qu [20], reprinted with permission of ASME.)

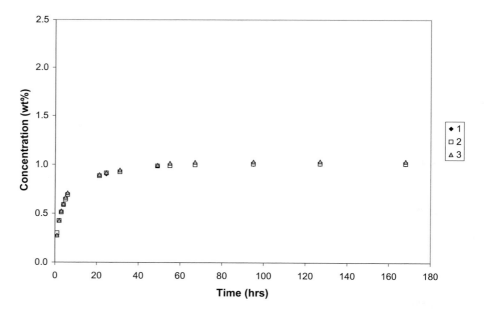

FIGURE 17.2. Moisture uptake profiles for UR-B at 85°C/85%RH. (Ferguson and Qu [20], reprinted with permission of ASME.)

found some of their epoxies to exhibit a gradual increase in moisture content with time, attributing this increase to void growth in the epoxy network caused by swelling.

The diffusivity of moisture through the thickness of the underfill resin is needed to apply an analytical, Fickian solution for modeling the moisture diffusion into the in-

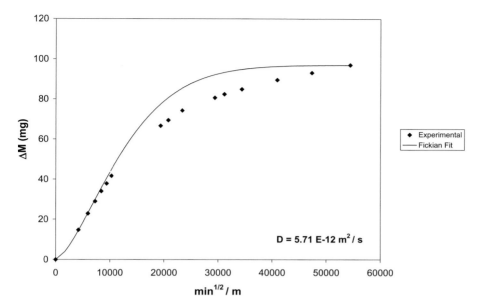

FIGURE 17.3. Fickian curve fit at 85°C/85%RH for UR-A. (Ferguson and Qu [20], reprinted with permission of ASME.)

terfacial fracture test specimens. The diffusion coefficient, D, can be experimentally determined using a test specimen that promotes predominantly one-dimensional diffusion into the test specimen. Using Equation (17.7) with absorption data, the diffusion coefficients were experimentally determined for both UR-A ($D = 5.71E{-}12$ m^2/s) and UR-B ($D = 1.47E{-}11$ m^2/s). Since the diffusion coefficient is a measure of how quickly a material responds to mass concentration changes in its environment, the larger value of diffusivity for UR-B indicates it will respond more quickly to those changes. As a result, UR-B test specimens will approach saturation more rapidly than UR-A test specimens, which quantitatively supports what was already qualitatively observed (Figures 17.1 and 17.2).

A Fickian curve was generated for each underfill to examine the extent that the moisture uptake of the specimens demonstrated Fickian behavior at conditions of 85°C/85%RH. The following relation developed by Shen and Springer [44] was implemented to generate the Fickian profile since it simplifies the infinite series of Equation (17.5):

$$\frac{M_t}{M_\infty} = 1 - \exp\left[-7.3\left(\frac{Dt}{h^2}\right)^{0.75}\right]. \tag{17.8}$$

A Fickian curve for each data set at 85°C/85%RH is shown in Figures 17.3 and 17.4.

It is clear from Figures 17.3 and 17.4 that neither UR-A nor UR-B exhibited true Fickian behavior at 85°C/85%RH, although UR-B appeared to obtain a better curve fit than UR-A. Since test specimens promoted predominately one-dimensional diffusion and exhibited non-Fickian absorption behavior, it is evident that the diffusion coefficients of both UR-A and UR-B were dependent on moisture concentration rather than being constant throughout the entire diffusion process at 85°C/85%RH.

Wong et al. [56] found varied diffusion behavior in the epoxy resins they evaluated at 85°C/85%RH, with some resins exhibiting Fickian diffusion while others did not. They

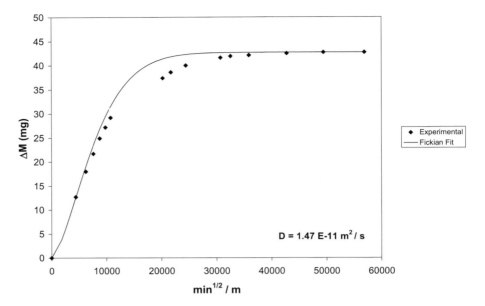

FIGURE 17.4. Fickian curve fit at 85°C/85%RH for UR-B. (Ferguson and Qu [20], reprinted with permission of ASME.)

postulated that diffusivity is constant and moisture diffusion exhibits Fickian behavior for epoxy resins at lower temperature and humidity levels such as 30°C/60%RH. Increasing the humidity level results in a corresponding amplification of the saturation level, while increasing the temperature level produces more prominent non-Fickian behavior [54]. Although test specimens will absorb more moisture in less time at higher temperatures and relative humidity levels, the trade-off is that the specimens will also exhibit an increased likelihood of non-Fickian diffusion behavior. The concentration dependence of the diffusivity in non-Fickian diffusion behavior complicates the modeling of the moisture ingress; however, numerical algorithms have been published that demonstrate how to model the non-Fickian diffusion process [56].

17.2.4. Moisture Absorption Modeling

To illustrate the moisture distribution graphically in interfacial fracture test specimens, a transient, finite element analysis was implemented to model the associated moisture concentration distribution in test specimens for small times of exposure. Since the substrates of the interfacial fracture test specimens were metallic and impermeable to moisture, it should be noted that only the moisture distribution in each underfill was modeled. Results of the finite element model illustrating the transient moisture distribution in the underfill resins are shown in Figures 17.5 and 17.6. Both figures refer to the interfacial fracture test specimens as unmodified, which indicates that this is the moisture absorption behavior exhibited by the test specimens if placed in 85°C/85%RH conditions immediately after test specimen manufacture without consideration to how the moisture uptake could influence fracture results.

It is apparent from the model of the transient moisture ingress that edge effects are significant. This can be clearly seen by examining the interface of the test specimens in

EFFECT OF MOISTURE ON THE ADHESION AND FRACTURE OF INTERFACES   439

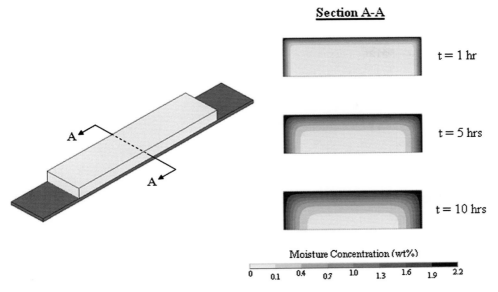

FIGURE 17.5. Moisture concentration distribution for unmodified UR-A interfacial fracture test specimen at 85°C/85%RH after 1, 5, and 10 hours of exposure. (Ferguson and Qu [20], reprinted with permission of ASME.)

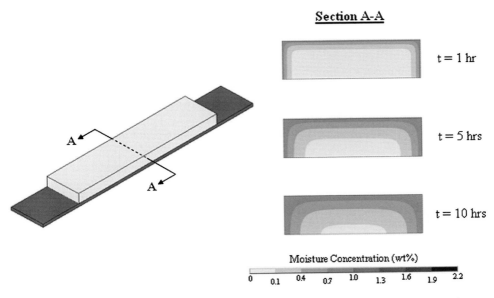

FIGURE 17.6. Moisture concentration distribution for unmodified UR-B interfacial fracture test specimen at 85°C/85%RH after 1, 5, and 10 hours of exposure. (Ferguson and Qu [20], reprinted with permission of ASME.)

Figures 17.5 and 17.6 (bottom of each cross section A–A), where it is evident that a gradient of moisture exists at the interface until saturation is reached. This is undesirable since the interface will experience different levels of moisture spatially relative to the exposure time, which will not allow a fracture toughness measurement to be identified with a par-

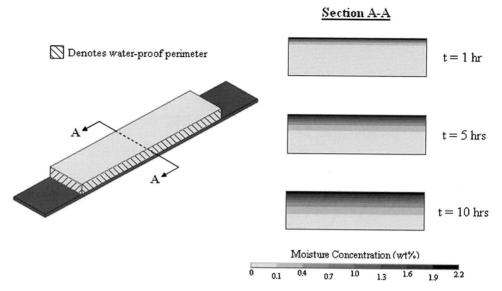

FIGURE 17.7. Moisture concentration distribution for modified UR-A interfacial fracture test specimen at 85°C/85%RH after 1, 5, and 10 hours of exposure.

ticular level of interfacial moisture concentration until saturation is reached. Furthermore, it is also possible that the non-uniform moisture gradient at the interface could influence interfacial fracture toughness results even if saturation is reached in a test specimen. This is attributed to different areas of the interface being exposed to varying degrees of moisture for different periods of time, which could have an effect on fracture toughness results even if test specimens are in a saturated state. Last, wicking of moisture along the interface could also introduce moisture concentration levels that remain unidentified through modeling of the absorption process alone. This would make it difficult to attribute a particular fracture toughness measurement with an associated interfacial moisture concentration level.

In view of these observations, the interfacial fracture test specimen design should be modified with a water-proof perimeter applied to test specimens prior to moisture preconditioning. The application of the water-proof perimeter forces 1D moisture uptake through the top surface of the test specimens and prevents wicking along the interface. Not only does this yield uniform concentrations spatially at the interface, but it also aids in the identification of an interfacial moisture concentration level by utilizing the inherent moisture absorption characteristics of the adhesive to restrict the amount of moisture arriving to the interface. Figures 17.7 and 17.8 depict the moisture concentration distribution in the modified interfacial test specimens.

Although percent weight is dependent on both the specimen volume and density, a comparison between the moisture concentration distributions can be made as a result of both underfills having similar densities (UR-A, $\rho = 1.14 \times 10^{-3}$ g/mm^3 and UR-B, $\rho = 1.16 \times 10^{-3}$ g/mm^3) and volumes. Figures 17.7 and 17.8 illustrate that although UR-A test specimens contain a significantly higher concentration of moisture near the underfill surface, the moisture will actually penetrate the interface first for comparably sized UR-B test specimens. It is clear from the progression of the constant-concentration lines depicted in Figures 17.7 and 17.8 that the moisture traversed much more easily through the UR-B

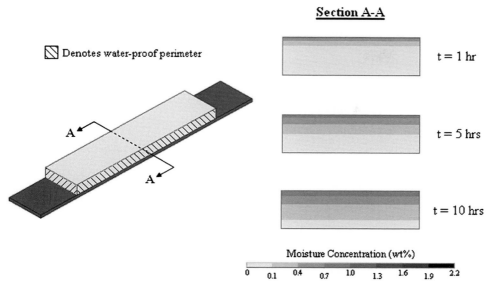

FIGURE 17.8. Moisture concentration distribution for modified UR-B interfacial fracture test specimen at 85°C/85%RH after 1, 5, and 10 hours of exposure.

test specimens. An explanation for this behavior lies in the particular chemistry of each underfill epoxy with respect to the polarity of water molecules. As previously noted, amine functional groups regulate transport through the nanopore channels of the epoxy by either blocking or allowing moisture to traverse the channels depending on the orientation of the resin with respect to nanopore position [48]. UR-A contains amine functional groups, while UR-B is a non-amine containing underfill [20]. Consequently, it would be anticipated that UR-B would have an enhanced diffusion coefficient than UR-A, which was found to be true based on experimental results. As demonstrated in Figures 17.7 and 17.8, the amine functional groups present in UR-A contributed to retard moisture penetration into the amine containing epoxy resin, whereas the moisture diffused more easily through the non-amine epoxy resin, UR-B. From this observation, there are three primary conclusions to consider:

1. Degradation of interfacial adhesion over the entire interface due to the presence of moisture will initially occur in UR-B test specimens prior to comparably sized UR-A test specimens.
2. Further degradation of interfacial adhesion will occur in UR-A test specimens than comparably sized UR-B test specimens for longer exposure times. This is due to UR-A absorbing more aggregate moisture than UR-B at longer durations (Figures 17.1 and 17.2).
3. Amine functional groups present in UR-A retard the rate by which moisture exits the underfill upon redrying. Consequently, UR-A test specimens will take longer to recover the reversible component of adhesion loss upon removal of moisture from the interface than comparably sized UR-B test specimens.

The absorption characteristics, exposure time, and adhesive performance from moisture dictate whether UR-A or UR-B is a more robust product in humid environments. For short exposure times to moist environments and considering only the absorption characteristics, UR-A represents a better encapsulant by retarding the rate of moisture ingress to the interface through the presence of amine functional groups in its chemistry. Conversely, UR-B represents a better product for longer exposure times to moist environments by absorbing less aggregate moisture than UR-A. Bare in mind neither of the aforementioned statements considers the relative adhesion performance in the presence of moisture nor the extent by which each underfill adhesive bond recovers after multiple exposures to moist environments. These are additional considerations when evaluating the long term reliability of a particular underfill to moist conditions.

17.3. ELASTIC MODULUS VARIATION DUE TO MOISTURE ABSORPTION

The deleterious effect of moisture not only damages interfacial adhesion by being physically present at the interface, but also through the degradation of the elastic modulus of the adhesive and substrate due to moisture uptake. The change in the elastic modulus after moisture uptake can be substantial, which can significantly affect material performance and adhesion results. Consequently, the variation in the elastic modulus of the adhesive and substrate as a function of moisture concentration should be determined to completely characterize the loss in interfacial adhesion due to moisture absorption. Since many of the substrates used in electronic packaging are impermeable to moisture (i.e., copper, aluminum, and silicon), it is often only necessary to characterize the change in the elastic modulus as a function of moisture concentration for the adhesives, which are typically epoxy based and highly susceptible to moisture uptake.

17.3.1. Background

Epoxy adhesives are found in many microelectronic packaging applications and widely used throughout the industry. One of the more substantial developments within the last ten years is underfill, which is an epoxy based encapsulant that mechanically couples the chip to the board. Underfill drastically enhances the fatigue life of microelectronic assemblies when compared to unencapsulated devices [51]; however, since underfills are epoxy based, they are also particularly vulnerable to moisture ingress [20,22,53,56]. Although the absorbed moisture can significantly alter its mechanical performance and the overall microelectronic assembly reliability, very few studies in the electronic packaging literature have addressed the issue of moisture on the mechanical properties of epoxies.

Throughout the literature, the availability of information regarding the effect of moisture on the mechanical properties of epoxy adhesives is in general limited and more work is needed to adequately characterize this response [12,24]. From the work that has been published, it has been found that water absorption can severely modify the mechanical properties of epoxies by decreasing the elastic modulus [36,63], shear modulus [27,64], yield stress [55], and ultimate stress [55] while increasing the failure strain [12,55] as water concentration increases. A representative stress/strain diagram is shown in Figure 17.9 illustrating these effects.

Moisture primarily affects the mechanical properties of epoxy adhesives through three mechanisms: plasticization, crazing, and hydrolysis. The first is considered reversible upon drying, while the latter two are irreversible. Several studies attribute the decrease in modulus due to the plasticizing action of the water on the adhesive [3,12,15,27,49,55,63,64]. By acting as an external plasticizer to the adhesive, the water spreads the polymer molecules apart and reduces the polymer–polymer chain secondary bonding. This provides more room for the polymer molecules to untangle and move, which results in a softer,

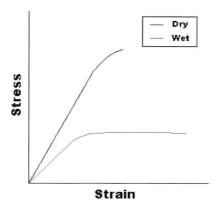

FIGURE 17.9. Representative stress/strain diagram depicting the effect of moisture on the mechanical properties of bulk epoxy adhesives.

more easily deformable mass [42]. Other studies show the decrease in epoxy modulus after moisture absorption resulting from crazing [34,36,37], where the absorbed water can act as a crazing agent continuously decreasing the mechanical strength of epoxies with exposure time in water [34]. This is supported by scanning electron micrographs of epoxies, which show cavities and fractured fibrils that could only be explained by a moisture-induced crazing mechanism [37]. The moisture-induced swelling creates dimensional changes and internal stresses that can ultimately craze and/or crack the material. As a result, lightly cross-linked networks will be more susceptible to crazing than highly cross-linked networks [36]. Last, moisture can also affect the mechanical properties of epoxy adhesives by causing hydrolysis leading to chain scission. Short term exposure to moisture results in chain scission with a chemical addition of water that remains permanently in the epoxy system even after subsequent drying. Long term exposure to moisture can result in an increased probability of chain scission detaching segments from the polymer network, yielding a permanent loss in weight after subsequent drying [59].

Studies by Zanni-Deffarges and Shanahan [63,64] and DeNeve and Shanahan [15] depict the decrease in elastic and shear modulus of an epoxy as a function of time exposure to moisture. Although this information is useful in evaluating the effect of exposure time to moisture on the modulus, it does not depict how the inherent wet modulus values change as a function of concentration. This is due to a gradient of mechanical properties that will exist in the adhesive until saturation is reached, where water concentrations become steady and uniform. Other studies have evaluated the effect of moisture on epoxy adhesives after saturation is established for a given level of moisture preconditioning. These studies have shown a decrease in the elastic modulus of epoxy adhesives of 24% [64], 29% [49], and 86% [49] for saturation concentrations of 4 wt%, 0.9 wt%, and 3.1 wt% respectively; however, they only tested one level of moisture preconditioning to compare to fully dried test results. Consequently, information regarding the mechanical response of epoxy adhesives to different levels of moisture concentrations is incomplete and fundamental insight into the intrinsic response of the adhesives to increasing saturation concentrations of moisture cannot be ascertained.

Even fewer investigations have evaluated the recovery of epoxies upon drying after moisture absorption with little information available regarding the extent of the reversible and irreversible nature of moisture uptake in epoxies. Netravali et al. [38] have shown for epoxy samples soaked in water at 25°C for 820 hours that much of the loss from moisture results from plasticization and is recoverable upon drying at 30°C for 400 hours; however, samples soaked in water at 70°C for 775 hours were highly irreversible after drying at 70°C for 125 hours. The irreversibility was attributed to water reacting with unreacted epoxide groups. It should be noted that neither groups of samples were completely dry at the time of testing after exposure to water and subsequent drying. Buehler and Seferis [4] found epoxy prepregs soaked in water at 71°C for 1200 hours exhibited varying degrees of reversible and irreversible damage to both the flexural modulus and flexural strength upon drying at 50°C for 450 hours. However, more time was needed to fully dry the specimens in this study as well, with 3% weight concentrations of moisture still existing in the specimens at the time of testing after drying. Wright [57] proposes that the permanent loss of properties that occur due to moisture uptake is most probably due to swelling of the matrix and the production of voids, while Xiao and Shanahan [59] suggest based on absorption behavior that the irreversible damage component of hydrolysis can play a significant role in the degradation process depending on the duration of exposure. Undoubtedly the mechanisms responsible for the observed losses in epoxies from moisture uptake are complex, and the material constitutive damage behavior is not entirely understood.

17.3.2. Effect of Moisture Preconditioning

To help characterize the elastic response of an epoxy adhesive to increasing moisture concentrations, an evaluation of the effect of moisture on the elastic modulus of a no-flow underfill was performed. The particular underfill evaluated was UR-B, which was determined to be ideal for studying the fundamental effect of moisture on interfacial adhesion due to its moisture diffusion kinetics and saturation behavior established from the moisture absorption portion of this research. Flexural specimens were tested in a three-point bend test according to ASTM D790 [2] to determine the effect of moisture on the elastic modulus.

Test specimens were divided into six test groups and subjected to five different levels of moisture preconditioning to ascertain the effect of moisture on the elastic modulus of the underfill. The test groups included fully dry, 85°C only, 85°C/50%RH, 85°C/65%RH, 85°C/85%RH, and 85°C/95%RH, with the latter five test groups being environmentally preconditioned for 168 hours. All test specimens were baked at 115°C for at least 12 hours to remove any moisture that may have been introduced during sample preparation prior to environmental aging, which was performed in a humidity chamber in an atmosphere maintained at a constant temperature (± 1°C), humidity (± 1%), and pressure (P_{atm}). In addition, all flexural tests were performed with both the surrounding environment and test specimens being at room temperature after environmental preconditioning. No measurable loss in moisture uptake occurred in the test specimens from the time they were removed from the environmental chamber, allowed to cool to room temperature, and experimentally tested.

Figure 17.10 illustrates the effect of moisture preconditioning on the underfill elastic modulus for several different temperature/humidity levels. All moisture preconditioned test specimens were fully saturated at the conclusion of the 168 hour exposure time, hence a gradient of moisture concentration did not occur within the specimens so that the inherent wet modulus was identified. In addition, Differential Scanning Calorimetry (DSC) test results demonstrate that the underfill was fully cured in the flexural specimens for the curing conditions and test specimen size and geometry used [18]. Therefore, incomplete curing

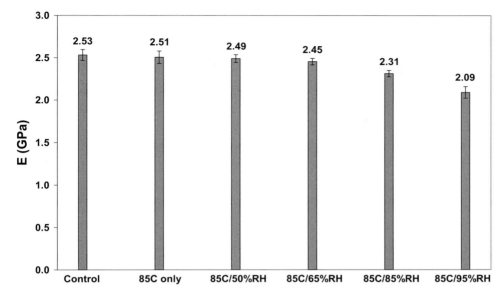

FIGURE 17.10. Effect of moisture preconditioning on the underfill elastic modulus. (Ferguson and Qu [18], reprinted with permission of IEEE.)

of the underfill in the flexural specimens did not influence any observed changes to the elastic modulus of the underfill after moisture preconditioning. Further information on the fundamentals and use of differential scanning calorimetry may be found in the works of Pasztor [40] and Prime [41].

When compared to unaged, control test specimen values, moisture preconditioning at 85°C/50%RH and 85°C/65%RH was found to have little to no effect on the elastic modulus of the underfill. A more noticeable effect occurs at 85°C/85%RH, while conditions of 85°C/95%RH yielded a significant decrease in modulus. To isolate the possible effect of thermal aging at 85°C from moisture preconditioning contributing to the observed changes in the elastic modulus of the underfill, flexural specimens were exposed to conditions of 85°C only for 168 hours and compared to unaged, control test specimen values. As shown in Figure 17.10, thermal aging at 85°C for 168 hours was found to have no effect on the elastic modulus with similar values obtained when compared to the control test group results. Again, it is important to note that all tests were performed at room temperature, hence only the effects of thermal aging were evaluated rather than the effect of testing at elevated temperatures on the elastic modulus. Since all environmental preconditioned test groups were exposed to the same temperature of 85°C and to the same duration of 168 hours, the observed changes in modulus from moisture preconditioning given in Figure 17.10 can be attributed to the effect of moisture and moisture alone.

A summary of the effect of moisture preconditioning on the elastic modulus of the underfill is given in Table 17.1, where C_{sat} represents the saturation concentration of moisture in the test specimens for each respective level of moisture preconditioning and given as both a percent weight change (wt%) and mg H_2O/mm^3.

Since saturation was reached in all moisture preconditioned test specimens prior to removal from the humidity chamber and thermal aging from the 85°C temperature component of moisture preconditioning was found to have no effect on the elastic modulus,

TABLE 17.1.
Change in the underfill elastic modulus from moisture uptake. (Ferguson and Qu [18], reprinted with permission of IEEE.)

T (°C)	RH (%)	C_{sat} (wt%)	C_{sat} (mg H_2O/mm^3)	E (GPa)	Modulus change (%)
Control	—	0	0.0000	2.53 ± 0.06	—
85	50	0.65	0.0075	2.49 ± 0.05	1.6
85	65	0.77	0.0089	2.45 ± 0.04	3.2
85	85	1.02	0.0118	2.31 ± 0.04	8.7
85	95	1.19	0.0138	2.09 ± 0.07	17.4

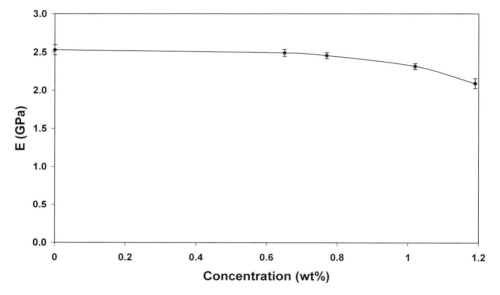

FIGURE 17.11. Underfill elastic modulus variation as a function of moisture concentration (wt%). (Ferguson and Qu [18], reprinted with permission of IEEE.)

the inherent wet modulus was identified and all observed changes in the modulus occurred solely from the influence of moisture. This allows the characterization of the change in modulus of the underfill from moisture uptake as a function of moisture concentration as shown in Figures 17.11 and 17.12.

Figures 17.11 and 17.12 depict the inherent change in the elastic modulus of an epoxy-based adhesive as a function of moisture concentration. Time dependent variation in the elastic modulus after saturation is assumed to be negligible, although it could be a consideration for longer durations of exposure at higher concentrations of moisture as a result of hydrolysis [59]. Previous studies on epoxy adhesives have shown the variation in modulus as a function of the square root of time corrected for specimen thickness [63]; however, this information depicts the change in modulus resulting from a transient, gradient of moisture concentration rather than demonstrating how the inherent wet modulus changes with increasing moisture content. Other studies have identified the inherent wet modulus for a single saturation level and compared to fully dry results [3,49,63]; however, these studies do not show the inherent wet modulus of the same adhesive for several dif-

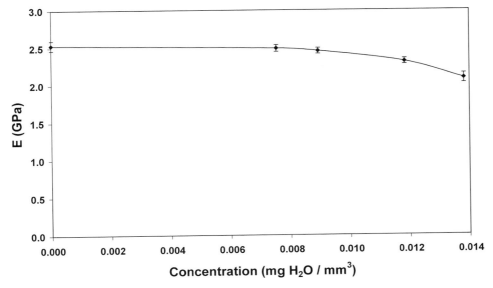

FIGURE 17.12. Underfill elastic modulus variation as a function of moisture concentration (mg H_2O/mm^3). (Ferguson and Qu [18], reprinted with permission of IEEE.)

ferent saturation levels and thus do not show the characteristic response of the adhesive as a function of increasing moisture concentration as given in Figures 17.11 and 17.12. Such information is extremely useful in predictive modeling efforts, where the intrinsic response of the elastic modulus as a function of increasing moisture concentration can be used in a coupled mechanical-diffusion analysis [55] to incorporate the transient effect of the continual variation of elastic modulus as moisture diffuses into the adhesive. These data are not only significant when modeling the effect of moisture on the bulk material behavior, but also on interfacial adhesion, where changes in the mechanical properties of the adhesive due to moisture uptake can play a significant role in the onset of package delamination.

17.3.3. Elastic Modulus Recovery from Moisture Uptake

To further characterize the response of the underfill from moisture uptake and identify the mechanisms responsible for the observed losses in the elastic modulus from moisture absorption, test specimens were moisture preconditioned for 168 hours followed by baking at 95°C until fully dry. A fully dried state was established when there was no measurable change in the weight of a specimen for a period of 24 hours. Since 85°C/85%RH and 85°C/95%RH moisture preconditioning conditions were found to noticeably decrease the elastic modulus of the underfill, only those conditions were evaluated for recovery of the elastic modulus from moisture uptake upon redrying. Figure 17.13 provides a graphical depiction of the recovery results for the underfill elastic modulus.

As shown in Figure 17.13, much of the observed loss in the elastic modulus from moisture uptake was recoverable upon subsequent drying. Since plasticization is the only primary degradation mechanism attributed to moisture that is regarded as a reversible process, the recovery results demonstrate that the majority of the loss in modulus resulted from plasticization of the underfill from moisture uptake. To further evaluate the change

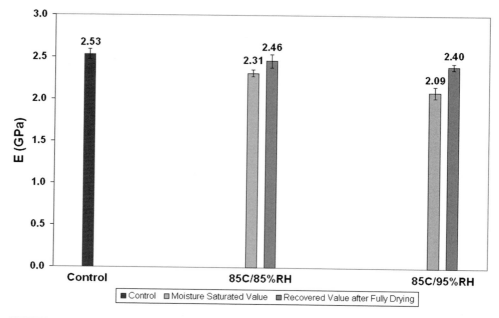

FIGURE 17.13. Recovery of the underfill elastic modulus on removal of moisture. (Ferguson and Qu [18], reprinted with permission of IEEE.)

TABLE 17.2.
Recoverability of the underfill elastic modulus from moisture uptake after subsequent drying. (Ferguson and Qu [18], reprinted with permission of IEEE.)

T (°C)	RH (%)	C_{sat} (wt%)	E_{sat} (GPa)	$E_{recovery}$ (GPa)	*Recoverability* (%)
Control	—	0.00	2.53 ± 0.06	—	—
85	50	0.65	2.49 ± 0.05	—	—
85	65	0.77	2.45 ± 0.04	—	—
85	85	1.02	2.31 ± 0.04	2.46 ± 0.08	68.2
85	95	1.19	2.09 ± 0.07	2.40 ± 0.05	70.5

in elastic modulus from moisture uptake, the recoverability for the elastic modulus will be defined as follows:

$$Recoverability\ (\%) = \frac{E_{recovery} - E_{sat}}{E_{dry} - E_{sat}} \cdot 100, \qquad (17.9)$$

where $E_{recovery}$ is the value of the elastic modulus upon fully drying from the moisture saturated state, E_{sat} is the saturated value of the elastic modulus after moisture absorption, and E_{dry} is the unaged, control value of the elastic modulus. The recoverability of the elastic modulus is given in Table 17.2.

Although a significant portion of the elastic modulus was recoverable after fully drying, some irreversible, permanent damage did occur. The average recoverable value of the elastic modulus suggests slightly more irreversible damage occurred at higher humidity levels, but it cannot be concluded unequivocally solely based on the modulus results due to the uncertainty associated within the two measurements. However, it can be concluded when considering the results from moisture uptake data. After fully drying, there was a slight net permanent weight increase in the test specimens, with specimens moisture preconditioned at 85°C/85%RH retaining $1.3 \pm 0.5\%$ of the total absorbed water while specimens moisture preconditioned at 85°C/95%RH retaining $2.3 \pm 0.4\%$ of the total absorbed water. The permanent weight increase in the test specimens after subsequent fully drying suggests that at least part of the irreversible damage resulted from hydrolysis with a greater extent occurring at higher humidity levels. In addition to hydrolysis, it is also possible that moisture-induced crazing also contributed to the irreversible damage to the elastic modulus. Overall, the irreversible damage was small with the majority of the loss in the elastic modulus from moisture uptake being fully recoverable after subsequent drying.

17.4. EFFECT OF MOISTURE ON INTERFACIAL ADHESION

The effect of moisture on interfacial adhesion is governed by two fundamental mechanisms. The first is the rate at which moisture is delivered to the interface, and the second is the change in adhesion performance as a consequence of moisture being present in the adhesive structure. This includes not only the primary effect of moisture being directly present at the interface itself, but also the secondary effect of moisture altering the mechanical performance of the two materials that constitute the bimaterial interface. Having previously quantified both the rate at which moisture is delivered to the interface and the degrading effect of moisture on the elastic modulus of the materials that constitute the bimaterial interface, a model depicting the intrinsic change in interfacial adhesion as a function of moisture concentration is developed. Interfacial fracture mechanics is used to characterize this change to develop relationships that are independent of test specimen geometry.

17.4.1. Background

With interconnect density increasing and package size decreasing, several adaptations to microelectronic assemblies have been developed to accommodate the increasing demand in both cost and performance requirements. In particular, epoxy-based encapsulants have been extensively used in microelectronic devices to enhance package reliability, provide environmental protection, and improve manufacturing yields. To insure these benefits are not compromised, the structural integrity of the adhesive bond must be maintained. Characterizing the primary adhesion mechanisms and identifying the factors that affect the strength and durability of encapsulants are critical to their success.

Traditional encapsulation processes, such as transfer molding, cavity filling, and glob-topping, are used to protect the IC device from environmental pollutants and provide mechanical support. In these devices, copper alloys are typically used as a lead frame material due to their low cost in conjunction with their high electrical and thermal conductivity. However, the adhesion at the epoxy/copper interface is poor [7,9,29,32]. In addition, the copper surface is highly susceptible to oxidation. This is an additional consideration when evaluating the interfacial adhesion of adhesives with copper.

A more recent encapsulant developed within the last ten years is underfill, which is an epoxy-based encapsulant that mechanically couples the chip to the board. Underfill drastically enhances the fatigue life of microelectronic assembles when compared to unencapsulated devices [51], provided the adhesive bond between the underfill and the printed wiring board, solder mask, copper, silicon, passivation, and solder is maintained. Characterizing the adhesion of underfill to these substrates has been the focus of several studies in adhesion and reliability research [13,14,17,19,60].

Although epoxy encapsulants have many benefits, they are susceptible to moisture uptake. A typical epoxy formulation can absorb between 1 and 7 wt% moisture [47], which can have a detrimental effect on interfacial adhesion and drastically reduce the reliability of encapsulated devices. While it has been shown that moisture can significantly alter adhesive performance in microelectronic packaging [21,34], the interfacial and material constitutive damage behavior from moisture exposure is not well understood. This largely arises due to the difficulty of the problem, which is governed by two fundamental mechanisms. The first is the rate at which moisture is delivered to the interface. The second is the response of the interfacial adhesion to varying levels of moisture concentration, where the deleterious effect of moisture not only affects interfacial adhesion by being physically present at the interface, but also through the degradation of the mechanical properties of the epoxy adhesive due to moisture uptake. Mass transport and in particular the diffusion of moisture in epoxy adhesives has been studied by several sources and is fairly well established [47,48,52,54,56]; however, the response of interfacial adhesion to moisture is much less understood. Although several studies have addressed the issue of moisture, much more work needs to be completed and there currently exists a lag in fundamental empirical data depicting the loss in interfacial adhesion as a function of interfacial moisture concentration. Since there exists this lag in experimental data, even less effort has been spent on developing predictive models that account for the effect of moisture on interfacial adhesion.

Of particular interest to the long-term reliability of an adhesive bond is ascertaining the permanent damage to the bond from exposure to moisture. Very few studies have examined the reversible and irreversible components of the loss in adhesion from moisture after subsequent drying. This has significant practical aspects, as the recoverability of the interface from moisture will identify the severity of the moisture damage. If the loss in adhesion from moisture is largely unrecoverable and irreversible, then the service life of the adhesive joint will be severely, permanently compromised as a result of exposure to moisture. Such consequences would bring added emphasis to protecting the encapsulated package from moisture ingress and developing more robust, moisture-resistant adhesives.

When evaluating the moisture recovery of an adhesive joint, there are two aspects to consider. The first is the recovery of the materials that constitute the adhesive joint, as absorbed moisture can alter the mechanical performance of those materials and indirectly affect adhesion [18]. The second aspect is the recovery of the interfacial bonding itself, as the direct presence of moisture at the interface can significantly alter adhesion. Butkus [5] examined the permanent change in Mode I fracture toughness of Aluminum/FM73M/Aluminum and Aluminum/FM73M/Boron-Epoxy joints after 5000 hours at 71°C and >90%RH followed by 5000 hours of desiccation at 22°C/10%RH prior to testing. Both the Al/FM73M/Al joints and the Al/FM73M/Boron-Epoxy joints recovered very little of their fracture toughness on subsequent drying, demonstrating large, permanent losses in toughness after exposure to moisture. Orman and Kerr [39] have shown that although some of the strength lost in the epoxy-bonded aluminum joints they studied was recovered, there was noticeable permanent damage from moisture suggesting an irreversible disruption at the interface as a result of attack by water. Contrary to this claim, Shaw et al. [43] found that nearly all of the strength lost after immersing steel/epoxy lap shear joints in distilled water for three weeks was recovered after drying. They attributed the loss in strength after moisture preconditioning to plasticization of the epoxy adhesive, which is generally regarded as a reversible process. Dodiuk et al. [16] found exposure to moisture of their epoxy/aluminum joints caused a reduction in lap shear strength; however, if the moisture concentration was below 0.3 wt%, the strength was fully recoverable after drying indicating a completely reversible process. The authors gave no explanation to this observed behavior other than to state that moisture concentrations exceeding 0.3 wt% would result in an irreversible process and permanent loss of adhesion at the interface. Undoubtedly the mechanisms responsible for the observed losses in both material behavior and interfacial adhesion from moisture uptake are complex, and the material constitutive damage behavior is not entirely understood.

17.4.2. Interfacial Fracture Testing

Interfacial fracture toughness is defined as the critical value of the energy release rate, G_c, at which a bimaterial interface will begin to delaminate. It is a property that characterizes the adhesion of a bimaterial interface, independent of the size and geometry of the cracked body. For a bimaterial interface loaded in four point bending under plane strain conditions, it can be shown that the critical value of the energy release rate, G_c, can be determined using the following equation [26]:

$$G = \frac{1}{2\overline{E_1}}\left(\frac{12M^2}{h^3}\right) - \frac{1}{2\overline{E_2}}\left(\frac{M^2}{Ih^3}\right), \quad (17.10)$$

where

$$\overline{E_i} \equiv \frac{E_i}{1 - v_i^2}, \quad (17.11)$$

M is the moment, v is Poisson's ratio, E is the elastic modulus, subscript 1 refers to material 1, subscript 2 refers to material 2, h is the height of material 1, and I is the dimensionless moment of inertia.

Since the interfacial fracture toughness only specifies the magnitude of the crack tip singularity, the mode mixity, ψ, must be determined from the complex stress intensity factor K. For a two-dimensional system, the complex stress intensity factor, K, is given by:

$$K = K_1 + iK_2. \quad (17.12)$$

For four-point loading conditions it can be shown [26]:

$$K = h^{-i\varepsilon}\sqrt{\frac{1-\alpha}{1-\beta^2}}\left(\frac{P}{\sqrt{2hU}} - ie^{i\gamma}\frac{M}{\sqrt{2h^3V}}\right)e^{i\omega} \quad (17.13)$$

with the mode mixity given by:

$$(K_1 + iK_2)L^{i\varepsilon} = |(K_1 + iK_2)|e^{i\psi}, \quad (17.14)$$

$$\psi = \tan^{-1}\left(\frac{\text{Im}(KL^{i\varepsilon})}{\text{Re}(KL^{i\varepsilon})}\right), \quad (17.15)$$

where L is the characteristic length and ε is a dimensionless quantity given by Hutchinson and Suo [26]. As shown in Equation (17.15), the mode mixity for a test specimen requires the specification of some length quantity, L. The choice for L is arbitrary, but it should be selected as a fixed length and reported with the calculated values for the mode mixity.

The flexural beam test for interfacial fracture testing has three primary benefits. First, it yields intermediate values for mode mixity, which is representative of the values experienced by electronic devices during actual application. Second, it provides a means for successful interfacial fracture test specimen construction utilizing substrates and adhesives common to microelectronic packaging. Last, the flexural beam test configuration yields an open-faced test specimen design, which allows saturated, steady state conditions to be reached in the test specimens in a relatively short amount of time. This is due to the large surface area for moisture uptake relative to the short diffusion path to the interface.

17.4.3. Effect of Moisture Preconditioning on Adhesion

Interfacial fracture mechanics was used to characterize the intrinsic effect of moisture on adhesion. The adhesive used was an epoxy-based underfill developed for no-flow assembly, designated as UR-B in this research. This particular underfill was determined to be ideal for studying the fundamental effect of moisture on interfacial adhesion due to its moisture diffusion kinetics and saturation behavior established from the moisture absorption portion of this research. The substrate used was oxygen-free electronic grade copper, alloy 101. The copper substrates were polished to a mirror finish and cleaned using the routine procedure given by Shi and Wong [45] prior to bonding. This was done to isolate the intrinsic effect of moisture on adhesion without mechanical interlocking and/or surface contamination influencing the results. Symmetric interface cracks were introduced into the underfill/copper bilayer test specimens by using a molding compound release agent [19]. Based on the results from the moisture absorption analysis, a water-proof perimeter was applied to the interfacial fracture test specimens during moisture preconditioning and removed before fracture testing. This perimeter served two purposes. First, the application of the perimeter forced 1D diffusion through the top, open surface of the underfill, yielding uniform concentrations of moisture spatially across the entire interface for the full duration of exposure to the humid preconditioning environment. Second, the water-proof perimeter prevented moisture wicking at the interface, which allowed identification of the test specimen moisture concentration by utilizing the inherent moisture absorption characteristics of the adhesive. Completed specimens were tested in a four-point bend test at room temperature to measure the critical load of fracture for the interface. A completed representative interfacial fracture toughness test specimen is shown in Figure 17.14.

Test specimens were divided into five test groups and subjected to four different levels of moisture preconditioning to ascertain the effect of moisture on interfacial fracture toughness. The test groups included fully dry, 85°C only, 85°C/50%RH, 85°C/65%RH, and 85°C/85%RH, with the latter four test groups being environmentally preconditioned for 168 hours. The 85°C temperature component in each moisture preconditioning environment will enhance diffusion rates and drive more moisture into test specimens over a smaller timeframe. In addition, as temperature increases, the moisture capacity of air increases. Consequently, more moisture will be available to diffuse into test specimens at higher relative humidity levels compared to similar, high relative humidity levels at lower temperatures. By gradually increasing the relative humidity while maintaining the temperature constant, the change in interfacial fracture toughness as a function of increasing

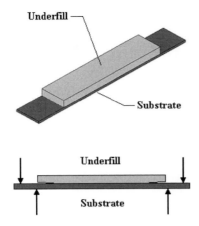

FIGURE 17.14. Interfacial fracture toughness test specimen.

moisture content can be identified. For additional information on psychometrics, refer to Thermodynamics: An Engineering Approach by Cengel and Boles [6].

All test specimens were baked at 115°C for at least 12 hours to remove any moisture that may have been introduced during sample preparation prior to environmental aging, which was performed in a humidity chamber in an atmosphere maintained at a constant temperature (± 1°C), humidity ($\pm 1\%$), and pressure (P_{atm}). All interface fracture tests were performed with both the surrounding environment and test specimens being at room temperature after environmental preconditioning. No measurable loss in moisture uptake occurred in the test specimens from the time they were removed from the environmental chamber, allowed to cool to room temperature, and experimentally tested.

Using the experimentally measured value for the critical load of fracture in conjunction with previously identified elastic modulus results, the interfacial fracture toughness of the underfill/copper test specimens was determined using Equation (17.10) for each particular level of moisture preconditioning. Figure 17.15 provides a graphical depiction of the results depicting the effect of environmental preconditioning on the underfill/copper interfacial fracture toughness.

The entire range of mode mixity for all interfacial test specimens fell between $-37.41°$ to $-37.64°$. The substrate height was used to define the characteristic length for all reported toughness values when evaluating the mode mixity. Since the variation in mode mixity was negligible, the effect of this variation affecting interfacial fracture toughness results between different test groups is insignificant. Consequently, interfacial fracture toughness results for different moisture preconditioned test groups can be compared to one another to ascertain the effect of increasing moisture content on toughness values. In addition, saturation was reached in each moisture preconditioning environment prior to fracture testing. As a result, a gradient of moisture concentration did not exist in the interfacial fracture toughness test specimens during testing. As shown in Figure 17.15, it is clear that the contribution of thermal aging at 85°C did not significantly affect the interfacial fracture toughness of the underfill/copper interface. It is important to remember that all tests were performed at room temperature, hence only the effects of thermal aging were evaluated rather than the effect of testing at elevated temperatures. Since all environmental preconditioned test groups were exposed to the same temperature component of 85°C and duration of 168 hours, any observed changes in the fracture toughness after

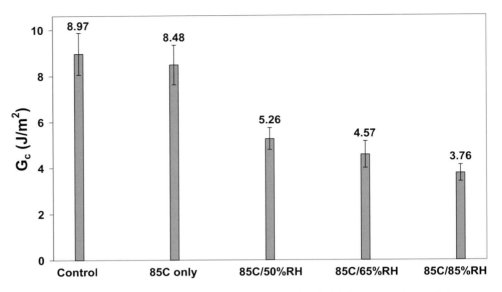

FIGURE 17.15. Effect of environmental preconditioning on the interfacial fracture toughness of the underfill/copper interface.

TABLE 17.3.
Change in the underfill/copper test specimen interfacial fracture toughness from moisture uptake.

T (°C)	RH (%)	C_{sat} (wt%)	C_{sat} (mg H_2O/mm^3)	G_c (J/m^2)	Toughness change (%)
Control	—	0	0.0000	8.97 ± 0.91	—
85	50	0.65	0.0075	5.26 ± 0.47	41.4
85	65	0.77	0.0089	4.57 ± 0.58	49.1
85	85	1.02	0.0118	3.76 ± 0.36	58.1

moisture preconditioning can be attributed to the contribution of moisture. Moisture preconditioning at 85°C/50%RH, 85°C/65%RH, and 85°C/85%RH had a substantial effect on the interfacial fracture toughness and yielded decreases of 41.4%, 49.1%, and 58.1% respectively. A summary of the effect of moisture preconditioning on the interfacial fracture toughness is provided in Table 17.3, where C_{sat} represents the saturation concentration of moisture for each respective level of moisture preconditioning and given as a percent weight change (wt%).

Figures 17.16 and 17.17 depict the inherent change in the underfill/copper interfacial fracture toughness as a function of moisture concentration.

Based on Figures 17.16 and 17.17, it is clear that the change in the interfacial fracture toughness is sensitive to small amounts of moisture. A significant reduction in interfacial adhesion was observed for concentrations as low as 0.65 wt%. Since the moisture did not significantly alter the elastic modulus of the underfill adhesive for the moisture conditions evaluated for the interfacial fracture toughness, plasticization of the underfill from moisture contributed little to the change in the interfacial fracture toughness. As a result, the reduction in toughness is primarily attributed to the weakening of the underfill/copper interface due to the direct presence of moisture at the interface. The moisture at the interface could decrease the adhesion through displacement of the underfill reducing Van der Waals forces

EFFECT OF MOISTURE ON THE ADHESION AND FRACTURE OF INTERFACES

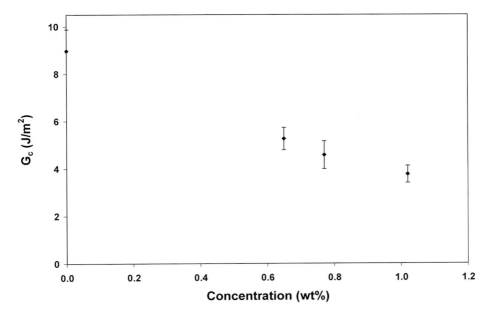

FIGURE 17.16. Underfill/copper interfacial fracture toughness variation as a function of moisture concentration (wt%).

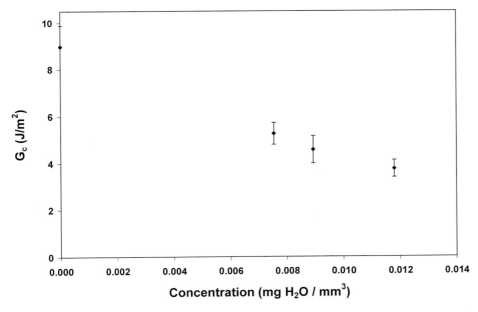

FIGURE 17.17. Underfill/copper interfacial fracture toughness variation as a function of moisture concentration (mg H_2O/mm^3).

as well as possible chemical degradation of adhesive bonds. Further investigations into the exact failure mechanism from moisture at the interface are provided in detail in subsequent sections of this chapter.

17.4.3.1. Moisture Induced Swelling In addition to the mechanical load applied to test specimens during interfacial fracture testing, the interface is also subjected to hygro-swelling and thermal contraction mismatch effects between the adhesive and substrate. These two effects have opposite outcomes on the interface, as the contribution from the hygro-swelling mismatch will cause the underfill to be in compression, while the contribution from the thermal contraction mismatch will cause the underfill to be in tension. This is attributed to the different stress free environments for each case. For the case of the hygro-swelling mismatch, fully dry conditions represent a stress-free state for the interface. As moisture is absorbed in the underfill, it will cause the underfill to expand, while the moisture impermeable substrate will retain its original dimensions. Since the moisture expansion in the underfill will be constrained by the substrate, the expansion in the underfill will yield compressive stresses within the underfill. For the case of the thermal contraction mismatch, the curing temperature of the underfill represents a stress-free state for the interface. Once test specimens are removed from the oven and allowed to cool to room temperature, the thermal mismatch between the copper and the underfill will cause the underfill to be in tension due to it wanting to shrink more than the copper substrate (CTE of experimental materials: underfill = 75 ppm/°C, copper = 17 ppm/°C). Whether the interface is dominated by the hygro-swelling mismatch, thermal contraction mismatch, or possibly neither due to the effects of one another canceling each other out for a particular moisture saturation level will depend on the characteristics of the materials that constitute each bimaterial interface relative to their moisture preconditioning environment.

To investigate the effect of hygro-swelling on interfacial fracture test results, the moisture swelling coefficient, β, of the underfill was experimentally determined for each moisture preconditioning environment. The moisture swelling coefficient is defined as

$$\beta = \frac{\Delta \ell / \ell_o}{C_{sat}}, \qquad (17.16)$$

where $\Delta \ell$ is the change in length of the specimen due to moisture absorption, ℓ_o is the initial dry length of the specimen, and C_{sat} is the saturation moisture concentration. Using Equation (17.16) with experimental test data, the moisture swelling coefficient was determined for conditions of 85°C/50%RH ($\beta = 1987$ ppm/wt%), 85°C/65%RH ($\beta = 1907$ ppm/wt%), 85°C/85%RH ($\beta = 1808$ ppm/wt%). Having identified the moisture swelling coefficient for each moisture preconditioning environment, a comparison can be made between the hygro-swelling and thermal mismatch strains for the underfill/copper interface. The hygro-swelling mismatch strain, ε_h, and thermal mismatch strain, ε_t, are defined as follows:

$$\varepsilon_h = \beta_1 C_{sat,1} - \beta_2 C_{sat,2}, \qquad (17.17)$$

$$\varepsilon_t = (\alpha_1 - \alpha_2)(T_f - T_i), \qquad (17.18)$$

where β is the moisture swelling coefficient, C_{sat} is the equilibrium moisture saturation concentration, α is the coefficient of thermal expansion, T is the temperature, and subscripts 1 and 2 refer to the two materials that constitute the bimaterial interface. The hygro-swelling mismatch strain and thermal expansion mismatch strain were calculated using Equations (17.17) and (17.18) respectively for each moisture preconditioning environment. Since the cooling of the interfacial fracture test specimens from the cure temperature to

TABLE 17.4.
Comparison of hygro-swelling and thermal mismatch strains for the underfill/copper interfacial fracture test specimens.

Environment	β (ppm/wt%)	C_{sat} (wt%)	ε_h	α_{uf} (ppm/°C)	α_{Cu} (ppm/°C)	T_i (°C)	T_f (°C)	ε_t
85°C/50%RH	1987	0.65	0.0013	75	17	190	25	0.0096
85°C/65%RH	1907	0.77	0.0015	75	17	190	25	0.0096
85°C/85%RH	1808	1.02	0.0018	75	17	190	25	0.0096

room temperature will result in a thermal contraction, while the uptake of moisture will result in an expansion from swelling, it should be noted that the hygro-swelling and thermal expansion mismatch strains act in opposite directions. The results are given in Table 17.4.

As shown in Table 17.4, the thermal mismatch strains were significantly greater than the hygro-swelling mismatch strains for all moisture preconditioning environments by roughly an order of magnitude. It is clear that the thermal mismatch strain dominated the interaction at the interface and was only slightly offset by a small contribution from the hygro-swelling mismatch strain for this particular bimaterial interface. As a result, the underfill will be in tension during interfacial fracture testing, effectively preloading the interface and requiring a lower critical load of fracture, P_c, from mechanical testing to advance the interface crack. Consequently, interfacial fracture toughness values will represent a conservative estimate of the interfacial fracture toughness of the interface. In addition, it is clear that increasing the saturation concentration did not significantly increase the hygro-swelling mismatch strain. All interfaces for all environments experienced similar hygro-swelling mismatch strains for the materials and moisture preconditioning environments tested in this study. Consequently, the trends exhibited in the interfacial fracture toughness as moisture concentration increases are essentially independent of the hygro-swelling mismatch relative to one another, and the observed changes between the different moisture preconditioning environments can be predominately attributed to more moisture being present at the interface resulting in a greater loss of adhesion.

17.4.3.2. Interfacial Hydrophobicity The polarity of the water molecule will affect its behavior at the interface, which can influence the extent of environmental degradation of an adhesive joint due to the presence of moisture [33]. The polar behavior of water arises from its structure, which is composed of a single oxygen atom bonded to two hydrogen atoms. The hydrogen atoms are covalently bonded to the oxygen atom through shared electrons. Two pairs of electrons surrounding the oxygen atom are involved in covalent bonds with hydrogen; however, there are also two unshared pairs of electrons (lone-pair) on the other side of the oxygen atom, which shift the electron cloud of the water molecule over to the oxygen atom as shown in Figure 17.18.

This uneven distribution of electron density in the water molecule yields a partial negative charge (δ^-) on the oxygen atom and a partial positive charge (δ^+) on the hydrogen atoms, giving rise to the polarity of the water molecule. Polarity allows water molecules to bond with each other, and hydrogen bonds will form between two oppositely charged ends of a water molecule as shown in Figure 17.19.

The hydrogen bonds have about a tenth of the strength of an average covalent bond, and are being constantly broken and reformed in liquid water. The polarity will also allow water to molecules to bond with other polar molecules, which will affect how the water will wet on different surfaces. Surfaces that contain polar molecules are hydrophilic. They

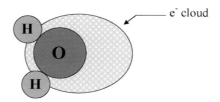

FIGURE 17.18. Electron cloud distribution on a water molecule.

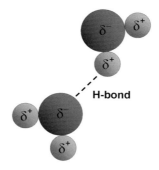

FIGURE 17.19. Hydrogen bonding between water molecules.

interact with the water molecules to enhance wetting, causing the water to smear flat. If a surface contains alcohols, O, or N, it will probably be hydrophilic. Conversely, surfaces that contain nonpolar substances are hydrophobic. They cannot interact with the water molecules, causing it to form a bubble on the surface. In general, if a surface contains C, H, or F, it will probably be hydrophobic.

Most materials will not be purely hydrophobic or hydrophilic, but will have varying degrees to which they are considered one or the other. This is addressed in Hydrophobicity, which is the study of the wetting characteristics of water on surfaces. One method used to test the hydrophobicity of a surface is through measurement of the contact angle, θ, using water as the probe liquid. The contact angle represents a balance between the adhesive forces between the liquid and solid and cohesive forces in the liquid. The adhesive forces cause the liquid drop to spread, while the cohesive forces cause the liquid drop to retain the shape of a sphere. The contact angle is a direct measure of wettability and provides an effective means to evaluate many surface properties such as surface contamination, surface hydrophobicity, surface energetics, and surface heterogeneity. When $\theta > 0$, the liquid is nonspreading and reaches an equilibrium position between the liquid-fluid and solid-liquid interfaces. When $\theta = 0$ the liquid wets without limit and spontaneously spreads freely over the surface. Hydrophobic surfaces repel water and produce high contact angles. Hydrophilic surfaces attract water and produce low contact angles. Figure 17.20 illustrates the contact angle behavior of water on both hydrophobic and hydrophilic surfaces.

By utilizing water as the probe liquid, the interfacial hydrophobicity can be ascertained by measuring the water contact angle of both the adhesive and substrate. To determine the hydrophobicity of interfacial fracture test specimens, contact angle measurements were made for the adhesive and substrate evaluated in this study. Both the clean copper substrate and underfill adhesive exhibited fairly hydrophobic behavior with contact angles of 74° and 83° respectively. Having established the hydrophobicity of the substrate

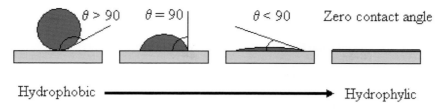

FIGURE 17.20. Hydrophobic and hydrophilic water contact angle behavior.

and adhesive, the interfacial hydrophobicity of the underfill/copper interfacial fracture test specimens can be evaluated. When addressing the relative hydrophobicity of the substrate and adhesive to moisture behavior at the interface, the interaction can become complex. The surface with the most dominant degree of hydrophobicity will govern the shape and response of the water at the interface. For example, if a hydrophobic substrate is bonded with a hydrophilic adhesive, then the water at the interface will want to minimize contact with the substrate and maximize contact with the adhesive. Depending on imperfections in the bonding, surface roughness, and the relative degree of hydrophobicity of the substrate to the adhesive, water at the interface will more or less form a somewhat hemi-spherical shape at the interface, with the spherical end minimizing contact on the substrate and the open end maximizing contact on the adhesive. Naturally, the shape of the water at the interface can have various permutations of the aforementioned shape depending on the degree of hydrophobic behavior of the substrate relative to the hydrophilic behavior of the adhesive, but the general idea remains the same. For other systems with varying degrees of hydrophobicity, the shape of the water at the interface relative to the hydrophobicity of the substrate and adhesive can be extremely difficult to characterize; however, qualitative conclusions can be made. For the case of the underfill/copper interfacial fracture test specimens, the relative hydrophobicity of the adhesive to the substrate was similar; consequently, the wetting behavior of the moisture at the interface would not be significantly dominated by either the adhesive or substrate.

An additional consideration unique to environmental preconditioning is the growth of oxides affecting the interfacial hydrophobicity. Copper has a strong affinity to oxygen, and the development of an oxidation layer between the substrate and adhesive after bonding is inevitable. Initially, cuprous oxide, Cu_2O, will form followed by the formation of a layer of cupric oxide, CuO. The oxidation of copper substrates can be significant, and previous studies have shown that the water contact angle on copper is affected by oxidation [7,25,28,61]. Due to oxidation growth on the copper substrates, contact angle measurements were made for each preconditioning environment to monitor any change in the hydrophobicity of the copper surface.

Since the copper bonding surface of the interfacial fracture test specimen will be shielded by the underfill adhesive, the oxidation growth rate will be different than for bare copper environmentally aged for a similar duration of time. Therefore, water contact angles for each environmental test group were measured using special test specimens that mimicked the exposure of the copper bonding surface to similar amounts of oxygen and moisture as the interfacial fracture test specimens. These specimens used the same geometry as the interfacial fracture test specimens, but the underfill adhesive was cured separately in an individual mold. After curing the adhesive, the underfill was placed on top of the copper substrate and held in place by c-clamps. Similar to the interfacial fracture test specimens, a water-proof sealant was applied around the perimeter of the test specimen to eliminate wicking of moisture at the interface and force 1D diffusion through the top surface of the underfill. After environmental preconditioning, the water-proof perimeter, c-clamps, and underfill were removed from the test specimen for contact angle measurement of the copper surface.

Experimentally measured water contact angle results were as follows: 76° for 85°C thermal aging, 76° for 85°C/50%RH moisture preconditioning, 77° for 85°C/65%RH moisture preconditioning, and 77° for 85°C/85%RH moisture preconditioning. All test groups were preconditioning for the same duration of 168 hours, which was the same criteria used in the evaluation of the effect of moisture on interfacial adhesion. Based on these results, it is evident that all levels of environmental preconditioning did not significantly alter the water contact angle and associated hydrophobicity of the interface. As a result, similar interfacial wetting characteristics of moisture at the interface will occur for all preconditioning environments.

Although the contact angle did not significantly change, there did appear to be a slight increase in the water contact angle with moisture preconditioning. Previous studies have shown both an increase [28,61] and decrease [7,25] in the water contact angle of copper with oxidation. The oxidation–reduction chemistry occurring at the interface relative to environmental preconditioning is complex, and the differences in trends could be attributed to the degree of oxidation altering the surface chemistry [7], change in surface roughness of the substrate from oxidation growth [25], and contamination of the surface by hydrocarbons from the environment [33]. In addition, Yi et al. [61] has provided data correlating the oxide layer thickness on copper leadframes to water contact angles. These data shows a slow, gradual increase in oxide thickness from water contact angles ranging from 72°–78°, but depicts a sharp increase in oxide layer thickness for contact angles exceeding 80°. Based on results for the water contact angle on copper in this study, all measurements yielded average contact angles less than 78° with vary little variation with each other. This indicates a similar level of interfacial hydrophobicity and oxide layer thickness for all environmentally preconditioned test groups. Both Mino et al. [35] and Chong et al. [8] have shown that the development of the copper oxide layer thickness is significantly slower and minimal for temperatures below 100°C and 120°C. Since the test specimens in this study had a temperature component of only 85°C, it is anticipated that the oxide layer thickness that developed on test specimens would have a minimal effect on toughness results. This is also supported by X-ray Photoelectron Spectroscopy (XPS) results. XPS showed the presence of cupric oxide not only in the 85°C/50%RH, 85°C/65%RH, and 85°C/85%RH test groups, but also in the 85°C thermal aging test group. As a result, identical oxide chemical formations existed at the interface for all environmentally preconditioned test groups. In addition, similar atomic percentages of cupric oxide were obtained when comparing thermal aging at 85°C to the moisture preconditioning environments of 85°C/50%RH, 85°C/65%RH, and 85°C/85%RH, indicating that the moisture component had a minimal contribution to oxidation growth rates on the copper compared to the available oxygen in the air common to all environmental preconditioned environments. Consequently, a similar level of oxidation thickness existed on all environmentally preconditioned test specimens, which supports the results from the water contact angle measurements.

Since oxides were removed from the copper surface before adhesive bonding and the flux present in the no-flow underfill would have removed any oxides that developed during adhesive curing, it is possible that the oxidation growth from environmental preconditioning would have an effect on the interfacial fracture toughness results. This oxide growth could displace the underfill from the copper substrate after bonding to contribute to the observed loss in adhesion after moisture preconditioning shown in Figure 17.15. Since both water contact angle measurements and XPS results demonstrate a similar oxidation thickness existed on all environmentally preconditioned test specimens, the 85°C thermal aging results can be compared to the control test results to ascertain the effect of oxidation growth on the loss in adhesion without the contribution from moisture. As shown in Figure 17.15, thermal aging at 85°C produced little to no effect on interfacial fracture toughness results, thus oxidation growth displacing the underfill after adhesive bonding had an insignificant effect on the adhesion loss compared to the effect of moisture from moisture preconditioning.

17.4.4. Interfacial Fracture Toughness Recovery from Moisture Uptake

The underfill/copper interface was found to be very sensitive to moisture, with large decreases in interfacial fracture toughness occurring for moisture preconditioning environments of 85°C/50%RH, 85°C/65%RH, and 85°C/85%RH (Figure 17.15). To further investigate the reversible and irreversible nature of moisture on the interfacial adhesion of the underfill/copper interface, additional test specimens were moisture preconditioned for each condition for 168 hours followed by baking at 95°C until fully dry. A fully dried state was established when there was no measurable change in the weight of a specimen for a period of 24 hours. Upon reaching a dry state, specimens were fracture tested to ascertain the interfacial fracture toughness. The entire range of mode mixity for all interfacial test specimens fell between $-37.43°$ to $-37.48°$. The substrate height was used to define the characteristic length for all reported toughness values when evaluating the mode mixity. Since the variation in mode mixity was negligible, the effect of this variation influencing interfacial fracture toughness results between different test groups is insignificant. Consequently, toughness recovery results for different moisture preconditioned test groups can be compared to one another to ascertain the effect of increasing moisture content on toughness values. Figure 17.21 provides a graphical depiction of the effect of environmental preconditioning and recovery of the underfill/copper interfacial fracture toughness.

As shown in Figure 17.21, most of the loss in interfacial fracture toughness from moisture was not recovered upon fully drying. Since the small change in the underfill elastic modulus from moisture was recoverable upon fully drying, the permanent reduction in the toughness of the underfill/copper interface is attributed to the direct presence of moisture at the interface debonding the underfill adhesive to the copper substrate. Similar in form to the recoverability of the elastic modulus given by Equation (17.9), the recoverability for the interfacial fracture toughness will be defined as follows:

$$Recoverability\ (\%) = \frac{G_{c,recovery} - G_{c,sat}}{G_{c,dry} - G_{c,sat}} \cdot 100, \qquad (17.19)$$

where $G_{c,recovery}$ is value of the interfacial fracture toughness upon fully drying from the moisture saturated state, $G_{c,sat}$ is the saturated value of the interfacial fracture toughness after moisture absorption, and $G_{c,dry}$ is the unaged, control value of the interfacial fracture toughness. Equation (17.19) only applies when the mode mixity of the interfacial fracture toughness before and after moisture preconditioning remains relatively unchanged, otherwise changes in the toughness due to a contribution from a change in the mode mixity will introduce error in the recoverability results. The recoverability of the underfill/copper interfacial fracture toughness is given in Table 17.5.

As shown by Table 17.5, the irreversible damage on interfacial fracture toughness from exposure to moisture was substantial for the underfill/copper interface. Very little

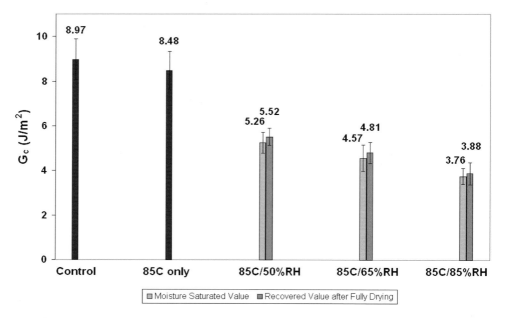

FIGURE 17.21. Recovery of the underfill/copper interfacial fracture toughness on removal of moisture.

TABLE 17.5.
Recoverability of the underfill/copper interfacial fracture toughness from moisture uptake after subsequent drying.

T (°C)	RH (%)	C_{sat} (wt%)	$G_{c,sat}$ (J/m^2)	$G_{c,recovery}$ (J/m^2)	Recoverability (%)
Control	—	0.00	8.97 ± 0.91	—	—
85	50	0.65	5.26 ± 0.47	5.52 ± 0.38	7.0
85	65	0.77	4.57 ± 0.58	4.81 ± 0.47	5.5
85	85	1.02	3.76 ± 0.36	3.88 ± 0.50	2.3

of the underfill/copper interfacial fracture toughness was recoverable after fully drying, with recoverability values for all moisture preconditioning environments less than 7%. It is also evident that a relatively small amount of moisture reaching the interface causes the structural integrity of the adhesive bond to be noticeably, permanently compromised.

17.4.5. Interfacial Fracture Toughness Moisture Degradation Model

Having implemented an extensive experimental program to ascertain the role of moisture in adhesion degradation and the physical mechanisms responsible for the change in interfacial adhesion, the focus now shifts to developing a model depicting the intrinsic loss in interfacial fracture toughness as a function of the critical parameters relevant to moisture. At the root of this model is characterizing the dominant mechanism for adhesion between the adhesive and substrate. There are four primary mechanisms for adhesion which have been proposed. They include mechanical interlocking, diffusion theory, electronic theory, and adsorption theory [30]. For the underfill/copper interface, the contributions of interfacial diffusion and electrostatic forces between the adhesive and substrate causing adhesion is far lower than the effects of mechanical interlocking and adsorption. Since the copper substrates in this study were polished to a mirror finish, the effects from mechanical interlocking of the adhesive into irregularities present on the substrate surface will be small compared to the effects from intermolecular secondary forces (i.e., Van der Waals) between the atoms and molecules in the surfaces of the adhesive and substrate. Consequently, adsorption theory will dominate the adhesive bonding at the underfill/copper interface of our test specimens.

Provided adsorption theory governs adhesion and only secondary forces are acting across an interface, the stability of an adhesive/substrate interface in the presence of moisture can be ascertained from thermodynamic arguments. The thermodynamic work of adhesion, W_A, in an inert medium is given by [30]:

$$W_A = \gamma_a + \gamma_s - \gamma_{as}, \quad (17.20)$$

where γ_a is the surface free energy of the adhesive, γ_s is the surface free energy of the substrate, and γ_{as} is the interfacial free energy. In the presence of a liquid, the thermodynamic work of adhesion, W_{Al}, is given by:

$$W_{Al} = \gamma_{al} + \gamma_{sl} - \gamma_{as}, \quad (17.21)$$

where γ_{al} and γ_{sl} are the interfacial free energies between the adhesive/liquid and substrate/liquid interfaces, respectively. Typically the thermodynamic work of adhesion of an adhesive/substrate interface in an inert medium, W_A, is positive, which indicates the amount of energy required to separate a unit area of the interface. However, the thermodynamic work of adhesion in the presence of a liquid, W_{Al}, can be negative, which indicates the interface is unstable and will separate when it comes in contact with the liquid. Thus, the calculation of W_A and W_{Al} can indicate the environmental stability of the adhesive/substrate interface. Kinloch [30] has shown that W_A and W_{Al} may be calculated from the following expressions:

$$W_A = 2\sqrt{\gamma_a^D \gamma_s^D} + 2\sqrt{\gamma_a^P \gamma_s^P}, \quad (17.22)$$

$$W_{Al} = 2(\gamma_{lv} - \sqrt{\gamma_a^D \gamma_{lv}^D} - \sqrt{\gamma_a^P \gamma_{lv}^P} - \sqrt{\gamma_s^D \gamma_{lv}^D} - \sqrt{\gamma_s^P \gamma_{lv}^P} + \sqrt{\gamma_a^D \gamma_s^D} + \sqrt{\gamma_a^P \gamma_s^P}), \quad (17.23)$$

where γ^D is the dispersion component of surface free energy, γ^P is the polar component of surface free energy, and γ_{lv} is the surface free energy of the liquid. Table 17.6 gives the polar and dispersion surface free energies of epoxy, copper, and water.

TABLE 17.6.
Polar and dispersion surface free energies of epoxy, copper, and water [30].

Substance	γ (mJ/m^2)	γ^D (mJ/m^2)	γ^P (mJ/m^2)
Epoxy	46.2	41.2	5.0
Copper	1360	60	1300
Water	72.2	22.0	50.2

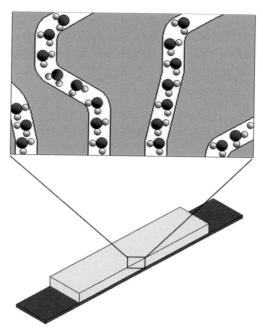

FIGURE 17.22. Moisture transport through the bulk epoxy of an interfacial fracture test specimen.

Using the values given in Table 17.6 and substituting into Equation (17.22), the thermodynamic work of adhesion of the epoxy/copper interface is 260.7 mJ/m^2. If water is present at the epoxy/copper interface, the thermodynamic work of adhesion given by Equation (17.23) is -270.4 mJ/m^2. Therefore, since the work of adhesion is positive before exposure to moisture and negative after exposure, all adhesion of the epoxy/copper interface is lost if water comes in contact with the interface. This is supported by the recovery interfacial fracture toughness results presented in Section 17.4.4, where virtually none of the observed loss in adhesion from moisture exposure was recovered upon fully drying.

Using adsorption theory as the physical basis for the loss in adhesion from moisture, expressions are now developed depicting the amount of moisture delivered to the underfill/copper interface. Since the interfacial fracture test specimens were designed to prevent wicking of moisture at the interface and the copper substrate provides a barrier for moisture transport, the moisture transport to the interface is governed by the epoxy network of the underfill. Soles and Yee [48] have shown that water traverses within the epoxy through the network of nanopores inherent in the epoxy structure. A typical nanopore ranges from 5.0 to 6.1 Å in diameter. Figure 17.22 illustrates the transport of moisture through the bulk epoxy of an interfacial fracture test specimen.

Assuming that the nanopore channels are the only mechanism by which moisture can be delivered to the interface, the saturation concentration in the epoxy expressed in mg H$_2$O/mm^3 is given by:

$$C_{sat} = \frac{\rho(NV)}{V_{tot}}, \qquad (17.24)$$

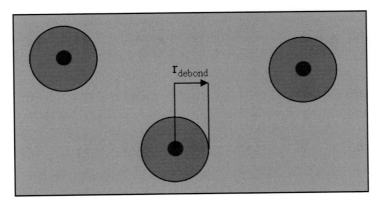

FIGURE 17.23. Graphical illustration of the parameter, r_{debond}, at the interface.

where ρ is the density of water measured in milligram per cubic millimeter (mg/mm^3), N is the number of nanopores actively participating within the epoxy network, V is the volume occupied by a single nanopore in the epoxy network, and V_{tot} is the total volume of the epoxy. After rearrangement of Equation (17.24), the number of nanopores actively participating within an epoxy system for a given saturation concentration is as follows:

$$N = \frac{4AC_{sat}}{\pi \rho D^2}, \tag{17.25}$$

where A is the total area of the interface and D is the nanopore diameter. Assuming adsorption theory holds, the adhesive bond area, A_{bond}, that remains intact after exposure to moisture will depend on the area occupied by the moisture at the interface, A_{H_2O}:

$$A_{bond} = A - A_{H_2O}. \tag{17.26}$$

Relating this adhesive bond area to the number of nanopores actively participating in transport yields:

$$A_{bond} = A - \pi N r_{debond}^2, \tag{17.27}$$

where r_{debond} represents the debond radius of moisture at the interface that occurs at each nanopore. The debond radius must be greater or equal to the nanopore radius and is governed by the interfacial hydrophobicity of the adhesive/substrate interface. Figure 17.23 provides a graphical depiction of the parameter, r_{debond}, at the interface.

Substituting Equation (17.25) into (17.27) provides an expression for the adhesive bond area that remains intact after exposure to a particular moisture saturation concentration:

$$A_{bond} = A - \frac{4AC_{sat}r_{debond}^2}{\rho D^2}. \tag{17.28}$$

We now want to employ a fracture mechanics development to relate the change in bond area due to the presence of moisture at the interface. Recall from fracture mechanics the general form of the stress intensity factor:

$$K = S\sigma\sqrt{\pi a}, \qquad (17.29)$$

where S is a dimensionless constant that depends on the geometry and mode of loading, σ is the remotely applied stress, an a is the crack length. The stress intensity factor is related to the fracture toughness, G_c, by the following expression:

$$G_c = Z\sigma^2, \qquad (17.30)$$

where

$$Z = \frac{\pi a S^2 (1-\upsilon^2)}{E}.$$

Based on the thermodynamic work of adhesion for the epoxy/copper interface, the interface will become unstable and debond in the presence of moisture; however, since interfacial fracture toughness is a material property that characterizes the adhesion of the interface, the toughness must be the same in all areas that remain bonded after exposure to moisture. Using mode I loading and making the following three assumptions: (1) Adsorption theory dominates the interfacial bonding; (2) The change in the mechanical properties of both the adhesive and substrate from moisture is small relative to the change in bond area from moisture, and (3) The relative change in fracture toughness from moisture remains constant irrespective to the means of measuring the toughness for a given moisture saturation concentration, an expression is obtained relating the change in bond area due to the presence of moisture to the change in the critical load of fracture:

$$\frac{P_{wet}}{A - \pi N r_{debond}^2} = \frac{P_{dry}}{A}. \qquad (17.31)$$

Rearranging Equation (17.31) to obtain an expression for P_{wet} and substituting that value into Equation (17.30) for the wet, saturated case yields the following expression:

$$G_{c,wet} = \left(1 - \frac{\pi N r_{debond}^2}{A}\right)^2 G_{c,dry}. \qquad (17.32)$$

As the saturation moisture concentration increases, so will the number of active nanopores participating. The incremental change in fracture toughness due to the participation of a single additional nanopore, $N+1$, is given by:

$$G_{c,wet} = \left(1 - \frac{\pi (N+1) r_{debond}^2}{A}\right)^2 G_{c,dry}. \qquad (17.33)$$

For convenience, define f such that for N nanopores participating:

$$f_N = \frac{\pi N r_{debond}^2}{A}. \qquad (17.34)$$

For $N+1$ nanopores participating:

$$f_{N+1} = \frac{\pi r_{debond}^2}{A}(N+1). \tag{17.35}$$

Restating Equations (17.32) and (17.33) in terms of f:

$$G_{c,wet}(f_N) = \left(1 - \frac{\pi N r_{debond}^2}{A}\right)^2 G_{c,dry}, \tag{17.36}$$

$$G_{c,wet}(f_{N+1}) = \left(1 - \frac{\pi r_{debond}^2}{A}\right)^2 G_{c,wet}(f_{N_N}). \tag{17.37}$$

Subtracting (17.36) from (17.37) and dividing by $f_{N+1} - f_N$ gives:

$$\frac{G_{c,wet}(f_{N+1}) - G_{c,wet}(f_N)}{f_{N+1} - f_N} = \frac{[1 - (\pi r_{debond}^2/A)]^2 G_{c,wet}(f_N) - G_{c,wet}(f_N)}{f_{N+1} - f_N}. \tag{17.38}$$

Utilizing a Taylor series expansion of f_N with first order accuracy and substituting Equations (17.34) and (17.35) into (17.38) yields:

$$\frac{dG_{c,wet}(f_N)}{df_N} = \frac{[1 - (\pi r_{debond}^2/A)]^2 G_{c,wet}(f_N) - G_{c,wet}(f_N)}{(\pi r_{debond}^2/A)}. \tag{17.39}$$

Simplification and elimination of higher order terms gives the following differential equation characterizing the loss in interfacial fracture toughness due to moisture:

$$\frac{dG_{c,wet}(f_N)}{df_N} = -2G_{c,wet}(f_N), \tag{17.40}$$

subject to the boundary condition:

$$G_{c,wet}(f_N = 0) = G_{c,dry}. \tag{17.41}$$

Solution of Equation (17.40) gives:

$$G_{c,wet} = G_{c,dry} \exp\left[\frac{-8C_{sat} r_{debond}^2}{\rho D^2}\right]. \tag{17.42}$$

Equation (17.42) characterizes the loss in interfacial fracture toughness from moisture in terms of key parameters relevant to moisture. Using the value for the density of water at room temperature (0.998 mg/mm^3), an average nanopore diameter of 5.5 Å, and the saturation concentration determined from the experimental portion of this study in conjunction with Equations (17.25) and (17.42), the number of active nanopores participating, N, and value of r_{debond} can be determined by the intrinsic response of the material system to each level of moisture preconditioning. The results are shown in Table 17.7.

TABLE 17.7.
Key parameters relevant to moisture for the underfill/copper interface.

Environment	Substrate	Adhesive	C_{sat} (mg H$_2$O/mm^3)	N	r_{debond} (mm)
85°C/50%RH	Copper	Underfill	0.0075	1.006×10^{13}	1.640×10^{-6}
85°C/65%RH	Copper	Underfill	0.0089	1.194×10^{13}	1.692×10^{-6}
85°C/85%RH	Copper	Underfill	0.0118	1.583×10^{13}	1.669×10^{-6}

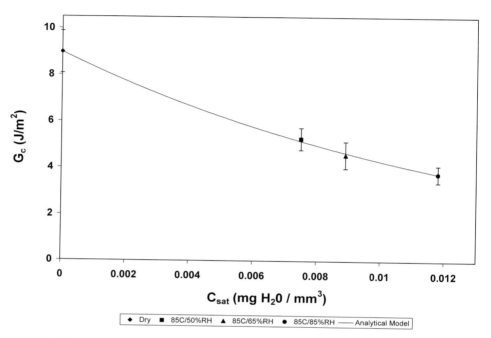

FIGURE 17.24. Analytical prediction of the loss in interfacial fracture toughness from moisture for the underfill/copper interface.

As shown in Table 17.7, the number of nanopores participating increases with saturation concentration. This is expected since an increase in saturation concentration would increase the available moisture for transport through the nanopores. In addition, the values for r_{debond} were similar for each moisture preconditioning environment for both respective interfaces, which is also expected since X-ray Photoelectron Spectroscopy and water contact angle results did not indicate a change in the interfacial hydrophobicity of the copper surface from moisture preconditioning. The slight variation in the values for r_{debond} could in part be attributed to experimental scatter. Since the results were similar, they were averaged to obtain a representative value for r_{debond} in the presence of moisture for each interface.

Using the moisture parameters identified for each interfacial material system, Equation (17.42) was used to predict the interfacial fracture toughness for the underfill/copper interface as a function of increasing saturation concentration.

As shown in Figure 17.24, Equation (17.42) accurately predicted the loss in interfacial fracture toughness as a function of increasing moisture concentration. Since Equation (17.42) was based on the physics of adsorption theory, it will yield a loss in interfacial fracture toughness provided there is moisture at the interface, no matter how small the concentration. This contradicts the results of previous studies, who have reported a critical concentration of water may exist below which there is no measurable loss in adhesion [10,23,31]. Based on the results of adsorption theory, it does not appear possible that a critical concentration of water could exist in theory. It is possible in those studies that other mechanisms for adhesion in addition to adsorption theory governed the adhesion at the interface, which could explain why a critical concentration of water was observed. An additional consideration is the method of testing used to obtain adhesion results. The aforementioned studies used lap shear test specimens to determine the interfacial strength after moisture preconditioning. Due to lacking a precrack at the interface and the applied load being distributed over the entire bonding area, these test specimens are not as sensitive to interfacial failure; consequently, possibly also explaining why in part a critical concentration of water appeared to exist for low concentrations of moisture. Conversely, interfacial fracture toughness test specimens are designed for interfacial failure through the use of a precrack at the interface, making them more sensitive to subtle changes in adhesion at the interface. The work of Wylde and Spelt [58] supports this observation. Using interfacial fracture toughness test specimens with a similar material system previously reported to exhibit a critical concentration of water from lap shear results, they found a decrease in the interfacial toughness from moisture for all concentrations of moisture, including those lower than the previously reported critical concentration of water. Consequently, provided adsorption theory dominates the adhesive bonding at the adhesive/substrate interface and the assumptions in the development of the model are satisfied, Equation (17.42) should accurately predict the loss in interfacial fracture toughness for a given moisture concentration.

REFERENCES

1. H. Ardebili, E.H. Wong, and M. Pecht, Hydroscopic swelling and sorption characteristics of epoxy molding compounds used in electronic packaging, IEEE Transactions on Components and Packaging Technologies, 26(1), pp. 206–214 (2003).
2. ASTM D790, Standard Test Methods for Flexural Properties of Unreinforced and Reinforced Plastics and Electrical Insulating Materials, Annual Book of ASTM Standards, Vol. 08.01, 1999.
3. D. Brewis, J. Comyn, A. Raval, and A. Kinloch, The effect of humidity on the durability of aluminum-epoxide joints, International Journal of Adhesion and Adhesives, 10, pp. 247–253 (1990).
4. F. Buehler and J. Seferis, Effect of reinforcement and solvent content on moisture absorption in epoxy composite materials, Composites: Part A: Applied Science and Manufacturing, 31, pp. 741–748 (2000).
5. L. Butkus, Environmental durability of adhesively bonded joints, Doctoral Thesis, Georgia Institute of Technology, Woodruff School of Mechanical Engineering, Atlanta, GA, 1997.
6. Y. Cengel and M. Boles, Thermodynamics: An Engineering Approach, McGraw-Hill, Inc., New York, 1994.
7. K. Cho and E. Cho, Effect of the microstructure of copper oxide on the adhesion behavior or epoxy/copper leadframe joints, Journal of Adhesion Science and Technology, 14(11), pp. 1333–1353 (2000).
8. C. Chong, A. Leslie, L. Beng, and C. Lee, Investigation on the effect of copper leadframe oxidation on package delamination, Proceedings of the 45th Electronic Components and Technology Conference, 1995, pp. 463–469.
9. P. Chung, M. Yuen, P. Chan, N. Ho, and D. Lam, Effect of copper oxide on the adhesion behavior of epoxy molding compound-copper interface, Proceedings of the 52nd Electronic Components and Technology Conference, 2002, pp. 1665–1670.
10. J. Comyn, C. Groves, and R. Saville, Durability in high humidity of glass-to-lead alloy joints bonded with and epoxide adhesive, International Journal of Adhesion and Adhesives, 14, pp. 15–20 (1994).
11. J. Crank, The Mathematics of Diffusion, Clarendon Press, Oxford, 1956.

12. A. Crocombe, Durability modeling concepts and tools for the cohesive environmental degradation of bonded structures, International Journal of Adhesion and Adhesives, 17, pp. 229–238 (1997).
13. X. Dai, M. Brillhart, and P. Ho, Adhesion measurement for electronic packaging applications using double cantilever beam method, IEEE Transactions on Components and Packaging Technology, 23, pp. 101–116 (2000).
14. X. Dai, M. Brillhart, M. Roesch, and P. Ho, Adhesion and toughening mechanisms at underfill interfaces for flip-chip-on-organic-substrate packaging, IEEE Transactions on Components and Packaging Technology, 23(1), pp. 117–127 (2000).
15. B. DeNeve and M. Shanahan, Effects of humidity on an epoxy adhesive, International Journal of Adhesion and Adhesives, 12, pp. 191–196 (1992).
16. H. Dodiuk, L. Drori, and J. Miller, The effect of moisture in epoxy film adhesives on their performance: I. Lap shear strength, Journal of Adhesion, 17, pp. 33–44 (1984).
17. L. Fan, C. Tison, and C. Wong, Study on underfill/solder adhesion in flip-chip encapsulation, IEEE Trans. on Advanced Packaging, 25(4), pp. 473–480 (2002).
18. T. Ferguson and J. Qu, Elastic modulus variation due to moisture absorption and permanent changes upon redrying in an epoxy based underfill, IEEE Transactions on Components and Packaging Technologies, 29(1), pp. 105–111 (2006).
19. T. Ferguson and J. Qu, Moisture and temperature effects on the reliability of interfacial adhesion of a polymer/metal interface, Proceedings of the 54th Electronic Components and Technology Conference, 2004.
20. T. Ferguson and J. Qu, Moisture absorption analysis of interfacial fracture test specimens composed of no-flow underfill materials, ASME Journal of Electronic Packaging, 125, pp. 24–30 (2003).
21. T. Ferguson and J. Qu, Effect of moisture on the interfacial adhesion of the underfill/soldermask interface, ASME Journal of Electronic Packaging, 124, pp. 106–110 (2002).
22. T. Ferguson and J. Qu, Moisture absorption in no-flow underfills and its effect on interfacial adhesion to solder mask coated FR-4 printed wiring board, International Symposium and Exhibition on Advanced Packaging Materials, Processes, Properties, and Interfaces, 2001, pp. 327–332.
23. R. Gledhill, A. Kinloch, and J. Shaw, A model for predicting joint durability, Journal of Ahdesion, 11, pp. 3–15 (1980).
24. B. Harper and V. Kenner, Effects of temperature and moisture upon the mechanical behavior of an epoxy molding compound, ASME Advances in Electronic Packaging, 1, pp. 1207–1212 (1997).
25. K. Hong, H. Imadojemu, and R. Webb, Effects of oxidation and surface roughness on contact angle, Experimental Thermal and Fluid Science, 8, pp. 279–285 (1994).
26. J. Hutchinson and Z. Suo, Mixed mode cracking in layered materials, Advances in Applied Mechanics, Vol. 29, Academic Press, New York, 1992.
27. R. Jurf and J. Vinson, Effect of moisture on the static and viscoelastic shear properties of epoxy adhesives, Journal of Materials Science, 20, pp. 2979–2989 (1985).
28. S. Kim, The role of plastic package adhesion in IC performance, Proceedings of the 41st Electronic Components and Technology Conference, 1991, pp. 750–758.
29. J.K. Kim, M. Lebbai, J. Liu, J.H. Kim, and M. Yuen, Interface adhesion between copper lead frame and epoxy molding compound: effects of surface finish, oxidation, and dimples, Proceedings of the 50th Electronic Components and Technology Conference, 2000, pp. 601–608.
30. A.J. Kinloch, Adhesion and Adhesives Science and Technology, Chapman and Hall, London, 1987.
31. A. Kinloch, Interfacial fracture mechanical aspects of adhesive bonded joints—a review, Journal of Adhesion, 10, pp. 193–219 (1979).
32. H. Lee and J. Qu, Microstructure, adhesion strength and failure path at a polymer/roughened metal interface, Journal of Adhesion and Science Technology, 17(2), pp. 195–215 (2003).
33. S. Luo, Study on adhesion of underfill materials for flip chip packaging, Doctoral Thesis, Georgia Institute of Technology, School of Textile and Fibers Engineering, Atlanta, GA, 2003.
34. S. Luo and C.P. Wong, Influence of temperature and humidity on adhesion of underfills for flip chip packaging, Proceedings of the 51st Electronic Components and Technology Conference, 2001, pp. 155–162.
35. T. Mino, K. Sawada, A. Kurosu, M. Otsuka, N. Kawamura, and H. Yoo, Development of moisture-proof thin and large QFP with copper lead frame, Proceedings of the 48th Electeonic Components and Technology Conference, 1998, pp. 1125–1131.
36. R. Morgan, J. O'Neal, and D. Fanter, The effect of moisture on the physical and mechanical integrity of epoxies, Journal of Materials Science, 15, pp. 751–764 (1980).
37. R. Morgan, J. O'Neal, and D. Miller, The structure, modes of deformation and failure, and mechanical properties of diaminodiphenyl sulphone-cured tetragylcidyl 4,4′ diaminodiphenyl methane epoxy, Journal of Materials Science, 14, pp. 109–124 (1979).

38. A. Netravali, R. Fornes, R. Gilbert, and J. Memory, Effects of water sorption at different temperatures on permanent changes in an epoxy, Journal of Applied Polymer Science, 30, pp. 1573–1578 (1985).
39. S. Orman and C. Kerr, in D.J. Alner, Ed., Aspects of Adhesion, University of London Press, London, 1971.
40. A. Pasztor, in F. Settle, Ed., Handbook of Instrumental Techniques for Analytical Chemistry, Chapter 50, Prentice Hall, New Jersey, 1997.
41. B. Prime, in E. Turi, Ed., Thermal Characterization of Polymeric Materials, Vol. 2, Chapter 10, San Diego, Academic Press, San Diego, 1997.
42. S. Rosen, Fundamental Principles of Polymeric Materials, John Wiley and Sons, New York, 1993.
43. G. Shaw, C. Rogers, and J. Payer, The effect of immersion on the breaking force and failure locus in an epoxy/mild steel system, Journal of Adhesion, 38, pp. 255–268 (1992).
44. C.H. Shen and G.S. Springer, Moisture absorption and desorption of composite materials, Journal of Composite Materials, 10, pp. 2–20 (1976).
45. S. Shi and C.P. Wong, Study of the fluxing agent effects on the properties of no-flow underfill materials for flip-chip applications, Proceedings of the 48th Electronic Components and Technology Conference, 1998, pp. 117–124.
46. R. Shook and J. Goodelle, Handling of highly-moisture sensitive components—an analysis of low-humidity containment and baking schedules, Proceedings of the 49th Electronic Components and Technology Conference, 1999, pp. 809–815.
47. C. Soles, F. Chang, D. Gidley, and A. Yee, Contributions of the nanovoid structure to the kinetics of moisture transport in epoxy resins, Journal of Polymer Science: Part B: Polymer Physics, 38, pp. 776–791 (2000).
48. C. Soles and A. Yee, A discussion of the molecular mechanisms of moisture transport in epoxy resins, Journal of Polymer Science: Part B: Polymer Physics, 38, pp. 792–802 (2000).
49. N. Su, R. Mackie, and W. Harvey, The effects of aging and environment on the fatigue life of adhesive joints, International Journal of Adhesion and Adhesives, 9, pp. 85–91 (1992).
50. E. Suhir, Failure criterion for moisture-sensitive plastic packages of integrated circuit (IC) devices: application of von-Karman's equations with consideration of thermoelastic strains, International Journal of Solids and Structures, 34(23), pp. 2991–3019 (1997).
51. D. Suryanarayana, R. Hsiao, T. Gall, and J. McCreary, Enhancement of flip-chip fatigue life by encapsulation, IEEE Transactions on Components, Hybrids, and Manufacturing Technology, 14(1), pp. 218–223 (1991).
52. M. Uschitsky and E. Suhir, Moisture diffusion in epoxy molding compounds filled with particles, Journal of Electronic Packaging, 123, pp. 47–51 (2001).
53. M. Uschitsky and E. Suhir, Moisture diffusion in epoxy molding compounds filled with silica particles, ASME Structural Analysis in Microelectronics and Fiber Optics, 21, pp. 141–170 (1997).
54. M. Vanlandingham, R. Eduljee, and J. Gillespie, Moisture diffusion in epoxy systems, Journal of Applied Polymer Science, 71, pp. 787–798 (1999).
55. M. Wahab, A. Crocombe, A. Beevers, and K. Ebtehaj, Coupled stress-diffusion analysis for durability study in adhesively bonded joints, International Journal of Adhesion and Adhesives, 22, pp. 61–73 (2002).
56. E. Wong, K. Chan, T. Lim, and T. Lam, Non-fickian moisture properties characterization and diffusion modeling for electronic packages, Proceedings of the 49th IEEE Electronic Components and Technology Conference, 1999, pp. 302–306.
57. W. Wright, The effect of diffusion of water into epoxy resins and their carbon-fibre reinforced composites, Journal of Composites, 12, pp. 201–205 (1981).
58. J. Wylde and J. Spelt, Measurement of adhesive joint fracture properties as a function of environmental degradation, International Journal of Adhesion and Adhesives, 18, pp. 237–246 (1998).
59. G.Z. Xiao and M. Shanahan, Water absorption and desorption in an epoxy resin with degradation, Journal of Polymer Science: Part B: Polymer Physics, 35, pp. 2659–2670 (1997).
60. D. Yeung, M. Yuen, D. Lam, and P. Chan, Measurement of interfacial fracture toughness for microelectronic packages, Journal of Electronics Manufacturing, 10, pp. 139–145 (2000).
61. S. Yi, C. Yue, J. Hsieh, L. Fong, and S. Lahiri, Effects of oxidation and plasma cleaning on the adhesion strength of molding compounds to copper leadframes, Journal of Adhesion Science and Technology, 13, pp. 789–804 (1999).
62. O. Yoshioka, N. Okabe, S. Nagayama, R. Yamagishi, and G. Murakami, Improvement of moisture resistance in plastic encapsulants MOS-IC by surface finishing copper leadframe, Proceedings of the 39th Electronic Components and Technology Conference, 1989, pp. 464–471.
63. M. Zanni-Deffarges and M. Shanahan, Diffusion of water into an epoxy adhesive: comparison between bulk behaviour and adhesive joints, International Journal of Adhesion and Adhesives, 15, pp. 137–142 (1995).
64. M. Zanni-Deffarges and M. Shanahan, Bulk and interphase effects in aged structural joints, Journal of Adhesion, 45, pp. 245–257 (1994).

18

Highly Compliant Bonding Material for Micro- and Opto-Electronic Applications

E. Suhir[a] and D. Ingman[b]

[a]University of California, Santa Cruz, CA, University of Maryland, College Park, MD, USA
[b]Technion, Israel

18.1. INTRODUCTION

Bonded assemblies (joints) that experience thermal and/or mechanical loading are widely used in micro- and opto-electronics. These assemblies are typically subjected to thermal stresses due to the thermal expansion (contraction) mismatch of the dissimilar materials of the adherends and/or because of temperature gradients. In other cases, bonded assemblies experience mechanical loading. It has been established that the most reliable adhesively bonded or soldered assemblies are characterized by stiff adherends and a compliant adhesive. It has been established also that the employment of low modulus bonding materials and thick bonding layers can lead to an appreciable stress relief [1–7]. This is true for both the interfacial stresses (which are responsible for the adhesive and the cohesive strength of the bonding material) and the stresses acting in the cross-sections of the adherends (these stresses are responsible for the strength of the bonded components) [8–10]. Since the interfacial stresses concentrate at the assembly ends, a substantial stress relief could be expected, if a low modulus material is used at the peripheral portions of the joint [11,12]. Low modulus and thick (up to 4 mils or even thicker) bonding layers are currently employed in micro- and opto-electronic packaging in order to provide a desirable strain buffer between the adherend materials. Certainly, there is always a need in the art for better bonding materials that combine good adhesive properties with high interfacial compliance.

In this connection it should be point out that the reliability of a bonded assembly is due to both high bearing capacity of the joint (i.e., its ability to withstand high stresses of any nature, as well as repetitive loading) and to the low enough level of loading. High compliance of the bonding layer is aimed at reducing the level of the loading, thereby making sufficiently reliable even those joints whose bearing capacity might not be very high. It is also noteworthy that in "conventional" assemblies (which are characterized by moderately compliant bonding layers), the interfacial compliance is due to both the bonding layer and the bonded components. It is not advisable, however, and, in many cases, is not even possible, to increase the interfacial compliance of the adherends. The most attractive way to increase the interfacial compliance and, hence, the reliability of a bonded joint is to employ a highly compliant, yet thin enough (say, for lower thermal resistance) bonding layer. In assemblies with highly compliant bonds, the role of the adherends, as far as the interfacial compliance is concerned, is insignificant. It is the bonding material only that is responsible for the high and favorable interfacial compliance.

In this chapter we suggest an analytical stress model that enables one to quantitatively assess the expected relief in the interfacial shearing stress due to the application of a highly compliant bonding layer. We use this model to demonstrate that the application of a wire array (WA) and/or a suitable modification of a newly developed nano-particle material (NPM) and structure can improve dramatically the compliance of a bonding layer. This NPM has extraordinary mechanical and environmental properties [13–15]. It has been recently tested in application to a new generation of cladding and coatings for fiber optic systems [16–18]. We expect that this material will find a wide application in micro- and opto-electronics, and beyond.

18.2. EFFECT OF THE INTERFACIAL COMPLIANCE ON THE INTERFACIAL SHEARING STRESS

The objective of the developed analytical stress model (see Appendix 18.A) is to evaluate the effect of the interfacial compliance on the interfacial shearing stress. We use this model to demonstrate the importance of highly compliant bonds. This could certainly be done in addition to, or sometime even instead of, employing adherends with a good thermal extension (contraction) match. The following major conclusions can be drawn from the analysis based on the developed model.

The interfacial shearing stress is proportional to the parameter of the interfacial compliance. This parameter is defined as a square root of the ratio of the axial compliance of the assembly (which, in typical assemblies with thin and/or low modulus bonds, is due to the adherends only and should be made as low as possible) to the assembly's interfacial compliance (which, in typical assemblies is due to both the adherends and the bonding layer). In "conventional" assemblies, i.e., in assemblies with a moderately compliant bonding layer, the interfacial compliance is due to both the adherends and the adhesive: thick and high-modulus adherends and a thin and low-modulus adhesives typically contribute more or less equally to the total interfacial compliance. If, however, a highly compliant bonding layer is employed, it is only this layer, and not the adherends, that is responsible for the interfacial compliance of the bond, and, hence, for the structural reliability of the joint.

Although the predictive model was developed for a "flat" (planar) joint, the above conclusions are valid for any assembly, whether circular (e.g., in fiber optics applications), flat (as in microelectronics applications), elliptical, etc.

The level of stress relief that could be expected by employing a WA as a compliant attachment can be assessed based on the following simple reasoning. Examine a bonded structure, in which the bonding layer is constructed of an array of wires, particularly, nano-wires. Let the wires have circular cross-sections and can be considered rigidly clamped at the ends. Treating each wire as a beam clamped at its ends and experiencing ends offset of the magnitude δ, we find that the lateral forces at the wire ends are

$$N = 12\frac{EI\delta}{h_0^3} = \frac{3\pi}{16}\frac{Ed^4\delta}{h_0^3}, \qquad (18.1)$$

where E is the Young's modulus of the wire material, d is the wire diameter, and h_0 is the wires' height, i.e., the thickness of the WA (bonding layer). In the Equation (18.1),

$$I = \frac{\pi \cdot d^2}{64}$$

is the moment of inertia of the wire cross-section. The interfacial compliance of the bonding layer can be found by multiplying the δ/N ratio by the wire cross-sectional area

$$A = \frac{\pi \cdot d^2}{4}. \qquad (18.2)$$

This results in the following formula for the interfacial compliance:

$$\kappa_w = \frac{16h_0^3}{3\pi Ed^4}\frac{\pi d^2}{4} = \frac{4}{3}\frac{h_0^3}{Ed^2}. \qquad (18.3)$$

Using the Equation (18.A7) for the interfacial compliance of a "conventional" (adhesively bonded or soldered) assembly, we conclude that the ratio

$$\eta = \frac{4}{3}\frac{G_0}{E}\left(\frac{h_0}{d}\right)^2 \qquad (18.4)$$

could be used to assess the advantage of a WA-based bonding layer in comparison with a "conventional" layer. Let, for instance, a conventional bonding material has a shear modulus of 5000 psi, the Young's modulus of the WA material be $E = 15 \times 10^6$ psi, the thickness of the bonding layer be $h_0 = 50$ μm, and the diameter of a single wire be 100 nm. Then the Equation (18.4) yields: $\eta = 111$. Hence, a significant increase in the interfacial compliance could be expected by using a WA-based bonding structure, and, because the interfacial shearing stress is inversely proportional to the square root of the interfacial compliance, one could expect an order of the magnitude decrease in the interfacial shearing stress.

Although the above analysis was carried out for a "clamped-clamped" wire, the results are equally applicable to wires with other boundary conditions at the ends. Additional increase in the interfacial compliance could be achieved, if the wires experience axial compression. In order to assess the expected effect of such a compression, we have examined, as an illustration, in Appendix 18.B, a cantilever wire (beam) subjected at its free end to a lateral (bending) force, P, and an axial compressive force, T. The force P can produce a significantly larger lateral deflection, if the wire is subjected to an external axial compression. Such an effect should be expected, of course, since a compressed wire (beam) experiences lateral deflections even in the absence of any lateral force, provided that the axial force reaches and exceeds its critical value.

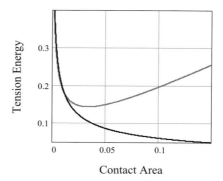

FIGURE 18.1. Surface potential energy vs. "bonded" surface-compound (adhesion) area. Red line — $\sigma_{SC} > \sigma_{SA}$ Blue line — $\sigma_{SC} < \sigma_{SA}$ (see Appendix 18.C).

18.3. INTERNAL COMPRESSIVE FORCES

In addition to the external compression that can be imposed deliberately to enhance the lateral compliance of the assembly, there exist also "internal" compressive forces caused by surface tension.

Consider two bonding surfaces (e.g., parallel surfaces, although these surfaces do not have to be necessarily parallel) and a certain amount of the bonding material (compound) that touches the bonded surfaces. Let us suppose, for the sake of simplicity, that this bonding compound has a shape of a circular cylinder, whose bases touch the bonded surfaces. The lateral surface of the cylinder is "free," i.e., is in contact with air. Let us address a situation, when the area of each of the two bonded surfaces is greater than the bonding area, i.e., the area of the contact with the bonding material. If the surface tension characteristics of the bonding and the bonded materials are properly chosen, then the attraction forces will arise between the bonding surfaces. This phenomenon is similar to the phenomenon observed during the formation of solder balls. So, the bonding compound material will tend to decrease its "free" area and increase its contact area with the bonded surfaces. The total volume of the cylinder remains unchanged. Such a "self-adjusting" process stops when the potential energy of the system (which is as a function of the distance between the bonded surfaces) becomes minimal. A typical plot for this potential energy vs. the contact area is shown in Figure 18.1. The development of the expression for the potential energy is given in the Appendix 18.C.

If the distance between the bonded surfaces becomes larger than the minimum distance that corresponds to the minimum strain energy of the system, then the bonding material behaves as a, sort of, a stretched spring, trying to bring the bonded surfaces closer, i.e., providing axial compression. This compression is proportional to the derivative of the elastic energy with respect to the distance between the bonded surfaces.

18.4. ADVANCED NANO-PARTICLE MATERIAL (NPM)

NPM is an unconventional inhomogeneous "smart" composite material that is equivalent to a "hypothetical" homogeneous material with the following major properties:

- Low Young's modulus.
- Immunity to corrosion.

- Good adhesion to the adjacent material.
- Non-volatile.
- Stable properties at temperature extremes (from $+350°C$ to as low as $-220°C$).
- Very long (practically infinite) lifetime.
- Strong hydrophobicity (against both water and water vapor).
- Ability for "self-healing:" ability to restore its dimensions and initial structure when damaged.
- Ability for "healing" the surfaces of the adjacent materials, i.e., to fill in the existing and/or the developed defects (surface cracks, flaws, and other imperfections) and to slow down their propagation and/or even to "arrest" them completely.
- Very low effective refractive index (if needed).
- High dielectric constant (if needed).

The NPM can be designed, depending on the particular application, in such a way that its important particular properties are enhanced. The conducted tests have confirmed these properties. In general, it is desirable to provide application-specific modifications of the NPM to master (optimize) its properties and performance. Because it is a nano-material, its surface chemistry and its performance depend a lot upon the contact materials and surfaces. The NPM applications include, but are not limited, to the following ones:

- Hermetic sealing of packages, components and devices, such as laser packages, MEMS, displays, plastic LEDs, etc.
- Effective protection (coating) for various metal and non-metal surfaces, well beyond the area of micro- and opto-electronic packaging: (cars, aerospace structures, offshore and ocean structures, marine vehicles, bridges, towers, tubes, pipes and pipelines, etc.). These applications benefit because the NPM is actively hydrophobic, does not induce additional stresses (owing to its low modulus), is inexpensive, is easy-to-apply, has practically infinite lifetime, and is self-healing. Application of the NPM can result in a significant resistance of a metal surface to corrosion, and, in addition, in substantial increase in the fracture toughness of the material, both initially and during the system's operation (use).
- The NPM can be added in the formulation of various coatings such as paints, thereby providing protective benefits without changing the application techniques.
- Because of a low refractive index, the NPM can be used, if necessary, as an effective cladding of optical silica fibers. The use of the NPM cladding eliminates the need to dope silica for obtaining light-guide cores. The new preform will consist of a single (undoped and, hence, less expensive) silica material.
- A derivative application is flexible light-guides. Multicore flexible fiber cables employing NPM are able to provide high spatial image resolution. As such, they might find important applications, when there is a need to provide direct high-resolution image transmission from secluded areas. Possible applications can be found in biomedicine, nondestructive evaluations, oil and other geological explorations, in ocean engineering, or in other situations when an image needs to be obtained and transmitted from relatively inaccessible locations.
- Another derivative application is a multicore fiber cable. Ultra-small diameter glass fibers with an NPM-based cladding/coating can be placed in large quantities within a NPM medium ("multiple cores in a single cladding"). In addition, owing to a much better inner-outer refractive index ratio in the NPM-based fibers, such cables will be characterized by very low signal attenuation.

- Yet another derivative application is sensor systems. The NPM-based fibers can be used in optical sensor systems that employ optical fibers embedded in a laminar or a cast material. Such systems are used, for instance, in composite airframes. With the NPM used as a cladding or, at least, as a coating of the silica optical fiber, the optical performance, the mechanical reliability and the environmental durability of the light-guide will be improved dramatically compared to the conventional systems.
- Ultra-thin planar light-guides are yet another derivative application of the NPM. In the new generation of the planar light-guides, NPM can be used as the top cladding material. It will replace silicon or polymer claddings, which are considered in today's planar light-guides. All the advantages of the NPM cladding material discussed above for optical fibers are equally applicable to planar light-guides. These are thought to have a "bright" future in the next generation of computers and other photonic devices.

18.5. HIGHLY-COMPLIANT NANO-SYSTEMS

In the current application, intended primarily for micro- and opto-electronic assemblies, we have developed a modification of the NPM for use as a highly compliant bonding structure. Here are some major characteristics of this material (additional to those described in the previous section).

The NPM for the application in question is highly inhomogeneous, anisotropic and thixotropic. It allows for an elevated, but still limited displacement in the in-plane direction, i.e., in the direction of the interfacial shearing stress. The material includes a filler and nanoparticles. The bonding elements are of a "strings-like" structural type. These elements are slightly compressed in the through-thickness direction. This enhances their compliance (flexibility) in the in-plane direction. The thixotropic matrix is quasi-solid in the absence of the applied stress field (disturbance) and becomes much less viscous and even quasi-liquid, when thermal or mechanical stress is applied. In other words, the bonding structure becomes much more compliant as a result of experiencing mechanical or thermal loading. The structure provides a highly effective strain buffer between the adherends.

The role of such a plurality of links (strings) are played by numerous nanotubes and/or nanowires embedded into the bonding matrix. Nanotubes and/or nanowires could be subjected to externally and/or internally deliberately induced compression in the through-thickness direction for even higher lateral compliance. In bonded structures of a circular (such as coated optical fibers or cables) or elliptic shape, the radial compression (hoop stresses) arises because of the thermal contraction mismatch in the radial direction of the high CTE coating and low CTE cladding. In flat (planar) bonded joints the compression in the through-thickness direction could be introduced deliberately (externally or internally).

The filler of the bonding compound has to have a high wetting ability with respect to the bonded surfaces. This means that the surface tension at the interface between the bonding compound (structure) and the bonded surface has to be much lower, than at the "free" surface of the bond. This allows for creating the desirable distance-depending attraction forces

In order to keep the system at the near-optimal distance between the bonded surfaces, the NPM-based compound has to maintain the adequate gap between the surfaces. This can be achieved by introducing microspheres into the compound. Such micro-spheres could serve as, sort of, bearing rolls that increase the system's stability and provide the desirable anisotropy of bonding matrix.

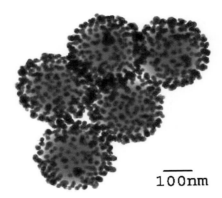

FIGURE 18.2. Microspheres with "protrusions."

The highly-compliant nano-systems in question are extremely heterogeneous. They consist of a bonding matrix with numerous embedded nanoparticles, microspheres and nanotubes/nanowires (whiskers). All these nano- and micro-particles embedded into a thixotropic matrix enable one to create a multi-purpose thixotropic structure. Figure 18.2 illustrates a possible combination of microspheres with "protrusions." These "protrusions" are formed by surface nanoparticles' clusters attracted to the microspheres.

18.6. CONCLUSIONS

The following conclusions can be drawn from the carried out analysis:

- Based on the developed simplified analytical stress model, we demonstrated that the employment of highly compliant materials and structures as bonding layers in bimaterial assemblies (joints) can lead to a significant relief in both the thermally induced and mechanical stresses. The model indicates that the interfacial shearing stress in an adhesively bonded or a soldered assembly is inversely proportional to the square root of the interfacial compliance, and that in "conventional" bimaterial assemblies (which are characterized by not-highly-compliant bonding layers), the interfacial compliance is due to both the bonding layer and the bonded components themselves. However, in assemblies with highly compliant bonds, it is the bonding material only that provides the high and favorable interfacial compliance.
- An appropriate wire array (WA) fabricated on one or both bonded components can be used as a suitable compliant bond. Based on the developed stress model, we demonstrate that its application can lead to a significant, about two orders of magnitude, increase in the interfacial compliance, thereby leading to a reduction in the maximum interfacial shearing stress of about an order of magnitude (compared to the bonded joints using "conventional" adhesives or solders).
- A modification of the newly developed nano-particle material (NPM) be used to increase dramatically the compliance of the bonding layer.
- The combination of both approaches could be effectively used to provide a highly compliant and a highly reliable bonding material and structure. In this case the NPM is employed as a suitable embedding material for the WA.
- Since the NPM has extraordinary mechanical and environmental properties, and, in combination with the appropriate WA, can make an extremely highly compliant

bond, we expect that the NPM and WA, used independently or in combination, might find a wide application in assemblies employed in micro- and opto-electronics, and beyond.

REFERENCES

1. E. Suhir, Thermal stress failures in microelectronics and photonics: prediction and prevention, Future Circuits International, (5) (1999).
2. E. Suhir, Buffering effect of fiber coating and its influence on the proof-test load in optical fibers, Applied Optics, 29(18) (1990).
3. Z. Kovac, et al., Compliant interface for semiconductor chip and method therefor, U.S. Patent #6,133,639, 2000.
4. T.H. DiStefano, et al., Compliant microelectronic mounting device, U.S. Patent #6,370,032, 2002.
5. Z. Kovac, et al., Methods for making electronic assemblies including compliant interfaces, U.S. Patent #6,525,429, 2003.
6. E. Paterson, et al., Mechanical highly compliant thermal interface pad, U.S. Patent #6,910,271, 2005.
7. Z. Kovac, et al., Methods of making microelectronic assemblies including compliant interfaces, U.S. Patent #6,870,272, 2005.
8. E. Suhir, Stresses in bi-metal thermostats, ASME Journal of Applied Mechanics, 53(3) (1986).
9. E. Suhir, Calculated thermally induced stresses in adhesively bonded and soldered assemblies, Proc. of the Int. Symp. on Microelectronics, ISHM, 1986, Atlanta, Georgia, Oct. 1986.
10. E. Suhir, Interfacial stresses in bi-metal thermostats, ASME Journal of Applied Mechanics, 56(3) (1989).
11. E. Suhir, Thermal stress in a polymer coated optical glass fiber with a low modulus coating at the ends, Journal of Materials Research, 16(10) (2001).
12. E. Suhir, Thermal stress in an adhesively bonded joint with a low modulus adhesive layer at the ends, Applied Physics Journal, (April) (2003).
13. E. Suhir, New nano-particle material (NPM) for micro- and opto-electronic packaging applications, IEEE Workshop on Advanced Packaging Materials, Irvine, March 2005.
14. D. Ingman and E. Suhir, Optical fiber with nanoparticle cladding, Patent Application, 2001 (allowed).
15. D. Ingman, V. Ogenko, and E. Suhir, Moisture-resistant nano-particle material and its applications, Patent Application, 2003 (pending).
16. E. Suhir, New hermetic coating for optical fiber dramatically improves strength, DARPA and Navair Workshop, St. Louis, MO, August 2004.
17. E. Suhir, Polymer coated optical glass fiber reliability: could nano-technology make a difference? Polytronic'04, Portland, OR, September 13–15, 2004.
18. E. Suhir, Mechanics of coated optical fibers: review and extension, ECTC'2005, Orlando, Florida, 2005.

APPENDIX 18.A. BIMATERIAL ASSEMBLY SUBJECTED TO AN EXTERNAL SHEARING LOAD AND CHANGE IN TEMPERATURE: EXPECTED STRESS RELIEF DUE TO THE ELEVATED INTERFACIAL COMPLIANCE

The following major assumptions are used in the analysis:

- Approximate methods of structural analysis (strength-of-materials) and materials science, rather than methods of elasticity, can be used to evaluate stresses and displacements.
- The bonded components can be treated, from the standpoint of structural analysis, as elongated rectangular strips that experience linear in-plane elastic deformations.
- At least one of the components ("substrate") is thick (stiff) enough so that bending deformations of the assembly as a whole do not occur and need not be considered.
- All the materials can be treated as linearly elastic.

- The interfacial shearing stresses can be evaluated based on the concept of the interfacial compliance, without considering the effect of the "peeling" stresses (normal interfacial stresses acting in the through-thickness direction).

Let a bimaterial assembly be subjected to both an external shearing force, $\widehat{T}$, and change, Δt, in temperature. Assuming, for instance, that the assembly was manufactured at an elevated temperature and subsequently cooled down to a low (say, room) temperature, one can seek, in an approximate analysis, the longitudinal interfacial displacements, $u_1(x)$ and $u_2(x)$, of the assembly components #1 and #2, respectively, in the form:

$$u_1(x) = -\alpha_1 \Delta t x + \lambda_1 \int_0^x T(\xi)d\xi - \kappa_1 \tau(x),$$

$$u_2(x) = -\alpha_2 \Delta t x - \lambda_2 \int_0^x T(\xi)d\xi + \kappa_{21} \tau(x). \tag{18.A1}$$

In this equations, α_1 and α_2 are the coefficients of thermal expansion (CTEs) of the dissimilar materials,

$$\lambda_1 = \frac{1-\nu_1}{E_1 h_1}, \quad \lambda_2 = \frac{1-\nu_2}{E_2 h_2} \tag{18.A2}$$

are the axial compliances of the assembly components, E_1 and E_2 are the Young's moduli of the materials, ν_1 and ν_2 are their Poisson's ratios, h_1 and h_2 are the thicknesses of the assembly components,

$$T(x) = x \int_{-l}^x \tau(\xi)d\xi \tag{18.A3}$$

are the thermally induced forces acting in the cross-sections of the assembly components, $\tau(x)$ is the thus far unknown interfacial shearing stress, l is half the assembly length,

$$\kappa_1 = \frac{h_1}{3G_1}, \quad \kappa_2 = \frac{h_2}{3G_2} \tag{18.A4}$$

are the interfacial compliances of the assembly components, and

$$G_1 = \frac{E_1}{2(1+\nu_1)}, \quad G_2 = \frac{E_2}{2(1+\nu_2)} \tag{18.A5}$$

are the shear moduli of the adherend materials. The origin, O, of the coordinate, x, is at the mid-cross-section of the assembly. The condition of the compatibility of the longitudinal interfacial displacements (18.A1) can be written as follows:

$$u_1(x) = u_2(x) - \kappa_0 \tau(x), \tag{18.A6}$$

where

$$\kappa_0 = \frac{h_0}{G_0} = 2(1+\nu_0)\frac{h_0}{E_0} \tag{18.A7}$$

is the interfacial compliance of the bonding layer, h_0 is the thickness of this layer, E_0 and ν_0 are the elastic constants of the bonding material, and G_0 is its shear modulus. Introducing the Equation (18.A1) into the compatibility condition (18.A6), we obtain the following equation for the sought interfacial shearing stress function, $\tau(x)$:

$$\kappa\tau(x) - \lambda \int_0^x T(\xi)d\xi = \Delta\alpha\Delta t x, \tag{18.A8}$$

where

$$\kappa = \kappa_0 + \kappa_1 + \kappa_2 \tag{18.A9}$$

is the total interfacial compliance of the assembly,

$$\lambda = \lambda_1 + \lambda_2 \tag{18.A10}$$

is the total axial compliance, and $\Delta\alpha = \alpha_1 - \alpha_2$ is the thermal contraction mismatch of the adherend materials.

Differentiating the Equation (18.A8) with respect to the coordinate, x, we obtain:

$$\kappa\tau'(x) - \lambda T(x) = \Delta\alpha\Delta t. \tag{18.A11}$$

The next differentiation, taking into account the Equation (18.A3), yields:

$$\kappa\tau''(x) - \lambda T(x) = 0. \tag{18.A12}$$

The boundary conditions

$$T(-l) = 0, \qquad T(l) = \widehat{T} \tag{18.A13}$$

for the induced force can be translated, using the Equation (18.A11), into the boundary conditions for the interfacial shearing stress, $\tau(x)$, as follows:

$$\tau'(-l) = \frac{\Delta\alpha\Delta t}{\kappa}, \qquad \tau'(l) = \frac{\Delta\alpha\Delta t + \lambda \widehat{T}}{\kappa}. \tag{18.A14}$$

Equation (18.A12) has the following solution:

$$\tau(x) = C_1 \sinh kx + C_2 \cosh kx, \tag{18.A15}$$

where

$$k = \sqrt{\frac{\lambda}{\kappa}} \tag{18.A16}$$

is the parameter of the interfacial sharing stress. Introducing the solution (18.A15) into the boundary conditions (18.A14), solving the obtained equations for the constants C_1

and C_2 of integration, and substituting the formulas for the constants of integration into the solution (18.A15), we obtain the following expression for the interfacial shearing stress:

$$\tau(x) = k\left\{\frac{\Delta\alpha\Delta t}{\lambda}\frac{\sinh kx}{\cosh kl} + \hat{T}\frac{\cosh[k(x+l)]}{\sinh 2kl}\right\}. \tag{18.A17}$$

The maximum value of this stress occurs, in this simplified stress model, at the end $x = l$:

$$\tau_{\max} = \tau(l) = k\left(\frac{\Delta\alpha\Delta t}{\lambda}\tanh kl + \hat{T}\coth 2kl\right). \tag{18.A18}$$

In sufficiently long and/or stiff assemblies, this formula yields:

$$\tau_{\max} = \tau(l) = k\left(\frac{\Delta\alpha\Delta t}{\lambda} + \hat{T}\right). \tag{18.A19}$$

Thus, it is the parameter, k, of the interfacial shearing stress that is responsible for both the level of this stress and the length of the end portions of the assembly that experience elevated interfacial stresses. As evident from the Equation (18.A16), this parameter decreases with a decrease in the axial compliance, λ, of the assembly and with an increase in the assembly's interfacial compliance κ. The axial compliance, λ, as evident from the Equations (18.A2), is due to the adherends only, and it is desirable, as evident from the Equation (18.A6), that these adherends are stiff, i.e., are characterized by a low axial compliance. As to the interfacial compliance, κ, this compliance, for conventional assemblies with not a very compliant bonding layer, is due to both the interfacial compliance of the bonding layer and the adherends themselves: the bonding layer is typically thin and low modulus, while the adherends are thick and high modulus, so, as evident from the Equations (18.A4) and (18.A7), the contribution of the compliances of the adherends and the bonding layer to the total compliance expressed by the Equation (18.A9), might be quite comparable. From the obtained formulas it is clear also that there is no incentive to increase the interfacial compliance of the adherends: by doing so, one would inevitably increase the axial compliance of the assembly as well, which is highly undesirable. Hence, for lower interfacial shearing stress, one should design a joint with a highly compliant bonding layer. This will make the factor k lower, will bring down the maximum interfacial shearing stress (which is inversely proportional to the square root of the interfacial compliance) and will spread the interfacial shearing loading over larger areas at the assembly ends.

APPENDIX 18.B. CANTILEVER WIRE ("BEAM") SUBJECTED AT ITS FREE END TO A LATERAL (BENDING) AND AN AXIAL (COMPRESSIVE) FORCE

Let a cantilever wire ("beam") be subjected at its free end to a lateral (bending) force, P, and an axial compressive force, T. The equation of bending of such a beam is as follows:

$$EIw''(x) + Tw(x) = 0. \tag{18.B1}$$

Here $w(x)$ is the deflection function of the beam, and EI is its flexural rigidity. The origin, O, of the coordinate x is at the clamped end of the wire. The following boundary conditions should be satisfied:

$$w(0) = 0, \ w'(0) = 0, \ w''(l) = 0,$$
$$EIw'''(l) + Tw'(l) + P = 0. \tag{18.B2}$$

In order to satisfy the four conditions (18.B2), the Equation (18.B1) is differentiated twice, and the solution to the obtained differential equation of the fourth order is as follows:

$$w(x) = C_0 + C_1 kx + C_2 \cos kx + C_3 \sin kx, \tag{18.B3}$$

where

$$k = \sqrt{\frac{T}{EI}} \tag{18.B4}$$

is the parameter of the compressive force. Introducing the solution (18.B3) into the boundary conditions (18.B2), we find:

$$C_0 = -C_2 = \frac{P}{kT} \tan kl,$$
$$C_1 = -C_3 = -\frac{P}{kT}, \tag{18.B5}$$

and the solution (18.B3) results in the following expression for the deflection function:

$$w(x) = \frac{P}{kT} \left\{ \tan kl - kx - \frac{\sinh[k(l-x)]}{\cosh kl} \right\}. \tag{18.B6}$$

The maximum deflection at the wire end is

$$w(l) = \frac{P}{kT} (\tan kl - kl). \tag{18.B7}$$

The maximum deflection at the free end of a cantilever beam subjected to a lateral force P applied at this end is

$$\delta = \frac{Pl^3}{3EI}. \tag{18.B8}$$

Comparing the Equations (18.B5) and (18.B6), we conclude that the parameter

$$\varsigma = 3 \frac{\tan kl - kl}{(kl)^3} \tag{18.B9}$$

considers the effect of the axial compression on the lateral compliance of a cantilever wire ("beam"). Within the framework of the linear approximation, the parameter ς is infinitely large, when the compressive force, T, reaches its critical (Euler) value. Indeed, this parameter is infinitely large, when $kl \to \pi/2 = 1.571$. In this case the Equation (18.B3) yields

$$T = \frac{\pi^2 \cdot EI}{4l^2},$$

which is a well known expression for a critical force for a cantilever beam. But even when the parameter kl is only $kl = 0.866$, the parameter ς is as high as about 6.4, and in such a case a 2.5 fold decrease in the interfacial shearing stress could be expected. Thus, there is a definite incentive, as far as the interfacial shearing stress is concerned, for using nanowire-based bonding structures subjected to compression.

APPENDIX 18.C. COMPRESSIVE FORCES IN THE NPM-BASED COMPOUND STRUCTURE

The desirable compression in the compound structure could be achieved by application of external forces, but it certainly can stem from the bonding compound features, as quantitatively shown below.

Let us define the system parameters as following:
S_S overall are of surfaces to be bonded,
S_B interfacial area of contact between the bonded surface and bonding compound,
S_C side area of the compound (in contact with air),
σ_{SA} interfacial tension coefficient between the bonded surface and air,
σ_{SC} interfacial tension coefficient between the bonded surface and the compound,
σ_{CA} interfacial tension coefficient between the compound and air,
d gap between the bonded surfaces,
R_C radius of the contact area,
and
V_C volume of the compound located between the bonded surfaces.

Constraint of the compound volume conservation during any spread/wetting process should be considered in the expression for the system surface tension energy:

$$E = 2 \cdot \sigma_{SA} \cdot (S_S - S_B) + \sigma_{CA} \cdot S_C + 2 \cdot \sigma_{SC} \cdot S_B$$

provided

$$S_B \cdot d = V_C. \tag{18.C1}$$

It is clear that

$$S_B = \pi \cdot R_C^2,$$

and

$$S_C = 2 \cdot \pi \cdot R_C \cdot d. \tag{18.C2}$$

Then, the system energy can be easily expressed vs. d or S_B. Let us bring the dependence of this energy on the contact area S_B:

$$E(S_B) = 2 \cdot \left[\sigma_{SA} \cdot S_S + \sigma_{CA} \cdot \frac{V_C \cdot \sqrt{\pi}}{\sqrt{S_B}} + (\sigma_{SC} - \sigma_{SA}) \cdot S_B \right]. \tag{18.C3}$$

In a case of $\sigma_{SC} < \sigma_{SA}$ there the energy function does not have any minimum, and the system tends to the complete bonded surface coverage with the compound, and the equilibrium is achieved only with the help of repulsive forces provided by the nanowires, microspheres and other embedded "elastic" elements. On the other hand, in an opposite case, when $\sigma_{SC} > \sigma_{SA}$, the energy has a minimum at

$$S_B = \sqrt[3]{\frac{\pi \cdot V_C^2}{4} \cdot \frac{\sigma_{SC}^2}{(\sigma_{SC} - \sigma_{SA})^2}},$$

or

$$d = \sqrt[3]{\frac{4 \cdot V_C}{\pi} \cdot \frac{(\sigma_{SC} - \sigma_{SA})^2}{\sigma_{SC}^2}}. \tag{18.C4}$$

As a quasi-elastic system (actually, visco-elastic system), this system tries to achieve a state of equilibrium that corresponds to the minimum of the elastic energy associated with surface tension and the nanowires buckling and microspheres compression. Then the system will operate in the condition of a dynamic steady-state (stable) equilibrium with the "distractive" reactive forces caused by the compressed nano-wires and the compression forces of the surfaces tension.

19

Adhesive Bonding of Passive Optical Components

Anne-Claire Pliska and Christian Bosshard

CSEM SA, Untere Grundlistrasse 1, 6550 Alpnach Dorf, Switzerland

19.1. INTRODUCTION

Fiber-optic communication involves generation, transmission, amplification, detection and processing of optical signals. For each of these functions, a multiplicity of optical components is needed: transmitters, receivers, modulators, splitters, fibers, filters, couplers, optical isolators, pump lasers, amplifiers. The proliferation of wavelength channels and passive components has triggered a dramatic increase in the amount of parts found in optical networks. As 60 to 80% of the cost of an optical component is in the packaging, the prospect of housing several components in the same package becomes attractive, especially when this strategy can eliminate several fiber pigtailing operations and lossy transitions between discrete components. The integration of electronic and optoelectronic chips on the same platform is expected to be the next breakthrough in optoelectronic packaging and to reduce the overall device cost. The benefits of squeezing more components into a single optical module are: reduced inventory costs, space savings, simplified mechanical design, reduction in fiber splicing. On the electronic side, as data rate continues to increase, there is a need to bring electronic circuitry (e.g., drivers, pre-amplifiers) and active optical components (e.g., lasers, amplifiers, photodiodes . . .) closer together to avoid parasitic capacitances.

Optics has not found an equivalent of the transistor, i.e., a single building block from which all other functions can be constructed in a single process. As a result, almost every optical function needs its own discrete element requiring its own fabrication process. In order to increase functionality into a single housing, two main strategies are being currently pursued:

- Monolithic integration uses a single-material to fabricate several components on a common substrate. So far, it is not clear which technology platform can handle fully integrated products and whether this platform can do it cost effectively.
- Hybridization relies on the assembly on the same board of specialty subcomponents, each optimized in its own material system.

The hybridization scheme and the need for reduced assembly costs are driving the development of new assembly technologies and new optical designs providing relaxed positioning tolerances:

- Optical building blocks with a collimated beam output [61].
- New passive waveguide technology: silicon waveguides, polymer waveguides [56], hollow waveguides [29].
- Disruptive fixing technologies including adhesive bonding or clipping mechanisms [37,48,55].
- MEMS-based alignment structures.
- Silicon-based heat pipes for improved thermal management [35].
- Local heating options or local curing options for multi-chip assembly [3].
- Development of dedicated automation tools [14].

To assemble the bare dies and optical fibers onto a common substrate or into housings, the following packaging processes are available today: soldering, laser welding, thermo-compression, mechanical interlocking, resistance welding and adhesive bonding.

Although adhesive bonding has been used for many years in the microelectronics industry, there is still a mental barrier for using adhesives in optical assemblies and packages. Historically, the reliability of adhesives in opto-electronic assemblies has often been questioned (outgassing, mechanical stability, photostability). With the release of new adhesives specifically developed for the assembly of fiber-optic components and a better understanding of the assembly processes, adhesives are nowadays increasingly used in optical packages, e.g., to fix optical fibers or lenses. Adhesives are applicable to a wide variety of materials and optical assembly tasks. Other advantages of adhesive bonding over alternative fixing techniques include low processing temperature and low-cost equipment. Additionally, a suitable adhesive can be selected for each optical assembly application from the large variety of products available on the market, making adhesive bonding a versatile fixing technology.

In order to better understand the requirements for a bonding process in opto-electronics packaging, we will first discuss the typical positioning tolerances when an optical connection is involved. Then, the influence of the thermal stress at a chip and package level, as well as the impact of creep on the performances of optical assemblies, will be evaluated. A comparison between adhesive and solder materials will be carried out.

In a next section, we will review the existing adhesive bonding theories and the requirements that must be met to create a reliable and low-stress joint. A special emphasis will be put on the surface energy concept and the surface preparation of adherends. Additionally, the influence of the glass transition temperature (T_g) and the coefficient of thermal expansion (CTE) on the relative displacement between optical devices will be discussed.

Ultimately, we will describe two examples of optical assemblies where an adhesive material is used to fix optical fibers. The first application consists in a laser pigtailing operation: we describe the alignment and fixing process and we present thermal cycling results. The second application concerns a channel waveguide where a fiber has to be aligned and fixed on both sides of the device.

19.2. OPTICAL DEVICES AND ASSEMBLIES

19.2.1. Optical Components

Through the exploitation of the unique properties of integrated, free space and fiber optics, a wide variety of active and passive optical devices are currently offered to supply the telecom and datacom market. Active devices require driving electronics using typical wiring technologies or external optical pump signals. These devices are used for:

- Opto-electronic conversion (optical-to-electrical, or vice-versa): diode laser (edge emitters, VCSEL,[1] LED,[2]) photodiode (surface or edge illuminated).
- Manipulation of the signal: modulators, amplifiers (SOA,[3] EDFA[4]), attenuators, switches.

Passive devices do not have driving electronics. They simply filter or route the signal based on wavelength, intensity or polarization. They include lenses, mirrors, prisms, isolators, couplers, multiplexers, and demultiplexers.

Table 19.1 and 19.2 list a variety of active and passive optical components along with the corresponding key material systems addressing the 1.55 μm telecommunication window. The applications of optoelectronic devices, ranging from long-haul to metro and access devices, include a large mix of material classes and properties. Thus, assembling several devices out of different materials in the same module requires special attention to packaging design (substrate, bonding material, etc.). So far, with an exception of low-cost TO[5]-based packages, the assembly processes remain proprietary and the standardization of packaging techniques are yet to come.

19.2.2. Opto-electronics Assemblies: Specific Requirements

In photonics, packaging must provide not only electrical connections and mechanical support but also thermal management and, more critically, optical connections. Figure 19.1 and 19.2 presents a "butterfly" housing used, e.g., in laser diodes or in high-speed photodetector packaging where the light from a semiconductor laser diode is coupled into an optical fiber through microlenses in free space propagation. Optical interconnects are highly directional. In order to achieve optimal signal transmission, they require precise control of the relative location between components and an attachment process that maintains the alignment over time. In single-mode fiber applications, the positional tolerance is typically in the sub-micron range. Therefore, the assembly of optoelectronic components provides unique requirements and challenges.

19.2.2.1. Optical Connection Optimizing optical connection is the most critical step in the assembly of an optoelectronic package. In a laser to fiber coupling configuration, the optical coupling efficiency η of the laser beam ψ_i in the optical fiber characterized by its

---

[1] VCSEL: vertical cavity surface emitting laser.
[2] LED: light emitting diode.
[3] SOA: semiconductor optical amplifier.
[4] EDFA: erbium-doped fiber amplifier.
[5] TO: transistor outline.

TABLE 19.1.
Examples of active functions and the corresponding material classes.

Active components	Material	References
Transmitters	III-V semiconductor	[4,15]
Receivers–photodetectors	InGaAsP	[1,22]
Modulators	III-V semiconductor,	[17,60]
	LiNbO$_3$, polymer, silicon	[18,36]
Amplifiers	III-V semiconductor,	[65]
	doped polymer, doped	[8]
	glass fiber	[2]
Pump lasers	III-V semiconductor	[43]
Switch	Polymer, silicone	[45]
	III-V semiconductor	[58]

TABLE 19.2.
Examples of passive functions and the corresponding material classes.

Passive components	Material	References
Arrayed waveguide grating	Silica, SiON, InP, polymer	[26,30]
Passive optical waveguides	Silica, SOI, SiON	[29,56]
	polymer, sol-gel, hollow waveguide	
Coupler, splitter	Silica, SOI, SiON	[23]
	polymer, sol-gel, hollow waveguide	
Lenses	Silica, silicon	[61]

FIGURE 19.1. Butterfly package used for the packaging of transmitters, pump lasers, edge detectors. Internal package dimensions are: 18.9 mm × 10.2 mm × 6.4 mm.

fundamental optical mode ψ_f is related to the overlap integral between the two electric fields and is defined as:

$$\eta = \frac{\left|\iint \psi_i \psi_f^* dA_f\right|^2}{\left(\iint |\psi_i|^2 dA_i\right)\left(\iint |\psi_f|^2 dA_f\right)}. \tag{19.1}$$

In Equation (19.1), A_i and A_f refer to the integration surface in the plane of the incident field and the optical fiber fundamental field, respectively. The small refractive

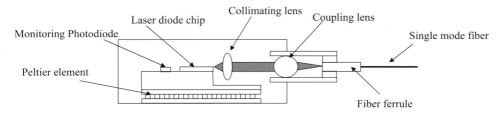

FIGURE 19.2. Typical structure of a laser module: the laser chip, a monitoring photodiode and a thermistor (not represented) are soldered on a submount. The laser beam is collimated by a first lens (collimating lens) and is then focused by the second lens (coupling lens) on the facet of the single mode optical fiber. A Peltier element provides active cooling of the laser chip during operation.

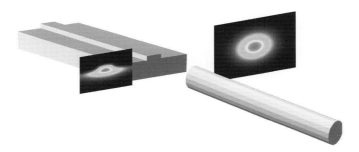

FIGURE 19.3. End-fire coupling between a laser diode chip and a single mode optical fiber. The mode shape of the ridge laser diode is highly asymmetric and smaller than the fundamental mode in the optical fiber.

index difference in a glass fiber of $\Delta n \approx 5 \times 10^{-3}$ results in a weakly guided optical mode with a typical mode size of 8–10 μm. In planar waveguides, including semiconductor integrated optical devices, Δn is often larger than 10^{-2}, leading to a mode size smaller than 3 μm. Moreover, unlike the circular mode in a fiber, the mode shape in a planar device is elliptical, resulting in an additional mode mismatch.

In order to estimate the coupling efficiencies and alignment tolerances, the Gaussian field approximation can be used in most cases [53]. In addition, the integration of intermediate microlenses or the use of lensed fibers can be simulated with the ABCD propagation matrix theory.

In the following example, the computation of the coupling efficiency of a 1550 nm laser diode into a single mode fiber is considered. We assume Gaussian profiles for the laser diode field distributions in the directions parallel and orthogonal to the junction plane with beam spot sizes $2\omega_{//} = 3$ μm and $2\omega_{\perp} = 1$ μm (divergence: $\theta_{//} = 9.4°$, $\theta_{\perp} = 28.3°$). The fundamental field HE_{11} in the single mode optical fiber is approximated by a Gaussian beam having a beam spot size $2\omega_f = 9.8$ μm. In an end-fire coupling configuration (direct fiber coupling without using intermediate lenses, see in Figure 19.3), the calculated optical coupling efficiency is roughly 16%. This percentage decreases if the alignment is not optimal. The solid line in Figure 19.4 shows coupling efficiency variations when a lateral (orthogonal) misalignment is introduced in the end-fire coupling situation. The coupling tolerance, defined as the lateral misalignment yielding 1 dB of additional coupling losses, is here 1.7 μm in the direction orthogonal to the junction plane.

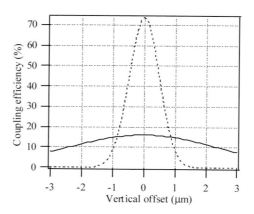

FIGURE 19.4. Evaluation of the coupling of a 1.55 μm laser diode with beam dimensions of 2 × 3 μm into a single mode fiber using a cleaved fiber (solid line) and a ball lens fiber (dash line). With a cleaved fiber, the optical coupling efficiency is roughly 16% and the "orthogonal" positioning tolerance for 1 dB additional coupling losses is 1.7 μm. With a ball lens fiber, the optical coupling efficiency rises up to 73% but the "orthogonal" positioning tolerance for 1 dB additional coupling losses becomes much tighter (0.4 μm).

Lensed fibers provide a highly effective way to improve the coupling efficiency between fibers and optical devices [42]. Various tip designs eliminate the need for a separate lens thereby reducing the return loss and the assembly costs. In the above example of the laser diode pigtailing, the coupling efficiency can be dramatically increased if a ball lens fiber is used. Indeed, an optimal coupling efficiency of 73% is found when the fiber lens radius is 4 μm and the distance between the laser facet and the fiber tip is 8 μm. Unfortunately, this optimized coupling configuration is achieved at the expense of reduced alignment tolerances. The dash curve in Figure 19.4 shows the coupling efficiency variations when a lateral (orthogonal) misalignment is introduced in the lensed fiber pigtailing situation. The coupling tolerance is here as low as 0.4 μm in the direction orthogonal to the junction plane.

In practice, in order to achieve optimal alignment, the fiber is held in a gripper mounted on an actuator-controlled stage in front of the powered laser diode endface. Stray light ("first light") is coupled into the optical fiber and detected with an optical power meter connected at the output of the fiber. Using dedicated alignment algorithms (hill-climb, triangulation, raster scan, spiral scan), the optical fiber stage is moved until maximum optical power is detected.

Fiber loading, alignment and fixing are still the bottleneck of an optoelectronic assembly. Indeed, it usually takes more than 4 minutes to align an optical fiber to a semi-conductor laser [62]. Using machine vision and pattern recognition, stray light detection is sped up and the alignment time can be shortened down to less than 1.5 minutes [47].

Passive alignment is an alternative approach to increase the integration of optoelectronic components and to drive down the packaging costs. This technique does not rely on emitted light for accurate coupling to single mode fibers and allows the alignment of optical fibers with passive devices, such as switches, array waveguide grating. In this assembly technique, optoelectronic chips, lenses and optical fibers have to be passively aligned. The flip-chip process on a structured substrate is the common tool for micropositioning the components [64] provided that:

- The alignment process of the chip is either based on vision if the machine specifications allow for a pre-bond accuracy below 1 μm [16] or/and on the C4[6] process first developed by IBM in the 60's [41].
- The fiber is, e.g., passively aligned in a micro-machined groove in front of the opto-electronic chip.

Chip to fiber alignment performance achievable with this technology is still somewhat reduced compared with standard active alignment techniques and cannot be used where sub-micron positioning accuracy is required.

19.2.2.2. Substrate Materials The choice of the submount material is mainly driven by the need of matching the CTE of both the optical components and the submount [44,57] as well as providing an efficient thermal pathway in uncooled devices or in high power applications [43].

19.2.2.2.1. Thermal Stresses. In this section, the thermal stresses in a tri-material assembly are considered. We evaluate the influence of the substrate material for a given GaAs laser chip and we compare the ability of the intermediate solder or adhesive material to accommodate the CTE-mismatch and to limit the thermal stresses in the assembly.

A GaAs laser chip soldered onto a heat spreading copper substrate with a AuSn (80 wt% Au) solder represents an extreme case of the CTE mismatch. The CTE of GaAs and Copper are, respectively, 6.5×10^{-6} K^{-1} and 17.8×10^{-6} K^{-1}. During the heating step of the soldering process, the semiconductor laser die and the copper substrate are free to expand. The copper substrate, due to its larger CTE, will expand more than the GaAs die. During the cooling phase of the process, the stress-free displacements between the die and the submount are prevented by the solder layer, and thermal stresses are induced in the assembly [57]. These include:

- Normal stresses acting over the cross sections of the components. These stresses are responsible for the strength of the components (die or substrate) themselves.
- Shearing and transverse normal ("peeling") stresses at the interfaces. These stresses are responsible for the adhesion or cohesion strength of the die attach material.

Using 2D modeling and assuming that there is no bending, the CTE mismatch induced shear stress $\tau_{\alpha,s}$ at the substrate-solder and at the chip–solder interface is maximum at the edge of the chip. Its value is given by [12]:

$$\tau_{\alpha,s} = \Delta \alpha \Delta T G_d \frac{\tanh \beta L}{\beta t_d}, \qquad (19.2)$$

where β is defined as:

$$\beta = \sqrt{\frac{G_d}{t_d} \left(\frac{1}{E_c t_c} + \frac{1}{E_s t_s} \right)}. \qquad (19.3)$$

The letters c, s and d stand for chip, substrate and die attach materials, respectively, $\Delta \alpha$ is the CTE mismatch between the chip and the substrate, ΔT is the temperature change. E and G are the elastic moduli in tension and shear, respectively, t represents the materials thickness, and $2L$ is the chip length.

---

[6]C4: controlled chip collapse connection.

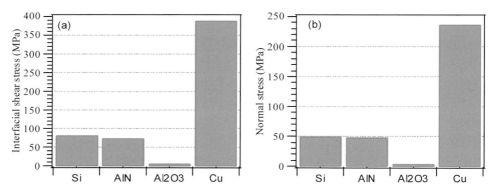

FIGURE 19.5. (a) Comparison of the CTE-mismatch induced interfacial shear stresses when a 2400 μm GaAs chip is soldered to different substrate materials. The intermediate solder material is AuSn (80%wt Au) and the temperature difference between the peak soldering temperature and the room temperature is 300°C. The shear stresses at the chip–solder and at the substrate–solder interfaces are minimum when an Alumina submount is used (7.4 MPa). The influence of the bow in the substrate and in the chip is not considered in these calculations. (b) Corresponding normal stresses in the GaAs die. The normal stress in the GaAs die soldered on a Copper substrate exceeds the GaAs strength fracture. Hence, the die is expected to crack upon a rapid cooling phase of the soldering process.

Similarly, the maximum normal stresses $\sigma_{\alpha,n}$ (at the center of the chip) are given by:

$$\sigma_{\alpha,n} = \frac{\Delta\alpha \Delta T G_d}{\beta^2 t_d}\left(1 - \frac{1}{\cosh \beta L}\right). \tag{19.4}$$

In Figure 19.5(a), we numerically evaluate the influence of the substrate on the maximum interfacial shear stress built-up during soldering of a 2400 μm long GaAs chip using a AuSn (80 wt% Au) solder. We assumed a solder material thickness of 20 μm in the calculations and a maximum temperature variation between room and soldering temperature of 300°C. The following submount materials are compared: copper, silicon, AlN and Al_2O_3. In many applications involving active optoelectronic semiconductor chips based on InP or GaAs, the preferred submount material is AlN due to high thermal dissipation and good CTE matching properties. On the other hand, Al_2O_3 and silicon are the materials of choice when cost savings are critical [7,64]. The substrates and solder materials properties involved in the calculations are summarized in Table 19.3.

The thermally induced shear stresses at the edge of the AuSn solder layer are approximately 10 times smaller when using an Al_2O_3 (7.4 MPa) instead of an AlN (73.5 MPa) or a silicon (80 MPa) substrate. For all the three substrates, the normal stresses displayed in Figure 19.5(b) remain below the GaAs fracture strength (85 MPa). On the other hand, the estimated interfacial shear stresses are 387 MPa when the GaAs chip is bonded onto a copper substrate. Assuming that the solder exhibits a linear elastic behavior at such stress values, the normal (tensile) stresses in the GaAs chip are estimated to be 236 MPa, i.e., above the fracture strength of GaAs. However, as the estimated interfacial shear stresses exceed the yield stress of the gold-tin solder (275 MPa), a plastic deformation is induced in the solder during the cooling phase of the soldering process. This plastic deformation redistributes the stresses in the solder layer and finally reduces the normal stresses in the GaAs chip [33]. Moreover, considering the creep-induced stress relaxation and slowing down the cooling process, once could further reduce the interfacial stress. This effect has proven to

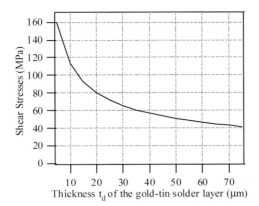

FIGURE 19.6. Dependence of the CTE mismatch induced shear stresses in a gold-tin solder as a function of the solder thickness assuming a difference between room and soldering temperature of 300°C. The interfacial shear stresses are better accommodated with a thick solder layer.

TABLE 19.3.
Thermal and mechanical properties of common packaging substrates materials.

Material	CTE (ppm/K)	Thermal conductivity (W/m K)	Young modulus (GPa)	Shear modulus (GPa)
Si	4.2	150	130	52
GaAs	6.5	54	86	33
Cu	17.8	400	110	46
CuW	7	180	260	
Al	23.6	240	70	25
Kovar (Fe:Ni:Co)	5.8	15	138	52
AlN	4.5	170	350	140
Al_2O_3	6.7	21	390	125
AlSiC	8	200	188	76
Diamond	2	2000	800	
AuSn (80% Au wt)	16	58	68	25
AgSn (96.5% Ag wt)	22	36	50	19
PbSn (37% Pb wt)	21	21	40	14
Epoxy	50	0.3	3	1.2
Acrylate	220		1.1	0.4

reduce also normal stresses from 130 MPa to 80 MPa in the case of GaAs chips soldered on diamond substrates [9] and could eventually be applied to the GaAs chip on copper to avoid chip cracking.

One can decrease the interfacial shear stress by increasing the die attach material thickness t_d. Figure 19.6 represents the dependence of the maximum induced shear stress in the gold-tin solder layer in a GaAs chip on a silicon submount, when the maximum temperature change between the room and the soldering temperature is 300°C. The shear stresses at the substrate-solder and at the chip–solder interfaces are reduced by approximately 40%, when the solder layer thickness is increased from 20 μm to 60 μm. By this means, the normal stresses in the GaAs die are also reduced and the assembly integrity is ensured.

For a defined CTE-mismatch between the substrate and the chip, the interfacial shear stresses is limited, if a compliant and/or thick die attach material is used. Indeed, the use of a compliant epoxy material requiring a reduced temperature excursion during the bonding process decreases the shear stresses from approximately 80 MPa (with a AuSn solder) down to 8.7 MPa.

Based on these calculations, we conclude that a compliant epoxy material limits the thermo-mechanical stresses and reduce the likelihood of cracks generation in the die or the substrate compared with a solder alloy material. Additionally, the reduced stress in adhesive bonds limits the adverse effect of creep in adhesively bonded assemblies.

19.2.2.2.2. Thermal Conductivity and CTE Trade-Off. A good trade-off between the CTE matching and thermal conductivity of the submount material must be found. Al_2O_3 offers ideal CTE properties, however, its low thermal conductivity can be detrimental for a proper operation of an active GaAs device. AlN or silicon materials are good alternatives to Al_2O_3. Indeed, a stack with a 150 μm thick GaAs chip mounted on an AlN substrate and a CuW package base, both 500 μm thick, has a thermal resistance of approximately 22 K W^{-1}. This assembly compares favorably to the Al_2O_3 substrate and package base configuration where a thermal resistance of 44 K W^{-1} is estimated. A stack with a silicon substrate and a light AlSi package base give identical heat flow performances as the AlN/CuW combination.

In addition to good CTE matching with other semiconductor materials and high thermal conductivity, silicon offers an extensive hybridization potential. Standard IC photolithography and structuring processes not only allow the fabrication of electrical interconnects but also opens the way to the fabrication of high-precision alignment features needed to mount optical devices [20]:

- V-grooves or U-grooves for fiber alignment [50,55].
- Microlens mounting [61].
- Standoff and indentations for flip-chip mounting of semiconductor chips [64].
- Reservoirs for adhesive [32].

Increased functionality of the silicon bench can be achieved with the monolithic integration of resistors, capacitors or inductors. For example, miniature polysilicone heaters can be integrated on a silicon board to enable local heating for reflow soldering of multichip modules [50]. Excellent RF properties can be achieved using Polyimide or BCB intermediate layers [59]. Additionally, deposition of TaN thin film resistor on silicon microbench can provide electrical damping for 10 Gbits/s driving signals in transceiver modules [3]. Optical functions can also be monolithically integrated on the silicon platform using either silica or silicon waveguides and devices [36].

19.2.2.3. Die Attach Material Requirements The most commonly used die attach materials in opto-electronics are solder alloys, adhesives and glass solders. Advantages and disadvantages of these materials are briefly summarized in Table 19.4. Minimizing the thermal stresses in the assembly process or in device operation is a concern in both microelectronics and opto-electronics. We have seen in Section 19.2.2.2.1 that an epoxy material compares favorably with a gold-tin solder as far as CTE mismatch induced shear stresses are concerned. In addition to the minimization of the mechanical stresses in the assembly, bonding materials in optical assemblies must fulfill the following requirements: low

TABLE 19.4.
Advantages and limitations of the three main die attach materials used in the opto-electronics industry. Cycling times are similar for all three materials.

Bonding material	Advantages	Limitations
Adhesive	Low curing temperature Low Young modulus (reduced stresses) Ease of automation	Rework not possible Large CTE Outgassing Low thermal and electrical conductivity Environmental sensitivity (moisture)
Solder	Good thermal and electrical conductivity Low CTE Rework possible	Require metalized surfaces Processing temperatures >200°C Need an inert gas atmosphere
Glass	Good thermal and electrical conductivity Low CTE Limited stress relaxation	Processing temperatures >300°C Rework is difficult

creep to minimize components shift over time and low stress-induced birefringence in the attached component (fiber, microlens, prisms).

19.2.2.3.1. Effect of Creep. In optical assemblies, there are very tight alignment requirements that do not allow extended movements of the semiconductor chip with respect to the optical fiber when heated or exposed to mechanical shock. Therefore bonding materials are supposed to take up the CTE mismatch, vibration and other stresses that may cause a die or a fiber to move. Moreover, the die attach material should exhibit low creep properties to maintain the optical alignment over time and under stress.

Driving forces of the creep are the internal and the external stress built-up during the solder cooling or as a result of adhesive polymerization and during device operation. The creep strain–stress relationship is determined by the material properties, i.e., by the microstructure and the diffusion properties of the solder alloy [40], or the free-volume ratio in adhesives [51]. Following Andrade's work on metals, the creep strain is a function of the applied stress (internal or external), σ, the time t and the temperature T:

$$\varepsilon = f(\sigma)g(t)h(T). \tag{19.5}$$

Several expressions can be found in the literature for each of these functions. A detailed description of the generally accepted expressions can be found in [34]. For example, the steady-state creep strain can be defined as:

$$\varepsilon = C\sigma^n t \exp\left(-\frac{\Delta H}{kT}\right), \tag{19.6}$$

where n, ΔH, k and C are the stress factor, the creep activation energy, the Boltzmann's constant and a material constant.

The stress factor n is usually a function of the stress level and temperature. For low stress conditions (typ. $\sigma \ll 1$ MPa) and high temperature, $n \sim 1$, i.e., the strain rate is

TABLE 19.5.
Creep parameters assuming a Norton-type stress dependence $d\varepsilon/dt = C \cdot \sigma^n \cdot \exp(-\Delta H/kT)$ for various solder alloys. ΔH is the creep activation energy and n is the stress factor at room temperature. The data for Sn-3.5Ag, Sn-9Zn, Sn-37Pb, Sn-2Cu-0.8Sb-0.2Ag are taken from the NIST database[a] and the parameters for the Sn-80Au are derived from [9].

Solder	C (MPa^{-n} s^{-1})	Activation energy ΔH (eV)	Stress factor n	Melting temperature (°C)
Sn-3.5Ag	1.5×10^{-3}	0.825	11.3	221
Sn-9Zn	2.17×10^{-2}	0.677	5.7	199
Sn-37Pb	0.205	0.49	5.25	183
Sn-2Cu-0.8Sb-0.2Ag	3.031	0.85	8.9	285
Sn-80Au	5.29×10^6	1.24	7	280

[a] http://www.boulder.nist.gov.

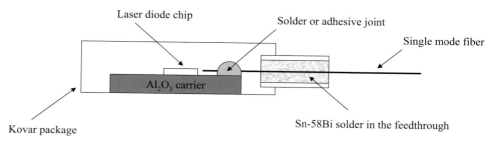

FIGURE 19.7. Assembly configuration used in the calculation of the creep in a solder and adhesive joint. The optical fiber is soldered or adhesively bonded on the laser carrier and is subsequently soldered in the fiber feedthrough with a bismuth-based solder.

proportional to the stress level. At higher stresses or lower temperatures, the stress exponent lies in the range of 3–20 depending on the material. This behavior is governed by a change in the dominant origin of the creep mechanism: diffusion-induced at low stresses or high temperature and dislocation induced at higher stresses or low temperature.

Table 19.5 lists the creep parameters for five common solders used in microelectronics and opto-electronics. These solders exhibit activation energies ranging from 0.6 eV to 1.3 eV, i.e., above the default activation energy specified in the Telcordia Generic Requirements GR 468[7] for module wear-out failure estimation (0.4 eV). It is noteworthy that the creep activation energy for all the listed binary solders increases linearly with the solder melting temperature.

In the following section, we compare the creep-induced displacement in a solder joint for different solder materials listed in Table 19.5 for the configuration sketched in Figure 19.7:

- An optical fiber is soldered on a planar substrate in front of a laser in a Kovar butterfly module. The mechanical properties of this solder are particularly critical as they

---

[7]GR468: generic reliability insurance requirements for optoelectronic devices used in telecommunications equipment, Bellcore, Issue 1, R4-62, 1998.

ADHESIVE BONDING OF PASSIVE OPTICAL COMPONENTS

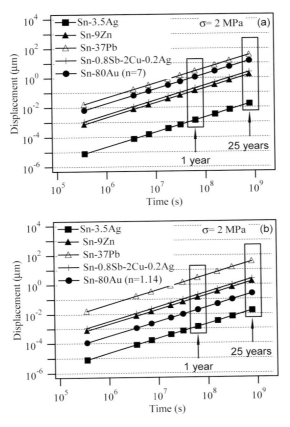

FIGURE 19.8. Solder joint displacement as a function of the time at room temperature using (a) $n = 7$ and (b) $n = 1.14$ for the gold-tin solder. From all solders listed in Table 19.5, Sn-3.5Ag exhibits the lowest creep compliance with an estimated displacement of approximately 20 nm after 25 years of device operation. Assuming a diffusion process as the dominant creep mechanism at room temperature for the Sn-80 Au, the displacement in the joint in the gold-tin solder is reduced from 14 µm down to 250 nm after 25 years of device operation.

define the evolution over time of the coupling efficiency between the laser beam and the optical fiber.

- Once properly aligned and soldered in front of the laser, the optical fiber is soldered at a second location in the fiber feedthrough of the butterfly module with a BiSn (58 wt% Bi) solder.

When the fiber is soldered in the module feedthrough, axial stresses are built up in the fiber when the assembly is cooled down from the manufacturing temperature ($T_m = 138°C$) and the room temperature T_a due to CTE mismatch between the housing, and the fiber substrate (the thermal expansion of the solder and the fiber itself is neglected). Assuming an alumina substrate and a fiber Young's modulus of 72 GPa, the initial axial stress in the fiber is estimated to be around 20 MPa. This corresponds to a force of 0.2 N acting on the solder joint. Based on a 2D modeling proposed by Suhir [57], the average shear stress in the solder is 2 MPa assuming a solder joint length of 500 µm.

The creep strain in the solder joint can be evaluated using Equation (19.6) for all the solders listed in Table 19.5. The results are displayed in Figure 19.8 as a function of time. The solder Sn-Ag (3.5%wt Ag) exhibits the lowest creep-induced displacement with an estimated fiber offset of approximately 20 nm after 25 years of device operation.[8] For all the others investigated solders, the creep compliances are expected to adversely impact device reliability after a few months of the device operation. Indeed, the expected fiber offset is already as large as 1.5 μm with the PbSn (37%wt Pb) after 1 year of service, which is already larger than the positioning tolerances in most single-mode fiber applications. The estimated behavior of the gold-tin solder is not as good as expected. The stress factor for the gold-tin solder at room temperature, extrapolated from the data at temperatures above 100°C found in [9] could be slightly overestimated. If we assume now that the predominant creep mechanism in the gold-tin solder remains atomic diffusion and that the stress factor keeps the value measured at 100°C (value of 1.14), the expected ultimate fiber displacement is 10 nm and 250 nm after 1 and 25 years of device operation, respectively. If the stress level is increased from 2 MPa to 5 MPa, this displacement remains below 1 μm after 25 years.

The strain developed over time in the solder fixing the fiber inside the module depends on the operating temperature, through the temperature dependence of the stress acting on the solder and the temperature dependence of the creep in the solder itself. Using the Norton model to describe the stress dependence of the creep, we can define a temperature acceleration factor as:

$$a(T) = \frac{\varepsilon(T)}{\varepsilon(T_a)} = \left(\frac{T_m - T}{T_m - T_a}\right)^n \exp\left[-\frac{\Delta H}{k}\left(\frac{1}{T} - \frac{1}{T_a}\right)\right]. \tag{19.7}$$

In this particular configuration where the stress experienced by the solder decreases with temperature and vanishes at $T = T_m$, the acceleration factor is not a monotonically increasing function of the temperature. Indeed, this function goes through a maximum for $T = T_c$ and then decreases for $T > T_c$. T_c can be easily evaluated for each of the solders under investigation. The calculations give 30°C for Sn-3.5Ag, 58°C for Sn-9Zn, 53°C for Sn-37Pb, 46°C for Sn-2Cu-0.8Sb-0.2Ag solder and 76°C for Sn-80Au. The acceleration factor is plotted in Figure 19.9 as a function of the package temperature T for the five solder materials. In fact, in the case of the Sn-3.5Ag solder, the acceleration factor assumes a maximum value of 1.1 at 28°C and for all temperature above 40°C, the acceleration factor is below 1: increase in the module temperature above 40°C decreases the creep compliance.

Note that, in these calculations, we assumed a single external stress. Other stresses in the solder will affect the true creep compliance and its dependence on temperature. The effect of internal stresses is not considered here. The presence of voids in the solder is not considered either.

As adhesive bonding is used in optical assemblies, it is necessary to evaluate the creep strain in adhesive joints and to compare it with the data obtained with solder joints. Polymeric materials do not necessarily follow the creep strain–stress relationship defined in Equation (19.6). While the creep mechanism in solders is governed by grain size and atomic diffusion or dislocation phenomenon, the creep in polymeric materials is associated with a viscoelastic behavior where the free volume ratio has a determinant effect. The intermolecular space between long polymer chains (defining the free volume) allows for chain mobility over time in response to an imposed mechanical deformation. A decrease

---

[8] The extrapolation of the fiber displacement after 25 years is only given for design rule purposes as the behavior of the solder material will become nonlinear over time.

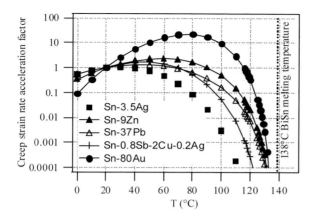

FIGURE 19.9. Steady-strain state acceleration factor as the function of the package temperature for Sn-3.5Ag, Sn-9Zn, Sn-37Pb, Sn-2Cu-0.8Sb-0.2Ag and Sn-80Au solders. The temperature dependence of the shear stress induced in the solder counteracts the Arrhenius term and strongly limits the usual adverse effect of the temperature on the creep compliances.

in the free volume yields a reduced chain mobility and a slower deformation under the imposed mechanical stress. In other words, the free volume directly impacts the time scale of a polymeric material strain–stress response. Moreover, the free volume and the T_g of a polymer are correlated: the larger the free volume, the smaller the T_g. Thus, the T_g and the viscoelastic properties of a polymer material are interrelated: the larger the T_g, the slower the viscoelastic deformation under a given mechanical load.

According to the Telcordia Generic Requirement GR1221,[9] adhesives used in optical assemblies should have a T_g above 95°C in order to limit joint deformations, i.e., components movements, under internal or external stress. However, published data on the creep strain of structural adhesive joints in optical assemblies are rather limited. Plitz et al. [46] have estimated the influence of post-curing aging on various mechanical properties, including CTE and creep compliances. Reith et al. [52] have evaluated the dimensional stability of adhesives in optical connectors ferrules and its influence on optical performance. They have shown that, although all the tested adhesives cannot restrain a permanent fiber pushback after a mechanical loading, fibers adhesively bonded into ferrules using high T_g adhesives ($T_g \sim 120°C$) exhibited a reduced fiber displacement of 10 nm after 400 hours testing. This displacement slightly increased to 15 nm when the test was performed at 65°C, however this pushback was not detrimental to the connector optical performances.

A rough estimation of the influence of the T_g on the creep compliance of optical adhesives can be performed using the data from [46]. They have evaluated the creep compliance at 40°C over time for two commercial UV adhesives having different T_g. After an initial time period where the creep-induced deformation steeply increases, both adhesives exhibited a steady-state creep compliance. The mechanical load applied on the bonded parts under investigation is small enough (0.3 MPa), so that a linear stress dependency model can be assumed. Assuming a similar assembly geometry as in the calculation of the creep strain in solders (Figure 19.7) and a similar strain–stress model than for solder materials, we can estimate the influence of the T_g on the creep displacement in an optical fiber bonded with an adhesive material. In order to account for the dependence of the creep properties on T_g,

---

[9]GR1221: generic requirements for passive optical components, R4-24, Issue 2 (1999).

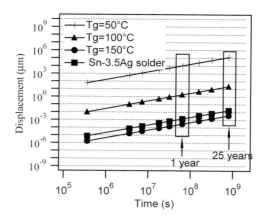

FIGURE 19.10. Adhesive joint displacement as a function of the adhesive glass transition temperature T_g. The results are given for an operating temperature of 40°C. In order to reach the performances of the Sn-3.5Ag material, the T_g of the adhesive should be above 150°C.

we assumed that the creep activation energy ΔH increased linearly with T_g and compared the displacement in the joint over time for adhesives having a $T_g = 50°C$, $T_g = 100°C$ and $T_g = 150°C$. The obtained results are shown in Figure 19.10. For comparison, the creep displacement in a Sn-3.5Ag solder at 40°C is also presented. We see that, in order to limit the displacement to an acceptable level (a fraction of a micron after 2 years), the T_g of the adhesive should be around 120–130°C. Again, the extrapolation of the fiber displacement after 25 years is only given as an illustration. It is expected that the behavior of the adhesive material becomes nonlinear over time.

19.2.2.3.2. Stress-Induced Birefringence. An adhesive or a solder joint can affect the optical properties of a bonded optical element through the build-up of a mechanical stress-induced birefringence in the optical component material. Whenever the polarization of the light propagating through the bonded element is critical, this stress should be controlled in order to obtain the desired state of propagation at the output of the element. For example, to achieve polarization-independence gain in semiconductor laser amplifier after flip-chip soldering, it has been shown that an additional bulk tensile strain needs to be preset into the semiconductor material when the component is soldered p-down on the substrate [19].

Other examples of mechanical stress-affected birefringence include:

- Optical fiber-based temperature sensor [25]. In this apparatus, the birefringence of the optical fiber, hence the state of polarization of the outcoming light, depends on the temperature-sensitive mechanical stress induced by the CTE mismatch between the glass fiber and an adhesive surrounding the fiber in a capillary tube. The sensitivity of the sensor depends on the CTE of the adhesive, the original birefringence of the fiber, as well as the length of the capillary tube.
- Reduction of birefringence in laser gyroscopes prisms using a soft indium solder as a bonding material [27]. The use of a soft indium solder reduces the optical birefringence at the basis of the prisms by a factor of 2–3 compared to an optical contact configuration.

19.3. ADHESIVE BONDING IN OPTICAL ASSEMBLIES

Common attach materials include solder alloy materials, adhesives and glass solders. Adhesive bonding provides advantages over other bonding techniques used in optoelectronic assemblies:

- ability to bond dissimilar materials,
- low processing temperature,
- refractive index-matching properties,
- no metalization required on the parts (as e.g., in the case of soldering or welding).

On the other hand, challenges in adhesive bonding include:

- limited operational temperature range,
- sensitivity to moisture,
- rework not possible,
- outgassing.

We will see in this paragraph that the main parameters that determine the performance of an adhesive bond are similar to the others bonding materials, i.e.,

- Surface preparation of the adherend.
- Physical and thermal properties of the adhesive (T_g, CTE, modulus).
- Joint design and shrinkage control.

We will not consider here electrically conductive adhesives. Their use and their properties have been described in [38].

19.3.1. Origin of Adhesion

Several theories of adhesion describe different types of adhesive bonding [28,31]: adsorption, chemical bonding, diffusion, mechanical interlocking, electrostatic bonding. However, no single model is able to explain a specific adhesive bond. The properties of an adhesive joint usually reflect the interplay of the effects described by various models. Nevertheless, it is clear that adherend–adhesive interface properties, in particular intermolecular forces at the interface, have a major contribution to the strength, as well as to the environment-induced degradation, of adhesive bonds.

19.3.1.1. Adsorption Adsorption is responsible for adhesion when intermolecular attractive forces, Van der Waals forces (electric dipole interactions) and/or hydrogen bonds [21], build up between the adhesive and the adherend. It is often believed that it is the most important adhesion mechanism and it has been experimentally demonstrated that the mechanism of adhesion in many adhesive joints only involves these interfacial secondary forces [5]. In most cases, adsorption is the relevant model when using adhesives in optical assemblies.

19.3.1.1.1. Wetting. According to this mechanism, the wetting of the adherend by the adhesive is a key factor in determining the bond strength. Wetting is defined as the tendency of a liquid to spread over a surface. The wetting process is controlled by three parameters: surface energy of the adherend γ_{SV} (solid–vapor interface), surface tension of the adhesive γ_{LV} (interface liquid–vapor) and the interfacial surface energy between the adherend and the adhesive γ_{SL}. The surface energy of a medium is defined as the free energy change when the surface area of this medium is increased by unit area. In other words, the surface

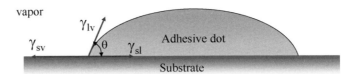

FIGURE 19.11. Adhesive drop resting at equilibrium on a solid surface. γ_{SV}, γ_{LV} and γ_{SL} are the surface energy of the adherend (solid–vapor interface), the surface tension of the adhesive (interface liquid–vapor) and the interfacial surface energy between the adherend and the adhesive. θ is the contact angle.

tension of the liquid γ_{LV} defines the energy required to create a new liquid surface area by moving molecules from the bulk liquid to the surface. The greater the surface tension, the greater is the energy needed to convert bulk molecules into surface molecules. Similarly, the surface energy of a solid γ_{SV} is equivalent to half the work of cohesion of this solid.

Young's equation [31] relates these three parameters at the three phase contacts to the equilibrium contact angle θ (Figure 19.11):

$$\gamma_{SV} = \gamma_{SL} + \gamma_{LV} \cos\theta. \tag{19.8}$$

When dispensing a liquid onto a substrate, the following equilibrium configurations can be reached:

- $\theta = 0°$: spontaneous spreading.
- $0° < \theta < 90°$: partial wetting.
- $90° < \theta < 180°$: partial non-wetting.
- $\theta = 180°$: negligible wetting.

For spontaneous spreading to occur, we need:

$$\gamma_{SV} > \gamma_{SL} + \gamma_{LV}. \tag{19.9}$$

By ignoring the interfacial free energy, Sharpe and Schornhorn [54] have proposed the following criteria:
For good wetting:

$$\gamma_{SV} > \gamma_{LV}. \tag{19.10}$$

For poor wetting:

$$\gamma_{SV} < \gamma_{LV}. \tag{19.11}$$

In other words, in order to wet the adherend, the surface tension of the adhesive must be lower than the surface energy of the adherend.

The surface tension of the adhesive is a given parameter in a dispensing process and typically ranges from 25 to 50 mN m^{-1}. It is not possible to change it without affecting the properties of the adhesive. Thus, in order to improve the adhesive wetting properties, it is advisable to look for an appropriate surface treatment to increase the surface energy of the substrate.

Table 19.6 presents the variation of the diameter of adhesive (acrylate) dots dispensed on a silicon substrate with the surface treatment. The raw silicon substrate exhibits a surface

TABLE 19.6.
Correlation between surface energy and dispensed dot size for various silicon surface preparations. The adhesive used in this experiment was an acrylate. The higher the surface energy of the silicon, the larger the adhesive dot due to improved wetting properties of the silicon surface.

Cleaning process	Surface energy (mJ m^{-2})	Dot size (μm)
Raw silicon substrate	42 ± 4	130 ± 10
2-propanol	52 ± 4	170 ± 10
O$_2$ plasma	>105	200

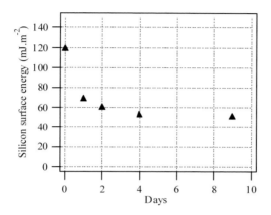

FIGURE 19.12. Evolution of the surface energy of an oxygen-plasma treated silicon substrate over time in ambient atmosphere.

energy as low as 42 ± 4 mJ m^{-2}. Two surface treatments were evaluated in these experiments: a wet cleaning with a 2-propanol solvent and a O$_2$ plasma etching. Using 2-propanol as a cleaning solution, the surface energy of the substrate and the diameter of the adhesive are increased by 25% and 40%, respectively. After an oxygen plasma treatment, the surface energy of the silicon substrate is higher than 105 mJ m^{-2}, and the diameters of the adhesive dot are increased by more than 50% compared to the uncleaned raw material. These results clearly demonstrate that the higher the surface energy of the silicon, the larger the adhesive dot due to improved wetting properties of the silicon surface.

Once properly treated, the substrates should be kept in a non-contaminating environment. Figure 19.12 shows the evolution of the surface energy of silicon substrates stored in ambient atmosphere. After two days, the substrate has lost half of its initial surface free energy through most probably the adsorption of the water molecules.

19.3.1.1.2. Kinetics of Wetting. The viscosity of a liquid arises from the intermolecular forces and steric-induced anchoring effects. The stronger the forces hindering the motion of the molecules, the higher the viscosity. Typical intermolecular forces include hydrogen bonding and dipole–dipole interactions. Hydrogen bonding accounts for the high viscosity of water compared to the aromatic benzene molecule, where there is no hydrogen bonding. Glycerol (C$_3$H$_7$O$_3$) as well is very viscous owing to the number of hydrogen bonds its molecule can form. Heavy hydrocarbon oils, which are not hydrogen bonded, are also

FIGURE 19.13. Description of the mechanical test assembly.

viscous. Their viscosity arises partially from the dipole interaction between molecules as well as steric effects (the long chainlike molecules get tangled to each other). On the other hand, the surface tension of a liquid is only determined by the intermolecular forces. These forces counteract the wetting of the liquid on a substrate. Steric-related effects have no influence on liquid surface tension.

There is no relationship between viscosity and surface tension of liquids. In other words, decreasing the viscosity will not decrease the contact angle nor affect adhesive dot size, but will speed up the kinetics of wetting [10].

The surface morphology of the substrate affects the kinetics of wetting as well. A liquid having a contact angle above 90° will spread along fine pores, scratches and other inhomogeneities by capillary action, even if it does not wet spontaneously a planar surface. It has been reported that random surface scratches can increase the spreading rate of liquids by as much as 50% [6]. This spreading is also observed when dispensing adhesive in V-grooves or U-grooves etched in silicon for passive alignment of optical fibers. The integration of larger sections along the main groove will prevent the adhesive flow down to the fiber tip owing to a decrease of the capillary pressure driving the adhesive flow and a reservoir-like functionality.

19.3.1.1.3. Effect of Surface Energies on Bond Strength in Optical Assemblies. The adhesive joint strength depends on the ability of the adhesive to spread spontaneously on the substrate [54]. Thus, the adhesive strength is reduced by the presence of contaminants, including hydrocarbons or moisture. The mechanical and environmental resistance of an adhesive joint will be improved through the prior application of dedicated surface treatments: solvent cleaning, wet chemical etching, plasma cleaning, UV radiation, silane adhesion promoter, ion-beam, laser surface treatment [31].

The adhesive bond strength can be assessed by a mechanical shear test. In order to evaluate the influence of adherend handling and surface preparation in passive optical components assemblies, stripped Corning SMF28 fibers have been bonded to silicon submounts and the joint strength has been measured through the application of a tensile stress on the free hanging part of the fiber until fracture. The tensile stress applied on the fiber translates into the shear stress in the adhesive. The test set-up is shown in Figure 19.13. To fulfill the Bellcore GR468,[10] such assemblies must be able to withstand 0.8 GPa tensile stress (120 kpsi or 1 kg load). The graph in Figure 19.14 shows three different failure cases of pull-tested adhesive-bonded fibers that are observed depending on adherend preparation:

- Adhesive failure of the adhesive joint on the silicon substrate [(Figure 19.15(a)]. Such failures occur for relatively low stress values (average value: 200 MPa tensile stress) when untreated silicon substrates are used.
- Fiber cohesive failure [(Figure 19.15(b)]. The applied tensile stress at failure is 500 MPa on average. This is typically observed when the silicon substrate has been

---

[10]GR468: generic reliability insurance requirements for optoelectronic devices used in telecommunications equipment, Bellcore, Issue 1, R4-40, 1998.

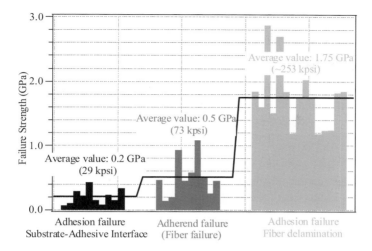

FIGURE 19.14. Tensile stress failure results on adhesive bonded fibers on silicon submounts. Results scattering is related to slight variations in adhesive dot size.

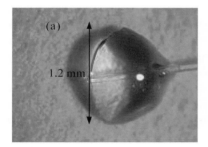 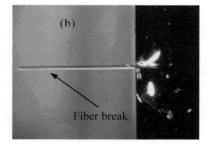

FIGURE 19.15. (a) Adhesive failure of the adhesive bond: the adhesive dot lifts off the silicon submount. (b) Fiber failure: the glass fiber broke at mid-span (here 1.5 mm away from the adhesive dot).

oxygen plasma treated and the stripping of the fiber has introduced micro-cracks at the bare fiber surface.
- Fiber delamination (adhesion failure at the adhesive–fiber interface). The applied tensile stress in this case reaches 1.75 GPa on average. This is typically observed with plasma treated silicon substrates and a proper fiber handling.

The Bellcore requirements are met only in the third case.

The low-stress adhesive failure case in Figure 19.15(a) is explained by a low surface energy value of the silicon submount (28 ± 5 mJ m^{-2}). This value is related to the presence of hydrocarbons or water on the surface of the substrate. After Oxygen plasma treatment of the substrates, surface energy of the silicon submount is above 105 mJ m^{-2} and subsequent pull test experiments with treated samples lead to failure cases 2 and 3 only (fiber cohesive failure or fiber delamination). Additional experiments showed that the transition between failure cases 1 and 2 or between 1 and 3 with 1 mm diameter dot occurs when the surface energy of the silicon submount is 48 ± 5 mJ m^{-2}. Fiber cohesive failure in Figure 19.15(b) is avoided with a proper handling of the fiber during and after stripping. In particular, we have seen that thermo-mechanical stripping compares favorably to standard mechanical stripping as micro-cracks at the surface of the glass appear using purely mechanical stripping techniques. The stripped length and the cleaning process must be carefully optimized to avoid subsequent mechanical contact on the bare fiber.

19.3.1.2. Mechanical Interlocking The mechanical interlocking model states that, when the surface of the adherend exhibits pores, holes and other irregularities, adhesive bonding can be enhanced through the mechanical interlocking of the adhesive and the adherend material. The adhesive should not only wet the substrate, but also have the right rheological properties to penetrate pores and openings. Since good adhesion can occur between smooth adherend surfaces as well, it is clear that while interlocking helps promote adhesion, it is not really a generally applicable adhesion mechanism. Pre-treatment techniques resulting in microroughness on the adherend surface can improve bond strength and durability [13]. Indeed, a larger contact area resulting from the roughening of the adherend surface contributes to the enhancement of the adhesive joint strength. Additionally, according to [11], the influence the bond strength depends not only on the contact angle and surface energies but also on the kinetics of wetting. The roughening of the substrate, speeding up the adhesive spreading, would then contribute to an increase of the bonding strength. However, this theory has not received a strong echo from adhesion scientists.

In most cases, the adhesive bonds found in the optical assemblies can be explained with the adsorption theory or the mechanical interlocking phenomenon or both. However, in order to explain all possible adhesive bond configurations, e.g., a plastic optical component on a plastic substrate, the models described in the following section may be necessary.

19.3.1.3. Other Models: Chemical Bonding, Diffusion Bonding and Electrostatic Bonding
Chemical bonding is responsible for adhesion when, in addition to an adsorption mechanism, there is a surface reaction, i.e., establishment of primary chemical bonds (covalent or ionic). Primary chemical forces have energies ranging between 60–1100 kJ/mol, which are considerably higher than the secondary bond energies have (0.08–5 kJ/mol) [31]. Chemical bonding will be the primary adhesion mechanism when silane-based adhesion promoters are used before application of the adhesive material. The chemical reactions occurring at the interface through the use of silane coupling agents have been reviewed by E. Plueddemann [49].

The diffusion bonding theory predicts a diffusion of molecules across the interface when the adherend and the adhesive have mutual solubility [63]. This theory may apply when both the adhesive and the adherend are polymers (e.g., when the optical element is bonded on a plastic substrate). The strength of the adhesive bond is related to the extent of the interdiffusion across the interface. The diffusion theory, however, is not justified where the adherend and adhesive are not soluble or when chain movement of the polymer materials is constrained by its highly crosslinked structure, or when it is below its glass transition temperature.

Electrostatic bonding is related to the formation of an electrical double layer of charges of opposite sign across the interface when the adherend and the adhesive have permanent electrical dipole moments or polar molecules. There are still some controversies around this theory because the electrical double layer cannot identified without separating the adhesive bond.

19.3.2. Adhesive Selection and Dispensing

19.3.2.1. Adhesives Detailed formulations of adhesives are usually proprietary information of the manufacturers and not available to the end-user. However, the active agent re-

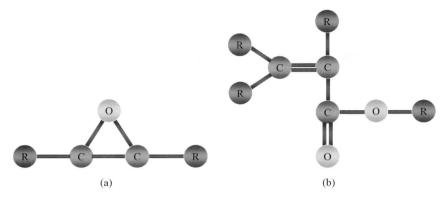

FIGURE 19.16. Chemical structures of (a) an epoxy functional group, (b) acrylate functional group.

sponsible for the chemical activity of the material is usually specified. In the fiber-optics industry, the adhesives are mostly based on either an *epoxy* or an *acrylate* group. The polymerization process is either light-based (UV or visible), heat-based or both.

The term "epoxy" refers to a chemical group consisting of an Oxygen atom bonded to two carbon atoms forming a ring structure. The simplest epoxy is a three-member ring structure known by the term "alpha-epoxy" or "1,2-epoxy" [Figure 19.6(a)]. *Thermal epoxies* are usually two component adhesives: the epoxy resin and the hardener (curing agent). The hardener, often an amine, is used to cure the epoxy by an addition reaction where two epoxy groups react with each amine site. This forms a complex three-dimensional molecular structure. Since the amine molecules co-react with the epoxy molecules in a fixed ratio, it is essential that the correct mix ratio is obtained between resin and hardener to ensure that a complete reaction takes place. If amine and epoxy are not mixed with the right stoichiometry, unreacted resin or hardener will remain within the matrix which will affect the final properties after cure. One component thermal epoxies are also available today. However, as they typically require 150°C curing temperature during one hour soaking time, they are cumbersome to use in low alignment tolerance optical assemblies.

The term acrylate refers to a chemical group consisting of a carbon–carbon double bond bonded to an ester functional group COOR [Figure 19.16(b)]. The resin base consists of a light molecular weight polymer (oligomer) having one or several acrylate functional groups. As opposed to epoxies, acrylates are polymerized with a catalyst rather than a hardener so that the curing proceeds as a chain reaction rather than an addition reaction. Typically, peroxides ROOR' are added to provide the resin with a source of free radicals. Upon curing, the peroxide in the resin base decomposes to yield free radicals RO. These radicals then initiate polymerization through the condensation of acrylate groups on the resin oligomers.

In UV curable adhesives, the polymerization proceeds in a chain mechanism involving cationic (epoxies) or free radical (acrylate) intermediates generated through the photolysis of a photoinitiator. The curing process of UV curable adhesives is fast, making them well suited for optical assemblies when the alignment between parts must be guaranteed during the polymerization process.

19.3.2.2. Critical Parameters When selecting an appropriate adhesive for an optical assembly application, the following parameters should be considered: T_g, CTE, propensity to creep, shrinkage, index-matching properties and photostability as well as moisture resis-

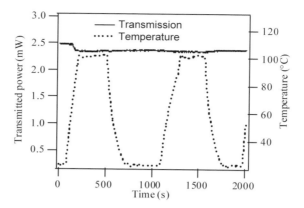

FIGURE 19.17. Thermal cycling test for a fiber-to-fiber assembly on a PEEK substrate. No change in coupling efficiency is observed up to 100°C.

tance and outgassing properties. In this section, we review the importance of each of these parameters.

19.3.2.2.1. CTE, Creep and Glass Transition Temperature. The glass transition temperature T_g is the temperature at which the material changes from a hard, glassy substance to a soft, rubbery one. In practice, this material transition occurs over a wide temperature range (up to 50°C) and only the central temperature is given on adhesive datasheets. CTE increases for $T > T_g$ whereas the Young's modulus and the hardness decrease for $T > T_g$.

To fulfill the Telcordia Generic Requirement GR1221, adhesives used in structural assemblies should have a T_g above 95°C measured by Differential Scanning Calorimetry (DSC). Indeed, the thermal expansion and the creep of adhesives, both T_g dependent,[11] can induce dimensional instabilities and ultimately misalignment between adhesive bonded optical devices.

The T_g of a UV adhesive is mainly defined by the adhesive temperature during curing: if the maximum temperature during curing is 60°C, the T_g of the UV-cured adhesive will be approximately 60°C. However, the T_g of this UV-cured adhesive can be further increased through, e.g., thermal post-curing. UV-curable adhesives were investigated with respect to thermal post-curing and accelerated aging [46]. Although the post-curing induced some shrinkage as well as some degradation, it was found that the probability that these devices exhibit dimensional instabilities was reduced due to an increase of the glass transition temperature. We have also performed thermal cycling tests on fiber-to-fiber assemblies. The results are displayed in Figure 19.17. The transmitted power from the incoming to the outcoming fiber is stable up to 100°C once the first temperature ramp-up is passed. This curve shows that a proper selection of the adhesive and a good bond design can lead to a stable optical coupling between room temperature and 100°C.

19.3.2.2.2. Shrinkage. The shrinkage of an adhesive joint is defined as the reduction of its linear dimensions and its volume during polymerization. Typically, the linear shrinkage of epoxies and acrylates is around 0.5% and 1.5% respectively. The shrinkage in optical assemblies results in a displacement between devices arising during the adhesive polymerization and a build-up of shear stresses in the adhesive joint leading to a creep-related misalignment over time.

---

[11] See Section 19.2.2.3.1 for the influence of T_g on the creep of adhesives.

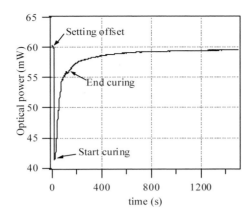

FIGURE 19.18. Shrinkage compensation using a fiber position offset.

The displacement occurring during polymerization can be corrected beforehand with an appropriate component offset. By this means, optimum alignment with minimized insertion losses is obtained after polymerization. Figure 19.18 presents the case of a laser to fiber pigtailing application where the adhesive bonded fiber was slightly offset from maximum optical coupling before starting the polymerization process. A UV epoxy was used in this experiment. The optical losses introduced by the offset are recovered during the epoxy polymerization. It is noteworthy that the shrinkage, or in other words the polymerization process, continues after switching off the UV light. Indeed, cationic species released during the UV curing have not reacted yet when the UV light is switched off but they remain available for further chain reactions.

In order to minimize shrinkage-related creep effects, it is necessary to limit the shrinkage-induced shear stress in the adhesive joint. Similarly to the CTE mismatch-induced thermo-mechanical stresses and assuming that the shrinkage-induced stress does not relax during the polymerization process, it is possible to introduce a maximum shrinkage-related shear stress σ_s:

$$\sigma_s = \frac{\Delta l}{l} G_D \frac{\tanh \beta L}{\beta t_D} = s G_D \frac{\tanh \beta L}{\beta t_D}, \qquad (19.12)$$

where β is defined in Equation (19.3) and s is the linear shrinkage (in %) of the adhesive upon polymerization.

As an example, we assume a chip length of 2400 μm and an adhesive thickness of 20 μm. The calculation of the maximum shrinkage-induced shear stresses in the bondline of this chip on submount assembly gives 126 MPa and 204 MPa in the epoxy and acrylate cases (see in Table 19.3 for adhesive material parameters). These stresses are not negligible compared to the CTE mismatch-induced stresses (e.g., 80 MPa for a GaAs chip on a silicon substrate upon soldering, see Section 19.2.2.2.1) and can readily influence the strength of the materials and the creep strain in adhesive bonded assemblies.

19.3.2.2.3. Outgassing. Outgassing of adhesive materials induce organic contamination on optically active parts, e.g., mirrors, lenses, fiber endfaces or laser facets and introduce a change in the response function of optical coating. The NASA has compiled outgassing data of adhesives intended for spacecraft applications. The method is based on the Total Mass Loss (TML) and Collected Volatile Condensable Materials (CVCM) measured in a

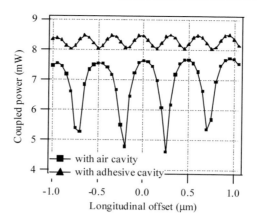

FIGURE 19.19. Fabry-Pérot interferences in the air cavity between a semiconductor laser and a fiber tip.

vacuum environment. Adhesive samples are heated to 398 K (125°C) for 24 hours. This causes the volatile materials to be driven out. The mass loss of the sample is determined from the weights before and after the 398 K exposure, and the percentage loss is calculated to provide the TML. The data can be used as a guide in selecting low-outgassing materials for optical applications.[12]

As a preliminary analysis, we verified that the facets of laser chips were not contaminated by the outgassing of the adhesive during the UV curing step and a subsequent burn-in procedure. Adhesive dots have been dispensed in front of the laser chips. The adhesive dots were UV cured and the assemblies were then submitted to a burn-in step. The analysis of the facet of the lasers was done using a Scanning Electron Microscope. None of the laser chips exhibited an organic contamination of the facets [48].

19.3.2.2.4. Refractive-Index Matching and Photostability. Adhesives usually have a refractive index between 1.4 and 1.6. Therefore, they can be used to reduce Fresnel reflection between optical devices, e.g., between a laser or a polymer waveguide and a glass optical fiber ($n = 1.46$ at a wavelength of 1.55 μm).

Figure 19.19 presents the variation of the coupled power in a laser-to-fiber coupling experiment as a function of the distance between the laser facet and the tip of the fiber. When the medium between the laser die and the optical fiber is air (squares data point), the reflection coefficient at the air–fiber interface and at the air–laser is as high as 4% and 13% respectively. Fabry-Perot interferences build up in this air cavity, generating coupled power variations as large as 35% when the longitudinal distance is varied. These oscillations limit the positioning tolerances of the optical fiber along the optical axis to less than 250 nm. When an adhesive is filling the gap between the laser die and the optical fiber, Fresnel reflections at the adhesive–optical fiber interface fall below 10^{-3} and the Fabry-Pérot interferences are strongly reduced (triangles data point). In this configuration, the influence of the thermal expansion of materials during device operation is not detrimental as far as the longitudinal displacement is concerned.

In addition to the minimization of Fresnel reflections, an adhesive, when used as a filling material between an optical component and an optical fiber, influences the coupling efficiencies and alignment tolerances due to improved mode matching. Figure 19.20 shows

---

[12] http://outgassing.nasa.gov.

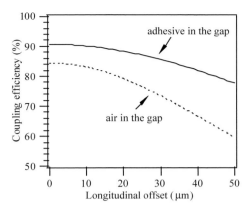

FIGURE 19.20. Comparison of the coupling efficiencies of a single-mode fiber to a 5 × 5 μm polymer channel waveguide with and without adhesive in the gap obtained in a end-fire coupling configuration.

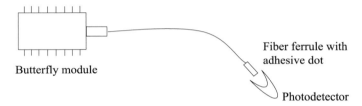

FIGURE 19.21. Schematic representation of the photostability measurement set-up. The light from the module passes through the adhesive dot on the fiber ferrule tip and is detected by the photodetector.

simulations of the optical coupling of a single-mode fiber to 5 × 5 μm channel waveguides fabricated on a silicon wafer from a fluorinated acrylate polymer with refractive indices of core and cladding of 1.47 and 1.46, respectively. We clearly see that the coupling efficiencies are larger when the gap between the device and the optical fiber is filled with an adhesive material: 91% vs 85% at a longitudinal distance of 1 μm and 77% vs 60% at a longitudinal distance of 50 μm. Therefore, if the photochemical and photomechanical stability is guaranteed, filling the gap between the waveguide and the fiber with an adhesive can be very favorable.

In order to assess the reliability of the gap filling approach, we have performed photostability experiments for various adhesives at 980 nm and under high intensity conditions (from 600 to 900 kW cm^{-2}):

- The organic material to be evaluated was directly deposited on the ferrule tip of a pigtailed module.
- The material is cured according to the polymerization conditions given by the manufacturer.

The module is powered on and light passes through the organic material dot. The transmitted light is detected with a photodetector and recorded versus time (Figure 19.21).

The results of the photostability experiments are summarized in Figure 19.22. After 1000 hours, the optical transmission is still above 0.98 for both adhesives A and B. Similar results were reproduced on 2 other samples for both adhesives. We can then conclude that the adhesives A and B exhibit good photostability properties at 980 nm at intensity levels

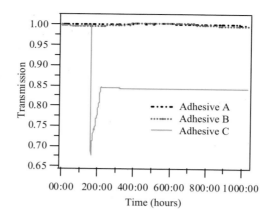

FIGURE 19.22. Evolution of the transmitted optical power over time for three different adhesives. The results are normalized relative to the value of the transmitted power at $t = 0$.

in the range of 600–900 kW cm^2 and could be used as gap filling material under similar stress conditions for low-end products. It is clear however that, in order to evaluate the risk of adhesive-related catastrophic optical damage on the laser chip itself, extended tests with the adhesive between the laser facet and the optical fiber should be performed.

The evolution of the optical transmission with adhesive C presents a steep decrease after 160 hours. The transmission value stabilizes then around 0.85 for the remaining lifetest time. A visual inspection of the adhesive dot at ferrule tip showed that this adhesive material presented cracks and "bubbles." As this behavior was confirmed on 3 other samples, we conclude that adhesive C cannot be used as an index matching material or as a fiber fixing material in 980 nm high power lasers applications.

It is clear that similar test experiments should be reconducted if the operating wavelength is different or if the intensity is higher.

19.3.2.2.5. Effect of Roughness on Moisture Resistance in Optical Assemblies. Moisture absorption is measured in terms of the percentage weight increase of the material caused by water absorption when placed under water or in a highly humid atmosphere for a given period of time. The effect of the adhesive moisture intake is twofold:

- decrease of the adhesive bonding strength,
- release of trapped moisture in sealed packages.

The danger of moisture absorption is significant in hermetically sealed packages when the moisture trapped in the adhesive is released during device operation, building a corroding atmosphere in the package. A corroding atmosphere is detrimental for the material strength of optical fibers [24,39] and electrical interconnects (wires bonds and metallic pads).

Ingress of moisture into the bondline (between the adhesive and the adherend) is the main source of moisture-related adhesive strength degradation and adhesive failure. In order to improve the moisture resistance of the adhesive bond, the properties of the adherend surface should be carefully investigated. In particular, adherend surfaces providing a high density of physical bonds (mechanical interlocks) show a better moisture resistance than smooth surfaces. Figure 19.23(a) and (b) shows the cases of two adhesive dots placed during 24 hours in deionized water. In both cases, the adhesive has been dispensed on silicon. The major difference between both adherends is their surface profile: polished smooth silicon surface in Figure 19.23(a) and non-polished silicon surface in Figure 19.23(b). The

ADHESIVE BONDING OF PASSIVE OPTICAL COMPONENTS

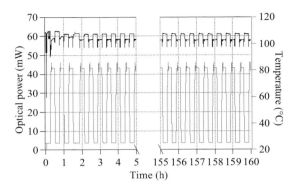

FIGURE 19.27. Thermal cycling tests performed during one week. The assembly went through more than 600 thermal cycles without degradation of the coupling efficiency over time. The upper and bottom lines represent the coupled optical power variation and the temperature cycles respectively. The variation of the coupling efficiency between 25°C and 80°C is 1.5%, resulting from the variation of the Fabry-Pérot air cavity length between the laser facet and the fiber tip as the temperature is increased.

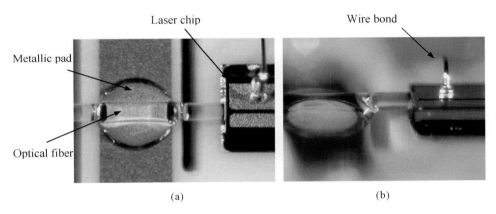

FIGURE 19.28. (a) Top view of the laser-to-fiber assembly. (b) Side-view of the laser-to-fiber assembly. The fiber is embedded in the adhesive dot owing to a planar assembly design.

Once the fiber was properly aligned, a position offset was introduced to compensate for the adhesive shrinkage upon curing (see Section 19.3.2.2.2). Once cured, the assembly went through a stability bake for a better thermal resistance. Figure 19.27 shows the thermal cycling results between 25°C and 80°C over one week (160 hours, more than 600 cycles). No degradation over time is observed during the thermal cycling tests. The variation of the coupling efficiency is 1.5% between 25°C and 80°C and mainly results from the variation of the distance between the fiber tip[13] and the laser endface as the temperature is increased, i.e., a variation of the Fabry-Perot cavity as described in Section 19.3.2.2.4. Indeed, due to the different CTE involved in this assembly, we estimate a cavity length increase of 200 nm. This represents a variation in coupling efficiency of 1% as, over this distance variation, the coupled power goes from a maximum to a minimum of the Fabry-Pérot interferences.

On the top view of the fiber-to laser assembly [Figure 19.28(a)], we see that the flow of the adhesive is well under control as the adhesive joint is limited to the metallic pad.

---

[13]There is no anti-reflection coating on the fiber tip.

On the side-view of Figure 19.28(b), the fiber is embedded in the adhesive dot owing to a planar chip on carrier assembly configuration.

19.4.2. Planar Lightwave Circuit (PLC) Pigtailing

In a second example, a PLC pigtailing process is described. The PLC's used are 5×5 μm channel waveguides fabricated on a silicon wafer from a fluorinated acrylate polymer with refractive indices of core and cladding of 1.47 and 1.46, respectively.

We first investigated the coupling efficiencies and alignment tolerances that can be achieved with different coupling methods (with/without lenses, with/without adhesive in the gap between waveguide and fiber). The simulation of coupling efficiencies showed that a coupling scheme using lenses gives the largest coupling efficiencies (95%) but tight alignment tolerances (e.g., a lateral fiber shift of 0.85 μm already leads to a decrease in coupling efficiency of 10% this in comparison to 1.4 μm for butt-coupling). Therefore, in order to ease the alignment, an end-fire coupling configuration with slightly lower coupling efficiencies was chosen.

The PLC pigtailing consisted of two main steps:

- The integrated optical polymer chip was first fixed on a submount with a thermal adhesive.
- Subsequently, the fibers were aligned for maximum throughput and finally fixed using a UV-curable adhesive.

In addition to the requirements given in Section 19.2.2.2, the submount should have a high operating temperature as the polymer chip is bonded with a thermal adhesive. Thus, the polymer PEEK (polyether ether ketone) was chosen. The chip bonding step proceeds as follows: three adhesive droplets were dispensed on the submount, the chip was placed onto the submount, and the assembly was put in an oven at a temperature of 120°C for 2 hours.

As the waveguide chip is fully passive, some 'external' light has to be first coupled into the waveguide in order to perform the fiber alignment. The fibers pigtailing setup is described in Figure 19.29. The chip holder was placed between two 3-axis fiber positioning stages having 0.1 μm resolution. The time-pressure adhesive dispenser was mounted on a 3-axis positioning stage for pinpoint adhesive dispensing. A 1550 nm external laser source and detector were used for the active alignment process of the fibers.

Some external light was coupled in the waveguide using the *first* fiber to provide an optical signal for the active alignment of the *second* fiber. The coarse fiber to waveguide in-plane alignment was done using a microscope. The *first* fiber alignment was further improved using an infrared card at the output of the waveguide. Once first light was observed at the output of the waveguide, the *second* fiber could be aligned for maximum throughput.

In order to ease the coarse alignment step of the *second* fiber, the alignment was first performed 100–300 μm away from the waveguide. This procedure was carried out as the laser beam was several tens of microns wide a few hundreds microns away from the waveguide facet due to the waveguide divergence. Once the fiber was properly positioned close to the waveguide optical axis, the fiber was brought 3 to 5 microns away from the waveguide facet and the fiber position was newly optimized. The *first* fiber was now realigned using the reference signal coupled in the waveguide from the *second* fiber in order to measure the maximum achievable optical throughput.

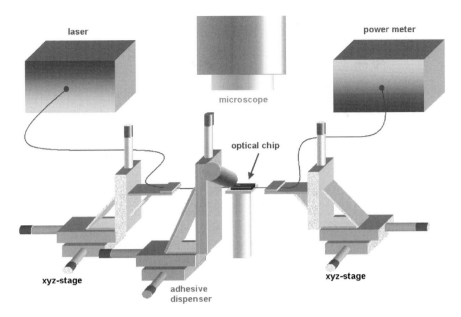

FIGURE 19.29. Setup used for the fiber pigtailing of PLC waveguides.

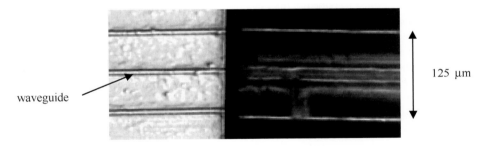

FIGURE 19.30. Image of waveguide (left) and single-mode fiber (right) prior to alignment. The core dimensions are 5×5 μm for the waveguide and 9 μm (diameter) for the fiber.

Figure 19.30 shows an image of the waveguide and the fiber after alignment. Figure 19.31 shows the nice comparison of the coupling efficiencies between simulations and actual experiments carried out before fixing.

Subsequently, the fibers were fixed with a UV-curable adhesive. The *second* fiber was moved away in longitudinal direction and the adhesive was dispensed on the submount. After this dispensing step, the fiber was moved back to its initial position and actively aligned. Finally the adhesive was UV cured. Using this approach, the adhesive also fills the gap between waveguide and fiber which leads to a decrease in the Fresnel losses according to the results of Figure 19.20 and an increase in the alignment tolerances.

The same procedure was applied to the *first* fiber. The current insertion losses (fiber-waveguide-fiber) after fixing were around 4.5 dB compared to a minimum expected loss of around 2.5 dB for the current configuration.

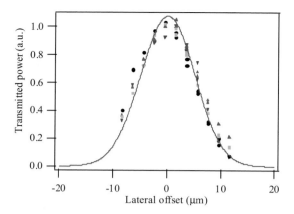

FIGURE 19.31. Influence of a lateral (horizontal) shift on the fiber-coupled power: comparison between simulations (solid line) and experiments (points) displaying a nice correspondence. The lateral alignment tolerance for 10% additional losses is 2.3 μm.

19.5. SUMMARY AND RECOMMENDATIONS

A large variety of active and passive optical devices are currently proposed to supply the telecom and datacom market. The devices are typically fitted into ceramic or metal alloy housings that must provide not only electrical connections and mechanical support but also thermal management and, more critically, optical connections. Optical interconnects are highly directional and, in order to achieve optimal signal transmission, they require extremely precise control of the relative location between components (typ. below 1 μm) and an attachment process that maintains the alignment over time.

Adhesive bonding of optical passive components is often seen as a cheap alternative to laser welding or gold-tin soldering. However, today, there is a mental barrier in using adhesives in optoelectronic packaging. In order to clarify the potential of such bonding approach, extensive reliability tests (thermo-mechanical stability, photostability) must be performed. In order to maximize the adhesion energy and to improve the reliability of adhesive-bonded assemblies, the parts must be free of contaminants and their surface energies should be above 100 mJ m^{-2}. Oxygen plasma cleaning is very effective in removing organic residues and should be done whenever possible. Optical fibers adhesively bonded on silicon substrates have shown to resist pulling load in excess of 2 kg when the silicon surface was oxygen plasma treated.

We have seen that the use of a compliant adhesive material strongly reduces CTE mismatch induced thermal stresses compared with typical solder alloy used in microelectronics and opto-electronics packaging. Provided that internal or external stresses remain at a low level during the entire assembly process, we have estimated that the creep-induced displacement can be limited to values below 1 μm using high T_g adhesives ($T_g > 120°C$). The T_g of a UV adhesive is mainly limited to the actual internal temperature during curing. It is very likely that this temperature does not exceed the required value given in the Telcordia GR1221. Thus, in order to increase the T_g and limit the creep compliance, UV cured adhesives must be thermally post-cured.

The inherent shrinkage of the adhesives during curing can be pre-compensated by applying a component offset before the adhesive polymerization. This offset is then recovered during the curing of the adhesive. The shrinkage is not a problem by itself as long as it is reproducible.

Finally, the refractive index-matching properties of adhesive materials are a unique advantage of adhesive bonding over other fixing technologies including soldering or laser welding. Providing the adhesive is photostable under given stress conditions, the use of an adhesive in the gap between an optical chip and a fiber can improve the optical coupling efficiency and increase positioning tolerances.

ACKNOWLEDGMENTS

The authors would like to thank Jens Kunde, Helmut Knapp and Rainer Bättig for stimulating discussions and valuable suggestions. The authors would also like to acknowledge Yanki Keles for the AFM pictures, Nicolaï Matuschek for the thermal resistivity calculations as well as Sylvain Grossman and Gustavo Aeppli for their technical assistance. Fundings for this work was provided by the Commission pour la Technologie et l'Innovation and by the Cantons of Central Switzerland. The company Bookham (Switzerland) AG provided the laser chips and the modules used in some of the experiments presented in this article. The authors are grateful for their material support.

REFERENCES

1. M. Achouche, V. Magnin, J. Harari, J.-L. Gentner, F. Lelarge, E. Derouin, C. Jany, D. Carpentier, F. Barthe, F. Blache, S. Demiguel, and D. Decoster, New high performance evanescent coupled waveguide UTC photodiodes for >40 Gb/s applications, Proceedings ECOC, Th 3.4.1, Rimini, Italy, 2003.
2. Y. Akasaka, Gain bandwith of optical amplifiers over 100 nm and beyond, Proceedings ECOC, Tu 3.7.1, Rimini, Italy, 2003.
3. T. Akashi, S. Higashiyama, H. Takemori, and T. Koizmi, A silicon optical bench incorporating a tantalum-nitride thin-film resistor, J. Micromech. and Microeng., 14, pp. 283–289 (2004).
4. M.C. Amann, Long-wavelength InP-based VCSELs, Proceedings ECOC, Tu 1.5.1, Rimini, Italy, 2003.
5. E.H. Andrews and A.J. Kinloch, Mechanics of elastomeric adhesion, J. Polymer Sci., 46, pp. 1–14 (1974).
6. W.D. Bascom, R.L. Cottington, and C.R. Singleterry, Dynamic surface phenomena in the spontaneous spreading of oils on solids, in R.F. Gould, Ed., Advances in Chemistry Series, Vol. 43, American Chemical Society, Washington, 1964, pp. 355–379.
7. R. Bättig, H.-U. Pfeiffer, N. Matuschek, and B. Valk, Reliability issues in pump laser packaging, Proc. Lasers and Electro-Optics Society Conference, WW1, Glasgow, 2002.
8. R. Brenot, S. Kerboeuf, N. Bouché, F. Mallecot, V. Colson, O. Gauthier-Lafaye, M. Picq, J.G. Provost, and B. Thédrez, New low-chirp and high power semiconductor amplifier for 10 Gbit/s metropolitan transmission, Proceedings ECOC, Th 3.5.3, Rimini, Italy, 2003.
9. B. Chandran, W.F. Schmidt, M.H. Gordon, and R. Djkaria, The determination and utilization of AuSn solder properties to bond GaAs dice to diamond substrates, ASME Proceedings of the Application of CAE/CAD to electronics systems congress, 1996, pp. 61–66.
10. B.W. Cherry and C.M. Holmes, Kinetics of wetting of surfaces by polymers, J. Colloïd Interf. Sci., 38, p. 174 (1969).
11. B.W. Cherry and S.E. Mudaris, Wetting kinetics and the strength of adhesive joints, J. Adhesion, 2, pp. 42–49 (1970).
12. W.T. Chen and C.W. Nelson, Thermal stresses in bonded joints, IBM J. Research and Development, 23, pp. 179–188 (1979).
13. H.M. Clearfield, D.K. McNamara, and G.D. Davis, Adherend surface preparation for adhesive bonding, in L.-H. Lee, Ed., Adhesive Bonding, Plenum Press, 1991.
14. A. Codourey, A.-C. Pliska, C. Bosshard, B. Sprenger, U. Gubler, A. Steinecker, M. Thurner, and M. Honegger, A robotic system for assembly and packaging of micro-optoelectronic Components, DTIP Conference, Montreux, 2004.

15. L.A. Coldren, Integrated tunable transmitters for WDM networks, Proceedings ECOC, Th 1.2.1, Rimini, Italy, 2003.
16. K.A. Cooper, R. Yang, J.S. Mottet, and G. Lecarpentier, Flip chip equipment for high end electro-optical modules, Proc. 48th ECTC, Seattle, 1998, pp. 176–181.
17. N. El Dahda, A. Shen, F. Devaux, G. Aubin, J.C. Harmand, A. Garreau, B.-E. Benkelfat, and A. Ramdane, Novel InGaAs/InGaAlAs MQW electroabsorption modulator for ultra-fast optical signal processing, Proceedings ECOC, We 2.5.3, Rimini, Italy, 2003.
18. L.R. Dalton, Novel polymer-based, high-speed electro-optic devices, Proceedings ECOC, Tu 3.5.1, Rimini, Italy, 2003.
19. F. Dorgeuille, S. Rabaron, F. Pommereau, C. Artigue, and P. Brosson, Optical amplifier device, United States Patent Application 20020154392 (2002).
20. J. Elwenspoek and H. Jansen, Silicon Micromachining, Cambridge University Press, 1998.
21. F.M. Fowkes, Role of acid-base interfacial bonding in adhesion, J. of Adhesion Science and Technology, 1, pp. 7–27 (1987).
22. K. Fukatsu, T. Takeuchi, K. Shiba, K. Makita, Y. Amamiya, Y. Susuki, and T. Kato, An extremely compact (0.3 cc) 40 Gb/s optical receiver module with ease of use receptacle interface and feedthrough launcher, Proceedings ECOC, Th 3.4.3, Rimini, Italy, 2003.
23. L. Guiziou, P. Ferm, J.M. Jouanno, and L. Shacklette, Low-loss extinction ratio 4×4 polymer thermo-optical switch, Proceedings ECOC, Paper TuL1.4, Amsterdam, 2001.
24. N. Gougeon, M. Poulain, and R.L. Abdi, Strength of silica fibers under various moisture conditions, Photonics Fabrication Europe Conference, Bruges, Proceedings SPIE, 4940 (2002).
25. C. Helming and J. Teunissen, Optical low-cost temperature point sensor, Proceedings SPIE, 4074 (2000).
26. B. Hvolbaek Larsen, L. Pleth Nielsen, K. Zenth, L. Leick, C. Laurent-Lund, L.-U. Aaen Andersen, and K.E. Mattsson, A low-loss silicon oxynitride process for compact optical devices, Proceedings ECOC, We 1.2.6, Rimini, Italy, 2003.
27. V.O. Indisov, V.N. Kuryatov, B.N. Semenov, I.M. Sokolov, and Ya.A. Fofanov, Polarization characteristics of total internal reflection lasers prisms, Part 1, Optics and Spectroscopy, 75, pp. 121–127 (1993). Polarization characteristics of total internal reflection lasers prisms, Part 2, Optics and Spectroscopy, 75, pp. 266–271 (1993).
28. J.N. Israelachvili, Intermolecular and Surface Forces, Academic Press, 1991.
29. R.J Jenkins, M.E. McNie, A.F. Blockley, N. Price, and J. McQuillan, Hollow waveguides for integrated optics, Proceedings ECOC, Tu 1.2.4, Rimini, Italy, 2003.
30. N. Keil, H. Yao, C. Zawadzki, O. Radmer, F. Beyer, M. Bauer, C. Dreyer, and J. Schneider, Polarization and temperature behavior of all-polymer arrayed-waveguide gratings, Proceedings ECOC, Tu 3.5.3, Rimini, Italy, 2003.
31. A.J. Kinloch, Adhesion and Adhesives, Science and Technology, Chapman and Hall, 1987.
32. J. Kunde, M. Thurner, A.-C. Pliska, Ch. Bosshard, A. Codourey, R. Bauknecht, and S. Egger, Comparison of microlens technologies for passive alignment platform applications, Proceedings MOC, V1.6, 2004.
33. J.-H. Kuang, M.-T. Sheen, C.-F. Chang, C.-C. Chen, G.-L. Wang, and W.-Hi. Cheng, Effect of temperature cycling on joint strength of PbSn and AuSn solders in lasers packages, IEEE Trans. Adv. Packaging, 24, pp. 563–568 (2001).
34. J. Lau, C.P. Wong, J.L. Prince, and W. Nakayama, Electronic Packaging: Design, Materials, Process and Reliability, McGraw-Hill, 1998.
35. M. Lee, M. Wong, and Y. Zohar, Characterization of an integrated micro heat pipe, J. Micromech. and Microeng., 13, pp. 58–64 (2003).
36. A. Liu, R. Jones, L. Liao, D. Samara-Rubio, D. Rubin, O. Cohen, R. Nicolaescu, and M. Panaccia, A high-speed silicon optical modulator based on a metal-oxide-semiconductor capacitor, Nature, 427, pp. 615–618 (2004).
37. T.J. Lu, D.F. Moore, and M.H. Chia, Mechanics of micromechanical clips for optical fibers, J. Micromech. and Microeng., 12, pp. 168–176 (2002).
38. M.A. Lyons and D. Dahringer, Electrically conductive adhesives, in A. Pizza and K.L. Mittal, Eds., Handbook of Adhesive Technology, 2nd edn, 2003.
39. M. Mattewson, Environmental effect on fatigue and lifetime prediction for silica optical fibers, Photonics Fabrication Europe Conference, Bruges, Proceedings SPIE, 4940 (2002).
40. M.A. Meyers, R.W. Armstrong, and H.O.K. Kirchner, Mechanics and Materials: Fundamentals and Linkages, John Wiley & Sons, New York, 1999.
41. L.F. Miller, Controlled collapse reflow chip joining, IBM J. Research and Development, 13, pp. 239–250 (1969).

42. R.A. Modavis and T.W. Webb, Anamorphic microlens for laser diode to single mode fiber coupling, IEEE Photon. Tech. Lett., 7, pp. 798–800 (1995).
43. S. Mohrdiek, T. Pliska, R. Bättig, N. Matuschek, B. Valk, J. Troger, P. Mauron, B.E. Schmidt, I.D. Jung, C.S. Harder, and S. Enochs, 400 mW uncooled MiniDIL pump modules, Elec. Let., 39, pp. 1105–1107 (2003).
44. H.S. Morgan, Thermal stresses in layered electrical assemblies bonded with solder, J. Elec. Packaging, 113, pp. 350–354 (1991).
45. A. Norris and J. DeGroot, Silicone materials for optical device applications, Proceedings ECOC, Tu 3.5.6, Rimini, Italy, 2003.
46. I.M. Plitz, O.S. Gebizlioglu, and M.P. Dugan, Reliability characterization of UV-curable adhesives used in optical devices, Proceedings SPIE, 2290, pp. 150–159 (1994).
47. A.C. Pliska, J. Kunde, S. Grossmann, C. Bosshard, T. Pliska, and B. Valk, Automated fiber alignment using machine vision and pattern recognition, Technology Leadership Day, Winterthur, 2003.
48. A.-C. Pliska, J. Kunde, S. Grossmann, Ch. Bosshard, R. Bättig, S. Pawlik, T. Pliska, S. Saintenoy, and B. Schmidt, Low-cost optoelectronic packages: development of a fast alignment technique and a stable bonding process of singlemode optical fibers, EMPC Conference, Brugges, 2005.
49. E.P. Plueddemann, Adhesion through silane coupling agents, in L.-H. Lee, Ed., Fundamentals of Adhesion, Plenum Press, 1991, pp. 279–290.
50. M. Pocha, O.T. Strand, and J.A. Kerns, A silicon microbench concept for optoelectronic packaging, Proc. Surface Mount Int., 1, pp. 377–382 (1996).
51. C.F. Popelar and K.M. Liechti, A distorsion-modified free volume theory for nonlinear viscoelastic behavior, Mechanics of Time Dependant Materials, 7, pp. 89–141 (2003).
52. L.A. Reith, O.S. Gebizlioglu, M. Koza, J. Mann, M. Ozgur, and T. Bowner, The dimensional stability of adhesives, zirconia and silica in optical connector ferrules and their impact on optical performance, Proceedings of Mat. Res. Soc. Symposium, 1998, pp. 65–76.
53. M. Saruwatari and K. Nawata, Semiconductor laser to single-mode fiber coupler, Appl. Opt., 18, pp. 1847–1856 (1979).
54. L.H Sharpe and H. Schonhorn, Advances in Chemistry Series, Vol. 8, American Chemical Society, Washington, 1964, p. 189.
55. C. Strandman and Y. Bäcklund, Passive and fixed alignment of devices using flexible silicon elements formed by selective etching, J. Micromech. and Microeng., 8, pp. 39–44 (1998).
56. A. Stump, U. Gubler, and C. Bosshard, Polymer optical waveguides structured by UV-exposure, EOS Topical Meeting, Engelberg, 2004.
57. E. Suhir, Calculated thermally induced stresses in adhesively bonded and soldered assemblies, Proc. Int. Microelectronics Symposium, Atlanta, 1986.
58. S. Tabata, T. Saito, K. Kawamura, T. Itoh, and T. Hatta, 32×32 bascule optical switch with polymer waveguide, Proceedings ECOC, Tu 3.5.2, Rimini, Italy, 2003.
59. T.G. Tessier, G. Ademon, and I. Turlik, Polymer dielectric options for thin film packaging applications, IEEE Proc. of 39th Elec. Comp. and Tech. Conference, 1989, pp. 127–132.
60. K. Tsuzuki, T. Ishibashi, T. Ito, S. Oku, Y. Shibata, R. Iga, Y. Kondo, and Y. Tohmori, A 40 Gbit/s InP-based n-i-n Mach-Zehnder modulator with a π-voltage of 2.2 V, Proceedings ECOC, We 2.5.2, Rimini, Italy, 2003.
61. M. Uekawa, H. Sasaki, D. Shimura, K. Kotani, Y. Maeno, and T. Takamori, Surface-mountable silicon microlens for low-cost laser modules, IEEE Photon. Tech. Lett., 15, pp. 945–947 (2003).
62. B. Valk, P. Müller, and R. Bättig, Fiber attachment for 980 pump lasers, Proc. Lasers and Electro-Optics Society Conference, ThV2, 2000, pp. 880–881.
63. S.S. Voyutski, Autohesion and Adhesion of High Polymers, Interscience, New York (1963).
64. E. Zielinski and H.P. Mayer, Optohybrids in high capacity communication systems, Symp. Opto-Microelectronics Devices and Circuits, Stuttgart, 2002.
65. W.H. Wong, E.Y.B. Pun, and K.S. Chan, Rare-earth doped polymer waveguide amplifiers, Proceedings ECOC, Th 4.2.7, Rimini, Italy, 2003.

20

Electrically Conductive Adhesives: A Research Status Review

James E. Morris[a] and Johan Liu[b,c]

[a] *Department of Electrical and Computer Engineering, Portland State University, Oregon, USA*
[b] *SMIT Center and Department of Microtechnology and Nanoscience, Chalmers University of Technology, Goteborg, Sweden*
[c] *SMIT Center, Shanghai University, China*

20.1. INTRODUCTION

20.1.1. Technology Drivers

There are two primary categories of Electrically Conductive Adhesive (ECA):

- Isotropic Conductive Adhesive (ICA).
- Anisotropic Conductive Adhesive (ACA).
- ACAs are available as paste (ACP) or film (ACF).

Both types conduct through metal filler particles in an adhesive polymer matrix.
ECAs have been used for electronics packaging applications for decades in hybrid, die-attach and display assembly. ICAs have been used extensively for die attach, and in automotive electronics. ACF technology is employed with almost every liquid crystal display. But there has been growing interest from the electronics industry over the past decade or so in other kinds of electronics packaging applications. While toxicity issues and environmental incompatibility of the lead in tin-lead solders triggered that greater interest at the outset, it has been the other evident advantages which continue to drive further research. ECAs can offer the following additional potential advantages:

- Fine-pitch capability, especially when using ACAs for flip-chip.
- Elimination of underfilling with ACA bonding.
- Low temperature processing capability.
- Flexible, simple processing and hence low cost.

In addition, solder failures due to the formation of voids or brittle intermetallics are controlled by diffusion rates at small dimensions, and time-to-failure scales as the square of the dimensions, limiting solder reliability lifetimes at ultra-fine pitch [1].
ACF has long been the interconnect of choice in the LCD display industry, and ACP is now finding applications in flex circuits and surface mount technology (SMT) for chip-

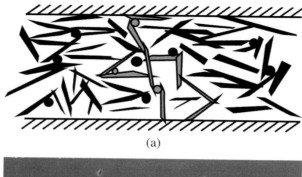

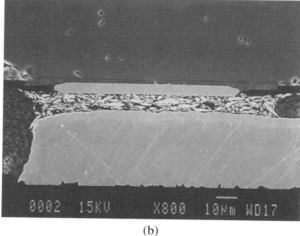

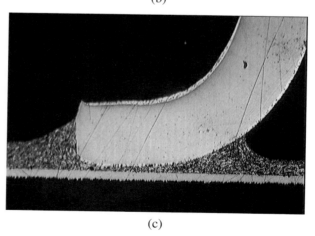

FIGURE 20.1. ICA contact joints: (a) schematic, (b) flip-chip on FR-4, (c) SMT on FR-4.

scale package (CSP), ASIC, and flip chip attachment for cell phones, radios, PDAs, laptop PCs, and cameras. ICAs are used extensively in die-attach, for many years for small passive chip attachment in automotive electronics, and in varied consumer products, e.g., in Matsushita/Panasonic's stud bump joints. More recent applications include RFID tags, potentially for the antenna as well as chip connections.

This chapter presents an overview of the current status of understanding of conductive adhesives in various electronics packaging applications, and of some fundamental issues relevant to their continuing development. It is organized with initial discussions of basic ECA concepts of structure-related properties, and how these are affected by material selection and processing, followed by general properties and reliability considerations. In each section, there is material common to both ICAs and ACAs, and parallel treatments of topics specific to each.

20.1.2. Isotropic Conductive Adhesives (ICAs)

Ag is usually used as the filler material due to its high conductivity and simple processing for ICA applications. Figure 20.1 shows the microstructure of an ICA joint, (a) in schematic form, (b) for a flip chip component, and (c) for SMT attachment (both on FR-4 substrates). The metallic filler content is high enough (25–30 volume percent) to cause direct metallic contact from bump or lead to circuit board pad.

20.1.3. Anisotropic Conductive Adhesives (ACAs)

Polymer based metal-plated spheres or Ni fillers are mainly used for ACA applications. In an ACA joint, the filler particles normally constitute between 5–10 volume percent, and do not cause any direct metallic contact. It is only after the application of pressure during curing that electrical conduction becomes possible in the normal (z-axis) direction, as is illustrated in Figure 20.2. The Ni particles are usually matched to soft Au contacts, and the contact deforms to define the contact area. By contrast, the polymer beads, usually Au-plated, deform to establish bump and pad contacts. As there is no direct contact between the particles, ACA technology is very suitable for fine pitch assembly, and is starting to find applications in flip-chip technology.

20.1.4. Non-Conductive Adhesive (NCA)

The complementary NCA technology is mentioned only briefly here, for completeness. In this case, the contacts themselves are in direct metal-to-metal contact, held together by the NCA. The concept relies upon surface asperities for the direct contacts.

20.2. STRUCTURE

20.2.1. ICA

As the proportion of metal in the ICA polymer matrix is increased (Figure 20.3), the resistance drops only slightly until the "percolation threshold" is reached, when the first continuous metal path is established through the composite material [2]. The resistance continues to drop more slowly as multiple parallel paths are developed with the continued addition of more metal filler. Ideally one would like to use the minimum quantity of filler necessary to pass the threshold, but in practice manufacturing tolerances require a design target composition significantly beyond the threshold. Very small contact volumes increase the statistical spread of resistivities already inherent in a percolation structure. Minimal filler content is a requirement for both economic reasons (since Ag is expensive) and to maximize the proportion of polymer adhesive. Both issues can be addressed by the use of

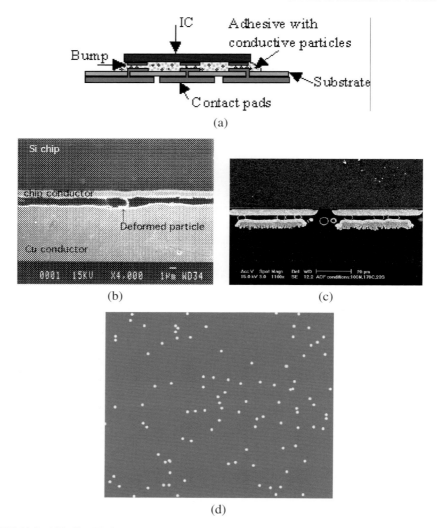

FIGURE 20.2. ACA flip-chip joining: (a) schematic, (b) Ni hard particle filler, (c) metal coated polymer filler, (d) vertical view showing random filler dispersion.

metal flakes (or rods) as filler instead of spheres. The benefits can be readily understood by considering the extreme case of flakes of zero thickness, which would clearly establish a percolation threshold at zero metal content, i.e., at 100% adhesive with zero filler cost. Practical commercial materials achieve substantial reduction in the threshold composition by the use of flakes, by virtue of the increased connectivity which accompanies the increased surface-to-volume ratio, compounded by bimodal particle size distributions (Figure 20.4). The efficiency of bimodal particle distributions has been demonstrated [3] and either flakes or powders can be used for the smaller particles [4].

Theoretical simulation has been carried out to optimize the electrical performance of a conductive adhesive joint using a bimodal distribution of metal fillers using computer modeling with finite-element analysis. Two classes of metal fillers were used, i.e., nano-scale and micro-scaled particles. The goal was to decrease the metal loading to improve

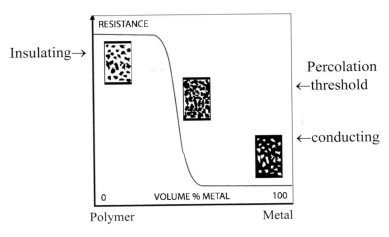

FIGURE 20.3. Percolation threshold [2].

FIGURE 20.4. ICA bimodal filler distribution (Ag flakes and powder) with surface layering [4,6].

the mechanical performance for specified electrical properties. It has been shown that it is possible to decrease the total metal loading with good electrical conductivity using a bimodal filler distribution [5], but that the nano-particles increase resistivity for a given total filler content, due to mean free path limits and increased numbers of contacts.

In addition, the flakes at the material surfaces seem to be aligned parallel to the surface to a depth of a few flake thicknesses (Figure 20.4). The effect seems to be universal, with squeegee or syringe dispensing, and also appears inside air bubbles, probably due to surface tension [6]. One also sees particle alignment inside the material, parallel to the contacts, if pressure is applied [7], as seen in Figure 20.1(b). Particulate orientations are important for predictions of both adhesive strength and electrical resistance. Flow modeling of the ICA under stencil, print, or dispensing, and/or positioning pressure is much more difficult than underfill flow modeling [8] due to filler sizes and shapes. Recently, a dynamic model of the effects of compression has demonstrated flake alignment quite dramatically [9].

There are various novel approaches to the improvement of electrical connectivity at a given metal content, including magnetic alignment of nickel filler rods [10], the use of polymer particles to force z-axis alignment of flakes [11], and electric fields [12]. Anything that increases the packing density/efficiency of the particles also increases the connectivity and reproducibility of the structure, permitting the design composition to move closer to the percolation threshold, and a reduction in metal filler.

Modeling of the material structure for prediction of the electrical properties has included

- Distributions of uniformly sized spheres [13].
- 2D [14] and 3D [15] rectangular particles with limited size distributions and x, y, z orientations.
- Rectangular particles rotated about the x, y, z axes in 1° increments [16].

The extension to bimodal flake representations requires a dramatic increase in computing power. The difficulty lies in the development of an efficient algorithm for the random placement of the particles, which must not impinge upon the space occupied by another. As the structure fills up, this process becomes more and more time consuming. A potential energy technique has proved effective [16], but the most recent advances have been by the use of compression algorithms applied to initially well separated particles [9,17,18].

20.2.2. ACA

The first structural issue for ACAs is the distribution of particles, and how many are captured and compressed between the contacts. A software package has been developed to calculate the average resistance of the ACA contacts for pastes or randomly loaded ACFs [19], with statistical distribution. Input parameters include particle size and distribution, particle loading, pad size, and pad spacing. The program also performs the inverse calculations, i.e., it determines the particle loading requirements for a specified minimum number of particles per pad (i.e., minimum conductance).

It is implicit in the program above that the particle distribution is well behaved, but for pastes, the realities of flow around and over pads as pressure is applied will dramatically alter the distribution. So the program has only been validated against ACFs, and studies on the entrapment of particles between pads continue. There have been efforts made to model ACP flow [20], and the problem should be more tractable than underfill flow modeling [8], due to the smaller particles, but there appears to be no result yet suitable for inclusion into a particle distribution model.

There is, of course, a finite probability that there are no particles in the joint. Williams et al. tried to estimate this by assuming the particles are placed in the bonding area according to the Poisson distribution [21]:

$$P(n) = \frac{e^{-\mu}\mu^n}{n!}, \qquad (20.1)$$

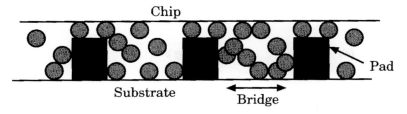

FIGURE 20.5. Bridging during ACA bonding (Courtesy of Mannan et al.) [24].

where P is the probability of an open joint, n is the number of particles per pad, and μ is the average number particles on a pad, given by:

$$\mu = \frac{3Af}{2\pi r^2}, \tag{20.2}$$

where A is the total bonding area, f is the volume fraction of the particles, and r is the radius of the particle. So, for a typical volume fraction ranging from 3 to 15 vol%, chip area (100 mm^2) and pad size (100 µm^2), the probability for an ACA open circuit is

$$P(0) = e^{-\mu} = e^{-\frac{3Af}{2\pi r^2}}, \tag{20.4}$$

which gives 10^{-13} to 10^{-3} for typical parameters, i.e., extremely small. However, in reality, there is always a crowding effect which must be taken into account. In this case, the particle distribution can be described using a binominal distribution model [21]:

$$P(n) = C_n^N (1-s)^{N-n} s^n, \tag{20.5}$$

where N is the maximum number of particles that can be contained in an area A, C_n^N is the binominal coefficient, and s is equal to f/f_m where f_m is the volume fraction corresponding to maximum packing [22]. In the limit when $f \ll 1$, Equations (20.4) and (20.5) give identical results for $P(0)$.

Bridging is possible due to there being too many particles between contacts, or insufficient contact separation, as illustrated in Figure 20.5 [23,24]. The probability for bridging is given as

$$p = 1 - \left[1 - \left(\frac{6f}{\pi}\right)^{\frac{d}{4r^2}}\right]^{\frac{hl}{4r^2}}, \tag{20.6}$$

where f is the volume fraction of the particles, h is the pad height, l the pad length, d the spacing between the pads, and r is the radius of the particle. Bridging probabilities are also very small, as seen for various cases in Figure 20.6, where it is clearly shown that the lowest combined probability for bridging and skipping occurs in the volume fraction between 7 and 15% depending on what model we use. This volume fraction range is also generally used for commercial ACA materials today.

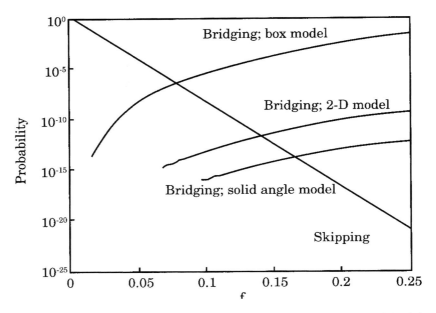

FIGURE 20.6. Probability of particles bridging gap as a function of volume fraction of particles during ACA bonding (Courtesy of Mannan et al.) [24].

20.2.3. Modeling

The goal of any physical modeling exercise is to demonstrate that the physical processes are well understood. Validation is achieved by agreement of the model predictions with experimental data. It is clear for both ICAs and ACAs, that successful comparisons of electrical modeling with experimental results require accurate structural modeling first. For ICAs, there have been some superficial efforts at structural modeling in the past, and as a result, comparisons of electrical models with experiment were either strictly qualitative [14,15] or subject to parameter adjustment to achieve a fit [16]. For ACAs, one would expect less difficulty in structural characterization, and the general form of the match of theory to experiment for the resistance variation with pressure [25] is convincing, even recognizing the need for fitting parameters. For ICAs, similar results are achievable for "dry" test systems [10], i.e., without epoxy, but cannot be correlated with actual ICA internal pressures because these are unknown. Real progress in realistic ICA structural modeling has bee made only recently [9,18].

20.3. MATERIALS AND PROCESSING

20.3.1. Polymers

The common use polymers in ECAs are thermosetting epoxies, with sufficient thermoplastic mixed in to allow for softening and release for rework under moderate heat. Up until now, the polymers used in ICAs have been adapted from those developed for other purposes. Recent research has provided clear pointers to the properties required of the new generation of materials, but the engineering trade-offs make implementation difficult, and the universal solution has yet to appear on the market. Certainly, the need for mechanical

ELECTRICALLY CONDUCTIVE ADHESIVES: A RESEARCH STATUS REVIEW

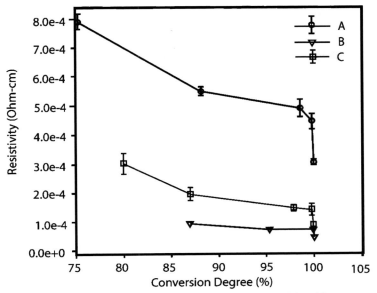

Resistivity versus conversion degree for Adhesives A, B and C.

FIGURE 20.7. Resistance dependence on cure as modeled by Equation (20.7) [4,6,27].

energy dissipative materials is well known, and the search for new materials must focus on this property, but without compromising on others. Polymer adhesion is a fundamental property which must be understood, with fundamental contact angle wetting experiments for ICA base polymers a first step.

The cure process has been modeled successfully by very simple mathematical expressions [26,27], which do however require accurately determined parameters from experimental Differential Scanning Calorimetry (DSC) data. For reaction rate

$$d\alpha/dt = kf(\alpha), \tag{20.7}$$

where α = degree of cure, $k = A \exp -(E/kT)$ is the chemical rate constant, and $f(\alpha)$ is a function of the reactant concentration, given by

$$f(\alpha) = (1 - \alpha)^n \tag{20.8}$$

for an n-th order model, the degree of cure is calculated as:

1st order $\quad d\alpha/dt = k(1 - \alpha), \quad \alpha = 1 - \exp(-kt),$ (20.9)

2nd order $\quad d\alpha/dt = k(1 - \alpha)^2, \quad \alpha = 1 - (1 + kt)^{-1},$ (20.10)

etc. More complex treatments employ a linear combination with the auto-catalyzed model

$$d\alpha/dt = kf(\alpha) = (k_1 + k_2 \alpha^m)(1 - \alpha)^n. \tag{20.11}$$

The success is demonstrated by the observation of the sudden decrease in ICA resistance (Figure 20.7) at the predicted point of 100% cure [6,27] according to Equation (20.9).

The assumption is that the resistance drop is due to physical shrinkage of the polymer matrix approaching complete cure, but this particular property, i.e., physical shrinkage and the development of an internal pressure to squeeze filler particles together, has not been measured, and does not appear to have been modeled. The measurement of a dimensional shrinkage with cure should not be a major problem, nor should the measurement of internal pressures.

Prior to the initiation of cure, however, the carrier solvents which permit the ICA paste (or ACP) to be printed must be expelled. Failure to do so degrades reliability, adhesion, and impact strength. The problem arises when partially cured polymer traps the evaporating solvent in bubbles. Brief exposure to vacuum prior to cure visibly decreases ICA paste volume as gas escapes from the surface [28], but the more practical technique is a pre-cure heat soak, e.g., for about 20–30 minutes at 100–120°C [29] which achieves the same result and marked reliability improvement.

ICAs typically cure at around 150°C for 20–30 minutes. This is considered too long to be competitive with solders [30], but there are much faster "snap-cure" alternatives.

20.3.2. ICA Filler

The metal filler of choice has been Ag, but the warning that Ag is accompanied by electromigration problems keeps on coming up. Certainly the existence of the electromigration problem has been documented [31], but it does not seem to be a problem in practice. It is evident in uncured material [12], and it has been suggested that commercial additives to the polymer seal the silver surface, defeating migration tendencies. There also appears to be a field threshold, and moisture is a requisite, but systematic study is required to establish the boundaries to the effect. In addition, the diffusion and clustering of metals in polymers is well established [32,33], and should be investigated for Ag in appropriate polymers as the limiting zero-field phenomenon. Sencaktar et al. have recently correlated electromigration with Ag surface pitting [34].

There have also been materials reported using low melting point alloys [35,36], or Sn-coated Ag particles [37]. The intent is for the particles to form metallurgical bonds during the polymer cure, to achieve lower contact resistances. The greater rigidity of the metallic network could be a problem if the contacts fracture under mechanical stress, but apparently the Sn inhibits Ag migration [37].

20.3.3. ACA Processing

ACA assembly typically requires a thermode for the bonding process, which takes place in about 30 seconds. Many parameters can affect the bonding quality during the ACA bonding process:

- Curing temperature and time of the ACA.
- Bonding temperature and time.
- Temperature ramp rate.
- Alignment accuracy.
- Pressure value, pressure distribution and pressure application rate.
- Bump height and uniformity.
- Board planarity and stiffness.

If the thermal ramp rate is too high, no particle contact will occur as the adhesive will be cured during the bonding process, before the pad and bump can both reach the particles.

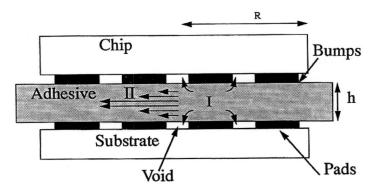

FIGURE 20.8. Schematic drawing of the ACA flow during bonding (Courtesy of Mannan et al.) [24].

The effect of bump to pad alignment accuracy is mainly related to the performance of the bonder. A flip-chip bonder that offers a 5 μm alignment accuracy is usually good enough. The question is how much tolerance is acceptable for the ACA bonding. As has been discussed earlier, the particle location will affect the electrical conduction performance of the joint, and bad alignment will have the same effect.

ACA flow during bonding has also been modeled by Mannan et al. [24]. A typical situation of the particle flow is shown in Figure 20.8. Assuming Newtonian flow

$$\tau_{xy} = \eta \delta\gamma/\delta t, \tag{20.12}$$

where τ_{xy} is the shear stress, $\delta\gamma/\delta t$ is the strain rate, and η is the viscosity. By solving the Navier-Stokes equation in this condition, one can obtain the following equation that describes the pressure under the chip during the bonding due to the flow of the ACA:

$$P = 2F/(\pi R^2)(1 - r^2/R^2), \tag{20.13}$$

where F is the bonding force, r is the distance from the chip center, and R is half of the edge length of the chip. Pressure variations cause planarity problems during the bonding. In reality, the ACA resin probably follows a power law type of flow as

$$\tau_{xy} = \eta_0 (d\gamma/dt)^n, \tag{20.14}$$

where n is the power law index and η_0 is the viscosity of the fluid. In this case, Mannan et al. have successfully obtained the following equation to predict the time for the ACA to be squeezed out between the chip and substrate:

$$Tp = (2n+1)/(n+1)\left\{2\pi\eta_0^{(n+3)}/[F(n+3)h_0^{(n+1)}]\right\}^{1/n}[(h_0/h)^{(n+1)/n} - 1], \tag{20.15}$$

where h is the ACA height. One of the important issues here is that the viscosity of the ACA will increase dramatically when it starts to get cured. This can be described as

$$\eta_0 = \eta_{00} e^{-b(T-T_0)}, \tag{20.16}$$

where b and η_{00} are constants and T_0 is an arbitrary reference temperature. However, measuring the viscosity of the ACA material close to the curing point remains a problem.

20.4. ELECTRICAL PROPERTIES

20.4.1. ICA

The electrical resistance of the ICA has four distinct components [2]. Three of these are obvious; they are the metal "intra-particle" resistance, the "inter-particle" contact resistance, and the "contact" resistance between the surface particles and the lead or contact pad. The fourth component comes from the meandering "percolation" path of the continuous metallic connection(s) through the material. Figure 20.9 illustrates the principle, showing how some metal particles carry current, while others do not.

(1) *Percolation theory* is well developed for the elementary system of uniform conducting spheres (or cubes) in a perfectly insulating medium [38]. There are no analytical solutions, and the theoretical results are deduced by the averaging of multiple Monte Carlo randomized simulations. Conducting particles are randomly assigned to sites on a specified regular array. For any realistic random system, the site separations must be much less than the particle sizes, and the problem becomes how to fit new particles into the structure at high particle concentrations (as required here) with reasonable efficiency.

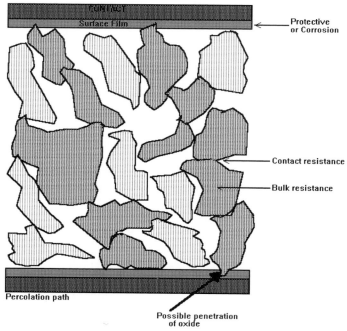

FIGURE 20.9. Schematic of metal particles/flakes in an ICA joint, showing sources of resistance [2].

The percolation modeling literature commonly includes finite intra-particle resistivities, but inter-particle resistances are seen less often. This is partly because the most common systems of interest would assume electron tunneling between particles, and the exponential dependence of the tunneling probability on separation introduces a strong parametric dependence on a poorly characterized variable. Percolation models of the electrical resistance of ICA systems have included both intra-particle and inter-particle resistances, but so far with only gross approximations, e.g., with the simplifying assumption of uniform tunneling thickness [13–16]. (To accommodate 1 nm variations in a 1–10 nm tunneling separation range between particles requires an underlying simulation grid with a 1 nm pitch in a brute force approach, i.e., a substantial increase in resolution over that otherwise required for micron sized particles. It would be more efficient, however, and just as valid to superimpose the tunnel gap distribution on contacting islands distributed on a coarser grid.) The structural modeling requirements have been outlined above.

Electrical modeling requires the addition of the conduction processes discussed below to each of the elements: intra-particle, inter-particle, and contact, with the structural model itself providing the percolation component. Existing models confirm the effects of surface layering at a qualitative level [15], and size effects [13,39], i.e., the increase and decrease in effective conductivity respectively for limited ICA sample dimensions parallel and perpendicular to current flow. In particular, the "short sample" percolation threshold is less than the bulk isotropic value, if the dimension parallel to current flow approaches or falls below the percolation "coherence length," as for practical flip-chip bump-to-pad interconnect. As contact separation decreases to filler particle size (the ACA situation), the threshold obviously tends to 0%. On the other hand, for a short dimension perpendicular to current flow, as, for example, in many experimental test "tracks" along a substrate surface, the percolation threshold increases due to the intersection of percolation paths with the surface. So pad-to-pad "contact" structures tend to possess lower resistivities than the nominal bulk value, and "track" configurations higher resistivities. Flake alignment along the surface tends to produce the opposite effects, increasing the number of inter-particle contacts in z-axis current flow, and short-circuiting the isotropic interior of the "track." Figure 20.10 shows the track conductance variation with thickness. At low thicknesses the track flakes are all layered, but as thickness increases, the proportion of internal disorder increases [40]. Figure 20.11 shows the resistance of a z-axis contact as it is mechanically thinned. The sharp drop in resistance corresponds to the removal of the aligned surface layer [41].

As models improve, one expects eventual quantitative agreement. It is probably satisfactory to continue to ignore the effect of finite polymer conductivity, but this assumption should be checked for new materials. One prediction of percolation theory which has never been validated in these systems is the frequency dependence of the conductivity in terms of the coherence length [42], which appears to be masked by the particle skin effect (see below).

It is noted here that there is evidence of percolation chains dropping in and out of the conduction paths during thermal cycling. This effect is demonstrated by reproducible hysteresis in resistance versus temperature plots during thermal cycling of ICA structures where only a limited number of percolation paths are expected to exist, i.e., for very small contact areas [43].

FIGURE 20.10. Variation of ICA resistance with track thickness [40].

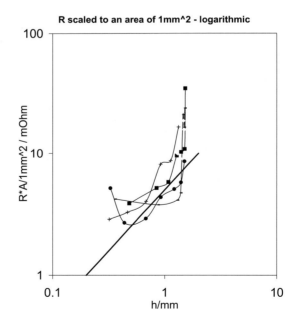

FIGURE 20.11. Log-log resistance variation with thinned thickness. The line of slope for constant bulk resistivity is shown [41].

(2) *Inter-particle conduction* is generally assumed to take place directly from metal to metal, or by tunneling through an insulating layer, whether of intervening polymer or of surface contaminants, or by conduction through a surface oxide film, which

for Ag would be a conductive degenerate semiconductor. There is a range of tools available to characterize conduction mechanisms, including frequency effects, non-ohmic high field behavior, temperature dependencies, etc., and the absence of any observation of negative temperature coefficient of resistance (TCR) or non-ohmic behavior is sufficient to eliminate most other conduction mechanisms from contention. Silver is typically tarnished, and presumably would oxidize within the polymer, even if initially "clean," but it is not clear what effect the surface lubricants identified on flake surfaces [44] would have on this process. An XPS surface analysis of the flake surfaces should be carried out to distinguish between the presence of Ag oxide and the oxygen content of the lubricant and/or polymer residue.

No matter which of the mechanisms apply at the gap itself, the contact area is accepted as being typically small, of diameter, d, 10 nm or less [6]. Clearly there will be some constriction of current flow between particles at the contact, and the Holm theory specifies this contact resistance to be $\rho/2d$ [2]. With the constriction resistance proportional to d^{-1}, and the tunneling or oxide resistances proportional to d^{-2}, one should be able to identify the dominant contribution from a pressure dependence, which has been theoretically matched to experimental data for a d^{-2} dependence, but not for an ICA [45]. The internal pressure exerted by the curing process must be quantified for further progress to be made. Returning to the differences between the contact conduction mechanisms, the TCR will be zero (or slightly negative) for tunneling and positive for the oxide. Unfortunately, it is difficult to find data on the exact electrical properties of the oxide, which are subject to local formation conditions, so it is not known whether the TCR would be less or greater than the metal particle TCR. It would appear from the formula that the constriction resistance would have the same TCR as the metal particles', but the derivation does not include the mean free path reduction which will be associated with contact dimensions less than the bulk value in the particles; this extension to the theory is necessary.

The inter-particle conduction mechanism remains undetermined, and the points made above, which reflect those in the literature, are very general. What is needed is a comprehensive basic study of the metal-polymer interface, to investigate charge transfer and band effects in the polymer(s), time and temperature effects, and how the process proceeds during curing. In addition, actual contact points need to be located and isolated, and the potential distribution plotted across the boundary from one particle to the other. Electrical noise measurements are also often a useful diagnostic tool, and there is noise data in the literature [46,47], but as is often the case, the interpretation is ambiguous. Wong has studied the role of the flakes' surface lubricant, and has achieved reduction of overall resistance by replacing the traditional stearic acid with shorter chain alternatives [36,48]. Benson, on the other hand, has shown that the lubricant breaks down during cure, and leaves a carbon residue on the flake surface [49], which is expected to control the inter-particle resistance.

Frequency dependencies are easily determined and can be definitive in the identification of some conduction mechanisms. In the ICA case, ac measurements were expected to short circuit the tunnel gaps between particles, with the corner frequencies providing the means to separate out the particle resistance from the inter-particle gap contribution. The method was validated by ac measurements before cure. Resistivities below the percolation threshold decreased with frequency to limiting values similar to those above it [50]. (This experiment suggests the use of impedance spectroscopy as a manufacturing quality test for ICAs, as for solder paste.) When applied to cured ICAs, however, no such effect was observed (see below), so either there is no tunneling gap (i.e., conducting oxide or metal–metal contact) or the tunnel resistance is just much less than the particle resistance [2,4].

(3) *Intra-particle resistance* accounts for a substantial proportion of the measured ICA resistance, and in some cases essentially all of it. This conclusion is based on temperature coefficient of resistance TCR measurements on a variety of commercial ICAs, where the TCR values range downward from the bulk metallic value, but are always positive [2,4,42], and hence clearly indicative of the dominance of the metallic component of the resistivity. (The other possible contributions to resistance all have zero or negative TCRs.) Within the limits of experimental accuracy, the data are consistent with a model of the intra-particle metallic resistance in series with a zero TCR contact resistance, but cannot be totally conclusive (Figure 20.12). Thermal testing needs to be extended to much lower temperatures to resolve this point.

For ten micron diameter flakes one micron thick, and micron-sized smaller particles, the electron mean free path (mfp) is not going to be reduced significantly from the bulk value, and no accounting is needed for size effects in the particles. (Note that this would not apply to the nano-particle ICA variant [51,52], where the mfp is limited by the particle dimensions.) But the nature of the surface could be important for the assessment of mfp limitation for constriction resistance (with rough surfaces limiting the mfp by random "diffuse" scattering, and with "specular" reflections from smooth surfaces having no such mfp effect).

The ac measurements mentioned above were run on the same ICA materials which gave TCRs identical to the bulk value for Ag, and so it is not surprising that the ac characteristics were in total accordance with the predictions of skin effect resistance and inductance for Ag [4]. These experiments should be duplicated for materials with greater inter-particle resistances.

At frequencies where skin effect is dominant, the lower resistance advantage enjoyed by solder disappears, as the effective cross-sectional area shrinks with the skin depth for solder and ICA alike [41].

(4) *Contact resistance* can be isolated from the bulk composite resistivity by the combination of three-terminal measurement with the more common four-terminal (see below) [26]. It is the contact resistance which has been shown to be the source of electrical reliability problems [53–55]. The oxidation (corrosion) of Cu or the Sn in Sn/Pb contact pads or Pb coatings has been demonstrated, and explains the greater long term stability of noble metal contacts (Au or Pd under thermal cycling and 85/85 stress testing). More recently galvanic corrosion has been identified between dissimilar metals [56,57]. One would fully expect to see the effects of interfacial diffusion in the longer term (albeit limited by low process temperatures), and the formation of brittle inter-metallics; although these would probably have no discernible effect on mechanical reliability, the electrical impact could be significant, given the limited number of low diameter contact points to the percolation paths. The solution to these problems would be to eliminate the dissimilar materials, and the use of silvered contacts with Ag-based ICAs would seem logical. However, Ag on Cu introduces much the same problems, and requires a barrier layer (of Ni, for example).

ELECTRICALLY CONDUCTIVE ADHESIVES: A RESEARCH STATUS REVIEW 543

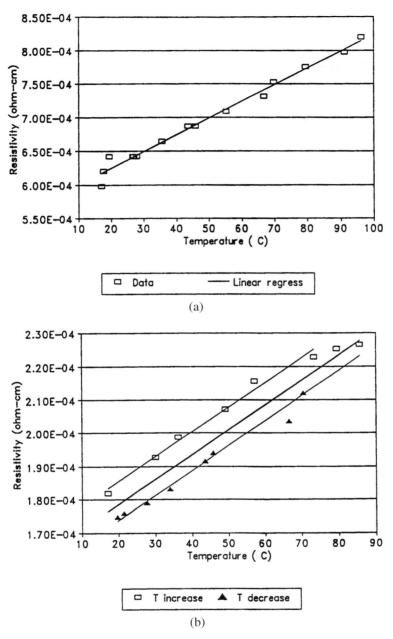

FIGURE 20.12. ICA resistance versus temperature. The slopes match the TCR of Ag [4]. (a) Brass stenciled CT-5047-02 thermoset sample (TCR = 0.0039/°C). (b) CSM-933-65-1 screen printed thermoplastic sample (TCR = 0.0038/°C).

The high-frequency ICA data of Li et al. [4] have been extended by Wu et al. [58] and by Dernevik et al. [59,60] Li and Wu focused mainly on the MHz region, and Dernevik on the GHz region. Wu reported that ICA joints can change their high frequency properties during bending.

20.4.2. Electrical Measurements

Resistivity measurements must be made on genuinely isotropic samples of suitable size, unless size or layering effects are the actual object of the measurement. On the other hand, the interconnect application will actually include both, as described above, so will measurements along a long thin sample, but with the opposite effects [2]. The measurement of small resistances with sufficient sensitivity to detect early corrosion, etc., is difficult and usually accomplished by depositing the long specimen just mentioned or by daisy-chaining multiple interconnect samples. In future, the impedance transformer should see increased application to solve this problem [2].

Printed ICA "tracks" are also commonly used for 4-terminal and 3-terminal measurements (Figure 20.13) which provide separation of bulk and contact resistances, an essential step to interpretation of any resistance data. Finite geometries can lead to errors, however, if care is not exercised (Figure 20.14). The same sort of problem can be experienced with z-axis samples (and with ACA testing) due to finite track resistances comparable to the sample's (Figure 20.15) [41]. However, self-consistent results have been demonstrated [61].

20.4.3. ACA

We move now to conduction through an individual particle between two contact pads.

The electrical conductivity in an ACA joint has been estimated by Williams et al. [23]. Assuming both hard and soft spheres, plastic deforming lands and spheres, the joint contact resistivity (ρ) has been estimated as:

$$\rho = \frac{A\rho_B \left(\sqrt{\frac{6\pi n \kappa}{\sigma A}} - \frac{1}{R_B} \right)}{4\pi n R_B} \tag{20.17}$$

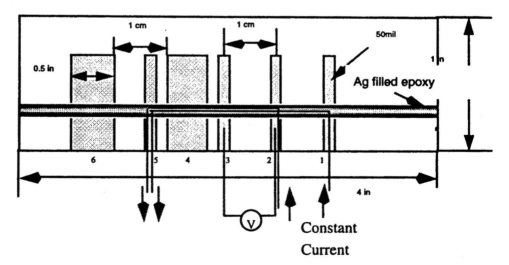

FIGURE 20.13. Four-terminal and three-terminal measurements [2,26]. The test current is injected at the right hand terminal for the usual 4-terminal measurement. Injection into the voltage test terminal includes that contact to the sample. The difference of the two measurements yields the contact resistance.

ELECTRICALLY CONDUCTIVE ADHESIVES: A RESEARCH STATUS REVIEW 545

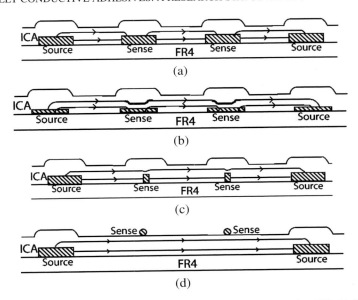

FIGURE 20.14. 4-point measurement using ICA track across proto-board current lines [41]. (a) Sense lines short ICA, (b) thinned contacts, (c) trimmed contacts, (d) surface point contacts.

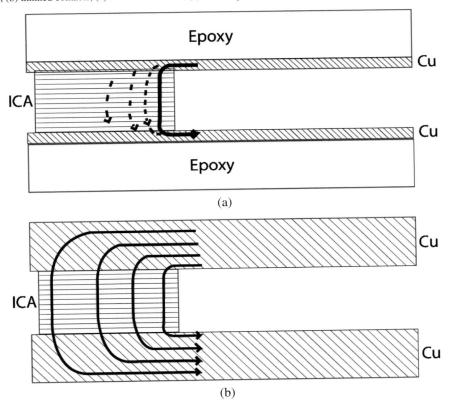

FIGURE 20.15. Current crowding effects due to contact track thickness [41]; (a) thin PWB track, (b) thick Cu plates.

for a single particle, where ρ_B is the resistivity of the sphere material, n is the number of contacts within the contact area A, κ is the shear yield stress on a single contact sphere of radius R_B, and σ is the pressure applied to the joint. Using Equation (20.7) one can calculate the electrical conductivity through an ACA joint. For instance, for 25 micron radius Ni alloy particles with $A = 0.1$ mm^2, $\kappa = 1$ GPa, $\rho_B = 6.1 \times 10^{-8}$ Ω m and $\sigma = 1$ MPa we obtain joint conductivity of 4×10^5 Ω^{-1} m^{-1} with 50 conducting particles and 1×10^5 Ω^{-1} m^{-1} for 2 particles.

Shi et al. [25] have calculated the resistance of a solid spherical contact particle analytically as the sum of the bulk resistance and constriction resistance components, and then extended the result to include distributed particle sizes and the influence of pressure on the particle contact area. Both elastic and plastic deformation models were included. Yim and Paik [62] also developed an analytical model, but also, like Oguibe et al. [63], performed finite element modeling of both solid sphere and coated polymer particle systems. There is direct comparison with experiment and good agreement is claimed. Meanwhile, Fu et al. have found that the particle location in an ACA joint can affect the electrical conductivity [64]. Generally speaking, a particle in the center of the joint contributes much more to the electrical performance than a particle close to the edge of the joint, which helps to explain the large scatter observed in single joint resistance values.

The high frequency properties of ACAs in packaging have also been studied by several groups [65–67]. ACA can offer similar high frequency flip-chip packaging performance up to 20 GHz [65] as solder. Significant transmission loss is due to the silicon material itself. It is characteristic of experimental comparisons of ECA joints with solder at high frequencies that the results owe more to the board lines and connections than to either the solder or ACA joints, which both contribute comparably negligible loss to the overall system. Kim [66] has used electrical equivalent circuit modeling of the ACA flip-chip interconnect in conjunction with experiments to extract the high frequency effect of the ACA interconnecting material itself. Fu et al. [67] used a physical modeling approach to understand the high frequency performance of the ACA joint. They found that at extremely high frequency, the capacitive displacement bypasses the filler contacts, and the high frequency performance is largely independent of the particle size and number. It is therefore doubtful if arrayed structures can improve high frequency ACF performance.

20.5. MECHANICAL PROPERTIES

20.5.1. ICA

The vast majority of the published ICA data is comprised of electrical resistance and adhesive strength measurements, both presented in the context of reliability testing. But while there has been some parallel effort to interpret the electrical properties in terms of both structure related and physical failure models, there has been little similar effort to understand the mechanisms of adhesion at a comparably fundamental level. Obviously, such a study requires a systematic approach to the measurement of adhesive strengths for a matrix of metal surface and polymer combinations. It is clear that surface cleanliness plays a crucial role in effective adhesion (and one which may be overlooked in the desire to apply ICAs as drop-in replacements for solder), so surface treatments must be included as a secondary variable, with consideration of roughening effects also included. While this could be regarded as an empirical study, the fundamental goal of determining the relative contributions of various adhesion mechanisms (e.g., chemical, mechanical, electrical) should not be lost sight of.

Plasma cleaning of the adherent surfaces would seem to be a logical step, but so far preliminary data shows no improvement in adhesive strength with either Ar or O_2 plasma treatments, despite the demonstrated removal of organic contaminants and oxides [12,59]. Wolter et al. [68,69] have demonstrated that it is the polar component of surface energy which is increased by plasma treatments. Experimental studies consistently show that the mechanical component of adhesion dominates [70,71], with best results from surface roughening (which may be accomplished by high energy plasmas). (A simple NCA shows good electrical stability, provided the contact surfaces are roughened [40].)

Published ICA adhesive and shear strengths are on the same order as those of solder, usually a little less [72], occasionally higher [69], but anyway adequate. The problem has been the drop test failure rate, leading to the widespread adoption of the NCMS (National Center for Manufacturing Science) criteria [73] as a de facto standard. In general, it is the larger devices which are most at risk, and indeed current commercial materials seem adequate for smaller devices such as SMT passives, which have been in mass production with ICA attachment for some years, e.g., in automotive electronics. Improved understanding of this particular phenomenon is currently leading to the development of ICA materials specifically designed to address the drop test problem. The key lies in the imaginary component of the complex Young's modulus, which represents energy dissipation in the material, as opposed to the energy storage of the simple deformation represented by the conventional form of Hooke's Law. The complex modulus is therefore directly analogous to the complex dielectric constant, and dynamic stress–strain relationship actually includes a phase shift. When one examines drop test survival data, the success rate correlates with the (imaginary) dissipation modulus rather than with adhesive strength [74,75]. One way to design materials with high dissipation modulus is to select polymers with glass transition temperature T_g below the operating range, i.e., below room temperature, in general [76]. However, operating polymers above T_g carries its own penalties, e.g., higher temperature coefficients of expansion, so the next step is to develop adhesive polymer blends with high mechanical absorption without those disadvantages.

Figure 20.16 shows the improvement in drop test results for a commercial ICA with the addition of a pre-cure heat soak to the processing schedule [77]. The additional step enables the material to survive an additional 20 4-foot drops and at least 10 more from 5 feet, greatly exceeding the NCMS standard. Improvement in resistance, adhesion, and drop test results are all tentatively attributed to the elimination of bubbles at the contact interfaces, where they reduce the effective contact area.

20.5.2. ACA

Several research groups have now studied the deformation effect on the electrical conduction development during the ACA assembly [58,62,64,78]. The key work has been done by Wu et al. [58] where they consider two types of particle situations as shown in Figure 20.17 (rigid and hard particle) and in Figure 20.18 (soft particle).

In addition, Wu et al. have also successfully deduced an equation that describes the relationship between the resistance and bonding pressure. These are shown in Figures 20.19 and 20.20 for the rigid and deformable particles respectively.

Wang et al. also focused on quantification of criteria for a good ACA flip-chip joint, as this is one of the most important issues during the ACA bonding process [79]. The

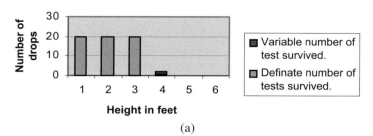

(a)

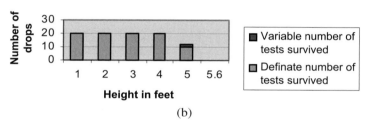

(b)

FIGURE 20.16. ICA drop test improvement from (a) to (b) with pre-heating before cure [77]. Each sample device was dropped 20 times from 1 foot, then 20 times from 2 feet, then 20 times from 3 feet, etc., until failure.

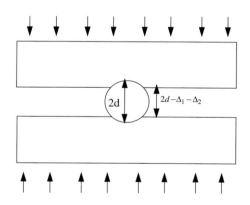

FIGURE 20.17. Schematics of a rigid particle system (courtesy of Wu et al.) [58].

purpose of their work was to study the relationship between bonding pressure and the deformation of conducting particles in ACAs. It has been shown that the deformation of filler particles plays an important role in both bonding quality and the reliability of ACA joints. The deformation degree as a function of bonding pressure during ACA film flip-

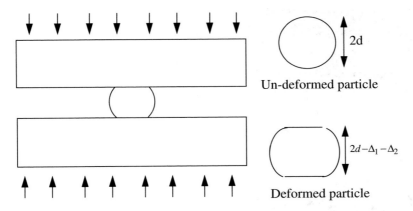

FIGURE 20.18. Schematics of a deformable particle system (courtesy of Wu et al.) [58].

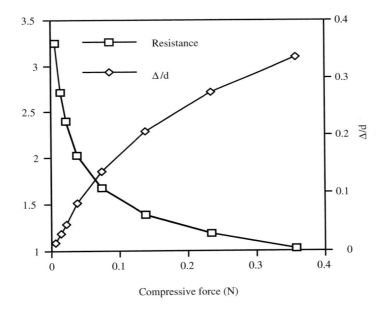

FIGURE 20.19. Force–resistance–deformation relationship for a deformable particle system (courtesy of Wu et al.) [58].

chip bonding is shown in Figure 20.21. As can be seen, deformation and bonding pressure exhibit a linear relationship.

As the mechanical properties of a thin layer adhesive material differs from its bulk material properties, it is important to know the ACA material properties for mechanical model and simulation purposes. Because of this, Young's modulus and Poisson's ratio have been determined by Zribi et al. [80] for an ACF over the temperature range of 15 to 60°C (Figure 20.22).

A group at Chalmers University has recently carried out theoretical simulation in order to explain the microscopic mechanism of the electrical contact conduction through the metal fillers for ACAs. By comparing with experiments performed by Wang et al. [79], it was concluded that the deformation of the metal filler is plastic even at rather low external

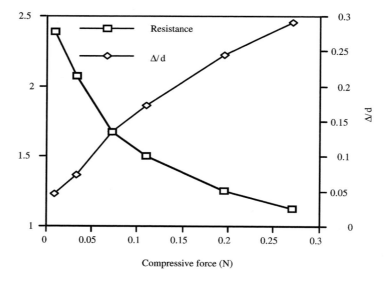

FIGURE 20.20. Force–resistance–deformation relationship for a rigid particle system (courtesy of Wu et al.) [58].

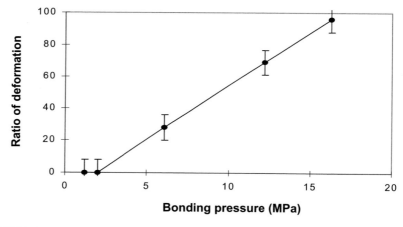

FIGURE 20.21. The relationship between nickel particle deformation and bonding pressure during ACA flip-chip bonding [79].

loads. Further theoretical simulation reveals two aspects of the conductance characteristics: the conductance is improved by increasing the external load, but the dependence of the conductance becomes stronger on the spatial position of the metal filler [65].

The consequence of the bonding pressure during the ACA bonding on the stress generation is evident. This stress generation is probably the reason for catastrophic failure. Wu et al. have simulated this using finite element modeling. They have found that both for rigid and deformable particles, significant stress is built up in the interface between the two contacts [58]. Wang et al. used a similar approach to calculate the stress after various degrees of deformation [65,69]. It is clearly shown in Figure 20.23 that the highest stress point is in the edge of the ACA particle.

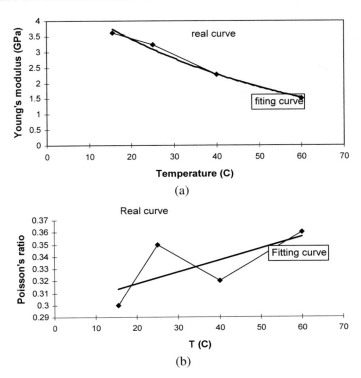

FIGURE 20.22. (a) Young's modulus, and (b) Poisson's ratio of ACA film vs. temperature after cure [80].

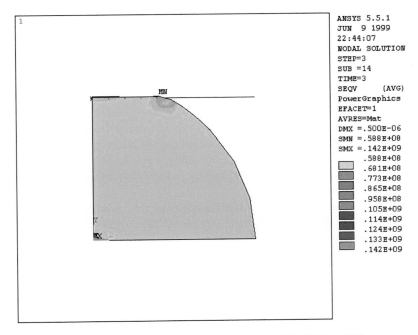

FIGURE 20.23. Stress profile at the reduction in height of 0.5 μm [20].

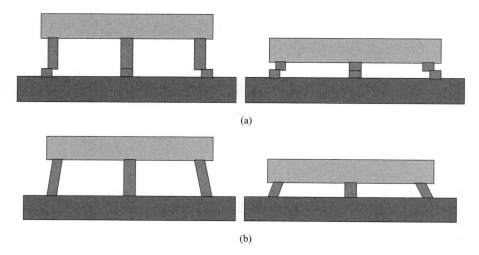

FIGURE 20.24. Deformation mechanism of the ACA joint due to thermally induced stress [81]. (a) Strain caused by thermal mismatch cannot change with bump height in ACA joint. (b) Schematic diagram showing that thermal mismatch strain reduces as bump height increases for solder joints.

Lai et al. have proposed a similar failure mechanism of the ACA flip-chip joint structure [81] as can be seen in Figure 20.24(a). In this case, the particles are exposed to large shear stress due to the expansion of the substrate. Therefore, it seems that the stress concentration is the most critical parameter that governs the reliability of the ACA joint. Therefore the failure mechanism of ACA joint differs from the traditional solder joint failure mechanism where plastic strain is the most critical parameter that governs the joint reliability, see Figure 20.24(b).

The most complete ACA joint stress analysis has been done by Wu et al. [82]. It is shown that the residual stress is larger on a rigid substrate than on a flexible substrate after bonding.

As an ACA bonded module may be used in connection with soldering technology for a final product, we need to know if the ACA material can withstand the soldering temperature. Sugiyama et al. reported that the ACA material from Sony can withstand the soldering profile three times and without causing reliability problems [83]. Törnvall reported different results in terms of soldering effect where he showed that an ACA flip-chip module cannot withstand a normal soldering process [84]. More research is therefore needed to clarify this matter further.

Oxidation is an ACA joint failure mechanism that was identified at an early stage [85]. The idea was that the particle is oxidized causing electrical performance decrease. For a parabolic oxide growth law, and assuming the resistance change, ΔR, is ohmic, then the change with time, t, is given for contact area, A, by

$$\Delta R = (2Dt)^{1/2} \rho_{\text{oxide}}/A, \tag{20.18}$$

where D is the diffusion constant of oxygen in the metal filler, and ρ_{oxide} is the volume resistivity of the oxide.

ELECTRICALLY CONDUCTIVE ADHESIVES: A RESEARCH STATUS REVIEW   553

20.6. THERMAL PROPERTIES

20.6.1. Thermal Characteristics

The thermal performance of an adhesively assembled chip is of vital interest as power dissipation in the chip increases. Power dissipations have been simulated by Sihlbom et al. for both ICA and ACA flip-chip joints [86]. They concluded that the ACA flip-chip joint is more effective in transferring heat to the substrate from the powered chip than the ICA joint, because the adhesive thickness is so much thinner than the ICA joint.

20.6.2. Maximum Current Carrying Capacity

Dernvik et al. have also studied the effect of maximum current carrying transmission capacity through the ACA adhesive joints at 3.2 GHz [59]. The copper bridge structure was subjected to a maximum transmission of 25 W of average pulsed power for 10 minutes, with a pulse length of 10 µs and peak power of 250 W at duty cycles of 1, 5 and 10%. The result indicated that bonding pressure has a strong influence of the joint quality. At 150 N, no electrical transmission loss is observed, but for 75 N bonding force, some deterioration was observed after the final power exposure; however, the effect was small, at about 0.5 dB.

In a similar dc study of three ICAs, Morris et al. [12] found that failure correlated directly with temperature rise, which in turn correlated with the joint resistance, and hence to power dissipation. The breakdown mechanism was clearly polymer degradation, accompanied by the emission of noxious fumes. The onset of breakdown shows thermal runaway (Figure 20.25), as measured at the adhesive surface. Comparisons of surface temperatures with estimates of internal temperatures (from resistance values and the previously mea-

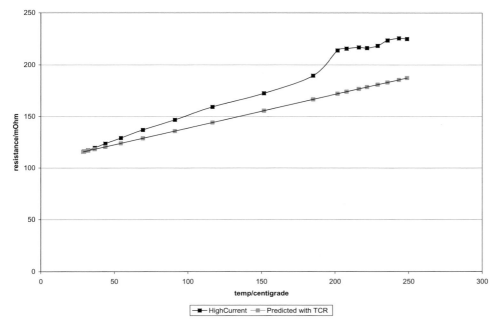

FIGURE 20.25. ICA resistance and surface temperature, as current is increased to thermal runaway and destruction [40]. The straight line corresponds to internal temperatures estimated from the resistance and the TCR.

sured temperature coefficient of resistance (TCR)) suggest that the internal temperature at this point is a good 60–80°C higher than the 200°C shown. While two of the ICAs failed like this at 9 A/mm^2, the other showed no ill effects whatsoever.

Bauer et al. has studied the power dissipation performance of NCA joints under dc conditions. It was found that the NCA joints still show ohmic behavior up to 25 A/mm^2 [87].

20.7. RELIABILITY

20.7.1. ICA

(1) *Cure schedule* control is undoubtedly very important for joint reliability. It seems that the electrical resistance of the joint is related to the curing degree, especially for non-noble metal surfaces, as can be seen in Figure 20.26. Figure 20.26(a) shows the contact resistance vs. curing time for an epoxy conductive adhesive cured at 150°C, and following 1000 hours of damp heat treatment at 85°C, 85%RH [88]. The corresponding curing degree varies between 65% and 90%, determined by DSC. Below a critical curing degree (77% for this adhesive), the electrical resistance of the joint increases significantly, because an incompletely cured epoxy can absorb a significant amount of moisture, which in turn causes oxidation/hydration of the Sn37Pb bonding surface in Figure 20.26(a) [89] and less crosslinking/shrinking of the polymeric matrix. If a noble metal, for instance Au or Pd is used as the bonding surface, no electrical resistance change is observed, despite the fact that the curing degree can be very low [Figure 20.26(b)]. Once a critical curing degree is reached (72%), it seems that the shear strength of the joint on the Sn37Pb bonding surface can be maintained at a constant level as is illustrated in Figure 20.27(a). However, on the noble metal bonding surface, the shear strength of the joint is almost independent on the curing degree in the range from 67% to 92% [Figure 20.27(b)]. In summary, it can be said that a minimum curing degree appears to be required to provide a certain level of mechanical and electrical performance in the adhesive system. Once this is achieved, increasing curing times do not result in significant improvement. At full cure conditions, however, the electrical resistance and the mechanical strength of conductive adhesives are

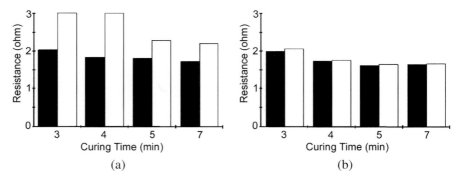

FIGURE 20.26. Series contact resistance of 10 epoxy-based ICA mounted chip components in series, before and after 1000 hours of 85°C and 85%RH [88]. (a) Sn37Pb plated chips and boards; (b) Ag/Pd plated chips and Au plated boards.

also guaranteed [90]. These above results also indicate that for conductive adhesive joining, noble metal surfaces are preferred to non-noble metal surface, due to the absence of galvanic corrosion.

(2) *Moisture effects* on the polymer degradation in conductive adhesives have been studied by Khoo and Liu [91]. Moisture sorption effects may be reversible or irreversible, and are usually small enough to make detection of the molecular changes during absorption/adsorption very difficult. Figure 20.28(i) shows the FTIR spectra of an ACA (but ICAs would be similar) after curing and subsequent 41 hours conditioning at 85°C and 85%RH. The difference spectrum (a)–(b) represents the changes occurring due solely to exposure to the moisture conditions at 85°C and 85%RH. The most obvious real changes are the negative bands at 868, 916, 1345, 3005 and 3058 cm^{-1}, implying decreasing epoxy functionality, and thus further progress of the cure reaction. The new bands at 3560 and 3350 cm^{-1} may both be attributed to hydroxyl groups, of which the former are free groups, which could be formed on further curing, or as an oxidation product resulting from thermo-oxidation/degradation processes. The latter are attributed to hydrogen-bonded hydroxyl groups, indicating the type of bonding of the adsorbed water to the epoxy resin. The slight rise at about 1640 cm^{-1} indicates the presence of absorbed water in the epoxy resin. Finally, new ester linkages indicative again of further curing are indicated by the presence of a broad absorption region between 1000 and 1300 cm^{-1}.

Figure 20.28(ii) compares the molecular events happening on further exposure to these same conditions. The figure shows the difference spectra (a) after 41 hours conditioning, (b) after 162 hours conditioning, (c) after 821 hours conditioning, all at 85°C and 85%RH. It is observed firstly that the subtracted spectra all show the same profile, indicating that the subtraction procedure has been consistent and that the subtraction spectra are valid, but a closer scrutiny also shows that the bands at 3560, 3350, 1640, 1573 cm^{-1} are increasing in intensity with increasing conditioning time. Both the increases at 3350 and 1640 cm^{-1} follow the increasing adsorption of water in the epoxy. The steadily growing absorption at 1573 cm^{-1} is tentatively attributed to unsaturated vinyl structures ("C=C") which are formed as a result of degradation actions. Moisture degradation is felt to occur by hydrolysis of the ester linkages ("R−(C=O)−OR"). Such hydrolytic attack breaks the polymer chain creating two new end groups, a hydroxyl and a carbonyl. Although

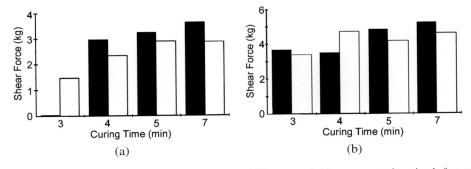

FIGURE 20.27. Average shear strength of 10 epoxy-based ICA-mounted chip components in series, before and after 1000 hours of 85°C and 85%RH [89]. (a) Sn37Pb plated chips and boards; (b) Ag/Pd plated chips and Au plated boards.

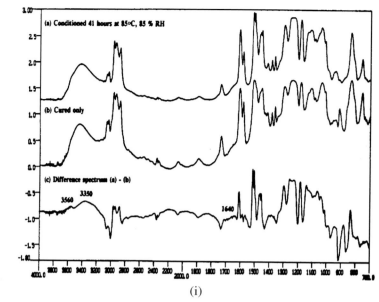
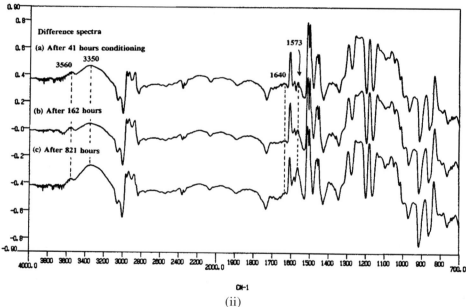

FIGURE 20.28. FTIR results for cured ACAs [91]. (i) FTIR Spectra (a) after 41 hours of 85/85 testing, (b) before the testing, and (c) the difference spectrum. (ii) Difference spectra for (a) 41 hours, (b) 162 hours, and (c) 821 hours at 85/85.

it is difficult to see a new emerging carbonyl group in this figure, the presence of the band at 3560 cm^{-1}, which indicates free hydroxyls, supports the suggestion of degradation reactions occurring with increasing exposure to heat and humidity at 85°C and 85%RH. Hence, in conclusion, it can be said that on exposure of the cured adhesive to 85°C and 85%RH, both moisture adsorption and further curing can be observed. After a certain time, however, further curing will not be observed, but instead, degradation effects may be seen.

(3) *Galvanic corrosion* at the contact interfaces has been demonstrated by Lu et al. [56,57]. They correlated the degree of contact corrosion (as indicated by resistance increases) to the electrochemical series, and demonstrated the requirement for moisture in the process. With this understanding of the process, it was shown that resistance drift could be inhibited by the addition of corrosion inhibitors, oxygen scavengers, and/or sacrificial anode material to the polymer matrix [92–94]. Figure 20.29 illustrates the classic result. The bulk ICA resistance varies little, decreasing somewhat with further cure. The contact resistance changes little for the Ag filler on a Au surface, but corrosion produces major increases for the Ag/Cu combination.

(4) *Thermomechanical* cycling is the litmus test for all package interconnect, especially for soldered flip-chip, and there are many examples of ICA data in the literature. Polymer cure temperatures are usually less than even the lowest solder reflow temperatures, so initial thermomechanical stress is lower at room temperature. In addition, the ICA's polymer base provides greatly increased creep properties in comparison to its solder competitors [95], so it may relax more readily to a zero stress state. It is therefore not surprising that ICAs out-perform solder on mechanical cycling tests by an order of magnitude (Figure 20.30) [96]. In an SMT application, however, the thermoplastic properties of the polymer lead to the accumulation of plastic strain, which initiates cracking [29].

20.7.2. ACA

The bonding pressure, if not homogeneously distributed, can cause uneven deformation of the particles. There are many possible reasons for uneven pressures, but the main reasons are improper electrical routing on the substrate, unadjusted pressure bonding head, etc. It is also clear that on rigid and thick substrates, electrical routing is not so critical, but it is crucial on thin and flexible circuitry.

Electrical failures during thermal cycling are observed at both low and high temperatures, and the different types of failure (Figure 20.31) can be attributed to different combinations of particle size and chip geometry, as can be seen in Figure 20.32 [97]. The best case is that the particles are deformed uniformly and metallurgical bonding between the particles and contacts is achieved (Type 1). If the particles are just in contact with the bonding surface, we will have a situation that electrical reliability is not good at high temperature due to the fact that epoxy will expand more than the metal particles (Type 2). If the particles are not the same in size and for unbumped dies and for chips with non-uniform bump heights, one obtains a situation that the smaller particles are not deformed and will shrink more than the big particles causing problems at low temperature (Type 3). The final case is when there is a uniform height of the bump and chip surface but very a large particle size variation. This will cause electrical opens at both low and high temperature (Type 4).

The dynamic pressure on the particle necessary to cause separation from the contact is, for the deformed (polymer particle) case, given by [97]:

$$P_i = N\left(\frac{1}{b^2}\right)\sqrt{\frac{\rho}{AB}\frac{l}{\cos\alpha}}, \tag{20.19}$$

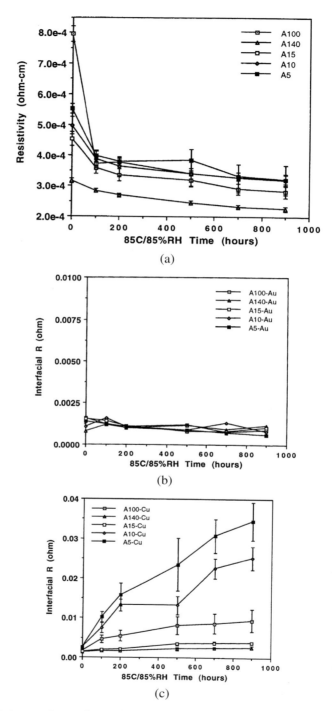

FIGURE 20.29. Post-cure resistance changes for an Ag-filled ICA [4,6,27]. (a) Bulk resistance; (b) contact resistance to Au; (c) contact resistance to Cu.

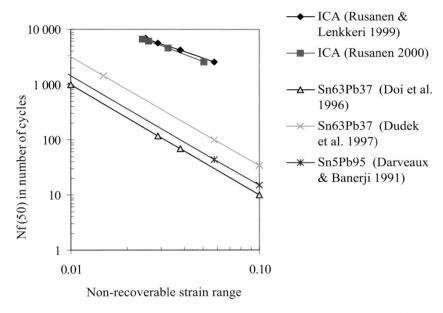

FIGURE 20.30. Strain cycling joint lifetimes according to the Coffin-Manson equation [96].

where P_i is the stress necessary for decohesion, N is the bonding force, ρ is the curvature of the particle, b, l and AB are geometric length, and α is the decohesion angle. For the undeformed (Ni sphere) case, a similar equation can be obtained [97]:

$$P'_i = N' \left(\frac{1}{b^2}\right) \sqrt{\frac{\rho'}{A'B'}} \frac{l}{\cos\alpha'}. \tag{20.20}$$

From Equations (20.19) and (20.20) it is clearly seen that $P_i > P'_i$. Therefore, for the deformed case, a larger stress is necessary to cause decohesion and, for the same stress level, it is easier to obtain failure with an undeformed particle.

As shown analytically by Hu et al., the electrical resistance decreases with increasing bonding pressure, until the polymer coating fractures. This has been demonstrated experimentally [14,98].

Another important issue is the substrate hardness, geometry and material. A soft substrate material may deform during the bonding due to the softening of epoxy at the bonding temperature, which is above the glass transition temperature (T_g) of the epoxy matrix. On an FR-4 substrate, it has been observed that the electrical conductivity and reliability of a joint depends on the depth of the glass fiber in the substrate where the bump exerts pressure on the pad [99]. Deeper fibers mean thicker layers of soft epoxy that deform more readily during the bonding. Therefore, insufficient particle deformation will be obtained at that point. Shorter distances to the fiber bundles resist substrate deformation better, and hence yield better electrical conductivity and reliability. See Figures 20.33 and 20.34. One approach to reduce the amount the pad can sink into the substrate is to use a relatively large pad area in comparison to the bump area to reduce the pad pressure for a given bonding force.

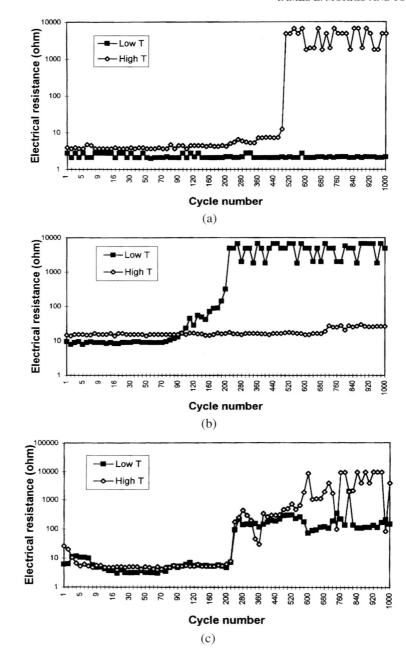

FIGURE 20.31. High temperature cycling [97]. (a) Type 2 joint failures occur at higher temperatures; (b) Type 3 joint failures occur at lower temperatures; (c) Type 4 joint failure occur in both high and low temperature regions.

Despite the complex processing and manufacturing conditions, under optimum conditions, good reliability data on ACA flip-chip joint has been reported [100]. Cumulative fails after the temperature cycling test from −40 to +125°C for 3000 cycles with a hold

ELECTRICALLY CONDUCTIVE ADHESIVES: A RESEARCH STATUS REVIEW

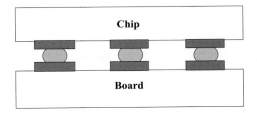

Type 1, the most reliable structure.

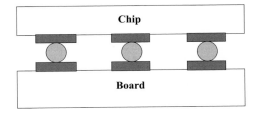

Type 2, unstable at high temperature.

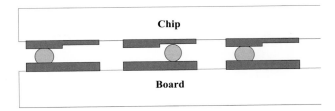

Type 3, unstable at low temperature.

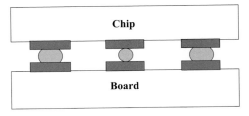

Type 4, unstable when temperature changes.

FIGURE 20.32. Schematics of four types of ACA joints caused by different bump geometries, variations in filler size, and differences in bonding pressure [97]. The classifications correspond to Figure 20.31.

time of 15 minutes at hold temperature are shown in Figure 20.35. However, the number of fails is dependent on the definition of the failure. Figure 20.35 shows three interpretations of the same cumulative failure data, based respectively on the different criteria: >20% of contact resistance increase; >50 mΩ; >100 mΩ. When the criterion was defined at 20% of resistance increase, all joints failed after 2000 cycles. This definition might be too harsh for those joints having initial contact resistances of only several mΩ. The 20% increase means only variations of a few milliohms are allowed. In some cases, the limitation is still within the margin of error of the measurement. Therefore, it is reasonable that the crite-

 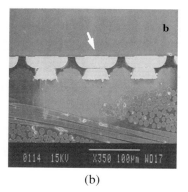

FIGURE 20.33. The case on the left is located close the glass fibers (a) and has an excellent resistance (5 mΩ) and reliability, whereas the latter (b) has a high contact resistance (14 mΩ) and poor reliability (49 mΩ after 1000 cycles of TC test) due to its location far from the fibers [99].

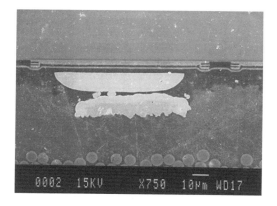

FIGURE 20.34. Pad sinking led to the bad bonding quality [99].

rion be defined according to the production requirements. However, if we define failure as 50 or 100 mΩ, the mean times between failures (MTBF) become 2500 and 3100 cycles, respectively.

In order to predict the real service life under different environmental conditions, a low-cycle fatigue testing machine was used to perform low-cycle fatigue experiments on ACA made joints in both dry and humid environments at different temperatures [80]. The final goal was to predict the real service life using the fatigue life data generated under different plastic strain loads. The plastic strain is a function of the temperature cycling interval, frequency and temperature ramp rate, etc. The daisy chain resistance curves versus the number of cycles are shown in Figure 20.36. A fairly significant decrease of the load level was noticed during the test. The load level ranged for both samples between 1 N and 0.7 N. According to the test data, failure occurred in the humid environment much faster than in the dry environment, which was physically expected. In fact in the humid atmosphere, the joints are subjected to very severe operating conditions (expansion of the adhesive due to moisture uptake, creep of the adhesive under exhaustive conditions).

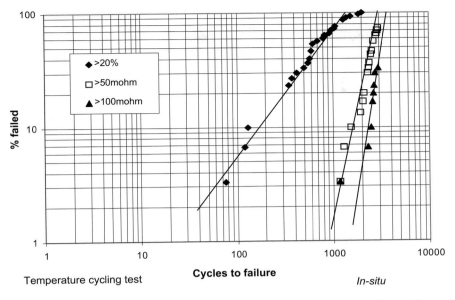

FIGURE 20.35. Cumulative failure of an epoxy-based ACA joint in after a temperature cycling test from −40 to +125°C with a dwell time of 15 min at hold temperatures [100].

There are two features of ACA flip-chip joining technology, which are different from other microelectronic interconnection techniques, such as flip-chip soldering, surface mounting, etc. These are:

- Application of bonding force.
- Simultaneous bonding of all bumps of a chip.

As the conductivity of ACA joints is directly determined by the mechanical contact between the terminals of chips and the electrodes on chip carriers, the bonding force plays a critical role in the electric performance. High bonding pressure is certainly favorable to an intimate contact, and thus to a low contact resistance. In addition, because the bonding of all bumps of a chip is performed simultaneously, uniform conductivity of all joints in the chip requires the bonding situation of every joint to be completely the same. In other words, every joint in the chip must have:

- Same bonding pressure.
- Same number of conducting particles.
- Same particle size.
- Same bump and pad geometry.

In practice, these requirements are hardly ever met. Many factors can affect the bonding situation, including:

- Distribution of bonding pressure on bonding tool.
- Alignment.
- Variation of conductive particle size.
- Distribution of particles in bonding area.
- Pad planarity.

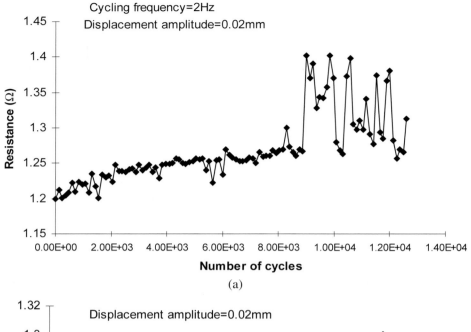

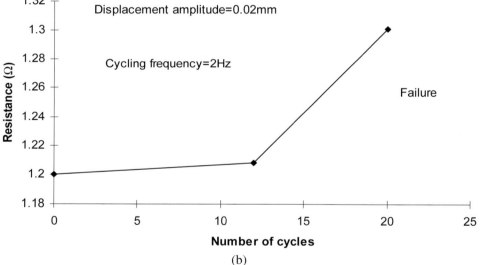

FIGURE 20.36. Electrical resistance of a daisy chain versus the number of cycles of low-cycle fatigue testing. (a) Dry environment; (b) 85°C and 85%RH [80].

- Bump planarity.
- Variation of substrate thickness.

Although we understand significant amount of the ACA joining technology, we still need to understand the following issues:

- High frequency behavior, its coupling with semiconductor devices, cross-talk between the particles especially at the range beyond 20 GHz.

- The maximum current carrying capability at high frequency and also after exposure to various environmental tests.
- The planarity effect of the substrate on the ACA joint reliability. This remains one of the most important issues to be understood before we can use ACA in real high volume for low-cost applications.
- Life time prediction models of ACA joints. For this, we need to understand further the effect of UV-degradation of the polymer chain, the corrosive gas effect and high temperature storage effect.

20.7.3. General Comments

As conductive adhesives are made of polymers and metal fillers, light may break the polymer chain. How this works is still unclear as there is no report on this work.

A similar situation is noted as to how various gases affect the polymer chain stability in conductive adhesive applications, although it is clear that polymer chains may be affected and broken by the corrosive gases.

One of the major important questions that need to be addressed is the estimation and prediction of the real service life of a conductive adhesive joint. As conductive adhesives consist of a metal part and an epoxy part, it is a composite material. Therefore, it is unlikely that the acceleration laws used for prediction of the real service lives of pure metals and pure polymers can be directly used for conductive adhesives.

20.8. ENVIRONMENTAL IMPACT

The environmental impact of ECAs has been studied by several research groups [101–103]. Segerberg et al. [102] compared use of conductive adhesive joining with soldering for SMT applications and concluded that it is really dependent on the mining condition of silver when determining the environmental load of the conductive adhesives. If silver is mined in a pure silver mine, then the environmental load is much smaller compared to soldering. On the other hand, if it is mined as a by-product in a copper mine, then conductive adhesive joining technology will have a much larger environmental load index compared to soldering.

Westphal et al. [101] came to the conclusion in their study that conductive adhesives are generally better in terms of environmental loading compared to solder. More work is needed to clarify this topic.

20.9. FURTHER STUDY

Readers are directed to references [104] and [105] for introductory concepts and a more comprehensive coverage respectively. Recent reviews of ACAs include references [106,107]. In addition, many of the points made here are expanded on in the on-line course at www.cpmt.org [108]. This work has been an update of an earlier review [109].

REFERENCES

1. J. Kivilahti (personal communication).
2. J.E. Morris, Conduction mechanisms and microstructure development in isotropic, electrically conductive adhesives, in J. Liu, Ed., Chapter 3 of Conductive Adhesives for Electronics Packaging, Electrochemical Publications, Ltd., UK, 1999, pp. 36–77.

3. R.P. Kusy, Influence of Particle size ratio on the continuity of aggregates, J. Appl. Phys., 48(12), pp. 5301–5305 (1977).
4. L. Li, H. Kim, C. Lizzul, I. Sacolick, and J.E. Morris, Electrical, structure and processing properties of electrically conductive adhesives, IEEE Trans. Compon. Packag. Manuf. Technol., 16(8), pp. 843–851 (1993).
5. Y. Fu, J. Liu, and M. Willander, Conduction modeling of conductive adhesive with bimodal distribution of conducting element, International Journal of Adhesion and Adhesives, (Dec.), pp. 281–286 (1998).
6. L. Li, Ph.D. dissertation, Basics and applied studies of electrically conductive adhesives, State university of New York at Binghamton, 1995.
7. J. Constable, T. Kache, S. Muehle, H. Teichmann, and M. Gaynes, Continuous electrical resistance monitoring, pull strength, and fatigue life of isotropically conductive adhesive joints, IEEE Trans. Compon. Packag. Technol., 22(2), pp. 191–199 (1999).
8. P. Li, E. Cotts, Y. Guo, and G. Lehmann, Viscosity measurements and models of underfill mixtures, Proc. 3rd International Conference on Adhesive Joining and Coating Technology in Electronics Manufacturing, Binghamton, NY (Adhesives'98), Sept., 1998, pp. 328–333.
9. M. Mundlein and J. Nicolics, Modeling of particle arrangement in an isotropically conductive adhesive joint, Proc. 4th International Conference on Polymers and Adhesives in Microelectronics and Photonics, Portland, Oregon, 2004, PP-3.
10. E. Sancaktar and N. Dilsiz, Anisotropic alignment of nickel particles in magnetic field for electrically conductive adhesives application, J. Adhesion Sci. Technol., 11(2), pp. 155–166 (1997).
11. T. Inada and C.P. Wong, Fundamental study on adhesive strength of electrical conductive adhesives (ECAs), Proc. 3rd International Conference on Adhesive Joining and Coating Technology in Electronics Manufacturing, Binghamton NY (Adhesives'98), Sept., 1998, pp. 156–159.
12. J.E. Morris, C. Cook, M. Armann, A. Kleye, and P. Fruehauf, Electrical conduction models for electrically conductive adhesives, Proc. 2nd IEEE International Symposium Polymeric Electronics Packaging (PEP' 99), Gothenburg, Sweden, 1999, pp. 15–25.
13. E. Sancaktar and N. Dilsiz, Thickness dependent conduction behavior of various particles for conductive adhesive applications, Proc. 3rd International Conference on Adhesive Joining and Coating Technology in Electronics Manufacturing, Binghamton, NY (Adhesives'98), Sept., 1998, pp. 90–95.
14. L. Li and J.E. Morris, Electrical conduction models for isotropically conductive adhesives, J. Electronics Manuf., 5(4), pp. 289–298 (1996).
15. L. Li and J.E. Morris, Electrical conduction models for isotropically conductive adhesive joints, IEEE Trans. Compon. Packag. Manuf. Technol. Part A., 20(1), pp. 3–8 (1997).
16. P. McCluskey, J.E. Morris, V. Verneker, P. Kondracki, and D. Finello, Models of electrical conduction in nanoparticle filled polymers, Proc. 3rd International Conference on Adhesive Joining and Coating Technology in Electronics Manufacturing, Binghamton, NY (Adhesives'98), Sept., 1998, pp. 84–89.
17. G.G.W. Mustoe, M. Nakagawa, X. Lin, and N. Iwamoto, Simulation of particle compaction for conductive adhesives using discrete element modeling, Proc. 49th IEEE Electronic Components and Technology Conference, San Diego, 1999, pp. 353–359.
18. M. Mundlein, G. Hanreich, and J. Nicolics, Simulation of the aging behavior of isotropic conductive adhesives, Proc. 2nd International IEEE Confer. Polymers and Adhesives in Microelectronics and Photonics (Polytronic 2002), Zalaegerszeg, Hungary, 2002, pp. 68–72.
19. L. Li and J.E. Morris, Structure and selection models for anisotropic conductive adhesives films, Proc. 1st International Conference on Adhesive Joining and Coating Technology in Electronics Manufacturing (Adhesives'94), Berlin, Germany, 1994.
20. A.O. Ogunjimi, S.H. Mannan, D.C. Whalley, and D.J. Williams, Assembly of planar array components using anisotropic conducting adhesives—A benchmark study, Proc. 2nd International Conference on Adhesive Joining and Coating Technology in Electronics Manufacturing (Adhesives in Electronics, '96), Stockholm, 1996, pp. 270–284.
21. D.J. Williams and D.C. Whalley, The effects of conducting particle distribution on the behavior of anisotropic conducting adhesives: non-uniform conductivity and shorting between connections, Journal of Electronics Manufacturing, 3, pp. 85–94 (1993).
22. R. Herczynski, Distribution function for random distribution of spheres, Nature, 255, pp. 540–541 (1975).
23. D.J. Williams, D.C. Whalley, O.A. Boyle, and A.O. Ogunjimi, Anisotropic conducting adhesives for electronic interconnection, Soldering and Surface Mount Technology, (14), pp. 4–8 (1993).
24. S.H. Mannan, D.J. Williams, D.C. Whalley, and A.O. Ogunjimi, Models to determine guidelines for the anisotropic conducting adhesives joining process, in J. Liu, Ed., Chapter 4 of Conductive Adhesives for Electronics Packaging, Electrochemical Publications Ltd., UK, 1999, pp. 78–98.

25. F. Shi, M. Abdulla, S. Chungpaiboonpatana, K. Okuyama, C. Davidson, and J. Adams, Electrically anisotropic conductive adhesives: a new model for conduction mechanism, Proc. 5th International Symposium Advanced Packaging Materials, Braselton, GA, 1999, pp. 163–168.
26. D. Klosterman, L. Li, and J.E. Morris, Materials characterization, conduction development, and curing effects on reliability of isotropically, IEEE Trans. Compon. Packag. Manuf. Technol. Part A, 21(1), pp. 23–31 (1998).
27. L. Li and J.E. Morris, Modeling cure schedules for electrically conductive adhesives, in J. Liu, Ed., Chapter 5 of Electrically Conductive Adhesives: A Comprehensive Review, Electrochemical Press, UK, 1999, pp. 99–116.
28. J.E. Morris and S. Probsthain, Investigations of plasma cleaning on the reliability of electrically conductive adhesives, Proc. 4th International Conference on Adhesive Joining and Coating Technology in Electronics Manufacturing (Adhesives in Electronics), Espoo, Finland, 2000, pp. 41–45.
29. M.G. Perichaud, J.Y. Deletage, D. Carboni, H. Fremont, Y. Danto, and C. Faure, Thermomechanical behavior of adhesive jointed SMT components, Proc. 3rd International Conference on Adhesive Joining and Coating Technology in Electronics Manufacturing, Binghamton, NY (Adhesives'98), Sept., 1998, pp. 55–61.
30. O. Rusanen and J. Laitinen, Reasons for using lead-free solders rather than isotropically conductive adhesives in mobile phone manufacturing, Proc. 4th International Confer. Polymers and Adhesives in Microelectronics and Photonics, Portland, Oregon, 2004, pp. 2–4.
31. Y. Wei, Ph.D. dissertation, Electronically conductive adhesives: conduction mechanisms, mechanical behavior and durability, Clarkson University, 1995.
32. J.E. Morris and J.H. Das, in J.E. Morris, Ed., Electronics Packaging Forum, Vol. 3, IEEE Press, 1994, pp. 41–71.
33. J.E. Morris and J. Das, Metal diffusion in polymers, IEEE Trans. Compon. Packag. Manuf. Technol.- Part B: Adv. Pkg., 17(4), pp. 620–625 (1994).
34. E. Sencaktar, P. Rajput, and A. Khanolkar, Correlation of silver migration to the pull-out strength of silver wire embedded in an adhesive matrix, Proc. 4th International Confer. Polymers and Adhesives in Microelectronics and Photonics, Portland, Oregon, 2004, RT2-1.
35. K.S. Moon, J. Wu and C.P. Wong, Improved stability of contact resistance of low meting point alloy incorporated isotropically conductive adhesives, IEEE Trans. Compon. Packag. Technol., 26(2), pp. 375–381 (2003).
36. C.P. Wong and Y. Li, Recent advances on Electrical Conductive Adhesives (ECAs), Proc. 4th International Conference Polymers and Adhesives in Microelectronics and Photonics, Portland, Oregon, 2004, PL-1.
37. K. Suzuki, Y. Shirai, N. Mizumura, and M. Konagata, Conductive adhesives containing Ag-Sn alloys as conductive filler, Proc. 4th International Conference Polymers and Adhesives in Microelectronics and Photonics, Portland, Oregon, 2004, MP3-1.
38. P. Smilauer, Thin metal films and percolation theory, Contemp. Phys., 32, pp. 89–102, (1991).
39. G.R. Ruschau, S. Yoshikawa, and R.E. Newnham, Percolation constraints in the use of conductor-filled polymers for interconnects, Proc. 42nd Electronic Components and Technology Conference, Atlanta GA, May 1992, pp. 481–486.
40. J.E. Morris, F. Anderssohn, E. Loos, and J. Liu, Low-tech studies of isotropic electrically conductive adhesives, Proc. 26th International Spring Seminar on Electronics Technology, Stara, Lesna, Slovak Republic, 2003, pp. 90–94.
41. J.E. Morris, F. Anderssohn, S. Kudtarkar, and E. Loos, Reliability studies of an isotropic electrically conductive adhesive, Proc. 1st International IEEE Conference Polymers and Adhesives in Microelectronics and Photonics, 2001, Potsdam, Germany, pp. 61–69.
42. R. Zallen, The Physics of Amorphous Solids, Chapter 4, Wiley, New York, 1983.
43. J.E. Morris, S. Youssof, and X. Feng, Electrically conductive adhesives for pin-through-hole applications, J. Electronics Manuf., 6(3), pp. 219–230 (1996).
44. C.P. Wong, D. Lu, and Q. Tong, Lubricants of silver fillers for conductive adhesive applications, Proc. 3rd International Conference on Adhesives Joining and Coating Technology in Electronics Manufacturing, Binghamton, NY (Adhesives'98), Sept., 1998, pp. 184–192.
45. E. Sancaktar and N. Dilsiz, Pressure dependent conduction behavior of various particles for conductive adhesive applications, Proc. 3rd International Conference on Adhesive Joining and Coating Technology in Electronics Manufacturing, Binghamton, NY (Adhesives'98), Sept., 1998, pp. 334–344.
46. U. Behner, R. Haug, R. Schutz, and H.L. Hartnagel, Characterization of anisotropically conductive adhesive interconnections by 1/f noise measurements, Polymers in Electronics Packaging Conference, Norrkoping, Sweden, 1997, pp. 243–248.

47. L.K.J. Vandamme, M. Perichaud, E. Noguera, Y. Danto, and U. Buehner, 1/f noise as a diagnostic tool to investigate the quality of isotropic conductive adhesive bonds, IEEE Trans. Compon. Packag. Technol., 22(3), pp. 446–454 (1999).
48. Y. Li, K. Moon, H. Li, and C.P. Wong, Conductive improvement of isotropically conductive adhesives, Proc. 6th IEEE CPMT Conference High Density Microsystem Design and Packaging and Component Failure Analysis (HDP'04), Shanghai, 2004, pp. 236–241.
49. J. Miragliotta, R.C. Benson, and T.E. Phillips, Measurements of electrical resistivity and raman scattering from conductive die attach adhesives, Proc. 8th International Symposium Advanced Packaging Materials, Stone Mountain, GA, 2002, pp. 132–138.
50. H. Kim, State University of New York at Binghamton, unpublished data (1992).
51. S. Kottaus, R. Haug, H. Schaefer, and B. Guenther, Investigation of isotropically conductive adhesives filled with aggregates of nano-sized Ag-particles, Proc. 2nd International Conference on Adhesive Joining and Coating Technology in Electronics Manufacturing (Adhesives in Electronics '96) Stockholm, 1996, pp. 14–17.
52. B. Guenther and H. Schaefer, Porous metal powders for conductive adhesives, Proc. 2nd International Conference on Adhesive Joining and Coating Technology in Electronics Manufacturing (Adhesives in Electronics, '96), Stockholm, 1996, pp. 55–59.
53. J. Liu, K. Gustafsson, Z. Lai, and C. Li, Surface characteristics, reliability and failure mechanisms of tin, copper and gold metallisations, Proc. 2nd International Conference on Adhesive Joining and Coating Technology in Electronics Manufacturing (Adhesives in Electronics, '96), Stockholm, 1996, pp. 141–153.
54. L. Li, J.E. Morris, J. Liu, Z. Lai, L. Ljungkrona, and C. Li, Reliability and failure mechanism of isotropically conductive adhesives, Proc. 45th Electronic Components and Technology Conference, Las Vegas, NV, May 1995, pp. 114–120.
55. H. Botter, R.B. Van Der Plas, and A.A. Junai, Factors that influence the electrical contact resistance of isotropic conductive adhesive joints during climate chamber testing, Int. J. Microelec. Pkg., 1(3), pp. 177–186 (1998).
56. D. Lu, C.P. Wong, and Q.K. Tong, Mechanisms underlying the unstable contact resistance of conductive adhesives, Proc. 49th Electronic Components and Technology Conference, San Diego, June 1999, pp. 324–346.
57. D. Lu and C.P. Wong, Conductive adhesives with improved properties, Proc. 2nd International Symposium Polymeric Electronics Packaging (PEP'99), Gothenburg, Sweden, 1999, pp. 1–8.
58. S. Wu, K. Hu, and C.P. Yeh, Contact reliability modeling and material behavior of conductive adhesives under thermomechanical loads, in J. Liu, Ed., Chapter 6 of Conductive Adhesives for Electronics Packaging, Electrochemical Publications Ltd., UK, 1999, pp. 117–150.
59. M. Dernevik, R Sihlbom, Z. Lai, P. Starski, and J. Liu, High frequency measurements and modeling of anisotropic, electrically conductive adhesive flip-chip joint, Advances in Electronics Packaging, 1997, EEP-Vol. 19-1, ASME, 1997, pp. 177–184.
60. R. Sihlbom, M. Dernevik, Z. Lai, P. Starski, and J. Liu, High-frequency measurements and simulation on the Sequential Build-Up Boards (SBU), Proc. 1st International Symposium on Polymeric Electronics Packaging (PEP'97), Norrköping, Sweden, 1997, pp. 131–139.
61. A. Kulkarni and J. Morris, Reliability life time studies on isotropic conductivity adhesives, Proc. 3rd International IEEE Conference Polymers and Adhesives in Microelectronics and Photonics, Montreux, Switzerland, 2003, pp. 333–336.
62. M.J. Yim and K.W. Paik, Design and understanding of Anisotropic Conductive Films (ACF's) for LCD packaging, IEEE Trans. Compon. Packag. Manuf. Technol., Part A, 21(2), pp. 226–234 (1998).
63. C. Oguibe, S. Mannan, D. Whalley, and D. Williams, Conduction mechanisms in anisotropic conducting adhesive assembly, IEEE Trans. Compon. Packag. Manuf. Technol., Part A, 21(2), pp. 235–242 (1998).
64. Y. Fu, Y.L. Wang, X. Wang, J. Liu, Z. Lai, G.L. Chen, and M. Willander, Experimental and theoretical characterization of electrical contact in anisotropically conductive adhesive, IEEE Trans. Compon. Packag. Manuf. Technol., Park B, Advance Packaging, 23(1), pp. 15–21 (2000).
65. R. Sihlbom, M. Dernevik, M. Lindgren, P. Starski, Z. Lai, and J. Liu, High frequency measurements and simulations on wire-bonded modules on the Sequential Build-Up boards (SBU's), IEEE Trans. Compon. Packag. Manuf. Technol., Part A, 21(3), pp. 478–491 (1998).
66. J. Kim, Proc. 2nd International Academic Conference, March 17–19, Atlanta, USA, 2000.
67. Y. Fu, J. Liu, and M. Willander, Electromagnetic wave transmission through lossless electrically conductive adhesive, Journal of Electronics Manufacturing, 9(4), pp. 275–281 (1999).
68. A. Paproth, K.-J. Wolter, T. Herzog, and T. Zerna, Influence of plasma treatment on the improvement of surface energy, Proc. 24th International Spring Seminar Electronics Technology, Romania, 2001, pp. 37–41.

69. T. Herzog, M. Koehler, and K.-J. Wolter, Improvement of the adhesion of new memory packages by surface engineering, Proc. 2004 Electronic Components and Technology Conference, Las Vegas, 2004, Vol. 1, pp. 1136–1141.
70. S. Liong, C.P. Wong, and W.F. Burgoyne, Jr., Adhesion improvement of thermoplastic isotropically conductive adhesives, Proc. 8th International Symposium on Advanced Packaging Materials, Braselton, GA, 2002, pp. 260–270.
71. L.L.W. Chow, J. Li, and M.M.F. Yuen, Development of low temperature processing thermoplastic intrinsically conductive polymer, Proc. 8th International Symposium on Advanced Packaging Materials, Braselton, GA, 2002, pp. 127–131.
72. R. Luchs, Application of electrically conductive adhesives in SMT, Proc. 2nd International Conference on Adhesive Joining and Coating Technology in Electronics Manufacturing (Adhesives in Electronics, '96), Stockholm, 1996, pp. 76–83.
73. M. Zwolinski, J. Hickman, H. Rubin, Y. Zaks, S. McCarthy, T. Hanlon, P. Arrowsmith, A. Chaudhuri, R. Hermansen, S. Lau, and D. Napp, Electrically conductive adhesives for surface mount solder replacement, Proc. 2nd International Conference on Adhesive Joining and Coating Technology in Electronics Manufacturing (Adhesives in Electronics, '96), Stockholm, 1996, pp. 333–340.
74. Q. Tong, S. Vona, R. Kuder, and D. Shenfield, Recent advances in surface mount conductive adhesives, Proc. 3rd International Conference on Adhesive Joining and Coating Technology in Electronics Manufacturing, Binghamton, NY (Adhesives'98), Sept., 1998, pp. 272–277.
75. S. Vona, Q. Tong, R. Kuder, and D. Shenfield, Surface mount conductive adhesives with superior impact resistance, Proc. 4th International Symposium and Exhibition on Advanced Packaging Materials, Processes, Properties and Interfaces, Braselton, GA, 1998, pp. 261–267.
76. S. Luo and C.P. Wong, Thermo-mechanical properties of epoxy formulations with low glass transition temperatures, Proc. 8th International Symposium on Advanced Packaging Materials, Braselton, GA, 2002, pp. 226–231.
77. S.A. Kudtarkar and J.E. Morris, Reliability studies of isotropic conductive adhesives: drop test for thermally cycled and 85/85 tested samples, Proc. 8th International Symposium on Advanced Packaging Materials, Braselton, GA, 2002, pp. 144–150.
78. Y.L. Wang, G.L. Chen, J. Liu, and Z. Lai, The contact characterization of conductive particles in anisotropically conductive adhesive using FEM, Proc. 2nd IEEE International Symposium on Polymeric Electronics Packaging, Göteborg, Sweden, 1999, pp. 199–206.
79. X. Wang, Y.L. Wang, G.L. Chen, J. Liu, and Z. Lai, Quantitative estimate of the characteristics of conductive particles in ACA by using nano indenter, IEEE Trans. Compon. Packag. Manuf. Technol. Part A, 21(2), pp. 248–251 (1998).
80. A. Zribi, K. Persson, Z. Lai, J. Liu, Y. Kang, S. Yu, and M. Willander, Effect of the bump height on the reliability of ACA made joints for flip-chip electronic packaging, 2nd International ATW on Flip-Chip Technology, Braselton, GA, 1998.
81. Z. Lai, R.Y. Lai, K. Persson, and J. Liu, Effect of bump height on the reliability of ACA flip-chip joining with FR4 rigid and polyimide flexible substrate, Journal of Electronics Manufacturing, 8(3&4), pp. 217–224 (1998).
82. C.M.L. Wu, J. Liu, and N.H. Yeung, Reliability of ACF in flip-chip with various bump height, Soldering and Surface Mount Technology, 13/1, pp. 25–30 (2001).
83. T. Sugiyama, The latest bare chip bonding by ACF, Sony Chemicals, seminar documentation at Chalmers University of Technology, 2000.
84. M. Törnvall, Proceedings of a Swedish National Seminar on Environmentally Compatible Materials Research for Electronics Packaging, IVF.
85. J. Liu and R. Rörgren, Joining of displays using thermo-setting anisotropically conductive adhesive joints, Journal of Electronics Manufacturing, 3, pp. 205–214 (1993).
86. A. Sihlbom, R. Sihlbom, and J. Liu, Thermal characterization of electrically conductive adhesive flip-chip joints, Proc. of the IEEE/CPMT Electronic Packaging Technology Conference, Dec. 8–10, Singapore, 1998, pp. 251–257.
87. A. Bauer and T. Gesang, Electrically conductive joints using Non-Conductive Adhesives (NCAs) in surface mount applications, in J. Liu, Ed., Chapter 12 of Conductive Adhesives for Electronics Packaging, Electrochemical Publications Ltd., UK, 1999, pp. 313–341.
88. C. Khoo, J. Liu, M. Ågren, and T. Hjertberg, Influence of curing on the electrical and mechanical reliability of conductive adhesive joints, Proc. 1996 IEPS Conference, Austin, Texas, 1996, pp. 483–501.
89. J. Liu, K. Gustafsson, Z. Lai, and C. Li, Surface characteristics, reliability, and failure mechanisms of tin/lead, copper, and gold metallizations, IEEE Trans. Compon. Packag. Manuf. Technol. Part A, 20(1), pp. 21–30 (1997).

90. J. Liu, P. Lundström, K. Gustafsson, and Z. Lai, Conductive adhesive joint reliability under full-cure conditions, EEP-Vol. 19-1, Advances in Electronics Packaging, Vol. 1, ASME 1997, pp. 193–199.
91. C. Khoo and J. Liu, Moisture sorption in some popular conductive adhesives, Circuit World, 22(4), pp. 9–15 (1996).
92. Q.K. Tong, D. Markley, G. Fredrickson, R. Kuder, and D. Lu, Conductive adhesives with stable contact resistance and superior impact performance, Proc. 49th Electronic Components and Technology Conference San Diego, 1999, pp. 347–352.
93. D. Lu and C.P. Wong, Development of conductive adhesives for solder replacement, IEEE Trans. Compon. Packag. Technol., 23(4), pp. 620–626 (2000).
94. H. Takezawa, T. Mitani, T. Kitae, H. Sogo, S. Kobayashi, and Y. Bessho, Effects of zinc on the reliability of conductive adhesives, Proc. 8th International Symposium on Advanced Packaging Materials, Braselton, GA, 2002, pp. 139–143.
95. O. Rusanen, Ph.D. dissertation, Adhesives in micromechanical sensor packaging, University of Oulu, VTT Publication 407, 2000.
96. O. Rusanen, Modeling of ICA creep properties, Proc. 4th International Conference on Adhesive Joining and Coating Technology in Electronics Manufacturing (Adhesives in Electronics), Espoo, Finland, 2000, pp. 194–198.
97. Z. Lai and J. Liu, Anisotropically conductive adhesive flip-chip bonding on rigid and flexible printed circuit substrates, IEEE Trans. Compon. Packag. Manuf. Technol., Part B: Advanced Packaging, 19(3), pp. 644–660 (1996).
98. Z. Lai and J. Liu, The effects of bonding force on ACA flip-chip reliability, IMAPS ATW on Flip-Chip Technology, Braselton, 1999.
99. J. Liu, A. Tolvgård, J. Malmodin and Z. Lai, A reliable and environmentally friendly packaging technology—flip-chip joining using anisotropically conductive adhesive, IEEE Trans. Compon. Packag. Technol., 22(2), pp. 186–190 (1999).
100. J. Liu and Z. Lai, Reliability of ACA flip-chip joints on FR-4 substrate, InterPACK'99, International, InterSociety, Electronic Packaging Technical/Business Conference and Exhibition, June 13–17, Hawaii, USA, Maui, 1999, pp. 1691–1697.
101. H. Westphal, Health and environmental aspects of conductive adhesives-the use of lead-based alloys compared with adhesives, in J. Liu, Ed., Chapter 18 of Conductive Adhesives for Electronic Packaging, Electrochemical Publications Ltd., UK, 1999, pp. 415–424.
102. T. Segerberg, Life Cycle Analysis—A comparison between conductive adhesives and lead containing solder for surface mount application, IVF report, 1997.
103. A. Tolvgård, J. Malmodin, J. Liu, and Z. Lai, A reliable and environmentally friendly packaging technology—flip chip joining using anisotropically conductive adhesive, Proc. 3rd International Conference on Adhesive Joining and Coating Technology in electronics Manufacturing, Binghamton, NY, 1998, pp. 19–26.
104. L. Li and J.E. Morris, An introduction to electrically conductive adhesives, Int. J. Microelectronic Packaging, 1(3), pp. 159–175 (1998).
105. J. Liu, Ed., Conductive Adhesives for Electronics Packaging, Electrochemical Publications Ltd., Port Erin, Isle of Man, UK, 1999, ISBN No. 0901150371.
106. G. Dou, D.C. Whalley, and C. Liu, Electrical conductive characteristics of ACA bonding: a review of the literature, current challenges and future prospects, Proc. 6th IEEE CPMT Confer. High Density Microsystem Design and Packaging and Component Failure Analysis (HDP'04), Shanghai, 2004, pp. 264–276.
107. J. Liu and Z. Mo, Reliability of interconnects with conductive adhesives, in S. Dongkai, Ed., Lead-free Solder Interconnect Reliability, ASM, Materials Park, OH, 2005, pp. 249–276.
108. J.E. Morris and J. Liu, An Internet course on conductive adhesives for electronics packaging, Proc. 50th Electronic Components and Technology Conference, Las Vegas, May, 2000, pp. 1016–1020.
109. J. Liu and J.E. Morris, State of the art in electrically conductive adhesives, Workshop Polymeric Materials for Microelectronics and Photonics Applications, EEP-Vol. 27, ASME, 1999, pp. 259–281.

21

Electrically Conductive Adhesives

Johann Nicolics and Martin Mündlein
Institute of Sensor and Actuator Systems, Gusshausstraße 27-29, A-1040 Wien, Austria

Abstract Electrically conductive adhesives are being used in electronic packaging for several decades. A brief review of the dynamic development of conductive adhesives under the influence of the miniaturization, the adaptation of environmental friendly manufacturing processes is presented.

With respect to the importance of isotropically conductive adhesives (ICA), a new contact model to analyze the principle influences of e.g., particle size, particle geometry, and filler content on the percolation threshold is introduced. With this model the arrangement of the particles within a contact is calculated by considering different types of forces (elastic, friction, adhesion, and inertia). Taking into account the electrical properties of the filler particles, the electrical contact behavior including its changes due to aging is investigated.

Finally, typical applications of isotropically conductive adhesives are presented. One example shows how the thermal requirements for attaching a GaAs heterojunction power transistor can be fulfilled using an adhesive with an extremely high filler content (thermal conductivity: >60 W/mK). In another case it is demonstrated how extreme thermomechanical requirements resulting from a thermal expansion mismatch of parts of a sealed IR sensor housing can be corresponded using an adhesive with a comparatively low glass transition temperature. A further example shows a packaging concept of a miniaturized, biocompatible multichip module. For mounting both, narrowly spaced SMDs and bare chips, an isotropically conductive adhesive has been applied.

21.1. INTRODUCTION AND HISTORICAL BACKGROUND

The industrial application of electrically conductive adhesives (ECA) for interconnecting and mounting electronic components on circuit carriers is comparatively new. However, a series of inventions related to various mixtures of conductive and non-conductive substances were made already in the first half of the last century. An early patent of an electrically conductive adhesive describes electromechanical applications already in 1926 [1]. In 1933 Hans Schuhmann claims a *conducting varnish* (consisting of a mixture of about 50% oil varnish, about 40% lithopone, and 10% of soot) which can be applied for shielding purposes in the radio industry [2]. Only two years later it turned out that conducting layers consisting of a binding agent mixed with a finely divided conductor lose their conductivity during operation or do not at all posses this conductivity even if applied in the form in which they are on the market [3]. In 1944 H.J. Loftis proposed a *molded electrically conductive body* consisting of Bakelite as the resin filled with graphite or metal powder and an additive of an alkaline earth metal salt providing hygroscopic properties to preserve a lubricating effect for the use as commutator brush [4]. An early description of an *electrically conductive adhesive* more similar to the modern ECA can be found in the patent of N.H. Collings et al. in 1948 on the base of a self-setting adhesive and, e.g., finely divided silver powder [5].

The first use of ECAs in electronic technology is known only in 1956: in the U.S. patent [6] *an electrically conducting cement comprising a thermosetting binding medium* is described *for fixing a semi-conducting crystal* (germanium) *to a metal base or holder*. Thereafter, the number of publications reporting experimental and theoretical investigations of electrically conductive adhesives increased continuously. Twenty years later C. Mitchel and H. Berg summarized the state-of-the-art in 1976. They reported that conductive adhesives due to their increasing reliability become more and more acceptable for many different applications. Also the most significant advantages compared to other interconnection techniques, such as low processing temperature and the possibility to use a large variety of low-cost surface fillers are already mentioned [7]. Especially for mounting chips with larger size on substrates with different coefficient of thermal expansion adhesives are superior to eutectic solder alloys due to their significantly higher flexibility. Conductive adhesive bonding started to become popular with the fabrication of plastic-foil keyboards and the use of flexible substrate materials [8].

More recently, in 2000, Alan J. Heeger, Alan G. MacDiarmid and Hideki Shirakawa are awarded the Nobel Prize in Chemistry for their discovery of electrically-conducting polymers whereby initial experiments with silvery shining films of polyacetylene lead back to the beginning of the 1970s [9].

Along with the development of liquid crystal displays (LCDs) in the 80s and their continuing miniaturization a special kind of ECA—the anisotropic conductive adhesive (ACA) has been invented for components with high-lead count and fine pitch interconnections [10,11]. ACAs allow to produce an excellent low-ohmic contact only perpendicular to the mating metalization planes whereby one-particle contact bridges between these metalizations are formed. However, due to their very low loading of conductive filler particles, ACAs remain insulating in the lateral direction. As such, no high-precision printing process is needed. In the meanwhile, ACAs are successfully applied especially for high-resolution displays in portable devices such as notebooks and PDAs (Personal Digital Assistant). Although, the capability of ACAs for comparatively high current load with even up to several Amperes per square millimeter were investigated [12], due to their low filler content their main application fields will probably remain the interconnection of components with low current consumption.

Whereas ACAs have a volume resistivity like an insulator, another type of ECA with a significantly higher degree of conductive filler has a low volume resistivity, comparable with that of conductors (Figure 21.1). They are called isotropic conductive adhesives (ICA, [13]). The goal of this chapter is to develop a deeper understanding of the contact formation process and the influence of application related parameters on the electrical contact resistance of ICAs.

The ICA technology became more interesting in the last years due to several reasons. One of the major driving forces is the increasing packaging density allowing to develop a huge manifold of complex but nevertheless light portable electronic equipments like e.g., cell phones, digital cameras, and notebooks. This development was mainly based on three key factors:

FIGURE 21.1. SEM cross-sectional view of an ICA.

- an inconceivable fineness of structures in the semiconductor level coming close to 0.1 μm and allowing a high integration level,
- a decrease of package size coming down from small outlined (SO) packages over quad flat pack (QFPs), ball-grid arrays (BGAs), and chip-scale packages (CSPs) to direct or bare chip attach techniques like flip-chips (FCs), and
- new printed circuit board manufacturing (PCB) techniques like microvia multilayer technology with plasma or laser structured vias with even less than 0.1 mm diameter allowing to realize a three-dimensional wiring net with extremely high density.

Although, soldering technology has matured through many decades, with the mentioned advances in microelectronics technology a limitation of solder paste become more and more evident in its use for fine pitch components [14]. With current solder paste technology very fine pitch interconnection become hard to handle due to soldering defects such as bridging and solder balling [15–17]. This means to connect the components on the circuit boards under these boundary conditions improved or new interconnection technologies are needed. One of these new techniques could be conductive adhesive bonding. Among these technical demands new legislative reforms make an adaptation of the existing technologies necessary. From the 1st July 2006 on member states of the European Union have to ensure that new electrical and electronic equipment does not contain lead, mercury, cadmium and other named materials. This is regulated in the directive 2002/95/EC of the European Parliament and of the Council on the restriction of the use of certain hazardous substances in electrical and electronic equipment [18]. Under these restriction the prohibition of lead is most serious for the packaging technology because the widely applied solder technology commonly uses solder pastes consisting of a tin lead alloy.

Due to their composition conductive adhesives are innately lead free. Additionally, they allow to avoid flux residues without cleaning. They need a low processing temperature but allow operation at high temperatures. Adhesives are preferable whenever high elasticity is needed to compensate for thermal expansion mismatch between substrate and component body [19–22]. However, compared to soldering, electrically conductive adhesive technology is still in its infancy [23,24]. Two critical issues of conductive adhesives for surface mount applications are the contact resistance shift during lifetime and a low impact performance [25]. As a long-term effect, the contact resistance of ECAs on ignoble metal finishes can increase dramatically during exposure to elevated temperature and humidity. The low impact performance is observed by separation of components from the board due to impact of significant shock during manufacturing and lifetime of the product. The ability of ECAs to withstand this stress can be determined with the standardized drop test [26]. Another drawback of one-component adhesives compared to soldering is also frequently cited: a lower productivity due to the long curing times. Results obtained by applying a new curing method which allows to reduce the duration of exposition to temperature to an extent comparable to that of soldering processes are also presented.

21.2. CONTACT FORMATION

Let us consider an insulating polymer mixed with a specific amount of conductive particles between a pair of parallel electrodes representing the mating pads of a contact to be formed. It is obvious that these mixture keeps non conductive as long as the particle content is small enough (below its percolation) and the particles are smaller than the gap between the electrodes. If the amount of particles is increased clusters of particles being in contact with each other are formed, whereby the mean size of these clusters increases with the growing amount of particles. From the moment when the first cluster touches both electrodes an electrical connection is formed. If the amount of filler is further increased additional clusters merge successively to one big cluster which leads to an improvement of the electric contact between the two mating pads (Figure 21.2).

The amount of filler particles needed to get the particle-adhesive mixture electrically conductive depends on microscopic geometrical parameters like shape, distribution, and orientation of the particles. The principle influences of these parameters can be analyzed using the percolation theory.

21.2.1. Percolation and Critical Filler Content

The percolation theory is a branch of probability theory dealing with the properties of randomly distributed media in general [27]. One main idea of the percolation theory

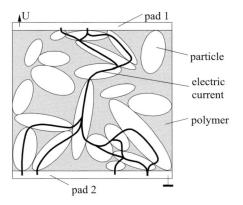

FIGURE 21.2. Schematic cross section of an isotropically conductive joint.

is to predict the macroscopic behavior of a randomly distributed composite material in dependence on its microscopic properties which are expressed purely by statistical data. In case of an isotropically conductive adhesive this composite material is the mixture of the non-conductive polymer and the conductive filler particles. In the simplest case (infinite conductivity of particles and infinite resistivity of the filler) the macroscopic behavior of the joint is described by the probability for a connection path between the two mating pads through the filler particles. Microscopic properties are amount, orientation, shape, size or size distribution of the filler particles. The percolation theory shows that there exists a distinct threshold particle content in excess to which the probability for the formation an *infinite* cluster of particles and therewith for an electrical conductance is bigger than zero and which rapidly increases with rising particle content. This threshold is called the percolation threshold [28].

The percolation theory is valid for infinite systems which means that the particle size is negligible with respect to the distance between the electrodes. Hence, this theory is unable to consider the influence of the particle size on the relation between filler content and percolation probability. However, how important the parameter "particle size" is can be seen by considering an ACA forming a contact with a filler content much below any percolation threshold. In many practical applications of ICAs the relation between the path length (distance between the mating pads in perpendicular direction) and the particle size is only one or two orders of magnitude which causes deviations between ideal (*infinite*) systems as considered by the percolation theory and real ones. For this reason it is important to distinguish between infinite and finite systems. The percolation threshold in an *infinite* system would correspond to the so called *critical filler content* in a real joint. In real (= *finite*) systems only clusters with a limited number of particles are necessary for an electrical connection between the mating borders. The smaller a contract system with respect to the particle size the mellower is the transition between the non conductive and the conductive state in the percolation curve or more precisely: in the conduction probability-versus-filler content function (a comparison with the ideal percolation curve can be seen in Figure 21.5). Nevertheless, here the terminology *percolation curve* is used defining the *percolation threshold* as the point on which the probability of an interconnection is 50% for real systems. (For definitions, symbols and terminology used in this chapter refer to the Section Notations and Definitions.)

21.2.2. ICA Contact Model

Using the percolation theory, it is possible to show principle influences on the properties of isotropically conductive adhesives depending on the geometric properties of the particles. But in order to get a deeper understanding of the processes inside a conductive adhesive joint additional factors have to be considered. Such factors are the volume electrical conductance of the filler particles, the contact resistance between the particles, and the particles and the pads, the influence of surface films on the contact resistance, the mechanical force between the particles, and the effects of oxidation and galvanic corrosion on the electric resistance of the joint [29].

For this purpose, a two-dimensional (2D) model of a complete joint including a numeric simulation was developed which is capable of calculating the particle alignment and density distribution as well as the voltage and current distribution inside the model area. On this base the total DC-resistance of the modeled joint was calculated. The model is based on the following considerations and assumptions [30]:

- the modeled system is two-dimensional,
- the considered area is rectangular,
- the pads are on the top and on the bottom borders of the model area,
- the conductivity of the polymer is neglected,
- the particles have elliptic shape with arbitrary aspect ratio (length ratio of major to minor axes),
- the particles may initially have an arbitrary orientation and can overlap whereby the condition for the final orientation and position of the particles is found by a simulation of the particles' alignment as described below by the assumption that the size of the overlapping area is directly related to a repulsion force between the particles.

A 2D simulation of a percolation problem shows a systematic difference to the three-dimensional (3D) problem. In a 2D percolation the threshold is shifted to higher filler contents. In the case of an ICA this can be observed by an increase of the contact resistance when the thickness of an adhesive layer is reduced [31–33]. Despite of this fundamental difference between the 2D and 3D model the percolation theory shows that the principle behavior of a random particle network obeys the same rules [34]. This means the absolute values derived from the 2D do not directly correspond to the 3D case but show the analog dependencies on the particle parameters like shape, orientation or distribution. However, the computational effort can be reduced drastically for a 2D numerical simulation and the computing capacity saved in this way can be used to investigate a higher number of different dependencies. For this reason, the presented simulation is restricted to the 2D case and can consider the change of size of the contact areas at the mating surfaces of the particles in dependence of the contact force. Moreover, the simulation considers the dependency of the contact resistance on particle shape, particle orientation, etc.

21.2.2.1. Simulation of the Particles' Alignment The simulation is capable of establishing the geometrical alignment of the elliptical particles based on mechanical effects like inertia, conservation of momentum, Hooke's law, and friction. This method is known as discrete element modeling (DEM) which is an explicit numerical method where individual separate elements (particles) react with their neighbors by their contact areas through friction and adhesion (DEM was first applied by Peter Cundall in 1971 to geotechnical and granular flow problems [35]).

For each time step an overlapping between the particles at their contact points is assumed in order to consider a pressure-dependent deformation of the particles. Figure 21.3 illustrates the forces of one particle in a non-equilibrium calculation step. The considered particle is in contact with another one and the upper pad. Elastic forces F_{e1} and F_{e2} are defined as functions of the size of the respective overlapping area whereby the direction of these forces is perpendicular to the chord between the intersection points S_{1a}, S_{1b}, S_{2a}, S_{2b}.

In order to define an electrical contact resistance between neighboring particles a contact force between them has to be present. In an ICA these contact forces are assumed to result from shrinking of the polymer during the curing process and have to be considered in order to simulate the DC-resistance of an ICA joint [36,37]. For this reason, an attracting force F_a between intersecting particles resulting from the compaction of the particle network during the curing of the polymer is introduced. The direction of this attracting force is antiparallel to the respective elastic force. The norm is proportional to the length of the chord between the intersecting points.

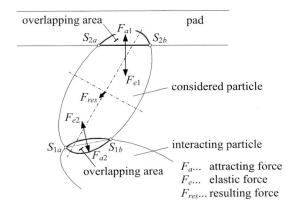

FIGURE 21.3. Schematic illustration of the overlapping area between adjacent particles, and mechanical forces leading to a rearrangement of the particles.

For each particle these attracting and repulsing elastic forces at different contact points can be added-up to one resulting force F_{Res} which causes a rearrangement. This rearrangement motion of the particles is controlled by inertia, a damping force and a friction force. Similarly, the resulting torsional moments are calculated for each particle. The distribution of the resulting rearrangement forces and moments are successively optimized in a looped calculation procedure leading to the final particle arrangement. Depending on the relation between the attracting force and the elastic force a defined overlapping between the particles remains at the equilibrium state.

21.2.2.2. Simulation of the Total Joint Resistance After the particle alignment is being calculated the electric resistance of the joint is simulated. This is done in several steps. The first step is to separate each single particle from the base model, whereby the voltage distribution inside the respective particle is calculated using a finite difference method [38]. As boundary condition, the potentials of the contact areas to the neighboring particles are defined. When the voltage distribution is calculated for $n - 1$ different sets of boundary conditions it is possible to calculate an admittance matrix describing the particle as $(n + 1)$-pole, where n denotes the number of neighbored particles. (The $(n+1)$ pole as an additional pole located in the particle center is introduced for simplifying indexing and set-up of the equation system.) This idea is depicted in Figures 21.4(a) and (b).

After an admittance matrix is being calculated for every particle, the resulting $(n+1)$-poles are connected to one network which contains the information on the particle distribution of the entire joint including the pad arrangement. The effort for calculating the contact resistance to the required extent is reduced by excluding particles which do not contribute to the current transportation from the network. In order to model more realistic contact properties a transition resistance between the particles is taken into account. For this purpose, additional resistors (R_{pp} and R_{ppad}) between the respective connections of the $(n + 1)$-poles are introduced [Figure 21.4(c)]. The value of each individual transition resistance depends on the contact force which controls the size of the overlapping area. As a measure for the size of the overlapping area and therewith for the contact resistance between the two considered particles the length of the intersection line (e.g., $w_s = \overline{S_{1a}S_{1b}}$ in Figure 21.3) is used. The interface conductivity between two particles is then obtained as product of a normalized specific conductivity λ_{pp} and the length of the intersection line w_s

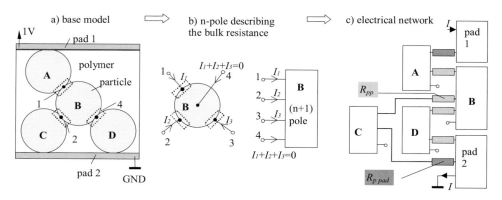

FIGURE 21.4. Basic principle for calculating the electric resistance.

of the involved particles. In turn, the normalized interface resistance R_{pp} between two particles or $R_{p\,pad}$ between a particle and a pad are calculated according to following relation, respectively:

$$R_{pp} = \frac{1}{\lambda_{pp} \cdot w_s} \quad \text{or} \quad R_{p\,pad} = \frac{1}{\lambda_{p\,pad} \cdot w_s}. \tag{21.1}$$

As final step the equations of this network are solved. Thereafter, the potential at each contact point is found and the total joint resistance is calculated.

21.2.3. Results

21.2.3.1. Percolation Behavior The established model allows us to analyze different fundamental properties of conductive adhesive joints. In the first step, only the percolation behavior is investigated. For this purpose, only the particle alignment is calculated and the probability of an interconnection p_p from one pad to the other is determined as a function of the filler content ϕ in % of area. For all simulations, a quadratic model area with a normalized edge length of 1 is chosen. Figure 21.5 shows the dependence of p_p for circular particles with different diameters d. It can be seen that the slope of the curve decreases for larger diameters in accordance with the theory. In the literature for the case of *hard-core* circles (neighboring particles only touch at one point), a critical filler content ϕ_c lies between 0.45 and 0.55 [34].

In Figure 21.5 the results of the simulation with the percolation theory is compared. At an interconnection probability of $p_p = p_c \equiv 0.5$ (which is defined as the percolation threshold for the finite system) the simulation provides filler content values of ϕ_c between 0.55 and 0.57. For the smaller particles the higher values are valid. The smaller the particles are, the more the infinite system is approached (which is described exactly by the percolation probability).

A further fact in this model is the deformability of the particles which corresponds to a combination of the *hard-core* and the *soft-core* model where the particles may touch along a line (not just a point) due to arbitrary overlapping. In the literature one can find the value $\phi_c = 0.68$ for the *soft-core* circles system [39].

In an ICA joint the electrical conductance is provided by the conductive particles, and the mechanical strength derives only from the polymer. This means: to preserve the

ELECTRICALLY CONDUCTIVE ADHESIVES

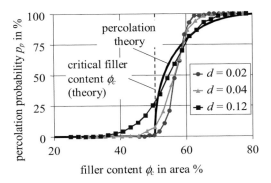

FIGURE 21.5. Comparison of percolation probability and probability of interconnection with circular particles as a function of the filler content.

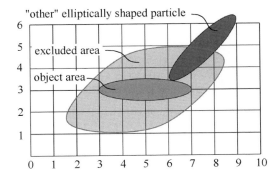

FIGURE 21.6. Excluded area for ellipses with the same shape but different orientation.

mechanical stability of the joint the particle content has to be as small as possible. This can be obtained with highly elongated particles like flakes, sticks, or tubes. However, for non-circular particles, a given filler content will lead to a different interconnection probability. In order to understand the influence of the particle shape on the filler content area which is necessary to obtain e.g., the 50% threshold to form an interconnection it is helpful to start at the definition of the so called *excluded volume* which is reduced to the *excluded area* in the two-dimensional case. The excluded area A_{ex} of an object (particle) is defined as the object area itself plus the area around this object into which the center of another similar object is not allowed to enter if overlapping of these two objects is to be avoided [34]. For example, for circles with the radius r the excluded area A_{ex} is $4r^2\pi$. For two ellipses of the same shape but different (constant) orientation the excluded area is shown in Figure 21.6. From a heuristic consideration it becomes obvious that the percolation threshold is lowered if the excluded volume is increased.

The excluded area is increased when circular particles are stretched to ellipses. The bigger the aspect ratio at a constant particle area the bigger is the excluded area. This tendency is shown in Figure 21.7 where three cases of excluded areas are formed with two ellipses of the same shape and size (relation between major and minor axis $X = 16$), respectively, but different mutual orientation ($\delta = 0°$, $\delta = 45°$, and $\delta = 90°$). One can see that the excluded area increases significantly with rising angle δ. The maximum possible value of the excluded area also depends on the shape of the particles. Figure 21.8 depicts

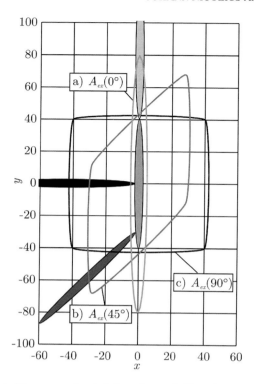

FIGURE 21.7. Comparison of three cases of excluded area for equal ellipses but different orientation: (a) $\delta = 0°$, $A_{ex} = 4 \cdot A$, (b) $\delta = 45°$, $A_{ex} = 16.5 \cdot A$, (c) $\delta = 90°$, $A_{ex} = 22.5 \cdot A$; relation between length of major and minor axis of ellipses $X = 16$.

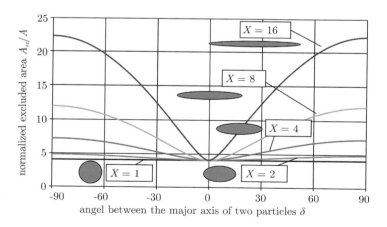

FIGURE 21.8. Normalized excluded area versus angle δ between elliptic particles with various aspect ratios ($1 \leq X \leq 16$).

the normalized excluded area versus angle δ between elliptic particles with various aspect ratios ($1 \leq X \leq 16$). It should be noted that the dependence of the excluded area on the orientation of the particles with respect to each other becomes stronger the more oblong the particles are.

ELECTRICALLY CONDUCTIVE ADHESIVES

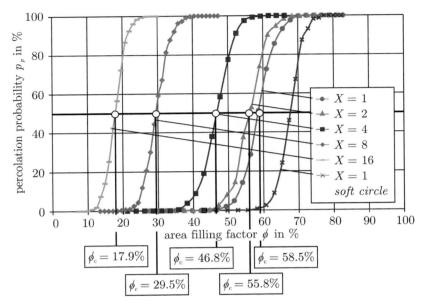

FIGURE 21.9. Probability of interconnection as a function of the filler content for elliptic particles with the same area but different aspect ratio.

The influence of the excluded area on the percolation behavior can be recognized in Figure 21.9 where the percolation probability p_p for elliptical particles with different aspect ratio is depicted as function of the filler content ϕ. It can clearly be seen the decreasing critical filler content (the respective filler content at a probability level of 50%) with the increasing aspect ratio of the particles. All particles in Figure 21.9 have the same area as a circular particle with the diameter $d = 0.04$.

A further important parameter is the mean number B of neighboring particles. With increasing filler content the number of particles being in contact with each other grows. On reaching the percolation threshold the following relation is valid:

$$B_c = \phi_c \frac{\langle A_{ex} \rangle}{\langle A \rangle}, \qquad (21.2)$$

where B_c is the critical mean number of neighboring particles, $\langle A \rangle$ is the mean particle area, and $\langle A_{ex} \rangle$ is the mean excluded area. For circular particles B_c varies in the range between 1.8 and 2.2 [34]. One can expect that this range should also be valid for elliptically shaped particles. However, the values of ϕ_c obtained from the simulation described here are consistently some percent bigger than those from the percolation theory. These differences derive from two factors: One as already mentioned above is that the percolation threshold derived from the percolation theory is defined differently as ϕ_c. The relation between the particles' size and the ICA layer thickness is considered in the simulation more realistically as a finite value which means that p_p is always higher than the theoretical percolation probability. The second factor is that a contact force was introduced in this model which equalizes the contraction forces (e.g., from the cured polymer binder) by a variable overlapping area between the particles. This leads to an increased mean value of neighboring particles B_c and changes the system in direction to the soft core model. Moreover, B_c also

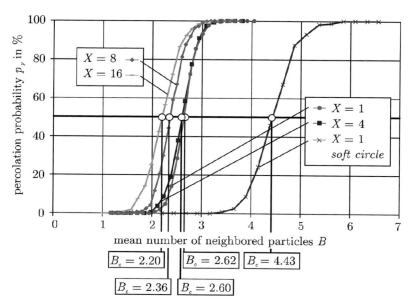

FIGURE 21.10. Percolation probability p_p versus the number of neighbor particles B for elliptical particles with different various aspect ratios ($1 \leq X \leq 16$).

varies with the particle geometry as can be seen in Figure 21.10 depicting the percolation probability versus the number of neighbor particles for elliptical particles with various aspect ratios ($1 \leq X \leq 16$). The diagram demonstrates that a higher number of neighboring particles with circular than with oblong form is needed to obtain the same percolation probability. For comparison, in Figure 21.10 the respective curve for soft circular particles is also displayed. In this case $B_c = 4.43$ which agrees with the respective value reported from Balberg et al. [34]. This comparison allows to understand the fundamental difference between *hard-core* and *soft-core* model.

Typical particle arrangements for illustration of the influence of the particle geometry on the respective filler content necessary to obtain an electrical connection between the pads are depicted in Figure 21.11. The particles have all the same area and in all cases the percolation probability is 50% ($\phi = \phi_c$). However, only arrangements are selected which provide at least one electrical connection between the mating pads. Particles contributing to an electrical path are displayed dark whereas the light particles are isolated or are linked to clusters which have electrical contact at most with only one pad.

In all subpictures it is conspicuous that only a comparatively small percentage of particles is involved in the current transportation. Frequently the current path is necked to a single chain of particles. This fact is directly related to the condition of a percolation probability far below 100%. The electrical consequences are discussed more thoroughly in the next subsection. In particular a strong dependence of the critical filler content on the particle geometry becomes obvious: ϕ_c varies from almost 55% for circular particles to less than 21% for elliptical particles with an aspect ratio of $X = 16$.

Until now we have considered the percolation probability under the condition that the particles may have an arbitrary orientation, i.e., δ was uniformly distributed in the interval from $-90°$ to $+90°$. However, oblong particles like silver flakes in an epoxy adhesive (in order to consider an example of the system most important for practical applications) may

ELECTRICALLY CONDUCTIVE ADHESIVES

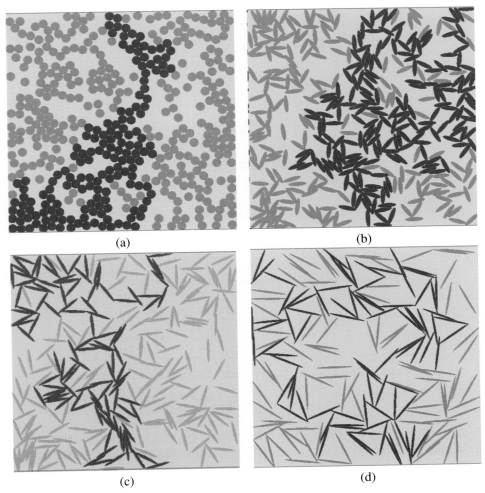

FIGURE 21.11. Particle arrangements with various filler content and various aspect ratio characterized by a percolation probability of 50%, respectively (all particles have the same area). (a) 445 particles, axis ratio 1, filler content 54.5%. (b) 380 particles, axis ratio 4, filler content 46.0%. (c) 255 particles, axis ratio 8, filler content 30.5%. (d) 170 particles, axis ratio 16, filler content 20.8%.

underlay constraints e.g., in the vicinity of the plane contact pads which, to some extent, force a parallel alignment of the particles as can be observed in microsections of ICA bonds. Therefore, a more realistic model should consider these constraints by limiting the freedom of orientation. In Figure 21.12 the effect of different degrees of freedom on the mean normalized excluded area $\langle A_{ex} \rangle$ is demonstrated. A restriction of the orientation to e.g., $-30° \leq \delta \leq +30°$ leads to a reduction of $\langle A_{ex} \rangle$ to about the half of the value for the unlimited freedom of orientation ($-90° \leq \delta \leq +90°$). If all particles are parallel ($\delta \approx 0$) the minimum of $\langle A_{ex} \rangle = 4$ is obtained. The practical meaning of this consideration is, that a higher filler content is needed in cases of a limited freedom of particle orientation.

Whereas in an infinite system the influence of the contact pads on the orientation of the particle does not exist, in the considered limited system this influence plays an impor-

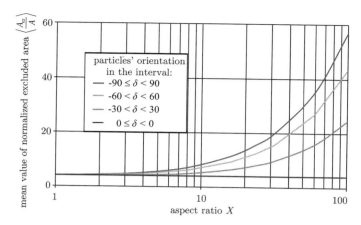

FIGURE 21.12. Mean normalized excluded area $\langle A_{ex} \rangle$ versus aspect ratio for different limitation degrees of possible particle orientation. Within the respective intervals δ is assumed to be uniformly distributed.

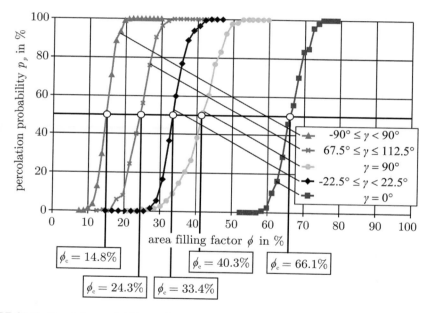

FIGURE 21.13. Percolation probability as function of the filler content for two cases of limited freedom of particle orientation ($\gamma = 0°$ and $\gamma = 90°$) and three different ranges of freedom (γ within $\pm 22.5°$ to the horizontal axis, $\pm 22.5°$ to the vertical axis, and no restriction: $-90° \leq \gamma \leq 90°$).

tant role. From a heuristic consideration one may expect a lower required filler content if the particles are oriented perpendicular to the pad surfaces. In order to analyze the influence of the particle orientation with respect to the contact pads the angle γ between the long particle axis and the pad plane (horizontal in this case) is introduced. The highest filler content is needed if all particles are aligned parallel to the pad plane ($\gamma = 0°$). On a first glance one could expect the lowest critical filler degree ϕ_c in the case when all particles are oriented perpendicular ($\gamma = 90°$). However, this is not the case. An analysis of the

ELECTRICALLY CONDUCTIVE ADHESIVES

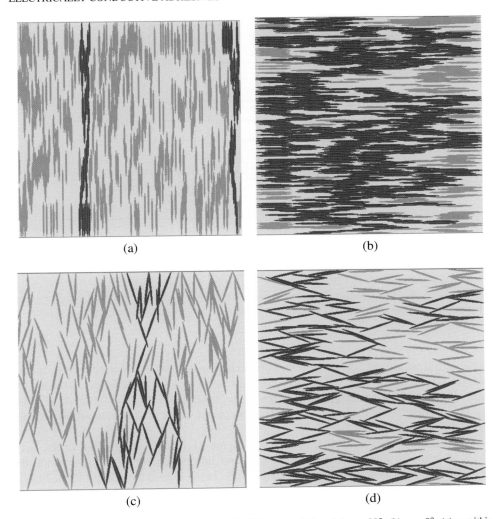

FIGURE 21.14. Accidental particle arrangements for different conditions (a) $\gamma = 90°$, (b) $\gamma = 0°$, (c) γ within $\pm 22.5°$ to the vertical axis, and (d) $\pm 22.5°$ to the horizontal axis. All particles have the same area $(0.02^2 \cdot \pi)$, the same aspect ratio $X = 16$, the percolation probability is 50%. Particles contribution to the interconnection between the pads are displayed in dark color. (a) 365 particles, $\gamma = 90°$, filler content 41.6%. (b) 609 particles, $\gamma = 0°$, filler content 66.1%. (c) 199 particles, $67.5° \leq \gamma \leq 112.5°$, filler content 24.2%. (d) 279 particles, $-22.5° \leq \gamma \leq +22.5°$, filler content 33.3%.

percolation probability as function of the filler content clearly shows a lower critical filler content in cases when the particle orientation may vary to some extent (γ within $\pm 22.5°$ to the horizontal axis, $\pm 22.5°$ to the vertical axis, and no restriction: $-90° \leq \gamma \leq +90°$) compared to cases without any freedom ($\gamma = 0°$ and $\gamma = 90°$, Figure 21.13). The lowest possible critical filler content ϕ_c is again reached with an unlimited fan out of the particle orientation.

The role of the particle orientation for the percolation probability is visualized in Figure 21.14 along with four accidental particle arrangements as calculated by this simulation for the different conditions (a) $\gamma = 90°$, (b) $\gamma = 0°$, (c) γ within $\pm 22.5°$ to the vertical

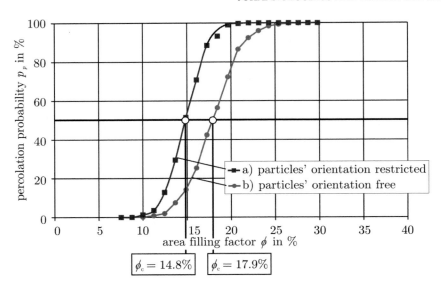

FIGURE 21.15. Percolation probability p_p versus particle filler content ϕ in case of (a) a model with purely translational particle rearrangement and (b) a model with translational and rotational rearrangement ($A = 0.02^2 \cdot \pi$, $X = 16$).

axis, and (d) within $\pm 22.5°$ to the horizontal axis. These simulations were carried out under following assumptions: All particles have the same area, the same aspect ratio, and the filler content is chosen such that the percolation probability is 50% ($A = 0.02^2 \cdot \pi$, $X = 16$, $p_p = p_c$). Again, only those arrangements are depicted where an interconnection between the pads is achieved.

From a comparison of Figure 21.9 and Figure 21.13 one can see that for the same cases of particles with elliptic shape $X = 16$ and no restrictions for the orientation the percolation threshold ϕ_c in one case is almost 18%, in the second case only 14.8%. The respective curves are shown in Figure 21.15. The difference of 3.2% results from the fact that the simulation of the particle arrangement has been carried out in different ways: In the first case the simulation started with the placement of particles with randomly and uniformly distributed orientation and only translational movements were allowed for the rearrangement of the particles. In this way the equidistribution of the orientation was constrained. By contrast, in case of the left curve of Figure 21.15 after placement of the particles besides the translational also a rotational rearrangement was permitted. These rotations allow the particles to increase the extent of parallel orientation of neighbored particles. This effect can counteract against the equidistribution of the orientation and in turn lead to an increase of the percolation threshold. The more the filler content is increased the more significant becomes this effect of paralleling of particles. This effect is clearly visible in Figure 21.16 showing a model with 400 particles being allowed to rotate after their placement. Due to the comparatively high filling degree of $\phi = 45\%$ a high degree of parallel clustering can be observed. Another aspect can also be recognized from this figure: not only the total number but also the percentage of electrically active particles (dark colored) is increased significantly when the percolation probability exceeds the threshold value. The number of parallel current paths per unit of contact area is of great practical importance for joints with high current loads. It must be noted that in any case the current is necked to the microscopic

FIGURE 21.16. Model of arrangement with 400 particles. A filler content significantly above the percolation threshold ($\phi = 45\%$) leads to paralleling of particles ($A = 0.02^2 \cdot \pi$, $X = 16$).

interfaces between the particles where an extremely high current density can be reached. In order to increase the number of interparticle contacts the filler content must be far enough beyond the percolation threshold.

In Figure 21.17 a microsection of a real ICA joint is shown. Some regimes with a distinct uniformity of orientation becomes visible. It can be assumed that the final particle arrangement in a real joint does not only depend on production process parameters but also varies e.g., during the printing process and during placing the components due to plastic deformation. As can be recognized in this figure the microhomogeneity is not ensured. The distribution of the particle orientation may vary significantly in a volume with an elongation of several ten to hundred particles. Although the properties of an ICA cannot be quantified accurately with this two-dimensional model, it just provides a deeper understanding of some phenomena and can be used to predict tendencies. The next section deals with the investigation of electric joint properties on the base of simulated particle arrangements.

21.2.3.2. Joint Resistance For a reliable isotropically conductive adhesive joint the percolation probability has to be 100%. In the simulation this means in the case of elliptical particles with an axis ratio of $X = 4$ a filler content of $>60\%$ is needed. In order to establish the percolation probability p_p and the average value of the total joint resistance as functions of the filler content ϕ, 600 simulations with accidental starting condition and identical parameters (but different particle arrangements) were performed for every ϕ value. Results are shown in Figure 21.18. In order to get the resistance curve also for filler contents below a percolation probability of 100% it was necessary to exclude particle arrangements without electrical connection between the contact pads from this consideration, since they would have led to an infinite resistance. However, the lowered number of evaluable simulation results is related to an increase of uncertainty in the resistance calculation which causes widening of the average deviation of the resistance curve.

For simplicity, in a first approximation the contact resistance between the particles is assumed to be zero (ideal contact) and the bulk conductivity σ_{pad} of the two pads is assumed to be 1000 times that of the particles' one ($\sigma_{pad} = 1000 \cdot \sigma$) so that the potential

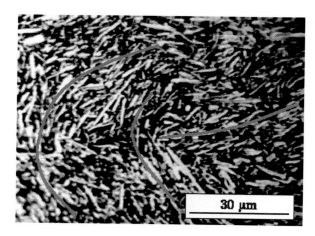

FIGURE 21.17. Microsection of an ICA joint demonstrating small areas with a distinct uniformity of orientation (along red lines). An orientation structure becomes visible.

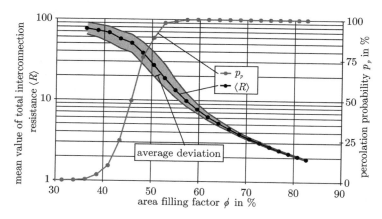

FIGURE 21.18. Probability of interconnection and the average joint resistance versus the filler content (elliptic particles with an axis ratio of $X = 4$).

in the pads can be considered as constant and the total joint resistance depends only on the particle resistance. In order to receive results which are independent from the system size a specific conductivity of the particles of $\sigma = 1$ is introduced. For the calculation of the joint resistance this means: If the quadratic model area would be completely filled with particle material ($\phi = 100\%$) the joint resistance would be 1. At a more realistic filler content of 77% the joint resistance reaches a value of 2.6. In the simulation the first contact occurred at $\phi = 38\%$ leading to a considerably higher joint resistance of 67. In between these two points, at $\phi = 60\%$, the interconnection probability just reaches the 100% value. At this level the normalized total joint resistance amounts 9.4. This shows that the major change of the contact resistance occurs in a range where the percolation probability is below 100%. By increasing the filler content above this point a strongly decreasing width of the average deviation of the joint resistance R can be observed indicating a regime of reliable interconnection.

ELECTRICALLY CONDUCTIVE ADHESIVES

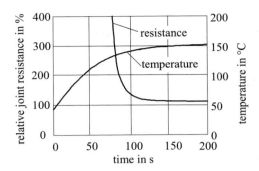

FIGURE 21.19. Change of the joint resistance during curing process.

Although, the filler content as a volume ratio is difficult to compare with that of a two-dimensional model, from a microsectional view of a real ICA joint one would obtain a value ϕ far beyond 60% which means definitely an interconnection probability of 100%. Nevertheless, an ICA just after being dispensed or printed on a substrate behaves like an insulator. The low joint resistance corresponding to an interconnection probability of 100% arises only successively during the curing process. In Figure 21.19 such a break-down of the joint resistance at the beginning of the curing process is depicted. This phenomenon is based on a strongly changing contact resistance R_{pp} between the particles which are brought into an intimate contact during the development of the contraction forces of the polymer binder as described in Section 21.2.2. It can be considered in this model e.g., by introducing a time and temperature dependent function describing the effect of the curing process. Such a simulation shows that the interface resistance between the particles and certainly also between particles and pads are the most significant parameters for modeling an ICA joint. In order to take into account different interface effects like corrosion at the pad-particle interfaces or oxidation between particles, both effects known as aging phenomena, the specific interface conductivity λ_{pp} between two particles and λ_{ppad} between particle and pad are introduced.

By contrast to Figure 21.18 where all transition resistances at interfaces where neglected let us assume the more realistic case of a finite conductivity at the interfaces. Taking the same particle arrangement but values of λ_{pp} and λ_{ppad} between 1 and 0.001 one can observe that in a logarithmic scale the total joint resistance as functions of the filler content have the same shape but are shifted to higher $\langle R \rangle$ values. This means that the effect of a change of transition conductivity is almost the same at any degree of filler. How strong this effect is, depends on the relation between the conductivity of the particles and their interfaces. The lesser the interface conductivity is, the more it controls the total joint resistance.

By keeping the interface conductivity between the particles constant at $\lambda_{pp} = 1$ and varying the one between particles and pad dependencies of the total joint resistance on ϕ are obtained (Figure 21.21). This corresponds to the practical case of a joint degradation e.g., due to galvanic corrosion at the interface between metalization and ICA.

It can be recognized that now the effect of λ_{ppad} on the total joint resistance is much weaker than in the case when $\lambda_{pp} = \lambda_{ppad}$ (Figure 21.20). A change of interface conductivity λ_{ppad} does not appear any more as resistance multiplier but rather acts as a supplement to the resistance of the ICA fill. This becomes clearer by considering the pad surfaces as equipontential planes and by defining a total transition resistance $\langle R_{tr} \rangle$ consisting of inter-

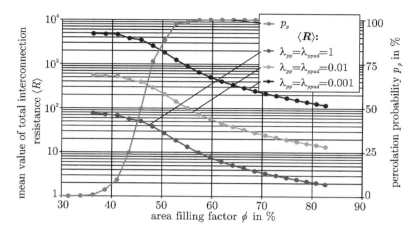

FIGURE 21.20. Normalized total joint resistance and percolation probability versus filler content for various values of normalized interface conductivity under the condition of $\lambda_{pp} = \lambda_{ppad}$.

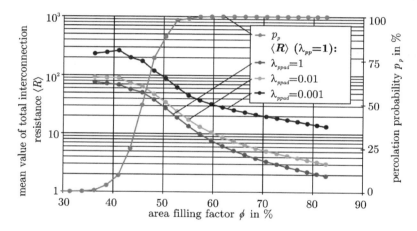

FIGURE 21.21. Normalized total joint resistance like in Figure 21.20 but for constant normalized particles' interface conductivity ($\lambda_{pp} = 1$).

face resistances of only those n particles touching the pad which are also involved in the current path according to the following relation:

$$\langle R_{tr} \rangle = \frac{1}{\sum_i^n \lambda_{ppad} \cdot w_{s,i}} = \frac{1}{\lambda_{ppad}} \cdot \frac{1}{\sum_i^n w_{s,i}}, \quad (21.3)$$

where $w_{s,i}$ is the length of the intersection line of the ith particle. The sum

$$w_s^* = \sum_i^n w_{s,i} \quad (21.4)$$

of all these lengths can be understood as *effective contact length* to which the current is necked at the respective pad interface. The total pad interface resistance is the sum of both

ELECTRICALLY CONDUCTIVE ADHESIVES

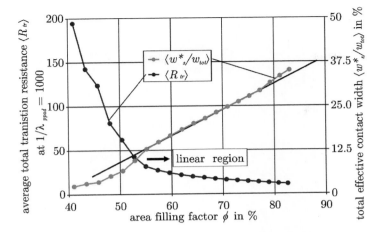

FIGURE 21.22. Total transition resistance at both pad-particle interfaces and normalized effective contact length as functions of the filler content.

interface effects (which are equal for both pads in this case but one could consider different interfaces as well):

$$\langle R_{tr} \rangle = \frac{2}{\lambda_{ppad} \cdot w_s^*}. \qquad (21.5)$$

This resistance can again be calculated as function from the filler content. A comparison with the results depicted in Figure 21.20 shows that it corresponds very well to the difference of the total joint resistance functions of the two cases $\lambda_{pp} = \lambda_{ppad} = 1$ and $\lambda_{pp} = \lambda_{ppad} = 0.001$. This is illustrated by Figure 21.22. This diagram also shows that the average effective contact width normalized to the total width w_{tot} of the modeled space demonstrates a linear increase with the filler content in a wide regime above the 100% percolation threshold. In Figure 21.23 the particle arrangement of a model with 70% filler content is depicted where the effective contact length amounts around 25% of the total model width.

For practical applications it is important to note that there is a distinct decrease of the interfacial resistance with increasing filler content. One could expect an improvement of quality of an ICA joint with rising filler content. However, in this case the bonding forces resulting from the adhesion of the polymer to the pad surface are reduced to the complement part of the effective contact area. This consideration allows to understand that a compromise between high mechanical strength and low total joint resistance has to be found which depends on the respective application field.

A further parameter influencing the joint resistance is the particle shape, since this is a controlling parameter of the percolation behavior as discussed before. For this purpose the joint resistance for various aspect ratios of elliptic particles is investigated. Figure 21.24 shows the results for the cases $X = 8$, $X = 4$, and circular particles, when $\lambda_{pp} = \lambda_{ppad} = 1$. The results clearly show qualitatively similar dependencies versus the filler content whereby the critical filler content is shifted in almost equal steps from 30%

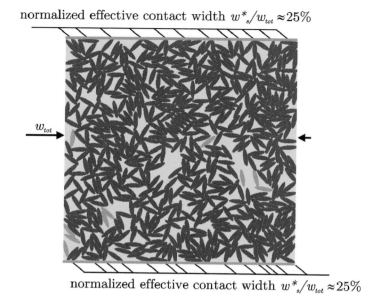

FIGURE 21.23. Particle arrangement of a model with 70% filler content providing a relation of around 25% effective contact width to total model width.

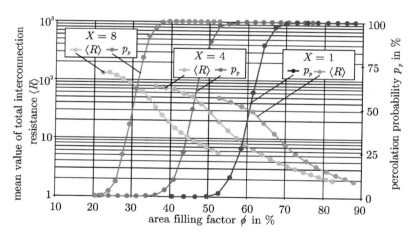

FIGURE 21.24. Total joint resistance and percolation probability versus filler content for particles with various aspect ratios.

for the oblong particles to around 60% for the circular ones. This comes out clearer by introducing the deviation $\Delta\phi$ of the filler factor ϕ from critical value ϕ_c as

$$\Delta\phi = \phi - \phi_c. \tag{21.6}$$

This transformation shifts the results of Figure 21.24 such that the point of percolation remains the same for all three cases and shows that also the minimum additional

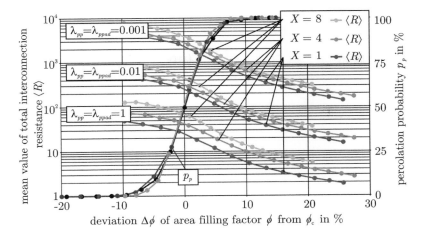

FIGURE 21.25. Results from Figure 21.24 but scaled as functions of the deviation $\Delta\phi$ of the filler content from the critical one for various cases of interfacial conductivity.

amount of filler content $\Delta\phi_{min}$ which is needed in excess to ϕ_c to allow to form a reliably operating joint is the same in all three cases ($\Delta\phi_{min} \approx 10\%$, Figure 21.25).

The practical meaning of this fact is that if the critical filler content of a system with a certain particle shape is known the necessary filler content of a system with a new particle shape can be found by increasing the filler content at least by the same percentage $\Delta\phi_{min}$ as in the known system.

Having in mind the afore mentioned relation between the filler content and the mechanical qualities one would undoubtedly prefer adhesives with long particles, since the lower the filler content is the bigger are the expectable bonding forces. However, it turns out that the joint resistance at the same $\Delta\phi$ is also shifted to higher values in case of the longer particles: In the case of $X = 8$ the joint resistance is between two and three times as high as in case of $X = 1$. Since the electrical parameters of all simulations ($X = 1$, $X = 4$, and $X = 8$) remained unchanged the shift of joint resistance can only be explained by geometrical effects such as the size of overlapping areas characterizing the intensity of the contacts, the number B of neighboring particles which is responsible for the density of the resistance network, and the geometry of the particles themselves. However, it can be shown that B doesn't change significantly in all discussed cases. Thus, B is not responsible for the big resistance changes. If it is further determined that the size of the overlapping contact areas and therewith the respective widths $w_{s,i}$ of the particles are only slightly depending on the filler content one can realize that the main part of the joint resistance change has to be assigned directly to the particle shape.

Besides the results of the case $\lambda_{pp} = \lambda_{ppad} = 1$ the joint resistance functions of the cases $\lambda_{pp} = \lambda_{ppad} = 0.01$ and $\lambda_{pp} = \lambda_{ppad} = 0.001$ are also demonstrated in Figure 21.25. A constant vertical shift of corresponding curves indicates that in all cases an increase of the axis relation X causes an increment of $\langle R \rangle$ of almost the same amount. This applies also for the dependence of the joint resistance functions on the interparticle conductivity.

For practical applications it can be concluded from this model that the particle shape (in the model represented by the aspect ratio for elliptical particles) has a certain influence on the joint resistance. However, the influence of the same relative change of filler content is much stronger. Moreover it can be observed that the joint resistance has a weaker depen-

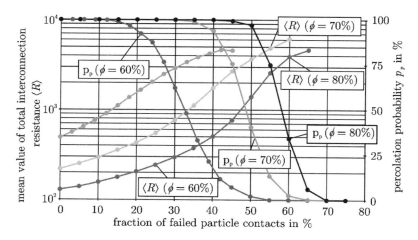

FIGURE 21.26. Joint resistance and percolation probability versus percentage κ_o of failed interparticle contacts for three values of filler content (particle parameters: $X = 4$, $A = 0.02^2 \cdot \pi$).

dency on the interparticle conductivity at high values of λ_{pp} and λ_{ppad} (a change of two orders of magnitude causes a resistance increase of only one order of magnitude). However, the more the interparticle conductivity decreases the more the joint resistance approaches to the inversely proportional dependence on the interparticle conductivity. Such a change could take effect if the molecular bonding forces of the polymer are weakened and oxides are formed at the particles interfaces.

As known from numerous investigations of the aging behavior under real operation conditions as well as under accelerated aging conditions the joint resistance can increase significantly [40–43]. By anticipating that changes do not occur in the volume of the conducting particles the resistance increase has to be assigned to degradation of the interfaces between particles and pad. The following part of the discussion is devoted to changes of the joint behavior due to aging.

In Figure 21.25 the shift of joint resistance with falling interparticle conductivity can be observed in cases when all interparticle contacts become homogeneously deteriorated to the same extent in the whole contact. However, this will remain the exception, since with growing thickness the contribution of tunneling becomes the dominant effect for the conductivity through the interface barriers. The initial specific tunneling conductivity e.g., on a noble metal surface is in the order of 10^{12} S/m^2 and with continuous growth of an oxide layer the conductivity decreases rapidly [44]. With respect to microscopic inhomogeneities in the ICA in a more realistic consideration, therefore, it should be assumed that due to e.g., oxidation more and more transitions between particles fail rather than that the interparticle conductivity decreases continuously and equally distributed. The results of simulations of such an aging behavior for joints with three different values of filler content ($\phi = 60\%$, $\phi = 70\%$, and $\phi = 80\%$) are demonstrated in Figure 21.26. For this purpose, the interparticle conductivity of a certain percentage of particle transitions are changed from $\lambda = 0.001$ to $\lambda = 0$ which means a total drop-out of the respective transitions. This percentage can be interpreted as the probability κ_o for such a drop-out. Thus, an increase of κ_o corresponds to an aging process in the ICA interconnection. At the beginning of the increase of κ_o only the joint resistance increases whereas the percolation probability remains constant at 100%. Only with continuing aging also the percolation probability starts to fall, indicating failing of the joint to the respective percentage. It is obvious, that joints with a higher filler content have a lower joint resistance and will withstand a longer aging process without failing which could be interpreted some how as a better long-term reliability. However, it is remarkable that the joint resistance in the moment when the percolation probability starts falling below 100% is almost independent from the filler content.

The change of contact resistance for various types of ICAs, different curing processes, and aging conditions were also experimentally investigated [45] and are summarized in the following section.

21.3. AGING BEHAVIOR AND QUALITY ASSESSMENT

21.3.1. Introduction

Besides the benefits mentioned in the introduction, ICAs provide an environmentally friendly alternative to solders for interconnections in electronic applications with the advantage of a superior fine pitch capability. However, unstable electrical conductivity under elevated temperature and humidity conditions and a low impact resistance of the interconnections were major obstacles preventing ICAs from becoming a general replacement for solders in SMT until now.

It is a well accepted opinion that corrosion is involved in the shift of contact resistance [20,24,46,47]. It is reported that galvanic corrosion rather than simple oxidation is the dominant interfacial degradation mechanism. However, galvanic corrosion doesn't only need an electrochemical potential difference but would also require a ions-containing aqueous phase which would have to be formed at e.g., 85% r.h.—that means at a humidity level where wet surfaces usually dry-up. On the other hand, galvanic corrosion would be an explanation why a much faster increase in resistance is attained under elevated humidity than under dry conditions at the same elevated temperature [25].

Recently developed ICAs promise an improved electrical and mechanical reliability on non noble metalizations as conventionally used in SMT and can be completely cured in the same short duration as known from a typical reflow soldering cycle. In order to investigate connections between processing parameters and the aging behavior eight different ICAs for solder replacement from four mayor manufactures were evaluated with respect to the behavior of the contact resistance and the shear force during a forced aging process. For this purpose test assemblies with ICA sample contacts were fabricated using PCBs with the four most common surface finishes. The test assemblies were then exposed to an elevated temperature and humidity environment (85°C/85% r.h.) for 1000 h, whereby the changes of the electrical contact resistance and the shear force were monitored.

The test assemblies consisted of SMD-chip components (1206 and 0805) which were mounted on FR4-printed circuit boards with the sample contacts between different surface finishes and the respective component metalization. Two different curing methods were investigated: curing in a conventional convection oven and curing in a vapor phase device.

The goal of this study was to experience advantages and drawbacks of ICAs presently available on the market and to present some recent progresses in ICA technology development.

21.3.2. Material Selection and Experimental Parameters

All selected adhesives are one component, epoxy based, thermosetting, silver filled and isotropically conductive. All are especially designed to replace solder pastes in SMT.

According to the manufactures data sheets, some of them promise a stable contact resistance even when applied on tin, tin/lead and OSP-coated copper. Table 21.1 gives an overview about the selected products and curing conditions as recommended by their manufacturers.

The test boards were pluggable and allowed to combine monitoring of contact resistance shift with the four-point probe method of 40 contacts simultaneously during forced aging (Figure 21.27), and after that mechanical quality testing.

In this way the voltage drop of the series connection of two joint resistances and the resistance of the component itself is measured. In order to keep uncertainties during evaluation of the joint resistance as small as possible chip resistors (1206 and 0805 package) with a low value and a small tolerance (51 mΩ ±5%) were used provided with lead free galvanic tin coated contacts. The joint resistance is defined as the resistance between contact pad and metalization of the component. The voltage drop within the copper pads and the metalization of the component can be neglected in this arrangement.

TABLE 21.1.
Chosen ICAs with the curing conditions as recommended by the manufactures.

Manufacturer	Adhesive	Recommended curing conditions
A	1	120 min @ 120°C
		30 min @ 140°C
	2	5 min @ 125°C-Reflow
		3 min @ 150°C-Reflow
		8 min @ 125°C-Convection
		5 min @ 150°C-Convection
B	1	30 min @ 125°C
		15 min @ 150°C
	2	120 min @ 125°C
		60 min @ 150°C
C	1	10 min @ 125°C
		6 min @ 150°C
		3 min @ 175°C
	2	6 min @ 130°C
		3 min @ 150°C
D	1	30 min @ 140°C
	2	N.A.

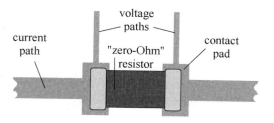

FIGURE 21.27. Conductor structure of one component.

ELECTRICALLY CONDUCTIVE ADHESIVES

With respect to the most commonly used surface finishes of PCBs test boards with the following types of surface treatments were manufactured:

- organic surface protection (OSP): ENTEK Plus Cu-106A
- electroless Ag-coating
- electroless Ni/Au-coating
- electroless Sn-coating.

The curing cycles were optimized by monitoring the electrical conductivity of the bondings during the curing process. This method is explained in the following subsection.

21.3.3. Curing Parameters and Definition of Curing Time

The results of quality testing of ICA joints depend strongly on how the adhesive has been cured. A comparison of quality, therefore, needs a thorough definition of curing conditions. Two different curing methods were applied in comparison. The first one as reference was curing in a conventional convection oven at a preselected temperature (two values were considered: 150°C or 200°C). The second method was curing in a saturated vapor phase of perfluorinated fluids with boiling points of 155°C and 200°C. The idea of using vapor phase condensation is to take advantage of the much higher heat transfer rates compared to gas convection allowing to shorten down the relatively long curing time of most ICAs to the length of a typical reflow soldering cycle by reducing the heating-up time.

In general ICAs have a high electrical resistance before curing. The conductivity develops during the curing process [48]. It is assumed that shrinkage of the resin matrix during curing causes the silver flakes to contact more intimately which leads to the dramatic contact resistance decreases [49]. If the relation between conductivity and curing degree of the adhesive is known for the respective combination of adhesive and curing method it can be used to determine the needed curing time. Figure 21.28 shows how the conductivity develops during the curing process. For this purpose, calibration measurements were performed with samples where the adhesive was printed over a spacing between two contact pads forming a quadratic resistor (length and width: 1 mm, thickness: 100 µm). The resistance between these contact pads was observed during the curing process. Reliable representa-

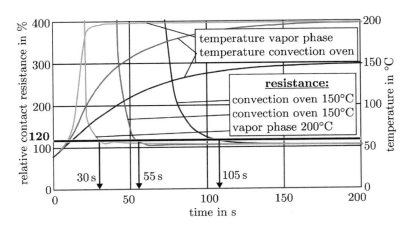

FIGURE 21.28. Change of contact resistance during curing process (adhesive type A1).

TABLE 21.2.
Applied curing times in minutes required to achieve 120% of the minimum possible resistivity.

Adhesive	Convection		Vapor phase	
	150°C	200°C	155°C	200°C
A1	8	3	2	2
A2	8	4	5	3
B1	20	4	15	3
B2	40	20	30	10
C1	8	2	3	2
C2	8	2	3	2
D1	8	5	6	3
D2	8	5	6	3

tive calibration curves were obtained by averaging results from six measurements at each combination. This kind of contact resistance measurement is an easy and reliable method of online monitoring of the curing progress allowing to closed-loop control the curing time and to optimize the curing condition for PCBs with different thermal properties.

Figure 21.28 shows the result of these measurements for curing the A1-type adhesive at 150°C and 200°C in the convection oven and for curing at 200°C in the vapor phase device. Curing at 200°C in the convection oven causes a faster decrease of the resistance as at 150°C due to the faster polymerization process. However, the resistance drop at curing the adhesive with 200°C in the vapor phase device is still much earlier and faster due to the significantly reduced heating-up time of the sample. As an indicator for the completion of the curing process the moment is used when the average resistance reaches the 120% mark of its final value (which was obtained from reference samples after a curing time of 1 hour). The time span until this moment multiplied by a safety factor was rounded up to at least 2 min. This time was defined as curing time for each evaluated adhesive/curing method combination and used for the further sample preparation. Table 21.2 gives a summary of the curing times determined in this way. In general the curing time as defined above is shorter than the one recommended in Table 21.1.

21.3.4. Testing Conditions, Typical Results, and Conclusions

For each particular combination of material and curing method the testing results from 80 adhesive joints were established. To evaluate the reliability of the contacts these samples were exposed to an elevated temperature and humidity (85°C/85% r.h.) environment for 1000 hours. During this forced aging process the electrical contact resistance was measured every 100 hours at room temperature and the results were averaged. The shear force was measured in the initial state, after 400, and after 1000 hours. At every measurement 6 to 7 components were sheared off. If either the contact resistance was higher than 5 Ω or the component dropped off from the test board the contact was counted as "failed" and not considered in further evaluations.

21.3.4.1. Influence of PCB finish on aging behavior The A1 type is a commercially applied ICA but not especially designed for the use on less noble metal surfaces. Therefore, it is not surprising that joints fabricated with this adhesive show a drastic increase of contact resistance during the testing period on PCB finishes like tin or copper [50,51]. Since the

ELECTRICALLY CONDUCTIVE ADHESIVES

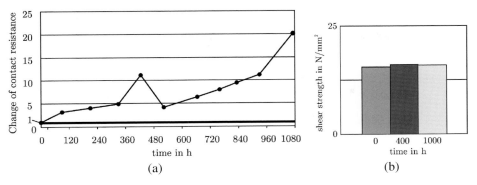

FIGURE 21.29. Adhesive A1 on Ag-coated PCB during forced aging test (85°C/85% r.h.); curing at 150°C in the convection oven. (a) Change of joint resistance (initial value: 10 mΩ). (b) Change of shear force.

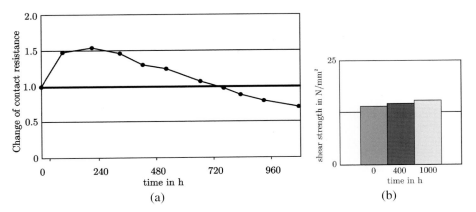

FIGURE 21.30. Adhesive A1 on Sn-coated PCB during forced aging test (85°C/85% r.h.); curing at 150°C in the convection oven. (a) Change of joint resistance (initial value: 37 mΩ). (b) Change of shear force.

increase of contact resistance is observable on each of the tested PCB-finishes it is assumed that the interface between the ignoble tin surface of the component and the adhesive is responsible for this effect. Despite of the high shift of contact resistance the shear strength did not decrease. Figure 21.29 shows the results measured at the Ag-coated PCB cured in the convection oven at 150°C.

The more recently developed product A2 is designed for solder replacement and suitable also for mounting components with tin surfaces. On all samples the average contact resistance after the forced aging test was even lower than the initial value. For the samples cured in the convection oven at 150°C only a slight increase of contact resistance in the first quarter of the testing time and a subsequent decrease of the resistance were observed. Figure 21.30 shows the results obtained from Sn-coated PCBs cured in the convection oven at 150°C. Also the shear force shows stable values over the entire testing period. Only at the NiAu coated PCB a decrease of the shear force was noticeable.

Another example is a product C2 offered as a comparatively quickly curing adhesive (3 minutes at 200°C) showing a strong increase of resistance only on the Sn-coated PCB. Rather high initial values decreased significantly after about 300 hours of aging (Figure 21.31). The used curing time of only 2 minutes was certainly insufficient.

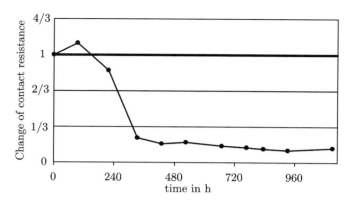

FIGURE 21.31. Change of contact resistance; adhesive C2 on OSP-coated PCB during forced aging test (85°C/85% r.h.); cured at 200°C in the convection oven.

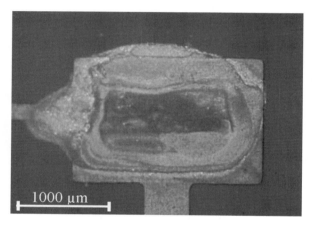

FIGURE 21.32. Area of fracture after shear test after 1000 hours forced aging (adhesive D2, curing condition: vapor phase 155°C, PCB OSP-coated).

Figure 21.32 depicts an area of fracture after a shear test of a sample joint manufactured with adhesive D2 after 1000 hours of aging. The fracture occurred at the PCB/adhesive–interface. A discoloration of the pad surface is visible resulting from a strong degradation of this interface. It should be noted that a low shear force can appear independent from a good conductivity.

By contrast, Figure 21.33 shows the area of fracture of a contact formed with adhesive D1 using the same parameters as in Figure 21.32. In this case the fracture occurred mainly at the component/adhesive interface. At the PCB pad where the fracture occurred a blank, shiny copper surface appeared indicating a low degradation. It is not surprising that this sample showed good mechanical and electrical properties.

21.3.4.2. Summary Eight different ICAs from four mayor manufactures were evaluated with respect to there suitability to serve as replacement of solder pastes. As a general result it should be noted that significantly different behavior of the tested adhesives could be observed in spite of consisting of quite similar compositions: silver-filled one-component epoxy resin.

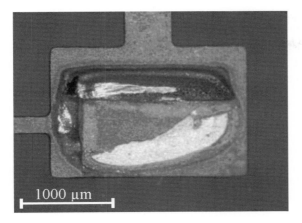

FIGURE 21.33. Area of fracture after shear testing a sample with a type D1 adhesive after 1000 hours forced aging (curing condition: in vapor phase at 155°C, PCB: OSP-coated).

Adhesives of the A1-type and D1-type are not recommended for oxygen rich systems or oxidizing surfaces from their manufacturers. Although, initial values of joint resistivity and shear strength were acceptable, the results during and after a forced aging process exhibit large shifts and an unstable values.

Using the vapor phase device generally the curing time of the adhesives can be decreased drastically due to the high heat transfer rate obtained by condensation. Although, with some adhesives perfect results were obtained with others the shear strength showed significantly lower values compared to equal samples cured in the convection oven. Further investigations would be required to fully understand the reasons for these different behaviors.

Using the NiAu-finished PCBs an increased failure rate was observable for the most adhesives. At the first glance this might be surprising since due to the low inclination to oxidation of noble metal surfaces one could expect a rather moderate tendency to changes of the contact resistance. However, the increase of resistance during aging might to a certain extent be assigned to the weaker adhesion between gold and the epoxy resin. The latter argument is in good agreement with the results of the shear tests frequently showing the area of fracture coinciding with the gold-adhesive interface.

The newer ICA types didn't show a significant increase of contact resistance during a forced aging under elevated temperature and humidity even on ignoble metal surfaces as well as on the OSP coated copper pads. Although, the protective layer could be assumed to impede an intimate contact between the filler particles and the metalization, rather the opposite turns out from the investigations as described afore. No retardation of the onset of contact formation during curing can be observed and a stable joint resistance during aging can be understood as an indicator for a reliable long-term behavior.

As a summarizing result of this investigation it can be concluded that the progress of the adhesive technology has reached a stage were a replacement of solder by an isotropic conductive adhesive can be performed in many cases without changing the involved surfaces treatments whereby a reliability level can be achieved comparable to solder joints.

21.4. ABOUT TYPICAL APPLICATIONS

Due to the large variety of filler content and chemical composition of the polymer the industrial use of ICAs is not restricted to the attachment of SMDs. The following examples demonstrate the numerous application fields.

21.4.1. ICA for Attachment of Power Devices

For optimizing device performance and reliability of power devices an accurate thermal design is a critical issue. The basic need is to remove the power loss produced in the device (junction) under operation condition at the lowest possible temperature drop with respect to the ambient. This quality is expressed as the junction-to-ambient thermal resistance $R_{th,j-a}$. It consists of the sum of the junction-to-case thermal resistance $R_{th,j-c}$ and the thermal resistance $R_{th,c-a}$ from the case to the ambient. Whereas $R_{th,j-c}$ is provided in the datasheet from the component's manufacturer $R_{th,c-a}$ highly depends on the mounting technique and can vary orders of magnitude. In power assemblies the device case is attached to a heat sink using a thermal grease, by adhesive bonding, or by soldering. Normally, heat dissipation from the case by radiation and convection can be neglected. $R_{th,c-a}$ is mainly controlled by the heat flow through the bond line between the device case and the heat sink. Therefore, $R_{th,j-a}$ can be expressed as

$$R_{th,j-a} = R_{th,j-c} + R_{th,c-a} = R_{th,j-c} + R_{th,bond} + R_{th,heatsink}, \quad (21.7)$$

where $R_{th,heatsink}$ is defined as the thermal resistance between the mounting surface of the heat sink and the ambient. The bond line thermal resistance $R_{th,bond}$ can in principle be calculated by dividing the expected or measured bond line thickness by the adhesive's intrinsic thermal conductivity λ_{bond} measured on a free-standing cured sample. However, at a typical bond line thickness of 15–75 μm, the interface thermal resistance $R_{th,if}$ between the adhesive and its adherents can be significant compared to the intrinsic thermal resistance of the adhesive and thus the bond line thermal resistance itself must be considered as a sum of the two components [52]

$$R_{th,bond} = \frac{t_{bond}}{A \cdot \lambda_{bond}} + R_{th,if}, \quad (21.8)$$

where A and t_{bond} are the bond line area and thickness. The practical importance can be seen on the example of a TO-247 package which is attached to a heat sink with a copper heat spreader as mounting surface (Figure 21.34). Two types of ICAs are compared: a conventional one with a low thermal conductivity and an ICA (type Diemat DM6030Hk) with a silver particle content of more than 95 weight percent.

For mounting the device on a heat sink, the ICA must flow as freely as grease to eliminate air voids and reduce the thermal resistance of the interface. However, a complete avoidance of pores and gas enclosures in the bond line is not always possible. In order to consider the thermal meaning of enclosures in the bond line Equation (21.8) has to be modified such that the thermally conducting cross section of the ICA is reduced by the area percentage p of the voids:

$$R_{th,bond} = \frac{1}{A} \cdot \frac{t_{bond}}{\lambda_{bond} \cdot \left(1 - \frac{p}{100}\right) + \lambda_{air} \cdot \frac{p}{100}} + R_{th,if}. \quad (21.9)$$

ELECTRICALLY CONDUCTIVE ADHESIVES

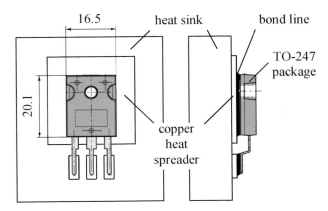

FIGURE 21.34. Power device assembly with TO-247 package (dimensions in millimeter).

TABLE 21.3.
Geometry and thermal data used for evaluation as demonstrated in Figure 21.35.

Bond line thickness	in μm	50	
Bonding area	in cm^2	2.31	
Thermal conductivity of voids (air)	in W/(m°C)	0,03	
Thermal conductivity of conventional ICA	in W/(m°C)	2	[53]
Thermal conductivity of DIEMAT ICA	in W/(m°C)	60	[53]
Thermal conductivity of Sn96Ag3.5Cu	in W/(m°C)	57	[54]
$R_{th,if}$ at bonding area of 2.31 cm^2	in °C/W	8.7×10^{-2}	[52]

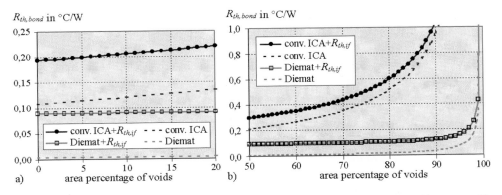

FIGURE 21.35. Bond line thermal resistance $R_{th,bond}$ versus percentage of gas enclosures; (a) at acceptable quantities of enclosure, (b) in an unacceptable range.

Using the values as listed in Table 21.3 Equation (21.9) is evaluated for different ICA types.

At low percentages of enclosures an almost linear increase of $R_{th,bond}$ can be observed in bond lines with conventional ICAs. No measurable thermal effect of enclosures exists in case of a high intrinsic thermal conductivity (DIEMAT or lead-free solder). In order to recognize the meaning of the interface thermal resistance for both ICA types $R_{th,bond}$

is also calculated as a reference with $R_{th,if}$ set to zero (dashed curves), respectively. In both cases the interface thermal resistance clearly predominates the thermal resistance increase due to enclosures [Figure 21.35(a)]. It is surprising that only if voids are unacceptably large from mechanical point of view they become thermally important [Figure 21.35(b)].

For $R_{th,j-c}$ of silicon devices in TO-247 packages frequently values below 1°C/W are provided from manufacturers. It should be noted that in cases of GaAs power devices due to the lower thermal conductivity of the substrate compared to Si the discussed phenomena are particularly important [53].

21.4.2. ICA for Interconnecting Parts with Dissimilar Thermal Expansion Coefficient

For many types of radiation sensors sealed housings are needed consisting of a metal cap and a base plate which carries the sensor substrate. The base plate including the bonding process for mounting the sensor substrate is dispensable if all functions of the base plate can be fulfilled by the sensor substrate itself. The cap can either be soldered to the substrate using a high-temperature solder in order to allow a second soldering process (e.g., for its attachment to a PCB) or an adhesive can be applied. The latter is favored whenever a low processing temperature is essential or when thermomechanical requirements cannot be fulfilled with a soldered joint. In the following example an alumina substrate with a wiring structure produced by thick film technology was required to be bonded to a cap of aluminum whereby soft soldering had to be excluded from the possible joining techniques, since due to very dissimilar coefficients of thermal expansion of base plate and cap thermal cycling caused microcracks in the solder layer. The basic structure of the sensor is schematically illustrated in Figure 21.36.

Adhesives with a high glass transition temperature T_g are reported as to be generally more susceptible to the formation of cracks than those with a low T_g which is understood as a different ability to stress relaxation [55]. However, due to the thermoelastic behavior of adhesives there are e.g., the possibility of directly attaching of 15 mm × 15 mm large silicon chips on FR-4 PCB without damage during temperature cycling as demonstrated in the DACTEL project [56].

A further essential parameter besides the thermoelastic behavior is the thickness of the ICA layer between the mating surfaces. The thicker the layer is, at a given thermal expansion mismatch, the lower is the shear plane angle and the inclination to formation of cracks. This is well known from reliability investigations of solder joints e.g., between chip components and PCBs where it was found that the number of thermal cycles until failure increases significantly with rising stand-off height [57].

Besides aspects of long-term reliability the housing of the sensor has to fulfill thermal requirements. An accurate calibration of the sensor frequently needs a constant temperature distribution in the housing. Therefore, the heat loss has to be dissipated from the substrate through a clamping part at the front face of the cap at the lowest possible temperature drop. From this consideration a thin bonding layer is needed which is in contradiction with the mechanical requirements. An optimum can be found by selecting an adhesive with a high thermal conductivity. Figure 21.37 depicts the result of a comparison of two axial temperature profiles in the cap obtained by thermal simulation. This study shows that the temperature drop perpendicular to an ICA layer can be reduced significantly by increasing the filler content. Whereas the thermal conductivity of adhesives for conventional surface mounting applications even if offered as thermally conductive adhesives is rather low (ranging in the order of some W/mK, [58], represented by type 1 ICA in Figure 21.37) there are some

ELECTRICALLY CONDUCTIVE ADHESIVES

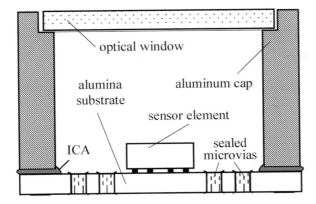

FIGURE 21.36. Schematic structure of the sensor housing.

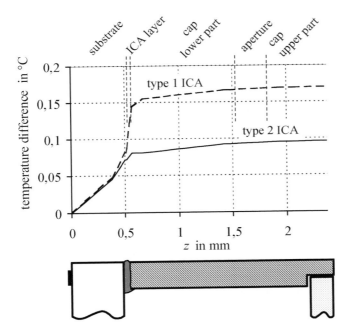

FIGURE 21.37. Influence of the thermal conductivity of the ICA layer on the temperature profile in axial direction (type 1 ICA: 3 W/mK, type 2 ICA: 30 W/mK).

types of ICA on the market with an extremely high silver content with a thermal conductivity of more than 50 W/mK (type 2 ICA, filler content: 93 weight % silver, 60 W/mK), allowing to achieve a negligible temperature drop across the ICA layer.

Due to the excellent thermoelastic behavior of the adhesive a lateral displacement of up to 20 μm due to different thermal expansions of substrate and cap could be handled with an adhesive layer thickness of only about 30 μm [59].

TABLE 21.4.
Comparison of number of process steps for SMD assembling either using soldering (left column) or adhesive bonding (right column).

...	...
Printing the solder (stencil) for SMDs and the transistor chip	Printing the adhesive (stencil) for all components
Placing SMDs and transistor	Placing all components
Reflow soldering	Curing the adhesive
Cleaning (flux removing)	–
Dispensing the adhesive for the remaining chips	–
Placing the chips	–
Curing	–
Wire bonding	Wire bonding
Applying glob top	Applying glob top
...	...

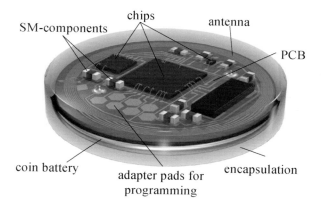

FIGURE 21.38. Schematic illustration of the temperature logger module.

21.4.3. ICA for Cost-Effective Assembling of Multichip Modules

Frequently multichip modules are the solution when a high degree of miniaturization is needed in an electronic assembly with complex functions. The application described here is a temperature logger system smaller than a one-Euro cent coin in diameter embedded in a removable brace to observe the wearing habits of the patient [60]. For this purpose the temperature is measured with a pn junction and recorded using a microcontroller which acts as a temperature logger and is capable of storing the thermal history from several months. This device is hermetically sealed using polymeric encapsulants. A wireless data exchange is performed using a radio frequency identification device (RFID). The electronic assembly contains a printed circuit board which is populated with SMDs in 0201 package (0.6 mm × 0.3 mm) and bare microprocessor and RFID chips (Figure 21.38).

The SMDs can be assembled by soldering or ICA bonding. The die components were not available for solder attachment as is frequently the case for small-batch production. Thus, wire bonding was necessary in the vicinity of the SMDs. In order to prevent the wire bonding pads from flux residues, the attachment of the SMDs by soldering would be related to additional process steps for protection and for cleaning of the bonding pads. The production process with the lowest possible number of process steps was found using ICA for the attachment of chips and SMDs [59]. The manufacturing processes are compared in Table 21.4.

21.5. SUMMARY

As a historical background the development of ICA technology is briefly discussed. Already decades ago some principle advantages were recognized such as the low processing temperature and the possibility to use non-wettable and base metal surfaces for interconnections. For this reason, the attachment of semiconductors was the first application of ICAs in electronic packaging. In the meanwhile, much progress has been achieved in the quality and stability of ICA materials. Nevertheless, the ICA technology is not widely seen as an equivalent replacement of soldering, although it is a lead-free and a flux-free technology and both of these facts are important environmental aspects. One reason might be the experience with the former ICA technology concerning a moderate long-term reliability due to the inclination of oxide formation and galvanic corrosion in humid environment at elevated temperature. One goal of this contribution is to provide a deeper understanding of silver filled epoxy-based adhesives. For this purpose a model of an ICA joint which considers the particle alignment and distribution as well as the voltage and current distribution within the adhesive is discussed. Using this model the influence of parameters like filler content, particle arrangement, and particle size on the joint resistance is estimated and tendencies of changes during aging are clarified. The behavior of real ICA joints under accelerated aging condition and the results of quality assessment is also discussed. Finally, some practical examples demonstrate advantages and drawbacks in manifold fields of applications.

NOTATIONS AND DEFINITIONS

A	particle area
A_{ex}	excluded area
$\langle A_{ex} \rangle$	mean value of excluded area
A_{ex}/A	normalized excluded area
B	mean number of neighbored particles
B_c	mean critical number of neighbored particles
δ	angle between major axes of two particles
ϕ	portion of area filled with particles (filler content in area %)
ϕ_c	critical portion of area filled with particles (critical area filling factor)
$\Delta\phi$	deviation of area filling factor ϕ from critical value ϕ_c
γ	angle between major axes of a particles and the (horizontal) pad plane
κ_o	probability for a nonconductive (failing) interconnection between two particles
λ_{pp}	specific interface conductivity between two particles, see Equation (21.1).
$\lambda_{p\,pad}$	specific interface conductivity between particle and pads
p_c	percolation probability of 50%
p_p	percolation probability
R	normalized total resistance of interconnection (joint resistance)
$\langle R \rangle$	mean value of the normalized total interconnection resistance
R_{pp}	transition resistance between two particles

$R_{p\,pad}$	transition resistance between a particle and a pad
$\langle R_{tr} \rangle$	total transition resistance at both pad-particle interfaces, average value
σ	specific conductivity of particle
σ_{pad}	specific conductivity of pad
w_{tot}	total width of the modeled space (= length of the contact pad)
$w_{s,i}$	contact width at the ith particle (length of intersection line between two considered particles or between particle and the pad as e.g., shown in Figure 21.3)
$w_s^* = \sum_i^n w_{s,i}$	total effective width of contact pad (= active width of pad involved in current flow), n is the number of particles in direct contact with the respective pad and contributing to current flow
$\langle w_s^*/w_{tot} \rangle$	percentage of total effective contact width with respect to the pad length, average value
X	length relation between major and minor axis of elliptical particles (axis relation)

REFERENCES

1. R.L. Henry, Improvements in commutator brushes, British Patent #246.972, 1926.
2. H. Schuhmann, Conductive varnish, U.S. Patent #1,913,214, June 6, 1933.
3. H. Bienfait and W.L. Carolus van Zwet, Electrical conductor and method of making the same, U.S. Patent #2,018,343, Oct. 22, 1935.
4. H.J. Loftis, Molded electrically conductive body, U.S. Patent #2,361,220, Oct. 24, 1944.
5. N.H. Collings and R.J. Heaphy Beverton, Electrically conductive adhesive, U.S. Patent #2,444,034, June 29, 1948.
6. H. Wolfson and G. Elliott, Electrically conducting cements containing epoxy resins and silver, U.S. Patent #2,774,747, Dec. 18, 1956.
7. C. Mitchel and H. Berg, Use of conductive epoxies for die-attach, Int. Symp. Microelectron, 1976.
8. D. Gerber and W. Scheel, Kleben elektronischer Baugruppen, in W. Scheel, Ed., Baugruppentechnologie der Elektronik, Verlag Technik Berlin, Eugen G. Leuze Verlag, Saulgau, 1997, pp. 393–394.
9. H. Shirakawa, Nobel lecture: The discovery of polyacetylene film—the dawning of an era of conducting polymers, Reviews of Modern Physics, 73(July), pp. 713–718 (2001).
10. P.M. Raj, J. Liu, and P. Öhlckers, Fundamentals of packaging materials and processes, in R.R. Tummala, Ed., Microsystems Packaging, McGraw-Hill, 2001, p. 739.
11. I. Watanabe and K. Takemura, Anisotropic conductive adhesive films for flip-chip interconnection, in J. Liu, Ed., Conductive Adhesives for Electronics Packaging, Electrochemical Publications Ltd., ISBN 0 901150 37 1, 1999, pp. 256–271.
12. A. Aghzout, Über die Leitfähigkeit von elektrisch anisotropen Klebern und deren Anwendungen in der Mikroelektronik, Dissertation thesis, Vienna University of Technology, June, 2002.
13. K.-i. Shinotani, J. Malmodin, and R. Trankell, Fundamentals of microsystems design for environment, in R.R. Tummala, Ed., Microsystems Packaging, McGraw-Hill, 2001, p. 859.
14. G. Hanreich, K.-J. Wolter, and J. Nicolics, Rework of flip-chip polpulated PCBS by laser desoldering, in H. Hauser, Ed., Sensors & Packaging, ÖVE-Schriftenreihe, Nr. 35, 2003, pp. 283–289.
15. L.M. Yu and W.C. Qing, Solder joints design attribute to no solder bridge for fine pitch device, Fifth International Conference on Electronic Packaging Technology, ICEPT2003, Shanghai, China, 28–30 Oct., 2003, pp. 70–75.
16. S. Kang, R.S. Rai, and S. Purushothaman, Development of high conductivity lead (Pb)-free conductive adhesives, IEEE Transactions on Components Packaging, and Manufacturing Technologies, 21(4), pp. 18–22 (1998).
17. J.S. Hwang, Fine pitch soldering and solder paste, in J.H. Lau, Ed., Handbook of Fine Pitch Surface Mount Technology, Van Nostrand Reinhold, New York, 1993, pp. 81–133.
18. Directive 2002/95/EC of the European Parliament and of the Council of 27 January 2003 on the Restriction of the Use of Certain Hazardous Substances in Electrical and Electronic Equipment, Official Journal of the European Union L37, 13.02.2003, pp. 19–23.

19. R. Dudek, H. Berek, T. Fritsch, and B. Michel, Reliability investigations on conductive adhesive joints with emphasis on the mechanics of the conduction mechanism, IEEE Transactions on Components and Packaging Technologies, 3(3), pp. 462–469 (2000).
20. C.P. Wong and D. Lu, Recent advances in electrically conductive adhesives for electronics applications, 4th International Conf. on Adhesive Joining and Coating Technology in Electronics Manufacturing, Proceedings, 2000, pp. 121–128.
21. S. Ganesan and M. Pecht, Lead-free electronics—2004 Edition, CALCE EPSC Press, ISBN 0-9707174-7-4, 2004.
22. J.H. Lau, C.P. Wong, N.-C. Lee, and R.S.W. Lee, Electronics Manufacturing—with Lead-Free, Halogen-Free, and Conductive-Adhesive Materials, McGraw-Hill, ISBN 0071386246, July 2002.
23. J.E. Morris, F. Anderssohn, S. Kudtarkar, and E. Loos, Reliability studies of an isotropic electrically conductive adhesive, 1st Int. IEEE Conference on Polymers and Adhesives in Microelectronics and Photonics, October 21–24, Potsdam, Germany, 2001, pp. 61–69.
24. D. Lu and C.P. Wong, Development of conductive adhesives for solder replacement, IEEE Transactions on Components and Packaging Technologies, 23(4), pp. 620–626 (2000).
25. D. Lu, C.P. Wong, and Q.K. Tong, Mechanisms underlying the unstable joint resistance of conductive adhesives, IEEE Transactions on Components, Packaging, and Manufacturing Technology, Part C, 22(3), pp. 228–232 (1999).
26. M. Zwolinski, J. Hickman, H. Rubin, Y. Zaks, S. McCarthy, T. Hanlon, P. Arrowsmith, A. Chaudhuri, R. Hermansen, S. Lau, and D. Napp, Electrically conductive adhesives for surface mount solder replacement, IEEE Transactions on Components, Packaging, and Manufacturing Technology, Part C, 19(4), pp. 251–256 (1996).
27. E. Stanley, J.S. Andrade, S. Havlin, H.A. Makse, and B. Suki, Percolation Phenomena: a broad-brush introduction with some recent applications to porous media, liquid water, and city growth, Physica A: Statistical and Theoretical Physics, 266(1-4), pp. 5–16 (1999).
28. A. Bunde, Percolation in composites, Journal of Electroceramics, 5(2), pp. 81–92 (2000).
29. J.E. Morris, Conduction mechanisms and microstructure development in isotropic, electrically conductive adhesives, in J. Liu, Ed., Conductive Adhesives for Electronics Packaging, Electrochemical Publications Ltd., ISBN 0 901150 37 1, 1999, pp. 37–77.
30. M. Mündlein, J. Nicolics, and G. Hanreich, Accelerated curing of isotropically conductive adhesivemns by vapor phase heating, Third International IEEE Conference on Polymers and Adhesives in Microelectronics and Photonics (Polytronic 2003), Montreaux, Switzerland, October 21–23, 2003, pp. 101–105.
31. L. Li and J.E. Morris, Electrical conduction models for isotropically conductive adhesive joints, IEEE Transactions on Components, Packaging, and Manufacturing Technology, Part A, 20(1), pp. 3–8 (1997).
32. X.-G. Liang and X. Ji, Thermal conductance of randomly oriented composites of thin layers, International Journal of Heat and Mass Transfer, 43, pp. 3633–3640 (2000).
33. E. Sancaktar and B. Lan, Modeling filler volume fraction and film thickness effects on conductive adhesive resistivity, 4th IEEE International Conference on Polymers and Adhesives in Microelectronics and Photonics (Polytronic 2004), Portland, USA, 12–15 Sept., 2004, pp. 38–49.
34. I. Balberg, C.H. Anderson, S.Alexander, and N. Wagner, Excluded volume and its relation to the onset of percolation, Physical Review B, 30(7), pp. 3933–3943 (1984).
35. P.A. Cundall, A computer model for simulating progressive, large-scale movements in blocky rock systems, Proc. Symp. Int. Rock Mech. 2(8), Nancy, Vol. 2, 1971, pp. 129–136.
36. B. Su and J. Qu, A micromechanics model for electrical conduction in isotropically conductive adhesives during curing, Electronic Components and Technology 2004, Volume 2, 1–4 June 2004, pp. 1766–1771.
37. G.G.W. Mustoe, M. Nakagawa, X. Lin, and N. Iwamoto, Simulation of particle compaction for conductive adhesives using discrete element modeling, Electronic Components and Technology Conference 1999, 1–4 June, 1999, pp. 353–359.
38. A. Drory, I. Balberg, and B. Berkowitz, Application of the central-particle-potential approximation for percolation in interacting systems, Physical Review E, 52, pp. 4482–4494 (1995).
39. B. Berkowitz, Percolation theory and network modeling applications in soil, Physics: Surveys in Geophysics, 19(1), pp. 23–72 (1998).
40. S. Xu, D.A. Dillard, and J.G. Dillard, Environmental aging effects on the durability of electrically conductive adhesive joints, International Journal of Adhesion and Adhesives, 23(3), pp. 235–250 (2003).
41. E. Suganuma and M. Yamashita, High temperature degradation mechanism of conductive adhesive/Sn alloy interface, International Symposium on Advanced Packaging Materials: Processes, Properties and Interfaces, Proceedings, 2001, March 11–14, 2001, pp. 19–22.

42. H. Li, K.-S. Moon, Y. Li, L. Fan, J. Xu, and C.P. Wong, Reliability enhancement of electrically conductive adhesives in thermal shock environment, Electronic Components and Technology, ECTC 2004, Vol. 1, June 1–4, Las Vegas, USA, 2004, pp. 165–169.
43. F. Kriebel, Leitkleben—eine Alternative zum Löten in der Oberflächenmontagetechnik, VTE, (4)(Aug.), pp. 182–191 (1998).
44. H. Hauser, Kontaktwerkstoffe, Anhang D, in G. Fasching, Ed., Werkstoffe für die Elektrotechnik, Springer, 4th edition, 2005, pp. 517–520.
45. M. Mündlein, J. Nicolics, and J.E. Morris, Reliability investigations of isotropic conductive adhesives, 25th International Spring Seminar on Electronics Technology, ISSE 2002, Prague, Czech Republic, May 11–14, ISBN 0-7803-9824-6, 2002, pp. 329–333.
46. K. Gilleo, Evaluating polymer solders for lead-free assembly part I, Circuits Assembly, (January), pp. 50–56 (1994).
47. J.C. Jagt, P.J.M. Beris, and G.F.C.M. Lijten, Electrically conductive adhesives: a prospective alternative for SMD soldering? IEEE Transactions on Components, Packaging, and Manufacturing Technology, Part C, 18(2), pp. 292–298 (1995).
48. D. Klosterman, L. Li, and J.E. Morris, Materials characterization, conduction development, and curing effects on reliability of isotropically conductive adhesives, IEEE Transactions on Components, Packaging, and Manufacturing Technology, Part A, 21(1), pp. 23–31 (1998).
49. D. Lu, Q.K. Tong, and C.P. Wong, Conductivity mechanisms of isotropic conductive adhesives (ICAs), IEEE Transactions on Electronics Packaging Manufacturing, 22(3), pp. 223–227 (1999).
50. J.C. Jagt, Reliability of electrically conductive adhesive joints for surface mount applications: a summary of the state of the art, IEEE Transactions on Components, Packaging, and Manufacturing Technology, Part A, 21(2), pp. 215–225 (1998).
51. M.G. Perichaud, J.Y. Deletage, H. Fremont, Y. Danto, C. Faure, and M. Salagorty, Evaluation of conductive adhesives for industrial SMT assembles, 23rd IEEE/CPMT Electronics Manufacturing Technology Symposium, 1998, pp. 377–385.
52. R.C. Campbell, S.E. Smith, and R.L. Dietz, Measurements of adhesive bondline effective thermal conductivity and thermal resistance using the laser flash method, 15th Annual IEEE Semiconductor Thermal Measurement and Management Symposium, 9–11 March, 1999, pp. 83–97.
53. M. Mayer, J. Nicolics, G. Hanreich, and M. Mündlein, Thermal aspects of GaAs power FET attachment using isotropically conductive adhesive, 2nd Int. IEEE Conf. on Polymers and Adhesives in Microelectronics and Photonics, Polytronic 2002, Zalaegerszeg, Hungary, June 23–26, 2002, pp. 38–43.
54. http://www.npl.co.uk/ei/iag/leadfree/propertiespbf.html, Lead-Free Alloy Properties, National Physical Laboratory.
55. D.W. Swanson and L.R. Enlow, Stress effects of epoxy adhesives on ceramic substrates and magnetics, Microelectronics Reliability, 41(4), pp. 499–510 (2001).
56. J. Vanfleteren, A. Vervaet, and A.V. Calster, Highlights of the DACTEL project: development of adhesive chip technologies for dedicated electronic applications, Proc. of 23rd Int. Spring Seminar on Electronics Technology, ISSE 2000, Balatonfüred, HU, May 6–10, 2000, pp. 315–318.
57. M. Sumikawa, T. Sato, C. Yoshioka, and T. Nukii, Reliability of soldered joints in CSPs of various designs and mounting conditions, IEEE Transactions on Components and Packaging Technology, 24(2), pp. 293–299 (2001).
58. R. Prasad, Adhesives and its applications, in G.R. Blackwell, Ed., The Electronic Packaging Handbook, CRC Press, IEEE Press, 2000, pp. 10-1–10-28.
59. M. Mündlein, G. Hanreich, and J. Nicolics, Application of an ICA for the production of a radiation sensor, Proc. of 3rd Int. IEEE Conf. on Polymers and Adhesives in Microelectronics and Photonics, Polytronic 2003, Montreux, Schweiz, October 21–23, 2003, pp. 123–127.
60. M. Mündlein, J. Nicolics, and M. Brandl, Packaging concept for a miniaturized wirelessly interrogable temperature logger, Proceedings of the 27th Int. Spring Seminar on Electronics Technology, Sofia, Bulgaria, May 14–16, 2004, pp. 68–73.

22

Recent Advances of Conductive Adhesives: A Lead-Free Alternative in Electronic Packaging

Grace Y. Li and C.P. Wong

School of Materials Science and Engineering, Georgia Institute of Technology, 771 Ferst Drive, Atlanta, GA 30332-0245, USA

22.1. INTRODUCTION

Although the electronics industry has made considerable advances over the past few decades, the essential requirements of interconnects among all types of components in all electronic systems remain unchanged. The components need to be electrically connected for power, ground and signal transmissions, where tin/lead (Sn/Pb) solder alloy has been the *de facto* interconnect material in most areas of electronic packaging. Such interconnection technologies include pin through hole (PTH), surface mount technology (SMT), ball grid array (BGA) package, chip scale package (CSP), and flip chip technology. Tin-lead alloy solder has been a primary means of all electronic systems [1,2]. Figure 22.1 shows the use of Sn-Pb solder in typical PTH, SMT and BGA configurations.

There are increasing concerns with the use of tin-lead alloy solders. First, tin-lead solders contain lead, a material hazardous to human and environment. Each year, thousands tons of lead are manufactured into various products especially consumer electronic prod-

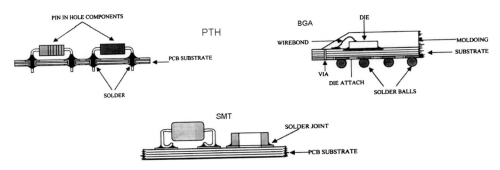

FIGURE 22.1. Schematic structures of PTH, SMT and BGA packages using solder interconnects.

ucts (e.g., cell phones, pagers, electronic toys, PDA, etc.) which tend to have a short (2–3 years) life cycle and millions of such lead-containing products simply end up in landfills. According to the latest report [U.S. Geological Survey, 2001], the total lead consumption by the U.S. industries was 52,400 metric tons in 2000. More than 10% of that (5,430 metric tons) was used to produce alloy solders. Worst of all, most of the electronic products have a very short service life. Recycling of lead-containing consumer electronic products has proven to be very difficult. Japan has banned the use of lead in all their electronic products in January of 2005, and European Union (E.U.) plans to ban all imports of lead-containing electronics in July, 2006. In response to the new legislation, most major electronic manufacturers all over the world, have stepped up their search for alternatives to lead-containing solders.

To date, these efforts have focused on two alternatives: lead-free solders and polymer based electrically conductive adhesives (ECAs) [3,4]. The most promising lead-free alloys contain tin as the primary element, because it melts at a relatively low temperature (232°C) and easily wets other metals. Depending on the applications, a number of lead-free solder alloys have found their way into commercial products [5,6]. However, most currently commercial lead-free solders, such as tin/silver (Sn/Ag), tin/silver/copper (Sn/Ag/Cu), have higher melting temperatures than that of tin-lead eutectic solder (183°C), typically at least 30°C higher. Therefore, the reflow temperature during electronic assembly must therefore be raised by 30°C to 40°C. This temperature increase reduces the integrity, reliability and functionality of printed wiring boards, components and other attachment, therefore severely limits the applicability of these metal alloys to organic/polymer packaged components and low-cost organic printed circuit boards. Although some low melting point alloys are available such as Sn/In (120°C), Sn/Bi (138°C), Sn/Zn/Ag/Al/Ga (189°C) [7], their material properties and processability in assembly are still of concern.

One the other hand, electrically conductive adhesives (ECA) can be processed at a much lower temperature. ECAs mainly consist of an organic/polymeric binder matrix and metal filler. These composite materials provide both physical adhesion and electrical conductivity. Compared to the solder technology, ECA offers numerous advantages, such as environmentally friendly, lower processing temperature, fewer processing steps (reducing processing cost), and especially, the fine pitch capability due to the availability of small-sized conductive fillers, especially for ACA [8–14].

ECA can be categorized with respect to conductive filler loading level into anisotropically conductive adhesives (ACA, with 3–5 μm sized conductive fillers, or sometimes in film form, ACF) and isotropically conductive adhesives (ICA, with 5–10 μm sized fillers), which are shown in Figure 22.2(a) and (b) [8]. The difference between ACA and ICA is based on the percolation theory (Figure 22.3). The percolation threshold depends on the shape and size of the fillers, but typically in the order of 15–25% volume fraction. For ICA, the loading level of conductive fillers is much more than the percolation threshold, providing electrical conductivity in x, y and z directions. For ACA, on the other hand, the loading level of conductive fillers is far below the percolation threshold, and the low volume loading is insufficient for inter-particle contact and prevents conductivity in the X–Y plane of the adhesive. Therefore, they provide a unidirectional electrical conductivity in the vertical or Z-axis. Both adhesive types are being adapted as interconnect materials for surface mount technology processes, such as chip on glass (COG), chip on flex (COF) and flip-chip bonding technologies in electronic packaging industries.

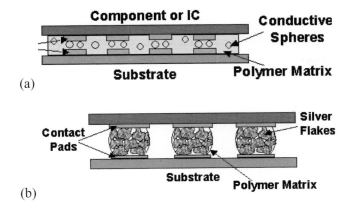

FIGURE 22.2. Schematic illustrations of (a) ACA and (b) ICA in flip-chip bonding.

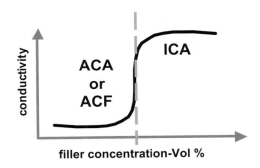

FIGURE 22.3. A typical percolation curve showing the abrupt drop in resistance at the percolation threshold.

22.2. ISOTROPIC CONDUCTIVE ADHESIVES (ICAs)

Isotropic conductive adhesives, also called as "polymer solder," are composites of polymer resin and conductive fillers. The adhesive matrix is used to form a mechanical bond at an interconnection. Both thermosetting and thermoplastic materials are used but thermoset epoxies are by far the most common binders due to the superior balanced properties, such as excellent adhesion, good chemical resistance and low cost. The conductive fillers provide the composite with electrical conductivity through tunneling and physical contact between the conductive particles. Silver flakes are the most commonly used conductive fillers for current commercial ICAs because of the unique properties of high conductivity of silver oxide and the maximum contact with flakes. ICAs have been used in the electronic packaging industry, primarily as die attach adhesives [15–17]. Recently, ICAs have been proposed as an alternative to tin/lead solders in surface mount technology (SMT) applications and a number of efforts have been conducted to improve the properties of ICAs in the past few years. Recent advances in material design and formulation have targeted to improved electrical conductivity, contact resistance stability, impact performance enhancement and metal migration control of ICAs. These will be reviewed in the following sections.

22.2.1. Improvement of Electrical Conductivity of ICAs

Polymer–conductive filler based electrically conductive adhesives (ECA) typically have lower electrical conductivity than Sn/Pb solders. To enhance the electrical conductivity of ECAs, various methods have been conducted and significant improvement of conductivity of ECA has been achieved.

22.2.1.1. Increase Polymer Matrix Shrinkage In general, ICA pastes exhibit insulative property before cure, but the conductivity increases dramatically after they are cured. ICAs achieve electrical conductivity during the polymer cure process caused by the shrinkage of polymer binder. Accordingly, ICAs with high cure shrinkage generally exhibit the best conductivity. Table 22.1 shows the relationship of shrinkage and conductivity for three different types of ECA, ECA1, ECA2 and ECA3 [18]. With increasing crosslinking density of ECA, the shrinkage of the polymer matrix increased, and subsequently, the obviously decreased resistivity of ECA was observed. Therefore, increasing the cure shrinkage of a polymer binder could improve electrical conductivity. For epoxy-based ICAs, a small amount of a multi-functional epoxy resin can be added into an ICA formulation to increase cross-linking density, shrinkage, and thus increase conductivity.

22.2.1.2. In situ Replacement of Lubricant on Ag Flakes An ICA is generally composed of a polymer binder and Ag flakes. There is a thin layer of organic lubricant on the Ag flake surface. This lubricant layer plays an important role for the performance of ICAs, including the dispersion of the Ag flakes in the adhesives and the rheology of the adhesive formulations [19–22]. This organic layer is a Ag salt formed between the Ag surface and the lubricant which typically is a fatty acid such as stearic acid [22,23]. This lubricant layer affects conductivity of an ICA because it is electrically insulating. To improve conductivity, the organic lubricant layer should be partially or fully removed or replaced. A suitable lubricant remover is a short chain dicarboxylic acid because of the strong affinity of carboxylic functional group ($-COOH$) to silver and stronger acidity of such short chain dicarboxylic acids. With the addition of only small amount of short chain single bond dicarboxylic acid, the conductivity of ICA can be improved significantly due to the easier electronic tunneling/transportation between Ag flakes and subsequently the intimate flake–flake contact [24].

22.2.1.3. Incorporation of Reducing Agent in Conductive Adhesives Silver flakes are by far the mostly used fillers for conductive adhesives due to the unique properties of high conductivity of silver oxide compared to other metal oxides, most of which are insulative. However, the conductivity of silver oxide is still inferior to metal itself. Therefore, incorporation of reducing agent would further improve the electrical conductivity of ICAs. Aldehydes were introduced into a typical ICA formulation and obviously improved con-

TABLE 22.1.
Relationship of shrinkage and electrical conductivity of ECAs.

Formulation	Cross-linking density (10^{-3} mol/cm^3)	Shrinkage (%)	Bulk resistivity (10^{-3} Ohm cm)
ECA1	4.50	2.98	3.0
ECA2	5.33	3.75	1.2
ECA3	5.85	4.33	0.58

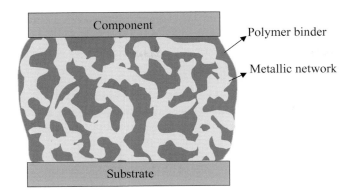

FIGURE 22.4. Diagram of transient liquid phase sintering conductive adhesives.

ductivity was achieved due to reaction between aldehydes and silver oxides that exist on the surface of metal fillers in ECAs during the curing process.

$$R-CHO + Ag_2O \rightarrow R-COOH + 2Ag. \tag{22.1}$$

The oxidation product of aldehydes, carboxylic acids, which are stronger acids and have shorter molecular length than stearic acid, can also partially replace or remove the stearic acid on Ag flakes and contribute to the improved electrical conductivity [25].

22.2.1.4. Low Temperature Transient Liquid Phase Fillers   Another approach for improving electrical conductivity is to incorporate transient liquid-phase sintering metallic fillers in ICA formulations. The filler used is a mixture of a high-melting-point metal powder (such as Cu) and a low-melting-point alloy powder (such as Sn-Pb). The low-melting-alloy filler melts when its melting point is achieved during the cure of the polymer matrix. The liquid phase dissolves the high melting point particles. The liquid exists only for a short period of time and then forms an alloy and solidifies. The electrical conduction is established through a plurality of metallurgical connections formed in situ from these two powders in a polymer binder (Figure 22.4). The polymer binder with acid functional ingredient fluxes both the metal powders and the metals to be joined and facilitates the transient liquid bonding of the powders to form a stable metallurgical network for electrical conduction, and also forms an interpenetrating polymer network providing adhesion. High electrical conductivity can be achieved using this method [26,27]. One critical limitation of this technology is that the numbers of combinations of low melt and high melt fillers are limited. Only certain combinations of two metallic fillers which are mutually soluble exist to form this type of metallurgical interconnections.

22.2.2. Stabilization of Contact Resistance on Non-Noble Metal Finishes

22.2.2.1. Mechanism of Unstable Contact Resistance   Contact resistance between an ICA and non-noble metal finished components increases dramatically during an elevated temperature and humidity aging. The National Center of Manufacturing and Science (NCMS) set a criterion for solder replacement conductive adhesives. The criterion is if the contact resistance shift after 500-hour at 85°C/85%RH aging is less than 20%, then the contact resistance is defined as "stable" [28]. Previous study in our laboratory has shown that galvanic corrosion rather than simple oxidation at the interface between an ICA and non-noble metal is the dominant mechanism for the unstable contact resistance. Under the aging conditions (for example, 85°C/85%RH), the non-noble acts as an anode, and is oxidized by losing electrons, and then turns into metal ion ($M - ne^- = M^{n+}$). The noble metal acts as a cathode, and its reaction generally is $2H_2O + O_2 + 4e^- = 4OH^-$. Then M^{n+} combines with OH^- to form a metal hydroxide or metal oxide. After corrosion, a layer of metal hydroxide or metal oxide is formed at the interface. Because this layer is electrically insulating, the contact resistance increases dramatically [29]. A galvanic corrosion process has several characteristics: (1) happens only under wet conditions, (2) an electrolyte must be present, and (3) oxygen generally accelerates the process. Based on the mechanism of unstable contact resistance of ECA, several methods could be applied to stabilize the contact resistance.

22.2.2.2. Oxygen Scavengers Since oxygen accelerates galvanic corrosion, oxygen scavengers could be added into ECAs to slow down the corrosion rate. When ambient oxygen molecules diffuse through the polymer binder, they react with the oxygen scavenger and are consumed. However, when the oxygen scavenger within the ECA is depleted, then oxygen can again diffuse onto the interface and accelerate the corrosion process. Therefore, oxygen scavengers can only delay the galvanic corrosion process, but does not solve the corrosion problem completely. Some commonly used oxygen scavengers include sulfates (such as Na_2SO_4), hydrazine ($H_2N=NH_2$), carbohydrazide ($H_2N-NH-CO-NH-NH_2$), diethylhydroxylamine (($C_2H_5)_2N-OH$), and hydroquinone ($HO-C_6H_4-OH$).

22.2.2.3. Corrosion Inhibitors Another method of preventing galvanic corrosion is to introduce organic corrosion inhibitors into ICA formulations [29–32]. In general, organic corrosion inhibitors act as a barrier layer between the metal and environment when forming a film over the metal surfaces [33]. Some chelating compounds are especially effective in preventing metal corrosion [34]. Chelating agents are organic molecules with at least two polar functional groups capable of ring closure with a metal cation. The functional groups may either be basic groups, such as $-NH_2$, which can form bonds by electron donation, or acidic groups, such as $-COOH$, which can coordinate after loss of the proton. Most organic corrosion inhibitors react with the epoxy resin at a specific temperature. Therefore, if an ICA is epoxy-based, the corrosion inhibitors must not react with the epoxy resin during curing which would cause them to be consumed and lose their effectiveness. Appropriate selection of corrosion inhibitors could be very effective in protecting the metal finishes from corrosion. However, the effectiveness of the corrosion inhibitors is highly dependent on the types of contact materials. As examples, Figure 22.5 and Figure 22.6 show the effect of different corrosion inhibitors on contact resistance of ICA on Sn/Pb and Sn surface, respectively. It can be seen that by using suitable corrosion inhibitors, stabilized contact resistance was achieved but the effective corrosion inhibitors were different for different metal finishes [31,32].

22.2.2.4. Sacrificial Anode To improve the contact resistance stability, applying a sacrificial anode is another method. For galvanic corrosion of ECAs during aging, the larger the different in potential, the faster the corrosion develops. Also, the self-corrosion rates of both metals will change: the comparably active metal (the anode) corrodes faster while the other (the cathode) corrodes slower than they would do without contact. When applying a third and more active metal (or alloy) on a dissimilar metal couple in electrical contact, the comparably active metal can be protected from galvanic corrosion by oxidation of the third metal. This corrosion control is very important in reliability issues of the conductive adhesive joints. The addition of low corrosion potential individual metals, metal mixtures

A LEAD-FREE ALTERNATIVE IN ELECTRONIC PACKAGING

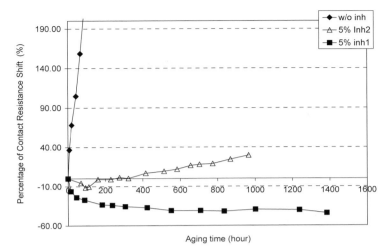

FIGURE 22.5. Shifts of contact resistance of conductive adhesives on Sn/Pb surface with and without corrosion inhibitors.

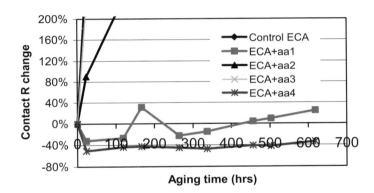

FIGURE 22.6. Shifts of contact resistance of conductive adhesives on Sn surface with and without corrosion inhibitors.

or metal alloys greatly reduces the electrode potential of ECAs, or in other words, narrows down the potential gap between the ECA and the metal finishes. Thus, these sacrificial anode materials act as an anode in this configuration and they are corroded first instead of the metal finishes, resulting in protecting the surfaces at the cathode. Comparing different types of sacrificial anodes, metal alloys are more effective than individual metals or metal mixtures due to the less easily oxidization characteristics for alloys [35].

22.2.2.5. Oxide-Penetrating Particles Another approach of improving contact resistance stability during aging is to incorporate some electrically conductive particles, which have sharp edges, into the ICA formulations. The particle is called oxide-penetrating filler. Force must be provided to drive the oxide-penetrating particles through oxide layer and hold them against the adherend materials. This can be accomplished by employing polymer binders that show high shrinkage when cured (Figure 22.7) [36]. This concept is used in polymer-

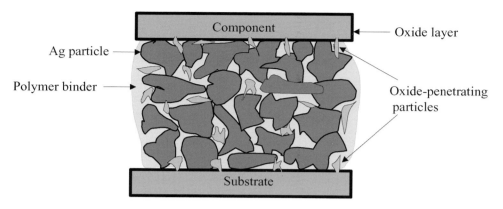

FIGURE 22.7. A Joint connected with an ICA containing oxide-penetrating particles and silver powders.

solder which has good contact resistance stability with standard surface-mounted devices (SMDs) on both solder-coated and bare circuit boards.

22.2.3. Silver Migration Control of ICA

Silver electromigration has long been a reliability concern in the electronic industry. The electromigration is an electrochemical process whereby, silver, in contact with an insulating material, in a humid environment and under an applied electric field, leaves its initial location in ionic form and deposits at another location [37]. Such migration may lead to a reduction in electrical spacing or cause a short circuit between interconnections. The migration process begins when a thin continuous film of water forms on an insulating material between oppositely charged electrodes. When a potential is applied across the electrodes, a chemical reaction takes place at the positively biased electrode where positive metal ions are formed. These ions, through ionic conduction, migrate toward the negatively charged cathode and over time, they accumulate to form metallic dendrites. As the dendrite growth increases, a reduction of electrical spacing occurs. Eventually, the silver growth reaches the anode and creates a metal bridge between the electrodes, resulting in an electrical short circuit [38]. The rates of silver migration is increased by: (1) an increase in the applied potential; (2) an increase in the time of the applied potentials; (3) an increase in the level of relative humidity; (4) an increase in the presence of ionic and hydroscopic contaminants on the surface of the substrate; and (5) a decrease in the distance between electrodes of the opposite polarity [37].

Several methods have been used to control the silver migration. The methods include: (1) alloying the silver with an anodically stable metal such as palladium [38] or platinum [39] or even tin [40]; (2) use of hydrophobic coating over the PWB to shield its surface from humidity and ionic contamination [41]; (3) plating of silver with metals such as tin, nickel or gold; and (4) coating the substrate with polymer [42]. Some other research is still going on for the migration prevention and high reliability ICAs.

22.2.4. Improvement of Reliability in Thermal Shock Environment

Besides the contact resistance increase due to galvanic corrosion, the poor thermal cycling (TC) performance of the ECA joints has been another serious issue for board level interconnects. It has been found rather difficult to control the TC failures of the ECA joints. Generally, the failure of the electrical interconnection during the TC test could be caused by many factors such as coefficient of the thermal expansion (CTE) mismatch between the IC component chip/the interconnection materials/the substrates, elastic modulus difference of these components, adhesion strength of the interconnect materials on the IC chip and the substrate, the mechanical properties of the IC chips, the glass transition or the softening point of the ECA materials, moisture uptake in the interface and the bulk ECAs, the surface or interface property change and so forth. Especially, the thermal stress in the ECA joints generated by a huge temperature difference during the TC and the interfacial delamination due to the adhesion degradation could be the critical reasons. In this aspect, a feasible solution to the TC failure problem is to introduce flexible molecules into the epoxy resin matrix. By releasing the thermal stress with the flexible molecules, the thermomechanical stresses could be dramatically reduced and the ECA/component joint interfaces could keep intact through the thermal cycling test.

22.2.5. *Improvement of Impact Performance of ICA*

Impact performance is a critical property of solder replacement ICAs. Effort has been continued in developing ICAs that have better impact strength and will pass the drop test, a standard test used to evaluate the impact strength of ICAs.

Nano-sized metal particles were used in ICAs to improve the electrical conduction and mechanical strength. Using nano-sized particles, agglomerates are formed due to surface tension effect [43]. Another approach is simply to decrease the filler loading to improve the impact strength [44]. However, such a process reduces the electrical properties of the conductive adhesives. A recent development was reported where conductive adhesives were developed using resins of low modulus so that this class of conductive adhesives could absorb the impact energy developed during the drop [45]. However, the electrical properties of these materials were not mentioned in the paper. Conformal coating of the surface-mounted devices was used to improve mechanical strength. It was demonstrated that conformal coating could improve the impact strength of conductive adhesives joints [46]. Furthermore, elastomer-modified epoxy resins were also used to enhance the damping properties and the impact performance of ICAs [47].

22.3. ANISOTROPIC CONDUCTIVE ADHESIVES (ACAs)/ANISOTROPIC CONDUCTIVE FILM (ACF)

Anisotropic conductive adhesives (ACAs) or anisotropic conductive films (ACFs) provide unidirectional electrical conductivity in the vertical or Z-axis. This directional conductivity is achieved by using a relatively low volume loading of conductive filler (5 to 20 volume percent) [48–50]. The low volume loading is insufficient for inter-particle contact and prevents conductivity in the X–Y plane of the adhesive. The Z-axis adhesive, in film or paste form, is interposed between the surfaces to be connected. Heat and pressure are simultaneously applied to this stack-up until the particles bridge the two conductor surfaces on the two adherends. Figure 22.8 shows the configuration of a component and a substrate bonded with ACF. Once the electrical continuity is produced, the polymer binder is hardened by thermally initiated chemical reaction (for thermosets) or by cooling (for thermoplastics). The hardened dielectric polymer matrix holds the two components together and helps maintain the pressure contact between component surfaces and conductive

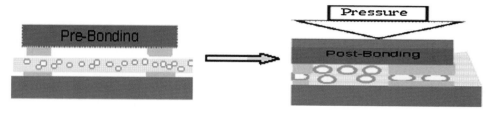

FIGURE 22.8. Schematic diagram of accomplishment of z-direction interconnects using ACF.

particles. Because of the anisotropy, ACA/ACF may be deposited over the entire contact region, greatly facilitating materials application. Also, an ultra fine pitch interconnection (<0.04 mm) could be achieved easily. The fine pitch capability of ACA/ACF would be limited by the particle size of the conductive filler, which can be a few microns or a few nanometers in diameter.

22.3.1. Materials

When designing materials to achieve fine pitch interconnections, several important variables must be considered and are application dependent. These variables include adhesive characteristics as well as particle type.

Two basic types of adhesives are available: thermosetting and thermoplastic materials. Thermoplastic adhesives are rigid materials at temperatures below the glass transition temperature (T_g) of a polymer. Above the T_g, polymers exhibit flow characteristics. When using this type of material, assembly temperatures must exceed the T_g to achieve good adhesion. The principal advantage of the thermoplastic adhesives is the relative ease with which the interconnection can be disassembled for repair operations. Thermosetting adhesives, such as epoxies and silicones, form a three-dimensional cross-linked structure when cured under specific conditions. Curing techniques include: heat, UV light, and added catalyst. As a result of this irreversible cure reaction, the initial uncross-linked material is transformed into a rigid solid. The curing reaction is not reversible. This fact may hinder disassembly and interconnection repair. The ability to maintain strength at high temperature and the deformation of robust adhesive bonds are the principal advantages of these materials. For the selection of the adhesive, the robust bonds should be formed to all surfaces involved in the interconnection. Numerous materials surfaces can be found in the interconnection region including: SiO_2, Si_3N_4, polyester, polyimide, FR-4, glass, gold, copper, and aluminum. Adhesion to these surfaces must be preserved after standard tests such as temperature-humidity-bias aging and temperature cycling. Some surfaces may require chemical treatments to achieve good adhesion. In addition, the adhesive must not contain ionic impurities that would degrade electrical performance of the interconnections.

The materials used as conductive particles must also be carefully selected. Silver offers moderate cost, high electrical conductivity, high current carrying ability, and low chemical reactivity, but problems with electromigration may occur. Nickel is a lower cost alternative, but corrosion of nickel surfaces has been found during accelerated aging tests. The material that offers the bests properties is gold; however, costs may be prohibitive for large-volume applications. Plated polymeric particles may offer the best combination of properties at moderate cost. Some ACA materials use solder particles to ensure electrical contacts with high reliability by creating a metallurgical bond.

22.3.2. Application of ACA/ACF in Flip Chip

Conventional flip-chip assembly involves two main steps: solder reflow and application of underfill, which is an organic adhesive placed between the IC chip and the substrate to improve mechanical reliability. In recent years, ACA flip-chip technology has been employed in many applications where flip chips are bonded to rigid chip carriers [51]. This includes bare chip assembly of ASICs in transistor radios, personal digital assistants (PDAs), sensor chip in digital cameras, and memory chip in lap-top computers. In all the applications, the common feature is that ACA flip-chip technology is used to assembly bare chips where the pitch is extremely fine, normally less than 120 μm. For those fine applications, it is apparently use of ACA flip chip instead of soldering is more cost effective.

ACA flip-chip bonding exhibits better reliability on flexible chip carriers because the ability of flex provides compliance to relieve stresses. For example, the internal stress generated during resin curing can be absorbed by the deformation of the chip carrier. ACA joint stress analysis conducted by Wu et al. indicated that the residual stress is larger on rigid substrates than on flexible substrates after bonding [52].

22.3.3. Improvement of Electrical Properties of ACAs

ACA flip chip technology has been employed in many applications because of the primary advantages such as fine pitch capability, lead-free, low processing temperature, absence of flux residue, and generally lower cost. Also, ACA flip chip technology does not require an additional underfilling process because the ACA resin acts as an underfill. However, ACAs have lower electrical current capability because of the restricted contact area and poor interfacial bonding of the ACAs and metal bond pad.

22.3.3.1. Self-Assembled Monolayer (SAM) In order to enhance the electrical performance of ACA materials, a class of chemicals that form a self-assembled monolayer (SAM) of conjugated molecular nano wires is introduced. These SAM molecules adhere to the metal surface and form physico-chemical bonds, which allow electrons to flow, as such, it reduces electrical resistance and enables a high current flow. The unique electrical properties of SAM originate from their linear chain structure and electron delocalization along the conjugated chain. Preliminary data have also shown that the current density of a SAM molecular wire such as 1,4-dithiol benzene can be up to 1×10^8 Å/cm^2, so the potential for conductivity improvement and high current carrying capability are significant.

An important consideration when examining the advantages of SAM compounds pertains to the affinity of SAM compounds to specific metal surfaces. Table 22.2 gives the examples of SAM molecules preferred for maximum interactions with specific metal finishes; although only molecules with symmetrical functionalities for both head and tail groups are shown, molecules and derivatives with different head and tail functional groups are possible for interfaces concerning different metal surfaces.

22.3.3.2. Electrical Properties Improvement of ACA with SAM Different SAM compounds, dicarboxylic acid and dithiol have been introduced into ACA joints for silver-filled and gold-filled ACAs, respectively [53,54]. Due to the strong affinity between those SAM compounds and metal fillers and metal bond pads, the physic-chemical bonding was introduced on the interface between the metal particles and between the ACAs/metal bond pads. The physico-chemical bonding could allow electrons to flow freely, as such, the molecules can reduce electrical resistance and achieve a high current density in the interface. For

TABLE 22.2.
Potential SAM interfacial modifier for different metal finishes.

Formula	Compound	Metal finish
H—S—R—S—H	Dithiol	Au, Ag, Sn, Zn
N≡C—R—C≡N	Dicyanide	Cu, Ni, Au
O=C=N—R—N=C=O	Diisocyanate	Pt, Pd, Rh, Ru
HO-C(=O)-R-C(=O)-OH	Dicarboxylate	Fe, Co, Ni, Al, Ag
imidazole structure	Imidazole and derivative	Cu
R = phenyl, cyclohexyl, —C=C—C=C—, —C—C—C—, etc.		

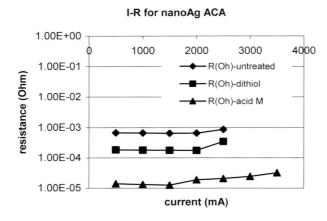

FIGURE 22.9. Electrical properties of nano Ag filled ACA with SAM.

dithiol incorporated ACA with micron-sized gold fillers, significantly lower joint resistance and higher maximum allowable current (highest current applied without inducing joint failure) was achieved for low temperature curable ACA (<100°C). For high curing temperature ACA (150°C), the improvement is not as significant as low curing temperature ACAs, due to the partial degradation of SAM compounds at the relatively high temperature. However, when dicarboxylic acid or dithiol was introduced into the interface of nano-silver filled ACA, significantly improved electrical properties could be achieved at high temperature curable ACA, suggesting de-bonding/degradation of SAM did not occur on silver nanoparticles at the curing temperature (150°C) (Figure 22.9). The enhanced bonding could be attribute to the larger surface area and higher surface energy of nano particles, which enabled the SAMs to be more readily coated and relatively thermally stable on the metal surfaces [54].

22.3.4. Thermal Conductivity of ACA

For the ACA interconnect joints to deliver high current, not only a low electrical resistance, but also a high thermal conductivity of the interconnect materials is required. The higher thermal conductivity can help dissipate heat more effectively from adhesive joints generated at high current. Therefore, a higher thermal conductivity could contribute to an improved current carrying capability. Reported shown that with the addition of high thermal conductivity fillers (e.g., SiC) into the ACA formulation, the higher thermal conductivity could be achieved, which also rendered a high current carrying capability [55] (Figure 22.10). By improving the interface properties of ACA joints with additives, a higher thermal conductivity could also be achieved.

22.4. FUTURE ADVANCES OF ECAs

There have been a significant number of studies to improve ECA technology. However, several critical issues associated with this technology must also be addressed before it can be used as a complete replacement for soldering technology.

22.4.1. Electrical Characteristics

The electrical conductivity must be further improved since their electrical conductivity is typically one or two orders of magnitudes lower than solders. Silver-filled ECAs achieve their electrical conductivity through a physical contact among the Ag flaks. Due to the high particle-particle contact resistance, ECAs exhibit a relatively higher bulk resistivity. Although some work has been conducted to improve electrical conductivity by in situ replacing or removing the lubricants on Ag flakes and helping electrons tunneling in ECA, additional work is still necessary to achieve the desired electrical conductivity.

Also, the electrical performance of ECAs on different metal surfaces under high temperature/humidity/bias conditions is not completely satisfactory. Although the mechanism for the reliability failure under aging environment has been well under-stood and some effective methods have been developed to stabilize the contact resistance on some metal finishes, further study is still needed to completely under-stand the reason for the reliability improvement. At the same time, it is necessary to look for new materials and methods to stabilize the contact resistance on different metal surfaces.

22.4.2. High Frequency Compatibility

The number of high-frequency applications and utilizations are increasing rap-idly, thus it is important to characterize the cross-linking between particles, coupling with semiconductor devices and other fundamental behavior of ECAs under high-frequency conditions. It is also necessary to maximize the current carrying capability of ECAs (especially ACAs) for high performance microprocessors and at high frequency range for wireless communication products, especially after exposure to various environment tests.

22.4.3. Reliability

It is necessary to understand the effect of the chip carrier material on ECA join reliability. This is a key issue before ECA technology is widely utilized in manufacturing (i.e.,

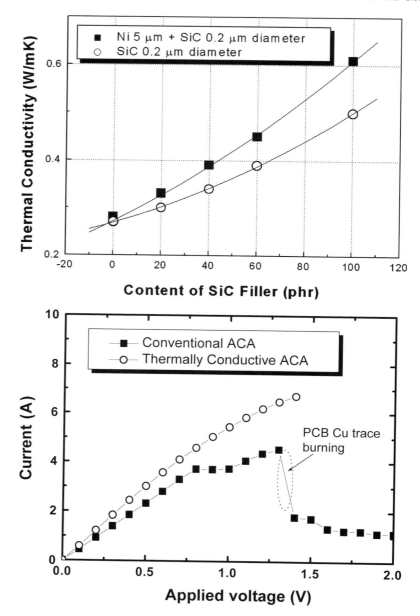

FIGURE 22.10. Thermal conductivity and I–V measurement of ACA with high thermal conductivity fillers.

in high-volume and low-cost applications). It is also necessary to establish failure rate prediction models for ECA joints for a wide variety of field conditions. It is essential to gain full understanding on effects of high current and high power on ECA joints, degradation and stress relaxation of polymeric matrices; and the effects of temperature, humidity, and other environments on matrix materials and the effects of fillers.

A LEAD-FREE ALTERNATIVE IN ELECTRONIC PACKAGING

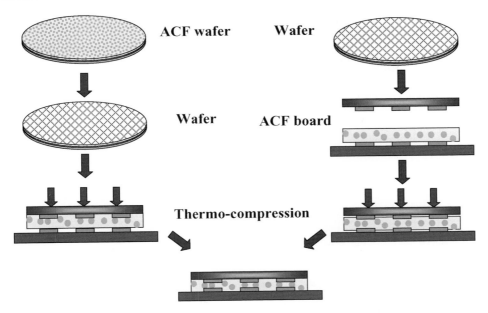

FIGURE 22.11. Schematic illustration of wafer level ACF process.

Silver-filled ICAs have a possibility for silver migration which causes electrical shorts especially in fine-pitch applications. Palladium treated silver filler exhibited much improved anti-migration characteristics compared to standard silver-filled ICAs. However, the preparation of the Pd-coated silver particles is expensive. Some low cost approaches must be developed. More comprehensive understanding of the fatigue resistance of ICA joints is required. Activities in this area have been limited and nonconclusive.

22.4.4. ECAs with Nano-filler for Wafer Level Application

For the next generation low cost, high efficiency ACA/ACF assembly, wafer level application might be a solution. Figure 22.11 shows the process of nano wafer level ACF (WLACF). Instead of using ACF in the assembly process, the ACF is applied on the wafer level before dicing. This eliminates the dispensing step in the component level and thus makes the ACF interconnect compatible with standard thermo-compression assembly process. The WLACF provides a lead-free and fine-pitch-capable interconnect, as well as a wafer level package to protect the wafer during test and burn-in. The cost of packaging can be dramatically reduced because it avoids the solder bumping process, combines interconnect and encapsulation, and enables wafer level test and burn-in. After the chip assembly, the ACF layer also acts as an underfill to redistribute the thermo-mechanical stress generated from the CTE mismatch between the chip and the substrate.

REFERENCES

1. M. Abet and G. Selvaduray, Lead-free solders in microelectronics, Materials Science & Engineering, 27, pp. 95–141 (2000).

2. J. Lau, C.P. Wong, N.C. Lee, and S.W.R. Lee, Electronics Manufacturing: With Lead-Free, Halogen-Free, and Conductive-Adhesive Materials, McGraw-Hill, New York, NY, 2002.
3. K.J. Puttlitz and K.A. Stalter, Eds., Handbook of Lead-free Solder Technology for Microelectronic Assemblies, Marcel Dekker, Inc., New York, NY, 2004, Chap. 1.
4. J. Lau, C.P. Wong, J. Prince, and W. Nakayama, Electronic Packaging; Design, Materials, Process and Reliability, McGraw-Hill, New York, 1998, Chap. 5, p. 393.
5. A.Z. Miric and A. Grusd, Lead-free alloys, Soldering and Surface Mount Technology, 10(1), p. 19 (1998).
6. J.S. Hwang, Solder materials, Surface Mount Technology, 18, p. 46 (2000).
7. K. Chen and K. Linz, Effects of gallium on wettability, microstructures and mechanical properties of the Sn-Zn-Ag-Ga and Sn-Zn-Ag-Al-Ga solder alloys, Int'l Symposium on Electronic Materials and Packaging, p. 49 (2002).
8. J. Liu, Ed., Conductive adhesives for Electronics Packaging, Electrochemical Publications Ltd. Isle of Man, British Isles, 1999.
9. E.P. Wood and K.L. Nimmo, In search of new lead-free electronic solders, J. Electron. Mater., 23(8), pp. 709–713 (1994).
10. E.R. Monsalve, Lead ingestion hazard in hand soldering environments, Proceedings of the 8th Annual SolderingTechnology and Product Assurance Seminar, Naval Weapons Center, China Lake, CA, February 1984.
11. Y. Li, K. Moon, and C.P. Wong, Electronics without lead, Science, 308(June 3), pp. 1419–1420 (2005).
12. S. Jin, D.R. Frear, and J.W. Morris, Jr., Foreword, J. Electron. Mater., 23(8), pp. 709–713 (1994).
13. Environmental Protection Agency, National Air Quality and Emission Trend Report, 1989, EPA-450/4-91-003, Research Triangle Park, NC, 1991.
14. E. Perrot, Electronic packaging for the 21st century, Advanced Packaging, (July-August) (1995).
15. J. Greaves, Evaluation of solder alternatives for surface mount technology, Proceedings of Nepcon West Technical Program, 1993, pp. 1479–1488.
16. M.A. Gaynes, R.H. Lewis, R.F. Saraf, and J.M. Roldan, Evaluation of contact resistance for isotropic conductive adhesives, IEEE Transactions on Components, Packaging, and Manufacturing Technology, Part B, 18(2), pp. 299–304 (1995).
17. G. Nguyen, J. Williams, F. Gibson, and T. Winster, Electrical reliability of conductive adhesives for surface mount applications, Proceedings of International Electronic Packaging Conference, 1993, pp. 479–486.
18. D. Lu, Q.K. Tong, and C.P. Wong, Conductivity mechanisms of isotropic conductive adhesives (ICAs), IEEE Transactions on Components, Packaging, and Manufacturing Technology, Part C, 22(3), p. 223 (1999).
19. L. Smith-Vargo, Adhesives that posses a science all their own, Electronic Packaging & Production, (August), pp. 48–49 (1986).
20. E.M. Jost and K. McNeilly, Silver flake production and optimization for use in conductive polymers, Proceedings of ISHM, 1987, pp. 548–553.
21. S.M. Pandiri, The behavior of silver flakes in conductive epoxy adhesives, Adhesives Age, Oct. 1987, pp. 31–35.
22. D. Lu, Q.K. Tong, and C.P. Wong, A study of lubricants on silver flakes for microelectronics conductive adhesives, IEEE Transactions on Components, Packaging and Manufacturing Technology, Part A, 22(3), pp. 365–371 (1999).
23. D. Lu, Q. Tong, and C.P. Wong, A fundamental study on silver flakes for conductive adhesives, 1998 International Symposium on Advanced Packaging Materials, Braselton, GA, 1998, pp. 256–260.
24. Y. Li, K. Moon, H. Li, and C.P. Wong, Conductivity improvement of isotropic conductive adhesives with short-chain dicarboxylic acids, Proceedings of 54th IEEE Electronic Components and Technology Conference, Las Vegas, Nevada, June 1–4, 2004, p. 1959.
25. Y. Li, A. Whitman, K. Moon, and C.P. Wong, High performance electrically conductive adhesives (ECAs) modified with novel aldehydes, Proceedings of 55th IEEE Electronic Components and Technology Conference, Lake Buena Vista, Florida, May 31–June 3, 2005, pp. 1648–1652.
26. C. Gallagher, G. Matijasevic, and J.F. Maguire, Transient liquid phase sintering conductive adhesives as solder replacement, Proceedings of 47th Electronic Components and Technology Conference, San Jose, CA, May 1997, p. 554.
27. J.W. Roman and T.W. Eagar, Low stress die attach by low temperature transient liquid phase bonding, Proceedings of ISHM, San Francisco, CA, Oct. 1992, p. 52.
28. M. Zwolinski, J. Hickman, H. Rubon, and Y. Zaks, Electrically conductive adhesives for surface mount solder replacement, Proceedings of the 2nd International Conference on Adhesive Joining & Coating Technology in Electronics Manufacturing, Stockholm, Sweden, June 1996, p. 333.
29. D. Lu, Q.K. Tong, and C.P. Wong, Mechanisms underlying the unstable contact resistance of conductive adhesives, IEEE Transactions on Components, Packaging, and Manufacturing Technology, Part C, 22(3), pp. 228–232 (1999).

30. Q.K. Tong, G. Fredrickson, R. Kuder, and D. Lu, Conductive adhesives with superior impact resistance and stable contact resistance, Proceedings of the 49th Electronic Components and Technology Conference, San Diego, CA, May 1999, p. 347.
31. D. Lu and C.P. Wong, Novel conductive adhesives for surface mount applications, Journal of Applied Polymer Science, 74, p. 399 (1999).
32. Y. Li, K. Moon, and C.P. Wong, Development of conductive adhesives with novel corrosion inhibitors for stabilizing contact resistance on non-noble lead-free finishes, Proceedings of 55th IEEE Electronic Components and Technology Conference, Lake Buena Vista, Florida, May 31–June 3, 2005, pp. 1462–1467.
33. H. Leidheiser, Jr., Mechanism of corrosion inhibition with special attention to inhibitors in organic coatings, Journal of Coatings Technology, 53(678), p. 29 (1981).
34. G. Trabanelli, Corrosion inhibitors, in F. Mansfeld, Ed., Corrosion Mechanisms, Marcel Dekker, Inc., New York, NY, 1987, p. 119.
35. H. Li, K. Moon, and C.P. Wong, A novel approach to stabilize contact resistance of electrically conductive adhesives on lead-free alloy surfaces, Journal of Electronic Materials, 33(2), p. 106 (2004).
36. D. Durand, D. Vieau, A.L. Chu, and T.S. Weiu, Electrically conductive cement containing agglomerates, flake and powder metal fillers, U.S. Patent 5,180,523, Nov. 1989.
37. G. Davies and J. Sandstrom, How to live with silver tarnishing/silver migration, Circuits Mfg., (Oct.), pp. 56–62 (1976).
38. G. Harsanyi and G. Ripka, Electrochemical migration in thick-film IC-S, Electrocomp. Science And Tech., 11, pp. 281–290 (1985).
39. R. Wassink, Notes on the effects of metalization of surface mounted components on soldering, Hybrid Circ., (13), pp. 9–13 (1987).
40. Y. Shirai, M. Komagata, and K. Suzuki, Non-migration conductive adhesives, 1st International IEEE Conference on Polymers and Adhesives in Microelectronics and Photonics, 21–24 Oct., 2001, pp. 79–83.
41. A. Der Marderosian, The Electrochemical Migration of Metal, Ratheon Co. Equipment Division, Equipment Development Laboratories, pp. 134–141.
42. H. Schonhorn and L.H. Sharpe, Prevention of surface mass migration by a polymeric surface coating, U.S. Patent #4,377,619, 1983.
43. S. Kotthaus, R. Haug, H. Schafer, and O.D. Hennemann, Proceedings of 1st IEEE International Symposium on Polymeric Electronics Packaging, Norrkoping, Sweden, Oct. 26–30, 1997, pp. 64–69.
44. S. Macathy, Proceedings of Surface Mount International, San Jose, August 27–31, 1995, pp. 562–567.
45. S.A. Vona and Q.K. Tong, Surface mount conductive adhesives with superior impact resistance, Proceedings of 4th International Symposium and Exhibition on Advanced Packaging Materials, Processes, Properties and Interfaces, Braselton, GA, March 14–16, 1998, pp. 261–267.
46. J. Liu and B. Weman, Modification of processes and design rules to achieve high reliable conductive adhesive joints for surface mount technology, Proceedings of the 2nd International Symposium on Electronics Packaging Technology, Dec. 9–12, 1996, pp. 313–319.
47. D. Lu and C.P. Wong, Conductive adhesives for solder replacement in electronics packaging, Proceedings—International Symposium on Advanced Packaging Materials: Processes, Properties and Interfaces, Braselton, GA, United States, Mar. 6–8, 2000, pp. 24–31.
48. A.O. Ogunjimi, O. Boyle, D.C. Whalley, and D.J. Williams, A review of the impact of conductive adhesive technology on interconnection, Journal of Electronics Manufacturing, 2, p. 109 (1992).
49. P.G. Harris, Conductive adhesives: a critical review of progress to date, Soldering & Surface Mount Technology, (20), p. 19 (1995).
50. K. Gilleo, Assembly with conductive adhesives, Soldering & Surface Mount Technology, (19), p. 12 (1995).
51. J. Liu, ACA bonding technology for low cost electronics packaging applications-current status and remaining challenges, Proceedings of 4th International Conference on Adhesive Joining and Coating Technology in Electronics manufacturing, Helsinki, Finland, June 2000, pp. 1–15.
52. C.M.L. Wu, J. Liu, and N.H. Yeung, Reliability of ACF in flip chip with various bump height, Proceedings of 4th International Conference on Adhesive Joining and Coating Technology in Electronics manufacturing, Helsinki, Finland, June 2000, pp. 101–106.
53. Y. Li, K. Moon, and C.P. Wong, Adherence of self-assembled monolayers on gold and their effects for high performance anisotropic conductive adhesives, Journal of Electronic Materials, 34(3), pp. 266–271 (2005).
54. Y. Li and C.P. Wong, Nano-Ag filled anisotropic conductive adhesives (ACA) with self-assembled monolayer for high performance fine pitch interconnect, Proceedings of 55th IEEE Electronic Components and Technology Conference, Lake Buena Vista, Florida, May 31–June 3, 2005, pp. 1147–1154.
55. M. Yim, J. Hwang, J. Kim, H. Kim, W. Kwon, K.W. Jang, and K.W. Paik, Anisotropic conductive adhesives with enhanced thermal conductivity for flip chip applications, Proceedings of 54th IEEE Electronic Components and Technology Conference, Las Vegas, Nevada, June 1–4, 2004, p. 159.

23

Die Attach Quality Testing by Structure Function Evaluation

Márta Rencz[a], Vladimir Székely[b], and Bernard Courtois[c]

[a]*MicReD Ltd., Hungary*
[b]*Budapest University of Technology and Economics, Hungary*
[c]*TIMA Laboratory Grenoble, France*

Abstract   In this chapter simulation and measurement experiments prove that the structure function evaluation of the thermal transient testing is capable to locate die attach failure(s), even in case of stacked die packages. Both the strength and the location of the die attach failure may be determined with the methodology of a fast thermal transient measurement and the subsequent computer evaluation. The paper presents first the theoretical background of the method. After this, application on single die packages is presented. In the rest of the paper first simulation experiments show the feasibility of locating die attach problem in stacked die structures with the presented algorithm and a large number of measured experiments prove that the methodology is applicable also in practice. At the end of the chapter, in the evaluation of the methodology the special advantage of the method, that normally it does not require any additional circuit elements on any of the possibly-stacked-dies is also presented.

NOMENCLATURE

A (m^2)	area, surface
c (Ws/Km^3)	volumetric heat capacity
C, C_{th} (Ws/K)	thermal capacitance
K (W^2 s/K^2)	value of the differential structure function
R, R_{th} (K/W)	thermal resistance
Z (K/W)	thermal impedance

Greek symbols

λ (W/mK)	thermal conductivity
τ (s)	time constant
ω (1/s)	angular frequency

Subscripts

Σ   cumulative values

23.1. INTRODUCTION

The reliability of packaged electronics strongly depends on the quality of the die attachment. Any void in it or a small delamination may cause instant temperature increase in the die, leading sooner or later to failure in the operation. Since die attach problems are built in time bombs, they have to be eliminated by any means. They usually can not be detected by the electronic test, only in special, very bad cases. They can be detected by steady state thermal measurements, but these measurements are on one hand time consuming on the other hand they can not locate the actual problem. In case of stacked die structures, that will dominate the packages in the near future the die attach quality control problem is multiplied: the structure contains as many die attach layers as the number of chips, which are very difficult to test, resulting in pronounced reliability concerns [1,2].

It has been presented in earlier papers that the structure functions [3], obtainable from the thermal transient measurements are applicable to indicate the die attach failures of single die packages [4]. In the paper of [4] die attach quality measurements on experimental packages of ST-Microelectronics were presented. It was demonstrated that the presence of die attach voids in different forms and locations influences the thermal resistance of the glue layer such, that the voids in the die attach material can be conveniently detected with the help of fast thermal transient measurements and the subsequent evaluation. A special advantage of the applied structure function evaluation methodology is that it delivers not only the value but also the location of the increased thermal resistance in the heat flow path.

Because of the extremely small thermal resistance value of the stacked dies it was not straightforward that the method can be used to uncover the locations of die attach failures of stacked dies as well. And even if the method is applicable up to a certain number of dies, it has to be experimented what are the limits in the resolution of the method. It had to be also investigated whether the methodology is applicable for stacked die structures of different die sizes as well; this was done in [5].

In this chapter we first give a short summary of the structure function evaluation methodology, presenting how it can be used to detect the thermal properties of various layers in the heat flow path. In the rest of the chapter we demonstrate the applicability of the method with simulation and measurement experiments.

23.2. THEORETICAL BACKGROUND

The simplest representation of a thermal system is presented in Figure 23.1: it is characterized by one thermal resistance and one thermal capacitance.

If we apply a P power step to this model the temperature will rise at its port according to

$$T(t) = PR_{th}(1 - \exp(-t/\tau)), \tag{23.1}$$

where $\tau = R_{th}C_{th}$ is the time constant of the system.

DIE ATTACH QUALITY TESTING BY STRUCTURE FUNCTION EVALUATION

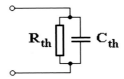

FIGURE 23.1. The simplest representation of a thermal system.

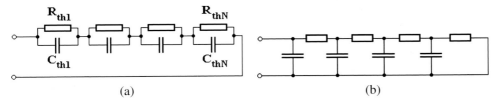

FIGURE 23.2. Foster (a) and Cauer (b) type representation of physical structures with finite number of time constants.

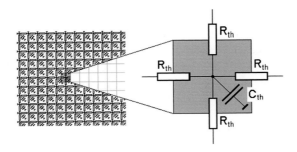

FIGURE 23.3. Real physical structures have infinite number of time constants.

Real physical structures have usually more time constants, in this case the temperature response function is the sum of the appropriate exponential functions, as

$$T(t) = P \sum_{i=1}^{N} R_{thi}(1 - \exp(-t/\tau_i)). \qquad (23.2)$$

To such response functions two different model networks may be ordered. These are not independent, one may be calculated from the other: a Foster and a Cauer type network, see Figure 23.2.

Thermal systems are to be represented by Cauer networks, since the thermal capacitances are always connected to the ground, but since every Cauer network has its Foster equivalent, the Foster approximation of thermal networks exists as well.

Real physical structures, however, are represented with an infinite number of time constants, since any infinitesimal cube of a matter shows a certain thermal resistance and capacitance value, see Figure 23.3 resulting in as many number of time constants in the response function as the number of its capacities.

These time constants may be represented only with their density function, see Figure 23.4.

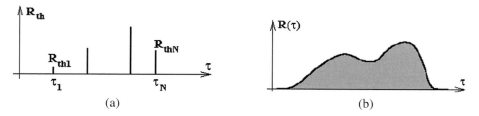

FIGURE 23.4. Time constants in a lumped element system (a) and in an infinite distributed parameter system (b). In the latter case the time constants form a continuous curve, the time constant density function.

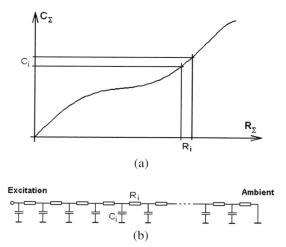

FIGURE 23.5. The cumulative structure function and the related Cauer equivalent circuit. (© 2002 IEEE SEMI-THERM Proceedings, "Determining partial thermal resistances with transient measurements and using the method to detect die attach discontinuities," M. Rencz, V. Szekely, A. Morelli, C. Villa, Fig. 1. © 2004 IEEE SEMI-THERM Proceedings, "Structure function evaluation of stacked dies," M. Rencz, V. Székely, Fig. 1.)

Knowing the time constant density of a system, an arbitrarily well approximating Foster equivalent circuit may be constructed by approximating the $R(\tau)$ function by infinitesimally narrow boxes with the height of $R(\tau)$, determining for each a parallel RC pair in the Foster chain. Finding the Cauer equivalent of this network by textbook transformations, we obtain a true physical equivalent of the heat transport in thermal systems. From this circuit we can draw up the so-called *Cumulative structure function* or PROTONOTARIOS-WING function [6], that gives the sum of the thermal capacitances C_Σ versus the sum of the thermal resistances R_Σ of the thermal system, measured from the point of excitation toward the ambient, see Figure 23.5.

On this monotonously increasing function, the plateaus represent the new materials with relatively low thermal capacity value and their widths give the related thermal resistances. The constant cross section intervals represent fields with heat propagation over linearly increasing areas.

The structure functions are calculated by direct mathematical transformations from the heating or cooling curves [3]. These latter curves may be obtained either from measurements or from the simulations of the detailed structural model of the heat flow path. In both cases a step function powering has to be applied on the structure, and the resulting increase (or decrease, in case of switching off) in the temperature at the location of the powering has to be measured, following the action of switching on or off has to be recorded.

The differential structure function is defined as the derivative of the cumulative thermal capacitance with respect to the cumulative thermal resistance, by

$$K(R_\Sigma) = \frac{dC_\Sigma}{dR_\Sigma}. \tag{23.3}$$

Considering a dx wide slice of a single matter of cross section A, we can calculate this value. Since for this case $dC_\Sigma = cAdx$, and the resistance is $dR_\Sigma = dx/\lambda A$, where c is the volumetric heat capacitance, λ is the thermal conductivity and A is the cross sectional area of the heat flow, the K value of the differential structure function is

$$K(R_\Sigma) = \frac{cAdx}{dx/\lambda A} = c\lambda A^2. \tag{23.4}$$

This value is proportional to the c and λ material parameters, and to the square of the cross sectional area of the heat flow, consequently it is related to the structure of the system. In other words: this function provides a map of the square of the cross section area of heat current-flow as a function of the cumulative resistance. In these functions the local peaks and valleys indicate reaching new materials in the heat flow path, and their distance on the horizontal axis gives the partial thermal resistances between them. More precisely, the peaks point usually to the middle of any new region where both the areas, perpendicular to the heat flow and the material are uniform. After identifying the peaks that refer to the chip and the chip carrier their distance on the horizontal axis can be read from the diagram. This value gives the die attach thermal resistance.

The method can be used for detecting die attach failures as follows:

1. The structure function of a good sample has to be created. This can be obtained by mathematical transformation [3] of the measured or simulated thermal transient curves.
2. The structure function of the sample package that we wish to evaluate has to be created similarly.
3. The structure function of the known good device has to be compared to the structure function of the device under test. The shift in the appropriate points in the structure functions gives the increased thermal resistance of the layer in question.

In the next section it is presented how to apply the methodology to single die packages. In the following sections it is first demonstrated with simulation experiments that the method is applicable also for quality testing of stacked die structures, after this we show how to use the method in practice. The presented measured results on real stacked die packages prove the applicability of the method for the detection of voids and delamination in the different die attach and soldering layers of stacked die packages as well.

23.3. DETECTING VOIDS IN THE DIE ATTACH OF SINGLE DIE PACKAGES

Fast detection of die attach problems is a crucial question in semiconductor packaging. In order to verify the applicability of the structure function evaluation for the detection and the analysis of die attach discontinuities in single die packages ST Microelectronics has fabricated experimental packages [4] with well defined die attach problems, see Figure 23.6. In these experimental samples voids were produced in the corners, at the sides and in the centers of the packages and a large number of experiments were carried out on a large number of samples with each type of failures to assure reliable results. Out of all these experimental data only a few are presented to demonstrate our findings: we selected to present here the curves that represent the averages of the centre void or corner void samples.

In Figure 23.7 average differential structure functions of samples with corner voids and central voids are presented together with the differential structure function of a reference device. These curves were obtained by carrying out thermal transient measurements on the structures, and calculating the structure functions, as described in [6]. We can see in this figure that the first peaks, representing the chips, are at the same R_{th} value, but on the right hand side of these, the R_{th} "distance" of the next peak is increased in case of corner voids, and highly increased in case of central voids.

Similarly large differences can be noticed in the dominant time constants of the samples in the time constant density functions. Figure 23.8 presents the time constant density functions of the samples, where the main time constants of the selected average samples are those which have the highest intensity values. It is observable in the figure that a dominant time constant appears with the value of 0.01 sec in case of the samples with central voids. Similar, though less dominant increase can be noticed in case of the samples with corner voids as well.

Figure 23.7 and Figure 23.8 demonstrate in one hand, that the highest increase in the R_{th} value is resulted in the cases when the voids are centrally located in the package. This result shows that in this package structure the heat flow path is such that the thermal properties of the package very strongly depend on the quality of the heat removal path at the middle of the structure. On the other hand it is also suggested from these figures that the heating curves of the samples have to be significantly different already after 0.01 sec. This can be in fact noticed on the measured Z_{th} curves of Figure 23.9 as well: at $t = 0.01$ sec

REFERENCE (NO VOID)   **CORNER VOID**   **CENTRAL VOID**

FIGURE 23.6. Experimental package samples with die attach voids prepared to verify the accuracy of the detection method (acoustic microscopic images). (© 2002 IEEE SEMI-THERM Proceedings, "Determining partial thermal resistances with transient measurements and using the method to detect die attach discontinuities," M. Rencz, V. Szekely, A. Morelli, C. Villa, Fig. 6.)

of the measurement the heating curves of the samples with die attach failures demonstrate significantly higher Z_{th} values than the perfect ones.

This last finding is extremely important from the point of view of in-production-line testing, because it means, that the method is applicable also for the in-line detection of die attach failures. In-production-line testing has to be done in milliseconds, and can not wait for the generation of the structure functions. For the generation of the structure function the whole thermal transient curve is needed, reaching the steady state, which may take minutes, or even longer, depending on the structure. To analyze however the origin of the Z_{th} increase, that is, for failure analysis the structure function method has to be used.

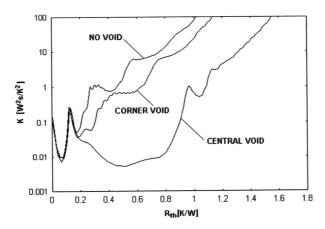

FIGURE 23.7. Differential structure functions of the experimental samples. The samples with central voids show highly increased R_{th} between the chip and the package. (© 2002 IEEE SEMI-THERM Proceedings, "Determining partial thermal resistances with transient measurements and using the method to detect die attach discontinuities," M. Rencz, V. Szekely, A. Morelli, C. Villa, Fig. 7.)

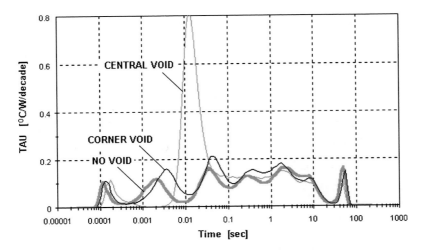

FIGURE 23.8. Main time constants of the experimental samples. (© 2002 IEEE SEMI-THERM Proceedings, "Determining partial thermal resistances with transient measurements and using the method to detect die attach discontinuities," M. Rencz, V. Szekely, A. Morelli, C. Villa, Fig. 8.)

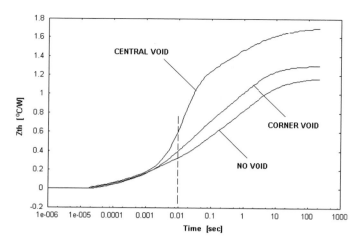

FIGURE 23.9. Measured Z_{th} curves of the average samples with no void, central void and corner voids in the die attach. At 0.01 sec the curves are with and without die attach voids are already clearly distinguishable. (© 2002 IEEE SEMI-THERM Proceedings, "Determining partial thermal resistances with transient measurements and using the method to detect die attach discontinuities," M. Rencz, V. Szekely, A. Morelli, C. Villa, Fig. 9.)

23.4. SIMULATION EXPERIMENTS FOR LOCATING THE DIE ATTACH FAILURE ON STACKED DIE PACKAGES

Today the major challenge to the packaging industry is the reliable increase of the packaging density with the help of stacking several dies in single packages. In case of stacked die packages the die attach quality control problem is multiplied: the structure contains as many die attach layers as the number of dies, and these are very difficult to test. Since any failure in the die attach may contribute to an early ruination of the whole circuit, testing the die attach quality of stacked dies is today a major task in assuring reliability.

The structure function evaluation methodology has demonstrated excellently the capability of detecting die attach problems of single die packages. It can be expected that the method will be applicable also to detect the location of the die attach problems also in the case of stacked die packages as well. The applicability can be proven first with the help of simulation tests. If these tests indicate the applicability of the method, we may start with experiments on real structures.

In order to calculate the structure function of the heat flow path by simulation first the thermal transient behavior has to be simulated, assuring one-dimensional heat flow through the structure. In the presented simulation experiments the simulation was done by the SUNRED program [8]. The one-dimensional heat flow was assured by supposing cold plate at the header of the simulated package, and constant and evenly distributed dissipation was switched on the top of the structure. The increasing temperature was observed on the location of the dissipation, on the top die.

The results of two different structures are presented: a 3D package with equal size dies, and a pyramidal structure, that is frequently used in case of stacked die structures in order to facilitate the bonding.

23.4.1. Simulation Tests Considering Stacked Dies of the Same Size

In the first simulation experiments 3-, 2- and 1-dimensional thermal simulations were applied for 3D stacked die structures of the same die size. Note, that 1-dimensional thermal simulation is very useful sometimes to get an immediate answer to some basic questions. 2- or 3-dimensional simulations need much more time to prepare, but to obtain accurate results we frequently need them. For all the simulation tests step function powering was applied on the top surface of the top layer silicon, and the resulting temperature transients were obtained in the structure by calculating the time dependent temperature in the middle of the top surface. In order to emulate measurement conditions random noise was added to the simulated results and the same resolution (bit-number) was used in the results that we use normally in the measurements.

From the results of the simulated and measurement emulated transient response function, we can calculate [3] the structure functions. This calculation involves supposing one-dimensional heat flow from the top surface of the stacked structure toward the cold plate. In the simulation experiments the dies were separated by die attach (glue) layers, having the same uniform thickness as the die thickness. The considered structure is presented in Figure 23.10.

From the simulated temperature transients the structure functions were created by the measurement evaluation software of the T3Ster tool [9]. In Figure 23.11 the differential structure function of the structure of Figure 23.10 is presented, calculated from the simulated time response. The arrows are pointing to the peaks, representing the silicon dies, their respective distances on the horizontal axis give the values of the respective thermal resistances of the die attach material between them.

In the presented first simulation experiment the thermal resistance of the die attach layer separating the 1st and the 2nd silicon dies was increased, modeling a die attach void in between these dies. In Figure 23.12 from the shift in the right hand side end of the second curve the increased total thermal resistance of the structure can be read, this is about 0.6 K/W.

In the second simulation experiment presented the same increased thermal resistance value was considered between the 2nd and the 3rd silicon dies. It is well observable from the comparison of the figures that the total thermal resistance of the structure is the same increased value in both cases, but the beginning of the displacement is different. In the

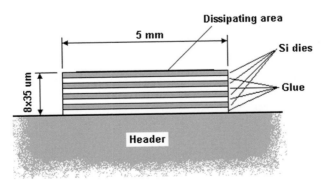

FIGURE 23.10. The simulated structure contained 4 silicon dies of 35 μm thickness, separated by die attach of the same thickness. The figure is not to scale. (© 2004 IEEE SEMI-THERM Proceedings, "Structure function evaluation of stacked dies," M. Rencz, V. Szekely, Fig. 2.)

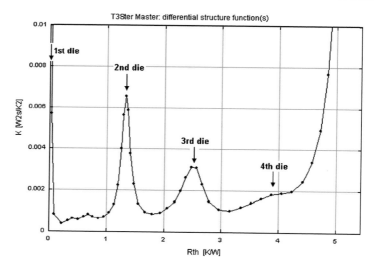

FIGURE 23.11. On the differential structure function of the structure the peaks represent the subsequent dies, their distances give the value of the die attach thermal resistance between them. (© 2004 IEEE SEMI-THERM Proceedings, "Structure function evaluation of stacked dies," M. Rencz, V. Szekely, Fig. 3.)

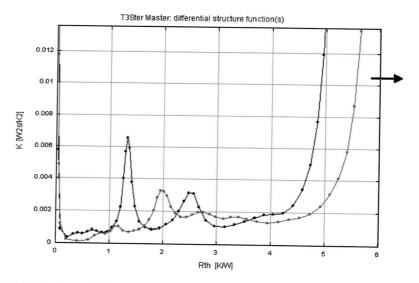

FIGURE 23.12. The increased thermal resistance of the die attach between the 1st and the 2nd silicon layer results in the displacement of the peak representing the 2nd die. (© 2004 IEEE SEMI-THERM Proceedings, "Structure function evaluation of stacked dies," M. Rencz, V. Szekely, Fig. 4.)

second experiment the source of the increased thermal resistance is the die attach between the 2nd and the 3rd dies, since the first two peaks are at the same location in the differential structure function as in the case of the nominal curve of Figure 23.13.

The simple simulation experiments presented demonstrate well the feasibility of the method for the detection of die attach problems in stacked die structures for at least 4 layers of dies.

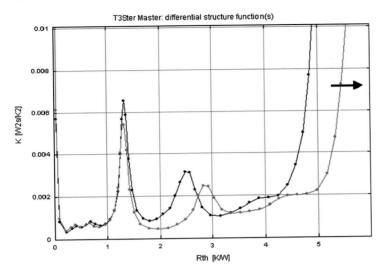

FIGURE 23.13. Inserting additional thermal resistance between the 2nd and the 3rd dies results in the shifting of the 3rd peak and from that on the whole right hand side of the curve. (© 2004 IEEE SEMI-THERM Proceedings, "Structure function evaluation of stacked dies," M. Rencz, V. Szekely, Fig. 6.)

23.4.2. Simulation Experiments on a Pyramidal Structure

Since today most of the stacked dies are of pyramidal structure to facilitate bonding it has to be examined if the method works appropriately also in such cases.

One quarter of the investigated pyramidal structure is presented in Figure 23.14. The dissipating area was supposed in the middle of the top layer. Note that the figure is not to scale.

The sizes of the simulated structure were as follows. The size of the bottom die was 14×14 mm (7×7 in the 1/4 simulation), the die size in the middle was 12×12 mm, and the size of the top die was 10×10 mm. The thickness of all the die and the glue layers were 0.035 mm. The considered thermal conductivity and volumetric thermal capacity values were 156 W/mK and 1.6×10^6 W s/m^3 K for the silicon, 1 W/mK and 10^6 W s/m^3 K for the glue layers. The considered heat transfer coefficients were 6000 W/m^2 K on the bottom, and 100 W/m^2 K on the top.

The simulated time range was between 1 µs–2 s, logarithmic time steps were applied. The differential structure function of the faultless structure shows the expected peaks, see Figure 23.15. The three chips and the glue under the structure can be well identified.

The thermal resistance and thermal capacitance values may be directly read from the cumulative structure functions. In the step function like cumulative structure function of Figure 23.16 the height values of the individual steps give the thermal capacitances, while the width values of the steps provide the related thermal resistances of the subsequent layers, measured from the top die top surface ($R_{th} = 0$) toward the bottom die and finally to the cold plate. The values that can be read from Figure 23.16 are good approximations of the values that can be calculated for the steady state from the geometrical and material parameters, but do not equal them. The explanation of the small difference is that as the structure is very shallow, the heat is not spreading out in the whole silicon area at once, and the one-dimensional heat flow approximation is not correct in the whole structure. The heat is reaching the cold plate under the structure rapidly, and for this reason practically only the

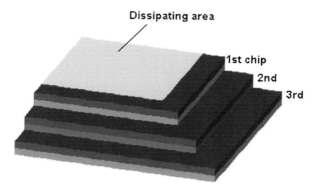

FIGURE 23.14. The investigated pyramidal structure. (© 2004 IEEE SEMI-THERM Proceedings, "Structure function evaluation of stacked dies," M. Rencz, V. Szekely, Fig. 7.)

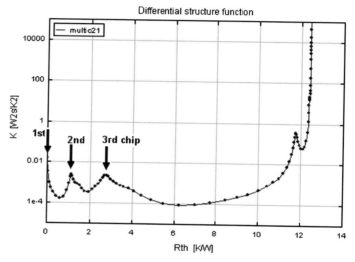

FIGURE 23.15. The structure function of the faultless pyramidal structure. (© 2004 IEEE SEMI-THERM Proceedings, "Structure function evaluation of stacked dies," M. Rencz, V. Szekely, Fig. 8.)

volume under the dissipating shape is participating in the heat conduction, showing smaller thermal capacitance values than expected in the steady state calculations.

Let us suppose now die-attach failure between the 2nd and the 3rd dies. This can be modeled with an increased thermal resistance of the die attachment between these dies. In the results of the simulation experiment presented in Figure 23.17 the thermal resistance of the die attach between the 2nd and 3rd dies was increased by 33%, that is, the thermal conductivity was decreased to 0.75 W/mK.

The curve denoted by A in Figure 23.17 refers to the faultless case, B refers to the case of die attach void between the 2nd and 3rd chips. As it is show in the figure, the peak referring to the 3rd chip is displaced now with the value of Δ, while the peak referring to the 2nd chip did not move from the original position. The value of Δ gives back the increased thermal resistance of the die attach layer, as expected.

DIE ATTACH QUALITY TESTING BY STRUCTURE FUNCTION EVALUATION

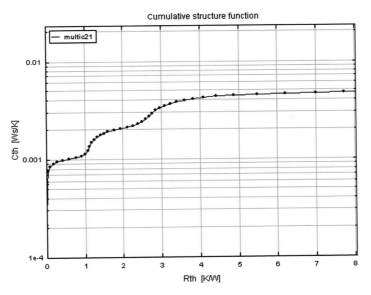

FIGURE 23.16. The cumulative structure function of the faultless structure. (© 2004 IEEE SEMI-THERM Proceedings, "Structure function evaluation of stacked dies," M. Rencz, V. Szekely, Fig. 9.)

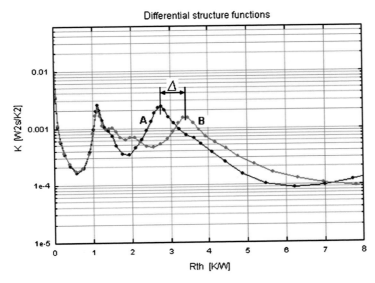

FIGURE 23.17. Die-attach voids are supposed between the 2nd and the 3rd chips. The displacement of the 3rd peak, that is, the distance between the A and B curves gives the increased thermal resistance. (© 2004 IEEE SEMI-THERM Proceedings, "Structure function evaluation of stacked dies," M. Rencz, V. Szekely, Fig. 10.)

In the next presented simulation experiment the die-attach failure was supposed to be between the 1st and the 2nd dies. This was modeled by an increased thermal resistance again, obtained by decreasing the value of the glue thermal conductivity to 0.75 W/mK. The new results are presented in Figure 23.18. In this figure A denotes the faultless curve,

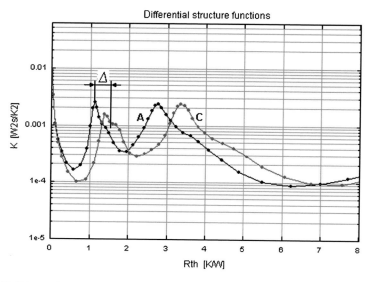

FIGURE 23.18. Die-attach voids are supposed between the 1st and the 2nd chips. The Δ displacement of the 2nd peak, that is, the distance on the A and C curves gives the increased thermal resistance. (© 2004 IEEE SEMI-THERM Proceedings, "Structure function evaluation of stacked dies," M. Rencz, V. Szekely, Fig. 11.)

C denotes the curve obtained by considering an increased thermal resistance between the 1st and 2nd dies. As it is shown in the figure, now the second peak is displaced, showing clearly the location of the increased glue thermal resistance, resulted from the die attach failure. The 3rd peak is displaced as well, with about the same value as the 2nd one.

With these simulation experiments it is demonstrated that the location of an increased thermal resistance, even if this increase is not larger than 20–30%, may be determined also in the case of pyramidal die structures.

23.5. VERIFICATION OF THE METHODOLOGY BY MEASUREMENTS

In order to verify the results obtained from simulations a large series of measurements were accomplished. In the presented first group of experiments the packages contained thermal test dies [10], enabling the verification of the measured results by steady state measurements as well. The thermal transient measurements and the subsequent evaluations were done with the T3Ster equipment [9].

23.5.1. Comparison of the Transient Behavior of Stacked Die Packages Containing Test Dies, Prior Subjected to Accelerated Moisture and Temperature Testing

In the first presented series of measurements we examined several samples that have been prior subjected to accelerated moisture and temperature shock testing, to induce different extent of interfacial integrity. The packages contained 2 test dies, and were of 44 lead QFN (Quad Flat No-lead) type [11].

The cross section of the pristine package is presented in Figure 23.19.

In the presented experiments the top die was used to heat and measure the temperature in the middle of the structure. Two samples were examined, denoted by EM2 and

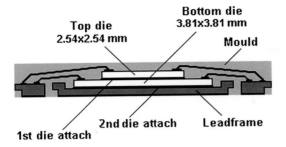

FIGURE 23.19. Cross section of the 44L QFN packages, denoted as EM2 series.

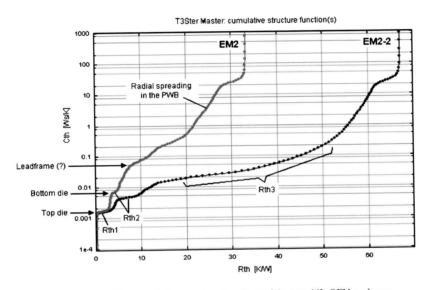

FIGURE 23.20. The cumulative structure functions of the two 44L QFN packages.

EM2-2. The cumulative structure functions calculated from the measured results are presented in Figure 23.20.

The first step in the structure function, denoted as "top die" thermal capacitance of Figure 23.20 indicates a 1.05 mm^3 volume of silicon. The volume of the top die calculated from the data of Figure 23.19 is $2.54 \times 2.54 \times 0.2 = 1.3$ mm^3. "Bottom die" in Figure 23.20 shows the thermal capacitance value of the bottom die. From this value the calculated volume of the bottom die is 3 mm^3, while $3.81 \times 3.81 \times 0.2 = 2.9$ mm^3 volume is expected from Figure 23.19. This difference between the real and the calculated volume is fairly small (∼3%). In case of the package denoted as EM2-2 this thermal capacitance value is smaller, which is may be originating from the detected anomalies, most probably delamination in the further regions of the heat flow path. The almost horizontal section under the top die denoted by R_{th1} is the thermal resistance of the die attach under the top die. It is slightly increased for the sample denoted by EM2-2, the values are: ∼2.7 K/W and ∼3.8 K/W, suggesting a small delamination in the package EM2-2 in the upper die attach layer.

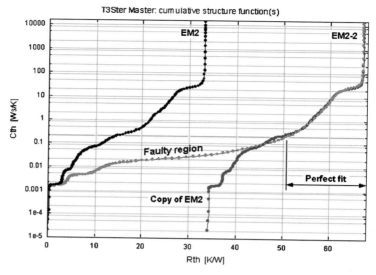

FIGURE 23.21. Cumulative structure functions of the two samples, a copy of the EM2 curve is shifted to match at the right end.

The second almost horizontal section in Figure 23.20, denoted as R_{th2} is the thermal resistance of the die attach under the bottom die. This is strongly different at the two samples: EM2 ∼1.6 K/W, EM2-2 ∼5 K/W, and before reaching the thermal capacitance of the leadframe we can identify a further large almost horizontal part, demonstrating a largely increased thermal resistance, shown by R_{th3} in Figure 23.20. This resistance can be even better seen from a figure where the structure functions of the two samples are drawn up in one figure, and the curve of the known good one is copied and shifted to demonstrate the fit on the right-hand side, see Figure 23.21. The section, where the curve denoted as EM2-2 travels alone, shows the increased thermal resistance on the EM2-2 sample. Since this section of the curve is running entirely under the capacitance value attributed to the leadframe this curve indicates the presence of an increased thermal resistance between the bottom die and the leadframe.

It can be stated from this large shift that there is probably a strong delamination at this sample between the lead-frame and the bottom die. This statement has been later verified also by C-mode Scanning Acoustic Microscope inspections, revealing in fact delamination of the bottom die. These pictures are so blurred however, that even the most experienced eyes have difficulties to evaluate them; this is why they are not presented here.

23.5.2. Comparison of the Transient Behavior of Stacked Die Packages Containing Real Functional Dies, Subjected Prior to Accelerated Moisture and Temperature Testing

The final test of the method is when the examined packages contain real functional stacked dies. In the presented experiments 48 lead QFN packages were examined, containing two stacked functional dies each. The measurements were done by using the substrate diode of the top die both for heating and temperature sensing. #789 and #630 are the codes of the two investigated sample packages. The measurements were blind, nothing was known about the reliability testing history of the packages. Figure 23.22 shows the

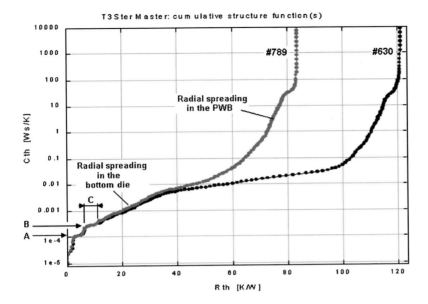

FIGURE 23.22. Cumulative structure functions of the measured two samples.

cumulative structure functions, calculated from the measured thermal transient results. The following train-of-thoughts shows how to draw inferences and conclusions by using the methodology.

The location of the top die in the structure function of Figure 23.22 needs some thinking. At the first glance the step marked with "*A*" with its value of $C_{th} = 1.3 \times 10^{-4}$ W s/K seems to be the thermal capacitance of the top die. Calculating however this C_{th} value from the geometrical data, using 1.6×10^{-3} W s/mm^3 K volumetric heat capacity of the Si we obtain $C_{th} = 1.555 \times 1.555 \times 0.076 \times 1.6 \times 10^{-3} \cong 2.9 \times 10^{-4}$ W s/K. This suggests that in fact the second step, marked with "*B*" corresponds to the top die, and the first one may be related to the transients inside the top die. The almost horizontal section denoted by "*C*" gives the ≈6 K/W thermal resistance of the die attach under the top die.

Ascendant linear regions of the structure function (in the lin/log representation) indicate always radial heat spreading. [12] Such a region is shown in Figure 23.22, denoted as "Radial spreading in the bottom die." This suggests that the heat moves from the top die first to the middle of the bottom die and continues to spread out laterally. From the slope of this region even the thickness of the bottom die may be deduced as follows:

$$w = \frac{1}{\lambda}\frac{1}{4\pi}\frac{\log(C_{th2}/C_{th1})}{R_{th2} - R_{th1}} = \frac{1}{120}\frac{1}{12.56}\frac{\log(10)}{30 - 11} = 83 \times 10^{-6} \text{ m} \qquad (23.3)$$

where R_{th1}, C_{th1} and R_{th2}, C_{th2} are representing two points in the linear section, the slope of which is calculated [12].

This calculation results in an about 83 μm thickness value for the bottom die, while the real thickness is 76.2 μm.

The "Radial spreading in the bottom die" section ends at $C_{th} \approx 0.002$ W s/K, the calculated heat capacity of the bottom die is $C_{th} = 3.467 \times 3.302 \times 0.1 \times 1.6 \times 10^{-3} = 0.00183$ W s/K, the fitting is acceptable. At the same location the differential structure

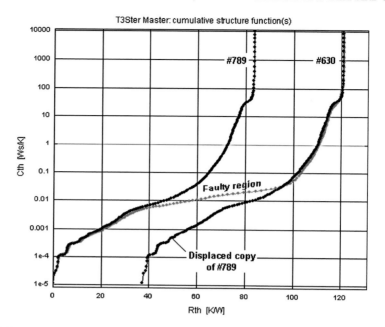

FIGURE 23.23. The copy of the structure function of the good device is displaced to fit at the ambient side to the structure function of the faulty device.

function (see Figure 23.24) gives $K = 3 \times 10^{-4}$ W s^2/K^2 value, from which, for Si material 1.25 mm^2 cross section is calculated by T3Ster. This is about the peripheral area of the bottom die. This means that at the right hand side of the "Radial spreading in the bottom die" section of Figure 23.22 the heat reaches exactly the periphery of the bottom die.

This calculation and the curves in Figure 23.23, where the copy of the structure function of the good device is displaced to fit at the ambient side to the structure function of the faulty device, suggest that the faulty region of the #630 sample is under the bottom die. If we calculate also the differential structure functions, we can identify also the location of the leadframe from the bump at the area referring to the leadframe.

Comparing Figure 23.23 and Figure 23.24 we can determine that the faulty section (from Figure 23.23) is divided by the "Cu leadframe" bump of Figure 23.24. This suggests that there is a strong delamination both in the die attach under the bottom die and in the soldering of the leadframe. In order to check this, the sample #630 was examined by using X-ray microscopy, see Figure 23.25. The X-ray images clearly show the deterioration of the soldering between the package and the PWB. The loss in the solder area is about 20%; this explains only partially the increase in Rth. The other part of the increase has to be attributed to the induced delamination of the die attach, that unfortunately can not be seen even by X-ray images.

The conclusions inferred from the structure functions may be cross checked with the help of the time constant density functions [3] and the complex loci of the $Z_{th}(\omega)$ curves as well [6]. These curves are generated by the same mathematical transformations that are used to create the structure functions and provided by the measuring equipment [9]. Figure 23.26 shows the calculated thermal time constant density of the measured samples, while Figure 23.27 shows the frequency dependence of their complex thermal impedances [6].

DIE ATTACH QUALITY TESTING BY STRUCTURE FUNCTION EVALUATION

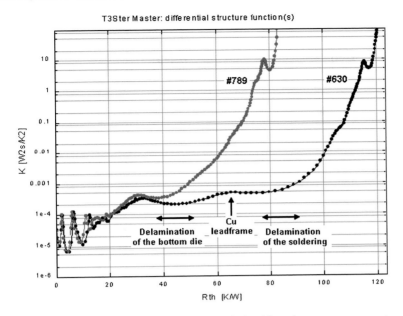

FIGURE 23.24. Differential structure functions calculated from the same measurements.

FIGURE 23.25. X-ray image shows the delamination in the soldering (dark area in the figure).

A characteristic feature of these curves is that the first parts of them show nice fitting, suggesting that there is no defect in the neighborhood of the top die, but the right hand side of the curves in both of the figures show large differences, suggesting a different material structure, that is delamination of the layers.

Although these functions are very useful for the experienced reliability engineer to locate die attach problems, we recommend to start the failure analysis with the structure function evaluation, since these functions can provide directly the data that can be used for characterizing the die attach problems.

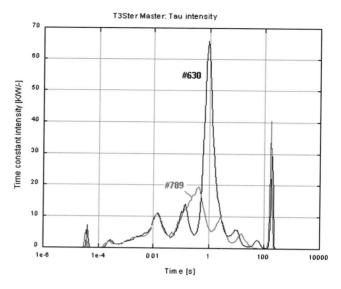

FIGURE 23.26. The calculated thermal time constant density of the measured samples.

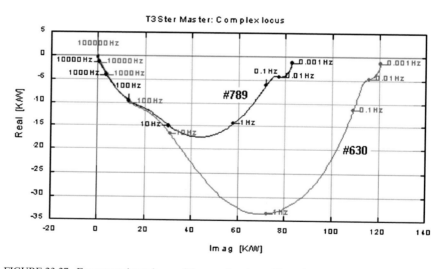

FIGURE 23.27. Frequency dependence of the complex thermal impedances of the measured samples.

A special advantage of the presented test technology is that thermal test dies are not needed to detect integrity problems in the packages. Usual working dies can be used both as heater and temperature sensor in the proposed methodology. For measuring the structure function we even do not need any extra elements on the dies if the substrate diodes of the individual dies are not connected electronically, since for the temperature sensing the substrate diode of the top die may be conveniently used. If for some reason the substrate diodes of the dies have to be connected electrically, other dedicated elements have to be selected on the dies for temperature sensing.

On the other hand, switching on either the whole dissipation on the top die, or just one dissipating element, may provide the constant source of heat during the time of the transient measurement. This means that obtaining the necessary transient curves is feasible normally without any special built in elements or test structures in any circuit. This way we may conclude that with the presented method the fast diagnosis of the die-attach problems of stacked dies is possible, and the method does not require any additional circuit elements in any of the stacked dies.

With the help of the method die attach problems affecting about 5% of a certain surface can be detected in the form of increased thermal resistance (shift in the structure functions). This value strongly depends on the actual structure. For usual package structures it can be even smaller, but sometimes it may be somewhat larger. The closer is the heat flow in the structure the 1-dimensional flow, the smaller is the die attach failure that can be detected.

Finding the exact locations of the material transitions is not always very simple. It strongly depends on the geometry and on the materials used in the structure. It can be stated however, that by finding the inflection points in the structure function, which can be easily automated in the structure function software, the material transition point can be easily approximated.

23.6. CONCLUSIONS

With the presented measurement experiments we have demonstrated the feasibility of locating die attach problems of packages containing either single or stacked dies. In the proposed methodology first the thermal transient response of the structure has to be obtained from measurement, and from this the structure functions have to be calculated. Comparing the structure functions of a measured sample with the appropriate structure functions of a known good stacked structure, which can be obtained, e.g., from the simulation of the structure, or from the measurement of a known good sample package, the differences can be easily observed. From the location of the shift in the structure function of the device under test, in case of structures with integrity problems, the location of the integrity problem can be determined.

The method works well both in the case of stacked dies of the same size, and in the case of pyramidal stacked die structures. The resolution of the method will be further investigated with different die sizes and layer numbers, but from the experiments up to now it can be expected that the method is well applicable for structures up to at least 3–4 layers of stacked dies.

The large number of experiments and cross verifications indicated that if we wish to use the methodology to determine the exact thermal resistance values of the individual die attach layers the measured values have to be corrected with the structure functions of the parallel heat flow paths [13]. The correction can be a built in function of the evaluation software of the thermal transient measurement equipment.

ACKNOWLEDGMENTS

This work was supported by the PATENT IST-2002-507255 Project of the EU and by the INFOTERM 2/018/2001 NKFP project of the Hungarian Government. The authors wish to thank G. Farkas his contribution by carrying out some of the measurements, and for R. Kovács for taking the X-ray image.

REFERENCES

1. L. Zhang, N. Howard, V. Gumaste, A. Poddar, and L. Nguyen, Thermal characterization of stacked dies, Proceedings of the XXth SEMI-THERM Symposium, San Jose, CA, USA, March 9–11, 2004, pp. 55–63.
2. C. Lin, S. Chiang, and A. Yang, 3D stackable packages with bumpless interconnect technology, 5th Electronics Packaging Technology Conference, Singapore, 10–12 December 2003.
3. V. Székely and T. van Bien, Fine structure of heat flow path in semiconductor devices: a measurement and identification method, Solid-State Electronics, 31, pp. 1363–1368 (1988).
4. M. Rencz, V. Székely, A. Morelli, and C. Villa, Determining partial thermal resistances with transient measurements and using the method to detect die attach discontinuities, Proceedings of the XVIIIth SEMI-THERM Symposium, San Jose, CA, USA, March 1–14, 2002, pp. 15–20.
5. M. Rencz and V. Székely, Structure function evaluation of stacked dies, Proceedings of the XXth SEMI-THERM Symposium, San Jose, CA, USA, March 9–11, 2004, pp. 50–54.
6. V. Székely, Distributed RC networks, The circuits and Filters Handbook, CRC Press, USA, 2003, pp. 1202–1221.
7. E.N. Protonotarios and O. Wing, Theory of nonuniform RC lines, IEEE Trans. on Circuit Theory, 14(1), pp. 2–12 (1967).
8. http://www.micred.com.
9. http://www.micred.com/t3ster.html.
10. http://www.delphi.com/pdf/techpapers/2004-01-1681.pdf.
11. M. Rencz, V. Székely, B. Courtois, L. Zhang, N. Howard, and L. Nguyen, Die attach quality control of 3D stacked dies, IEMT, 2004, pp. 78–84.
12. V. Székely, M. Rencz, S. Török, and S. Ress, Calculating effective board thermal parameters from transient measurements, IEEE Transactions on Components and Packaging Technology, 24(4), pp. 605–610 (2001).
13. M. Rencz, A. Poppe, E. Kollár, S. Ress, V. Székely, and B. Courtois, A procedure to correct the error in the structure function based thermal measuring methods, Proceedings of the XXth SEMI-THERM Symposium, San Jose, CA, USA, March 9–11, 2004, pp. 92–98.

24

Mechanical Behavior of Flip Chip Packages under Thermal Loading

Enboa Wu[a,e], Shoulung Chen[a,b], C.Z. Tsai[a,c] and Nicholas Kao[a,d]

[a]Institute of Applied Mechanics, National Taiwan University, Taipei 106, ROC
[b]Electronics Research and Service Organization, Industrial Technology Research Institute, Hsin-Chu 310, Taiwan, ROC
[c]Macronix International Co., Ltd., Hsin-Chu 300, Taiwan, ROC
[d]Siliconware Precision Industries Co., Ltd., Taichung 400, Taiwan, ROC
[e]Hong Kong Applied Science and Technology Research Institute, Hong Kong

Abstract   A complete report on mechanical behavior of large flip chip plastic ball grid array (FC-PBGA) packages under reflow condition is presented in this chapter. The coefficients of thermal expansion (CTE) of BT substrates were also measured using electronic speckle pattern interferometry (ESPI) and were found to change significantly at different processing stages. Careful selection of a substrate CTE is needed for accurate warpage prediction using the finite element method. A linear relationship between the temperature and the FC-PBGA warpage was observed from the data measured using both the phase-shifted shadow moiré and the ESPI. Zero warpage was measured at approximately 150°C regardless of the size of the chip composed of the package. An optimal warpage design for FC-PBGA was also suggested.

24.1. INTRODUCTION

The mechanical behavior of a flip chip package under temperature change is always important as warpage and stresses are generated due to mismatch of the mechanical properties of the materials used to make the package. A flip chip package is usually composed of a silicon chip, solder joints, an organic substrate, and underfill. In order to understand the behavior of flip chip packages under thermal loading, numerous numerical and theoretical methods have been employed. For theoretical analysis, bi-material and tri-material models have been developed to predict stresses and failure of flip chip packages [1–3]. Simple structural mechanics models, such as those based on beam or plate theories, have frequently been adopted. These models have also been used to predict stresses on solder joints [4,5], underfills [6], or UBM [7] of flip chip packages, and could possibly be applied under different environmental conditions [8,9]. For finite element analysis, two-dimensional and three-dimensional finite element models have been constructed to study packaging deformation [10,11], interfacial stresses [12], solder joint shapes and stresses [13–15], and underfill behaviors and stresses [16] to determine the reliability of packages [17–19]. Parametric studies have also been presented to determine the optimal properties and dimensions of the materials and devices in packages [10–22].

On the other hand, experimental tools have been adopted to directly measure the behavior of flip chip packages at different temperatures. Full field optics methods, such as the shadow moiré, moiré interferometry, reflection moiré, or electronic speckle pattern interferometry (ESPI) schemes, have frequently been adopted. The shadow moiré method has been employed to measure the out-of-plane deformation of flip chip packages [23–26], printed wiring boards [27,28], and organic substrates [29]. The resolution for shadow moiré can be enhanced by an order of magnitude if the phase shifting setup and subsequent algorithms for phase diagram construction are employed [30]. Moiré interferometry, on the other hand, has frequently been adopted to measure the in-plane deformation of the cross-sections of solder joints and flip chip packages after the temperature cools down [31]. Because a package subjected to moiré interferometry measurement has to be cross-sectioned, the deformation on the cross-section is disturbed. As a result, the measurement result may not reflect the true deformation conditions in the package. This disadvantage can be avoided by using ESPI, where the experimental setup can be arranged to measure either the in-plane or the out-of-plane deformation of the package [32–34]. On the other hand, as the sensitivity of ESPI is high, it is suitable only for deformation measurement at small scale. In addition, ESPI is fragile to the ambient environment. When the measurement is performed at elevated temperature, care must be taken to avoid drifting of the fringes due to the unstable hot air flow. When the measurement surface is reflective or shiny, such as a polished die surface in a flip chip package, reflection moiré is an ideal method for measuring the out-of-plane deformation [35]. The sensitivity is high because the slope of the out-of-plane deformation is directly measured.

In this chapter, we will discuss use of both the ESPI and phased shifted shadow moiré methods to measure the mechanical behavior of flip chip packages. We will also discuss how the coefficients of thermal expansion of flip chip BGA substrates at different fabrication stages are measured using ESPI, and how the out-of-plane deformation of different types of flip chip packages is measured using the phased-shifted shadow moiré and the ESPI methods. The use of two-dimensional and three-dimensional finite element models to perform parametric analyses of flip chip packages will also be discussed.

24.2. FLIP CHIP PACKAGES

A typical picture of a flip chip package is shown in Figure 24.1. The flip chip packaging process generally includes wafer bumping and flip chip assembly. In the wafer bumping process, the peripheral pads on each chip in a wafer are redistributed to form area array pads. Under bump metallurgy and solder bumps are then deposited on the redistributed pads. In the flip chip assembly process, bumped chips, after being diced from a wafer, are placed on a substrate and undergo a subsequent reflow process. Underfill is then deposited to reinforce the solder bumps and enhance the reliability of the flip chip package.

In this chapter, we will focus on the mechanical behavior of flip chip packages when they are subjected to different temperature conditions. Except for the flip chip packages discussed in Section 24.5.3 to study the underfill effect, flip chip packages used for warpage measurement can be divided into four groups, as listed in Table 24.1. The chip sizes used

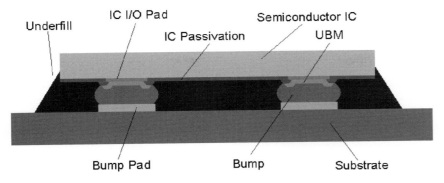

FIGURE 24.1. Schematic diagram of a typical flip-chip package.

TABLE 24.1.
Dimensions in mm of flip chip packages and the number of samples used.

Group	1	2	3	4
Chip size	10 × 10	20 × 20	26 × 26	26 × 26
Chip thickness	0.73	0.73	0.73	0.40
Solder bump pitch	0.5	0.5	0.5	0.5
Bump height	0.07	0.07	0.07	0.07
Substrate size	40 × 40	40 × 40	40 × 40	40 × 40
Substrate thickness	1.12	1.12	1.12	1.12
Sample No.	3	3	2	2

TABLE 24.2.
Mechanical properties of chip, solder bump and underfill.

	E (GPa)	CTE (ppm/°C)	Poisson ratio
Chip	156	2.7	0.278
Solder bump	57.4	24.5	0.4
Underfill	5.5	45	0.35

are 10 × 10 mm, 20 × 20 mm, and 26 × 26 mm. The thicknesses of the chips are 0.73 mm and 0.40 mm. The solder bump height and the pitch are 70 μm and 500 μm, respectively. The mechanical properties of the silicon chip, the solder bump and the underfill are listed in Table 24.2. For these flip chip packages, the build-up substrates used are 40 × 40 × 1.12 mm in size and are made by deposition of one top and one bottom build-up layers onto two 2-metal-layer cores (1 + 4 + 1). The material data are listed in Table 24.3 and were provided by manufacturers. The coefficient of thermal expansion (CTE) will be verified in Section 24.4. Figure 24.2 shows pictures of Group 1, 2, 3, and 4 flip chip packages. In this chapter, we will also discuss the use of other types of substrates for CTE measurement. The results will be given in Section 24.5.

TABLE 24.3.
Mechanical properties of the substrate used for the flip chip packages listed in Table 24.1.

	E (GPa)		CTE (ppm/°C)		Poisson ratio	
	In-plane	Out-of-plane	In-plane	Out-of-plane	In-plane	Out-of-plane
Substrate	24.5	9.81	12	50	0.143	0.02

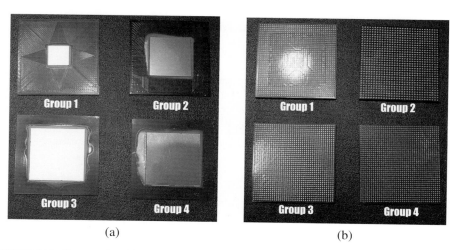

FIGURE 24.2. Photographs of Groups 1, 2, 3, and 4 flip chip packages. (a) Chip side; and (b) substrate side.

24.3. MEASUREMENT METHODS

24.3.1. Phase Shifted Shadow Moiré Method

For shadow moiré [23–29], the fringe patterns are formed through superposition of the reference gratings and the object gratings. The reference gratings are placed in front of a specimen whose out-of-place displacement or surface profile is to be measured. The object gratings are the shadows of the reference gratings, which are produced by shining parallel light onto the specimen surface through the reference gratings. The out-of-plane displacement can be expressed as

$$w = \frac{Np}{\tan\alpha + \tan\beta}, \qquad (24.1)$$

where N is the fringe order, p is the pitch of the object grating, α is the angle between the projected parallel light and the normal of the reference grating plane, and β is the angle between the axis of the CCD camera and the normal of the reference grating plane. In this chapter, α will always be equal to 45°, β will always be equal to 0°, and $p = 50$ μm, i.e., 20 lines/mm.

In order to enhance the sensitivity to warpage in shadow moiré measurement, the phase of the recorded moiré fringes is used instead of the intensity of the fringe patterns. The general expression for the fringe intensity can be expressed as in Equation (24.2) when the reference plane moves a distance equal to p/n along its normal direction. If we take $n = 4$, i.e., adopt the four-step phase shifting method, the phase of the recorded light intensity can be extracted by using Equation (24.3) [36]:

$$I_k(x, y) = I_0(x, y)\{1 + \gamma(x, y)\cos[\phi(x, y) + \alpha_k]\},$$
$$\alpha_k = (k-1)\alpha_r,$$
$$k = 1, 2, \ldots, n, \qquad (24.2)$$

$$\phi(x, y) = \tan^{-1}\left(\sqrt{3}\frac{I_1 - I_3}{2I_2 - I_1 - I_3}\right) + \frac{2\pi}{3}. \qquad (24.3)$$

All the moiré measurements discussed in this chapter are performed using the four-step phase shifting method. Measurement, which is frequently performed at an elevated temperature, can thus be accomplished in a relatively short period of time. Movement of the reference grating plane is controlled by a stepping motor.

24.3.2. *Electronic Speckle Pattern Interferometry (ESPI) Method*

When laser light is directed onto the surface of an object, speckles are formed because of the roughness of the surface. These speckles move as the object deforms. As a result, fringes are formed, and the corresponding deformation can be measured. In this study, ESPI was adopted for both in-plane and out-of-plane deformation measurements of flip chip packages [32–34].

The experimental set up for in-plane deformation measurement performed using ESPI is depicted in Figure 24.3. Laser light was split into two beams of equal intensity. These two beams traveled the same distance and, by means of two spatial filters, were uniformly distributed onto the surface of the object whose deformation was to be measured. The speckle patterns were recorded by a CCD camera when the surface of the body to be measured moved.

Because the two laser beams are symmetric along the CCD axis, any movement in the out-of-plane direction of the object will not change the traveling distance of the two optical paths. As a result, no fringe is formed by this out-of-plane movement. On the other hand, when the object moves in the in-plane direction, the difference in the optical paths of the

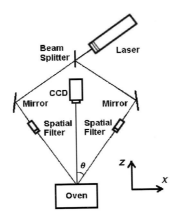

FIGURE 24.3. Schematic diagram of the experimental setup for in-plane ESPI displacement measurement.

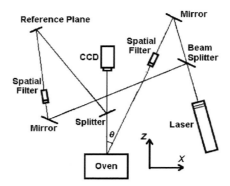

FIGURE 24.4. Schematic diagram of the ESPI experimental setup for out-of-plane deformation measurement.

two laser beams that hit a specific speckle changes. As a result, the in-plane displacement can be expressed as

$$U_x = \frac{n\lambda}{2\sin\theta}. \tag{24.4}$$

The ESPI experimental set up for out-of-plane displacement measurement is shown in Figure 24.4. A laser beam was split into object and reference beams by a beam splitter. Spatial filters were employed to produce uniformly distributed scattering light projected onto the object and reference planes. The path distances between these two beams were adjusted so as to be equal. A CCD camera was employed to record the fringe patterns when the object inside the oven deformed due to temperature change. The incident angle, θ, was kept as small as possible to reduce the effect of in-plane displacement on the out-of-plane displacement measurement. The out-of-plane displacement corresponding to the nth dark fringe is

$$U_z = \frac{n\lambda}{1+\cos\theta}. \tag{24.5}$$

In this study, the surface of the electronic package to be measured using either the shadow moiré or ESPI methods was always sprayed with white paint in order to enhance contrast. During the measurement process, the flip chip package was mounted on a tripod inside an oven. Because of the high sensitivity of the ESPI method, the temperature loading applied to the specimens was caused by a 5 to 10°C temperature difference, and the accuracy of the temperature adjustment was controlled so as to be ±0.1°C during the measurement process.

24.4. SUBSTRATE CTE MEASUREMENT

The in-plane ESPI experimental setup shown in Figure 24.3 was employed to perform in-plane deformation measurement of substrates. The specimen to be measured was supported by a small tripod inside an oven. Once the in-plane deformation had been recorded, with knowledge of the temperature change, the equivalent CTE value was calculated. In this experimental setup, the incident angle, θ, that is the angle between the laser beam

TABLE 24.4.
Specimens used for CTE measurement.

Specimen		Dimensions (mm)
A	Aluminum	$31.0 \times 30.1 \times 1.0$
B	Copper	$30.6 \times 29.7 \times 1.5$
C	BT Laminates	$35.0 \times 35.0 \times 0.45$
D	BT Laminates with via	$35.0 \times 35.0 \times 0.45$
E	BT Laminates with via and copper trace	$35.0 \times 35.0 \times 0.50$
F	BT substrate	$35.0 \times 35.0 \times 0.55$

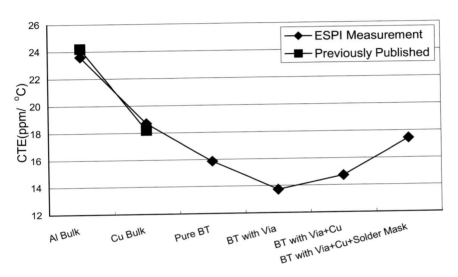

FIGURE 24.5. Measured CTE values for aluminum, copper, and BT substrates at different fabrication stages.

and the axis of the CCD camera shown in Figure 24.3, was 30°. Therefore, the in-plane displacement of the specimen was equal to $n\lambda$ and could be determined directly from the speckle fringe number.

In this study, four types of BT substrates were used to perform CTE measurement. In the first group of CTE measurements, six specimens were employed. The purposes of this group of measurements were (a) to verify the accuracy of the ESPI method by measuring the CTE values of aluminum and copper; and (b) to study the change of CTE values for BT substrates in different fabrication stages. Table 24.4 lists the specimens and their dimensions.

To fabricate a BT substrate for use in electronic packages, glass/BT composite laminates need to undergo drilling, via filling, trace forming, and solder mask forming processes. In this part of study, the BT substrate used in Table 24.4 was for PBGA packaging. The final thickness of the BT substrate was 0.55 mm. The in-plane dimensions of the BT substrate were 35×35 mm.

For each measurement, the temperature range was approximately 10.0°C. For each specimen, the temperature varied within the range of 50°C to 70°C. Four measurements were performed for each specimen. The results are plotted in Figure 24.5. The recorded values for aluminum and copper were 24.2 ppm/°C and 18.2 ppm/°C, respectively. The

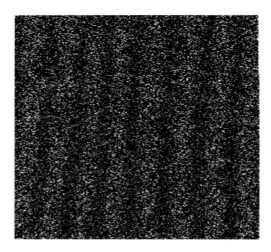

FIGURE 24.6. A typical ESPI fringe pattern for a BT laminate without via holes.

previously published values are 23.6 ppm/°C and 18.7 ppm/°C, respectively [37]. The deviation is less than 3%, which shows that the measurement method is accurate.

The measured average CTE for the BT/glass laminates was 15.86 ppm/°C. This value is very close to the value provided by the manufacturer, which is 15.0 ppm/°C (Mitsubishi Gas Chemical Co.). After the BT/glass laminates were drilled to form vias, the CTE value became 13.71 ppm/°C, a reduction of approximately 14%. This is because the drilled vias provide additional flexibility for BT/glass composite laminates to expand onto the neighborhood of these vias when the temperature is increased. As a result, the overall elongation for the laminates with vias becomes smaller as compared to the overall elongation for the laminates without these vias when the temperature is increased. Once the copper foil was plated, the CTE value increased to 14.71 ppm/°C, approximately a 7% increase. This increase of the CTE value occurred because the CTE value of the copper was approximately 18.2 ppm/°C, i.e., 33% higher than the recorded CTE of the BT/glass laminates. The last process in substrate fabrication was deposition of the solder mask. The CTE value was found to jump to 17.45 ppm/°C, a 19% increase. This significant increase of the CTE value indicated that the CTE value of the solder mask is much higher than that of the BT/glass laminates. As a result, significant stresses were induced at the interfaces among the BT/glass laminates, copper foil, and solder mask. The results for CTE measurement performed at different manufacturing stages for BT substrates are shown in Figure 24.5. For illustration purposes, in Figure 24.6 it shows a typical fringe pattern for a BT/glass composite laminate without via holes. The total pixel number for this 35 mm wide laminate was 286, and there existed six complete fringes and two additional fringes at two edges of the laminates that were not completely shown. For these six complete fringes the pixel number was 218, and the total spacing was 26.68 mm. In this measurement set-up each fringe is equal to an in-plane deformation of 0.6328 µm, and during the measurement the temperature was increased from 65.0°C to 73.9°C. As a result, the CTE is calculated as $6 \times 0.6328 \times 1000/(26.68 \times (73.9 - 65.0)) = 15.99$ ppm/°C, which is very closed to the data of 15.86 ppm/°C shown in Figure 24.5. The slight deviation is that in Figure 24.5 the CTE value was averaged from multiple measurements at different temperatures.

The second type of specimen used for CTE measurement was a BT substrate used in a two-chip PBGA module, i.e., an MCM module. The dimensions were $30 \times 12 \times 0.56$ mm.

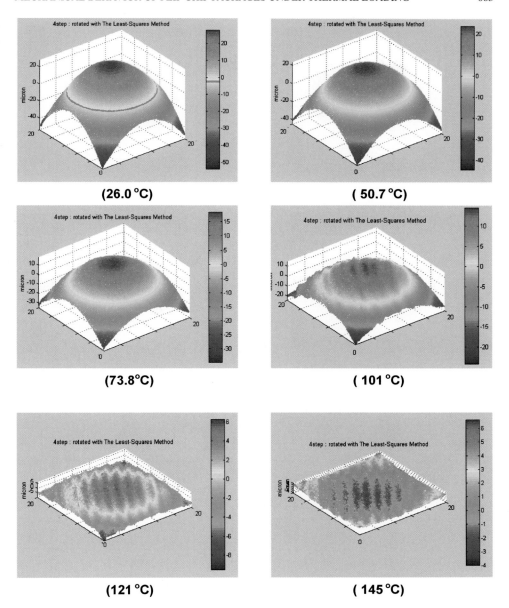

FIGURE 24.11. Recorded warpage profiles measured on the chip side of the FC package at 26.0°C, 50.7°C, 73.8°C, 101°C, 121°C, and 145°C. The chip size was 20 × 20 mm.

chip size increased. On the other hand, a comparison of the results for Groups 3 and 4 reveals that the warpage remained the same regardless of the change of the chip thickness.

A linear relationship was also observed between the warpage measured on the substrate side and the corresponding temperatures. For the flip chip packages with 10 × 10 mm and 20 × 20 mm chips (Groups 1 and 2), this linear relationship was found for a temperature

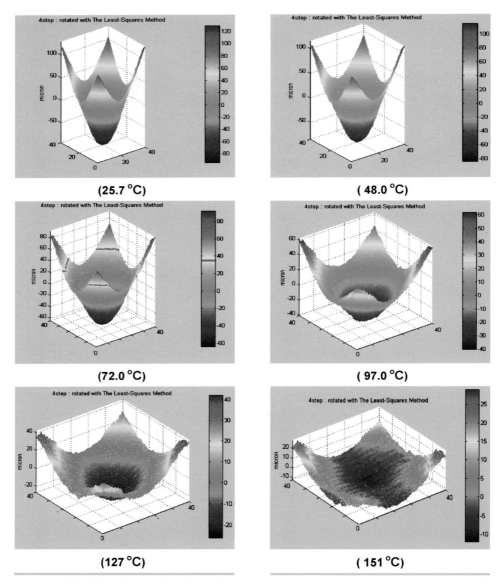

FIGURE 24.12. Recorded package warpage profiles at 25.7°C, 48.0°C, 72.0°C, 97.0°C, 127°C, 151°C, 176°C, 201°C, and 228°C. The chip size was 20 × 20 mm. Warpage measurements were performed on the substrate side.

range of from room temperature to 225°C. On the other hand, due to the constraint from the 26 × 26 mm chips on the flip chip packages in Groups 3 and 4, the substrate warpage remained unchanged as the temperature passed the glass transition temperature of the BT substrate, as shown in Figures 24.15 and 24.16. When the temperature dropped to room temperature, it was found that the warpage values measured on the substrate side became larger when the chip size increased and the chip thickness decreased.

MECHANICAL BEHAVIOR OF FLIP CHIP PACKAGES UNDER THERMAL LOADING   665

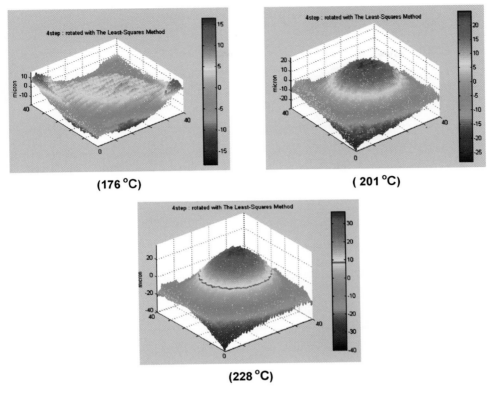

(176 °C)

(201 °C)

(228 °C)

FIGURE 24.12. (Continued.)

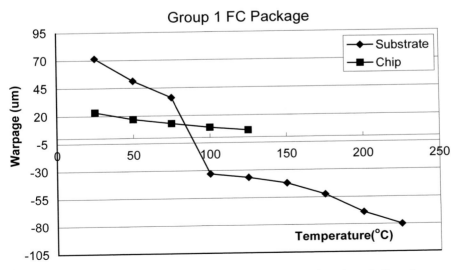

FIGURE 24.13. Warpage of the chip and substrate in the flip chip packages in Group 1.

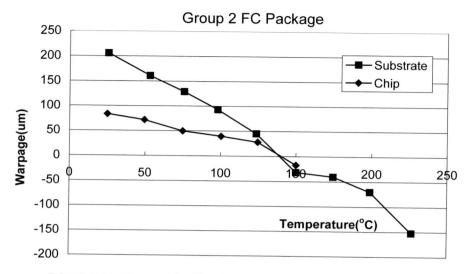

FIGURE 24.14. Warpage of the chip and substrate in the flip chip packages in Group 2.

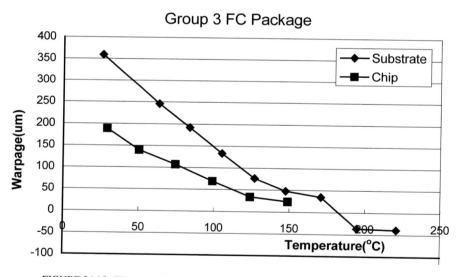

FIGURE 24.15. Warpage of the chip and substrate in the flip chip packages in Group 3.

24.5.3. Effect of Underfill on Warpage

In order to investigate the effect of underfill on the warpage behavior of flip chip packages, warpage in two 27 × 27 mm FC packages with 10 × 10 mm chips was measured using the developed out-of-plane ESPI method (Figure 24.4). The results are shown in Figures 24.17 and 24.18 for flip chip packages without and with underfill reinforcement, respectively. In both figures, two measurement results for the same packages are plotted. It was found that the relationship between the package warpage and the corresponding temperatures was not linear when the package was not reinforced by underfill. When the packaged was reinforced by underfill, the warpage to temperature relationship became linear,

MECHANICAL BEHAVIOR OF FLIP CHIP PACKAGES UNDER THERMAL LOADING

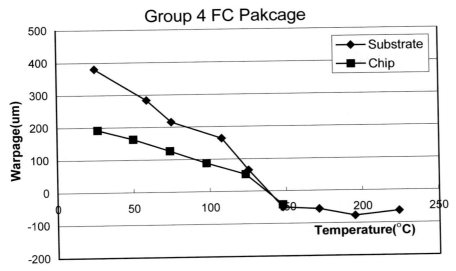

FIGURE 24.16. Warpage of the chip and substrate in the flip chip packages in Group 4.

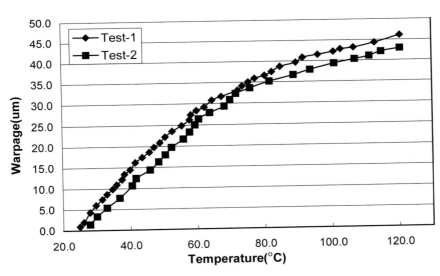

FIGURE 24.17. Warpage versus temperature for a 27 × 27 mm FC package without underfill reinforcement.

and the values also became larger over the same temperature range. For example, when the temperature increased from room temperature to 60°C, a linear relationship was observed between warpage and temperature for both packages, but the warpage in the package with underfill was 1.1 times larger than that in the package without underfill. When the temperature reached 120°C, this ratio increased to 2.3.

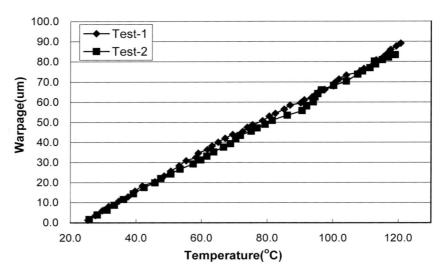

FIGURE 24.18. Warpage versus temperature for a 27 × 27 mm FC package with underfill reinforcement.

24.6. FINITE ELEMENT ANALYSIS OF FLIP CHIP PACKAGES UNDER THERMAL LOADING

Both two-dimensional and three-dimensional finite element models were constructed using ANSYS for warpage analysis of a flip chip package. The material properties adopted in these models are listed in Table 24.2 and Table 24.3. The CTE value and the Young's modulus for the 1 + 4 + 1 build-up substrate were the two most sensitive properties used in analysis. The CTE value was verified by measurement, as described in Section 24.4. The Young's modulus of the substrate was verified in [38]. For 3D analyses, a 1/8 model was constructed using a 10-node tetrahedral element (SOLID45) to avoid erroneous spurious shear strain [39]. The total number of elements used was 69,659. For 2D analyses, a 1/2 model was constructed. The element used was PLANE82. The total number of elements was 11,571. The finite element models discussed in this section were constructed based on the geometry of a flip chip package assembled using a 20 × 20 mm chip. The underfill at the edge of the chip, called the fillet, was modeled so that the height would be equal to a chip thickness of 730 μm and a tail length of 500 μm. According to the experimental results discussed in Section 24.5, the stress free temperature was 150°C. Therefore, the thermal loading applied in the finite element models was based on a temperature change from 150°C to 26°C. The material properties were assumed to remain constant during the change of temperature.

Figure 24.19(a) and (b) plot the constructed 3D and 2D finite element models, respectively. The corresponding warpage profiles of the two models are depicted in Figure 24.20(a) and (b).

A comparison of the experimental and numerical data for the maximum warpage values, i.e., at warpage 1 and warpage 2, is shown in Table 24.6. The definitions of warpage 1 and warpage 2 are the same as those given in Section 24.5.1. The correlation is considered to be satisfactory. Therefore, both the 2D and 3D finite element models are considered to be accurate when adopted for warpage analysis purposes.

MECHANICAL BEHAVIOR OF FLIP CHIP PACKAGES UNDER THERMAL LOADING

FIGURE 24.19. Plots of finite element models for (a) 3D and (b) 2D analyses. The flip chip package had a 20 × 20 mm chip.

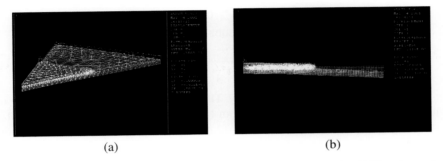

FIGURE 24.20. Warpage profiles of the models plotted in Figure 24.19 for (a) 3D and (b) 2D analyses when the temperature changed from 150°C to 26°C.

TABLE 24.6.
Comparison of the measured data and numerical warpage predictions obtained using 2D and 3D finite element models.

		Shadow moiré	3D FEM		2D FEM	
		Result	Result	Difference	Result	Difference
Chip	Warp. 1	41.9	45.7	8.84%	43.5	3.60%
	Warp. 2	79.0	87.5	10.7%	N/A	N/A
Substrate	Warp. 1	118	118	0%	109	−7.63%
	Warp. 2	205	203	−0.86%	N/A	N/A

24.7. PARAMETRIC STUDY OF WARPAGE FOR FLIP CHIP PACKAGES

In this section, the results of the parametric analysis will be used to evaluate the effects of the (a) chip thickness, (b) substrate thickness, (c) Young's modulus of the underfill, (d) CTE of the underfill, and (e) geometry of the underfill fillet on the warpage behavior of flip chip packages. The studied chip thickness ranged from 0.119 mm to 1.793 mm. The substrate thickness ranged from 0.3 mm to 1.5 mm. The Young's modulus and CTE of the underfill ranged from 5.0 GPa to 30 GPa and from 15.0 ppm/°C and 52.5 ppm/°C, respectively. The 2D finite element model described in Section 24.6 was employed in this study. The analyzed chip and the substrate had square in-plane dimensions, which were

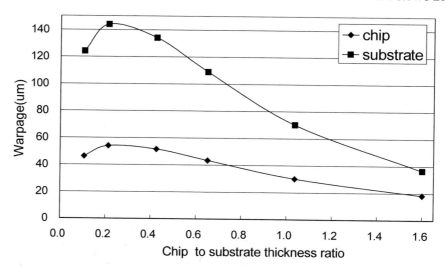

FIGURE 24.21. Effect of chip thickness change on flip chip warpage.

20 × 20 mm and 40 × 40 mm, respectively. The analyses were performed while the temperature of the flip chip package changed from 150°C to 26°C. The baseline model had chip and substrate thicknesses of 0.73 mm and 1.12 mm, respectively, and the Young's modulus and CTE of the underfill were 5 GPa and 45 ppm/°C, respectively. For each parametric analysis, only one parameter was changed.

24.7.1. Change of the Chip Thickness

The results of the warpage analyses of the chip and substrate in a flip chip package are plotted in Figure 24.21. The chip thickness for the baseline analysis was 0.73 mm. The analyses covered change of the chip thickness from 0.16 to 2.46 mm. When the chip to substrate ratio was reduced to 0.2, the maximum warpage values for both the chip and substrate were obtained. When the chip thickness increased, the warpage values for both the chip and substrate decreased due to the increase of the stiffness of the chip. On the other hand, when the ratio became smaller, the warpage values for both the chip and substrate dropped again. This was because the stiffness of the chip became small enough to comply with the deformation of the substrate.

24.7.2. Change of the Substrate Thickness

The substrate thickness for baseline analysis was 1.12 mm. The substrate thickness changed from 0.3 mm to 1.5 mm, which corresponded to ratios of 0.3 to 1.34, compared to the baseline substrate. The results are plotted in Figure 24.22. The change of warpage of the chip was found to be insignificant. On the substrate side, the maximum warpage occurred when the substrate to chip thickness ratio was 1.25. As the substrate thickness decreased from 1.12 mm to 0.3 mm, the warpage decreased progressively. This was because the stiffness of the substrate became so small that its influence on the change of warpage became progressively smaller. On the other hand, when the thickness of the substrate increased, the effect of the increase of the substrate stiffness on the decrease in the warpage was smaller

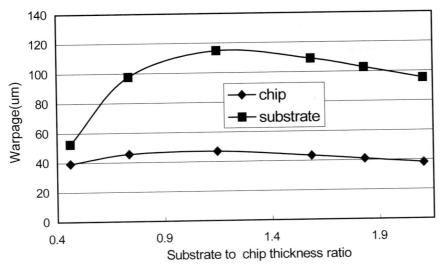

FIGURE 24.22. Effect of the substrate thickness change on the flip chip warpage.

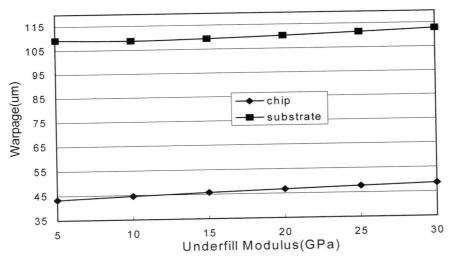

FIGURE 24.23. Effect of the change of the Young's modulus of the underfill on the flip chip warpage.

than the effect of the increase of the chip thickness (Figure 24.21). This was because the stiffness of the chip was much higher than that of the substrate.

24.7.3. Change of the Young's Modulus of the Underfill

The Young's modulus of the underfill used in the baseline analysis was 5 GPa. In this part of the study, the Young's modulus changed from 5 GPa to 30 GPa, i.e., a six-fold increase. The results are plotted in Figure 24.23. The warpage on both the chip and substrate sides increased almost linearly with the increase of the Young's modulus of the underfill. The experimental results plotted in Figures 24.17 and 24.18 confirmed this prediction. As

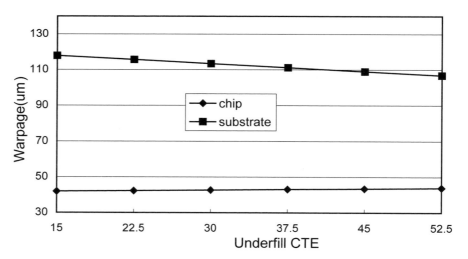

FIGURE 24.24. Effect of the change of the CTE of the underfill on the flip chip warpage.

shown in Figure 24.18, when the temperature increased, the warpage of the package with underfill increase more than the warpage of the package without underfill. This was because the solder joints functioned as cushions, absorbing the deformation due to the CTE mismatch between the substrate and chip. Once the underfill filled the gap between the chip and substrate, the deformation of the solder joints was constrained. As a result, the deformation induced by the CTE mismatch between the chip and substrate resulted in global warpage of the package because the deformation of the solder joints was now constrained by the surrounding underfill. Therefore, if the Young's modulus of the underfill increased, the warpage of both the chip and substrate in the flip chip package also had to increase, as evidenced by the results shown in Figure 24.23.

24.7.4. Change of the CTE of the Underfill

The CTE value for baseline analysis was 45 ppm/°C. The analysis covered CTE values ranging from 15 ppm/°C to 52.5 ppm/°C. The results are plotted in Figure 24.24. The change of the flip chip warpage was not significant. However, it is interesting that the warpage change trends of the chip and substrate were different. This was due to the underfill at the edge of the chip, i.e., at the fillet location. The fillet was modeled so as to have an inclined surface. The height and tail length of the fillet were 0.732 mm and 0.5 mm, respectively. When the CTE of the underfill increased, due to the temperature change from 150°C to room temperature, the tensile force induced by contraction at the fillet increased. As a result, the substrate warpage decreased and the chip warpage increased.

24.7.5. Effect of the Geometry of the Underfill Fillet

The height and tail of the fillet used in the baseline analysis were 0.732 mm and 0.5 mm, that is, 100% and 68.5% of the thickness of the chip, respectively. In this part of parametric analysis, the heights of the underfill fillet were selected to be 100%, 90%, 70%, and 50% of the flip chip thickness. On the other hand, the lengths of the underfill tail were selected to be 0.3 mm, 0.5 mm, 0.7 mm, and 0.9 mm. The results are plotted in

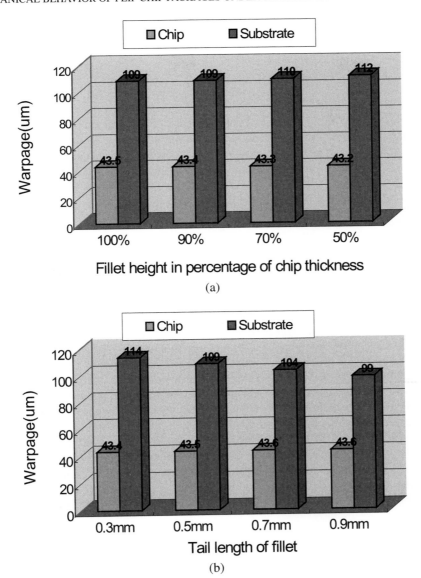

FIGURE 24.25. Effect of the change of the geometry of the underfill fillet on the flip chip warpage. (a) Change of the fillet height; and (b) change fillet the tail length.

Figure 24.25(a) and (b). It is found that the chip warpage remained essentially the same regardless of the geometric change of the fillet. On the other hand, when the fillet height increased from 50% to 100% of the chip thickness, the substrate warpage decreased only 3%. Also, when the tail length increased from 0.3 mm to 0.9 mm, the substrate warpage decreased by 15%. These results are considered to be reasonable because, when the height and tail length increased, the tensile force induced by fillet contraction, which, in turn, was induced by the decrease in temperature, also increased. As a result, the substrate warpage decreased. The insignificant change in chip warpage was mainly due to the fact that the chip stiffness was much larger than the substrate stiffness.

24.8. SUMMARY

In this chapter, the mechanical behavior of flip chip packages has been reported in detail. The phase shifted shadow moiré approach was adopted to measure the warpage of flip chip packages, and both the in-plane and out-of-plane ESPI methods were adopted to measure the CTE values of the substrate and the warpage of flip chip packages. Meanwhile, two-dimensional and three-dimensional finite elements models were constructed for numerical analysis of flip chip packages. It was found that:

(a) The substrate CTE values varied significantly due to the use of different BT cores, numbers of vias drilled in the substrate, and thicknesses of the copper and the solder mask. Therefore, the CTE of each substrate needs to be carefully measured so that the mechanical behavior of the corresponding flip chip package can be accurately determined.
(b) At room temperature, warpage of flip chip packages increased with increasing chip size and decreasing chip thickness. The warpage of the flip chip package also increased after the underfill reinforcement, and the relationship between the package warpage and the temperature changed from nonlinear to linear.
(c) The flip chip warpage was found to decrease to zero when the temperature increased to around 150°C. Linear relationships between the warpage values, for both the chip and the substrate, and the corresponding temperatures were observed. When the temperature increased further, the amount of substrate warpage gradually approached to the amount of chip warpage, especially when the chip size became large, as in the packages with 20 × 20 mm and 26 × 26 mm chips.
(d) The parametric analysis showed that a minimum warpage value for a flip chip package could be obtained by selecting a chip thickness that was much smaller than the thickness of the substrate, by choosing underfill materials with small Young's modulus and large CTE values, and by selecting an underfill fillet with a long tail.

REFERENCES

1. K. Wang, et al., Interfacial shear stress, peeling stress, and die cracking stress in trilayer electronics assembles, 2000 Inter Society Conference on Thermal Phenomena, 2000, pp. 56–64.
2. H.B. Fan, M.M.F. Yuen, and E. Suhir, Prediction of delamination in a bi-material system based on free-edge energy evaluation, IEEE Conference on Electronic Components and Technology Conference, 2003, pp. 1160–1164.
3. E. Suhir and J.D. Weld, Application of a surrogate layer for lower bending stress in a tri-material body, IEEE Conference on Electronic Components and Technology Conference, 1996, pp. 435–439.
4. T.M. Robert, Shear and normal stresses in adhesive joints, J. Eng. Mech., 115(11), pp. 2460–2479 (1989).
5. E. Suhir, Flip-chip solder joint interconnections and encapsulants in silicon-on-silicon MCM technology: thermally induced stresses and mechanical reliability, Multi-Chip Module Conference, 1993, pp. 92–99.
6. P. Palaniappan, et al., Correlation of flip chip underfill process parameters and material properties with in-process stress generation, Electronic Component and Technology Conference, 1998, pp. 838–847.
7. Y. Guo, S.M. Kuo, and L. Mercado, Stress/strength and reliability evaluations on UBM in different solder systems, 2000 Inter Society Conference on Thermal Phenomena, 2000, pp. 193–199.

8. W. Engelmaier, The use environments of electronic assemblies and their impact on surface mount solder attachment reliability, IEEE Transaction on Components, Hybrids, and Manufacturing Technology, 13(4), pp. 903–908 (1990).
9. E. Suhir, Predicted failure criterion (von-Mises stress) for moisture-sensitive plastic packages, IEEE Conference on Electronic Components and Technology Conference, 1995, pp. 266–284.
10. J.H. Zhao, X. Dai, and P.S. Ho, Analysis and modeling verification for thermal-mechanical deformation in flip-chip package, Electronic Components and Technology Conference, 1998, pp. 336–344.
11. K.S. Beh, A. Ourdjini, V.C. Venkatesh, and Y.L. Khong, Finite element analysis of substrate warpage during die attach process, Electronic Materials and Packaging, 2002, pp. 94–98.
12. A.O. Ayhan and H.F. Nied, Finite element modeling of interface fracture in semiconductor packages: issue and applications, 1998 Inter Society Conference on Phenomena, 1998, pp. 185–192.
13. T. Lee, J. Lee, and I. Jung, Finite element analysis for solder ball failures in chip scale package, Microelectronics Reliability, 38, pp. 1941–1947 (1998).
14. K.N. Chiang, Y.T. Lin, and H.C. Cheng, On enhancing eutectic solder joint reliability using a second-reflow-process approach, IEEE Transactions on Advanced Packaging, 23(1), pp. 9–14 (2000).
15. M.J. Pfeifer, Solder bump size and shape modeling and experimental validation, IEEE Transaction on Components, Packaging, and Manufacturing Technology, Part B, 20(4), pp. 452–457 (1997).
16. J.H. Lau, R. Lee, and C. Chang, Effect of underfill material properties on the reliability of solder bumped flip chip on board with imperfect underfill encapsulants, IEEE Transactions on Components and Packaging Technologies, 23(2), pp. 323–333 (2000).
17. C. Bailey, H. Lu, and D. Wheeler, Computer based modeling for predicting reliability of flip-chip components on printed circuit boards, 1999 IEEE/CPMT Int. Electronics Manufacturing Technology Symposium, 1999, pp. 42–49.
18. E.C. Ahn, T.J. Cho, J.B. Shim, H.J. Moon, J.H. Lyu, K.W. Choi, S.Y. Kang, and S.Y. Oh, Reliability of flip chip BGA package on organic substrate, Electronic Components and Technology Conference, 2000, pp. 1215–1220.
19. L. Leicht and A. Skipor, Mechanical cycling fatigue of PBGA package interconnects, Microelectronics Reliability, 40, pp. 1129–1133 (2000).
20. A. Mertol, Application of the Taguchi method to chip scale package (CSP) design, IEEE Transaction on Advanced Packaging, 23(2), pp. 266–276 (2000).
21. V.V. Calmidi and R.L. Mahajan, Optimization for thermal and electrical performance for a flip-chip package using physical-neural network modeling, Electronic Components and Technology Conference, 1997, pp. 1163–1169.
22. A.O. Cifuentes and I.A. Shareef, Modeling of multilevel structures: a general method for analyzing stress evolution during processing, IEEE Transaction on Semiconductor Manufacturing, 5(2), pp. 128–137 (1992).
23. K.S. Chen, T.Y.F. Chen, C.C. Chuang, and I.K. Lin, Full-field wafer level thin film stress measurement by phase-stepping shadow moiré, IEEE Transactions on Components and Packaging Technologies, 27(3), pp. 594–601 (2004).
24. Y.Y. Wang and P. Hassell, Measurement of thermally induced warpage of BGA packages/substrates using phase-stepping shadow moiré, Electronic Packaging Technology Conference, 1997, pp. 283–289.
25. S. Chen, C.Z. Tsai, E. Wu, and C.A. Shao, A study on effect of flip-chip BGA warpage and stresses using the different chip sizes, The 25th Conference on Theoretical and Applied Mechanics, 2001 (in Chinese).
26. Z.S. Chen, C.Z. Tsai, E. Wu, and C.A. Shao, Stress measurement and analysis of flip-chip BGA, SEMICON Taiwan 2001, Packaging & Testing Seminar, pp. 245–254.
27. M.R. Stiteler, I.C. Ume, and B. Leutz, In-process board warpage measurement in a lab scale wave soldering oven, IEEE Transaction on Components, Packaging, and Manufacturing Technology, Part A, 19(4), pp. 562–569 (1996).
28. D. Zwemer, et al., PWB warpage analysis and verification using an AP210 standards-based engineering framework and shadow moiré, Thermal and Mechanical Simulation and Experiments in Microelectronics and Microsystems, EuroSimE 2004, 2004, pp. 121–131.
29. K.S. Beh, A. Ourdjini, V.C. Venkatesh, and Y.L. Khong, Finite element analysis of substrate warpage during die attach process, Electronic Materials and Packaging, 2002, pp. 94–98.
30. I. Tsai, C.Z. Tsai, E. Wu, C.A. Shao, On accurate measurement of warpage for electronic packages, Proceedings of IMAPS Taiwan Technical Symposium, 2001, pp. 290–297.
31. Y. Guo and J.H. Zhao, A practical die stress model and its applications in flip-chip packages, 2000 Inter Society Conference on Thermal Phenomena, 2000, pp. 393–399.
32. J. Woosoon, et al., Evaluation of thermal shear strains in flip-chip package by electronic speckle pattern interferometry (ESPI), Electronic Materials and Packaging, 2001, pp. 310–314.

33. J.R. Huang, H.D. Ford, and R.P. Tatam, Speckle techniques for material testing, Optical Techniques for Structural Monitoring, IEE Colloquium, 1995, pp. 6/1–6/6.
34. N. Kao, Z.C. Tsai, and E. Wu, The application of electronic speckle pattern interferometry (ESPI) on the PBGA package, The Sixteenth National Conference on Mechanical Engineering, The Chinese Society of Mechanical Engineers, 1999, Vol. 5, pp. 254–261 (in Chinese).
35. E. Wu and A.J.D. Yang, Simultaneously determining Young's modulus, coefficient of thermal expansion, Poisson ratio and thickness of mult-layered thin films on silicon wafer, Electronic Components and Technology Conference, 2004, Vol. 1, pp. 901–905.
36. S. Liu, J. Wang, D. Zou, X. He, Z. Qian, and Y. Guo, Resolving displacement field of solder ball in flip-chip package by both phase shifting moire interferometry and FEM modeling, Electronic Components and Technology Conference, 1998, pp. 1345–1353.
37. F.F. Beer and E.R. Johnston, Mechanics of Materials, second edition, McGraw-Hill Publ., New York, 1992.
38. Y.K. Chung, G.Y. Tzeng, and B.C. Chen, Determining the Mechanical Properties of PBGA Substrates using Inverse Method, Report NSC-88-2815-002-038-E, National Taiwan University, 1999 (in Chinese).
39. R.D. Cook, D.S. Malkus, and M.E. Plesha, Concepts and Applications of Finite Element Analysis, John Wiley and Sons, 1981.

25

Stress Analysis for Processed Silicon Wafers and Packaged Micro-devices

Li Li[a], Yifan Guo[b], and Dawei Zheng[c]

[a]*Cisco Systems, Inc., USA*
[b]*Skyworks Solutions, Inc., USA*
[c]*Kotura, Inc., USA*

25.1. INTRINSIC STRESS DUE TO SEMICONDUCTOR WAFER PROCESSING

For quality control and reliability analysis in semiconductor product development, it is important to be able to determine the intrinsic stresses in the devices and interconnects due to various wafer processes. These processing steps include thin film deposition, plating, patterning, etching and heat treatment [1]. The local intrinsic stress built up in the devices, which is closely related to the wafer processes, is one of the major causes of low manufacturing yield and early failures in semiconductor products.

Currently, there are no effective tools that can characterize and monitor the local intrinsic stress induced by wafer processes. Average intrinsic stresses in thin films on a wafer are measured indirectly by the FLEXUS machine (KLA-Tencor Corporation, San Jose, CA) based on the Stoney's equation [5–7]. The FLEXUS machine is widely used in thin film stress analysis treating whole wafers as substrates (the thin film carrier). This approach can only provide information about the intrinsic stress in thin films of large dimensions. It is not capable of determining the intrinsic stresses in the thin films of small dimensions, such as bond pads and UBMs (under bump metallurgy). Other existing methods for intrinsic stress measurements are the X-ray diffraction and Raman microscopy [8–11]. The best spatial resolution of the X-ray diffraction is about 30 μm, which is not enough to determine the stress distribution at a local area especially when the strain gradient is high. The Raman microscopy has a spatial resolution of about 1 μm. However, it is only effective for single crystal materials. And it does not provide whole-field stress maps, which are very important in the analysis of stress distributions. Both methods need a stress free state as reference. This is not always available under actual manufacturing conditions. In addition, both methods require sophisticated measuring instruments and typically can not be used for on-site monitoring and measuring during wafer processing.

In this section, a novel testing methodology to evaluate the whole-field intrinsic stress is presented. Using this methodology, a comprehensive study of a Ni plating process is conducted. The testing structure consists of a Si membrane with SiO_2 and $Si_xN_yH_z$ buffer layer on top, which is used in the pressure sensor applications. The thin film material and the plating are processed on the top of the Si membrane using the regular device processing. The Si membrane deformation caused by the intrinsic stresses from the plating process is measured by an optical method. A finite element model is then used to calculate the intrinsic stress and the resulting full field stress distribution in the thin film and the Si membrane using the input from the measured membrane deformation.

25.1.1. Testing Device Structure

As shown in Figure 25.1, a sensing device wafer is, essentially, a pressure sensor wafer of specified membrane thickness. Either the whole pressure sensor wafer or a single sensing chip can be used for the monitoring and testing functions. The sensing wafer (or chip) is put into the processes which are under investigation or being monitored. During each material deposition or other process steps in the wafer process, the membrane will deform due to the process induced intrinsic stress.

In this study, an n-type (100) Si of 22 μm thick was grown onto a p-type (100) Si wafer of 380 μm thick as the starting substrate. 3500 Å thick SiO_2 and 4000 Å thick $Si_xN_yH_z$ were deposited on top of the Si wafer as insulating layers. The Si wafer was patterned using Si_3N_4 as the etching mask. The etching stopped automatically at the n-type Si layer where a Si membrane window of 2650 × 2650 μm^2 was formed at the center of a die of 4214 × 4214 μm^2 area. This process formed a membrane with a thickness of about 23 μm. This sensing device was used to determine the intrinsic stress developed during an electroless Ni plating process. In the process, Ni films with different sizes and thickness were plated at the central portion of the sensor membranes.

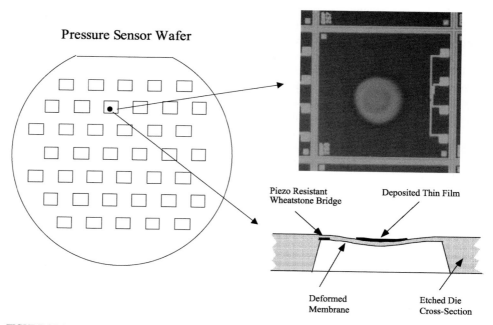

FIGURE 25.1. A sensing device wafer is essentially a Si wafer of specified membrane thickness. The membrane deformations are determined and monitored by an optical measurement technique.

25.1.2. Membrane Deformations

The deformations of the sensor membranes were determined using the Twyman-Green Interferometry technique as shown in Figure 25.2. The technique is a whole-field optical method with very high displacement sensitivity. It provides fringe patterns which are contour maps of out-of-plane displacements. The standard displacement resolution is 0.3165 μm which is one half of the wavelength of the laser used in the measurement. This optical technique has been used to measure static or dynamic deformations of Si membranes as a function of time and during progress of processes [12].

The membrane deformations were measured before and after the Ni plating process. Figure 25.3 shows the front and back view images of a membrane prior to the electroless Ni plating. The front side of the membrane is very flat. The back surface of the membrane has little variations caused by the thickness changes of the membrane as a result of the Si etching process. The Ni films are plated at the center of the Si membrane. In the fringe patterns, the out-of-plane displacement is determined from the fringe orders with a resolu-

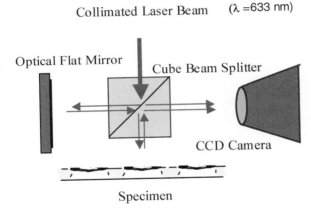

FIGURE 25.2. The deformation in the membranes of the sensor wafer is measured using the Twyman-Green interferometer.

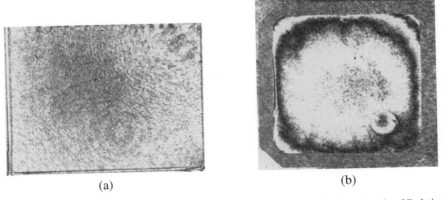

FIGURE 25.3. The front (a) and back side (b) images of a Si membrane before the electroless Ni plating.

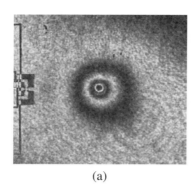 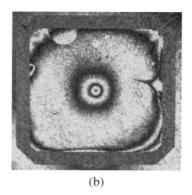

(a)   (b)

FIGURE 25.4. The membrane deformation produced by the Ni plating. Membrane deformations are measured from the front (a) and back side (b) of the membrane.

tion of 0.3165 μm. However, the resolution of this technique is not limited to 0.3165 μm. A phase-shift mechanism can be used to achieve a resolution of 1/10 of the fringe order.

The back surface of the membrane is optically reflective after the etching process. In practice, the membrane deformation can be measured from the backside of the membrane. The variation in the membrane thickness, which is measured before the plating process, as shown in Figure 25.3(b), is subtracted from the final deformation measured after the plating process. Thus, the final result is the deformation caused by the plating only. The measurement provides the whole-field deformation of the membrane including the area directly under the Ni film. However, sometimes, the etching process results in a rough surface with optical noises. In that case, the front side deformation can be used for the analysis. If the front side is used for deformation measurements, the thickness of the Ni film should be taken into consideration. If the data is taken from the backside of the membrane, the membrane thickness variation should be subtracted.

Figure 25.4 shows the typical interferometric fringe patterns obtained after the Ni plating process. The measurements were conducted at room temperature. In the experiment, membrane deformations were measured from the front and backsides of the membrane. The membrane size is 2600 × 2600 μm; membrane thickness is 23 μm; Ni film diameter is 200 μm; and Ni film thickness is 6 μm.

Since the plating temperature of the electroless Ni is at 85°C, in addition to the intrinsic stress, the measurements conducted at the room temperature will include the thermal stress caused by the Coefficient of Thermal Expansion (CTE) mismatch between the Ni and Si membrane. In order to determine the deformation induced by the intrinsic stress only, the measurements were also conducted at the plating temperature, 85°C. A hot plate with an enclosure and a glass window was used to heat the wafer and keep it at 85°C. Figure 25.5 shows the results of the measurements at the plating temperature. The images are membranes with the plated Ni films of 200, 500, 1000 and 1500 μm in diameters, respectively. The Ni thickness is 6 μm.

From the comparison of the membrane deformations obtained at the room temperature and those at the plating temperature, it was found that the deformation directions of the membranes were the same. When the measurements were conducted at room temperature, the thermal stress added into the intrinsic stress and the measurements showed that the magnitude of the deformation increased, but the deformation shape remained the same.

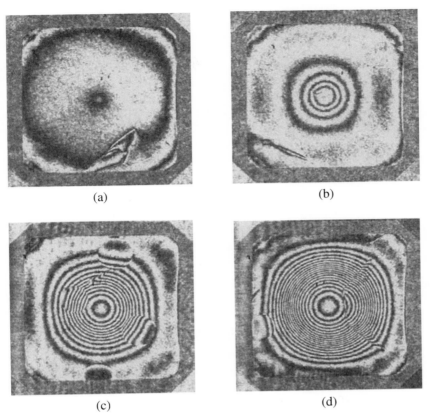

FIGURE 25.5. The membrane deformation produced by the Ni plating. The diameters of the Ni films are (a) 200 μm, (b) 500 μm, (c) 1000 μm, and (d) 1500 μm. Deformations are measured on the back surface of the membrane at 85°C, which is the plating temperature.

It means that the intrinsic stress in the Ni film has the same sign as the thermal stress in the Ni film caused by the negative ΔT (temperature decrease), which was a tensile stress.

25.1.3. Intrinsic Stress

In order to obtain the actual values of the intrinsic stresses, a finite element model (FEM) was used (Figure 25.6). With a parametric FEM model, we can model the geometry of the membrane sensing device exactly and can cover a range of wafer designs. In the process of the stress calculation, a pre-assumed intrinsic stress value was applied to the Ni film in the FEM model and the resulting membrane deformation was calculated. The membrane deformation from the FEM model was then compared with the experimentally measured membrane deformation. According to the comparison, the intrinsic stress value was adjusted in the FEM model until the deformation output from the model matched the experimental results. Figure 25.7 shows the final output of the FEM and the experimental results on the membrane deformations for the four Ni film sizes. With the right amount of intrinsic stress input in the model, the deformations output can be closely matched with the experimental results. Since all the boundary conditions are exact, the intrinsic stress in the FEM model, therefore, represents the intrinsic stress value in the Ni film.

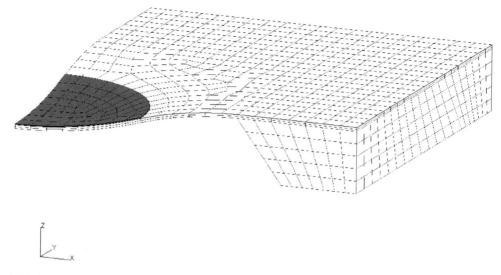

FIGURE 25.6. The FEM axisymmetric model simulated the membrane deformation with intrinsic stress input in the Ni film.

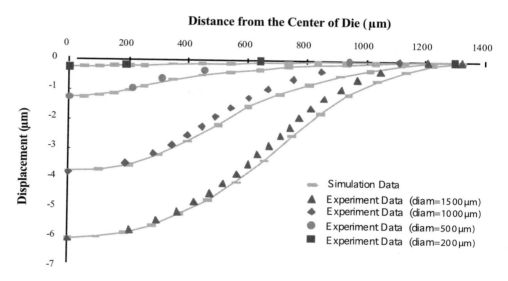

FIGURE 25.7. Experimental and FEM results on the membrane deformations for the four Ni film sizes at plating temperature.

Figure 25.8 gives the intrinsic stress values in the four Ni films with diameters of 200, 500, 1000 and 1500 μm. It is clear that the intrinsic stresses are dimensional sensitive. The intrinsic stresses increase as the Ni film dimension increases. In the real UBM, the Ni pad size is 120 μm, and the Ni intrinsic stress is around 50 MPa.

The resulting stress distributions in the Ni and the Si membrane can also be calculated from the FEM model. Figure 25.9 shows the normal and shear stress in the Si membrane at the edges of the Ni film. The resulting stress distributions are obtained based on the

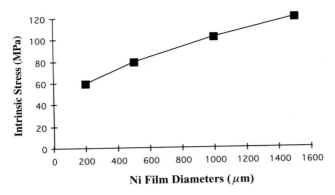

FIGURE 25.8. The intrinsic stress values in the four Ni films with different diameters. The intrinsic stress is a function of the Ni film size.

assumption that there is no Si membrane deformation before the metal was deposited. In practice, this assumption is not necessarily needed since the process induced Si membrane deformation can be obtained by subtracting the initial deformation from the final deformation after each process step.

25.1.4. Intrinsic Stress in Processed Wafer: Summary

The proposed technique uses a silicon membrane to detect and display the intrinsic stress, which is the stress accumulated during the plating processes. Deformations in thin reflective Si membrane are measured accurately with laser interferometry and the stress calculations are completed by finite element method (FEM).

In the study of the electroless Ni plating process, it is found that the intrinsic stress value in the Ni film is a function of the film dimension, and the dimension dependency is very strong. As the film dimension increases, the intrinsic stress should approach the value measured by the FLEXUS machine which provides the intrinsic stress values in a thin film with a dimension close to infinity. This hypothesis is currently being verified.

The intrinsic stress from the wafer process is a critical parameter affecting the reliability of the device, such as the bump strength, UBM (under bump metalization) strength, thin film adhesion, failure in the passivation layer and die cracking. The technique experimented should have impact on the following areas:

(a) Intrinsic stress measurements of different UBM.
(b) Process characterization, a simple technique to determine the local thin-film stress induced by metalization, deposition, plating, etching and heat treatment.
(c) In-situ monitoring of stress change in interconnect lines during electromigration and thermal migration in the accelerated testing.
(d) Local stress in active region after ion implantation (doping) and subsequent annealing.

The current technique used in the intrinsic stress measurement is the FLEXUS machine. Very often, it does not have enough sensitivity to measure the deformation if the thin film is patterned on a regular Si substrate. In practice, the Si substrates have to be thinned down to get enough sensitivity. In the FLEXUS machine, the wafer deformation is measured by the laser reflection, which can only obtain the average thin-film stress but not the

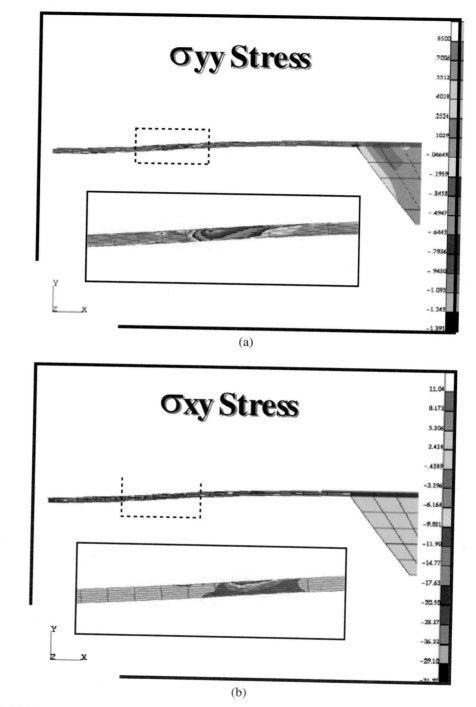

FIGURE 25.9. The normal (a) and shear stresses (b) in the silicon membrane at the interface with the Ni film by FEM analysis. High stresses are shown in the silicon membrane next to the edge of the Ni layer.

local stress variations. The proposed device with a thin membrane can provide extremely high sensitivity. By using a full field optical method, local deformations from small and various thin film geometry can be determined with very high resolution. With the aid of the FEM method, the intrinsic stress and the resulting stress field can be determined. This technique has its advantages to complement the FLEXUS machine for determinations of local stress fields caused by the intrinsic stress.

There is no practical limitation regarding the dimensions and shapes of the thin films studied. Currently, a great interest is in the area of determining the intrinsic stress in the UBM processes. In this study, the dimensional effect on intrinsic stress in the electroless Ni UBMs is analyzed. This Si membrane based testing sensor has great potentials in the wafer process characterization with high accuracy solution and a low cost. This new technology could also promise the capability to study the stress in much finer structures, such as interconnect line, stress change due to electromigration and thermal migration in interconnect lines and, more importantly, the stress at gate SiO_2/Si interface.

25.2. DIE STRESS RESULT FROM FLIP-CHIP ASSEMBLY

In flip chip plastic ball grid array (FC-PBGA) packages, silicon die is attached on a laminate substrate by solder joints. Underfill material is filled in the gap between the die and the substrate to protect the solder joints for better reliability. After the underfill process, the die and the substrate are rigidly bonded together and no interface delamination and separation should be present. A silicon die, with a coefficient of thermal expansion (CTE) of 2.6 ppm/°C, and a laminate substrate, with a CTE from 15 to 25 ppm/°C, are connected by underfill material. As a result of CTE mismatch, significant thermal stress occurs in the die and the substrate during thermal cycles. In component level reliability testing, this thermal stress is the major cause of many failure modes including die cracking. Lately in the development of FC-PBGA packages, reducing die stress and improving reliability has been a very serious issue.

25.2.1. Consistent Composite Plate Model

The flip chip package can be treated as a multilayer composite system. The consistent plate model treats the chip, the underfill and the carrier substrates, i.e., the chip/carrier module, as plies of an integrated laminated plate. The only assumptions made are those normal to thin and classical laminate plate theory [2].

The term consistent plate model comes from the assumption of consistent deformations in the chip and substrate. The strain in the system will therefore depend only on the strain on the reference plane somewhere in the middle of the plate, ε^0, and the curvature of the laminated plate, κ,

$$\begin{Bmatrix} \varepsilon_x \\ \varepsilon_y \\ \varepsilon_{xy} \end{Bmatrix} = \begin{Bmatrix} \varepsilon_x^0 \\ \varepsilon_y^0 \\ \varepsilon_{xy}^0 \end{Bmatrix} + z \begin{Bmatrix} \kappa_x \\ \kappa_y \\ \kappa_{xy} \end{Bmatrix}. \tag{25.1}$$

The constitutive relation for any ply of a laminated plate is

$$\sigma = \mathbf{Q}(\varepsilon - \mathbf{\Lambda}), \tag{25.2}$$

where the stress vector σ and induced strain vector Λ are

$$\sigma = \begin{Bmatrix} \sigma_x \\ \sigma_y \\ \sigma_{xy} \end{Bmatrix}, \quad \Lambda = \begin{Bmatrix} \Lambda_x \\ \Lambda_y \\ \Lambda_{xy} \end{Bmatrix} = \begin{Bmatrix} \alpha_x \Delta T \\ \alpha_y \Delta T \\ \alpha_{xy} \Delta T \end{Bmatrix}. \tag{25.3}$$

Here α is the coefficient of thermal expansion and ΔT is the temperature change.

The matrix $\mathbf{Q}$ is the transformed reduced stiffness of the lamina and is given by Ref. [4] as following

$$\mathbf{Q} = \begin{Bmatrix} \overline{Q}_{11} & \overline{Q}_{12} & \overline{Q}_{16} \\ \overline{Q}_{12} & \overline{Q}_{22} & \overline{Q}_{16} \\ \overline{Q}_{16} & \overline{Q}_{16} & \overline{Q}_{66} \end{Bmatrix}. \tag{25.4}$$

The load-deformation relationship for the consistent plate is given by

$$\begin{bmatrix} \mathbf{N} \\ \mathbf{M} \end{bmatrix} = \begin{bmatrix} \mathbf{A} & \mathbf{B} \\ \mathbf{B} & \mathbf{D} \end{bmatrix} \begin{Bmatrix} \varepsilon^0 \\ \kappa \end{Bmatrix} - \begin{bmatrix} \mathbf{N}^\Lambda \\ \mathbf{M}^\Lambda \end{bmatrix}, \tag{25.5}$$

where the conventional mechanical stress resultants, the mechanical forces and moments, are

$$\mathbf{N} = \int_t \sigma dz, \tag{25.6}$$

$$\mathbf{M} = \int_t \sigma z dz. \tag{25.7}$$

The matrices $\mathbf{A}$, $\mathbf{B}$, and $\mathbf{D}$ are the usual extensional stiffness, bending-stretching coupling stiffness and bending stiffness of the plate [4]. The integrations in the above equations are carried out through the composite plate thickness. The equivalent thermal forces and moments are

$$\mathbf{N}^\Lambda = \int_t \mathbf{Q}\Lambda dz, \tag{25.8}$$

$$\mathbf{M}^\Lambda = \int_t \mathbf{Q}\Lambda z dz. \tag{25.9}$$

The total potential energy stored in the plate is given by

$$U = \frac{1}{2} \iint_\Omega \begin{Bmatrix} \varepsilon^0 \\ \kappa \end{Bmatrix}^T \begin{bmatrix} \mathbf{A} & \mathbf{B} \\ \mathbf{B} & \mathbf{D} \end{bmatrix} \begin{Bmatrix} \varepsilon^0 \\ \kappa \end{Bmatrix} d\Omega - \iint_\Omega \begin{bmatrix} \mathbf{N}^\Lambda \\ \mathbf{M}^\Lambda \end{bmatrix}^T \begin{Bmatrix} \varepsilon^0 \\ \kappa \end{Bmatrix} d\Omega. \tag{25.10}$$

This strain energy equation together with a Ritz approximate solution method can be used to solve for the approximate strains and curvatures in the multilayer system.

25.2.2. Free Thermal Deformation

When there is no external mechanical load, Equation (25.5) can be reduced to

$$\begin{bmatrix} \mathbf{A} & \mathbf{B} \\ \mathbf{B} & \mathbf{D} \end{bmatrix} \begin{Bmatrix} \varepsilon^0 \\ \kappa \end{Bmatrix} = \begin{bmatrix} \mathbf{N}^\Lambda \\ \mathbf{M}^\Lambda \end{bmatrix}. \qquad (25.11)$$

Equation (25.11) can be solved directly for ε^0 and κ.

As shown in Figure 25.10, three special cases are of particular interests for electronics packaging analysis, namely, the beam bending, the cylindrical bending and the axisymmetrical bending. When material anisotropy for each layer is small, Equation (25.4) can be written as

$$\mathbf{Q} = \begin{Bmatrix} \dfrac{E_i}{1-v_i^2} & \dfrac{v_i E_i}{1-v_i^2} & 0 \\ \dfrac{v_i E_i}{1-v_i^2} & \dfrac{E_i}{1-v_i^2} & 0 \\ 0 & 0 & \dfrac{E_i}{2(1+v_i)} \end{Bmatrix}. \qquad (25.12)$$

Here E_i and v_i are the Young's modulus and Poisson's ratio for the ith layer lamina. Note Equation (25.12) is given for the axisymmetrical case. For the cylindrical and beam bending cases, we can simply replace the term $E_i/(1-v_i)$ with E_i.

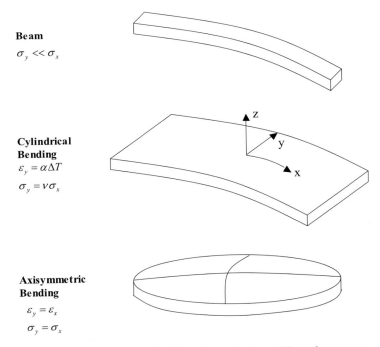

FIGURE 25.10. Three possible deformation modes for multilayered structures.

The deformation of the composite plate can be further simplified such that for beam bending,

$$\sigma_y \ll \sigma_x;$$

for cylindrical bending,

$$\begin{Bmatrix} \kappa_x \\ \kappa_y \\ \kappa_{xy} \end{Bmatrix} = \begin{Bmatrix} \kappa_x \\ 0 \\ 0 \end{Bmatrix}, \begin{Bmatrix} \varepsilon_x^0 \\ \varepsilon_y^0 \\ \varepsilon_{xy}^0 \end{Bmatrix} = \begin{Bmatrix} \varepsilon_x^0 \\ \varepsilon_y^0 \\ 0 \end{Bmatrix},$$

for axisymmetrical bending,

$$\begin{Bmatrix} \kappa_x \\ \kappa_y \\ \kappa_{xy} \end{Bmatrix} = \begin{Bmatrix} \kappa_x \\ \kappa_x \\ 0 \end{Bmatrix}, \begin{Bmatrix} \varepsilon_x^0 \\ \varepsilon_y^0 \\ \varepsilon_{xy}^0 \end{Bmatrix} = \begin{Bmatrix} \varepsilon_x^0 \\ \varepsilon_x^0 \\ 0 \end{Bmatrix}.$$

The force and the moment equations for the above three cases become

$$\begin{bmatrix} A_x & B_x \\ B_x & D_x \end{bmatrix} \begin{Bmatrix} \varepsilon_x^0 \\ \kappa_x \end{Bmatrix} = \begin{bmatrix} N_x^\Lambda \\ M_x^\Lambda \end{bmatrix}. \qquad (25.13)$$

Here coefficients A_x, B_x, D_x, N_x^Λ and M_x^Λ are given by the following equations [2,3]

$$\begin{cases} A_x = \int_t \frac{E_i}{1 - \nu_i} dz \\ B_x = \int_t \frac{E_i}{1 - \nu_i} z \, dz \\ D_x = \int_t \frac{E_i}{1 - \nu_i} z^2 \, dz \\ N_x^\Lambda = \int_t \frac{E_i \alpha_i \Delta T}{1 - \nu_i} z \, dz \\ M_x^\Lambda = \int_t \frac{E_i \alpha_i \Delta T}{1 - \nu_i} z \, dz. \end{cases} \qquad (25.14)$$

Again, Equation (25.14) is given for the axisymmetrical case. For cylindrical and beam bending cases, we simply replace the term $E_i/(1 - \nu_i)$ with E_i in the above equation.

25.2.3. Bimaterial Plate (BMP) Case

As an example, we consider a flip chip package shown in Figure 25.11. Using the consistent composite plate model, the flip-chip package can be treated as a system of two plates bonded together if only the die bending curvature and the stress on top of the die is interested. The underfill and solder joins can be treated as part of the substrate in this scenario. The Young's modulus, Poisson's ratio, CTE and thickness of upper plate (plate 1 or die) are E_1, ν_1, α_1, h_1, respectively (Figure 25.12). The corresponding quantities for

STRESS ANALYSIS FOR PROCESSED SILICON WAFERS AND PACKAGED MICRO-DEVICES 689

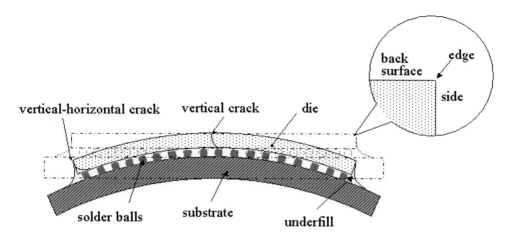

FIGURE 25.11. A schematic diagram of a flip-chip plastic ball grid array package and die failure modes due to die bending. The dashed lines represent the package at underfill curing temperature, which is stress free. The solid lines represent the package at room temperature, at which the package bends due to the CTE mismatch between the die and the substrate.

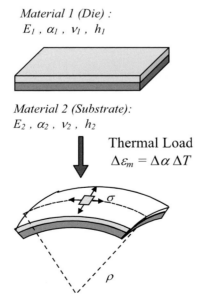

FIGURE 25.12. Bimaterial plate model for a flip chip module.

the lower plate (plate 2 or substrate) are E_2, v_2, α_2, h_2, respectively. In the flip chip package, the two plates were bonded together at a certain temperature T_0. The bimaterial plate (BMP) system will bend into a surface which is part of a sphere when temperature is reduced to T (Figure 25.12). Solving Equation (25.13) for the BMP system, we have

$$\begin{cases} \varepsilon_x^0 = \dfrac{D_x N_x^\Lambda - B_x M_x^\Lambda}{A_x D_x - B_x^2} \\ \kappa_x = \dfrac{A_x M_x^\Lambda - B_x N_x^\Lambda}{A_x D_x - B_x^2}. \end{cases} \quad (25.15)$$

Please note that ε_x^0 is dependent on the location of x–y plane while κ_x is consistent through the plate thickness. Substituting Equation (25.14) into Equation (25.15) and after some mathematical manipulations we have

$$\begin{cases} \varepsilon_x^0 = \dfrac{hm(4+3h+h^3m)(\alpha_2+\alpha_1)\Delta T}{(1+hm(4+6h+4h^2+h^3m))} \\ \kappa_x = \dfrac{6\varepsilon_m hm(1+h)}{h_1(1+hm(4+6h+4h^2+h^3m))}, \end{cases} \quad (25.16)$$

where $h = h_2/h_1$ is the thickness ratio of the lower plate to the upper plate, $m = M_2/M_1$ is the ratio of the biaxial modulus of the lower plate $[M_2 = E_2/(1-v_2)]$ respect to the upper plate $[M_1 = E_1/(1-v_1)]$, and $\varepsilon_m = (T-T_0)(\alpha_1-\alpha_2)$ is the thermal mismatch strain between the two plates.

In practice, it is convenient to express the curvature in a dimensionless quantity or, so called, characteristic curvature:

$$\bar{\kappa} = \frac{h_1}{6\varepsilon_m}\kappa_x = \frac{hm(1+h)}{1+hm(4+6h+4h^2+h^3m)}. \quad (25.17)$$

The characteristic curvature is a function of only the thickness ratio h and the biaxial modulus ratio m.

From Equation (25.2) and Equation (25.16), the stress on the top surface of plate 1, the die, can be expressed as

$$\sigma_{top} = \frac{\varepsilon_m M_1 hm(2+3h-h^3m)}{1+hm(4+6h+4h^2+h^3m)}. \quad (25.18)$$

The stress is uniform at any point on the top of the die which is a direct result of neglecting the edge effect. The dimensionless stress or characteristic stress is defined as

$$\bar{\sigma} = \frac{\sigma_{top}}{\varepsilon_m M_1} = \frac{hm(2+3h-h^3m)}{1+hm(4+6h+4h^2+h^3m)}. \quad (25.19)$$

It is instructive to express the stress on top of the die in terms of the curvature or characteristic curvature.

$$\sigma = \kappa_x \frac{h_1 M_1 (2+3h-h^3m)}{6(1+h)}, \quad (25.20)$$

$$\bar{\sigma} = \bar{\kappa}_x \frac{2 + 3h - h^3 m}{1 + h}. \tag{25.21}$$

According to Equations (25.20) and (25.21), the stress contains the curvature κ_x. Therefore, the curvature κ_x can provide direct information of the stress.

25.2.4. Validation of the Bimaterial Model

The mechanical behavior of the FC-PBGA package is similar to a mechanical system with two bonded plates. A bimaterial laminate model can be used to estimate the stress and bending curvature in each plate. The bimaterial plate model (BMP) has been widely used in the stress determinations of the wafer level process with thin films deposited on silicon wafers. With the thin film approximation, which is known as Stoney's equation [13], it provides an effective method to determine the stresses in thin films deposited on wafers [14]. However, this model can not be directly applied to the FC-PBGA packages because of the assumptions in the model that the film is much thinner than the wafer. In practice, the die stress in the FC-PBGA packages is usually determined from finite element method (FEM) models which require simulation tools and are comparatively time consuming when packages with different geometry are analyzed.

A typical FC-PBGA package is shown in Figure 25.11. In the assembly process, the die is attached to the substrate through solder joints. The die attach process does not produce high die stress because of the creep and relaxation in the solder joints. Most of the die stress is produced by the underfill process. When the underfill material is cured at high temperature (usually 150 to 170°C), the die and the substrate is connected. As the package cools down to room temperature, the thermal stress accumulates because of the CTE mismatch between the die and the substrate. A bending deformation is developed by cooling the package from the underfill curing temperature to room temperature. Further cooling to lower temperature in the reliability test will induce more package bending and higher die stress. During the package qualification, thermal cycles from −55 to 125°C are used for the package level test.

In a real package, the underfill material is adhered to the die and substrate, the degree of the creep and relaxation is not significant in the underfill material when it is fully cured. The stress free temperature of the system is near the underfill curing temperature. When the package is thermally cycled, the thermal stress in the die can cause several failure modes in die cracking at the low temperature. Figure 25.11 also shows the failure due to die cracking that have been identified during the reliability tests: (a) vertical crack, the crack initiated from the die back side and propagated vertically down into the die, (b) vertical-horizontal crack, the crack initiated from the back side propagated vertically first then turned to horizontal. The BMP model developed can be applied in the study of these two types of die cracks.

The tensile stress at the backside of the die is obviously the cause of the vertical cracks initiated from the backside. It is clear that the back surface of the die is in tension. The cracks are initiated from the back surface or the edges of the back surface and across the die from edge to edge. The depth of the crack depends on the stress level in the package. Simulation shows that the tensile stress at the backside of the die is distributed almost over entire back surface.

The BMP model is validated by experiments and finite element method (FEM). The curvature was directly validated by curvature measurement using an optical method. The stress calculated from the BMP model was compared with FEM results.

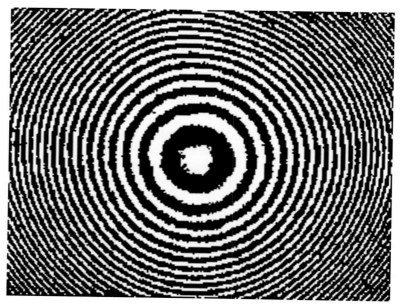

FIGURE 25.13. A typical interference fringe pattern obtained from the Si chip surface with the Twyman-Green interferometer.

In order to validate the curvature model for different thickness ratio and different materials, two sets of sample were prepared. The dice chosen in this study are single Si with (100) orientation. The dimensions are 6.78 mm × 5.50 mm × 0.54 mm. The dice are both-side polished. Substrates are printed circuit board (PCB) and Cu substrates. The dimension of the dice is kept unchanged for different samples. The thickness of the substrates varies from 0.25 times to 2 times of the die thickness for both PCB and Cu substrates. The length and the width of the substrates are kept the same as those of the dies. The die and substrate are bonded together by a thin layer of epoxy. After about 1 hour curing at 83°C, the epoxy layer firmly bonds the die and substrate. As the assemblies cool down to room temperature, bending and stress are generated due to the CTE mismatch. The thickness of the epoxy layer is less than 10 μm, which is negligible in the die bending and die stress calculation.

The curvature measurement was carried out using a Twyman-Green interferometer [15] with an *in situ* heating chamber. A typical fringe pattern of the die bending is shown in Figure 25.13. It is the interference pattern of a Si die (6.78 mm × 5.50 mm × 0.54 mm) on a 0.77 mm thick Cu substrate at 23.5°C. The package is flat at 78.5°C. The fringe pattern is a displacement contour of the sample surface. Each circular fringe corresponds to 316.4 nm out-of-plane deformation. Figure 25.13 clearly shows that the fringe pattern is consisted of concentric circles. The bimaterial plate theory treats the two plates as an axisymmetric mechanical system. The experiment proves that the deformation of the sample is indeed an axisymmetric spherical surface. It is noteworthy to point out that the zero bending temperature (78.5°C) is lower than the epoxy curing temperature 83°C in this case. The reason is that the thin epoxy layer had experienced a small stress relaxation before the test. It is crucial to check the zero bending temperature for each sample in order to get the accurate data.

The FEM calculation is carried out using the commercial software ANSYS 5.5. Two-dimensional (2D) 8-node axisymmetric solid elements were used to calculate the die bend-

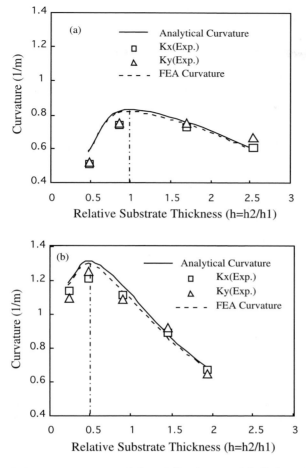

FIGURE 25.14. Bending curvatures (a) Si die on PCB substrate and (b) Si die on Cu substrate.

ing and the stress on top of the die. The FEM model treats the bimaterial plates as circular plates in order to be consistent with the analytical BMP model. Linear elastic deformation is assumed in the FEM model.

The bending curvatures from experiment measurement, BMP model and FEM are plotted in Figure 25.14. Figure 25.14(a) is the result of die on PCB substrate, and Figure 25.14(b) is the result of die on Cu substrate. The Young's modulus, Poison's ratio and CTE of Si used in the calculation are 131 GPa, 0.28 and 2.5 ppm/°C, respectively. The corresponding values are 21 GPa, 0.33, and 20 ppm/°C respectively, for PCB substrate, and 117 GPa, 0.35, and 17 ppm/°C, respectively, for Cu substrate. The experimental results of the curvature are in two orthogonal directions. The x-direction is along the length of the samples and the y-direction is along the width of the samples. The horizontal axis of the figure is the thickness ratio $h = h_2/h_1$, with the die thickness of $h_1 = 0.54$ mm fixed. For each substrate thickness, at least two samples were tested; the curvatures shown in the figure are the averaged values. Both Figure 25.14(a) and Figure 25.14(b) clearly show that the calculated curvatures from both BMP and FEM agree well with the experimentally mea-

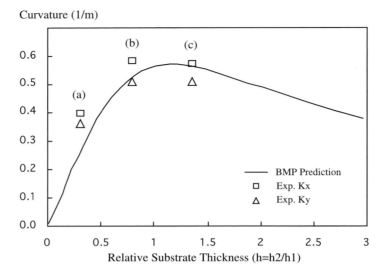

FIGURE 25.15. Curvature measured by Twyman-Green interferometer and calculated by BMP model for flip-chip packages.

sured curvatures. The curvatures of the two orthogonal directions are the same from the experimental measurement.

A very interesting feature shown in Figure 25.14(a) and Figure 25.14(b) is that there is a maximum bending curvature in both substrate systems. For PCB substrates, the maximum bending curvature occurs when the substrate to die thickness ratio equals to one, and the corresponding value for the Cu substrate occurs when the substrate to die thickness ratio equals to 0.5. This maximum bending curvature should be avoided in package design in order to reduce the package bending and die stress.

A set of real FC-PBGA packages was also tested. The die dimension is 7.0 mm × 6.5 mm × 0.61 mm. The underfill has a thickness of 0.13 mm. Figure 25.15 shows the curvature of this set of packages measured by Twyman-Green interferometer. The two orthogonal curvatures are plotted in the same figure. The curvature predicted by the BMP theory is plotted for comparison. An approximation made in the calculation of the curvature is that the underfill layer is treated as part of the substrate since the underfill layer has similar material properties as the substrate. It is clear that the predicted curvature by the BMP model is very close to the average curvature from experimental measurements in real packages.

The stresses on the top of the dies were calculated by BMP model and FEM for different samples. The stresses are shown in Figure 25.16. Although the BMP is based on plate approximation and the FEM uses 2D axisymmetric solid elements, the stress results are very close. The FEM calculation does show some edge effect at the die outer peripheral, this effect only affects the local area of about the dimension of order of the die thickness. The stress maintains constant in the rest part of the die. The FEM calculated stress shown in Figure 25.16 is the stress near the die center. The good agreement enables us to use BMP model to have a quick estimate of the stress on top of the die.

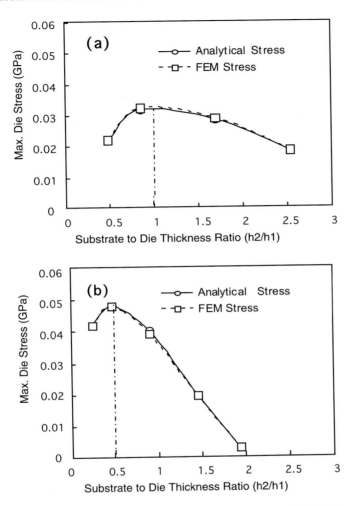

FIGURE 25.16. The stress on top of the die calculated from BMP model and FEM for (a) Si die on PCB substrate and (b) Si die on Cu substrate.

25.2.5. Flip-Chip Package Design

Since the BMP model gives closed-form solutions for bending curvature and die stress, it is very convenient to use the model to do parametric study in package design. The data can be tabulated or plotted as a handy reference. Figure 25.17 plots the characteristic curvature and characteristic die stress as a function of thickness ratio for different biaxial modulus ratios of the substrate to die. The reason to choose the characteristic curvature defined in Equation (25.17) is that it is dimensionless, thermal load independent and absolute thickness independent. This ensures that Figure 25.17 is universally applicable for arbitrary material combination, arbitrary geometry combination, and arbitrary thermal load. It is clearly shown that for every different material combination there is a maximum stress. As the ratio of the biaxial modulus of the substrate to die (m) increases, the maximum die stress (σ) increases and the position of the maximum die stress in terms of thickness

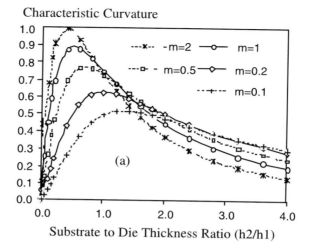

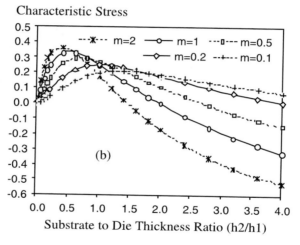

FIGURE 25.17. Dimensionless or characteristic curvature and stress in a bimaterial plate system as a function of thickness ratio and biaxial modulus ratio: (a) characteristic curvature; (b) characteristic stress.

ratio (h) tends to be smaller simultaneously. The position of maximum curvature is dependent on the material properties. Most of the FC-PBGA packages have the maximum curvature at $h \approx 1$, which is a coincident of the fact that most of packages are Si on PCB with $m \approx 0.2$. For Si on Cu ($m \approx 1$), the maximum curvature is located at $h \approx 0.5$. The characteristic curvature is approaching to zero as relative die thickness approaching zero or infinite. It is a reflection that the either the die or the substrate is too weak to let the system bend at these two extremes.

Figure 25.17(b) plots the stress on the top of the die predicted by the BMP theory. Although the die stress is related to the die bending, it is not simply related. The behaviors of the two quantities are very similar in certain range of thickness ratio h. This can be clearly seen in Figure 25.17(a) and Figure 25.17(b). Actually, if $h \to 0$, from Equation (25.21), $\bar{\sigma} \to 2\bar{\kappa}$. The behavior of stress is totally different from the curvature if h is large. From Equation (25.17), $\lim_{h \to \infty} \bar{\kappa} = 0$. On the other hand, Equation (25.19) leads to $\lim_{h \to \infty} \bar{\sigma} = -1$.

The physical meaning is that the system does not bend too much as the substrate is relatively thick or in other words the die is relatively thin. In this case, the curvature approaches to zero. The stress comes from two competing factors. One is the bending, and the other is the thermal mismatch strain. If the substrate is much thicker than the die, the bending of the bimaterial plate system is not significant and the die stress mainly comes from the mismatches in thermal deformations. Therefore, the die will be under a compressive stress, which is not critical to die cracking. If we are only interested in the lower thickness ratio range, a simple relation of the curvature and the die stress can be established. From Equation (25.21), one can have a relation $\bar{\sigma} = C\bar{\kappa}$ where $C = \bar{\kappa}/\bar{\sigma} = (1+h)/(2+3h-mh^3)$. Generally, C is not a constant. For Si on PCB with $h=1$, $C(m=0.2, h=1) = 0.41$. If we use this $C = 0.41$ for thickness ratio from 0 to 1.5, the correlation between the curvature and stress is very good.

Although the BMP model neglects the edge effect, it does not introduce large errors in the estimation the curvature and the die stress. Three dimensional FEM calculation results show that the package size effect and aspect ratio effect only make 5% difference [16]. Furthermore, the BMP model provides us more insight in the packaging design when the die size is becoming very large. From Equations (25.16) and (25.18), it can be clearly seen that the curvature and die stress are independent of the package size. The curvature is only the function of modulus ratio (m), thickness ratio (h), thermal load (ε_m), and die thickness (h_1), and the die stress is only the function of modulus ratio (m), thickness ratio (h), thermal load (ε_m), and die biaxial modulus (M_1). This indicates that the increase of die size will not cause intrinsic limiting factor coming from die stress as long as material combination and thickness ratio is properly chosen. In real packaging design, one does experience lower yield when package size increases. The reason is that the number of defects per die increases for larger package due to the larger area of die (assuming defect density is a constant).

Another important observation is that the die stress is independent of absolute die thickness when the thickness ratio is fixed. Equation (25.18) does not contain the factor of absolute die thickness. In real flip-chip packaging design, it was found that thinning and polishing the die can significantly enhance the yield. This can be understood by looking at the Figure 25.17. Thinning the die and keeping the substrate thickness unchanged is equivalent to increasing the thickness ratio (h). This actually offsets the maximum stress location to get a good yield if the die is originally thicker than the substrate. The other advantage is that polishing the die can reduce the defect density, which means higher die strength, and as a consequence, the yield is improved significantly.

25.2.6. Die Stress in Flip Chip Assembly: Summary

An analytical model and governing equations are established for describing the intrinsic behavior of flip-chip packages. The model is validated by experiment and FEM calculation. The curvature and the bending stress are independent of the die size. The bending stress is independent of absolute die thickness if substrate to die thickness ratio (h) is kept the same. Both the curvature and the die stress have maximum values as the thickness changes. For FC-PBGA packages, which correspond to biaxial modulus ratio ($m = 0.2$), the die stress is maximum when the thickness ratio (h) equals to one. The die curvature and the bending stress are approximately proportional in certain range of thickness ratio.

25.3. THERMAL STRESS DUE TO TEMPERATURE CYCLING

Due to thermal expansion mismatch between the microelectronics package and the printed circuit board on which the package is mounted, thermal stress is introduced in the cooling step of the industry standard solder assembly process. Dynamic stress is further produced in the completed package-board assembly, especially in the solder joints, when the assembly is subjected to power cycling, thermal cycling, or thermal shock. The dynamic stress/strain has a great impact on the long-term reliability of the solder joint.

25.3.1. Finite Element Analysis

Non-linear finite element simulation methodologies have been adopted to investigate the time-dependent thermal stress/strain in the solder joints. As an example, the finite element modeling of a FC-PBGA assembly under accelerated temperature cycling conditions is given here. ANSYS 6.0 is used for all aspects of the study: pre-processing, solution, and post-processing.

Due to symmetry, only one-quarter of the module-board assembly is modeled. Figure 25.18 shows one quarter of the module-board assembly for the FC-PBGA. The solder material is modeled as a viscoplastic solid. The PBGA substrate along with the printed circuit board are modeled as composite materials with their copper power and ground planes, and their orthotropic, temperature dependent dielectric materials explicitly modeled. The physical and mechanical properties of the silicon chip, underfill material and board are summarized in Table 25.1.

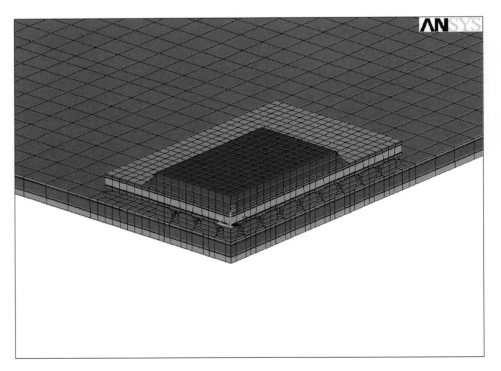

FIGURE 25.18. Finite element model of a FC-PBGA assembled on a board.

TABLE 25.1.
Material properties used in modeling of an FC-PBGA assembly.

Material	Modulus @ 23°C (GPa)	Poisson ratio @ 23°C	CTE (10^{-6}/°C) @ 23°C	T_g (°C)
Si Chip	186	0.28	3.2	NA
Underfill	3.4	0.33	24	145
PBGA substrate (Power Plane)	45.6	0.34	17.6	NA
PBGA substrate (Dielectric)	16.4 (x, y) 2.6 (z)	0.48 (x, y) 0.16 (z)	14.8 (x, y) 67 (z)	170
SnPb solder	Viscoplastic, see Ref. [21]	0.29	22	NA
Printed circuit board (Power Plane)	45.6	0.34	17.6	NA
Printed circuit board (Dielectric)	88.6 (x, y) 2.7 (z)	0.48 (x, y) 0.16 (z)	22.4 (x, y) 67 (z)	170

25.3.2. Constitutive Equation for Solder

Accurate constitutive modeling of the solder plays an important role in the simulation of solder joint reliability. ANSYS does have viscoplastic elements as a standard option, but they use Anand's constitutive model [17,18]. The use of these elements is convenient since the user does not have to modify the source code. Anand's model was developed for hot working metals, which unifying plasticity and creep via a set of flow and evolutionary equations:

Flow Equation

$$\frac{d\varepsilon_p}{dt} = A[\sinh(\xi\sigma/s)]^{1/m} \exp\left(-\frac{Q}{kT}\right). \tag{25.22}$$

Evolution Equations

$$\frac{ds}{dt} = \left[h_0(|B|)^a \frac{B}{|B|}\right]\frac{d\varepsilon_p}{dt}, \tag{25.23}$$

$$B = 1 - \frac{s}{s^*}, \tag{25.24}$$

$$s^* = \hat{s}\left[\frac{\frac{d\varepsilon_p}{dt}}{A}\exp\left(-\frac{Q}{kT}\right)\right]^n. \tag{25.25}$$

Darveaux et al. [19] were the first to modify the constants in Anand's constitutive relation to account for both time-dependent and time-independent phenomenon. Parameters for near-eutectic 62Sn/36Pb/2Ag are given in Table 25.2 (see Refs. [19,21]).

TABLE 25.2.
Anand's constants for 62Sn36Pb2Ag solder in ANSYS.

ANSYS	Parameter	Value	Definition
C1	S_o (psi)	1800	Initial value of deformation resistance
C2	Q/k (1/K)	9400	Activation energy/ Boltzmann's constant
C3	A (1/sec)	4.0×10^6	Pre-exponential factor
C4	ξ	1.5	Multiplier of stress
C5	m	0.303	Stain rate sensitivity of stress
C6	h_o (psi)	2.0×10^5	Hardening constant
C7	$s\char`\^$ (psi)	2.0×10^3	coefficient for deformation resistance saturation value
C8	n	0.07	Strain rate sensitivity of saturation (deformation resistance) value
C9	a	1.3	Strain rate sensitivity of hardening

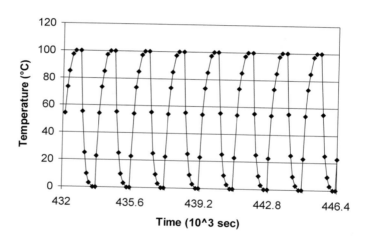

FIGURE 25.19. Cyclic temperature loading used in finite element modeling.

25.3.3. *Time-Dependent Thermal Stresses of Solder Joint*

The cyclic temperature loading imposed on the FC-PBGA/PCB assembly is closely matched to the temperature profile measured in accelerated temperature cycling test and is shown in Figure 25.19. It can be seen that for each cycle (30 minutes) the temperature is between 0 and +100°C, with 10 minutes ramp, 5 minutes hold at the two temperature extremes. In addition, pre-cycling, temperature-time history which corresponds to 5 days at 23°C is also included in the simulation.

Residual stresses in the solder joint after the FC-PBGA module was assembled to the board, cooled down to room temperature but before the temperature cycling test are shown in Figure 25.20. As expected the residual stresses relaxed very rapidly in the first hour after the assembly process. The stresses shown in Figure 25.20 are for the solder ball which is

STRESS ANALYSIS FOR PROCESSED SILICON WAFERS AND PACKAGED MICRO-DEVICES

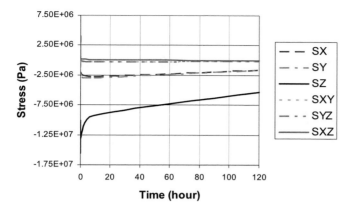

FIGURE 25.20. Residual stresses in the solder joint after assembly process.

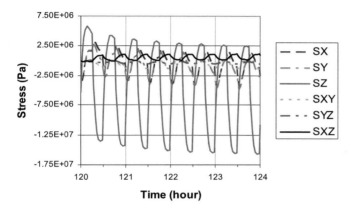

FIGURE 25.21. Time-dependent stresses in the solder joint due to cyclic temperatures.

under the corner of the chip as indicated in Figure 25.24. The dynamic stresses and strains in the same solder ball due to cyclic temperature variation are shown in Figure 25.21 and Figure 25.22, respectively. It can be seen from Figure 25.21 and Figure 25.22 the stresses and strains oscillate within certain ranges in accordance with the cyclic temperature load. Figure 25.23 shows the normal stress (σ_z) and the normal viscoplastic strain hysteresis loops for multiple cycles. For viscoplastic analysis, it is important to study the stress-strain responses for multiple cycles until the hysteresis loops become stabilized. Figure 25.23 indicates that the plastic strain is quite stabilized after the third cycle and the creep response converged after the fourth cycle.

25.3.4. Solder Joint Reliability Estimation

In addition to time-dependent thermal stress modeling, another goal of the simulation is to calculate the plastic work per unit volume (or viscoplastic strain energy density) accumulated per thermal cycle. Over the years, an energy based metric for predicting crack initiation and growth in solder joints was developed and refined [19–21]. The plastic work accumulated during the last cycle is used for crack growth and solder joint fatigue life cor-

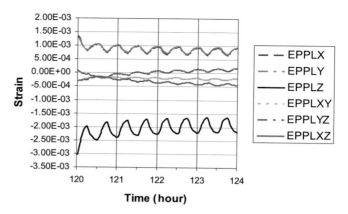

FIGURE 25.22. Time-dependent strains in the solder joint due to cyclic temperatures.

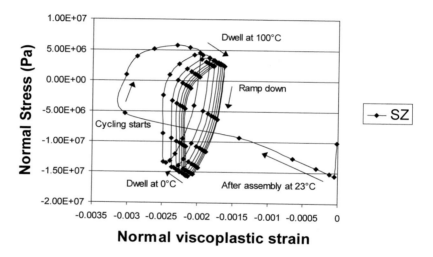

FIGURE 25.23. Hysteresis loops of stress and viscoplastic strain.

relations. The characteristic fatigue life (N_a) for the solder ball with a pad diameter of a can be written as

$$N_a = A(\Delta W_{ave})^B + C(\Delta W_{ave})^D, \tag{25.26}$$

where A, B, C, D are constants that depend on material and solder joint design and assembly process and ΔW_{ave} is the volume averaged viscoplastic strain energy density increment per cycle.

In practice, several complete thermal cycles are simulated in order to establish a stable stress–strain hysteresis loop. The A, B, C, D constants are determined by correlating the simulated results to the temperature cycling test results. Figure 25.24 shows the viscoplastic strain energy density increment calculated from the eighth to the seventh cycle. It is noticed that the solder ball with the highest accumulation in viscoplastic strain energy density is located just under the corner of the die edge. Experimental results show that the

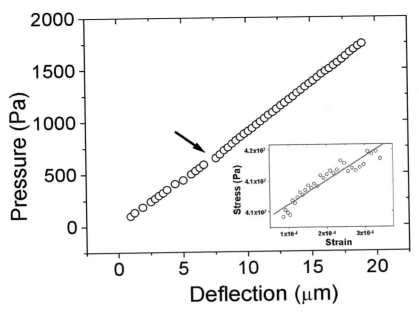

FIGURE 25.26. The total deflection of a 0.46-μm thick dense PAE membrane versus the applied differential pressure. Lower-right inset: the stress versus strain at the center of the membrane.

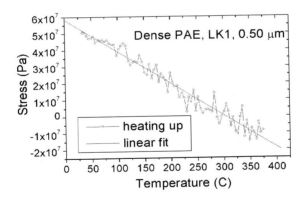

FIGURE 25.27. The stress versus temperature relationship of a blanket PAE film on a 4-inch (100) Si wafer.

the substrate Si was removed in the bulge testing. The slope of the stress-temperature curve is linked to the biaxial modulus and CTE of the film through the following equation:

$$\frac{d\sigma}{dT} = \left(\frac{E}{1-\gamma}\right)_f \cdot (\alpha_s - \alpha_f), \qquad (25.30)$$

where $(E/1-\gamma)_f$, α_s, and α_f are the biaxial modulus of the film, the substrate CTE and the film CTE, respectively.

Figure 25.27 shows a typical stress versus temperature curve, where the temperature profile was ramping up from room temperature (24°C) to 375°C in 1.75 hr, held at 375°C for 0.25 hr, and ramped down to room temperature in 2 hr. The slope of the PAE film ranges from 1.526×10^5 Pa/°C to 1.923×10^5 Pa/°C, and the corresponding CTE calculated range from 26.8–32.6 ppm/°C.

In summary, we have extended the bulge testing method to polymeric thin film of sub-micron thickness. The biaxial modulus and thermal expansion coefficient of the PAE film was obtained.

REFERENCES

1. F.M. D'Heuvle, Metallurgical topics in silicon device interconnections: Thin film stresses, International Materials Reviews, 34(2), p. 53 (1989).
2. J.E. Ashton and J.M. Whitney, Theory of Laminated Plates, Vol. IV, Progress in Materials Science Series, Stamford, CN: Technomic, 153, 1970.
3. E. Suhir, Predicted bow of plastic packages of integrated circuit (IC) device, Thermal Stress and Strain in Microelectronics Packaging, Van Nostrand Reinhold, New York, 1993.
4. R. Jones, Mechanics of Composite Materials, Hemisphere Publishing Corporation, 1975.
5. R.J. Jaccodine and W.A. Schlegel, Measurement of strains at Si-SiO$_2$ interface, J. Appl. Phys., 37(6), p. 2429 (1966).
6. P.A. Flinn, D.S. Gardner, and W.D. Nix, Measurement and interpretation of stress in aluminum-based metallization as a function of thermal history, IEEE Trans. ED, 34(3), p. 689 (1987).
7. R. Glang, R.A. Holmwood, and R.L. Rosenfeld, Determination of stress in films on single crystalline silicon substrates, Rev. Sci. Instru., 36(1), p. 7 (1965).
8. A. Segmüller, J. Angilelo, and S.J. LaPlaca, Automatic x-ray diffraction measurement of the lattice curvature of substrate wafers for the determination of linear strain patterns, J. Appl. Phys., 51(12), p. 6224 (1980).
9. S.G. Malhotra, Z.U. Rek, S.M. Yalisove, and J.C. Bilello, Depth dependence of residual strains in polycrystalline Mo thin films using high-resolution x-ray diffraction, J. Appl. Phys., 79(9), p. 6872 (1996).
10. K. Ajito, J.P.H. Sukamto, L.A. Nagahara, K.Hashimoto, and A. Fujishima, Strain imaging analysis of Si using Raman microscopy, J. Va. Sci. Technol. A, 13(3), p. 1234 (1995).
11. J. Groenen, A. Mlayah, and R. Carles, Strain in InAs islands grown on InP(001) analyzed by Raman spectroscopy, Applied Physics Letters, 69(7), p. 943 (1996).
12. Y. Guo, D. Mitchell, V. Sarihan, and T. Lee, A testing method and device for intrinsic stress measurement in wafer bumping process, IEEE Transactions On Components And packaging Technologies, 23(2) (2000).
13. G. G. Stoney, Proc. R. Soc. London, Ser. A, 82, p. 172 (1909).
14. R. W. Hoffman, The mechanical properties of thin condensed films, Phys. Thin Film, vol. 3, 1966, pp. 211–273.
15. D. Post, B. Han, and P. Ifju, High Sensitivity Moire: Experimental Analysis for Mechanics and Materials, Chap. 2, Springer-Verlag, New York, 1994.
16. L. Mercado, Effect of die size and gap height on die stress in flip-chip PBGA packages, Motorola Internal Report, 1999.
17. L. Anand, Constitutive equations for hot-working of metals, International Journal of Plasticity, 1, pp. 213–231 (1985).
18. S.B. Brown, K.H. Kim, and L. Anand, An internal variable constitutive model for hot working of metals, International Journal of Plasticity, 5, pp. 95–130 (1989).
19. R. Darveaux, K. Banerji, A. Mawer, and G. Dody, Reliability of plastic ball grid array assembly, in J. Lau, Ed., Ball Grid Array Technology, McGraw-Hill, NY, 1995.
20. R. Darveaux, Solder joint fatigue life model, Proc. TMS Annual Meeting, Orlando, FL, 1997, pp. 213–218.
21. R. Darveaux, Effect of simulation methodology on solder joint crack growth correlation, ECTC, Las Vegas, NV, 2000, pp. 1048–1063.
22. T. Tsakalakos, The bulge test: a comparison of theory and experiment for isotropic and anisotropic films, Thin Solid Films, 75, pp. 293–305 (1981).
23. E.I. Bromley, J.N. Randall, D.C. Flanders, and R.W. Mountain, A technique for the determination of stress in thin films, Journal of Vacuum Science and Technology, B1(4), pp. 1364–1366 (1983).
24. D. Maier-Schneider, A. Ersoy, J. Maibach, D. Schneider, and E. Obermeier, Influence of annealing on elastic properties of LPCVD silicon nitride and LPCVD polysilicon, Sensors and Materials, 7(2), pp. 121–129 (1995).

25. J.J. Vlassak and W.D. Nix, A new bulge test technique for the determination of Young's modulus and Poisson's ratio of thin films, Journal of Material Research, 7(12), pp. 3242–3249 (1992).
26. J.Y. Pan, P. Lin, F. Maseeh, and S.D. Senturia, Verification of FEM analysis of load-deflection methods for measuring mechanical properties of thin films, in Technical Digest IEEE Solid-state Sensors and Actuators Workshop, IEEE, New York, 1990, pp. 70–73.
27. G.F. Cardinale and R.W. Tustison, Fracture strength and biaxial modulus measurement of plasma silicon nitride films, Thin Solid Films, 207, pp. 126–130 (1992).
28. M.G. Allen, M. Mehregany, R.T. Howe, and S.D. Senturia, Microfabricated structures for the in-situ measurement of residual stress, Young's modulus, and ultimate strain of thin films, Applied Physics Letters, 51, pp. 241–243 (1987).
29. O. Tabata, K. Kawahata, S. Sugiyama, and I. Igarashi, Mechanical property measurement of thin films using load-deflection of composite rectangular membranes, Sensors and Actuators, 20, pp. 135–141 (1989).
30. D. Maier-Schneider, J. Maibach, and E. Obermeier, Computer-aided characterization of the elastic properties of thin films, Journal of Micromechanical and Microengineering, 2, pp. 173–175 (1992).
31. Y.H. Xu, Y.-P. Tsai, K.N. Tu, B. Zhao, Q.-Z. Liu, M. Brongo, G.T.T. Sheng, and C.H. Tung, Dielectric property and microstructure of a porous polymer material with ultra low dielectric constant, Applied Physics Letters, 75, pp. 853–855 (1999).
32. D.W. Zheng, Ph.D. dissertation, University of California at Los Angeles, USA, 1999.
33. R.E. Acosta, J.R. Maldonado, L.K. Towart, and J.M. Warlaumont, B-Si masks for storage ring X-ray lithography, Solid State Technology, 27, pp. 205–208 (1984).
34. D. Zheng, R.P. Hwang, X.Y. Dou, C.P. Yeh, M. Prakash, K. Boardman, and G. Ridsdale, Warpage analysis of 144-pin TQFP during reflow using image processing, Proceedings of 47th Electronic Components and Technology Conference, IEEE, New York, NY, 1997, pp. 1176–1181.

Index

Numbers in bold indicate the volume number.

abrasion 602 (**1**)
absorption coefficient 86 (**1**)
absorption rates 65 (**1**)
accelerated life test (ALT) 203 (**2**), 210 (**2**), 212 (**2**), 214 (**2**), 217 (**2**), 218 (**2**), 230 (**2**), 255 (**2**), 256 (**2**)
– model 223 (**2**)
accelerated stress testing (AST) 454 (**1**), 190 (**2**), 194 (**2**), 200 (**2**)
accelerated test conditions 215 (**2**)
accelerated testing 316 (**1**), 348 (**1**), 385 (**1**), 514 (**1**), 4 (**2**), 102 (**2**), 213 (**2**), 215 (**2**), 222 (**2**), 683 (**2**)
accelerated tests 454 (**1**), 301 (**2**)
accelerated thermal cycling (ATC) 485 (**1**), 257 (**2**), 291 (**2**)
acceleration factor 216 (**2**), 218 (**2**), 225 (**2**), 500 (**2**)
acceleration models 388 (**1**)
activation energy 44 (**1**), 215 (**1**), 367 (**1**), 371 (**1**), 576 (**1**), 127 (**2**), 130 (**2**)
adherend failure 6 (**2**)
adhesion 40 (**1**), 137 (**1**), 141 (**1**)–143 (**1**), 146 (**1**), 156 (**1**)–157 (**1**), 165 (**1**)–166 (**1**), 169 (**1**), 170 (**1**), 175 (**1**)–177 (**1**), 263 (**1**), 269 (**1**), 273 (**1**), 538 (**1**), 392 (**2**), 404 (**2**), 405 (**2**), 425 (**2**), 426 (**2**), 431 (**2**)–433 (**2**), 441 (**2**)–445 (**2**), 452 (**2**), 506 (**2**)–508 (**2**), 547 (**2**), 620 (**2**)
– failures 166 (**2**), 318 (**2**)
– promoter 392 (**2**), 506 (**2**)
– strength 404 (**2**), 409 (**2**), 414 (**2**), 419 (**2**), 426 (**2**), 432 (**2**), 619 (**2**)
adhesive 172 (**1**), 302 (**1**), 310 (**1**), 463 (**1**), 479 (**1**), 480 (**1**), 506 (**1**)–508 (**1**), 519 (**1**), 520 (**1**), 523 (**1**), 6 (**2**)–12 (**2**), 116 (**2**), 383 (**2**), 386 (**2**), 387 (**2**), 390 (**2**), 398 (**2**), 399 (**2**), 432 (**2**), 442 (**2**), 449 (**2**), 466 (**2**), 488 (**2**), 496 (**2**), 497 (**2**), 501 (**2**), 503 (**2**), 508 (**2**)–510 (**2**), 513 (**2**), 517 (**2**), 522 (**2**), 529 (**2**), 554 (**2**), 565 (**2**), 571 (**2**), 572 (**2**), 574 (**2**), 597 (**2**), 601 (**2**), 612 (**2**), 613 (**2**), 617 (**2**), 619 (**2**)
– bonding 302 (**1**), 506 (**1**), 7 (**2**), 116 (**2**), 397 (**2**), 460 (**2**), 469 (**2**), 488 (**2**), 503 (**2**), 514 (**2**), 515 (**2**), 572 (**2**), 573 (**2**), 602 (**2**), 606 (**2**)
– failure 310 (**1**), 6 (**2**)
– strength 172 (**1**), 8 (**2**), 260 (**2**), 405 (**2**), 514 (**2**), 532 (**2**), 546 (**2**)
adsorption theory 463 (**2**), 464 (**2**), 508 (**2**)
advanced configuration power interface (ACPI) 293 (**1**)
Al–Ni bond interfaces 124 (**2**)
Al–Ni interface 124 (**2**)
ALD-SAM coating 318 (**2**)
AlGaAs 139 (**1**)
Alloy 42 substrate 373 (**1**)
analysis of variance (ANOVA) 230 (**1**), 232 (**1**), 251 (**1**), 319 (**1**)
analytical, computational, and experimental solutions (ACES) 324 (**2**)
– methodology 324 (**2**)
analytical modeling 270 (**1**), 557 (**1**), 6 (**2**), 16 (**2**), 223 (**2**), 261 (**2**), 330 (**2**), 334 (**2**), 697 (**2**)
analytical stress model 272 (**1**), 13 (**2**), 16 (**2**), 474 (**2**)
analytical thermal model 78 (**1**)
Anand model 430 (**1**), 452 (**1**)
angular acceleration 328 (**2**)

anisotropic conductive adhesive (ACA) 527 (**2**), 529 (**2**), 533 (**2**), 536 (**2**), 572 (**2**), 619 (**2**)
anisotropic conductive film (ACF) 527 (**2**), 532 (**2**), 612 (**2**), 619 (**2**), 625 (**2**)
anisotropic conductive paste (ACP) 527 (**2**), 532 (**2**)
anisotropic elastic constants 156 (**1**)
anisotropic thermal conductivity 186 (**1**)
anisotropic thermal expansion 186 (**1**)
any layer inner via hole (ALIVH) 271 (**2**)
apparent Bragg wavelength shift 98 (**1**)
apparent elastic modulus 435 (**1**), 437 (**1**), 439 (**1**)
approximate fully cure dependent model 34 (**1**)
approximate method 11 (**1**)
approximation method 238 (**1**)
arc-discharge 186 (**1**)
area array packages (AAPs) 284 (**1**)
area array technology 283 (**1**), 398 (**1**)
Arrhenius equation 61 (**1**), 365 (**1**), 370 (**1**), 130 (**2**), 186 (**2**)
atomic force microscopy (AFM) 148 (**1**), 192 (**1**), 604 (**1**), 515 (**2**)
atomic layer deposition (ALD) 316 (**2**)
Au–Al interfaces 122 (**2**)
Au–Al wire bonds 122 (**2**)
Au–Au interfaces 123 (**2**)
axial elastic compression 560 (**1**)
axial free vibrations 674 (**1**)
axial power variation 75 (**1**)
axial spring constant 560 (**1**)
axial strength testing 616 (**1**)
axisymmetrical bending 687 (**2**)

ball bond shear test 89 (**2**), 102 (**2**), 117 (**2**)
ball grid array (BGA) 6 (**1**), 284 (**1**), 291 (**1**), 313 (**1**), 318 (**1**), 411 (**1**), 481 (**1**), 557 (**1**), 8 (**2**), 109 (**2**), 138 (**2**), 257 (**2**), 270 (**2**), 281 (**2**), 370 (**2**), 383 (**2**), 389 (**2**), 573 (**2**), 611 (**2**)
– reliability 401 (**1**)
– solder balls 398 (**1**)
– solder joint 400 (**1**)
ballistic electron transport 182 (**1**)
bare fiber 269 (**1**), 271 (**1**)
$BaTiO_3$ 112 (**1**)
beam bending 687 (**2**)
beam propagation method (BPM) 25 (**2**)

belt slide apparatus 605 (**1**)
benzocyclobutene (BCB) 142 (**1**), 79 (**2**), 108 (**2**), 137 (**2**), 313 (**2**)
BeO 484 (**1**)
BGO 112 (**1**)
bi-material assemblies 6 (**2**)
bi-material model 651 (**2**)
bias on the substrate 197 (**1**)
biaxial elastic modulus 147 (**1**)
bimaterial plate (BMP) 690 (**2**), 697 (**2**)
bimodality 577 (**1**)
binary phase diagram 354 (**1**)
BiTe 484 (**1**)
bithermal loading 481 (**1**)
board level model 322 (**1**)
body of knowledge (BOK) 283 (**1**)
Boltzmann-Arrhenius equation 225 (**2**)
bond failure 87 (**2**)
bond pad metallurgy 100 (**2**)
bond strength 103 (**2**), 108 (**2**), 396 (**2**), 503 (**2**), 508 (**2**)
bondability 79 (**2**), 101 (**2**), 110 (**2**), 185 (**2**)
bonding layer 14 (**2**), 475 (**2**), 604 (**2**)
bonding machine 77 (**2**), 80 (**2**), 90 (**2**), 110 (**2**)
bonding materials 6 (**2**), 15 (**2**), 184 (**2**), 218 (**2**), 384 (**2**), 473 (**2**)
bonding matrix 479 (**2**)
bonding mechanism 419 (**2**)
bonding pad 104 (**2**), 306 (**2**), 606 (**2**)
bonding strength 352 (**1**), 404 (**2**), 419 (**2**)
bonding wire 80 (**2**)
bondline 385 (**2**), 388 (**2**), 393 (**2**), 394 (**2**), 398 (**2**), 511 (**2**), 514 (**2**)
borosilicate glass 142 (**1**)
boundary, initial, and loading (BIL) 324 (**2**)
– conditions 324 (**2**)
boundary element method (BEM) 551 (**1**), 324 (**2**)
Bragg grating (BG) 65 (**1**), 4 (**2**), 16 (**2**)
Bragg wavelength 65 (**1**)
breaking strength 599 (**1**), 98 (**2**)
brittle fracture 418 (**1**), 454 (**1**), 204 (**2**)
brittle intermetallic compounds 313 (**1**), 122 (**2**), 225 (**2**)
brittle intermetallic layers 318 (**1**)
brittle materials 221 (**1**)
Bueche-Zhurkov equation 227 (**2**), 229 (**2**)
buffered oxide etchant (BOE) 704 (**2**)
bulk composite 185 (**1**)
bulk index grating 67 (**1**)
bulk relaxation modulus 21 (**1**)

INDEX

buoyancy   361 (**1**), 362 (**1**), 152 (**2**), 172 (**2**)
– force   361 (**1**)
– time   362 (**1**)
Burgers model   6 (**1**)
burn-in   310 (**1**), 145 (**2**), 192 (**2**), 205 (**2**), 220 (**2**), 301 (**2**), 625 (**2**)

camera stage   463 (**1**)
cantilever   192 (**1**), 305 (**2**), 307 (**2**), 318 (**2**), 483 (**2**), 484 (**2**)
carbon nanotube (CNT)   181 (**1**)
catalyst deposition   189 (**1**)
CCGA assembly   306 (**1**)
CdTe   112 (**1**)
central composite design (CCD)   233 (**1**)
ceramic ball grid array (CBGA)   283 (**1**), 294 (**1**)–297 (**1**), 299 (**1**), 303 (**1**), 305 (**1**), 309 (**1**), 386 (**1**), 489 (**1**)
ceramic column grid array (CCGA)   283 (**1**), 286 (**1**)
ceramic packages   293 (**1**)
channel dispersion   127 (**1**)
chaotic response   646 (**1**)
characteristic life times   342 (**1**)
chemical mechanical planarization (CMP)   141 (**1**), 155 (**1**), 192 (**1**)
chemical mechanical polishing (CMP)   193 (**1**), 240 (**2**)
chemical potential   324 (**1**)
chemical reaction   6 (**1**), 144 (**1**), 164 (**1**), 330 (**1**)
chemical vapor deposition (CVD)   144 (**1**), 186 (**1**), 187 (**1**)
chip carrier   310 (**1**), 483 (**1**), 501 (**1**), 263 (**2**), 563 (**2**), 623 (**2**)
chip density   71 (**2**)
chip on flex (COF)   612 (**2**)
chip on glass (COG)   612 (**2**)
chip scale package (CSP)   284 (**1**), 286 (**1**), 313 (**1**), 335 (**1**), 411 (**1**), 557 (**1**), 109 (**2**), 135 (**2**), 149 (**2**), 257 (**2**), 268 (**2**), 272 (**2**), 276 (**2**), 281 (**2**), 406 (**2**), 413 (**2**), 528 (**2**), 611 (**2**)
chip-on-board (COB)   94 (**2**), 108 (**2**)
chips per wafer (CPW)   136 (**2**)
chromatic mirror   115 (**1**)
cladding radius   120 (**1**)
cladding refractive index   69 (**1**), 84 (**1**), 120 (**1**)
Class I   206 (**2**)
Class II   207 (**2**)
Class III   207 (**2**)

classic wetting theory   358 (**1**)
coated fibers   272 (**1**)
coefficient of hygro expansion (CHE)   531 (**1**)
coefficient of hygroscopic swelling (CHS)   494 (**1**), 496 (**1**)
– mismatch   499 (**1**)
coefficient of thermal expansion (CTE)   5 (**1**), 186 (**1**), 199 (**1**), 212 (**1**), 215 (**1**), 253 (**1**), 271 (**1**), 294 (**1**), 380 (**1**), 430 (**1**), 459 (**1**), 483 (**1**), 524 (**1**), 530 (**1**), 532 (**1**), 635 (**1**), 3 (**2**), 11 (**2**), 14 (**2**), 32 (**2**), 38 (**2**), 39 (**2**), 75 (**2**), 87 (**2**), 97 (**2**), 122 (**2**), 129 (**2**), 136 (**2**), 160 (**2**), 234 (**2**), 241 (**2**), 273 (**2**), 305 (**2**), 350 (**2**), 373 (**2**), 456 (**2**), 478 (**2**), 481 (**2**), 488 (**2**), 619 (**2**), 653 (**2**), 670 (**2**), 680 (**2**), 685 (**2**), 686 (**2**), 706 (**2**)
– mismatch   219 (**1**), 289 (**1**), 294 (**1**), 395 (**1**), 405 (**1**), 499 (**1**), 9 (**2**), 127 (**2**), 130 (**2**), 139 (**2**), 149 (**2**), 185 (**2**), 273 (**2**), 275 (**2**), 307 (**2**), 373 (**2**), 496 (**2**), 497 (**2**), 502 (**2**), 522 (**2**), 689 (**2**), 691 (**2**)
Coffin-Manson equation   225 (**2**), 559 (**2**)
Coffin-Manson fatigue model   432 (**1**)
Coffin-Manson model   224 (**1**), 446 (**1**)
Coffin-Manson relationship   224 (**1**), 293 (**1**), 300 (**1**)
coherent beams   476 (**1**)
cohesive failure   416 (**1**), 6 (**2**), 386 (**2**), 506 (**2**)
collected volatile condensable materials (CVCM)   511 (**2**)
commercial-off-the-shelf (COTS)   283 (**1**)
compact tension (CT)   244 (**2**)
compatibility   129 (**1**), 377 (**1**), 131 (**2**), 302 (**2**)
complete failure   243 (**2**)
complete optical failure   4 (**2**)
compliance   10 (**1**), 15 (**1**), 16 (**1**), 22 (**1**), 24 (**1**)–26 (**1**), 42 (**1**), 147 (**1**), 162 (**1**), 294 (**1**), 408 (**1**), 688 (**1**), 5 (**2**), 59 (**2**), 77 (**2**), 136 (**2**), 149 (**2**), 293 (**2**), 473 (**2**), 500 (**2**)
– functions   10 (**1**)
compliant adhesive   473 (**2**), 522 (**2**)
component forward/backward compatibility   386 (**1**)
component level model   322 (**1**)
component thermal stability   386 (**1**)
compound   5 (**1**), 26 (**1**), 47 (**1**), 144 (**1**), 293 (**1**)
compressive strength   392 (**2**)
compressive stress   267 (**2**), 355 (**2**), 357 (**2**), 697 (**2**)
computer-aided design (CAD)   136 (**1**)

conducting failure analysis   177 (**2**), 180 (**2**), 269 (**2**)
conductive anodic filament (CAF)   264 (**2**), 267 (**2**)
consistent composite plate model   685 (**2**)
constant load effect   646 (**1**)
constitutive models   4 (**1**), 212 (**1**), 262 (**1**), 435 (**1**)
contact force   147 (**2**), 219 (**2**), 576 (**2**), 581 (**2**)
contact resistance   181 (**1**), 197 (**1**), 76 (**2**), 139 (**2**), 542 (**2**), 572 (**2**), 575 (**2**), 587 (**2**), 595 (**2**), 599 (**2**), 600 (**2**), 613 (**2**), 615 (**2**), 623 (**2**)
contemporary systems   463 (**1**)
continuous stiffness method (CSM)   163 (**1**)
conventional passive alignment methods   158 (**2**)
converse piezoelectric effect   138 (**1**)
convolution integral   11 (**1**)
cooling chamber   469 (**1**)
coplanar waveguide (CPW)   313 (**2**)
copper heat sink   349 (**2**), 350 (**2**), 354 (**2**)
copper pad   491 (**1**), 97 (**2**), 374 (**2**), 596 (**2**), 601 (**2**)
copper wire   97 (**2**)
core radius   84 (**1**), 120 (**1**)
core refractive index   84 (**1**), 120 (**1**)
corner adhesive bonds   302 (**1**)
corrosion   6 (**1**), 144 (**1**), 155 (**1**), 262 (**1**), 269 (**1**), 273 (**1**), 274 (**1**), 388 (**1**), 389 (**1**), 395 (**1**), 100 (**2**), 119 (**2**), 130 (**2**), 182 (**2**), 184 (**2**), 219 (**2**), 255 (**2**), 256 (**2**), 265 (**2**), 286 (**2**), 287 (**2**), 396 (**2**), 404 (**2**), 433 (**2**), 476 (**2**), 477 (**2**), 542 (**2**), 544 (**2**), 557 (**2**), 589 (**2**), 616 (**2**), 620 (**2**)
corrosion inhibitors   557 (**2**), 617 (**2**)
cost effectiveness   136 (**1**), 252 (**1**), 454 (**1**), 203 (**2**), 400 (**2**), 621 (**2**)
coupled axial-radial axisymmetric vibrations   677 (**1**)
coupling coefficient   69 (**1**), 83 (**1**), 122 (**1**)
coupling efficiency   23 (**2**), 28 (**2**), 30 (**2**), 42 (**2**), 48 (**2**), 50 (**2**), 64 (**2**), 152 (**2**), 168 (**2**), 204 (**2**), 373 (**2**), 379 (**2**), 383 (**2**), 489 (**2**), 492 (**2**), 499 (**2**), 510 (**2**), 518 (**2**), 520 (**2**), 523 (**2**)
coupling loss   121 (**1**), 122 (**1**), 27 (**2**), 380 (**2**), 492 (**2**)
crack   165 (**1**), 169 (**1**), 170 (**1**), 220 (**1**), 222 (**1**)–225 (**1**), 418 (**1**)–422 (**1**), 431 (**1**), 527 (**1**)–529 (**1**), 531 (**1**)–537 (**1**), 544 (**1**)–550 (**1**), 594 (**1**)–598 (**1**), 600 (**1**), 601 (**1**), 608 (**1**)–610 (**1**), 612 (**1**)–616 (**1**), 620 (**1**)–625 (**1**), 348 (**2**), 387 (**2**), 413 (**2**), 443 (**2**), 451 (**2**), 457 (**2**), 466 (**2**), 477 (**2**), 494 (**2**), 514 (**2**), 691 (**2**), 701 (**2**)
– detection   243 (**2**), 246 (**2**)
– geometry   610 (**1**)
– opening displacement (COD)   248 (**2**)
– propagation   165 (**1**), 169 (**1**), 223 (**1**), 419 (**1**), 529 (**1**)269 (**2**)
– surface displacement extrapolation method (CSDEM)   531 (**1**), 550 (**1**)
crack growth   222 (**1**), 224 (**1**), 405 (**1**), 433 (**1**), 596 (**1**), 600 (**1**), 614 (**1**), 623 (**1**)
– model   596 (**1**)
– rate   596 (**1**), 623 (**1**)
creep   4 (**1**), 16 (**1**), 22 (**1**), 24 (**1**)–26 (**1**), 42 (**1**), 156 (**1**), 213 (**1**), 252 (**1**), 272 (**1**), 300 (**1**), 301 (**1**), 389 (**1**), 390 (**1**), 393 (**1**)–395 (**1**), 405 (**1**), 454 (**1**), 4 (**2**), 51 (**2**), 219 (**2**), 274 (**2**), 701 (**2**)
– deformation   215 (**1**), 393 (**1**), 395 (**1**), 274 (**2**)
– description   10 (**1**)
– model   212 (**1**)–215 (**1**), 389 (**1**), 440 (**1**), 57 (**2**)
– rate   10 (**1**), 214 (**1**), 301 (**1**), 316 (**1**), 334 (**1**), 392 (**1**), 430 (**1**), 442 (**1**), 491 (**1**), 228 (**2**)
– strain   10 (**1**), 14 (**1**), 42 (**1**), 212 (**1**)–215 (**1**), 334 (**1**), 389 (**1**), 402 (**1**), 430 (**1**), 443 (**1**), 491 (**1**), 138 (**2**), 497 (**2**), 499 (**2**), 511 (**2**)
– strength   390 (**1**)
– test   440 (**1**), 53 (**2**)
– – program   434 (**1**)
creep-fatigue interaction model   432 (**1**)
critical stress   421 (**1**)
cross correlation algorithm   233 (**2**)–235 (**2**)
cross correlation coefficient   236 (**2**)
crystal elastic moduli   147 (**1**)
CT specimen   246 (**2**), 249 (**2**)
CU layers   464 (**1**)
cumulative distribution function (CDF)   388 (**1**), 583 (**1**), 47 (**2**), 180 (**2**), 182 (**2**), 184 (**2**)
cumulative structure function   632 (**2**), 639 (**2**), 641 (**2**)
cure dependent modeling   53 (**1**)
cure-temperature modeling   4 (**1**)
curing   4 (**1**), 5 (**1**), 124 (**1**), 281 (**2**), 384 (**2**), 398 (**2**), 597 (**2**), 606 (**2**)
– polymers   34 (**1**)
– profile   6 (**1**)

INDEX

– temperature 6 (**1**), 530 (**1**), 7 (**2**), 14 (**2**), 159 (**2**), 387 (**2**), 536 (**2**), 689 (**2**), 691 (**2**)
current density 193 (**1**), 389 (**1**), 220 (**2**)
curvature 146 (**1**), 157 (**1**), 164 (**1**), 182 (**1**), 233 (**1**), 238 (**1**), 240 (**1**), 385 (**1**), 435 (**1**), 483 (**1**), 59 (**2**), 411 (**2**), 412 (**2**), 685 (**2**), 688 (**2**), 690 (**2**), 697 (**2**)
– method 147 (**1**)
cycles-to-failure (CTF) 290 (**1**)–292 (**1**), 296 (**1**), 125 (**2**)
cyclic strain 405 (**1**), 8 (**2**)
cylindrical bending 687 (**2**)

D-optimal design 234 (**1**)
damage accumulation 572 (**1**), 580 (**1**), 593 (**1**)
damage mechanisms 274 (**2**)
dashpots 6 (**1**)
data acquisition 465 (**1**), 605 (**1**)
daughter distribution dynamics 583 (**1**)
debonding 179 (**1**), 263 (**1**), 403 (**2**), 432 (**2**), 461 (**2**), 622 (**2**)
deformed and resonant cantilever 165 (**1**)
degradation rate 223 (**2**)
delamination 6 (**1**), 140 (**1**), 146 (**1**), 164 (**1**)–170 (**1**), 179 (**1**), 263 (**1**), 272 (**1**), 381 (**1**), 525 (**1**), 540 (**1**)–544 (**1**), 547 (**1**), 12 (**2**), 130 (**2**), 185 (**2**), 219 (**2**), 242 (**2**), 260 (**2**), 283 (**2**), 350 (**2**), 353 (**2**), 394 (**2**), 399 (**2**), 403 (**2**), 404 (**2**), 406 (**2**), 414 (**2**), 419 (**2**), 423 (**2**), 427 (**2**), 432 (**2**), 447 (**2**), 507 (**2**), 619 (**2**), 630 (**2**), 633 (**2**), 643 (**2**), 646 (**2**), 685 (**2**)
delamination temperature 539 (**1**), 542 (**1**)
deleterious effect 143 (**1**), 148 (**1**), 281 (**2**), 442 (**2**)
dense wavelength division multiplexing (DWDM) 362 (**2**)
density measurement 48 (**1**)
dependability 270 (**1**), 206 (**2**), 211 (**2**), 213 (**2**)
deposition method 157 (**1**)
design for reliability (DFR) 155 (**1**), 270 (**1**), 429 (**1**), 432 (**1**), 433 (**1**)
design of experiments (DoE) 242 (**1**)
– method 225 (**1**), 226 (**1**)
– procedure 206 (**1**)
design-qualification AST 190 (**2**), 193 (**2**)
deviatoric creep compliance 15 (**1**)
device under test (DUT) 314 (**2**)
die fracture 276 (**2**)
die-substrate assemblies 6 (**2**)
dielectric deposition 189 (**1**)
dielectric dry-etch 189 (**1**)

dielectric film 121 (**1**), 137 (**1**), 140 (**1**)
differential scanning calorimetry (DSC) 42 (**1**), 385 (**2**), 388 (**2**), 444 (**2**), 535 (**2**)
differential structure function 633 (**2**), 634 (**2**), 637 (**2**)
differential thermal expansion 155 (**1**)
diffraction efficiency 120 (**1**)
diffusion 112 (**1**), 113 (**1**), 143 (**1**), 152 (**1**), 155 (**1**), 177 (**1**), 262 (**1**), 323 (**1**), 329 (**1**), 373 (**1**), 385 (**1**), 526 (**1**), 587 (**1**), 102 (**2**), 103 (**2**), 130 (**2**), 184 (**2**), 192 (**2**), 219 (**2**), 225 (**2**), 274 (**2**), 347 (**2**), 354 (**2**), 357 (**2**), 396 (**2**), 433 (**2**), 437 (**2**), 462 (**2**), 508 (**2**), 527 (**2**), 536 (**2**), 542 (**2**), 704 (**2**)
– barrier 143 (**1**), 177 (**1**), 124 (**2**), 284 (**2**)
– coefficient 526 (**1**), 587 (**1**), 434 (**2**)
– creep 274 (**2**)
– kinetics 324 (**1**)
– models 373 (**1**)
– reactions 177 (**1**), 323 (**1**), 324 (**1**), 330 (**1**), 385 (**1**), 130 (**2**)
– theory 434 (**2**), 508 (**2**)
diffusional creep 334 (**1**)
digital fringe multiplication (DFM) 478 (**1**)
digital image correlation (DIC) 233 (**2**), 234 (**2**), 238 (**2**), 241 (**2**), 250 (**2**)
digital light processor (TM) systems 136 (**1**)
digital mirror device (DMD) 301 (**2**)
dipping technique 517 (**2**)
direct chip attachment (DCA) 8 (**2**)
direct numerical method 82 (**1**)
direct piezoelectric effect 668 (**1**)
directional coupler (DC) 330 (**2**)
disclinations 182 (**1**)
discrete element modeling (DEM) 576 (**2**)
dislocation 8 (**2**), 13 (**2**), 185 (**2**), 344 (**2**), 348 (**2**)
– creep 274 (**2**)
– glide 274 (**2**)
– separation (DS) 342 (**2**), 344 (**2**)
dislocations 152 (**1**), 182 (**1**)
dispensing technologies 515 (**2**)
dispersion management filters 129 (**1**)
dispersion slope compensation 128 (**1**), 129 (**1**)
dissolution 145 (**1**), 330 (**1**)
distance from the neutral point (DNP) 219 (**1**), 300 (**1**), 491 (**1**)
distributed Bragg reflector (DBR) 67 (**1**), 113 (**1**), 140 (**1**), 363 (**2**)
– laser diodes (DBR LD) 67 (**1**)
distributed feedback (DFB) 362 (**2**)

divergence angle   26 (**2**), 27 (**2**), 37 (**2**)–40 (**2**)
double simple shear tool (DST)   39 (**1**)
double-sided mirror image assembly   291 (**1**)
downtime   180 (**2**)
drive gear   334 (**2**), 338 (**2**)
drop impact loading   454 (**1**)
drop testing   334 (**1**), 341 (**1**)
drop tests   348 (**1**)
drops-to-failure   319 (**1**), *see also* times-to-failure (TTF)
dual-coated fiber   269 (**1**), 3 (**2**), 15 (**2**)
ductile fracture   220 (**1**), 418 (**1**)
ductile materials   222 (**1**)
Duffing equation   630 (**1**), 633 (**1**), 635 (**1**)
dummy component   416 (**1**)
durability   263 (**1**), 270 (**1**), 302 (**1**), 387 (**1**), 206 (**2**), 253 (**2**), 261 (**2**), 341 (**2**), 392 (**2**), 432 (**2**), 478 (**2**)
DWDM systems   117 (**1**)
dwell time   316 (**1**)
dynamic fatigue   601 (**1**), 609 (**1**)
– test   604 (**1**), 621 (**1**), 624 (**1**)
dynamic loading   270 (**1**), 555 (**1**)
dynamic mechanical analysis (DMA)   19 (**1**), 27 (**1**), 54 (**2**), 385 (**2**)
dynamic mechanical elongation tests   56 (**1**)
dynamic model   163 (**1**)
dynamic physical damage   572 (**1**)
dynamic response   272 (**1**), 555 (**1**), 557 (**1**)
dynamic strength   556 (**1**), 261 (**2**)
dynamic stress   272 (**1**), 556 (**1**), 698 (**2**), 701 (**2**)
dynamic testing   19 (**1**), 27 (**1**)

E-modulus   5 (**1**)
edge dispensing method   163 (**2**)
edge-indentation method   405 (**2**), 428 (**2**)
EDX elemental map   288 (**2**)
effect of contact metalization dissolution   325 (**1**)
effect of damping   678 (**1**), 679 (**1**)
effective contact area   547 (**2**), 591 (**2**)
effective contact length   590 (**2**)
effective refractive index   68 (**1**), 270 (**1**), 273 (**1**)
elastic approximations   17 (**1**)
elastic constant   8 (**2**)
elastic constants   138 (**1**)
elastic energy distribution   572 (**1**)
elastic model   50 (**1**), 212 (**1**), 213 (**1**)
elastic modulus   17 (**1**), 400 (**1**), 442 (**2**), 448 (**2**)

elastic potential energy   145 (**1**)
elastic stability   270 (**1**), 272 (**1**), 16 (**2**)
elastic strain   222 (**1**), 317 (**1**), 430 (**1**), 410 (**2**)
elastic stress   492 (**1**), 7 (**2**)
elastic-plastic-creep analysis   430 (**1**)
elasto-plastic analyses   223 (**1**), 414 (**1**), 416 (**1**)
elastomer   460 (**1**)
electric enthalpy density   671 (**1**)
electrical failure   285 (**2**), 557 (**2**)
electrically conductive adhesive (ECA)   503 (**2**), 527 (**2**), 534 (**2**), 571 (**2**), 612 (**2**)
electro-optic (Pockels) effect   112 (**1**)
electrochemical migration (ECM)   264 (**2**), 265 (**2**), 286 (**2**), 288 (**2**), 290 (**2**)
electrodeposition   145 (**1**)
electroless gold plating   103 (**2**)
electroless nickel and immersion gold (ENIG)   303 (**1**), 310 (**1**), 381 (**1**), 270 (**2**), 272 (**2**), 283 (**2**)
electromagnetic interference (EMI)   193 (**2**)
electromigration   144 (**1**), 153 (**1**), 191 (**1**), 380 (**1**), 79 (**2**), 100 (**2**), 124 (**2**), 127 (**2**), 182 (**2**), 184 (**2**), 204 (**2**), 219 (**2**), 267 (**2**), 536 (**2**), 618 (**2**)
electron back scatter diffraction (EBSD)   149 (**1**)
electron transport characteristics   192 (**1**)
electronic manufacturing service (EMS)   371 (**2**)
electronic speckle pattern interferometry (ESPI)   158 (**1**), 652 (**2**), 655 (**2**), 674 (**2**)
electronics   102 (**1**), 103 (**1**), 283 (**1**)–285 (**1**), 315 (**1**), 351 (**1**), 555 (**1**), 556 (**1**), 629 (**1**), 136 (**2**), 177 (**2**), 234 (**2**), 243 (**2**), 255 (**2**), 270 (**2**), 294 (**2**), 304 (**2**), 311 (**2**), 323 (**2**), 364 (**2**), 366 (**2**), 389 (**2**), 392 (**2**), 397 (**2**), 403 (**2**), 404 (**2**), 432 (**2**), 442 (**2**), 462 (**2**), 487 (**2**), 493 (**2**), 527 (**2**), 529 (**2**), 571 (**2**), 595 (**2**), 606 (**2**), 607 (**2**), 611 (**2**), 613 (**2**), 618 (**2**), 630 (**2**), 656 (**2**), 703 (**2**)
– packaging   4 (**1**), 273 (**1**), 283 (**1**), 284 (**1**), 121 (**2**), 132 (**2**), 149 (**2**), 177 (**2**), 302 (**2**), 489 (**2**), 611 (**2**), 687 (**2**)
electrostatic discharge (ESD)   184 (**2**)
electrostatically driven microengines   324 (**2**)
elongation to failure   156 (**1**)
encapsulated package   139 (**2**)
end-contacted nanotubes   197 (**1**)
energy dispersive spectroscopy (EDS)   350 (**2**)
energy-based fatigue models   405 (**1**), 433 (**1**)

INDEX
717

engineering reliability   204 (**2**), *see also* reliability engineering
environmental chamber   463 (**1**)
environmental stress screening (ESS)   190 (**2**), 194 (**2**)
epitaxial film   137 (**1**)
epoxy adhesive   508 (**1**), 443 (**2**), 446 (**2**), 582 (**2**)
epoxy molding compound (EMC)   138 (**2**)
epoxy resin   6 (**1**), 34 (**1**), 44 (**1**), 390 (**2**), 433 (**2**), 600 (**2**), 601 (**2**), 614 (**2**), 616 (**2**), 619 (**2**)
epoxy viscosity   164 (**2**), 169 (**2**)
equation of motion   561 (**1**), 565 (**1**)
equilibrium moduli   17 (**1**)
equilibrium phase diagram   328 (**1**)
Er-doped fiber amplifiers (EDFA)   114 (**1**)
erbium-doped fiber amplifier (EDFA)   489 (**2**)
eutectic solder   294 (**1**), 298 (**1**), 351 (**1**), 354 (**1**), 360 (**1**), 382 (**1**), 397 (**1**), 401 (**1**), 411 (**1**), 412 (**1**), 418 (**1**), 453 (**1**), 491 (**1**), 572 (**2**)
evolution of microstructure   347 (**1**)
excluded area   579 (**2**), 607 (**2**)
excluded volume   579 (**2**)
exposure time   127 (**1**)
Eyring equation   227 (**2**)

fab integrated packaging (FIP)   138 (**2**)
Fabry-Perot filter with DBRs   116 (**1**)
Fabry-Pérot interferences   512 (**2**)
face-centered-cubic (FCC) structure   146 (**1**)
failure   3 (**1**), 136 (**1**), 144 (**1**), 156 (**1**), 157 (**1**), 185 (**1**), 205 (**1**), 215 (**1**), 223 (**1**), 253 (**1**), 262 (**1**), 270 (**1**), 290 (**1**), 293 (**1**), 309 (**1**), 314 (**1**), 317 (**1**), 318 (**1**), 380 (**1**), 386 (**1**), 411 (**1**)–414 (**1**), 431 (**1**)–433 (**1**), 524 (**1**), 573 (**1**), 595 (**1**), 631 (**1**), 3 (**2**), 6 (**2**), 12 (**2**), 47 (**2**), 87 (**2**), 104 (**2**), 123 (**2**), 124 (**2**), 130 (**2**), 177 (**2**), 178 (**2**), 191 (**2**), 203 (**2**), 205 (**2**), 241 (**2**), 255 (**2**), 256 (**2**), 258 (**2**), 262 (**2**), 274 (**2**), 279 (**2**), 282 (**2**), 294 (**2**), 301 (**2**), 315 (**2**), 348 (**2**), 357 (**2**), 394 (**2**), 401 (**2**), 403 (**2**), 546 (**2**), 630 (**2**), 683 (**2**)
– analysis   217 (**1**), 223 (**1**), 321 (**1**), 654 (**1**), 661 (**1**), 89 (**2**), 177 (**2**), 179 (**2**), 250 (**2**), 255 (**2**), 269 (**2**), 294 (**2**), 635 (**2**), 647 (**2**)
– criterion   221 (**1**), 405 (**1**), 573 (**1**), 268 (**2**)
– density function   217 (**1**)
– detection   302 (**1**), 269 (**2**), 635 (**2**)
– diagram   658 (**1**)
– distribution   217 (**1**), 218 (**1**), 220 (**1**), 299 (**1**), 318 (**1**), 319 (**1**), 389 (**1**), 524 (**1**), 572 (**1**), 595 (**1**), 605 (**1**), 655 (**1**), 47 (**2**), 177 (**2**), 268 (**2**), 350 (**2**), 604 (**2**)
– free life   319 (**1**), 342 (**1**)
– in time (FIT)   181 (**2**)
– intensity   220 (**1**), 388 (**1**), 524 (**1**), 551 (**1**), 623 (**1**), 661 (**1**), 180 (**2**), 204 (**2**), 226 (**2**)
– link   454 (**1**), 206 (**2**), 262 (**2**)
– mechanism   136 (**1**), 185 (**1**), 298 (**1**), 314 (**1**), 177 (**2**), 183 (**2**), 184 (**2**), 204 (**2**), 221 (**2**), 242 (**2**), 255 (**2**), 267 (**2**), 269 (**2**), 273 (**2**), 274 (**2**), 313 (**2**), 349 (**2**), 384 (**2**), 401 (**2**), 433 (**2**), 552 (**2**)
– mode   144 (**1**), 152 (**1**), 315 (**1**), 127 (**2**), 178 (**2**), 186 (**2**), 204 (**2**), 214 (**2**), 242 (**2**), 254 (**2**), 262 (**2**), 272 (**2**), 348 (**2**), 357 (**2**), 689 (**2**), 691 (**2**)
– probability   261 (**1**), 388 (**1**), 597 (**1**)
– rate   215 (**1**), 321 (**1**), 574 (**1**), 3 (**2**), 94 (**2**), 177 (**2**), 217 (**2**), 254 (**2**), 547 (**2**), 601 (**2**), 624 (**2**)
– site   153 (**1**), 299 (**1**), 204 (**2**), 262 (**2**)
– software   550 (**1**), 187 (**2**), 213 (**2**), 649 (**2**)
failure-time distribution   255 (**2**)
fall time   124 (**1**)
far infrared Fizeau interferometry (FIFI)   476 (**1**), 514 (**1**)
fast Fourier transform (FFT)   92 (**2**)
fatigue   156 (**1**), 220 (**1**), 222 (**1**), 294 (**1**), 337 (**1**), 390 (**1**), 400 (**1**), 412 (**1**), 431 (**1**)–435 (**1**), 443 (**1**)–448 (**1**), 453 (**1**), 454 (**1**), 595 (**1**), 599 (**1**), 614 (**1**), 620 (**1**)–625 (**1**), 3 (**2**), 9 (**2**), 75 (**2**), 125 (**2**), 132 (**2**), 204 (**2**), 219 (**2**), 221 (**2**), 274 (**2**), 286 (**2**), 303 (**2**), 562 (**2**)
– analysis   222 (**1**)
– – inelastic energy   222 (**1**)
– – steady-state crack growth   222 (**1**)
– – strain based   222 (**1**)
– – stress based   222 (**1**)
– ductility coefficient   432 (**1**), 446 (**1**), 226 (**2**)
– effect   595 (**1**)
– failure   215 (**1**)–217 (**1**), 220 (**1**)–223 (**1**), 300 (**1**), 388 (**1**), 395 (**1**), 411 (**1**)–414 (**1**), 422 (**1**), 454 (**1**), 595 (**1**), 608 (**1**)–613 (**1**), 616 (**1**), 623 (**1**)–625 (**1**), 104 (**2**), 192 (**2**), 204 (**2**), 268 (**2**), 315 (**2**), 562 (**2**)
– – mode   417 (**1**)
– fracture   342 (**1**), 396 (**1**), 412 (**1**)
– – mode   419 (**1**)

– life 213 (**1**), 223 (**1**), 240 (**1**), 294 (**1**), 395 (**1**), 413 (**1**), 415 (**1**), 424 (**1**), 431 (**1**), 451 (**1**), 292 (**2**)
– – model 390 (**1**), 398 (**1**), 402 (**1**), 405 (**1**)
– – – thermal 402 (**1**)
– reliability 398 (**1**), 401 (**1**), 413 (**1**), 424 (**1**)
– resistance 156 (**1**), 75 (**2**)
– strength 363 (**1**), 414 (**1**), 426 (**1**), 9 (**2**)
– test program 435 (**1**)
– tests 242 (**1**), 388 (**1**)–390 (**1**), 396 (**1**), 402 (**1**), 411 (**1**), 416 (**1**), 431 (**1**), 445 (**1**), 454 (**1**), 610 (**1**)–612 (**1**), 215 (**2**), 262 (**2**)
ferrules 272 (**1**)
fiber 11 (**1**), 65 (**1**)–71 (**1**), 81 (**1**), 102 (**1**), 112 (**1**), 119 (**1**), 133 (**1**), 150 (**1**), 170 (**1**), 182 (**1**), 194 (**1**), 269 (**1**), 272 (**1**), 274 (**1**), 277 (**1**), 380 (**1**), 389 (**1**), 459 (**1**), 468 (**1**), 483 (**1**), 502 (**1**), 557 (**1**), 571 (**1**), 577 (**1**), 580 (**1**), 585 (**1**), 592 (**1**), 595 (**1**), 605 (**1**), 23 (**2**), 109 (**2**), 152 (**2**), 158 (**2**), 164 (**2**), 166 (**2**), 185 (**2**), 186 (**2**), 332 (**2**), 367 (**2**), 368 (**2**), 395 (**2**), 477 (**2**), 487 (**2**), 488 (**2**), 497 (**2**), 520 (**2**), 523 (**2**)
– Bragg grating (FBG) 67 (**1**), 70 (**1**), 113 (**1**), 126 (**1**)
– lifetime 597 (**1**), 613 (**1**), 614 (**1**)
– – model 595 (**1**)
– optics 111 (**1**)–114 (**1**), 272 (**1**), 557 (**1**), 332 (**2**), 394 (**2**)
– – directional coupler 330 (**2**)
– strain 11 (**1**), 97 (**1**), 269 (**1**)–271 (**1**), 389 (**1**), 483 (**1**), 605 (**1**), 667 (**1**), 688 (**1**), 692 (**1**)–695 (**1**), 395 (**2**), 511 (**2**)
– strength 277 (**1**), 571 (**1**), 578 (**1**)
fiber-on-board (FOB) 368 (**2**)
fiber-optic system 23 (**2**), 27 (**2**), 37 (**2**), 67 (**2**)
fiber-optics structural mechanics (FOSM) 270 (**1**), 277 (**1**), 65 (**2**)
field failures 385 (**1**), 205 (**2**), 207 (**2**), 254 (**2**), 256 (**2**), 267 (**2**)
filler 25 (**1**), 44 (**1**), 48 (**1**), 347 (**1**), 349 (**1**), 385 (**2**), 392 (**2**), 435 (**2**), 478 (**2**), 527 (**2**), 529 (**2**)–531 (**2**), 536 (**2**), 549 (**2**), 557 (**2**), 565 (**2**), 572 (**2**), 574 (**2**), 578 (**2**), 579 (**2**), 582 (**2**), 583 (**2**), 591 (**2**), 612 (**2**)–614 (**2**), 619 (**2**), 625 (**2**)
filter 114 (**1**), 117 (**1**), 130 (**1**), 134 (**1**), 343 (**2**), 423 (**2**), 427 (**2**), 487 (**2**), 655 (**2**)
fine pitch bonding 105 (**2**)
fine pitch surface mount components 285 (**1**)
fine pitch wirebonding 107 (**2**)
fine tuning 275 (**1**)

finite difference method (FDM) 208 (**1**), 324 (**2**)
finite element analysis (FEA) 87 (**1**), 295 (**1**), 400 (**1**), 432 (**1**), 433 (**1**), 448 (**1**), 630 (**1**), 4 (**2**), 5 (**2**), 62 (**2**), 416 (**2**), 438 (**2**)
– modeling 448 (**1**)
– thermal model 80 (**1**)
finite element method (FEM) 10 (**1**), 35 (**1**), 163 (**1**), 206 (**1**), 208 (**1**), 209 (**1**), 294 (**1**), 321 (**1**), 322 (**1**), 508 (**1**), 524 (**1**), 25 (**2**), 324 (**2**), 681 (**2**), 683 (**2**), 691 (**2**)
– method 209 (**1**)
– model 243 (**1**), 263 (**1**), 508 (**1**), 681 (**2**)
– packages 10 (**1**), 50 (**1**)
first and second order second moment methods 44 (**2**)
first-time-to-failure (FTTF) 456 (**1**)
fixturing 397 (**2**)
flat-topped tunable filter 114 (**1**)
flex circuit 8 (**2**)
flexibilizers 392 (**2**)
flip chip (FC) 6 (**1**), 62 (**1**), 141 (**1**), 283 (**1**), 286 (**1**), 291 (**1**), 310 (**1**), 400 (**1**), 411 (**1**), 450 (**1**), 501 (**1**)–504 (**1**), 517 (**1**), 8 (**2**), 74 (**2**), 79 (**2**), 80 (**2**), 161 (**2**), 261 (**2**), 304 (**2**), 492 (**2**), 502 (**2**), 527 (**2**), 546 (**2**), 611 (**2**)
– assembly 74 (**2**), 304 (**2**), 320 (**2**), 652 (**2**), 697 (**2**)
– attachment 286 (**1**), 288 (**1**), 72 (**2**)
– BGA (FCBGA) 283 (**1**), 286 (**1**), 652 (**2**)
– components 313 (**1**)
– failures 6 (**1**)
– gold bump 235 (**2**)
– interconnection 283 (**1**), 310 (**1**), 313 (**1**), 294 (**2**)
– module 552 (**2**), 689 (**2**)
– on board (FCOB) 136 (**2**)
– package 284 (**1**), 286 (**1**)–288 (**1**), 291 (**1**), 310 (**1**), 501 (**1**), 519 (**1**), 145 (**2**), 651 (**2**), 652 (**2**), 656 (**2**), 659 (**2**), 660 (**2**), 662 (**2**), 668 (**2**), 670 (**2**), 672 (**2**), 674 (**2**), 685 (**2**), 688 (**2**)
– plastic ball grid array (FC-PBGA) 481 (**1**), 493 (**1**), 685 (**2**)
– – package assembly 516 (**1**)
– reflow soldering 77 (**2**)
– underfill inspection 282 (**2**)
flow soldering 412 (**1**)
flow stress 165 (**1**), 344 (**1**), 418 (**1**)
flux chemical reactions 360 (**1**)
focused ion beam imaging (FIB) 148 (**1**)

INDEX

four-beam moiré interferometry   476 (**1**)
four-wave mixing (FWM)   112 (**1**)
fractional factorial experiments (FFEs)   232 (**1**)
fracture   5 (**1**), 156 (**1**), 165 (**1**), 220 (**1**),
    325 (**1**), 342 (**1**), 347 (**1**), 412 (**1**), 419 (**1**),
    523 (**1**)–525 (**1**), 529 (**1**), 3 (**2**), 262 (**2**),
    284 (**2**), 303 (**2**), 395 (**2**), 506 (**2**), 600 (**2**)
– analysis   221 (**1**), 222 (**1**), 347 (**1**), 523 (**1**),
    529 (**1**), 5 (**2**), 184 (**2**), 247 (**2**), 452 (**2**)
– mechanics approach   524 (**1**)
– toughness testing   169 (**1**)
free space optical interconnect (FSOI)   366 (**2**)
free vibrations   690 (**1**)
frequency response   646 (**1**), 648 (**1**)
frequency scanning   30 (**1**)
frequency-modified Coffin-Manson model
    447 (**1**)
frequency-modified Morrow model   447 (**1**)
Fresnel reflection   373 (**2**), 512 (**2**)
fretting corrosion   287 (**2**)
fringe patterns   469 (**1**), 485 (**1**), 488 (**1**),
    498 (**1**), 334 (**2**), 337 (**2**), 654 (**2**)
fringe-locus function   331 (**2**)
full array configuration   289 (**1**)
full factorial design   229 (**1**)
full width at half maximum (FWHM)   71 (**1**),
    87 (**1**), 350 (**2**), 357 (**2**)
fully cure dependent modeling   50 (**1**), 51 (**1**)
fully cured polymer   18 (**1**)
fully state dependent modeling   10 (**1**), 34 (**1**),
    49 (**1**)
function variable incidence matrix   30 (**2**)
function-rich   253 (**2**)
functionality   3 (**1**), 53 (**1**)–57 (**1**), 132 (**1**),
    244 (**1**), 252 (**1**), 262 (**1**), 288 (**1**), 629 (**1**),
    253 (**2**), 336 (**2**), 348 (**2**), 391 (**2**), 424 (**2**),
    487 (**2**), 496 (**2**), 506 (**2**), 555 (**2**)
fundamental moisture absorption   435 (**2**)
fused biconical taper (FBT)   272 (**1**), 16 (**2**)
– couplers   686 (**1**), 16 (**2**)

G-Helix   139 (**2**)
GaAs   112 (**1**), 139 (**1**), 9 (**2**), 109 (**2**), 131 (**2**),
    350 (**2**), 357 (**2**), 363 (**2**)
– laser   493 (**2**)
galvanic corrosion   264 (**2**), 287 (**2**), 557 (**2**),
    575 (**2**), 595 (**2**), 607 (**2**), 615 (**2**), 616 (**2**),
    618 (**2**)
garbage in garbage out (GIGO)   208 (**1**)
Gatan image filter (GIF)   343 (**2**)
Gaussian beam   26 (**2**), 491 (**2**)

generalized Galerkin procedure   630 (**1**),
    673 (**1**), 677 (**1**)
geometrical nonlinearity   668 (**1**)
glass FBG   82 (**1**)
glass fiber   65 (**1**), 380 (**1**), 10 (**2**), 392 (**2**),
    477 (**2**), 491 (**2**), 559 (**2**), 562 (**2**)
glass optical fiber   68 (**1**), 81 (**1**), 10 (**2**)
glass solder   496 (**2**)
glass transition temperature   4 (**1**), 460 (**1**),
    468 (**1**), 530 (**1**), 7 (**2**), 85 (**2**), 109 (**2**),
    128 (**2**), 143 (**2**), 385 (**2**), 488 (**2**), 510 (**2**),
    547 (**2**), 559 (**2**), 571 (**2**), 620 (**2**), 662 (**2**),
    664 (**2**)
glassy modulus   17 (**1**)
glob-top manner   152 (**2**)
global mismatch   10 (**2**)
global stress   299 (**1**), 10 (**2**)
global-local modeling techniques   450 (**1**)
gold electroplating   103 (**2**)
gold thermosonic ball bond   113 (**2**)
gold-tin eutectics   9 (**2**)
Graham-Walles equation   228 (**2**)
grain   142 (**1**), 147 (**1**)–149 (**1**), 157 (**1**),
    175 (**1**), 191 (**1**), 212 (**1**), 261 (**1**), 315 (**1**),
    325 (**1**), 335 (**1**), 87 (**2**), 103 (**2**), 241 (**2**),
    274 (**2**), 352 (**2**), 500 (**2**)
– boundary (GB)   275 (**2**), 352 (**2**)
– boundary sliding   334 (**1**), 274 (**2**)
grain growth   307 (**1**), 335 (**1**)
graph conversion   31 (**2**)
graph partitioning   34 (**2**)
graph partitioning algorithms   34 (**2**)
graph partitioning analysis   37 (**2**)
graphene   182 (**1**), *see also* graphitic carbon
graphitic carbon   181 (**1**)
graphoepitaxy   137 (**1**)
grating   65 (**1**), 80 (**1**), 115 (**1**), 461 (**1**),
    476 (**1**), 510 (**1**), 384 (**2**), 654 (**2**)
– length   70 (**1**), 120 (**1**)
– period effect   81 (**1**)
– pitch   462 (**1**)
gravimetric method   146 (**1**)
group delay (GD)   127 (**1**)
growth   182 (**1**)–184 (**1**), 186 (**1**)–189 (**1**),
    197 (**1**), 222 (**1**)–224 (**1**), 335 (**1**),
    363 (**1**)–376 (**1**), 385 (**1**), 413 (**1**),
    595 (**1**)–601 (**1**), 621 (**1**)–625 (**1**), 8 (**2**),
    13 (**2**), 83 (**2**), 186 (**2**), 226 (**2**), 293 (**2**),
    317 (**2**), 348 (**2**), 354 (**2**), 460 (**2**), 618 (**2**)
– kinetics   365 (**1**)
– of the intermetallic compound   413 (**1**)

– temperature 197 (**1**)
– time 187 (**1**)

H-PDLC 123 (**1**)
Hall-Petch rule 172 (**1**)
Hamilton principle 671 (**1**)
hard solder 9 (**2**)
hard-core model 578 (**2**)
hardness 159 (**1**), 163 (**1**), 172 (**1**), 175 (**1**), 335 (**1**)
harmonic forcing 651 (**1**)
heat generation 77 (**1**)
heat loss 109 (**1**)
heat transfer 79 (**1**), 87 (**1**), 209 (**1**), 525 (**1**)
heavily alloyed gold 102 (**2**)
heteroepitaxy 137 (**1**)
high cycle fatigue 222 (**1**)
high density interconnect (HDI) 73 (**2**), 146 (**2**), 271 (**2**), 291 (**2**)
high performance dispersion compensators 126 (**1**)
high-cycle fatigue 454 (**1**)
high-speed testing 606 (**1**)
high-temperature-storage (HTS) 485 (**1**)
highly accelerated life test (HALT) 454 (**1**), 4 (**2**), 194 (**2**), 217 (**2**), 218 (**2**)
highly accelerated stress screening (HASS) 194 (**2**)
hinged micromirror 325 (**2**), 327 (**2**)
holographic grating 118 (**1**)
holographic memory assemblies 11 (**2**)
holographic polymer dispersed liquid crystal (H-PDLC) 113 (**1**)
homoepitaxy 139 (**1**)
hot air solder leveling (HASL) 297 (**1**), 302 (**1**), 310 (**1**), 381 (**1**), 272 (**2**)
Hsai-Wu failure criteria 654 (**1**)
humidity 155 (**1**), 262 (**1**), 275 (**1**), 388 (**1**), 500 (**1**), 523 (**1**), 572 (**1**), 576 (**1**), 587 (**1**), 595 (**1**), 606 (**1**), 178 (**2**), 212 (**2**), 220 (**2**), 254 (**2**), 255 (**2**), 303 (**2**), 318 (**2**), 386 (**2**), 393 (**2**), 396 (**2**), 398 (**2**), 435 (**2**), 444 (**2**), 453 (**2**), 556 (**2**), 574 (**2**), 595 (**2**), 601 (**2**), 615 (**2**), 620 (**2**), 623 (**2**)
– chamber 435 (**2**)
hybrid integration 304 (**2**)
hydrophilic 586 (**1**), 587 (**1**), 458 (**2**)
hydrophilic nanoparticle 586 (**1**)
hydrophobic 274 (**1**), 587 (**1**), 318 (**2**), 458 (**2**), 477 (**2**), 618 (**2**)
hydrophobic nanoparticle 586 (**1**)
hydrophobicity 273 (**1**), 458 (**2**), 477 (**2**)

hygroscopic deformation 494 (**1**)
hygroscopic strain 495 (**1**), 499 (**1**)
hygroscopic stress 496 (**1**)
hygroscopic swelling 500 (**1**)
– properties 494 (**1**)
hygrostrain 531 (**1**)
hygrostress 525 (**1**)
hyperbolic-sine model 442 (**1**)
hypercorrosion 286 (**2**)
hypergraph 33 (**2**)

imaging lens 516 (**1**)
IMC growth 364 (**1**)
IMC layer growth 366 (**1**)
IMC phase 274 (**2**), 275 (**2**)
immersion interfeometer 478 (**1**)
In composition fluctuation 343 (**2**), 344 (**2**), 347 (**2**)
in-circuit test (ICT) 380 (**1**)
in-plane thermal field 640 (**1**)
indentation 146 (**1**), 162 (**1**), 385 (**1**), 603 (**1**)
indentation-induced delamination test 404 (**2**)
independent plastic strain 430 (**1**)
indium tin oxide (ITO) 403 (**2**), 419 (**2**)
inelastic energy approach 224 (**1**)
inelastic strain 222 (**1**), 491 (**1**), 9 (**2**)
InGaAs 140 (**1**)
initial curvature 558 (**1**), 15 (**2**)
initial spring constant 561 (**1**)
initial stage of intermetallic growth 372 (**1**)
initial strength 575 (**1**), 597 (**1**), 599 (**1**)
initial stress 7 (**1**), 8 (**1**), 392 (**1**), 706 (**2**)
initial wetting time 361 (**1**), 363 (**1**)
ink-jet dispensing 517 (**2**)
inner lead bonding (ILB) 75 (**2**)
InP 112 (**1**)
insertion loss 120 (**1**), 372 (**2**)
integrated circuit (IC) 155 (**1**), 252 (**1**), 263 (**1**), 264 (**1**), 429 (**1**), 450 (**1**), 486 (**1**), 13 (**2**), 71 (**2**), 79 (**2**), 109 (**2**), 113 (**2**), 135 (**2**), 182 (**2**), 186 (**2**), 190 (**2**), 318 (**2**), 320 (**2**), 366 (**2**), 380 (**2**), 621 (**2**)
inter-particle conduction 540 (**2**)
interaction integral method 536 (**1**)
interconnect failure 273 (**2**)
interconnection 112 (**1**), 210 (**1**), 215 (**1**), 283 (**1**), 313 (**1**), 315 (**1**), 317 (**1**), 323 (**1**), 327 (**1**), 329 (**1**), 337 (**1**), 339 (**1**), 342 (**1**), 351 (**1**), 389 (**1**), 390 (**1**), 405 (**1**), 459 (**1**), 491 (**1**), 273 (**2**)
– electronics industries (IPC) 283 (**1**)
– level model 322 (**1**)

interdiffusion 81 (**2**), 103 (**2**), 114 (**2**), 219 (**2**), 508 (**2**)
interface failure mode 413 (**1**)
interface fatigue crack 426 (**1**)
interface fatigue mode 412 (**1**)
interface fracture mode 417 (**1**)
interface fracture toughness 529 (**1**)
interface metallurgy 76 (**2**), 349 (**2**), 357 (**2**)
interface strengths 262 (**1**)
interfacial adhesion 263 (**1**), 318 (**2**), 403 (**2**), 433 (**2**), 441 (**2**), 444 (**2**), 449 (**2**)
interfacial bond 271 (**2**)
interfacial compliance 7 (**2**), 12 (**2**), 59 (**2**), 473 (**2**)–475 (**2**), 479 (**2**), 481 (**2**), 483 (**2**)
interfacial conductivity 593 (**2**)
interfacial contact materials 196 (**1**)
interfacial failure 262 (**1**), 276 (**2**), 432 (**2**), 469 (**2**)
interfacial fracture 527 (**1**), 285 (**2**), 404 (**2**), 437 (**2**)–439 (**2**), 449 (**2**), 451 (**2**), 452 (**2**), 456 (**2**), 461 (**2**), 468 (**2**)
– toughness 461 (**2**), 462 (**2**), 469 (**2**)
interfacial fracture mechanics 527 (**1**)
interfacial hydrophobicity 457 (**2**), 458 (**2**)
interfacial metallurgical reactions 357 (**1**)
interfacial reactions 347 (**1**), 360 (**1**)
interfacial shearing stress 474 (**2**), 482 (**2**), 485 (**2**), 495 (**2**)
interfacial strength 263 (**1**), 260 (**2**), 263 (**2**), 403 (**2**), 411 (**2**), 414 (**2**), 423 (**2**), 425 (**2**), 426 (**2**), 469 (**2**)
interfacial stress 7 (**2**), 401 (**2**), 473 (**2**), 481 (**2**), 483 (**2**), 652 (**2**)
interfacial tension coefficient 485 (**2**)
interferometer (IT) 331 (**2**), 692 (**2**)
intermetallic bond 383 (**1**), 271 (**2**)
intermetallic compound 267 (**2**)
intermetallic compound (IMC) 330 (**1**), 412 (**1**)
– layer 417 (**1**)
intermetallic growth 262 (**1**), 363 (**1**), 83 (**2**), 93 (**2**)
– kinetics 363 (**1**), 376 (**1**)
intermetallic layer 330 (**1**), 339 (**1**), 360 (**1**), 365 (**1**), 375 (**1**), 415 (**1**)
internal compressive forces 476 (**2**)
interposer 294 (**1**)–296 (**1**), 305 (**1**), 309 (**1**), 519 (**1**), 81 (**2**), 278 (**2**), 371 (**2**)
intra-particle resistance 542 (**2**)
intrinsic stress 146 (**1**), 243 (**2**)
inverse piezoelectric effect 668 (**1**)
Ishikawa diagram 219 (**1**)

isothermal mechanical fatigue test 413 (**1**)
isotropic conductive adhesive (ICA) 527 (**2**), 529 (**2**), 538 (**2**), 572 (**2**), 575 (**2**), 595 (**2**), 604 (**2**), 613 (**2**)
isotropic laminate 631 (**1**)
isotropic materials 14 (**1**)
isotropic PFBG 80 (**1**)

J-integral 542 (**1**), 416 (**2**), 419 (**2**)

Kelvin element 6 (**1**), 52 (**2**)
kernel function 11 (**1**)
kickstand 310 (**2**), 313 (**2**)
$KNbO_3$ 112 (**1**)
Knoop hardness scale 97 (**2**)
– micro-hardness 354 (**2**)
known good die (KGD) 80 (**2**), 136 (**2**), 145 (**2**)
Kriging error estimation 245 (**1**)

laminated microstructure 660 (**1**)
land grid array (LGA) 138 (**2**), 260 (**2**)
large displacement 564 (**1**)
laser 67 (**1**), 71 (**1**), 85 (**1**), 94 (**1**)–101 (**1**), 118 (**1**), 128 (**1**), 139 (**1**), 147 (**1**), 186 (**1**), 273 (**1**), 298 (**1**), 483 (**1**), 514 (**1**), 23 (**2**), 73 (**2**), 144 (**2**), 216 (**2**), 221 (**2**), 229 (**2**), 234 (**2**), 271 (**2**), 272 (**2**), 314 (**2**), 330 (**2**), 332 (**2**), 477 (**2**), 487 (**2**), 489 (**2**), 518 (**2**), 523 (**2**), 573 (**2**), 655 (**2**), 679 (**2**)
– ablation 186 (**1**)
– bars 357 (**2**)
– beam 70 (**1**), 119 (**1**), 147 (**1**), 476 (**1**), 485 (**1**), 514 (**1**)–516 (**1**), 23 (**2**), 32 (**2**), 40 (**2**), 91 (**2**), 489 (**2**), 491 (**2**), 499 (**2**), 520 (**2**), 655 (**2**)
– DFB 67 (**1**), 363 (**2**)
– diode 71 (**1**), 85 (**1**), 95 (**1**)–101 (**1**), 25 (**2**), 222 (**2**), 330 (**2**), 343 (**2**), 348 (**2**), 354 (**2**), 362 (**2**), 489 (**2**)
– – high power 348 (**2**), 357 (**2**)
– emission 351 (**2**)
– Excimer 313 (**2**)
– Fabry-Perot (FP) 363 (**2**)
– interferometer 91 (**2**), 110 (**2**), 234 (**2**), 330 (**2**), 704 (**2**)
– interferometry 476 (**1**), 514 (**1**), 683 (**2**)
– power 70 (**1**), 86 (**1**), 93 (**1**), 97 (**1**)–99 (**1**), 101 (**1**), 119 (**1**), 485 (**1**), 28 (**2**), 514 (**2**)
– pump 487 (**2**)
– ridge 357 (**2**)
– welding 483 (**1**), 488 (**2**)
latin hypercube design 235 (**1**)

lattice-mismatch strain 8 (**2**)
lead-free flux 380 (**1**)
lead-free solder 283 (**1**), 324 (**1**), 353 (**1**), 411 (**1**), 435 (**1**), 557 (**1**), 261 (**2**), 273 (**2**), 603 (**2**), 612 (**2**)
– alloy compositions 378 (**1**)
– joint 424 (**1**)
– – interconnections 351 (**1**)
lead-free soldering processes 330 (**1**), 377 (**1**), 381 (**1**)
life distribution model 455 (**1**)
lifetime 6 (**1**), 165 (**1**), 218 (**1**), 269 (**1**), 273 (**1**), 319 (**1**), 586 (**1**), 587 (**1**), 595 (**1**), 8 (**2**), 75 (**2**), 94 (**2**), 122 (**2**), 195 (**2**), 208 (**2**), 214 (**2**), 216 (**2**), 221 (**2**), 225 (**2**), 226 (**2**), 229 (**2**), 315 (**2**), 477 (**2**), 574 (**2**)
– failure fraction 194 (**2**)
light emitting diode (LED) 6 (**1**), 71 (**1**), 216 (**2**), 343 (**2**), 489 (**2**)
lightly alloyed gold 102 (**2**)
linear acceleration 328 (**2**)
linear elastic fracture mechanics (LEFM) 220 (**1**), 248 (**2**)
linear state dependent viscoelasticity 13 (**1**)
linear thermally-dependent contributions 673 (**1**)
linear visco-elastic modeling 4 (**1**), 18 (**1**)
liquid crystal display (LCD) 225 (**2**), 258 (**2**), 527 (**2**), 572 (**2**)
liquid crystal (LC) 113 (**1**), 118 (**1**), 186 (**2**)
– polymer (LCP) 302 (**2**)
liquid encapsulant 6 (**1**)
liquid Pb-free solders 376 (**1**)
liquid solder 330 (**1**), 332 (**1**)
load 7 (**1**), 17 (**1**), 42 (**1**), 158 (**1**), 163 (**1**), 170 (**1**), 208 (**1**), 222 (**1**), 270 (**1**), 302 (**1**), 388 (**1**), 400 (**1**), 414 (**1**), 419 (**1**), 435 (**1**), 454 (**1**), 460 (**1**), 475 (**1**), 476 (**1**), 508 (**1**), 532 (**1**), 556 (**1**), 571 (**1**), 575 (**1**), 584 (**1**), 600 (**1**), 604 (**1**), 629 (**1**), 640 (**1**), 661 (**1**), 684 (**1**), 696 (**1**), 32 (**2**), 48 (**2**), 124 (**2**), 204 (**2**), 212 (**2**), 235 (**2**), 259 (**2**), 261 (**2**), 331 (**2**), 334 (**2**), 349 (**2**), 354 (**2**), 400 (**2**), 404 (**2**), 405 (**2**), 407 (**2**), 410 (**2**), 422 (**2**), 469 (**2**), 506 (**2**), 522 (**2**), 550 (**2**), 562 (**2**), 572 (**2**), 586 (**2**), 686 (**2**)
– application time 7 (**1**)
– sensor 616 (**1**)
loading conditions 17 (**1**), 206 (**1**), 215 (**1**), 270 (**1**), 314 (**1**), 398 (**1**), 451 (**2**)
local area networks (LAN) 363 (**2**)
local mismatch 10 (**2**)

local stress 10 (**2**), 241 (**2**), 685 (**2**)
localization length 184 (**1**)
localized galvanic corrosion 287 (**2**)
lognormal failure rate distribution 184 (**2**)
long-range thickness variation (LR-TV) 347 (**2**)
long-term reliability 272 (**1**), 379 (**1**), 483 (**1**), 6 (**2**), 172 (**2**), 182 (**2**), 186 (**2**), 209 (**2**), 215 (**2**), 223 (**2**), 432 (**2**), 450 (**2**), 595 (**2**), 604 (**2**), 607 (**2**), 698 (**2**), 703 (**2**)
longitudinal relaxation modulus 27 (**1**)
longitudinal strain 26 (**1**), 323 (**1**)
low cycle fatigue models 443 (**1**)
low-cycle fatigue 222 (**1**), 454 (**1**)
low-pressure chemical vapor deposition (LPCVD) 325 (**2**), 704 (**2**)

macrocrack 275 (**2**)
maintainability 206 (**2**)
maintenance 211 (**2**), 400 (**2**)
– preventive 211 (**2**)
– reactive 211 (**2**)
mandrel device 616 (**1**)
Manson-Coffin's law 412 (**1**)
manufacturability 144 (**1**), 285 (**1**)
manufacturing-qualification AST 193 (**2**)
mask 67 (**1**), 68 (**1**), 85 (**1**), 128 (**1**)–130 (**1**), 306 (**1**), 311 (**1**), 74 (**2**), 152 (**2**), 155 (**2**), 658 (**2**), 659 (**2**), 678 (**2**)
master curve 23 (**1**), 26 (**1**), 48 (**1**), 57 (**1**)
mathematical modeling 400 (**1**)
matrix creep fatigue model 432 (**1**)
maximum acceleration 273 (**1**), 562 (**1**)
maximum acceleration (deceleration) 566 (**1**)
maximum compressive force 561 (**1**), 565 (**1**)
maximum displacement 494 (**1**), 561 (**1**)
maximum stress criteria 221 (**1**), 524 (**1**), 654 (**1**), *see also* normal stress
maximum stress failure criteria 657 (**1**)
maximum wetting force 361 (**1**)
Maxwell element 6 (**1**)
mean downtime (MDT) 181 (**2**)
mean operating time (MOT) 181 (**2**)
mean-time-between failures (MTBF) 181 (**2**), 562 (**2**)
mean-time-to-failure (MTTF) 456 (**1**), 181 (**2**), 212 (**2**), 226 (**2**)
mechanical failure 144 (**1**), 215 (**1**), 314 (**1**), 524 (**1**), 551 (**1**), 105 (**2**), 206 (**2**), 315 (**2**)
mechanical fatigue test 414 (**1**), 416 (**1**)
mechanical interlocking 452 (**2**), 462 (**2**), 488 (**2**), 508 (**2**)

INDEX

mechanical load   388 (**1**), 400 (**1**), 454 (**1**), 508 (**1**), 627 (**1**), 631 (**1**), 633 (**1**)–635 (**1**), 640 (**1**), 657 (**1**), 254 (**2**), 456 (**2**), 501 (**2**)
mechanical loading   215 (**1**), 269 (**1**), 406 (**1**), 233 (**2**), 272 (**2**), 473 (**2**)
mechanical reliability   274 (**1**), 411 (**1**), 523 (**1**), 571 (**1**), 595 (**1**), 613 (**1**), 141 (**2**), 270 (**2**), 293 (**2**), 373 (**2**), 478 (**2**), 542 (**2**), *see also* structural reliability
mechanical shearing   390 (**1**)
mechanical shock   283 (**1**), 305 (**1**), 315 (**1**), 317 (**1**), 341 (**1**), 215 (**2**), 280 (**2**), 497 (**2**)
mechanical shock test   315 (**1**), 317 (**1**)
mechanical simulation   49 (**1**), 322 (**1**)
mechanical strength   141 (**1**), 198 (**1**), 275 (**1**), 357 (**1**), 388 (**1**), 191 (**2**), 443 (**2**), 554 (**2**), 578 (**2**)
median strength   607 (**1**), 608 (**1**)
melting point   5 (**1**), 145 (**1**), 342 (**1**), 352 (**1**), 367 (**1**), 411 (**1**), 75 (**2**), 97 (**2**), 123 (**2**), 127 (**2**), 311 (**2**), 615 (**2**)
melting temperature   5 (**1**), 149 (**1**), 212 (**1**), 298 (**1**), 351 (**1**), 383 (**1**), 430 (**1**), 75 (**2**), 498 (**2**), 612 (**2**)
membrane deflection   159 (**1**), 160 (**1**), 225 (**1**), 706 (**2**)
membrane deformation   678 (**2**), 681 (**2**), 683 (**2**)
mesh density   211 (**1**)
metal film   137 (**1**), 141 (**1**), 479 (**1**)
metal-oxide-silicon (MOS) transistor   140 (**1**)
metalization systems   331 (**1**)
metalized fiber   269 (**1**), 16 (**2**)
metallurgical interface   122 (**2**), 129 (**2**)
metallurgical structures   406 (**1**)
metastable solubility   331 (**1**)
metropolitan access networks (MAN)   362 (**2**)
micro-optics   310 (**2**)
micro-optoelectromechanical systems (MOEMS)   507 (**1**), 323 (**2**), 338 (**2**), 380 (**2**)
micro-Raman spectroscopy   350 (**2**)
micro-via technology   272 (**2**)
microbending   272 (**1**), 684 (**1**), 3 (**2**), 15 (**2**)
microcrack   307 (**1**), 579 (**1**), 580 (**1**), 592 (**1**), 110 (**2**), 185 (**2**), 275 (**2**), 507 (**2**)
– growth   580 (**1**), 186 (**2**)
microDAC   234 (**2**)
microelectromechanical systems (MEMS)   135 (**1**), 142 (**1**), 262 (**1**), 273 (**1**), 627 (**1**), 4 (**2**), 10 (**2**), 105 (**2**), 241 (**2**), 299 (**2**), 310 (**2**), 323 (**2**), 477 (**2**)
– device   301 (**2**), 311 (**2**)
– failures   303 (**2**)
– packaging   299 (**2**), 300 (**2**), 302 (**2**)
microelectronics   3 (**1**), 135 (**1**), 136 (**1**), 155 (**1**), 193 (**1**), 200 (**1**), 205 (**1**), 217 (**1**), 253 (**1**), 259 (**1**), 272 (**1**), 273 (**1**), 460 (**1**), 475 (**1**), 556 (**1**), 629 (**1**), 3 (**2**), 8 (**2**), 14 (**2**), 16 (**2**), 81 (**2**), 87 (**2**), 101 (**2**), 111 (**2**), 203 (**2**), 204 (**2**), 230 (**2**), 244 (**2**), 299 (**2**), 310 (**2**), 383 (**2**), 403 (**2**), 432 (**2**), 442 (**2**), 473 (**2**), 480 (**2**), 488 (**2**), 563 (**2**), 573 (**2**), 698 (**2**)
– bonding wire   95 (**2**)
microengine   328 (**2**), 334 (**2**), 335 (**2**)
micromachining   157 (**1**), 324 (**2**)
micromechanical model   572 (**1**)
micromirror   142 (**1**), 507 (**1**), 299 (**2**), 324 (**2**), 337 (**2**), 338 (**2**)
microscopic moiré interferometry   475 (**1**), 477 (**1**), 502 (**1**)
microspheres   479 (**2**)
microspring contact on silicon technology (MOST)   139 (**2**)
microstrain   461 (**1**), 670 (**1**)
microstructural characterization   321 (**1**)
microstructural effect   138 (**1**)
microstructure   137 (**1**), 173 (**1**), 212 (**1**), 261 (**1**), 313 (**1**), 334 (**1**), 351 (**1**), 273 (**2**), 275 (**2**)
microsystem   205 (**1**), 212 (**1**), 262 (**1**), 324 (**2**)
microtensile testing (MT)   157 (**1**)
mirror   65 (**1**), 291 (**1**), 462 (**1**), 483 (**1**), 506 (**1**), 302 (**2**), 363 (**2**), 370 (**2**), 452 (**2**)
– holders   463 (**1**)
– integrated   371 (**2**)
Mises equivalent stresses   419 (**1**)
mode field diameter   26 (**2**), 518 (**2**)
mode mixity phase angle   528 (**1**), 529 (**1**), 549 (**1**)
model molding compounds   44 (**1**)
modified crack surface displacement extrapolation method (MCSDEM)   531 (**1**)
modified J-integral (MJI) method   532 (**1**), 549 (**1**)
modified passive alignment method   162 (**2**)
modified virtual crack closure method (MVCCM)   533 (**1**), 550 (**1**)
modulator   330 (**2**), 333 (**2**), 487 (**2**)
moiré field   461 (**1**)
moiré grating   461 (**1**)

moiré interferometry 460 (**1**), 470 (**1**), 475 (**1**), 476 (**1**), 484 (**1**), 486 (**1**), 263 (**2**), 652 (**2**)
moiré pattern 461 (**1**), 463 (**1**), 469 (**1**), 477 (**1**)
moisture 5 (**1**), 169 (**1**), 262 (**1**), 269 (**1**), 508 (**1**), 525 (**1**), 571 (**1**), 595 (**1**), 13 (**2**), 16 (**2**), 82 (**2**), 186 (**2**), 267 (**2**), 294 (**2**), 303 (**2**), 386 (**2**), 392 (**2**), 395 (**2**), 396 (**2**), 401 (**2**), 432 (**2**), 433 (**2**), 450 (**2**), 464 (**2**), 503 (**2**), 515 (**2**), 555 (**2**), 642 (**2**)
– absorption 65 (**1**), 525 (**1**), 185 (**2**), 396 (**2**), 432 (**2**), 442 (**2**), 514 (**2**)
– concentration distribution 440 (**2**)
– diffusion 262 (**1**), 525 (**1**), 544 (**1**), 436 (**2**), 536 (**2**)
– diffusion coefficient 433 (**2**)
– ingress 6 (**1**)
– preconditioning 62 (**1**), 523 (**1**)–526 (**1**), 542 (**1**), 444 (**2**), 452 (**2**), 460 (**2**)
– sensitivity 386 (**1**), 523 (**1**)
– – level (MSL) 5 (**1**), 386 (**1**)
– swelling coefficient 456 (**2**)
mold compound 494 (**1**)
molding compound 5 (**1**), 252 (**1**), 524 (**1**), 8 (**2**), 13 (**2**), 14 (**2**), 432 (**2**)
molding plastic 389 (**1**)
molecular dynamic 172 (**1**)
Morrow's energy-based model 446 (**1**)
multi-beam optical stress sensor (MOSS) 385 (**1**)
multi-channel dispersion-slope compensator 126 (**1**)
multi-mode fiber (MMF) 27 (**2**), 167 (**2**), 365 (**2**)
multi-objective optimization 263 (**1**), 42 (**2**)
multi-points approach 11 (**1**)
multi-stage peel test (MPT) 405 (**2**), 407 (**2**), 410 (**2**), 427 (**2**)
multichip module (MCM) 87 (**2**), 571 (**2**), 606 (**2**), 659 (**2**)
multicore fiber cable 274 (**1**)
multiquantum well (MQW) 342 (**2**)
multiwalled carbon nanofibers (CNFs) 194 (**1**)
multiwalled CNT (MWCNT) 182 (**1**), 184 (**1**), 186 (**1**), 194 (**1**), 196 (**1**)
mutually-pumped phase conjugators (MPPC) 112 (**1**)

nano wafer level ACF (WLACF) 625 (**2**)
nano-electromechanical systems (NEMS) 136 (**1**), 247 (**2**), 320 (**2**)
nano-indentation 141 (**1**), 146 (**1**), 153 (**1**), 157 (**1**), 173 (**1**), 175 (**1**), 385 (**1**)
nano-machines 404 (**2**)
nano-particle material (NPM) 269 (**1**), 273 (**1**), 585 (**1**), 16 (**2**), 474 (**2**), 476 (**2**), 478 (**2**), 479 (**2**)
– cladding 274 (**1**)
NPM-based fibers 274 (**1**), 275 (**1**), 585 (**1**), 587 (**1**), 590 (**1**)
nanoDAC 239 (**2**), 242 (**2**), 246 (**2**)
nanofiber 182 (**1**), 188 (**1**), 192 (**1**)
nanostructures 181 (**1**)
National Electronics Manufacturing Initiative (NEMI) 429 (**1**)
Nickel substrate 373 (**1**)
non-conductive adhesive (NCA) 529 (**2**), 554 (**2**)
non-destructive evaluations (NDE) 222 (**2**), 477 (**2**)
non-linear deformation 351 (**1**), 489 (**1**)
non-linear plastic 430 (**1**)
non-solder mask defined (NSMD) 260 (**2**)
non-volatile 273 (**1**)
nondestructive pull test (NDPT) 89 (**2**)
nonhomogeneous Mathieu-Hill equation 675 (**1**)
nonlinear dynamics 630 (**1**)
– response 653 (**1**)
nonlinear thermo-elasticity theories 630 (**1**)
normal stress 221 (**1**), 440 (**1**), 442 (**1**), 527 (**1**), 14 (**2**), 61 (**2**), 63 (**2**), 395 (**2**), 494 (**2**), 684 (**2**)
Norton creep law 228 (**2**)
null field 465 (**1**)
– patterns 498 (**1**)
numerical aperture (NA) 23 (**2**), 26 (**2**)
numerical modeling 208 (**1**), 271 (**1**), 475 (**1**), 4 (**2**), 358 (**2**)

Oliver and Pharr method 162 (**1**)
on-wafer floating pad technology 139 (**2**)
ongoing AST 194 (**2**)
ongoing ESS 194 (**2**)
Optica Cross-connect (OXC) 371 (**2**)
optical beam 24 (**2**), 300 (**2**), 323 (**2**)
optical code division multiplexing 132 (**1**)
optical efficiency 342 (**2**), 373 (**2**)
optical fiber 67 (**1**), 68 (**1**), 79 (**1**), 86 (**1**), 112 (**1**), 114 (**1**), 142 (**1**), 269 (**1**)–272 (**1**), 274 (**1**), 275 (**1**), 483 (**1**), 572 (**1**), 576 (**1**), 609 (**1**), 683 (**1**), 4 (**2**), 23 (**2**), 24 (**2**), 37 (**2**), 63 (**2**), 151 (**2**), 157 (**2**), 160 (**2**), 166 (**2**), 226 (**2**), 365 (**2**), 367 (**2**), 478 (**2**),

488 (**2**), 490 (**2**), 498 (**2**), 502 (**2**), 512 (**2**), 518 (**2**)
– curvature   271 (**1**)
– reliability   575 (**1**)
– strength   582 (**1**)
– system   270 (**1**)
optical networking forum (OIF)   365 (**2**)
optical power loss   72 (**1**)
optimal fitting of a regression model objective   227 (**1**)
opto-electrical circuit board (OECB)   369 (**2**), 380 (**2**)
optoelectronic   139 (**1**), 273 (**1**), 557 (**1**), 3 (**2**), 14 (**2**), 151 (**2**), 165 (**2**), 203 (**2**), 204 (**2**), 225 (**2**), 230 (**2**), 299 (**2**), 311 (**2**), 330 (**2**), 338 (**2**), 383 (**2**), 403 (**2**), 473 (**2**), 480 (**2**), 487 (**2**), 498 (**2**), 503 (**2**), 522 (**2**)
optoelectronic holography (OEH)   330 (**2**)
optoelectronic laser interferometric microscopy (OELIM)   324 (**2**), 332 (**2**), 338 (**2**)
optoelectronics package   483 (**1**)
organic polymers   142 (**1**)
organic solderability preservative (OSP)   302 (**1**), 313 (**1**), 325 (**1**), 381 (**1**), 272 (**2**)
organic surface protection (OSP)   597 (**2**)
outer lead bonding (OLB)   76 (**2**)

package to board interconnection shear strength (PBISS)   263 (**2**)
packaging   3 (**1**), 4 (**1**), 35 (**1**), 261 (**1**), 262 (**1**), 135 (**2**), 177 (**2**), 255 (**2**), 294 (**2**), 302 (**2**), 311 (**2**), 320 (**2**), 348 (**2**), 358 (**2**), 383 (**2**), 390 (**2**), 432 (**2**), 433 (**2**), 442 (**2**), 450 (**2**), 473 (**2**), 477 (**2**), 487 (**2**), 489 (**2**), 495 (**2**), 522 (**2**), 527 (**2**), 529 (**2**), 546 (**2**), 572 (**2**), 607 (**2**), 611 (**2**), 613 (**2**), 625 (**2**), 634 (**2**), 636 (**2**), 652 (**2**), 657 (**2**), 660 (**2**), 697 (**2**)
pad design   302 (**1**)
pad metallurgy   74 (**2**), 273 (**2**)
parameterized polymer modeling (PPM)   53 (**1**), 60 (**1**), 62 (**1**)
Paris-Erdogan equation   226 (**2**)
partial differential equation (PDE)   207 (**1**), 324 (**2**)
partly state dependent modeling   10 (**1**), 34 (**1**), 35 (**1**)
passive alignment method   163 (**2**), 166 (**2**)
Pb-free HASL   381 (**1**)
Pb-free solder   351 (**1**), 267 (**2**), 274 (**2**)
– joint failure   388 (**1**)
– reliability   401 (**1**)

Pb-free solder alloys   353 (**1**), 380 (**1**), 384 (**1**), 401 (**1**), 404 (**1**), 429 (**1**), 273 (**2**)
Pb-Sn alloys   273 (**2**)
Peck and Black equations   227 (**2**)
peel test   405 (**2**), 426 (**2**)
peeling failure   412 (**1**)
peeling stress   7 (**2**), 61 (**2**)
percolation probability   581 (**2**), 585 (**2**), 586 (**2**), 594 (**2**)
percolation theory   538 (**2**), 574 (**2**), 578 (**2**), 581 (**2**), 612 (**2**)
peripheral array configuration   289 (**1**)
personal digital assistant (PDA)   253 (**2**), 259 (**2**), 572 (**2**), 621 (**2**)
phase fraction diagram   328 (**1**)
phase grating   461 (**1**), 479 (**1**)
phase separation   342 (**2**)
phase shifter   115 (**1**)
phase shifting technique   512 (**1**)
photo detector (PD)   366 (**2**), 489 (**2**), 513 (**2**)
photo-sensitive glasses   112 (**1**)
photoluminescence (PL)   346 (**2**)
photonic   65 (**1**), 114 (**1**), 136 (**1**), 270 (**1**), 272 (**1**), 475 (**1**), 555 (**1**), 4 (**2**), 8 (**2**), 16 (**2**), 23 (**2**), 32 (**2**), 151 (**2**), 204 (**2**), 221 (**2**)
– element   137 (**1**)
– reliability engineering   270 (**1**), 223 (**2**)
– system   111 (**1**), 23 (**2**), 44 (**2**), 67 (**2**)
photopolymers   112 (**1**)
photorefractive effect   112 (**1**)
photorefractive materials   112 (**1**)
photoresist   145 (**1**), 154 (**2**)
photosensitive optical fiber   67 (**1**)
photovoltaic effect   112 (**1**)
physical nonlinearity   668 (**1**), 678 (**1**)
physical vapor deposition (PVD)   144 (**1**)
physically nonlinear axial vibrations   683 (**1**)
piezoelectric coefficients   669 (**1**)
piezoelectric rod   671 (**1**), 672 (**1**), 683 (**1**)
– subject   678 (**1**)
piezoelectricity   668 (**1**)
pigtails   272 (**1**)
pin through hole (PTH)   611 (**2**)
pin-transfer dispensing   517 (**2**)
pitch   285 (**1**), 286 (**1**), 170 (**2**), 267 (**2**), 284 (**2**), 654 (**2**)
pitfall   204 (**2**), 219 (**2**), 255 (**2**), 268 (**2**)
pitting corrosion   287 (**2**)
placement machine   143 (**2**)
planar lightwave circuit (PLC)   225 (**2**), 520 (**2**)
– pigtailing   520 (**2**)

plasma enhanced CVD (PECVD) 187 (**1**)
plasma power 197 (**1**)
plastic ball grid array (PBGA) 263 (**2**)
plastic deformation 162 (**1**), 169 (**1**), 214 (**1**), 222 (**1**), 334 (**1**), 491 (**1**)
plastic encapsulated microcircuit (PEM) 494 (**1**)
plastic model 212 (**1**), 214 (**1**)
plastic package assembly reliability 289 (**1**)
plastic packages 288 (**1**)
plastic quad flat package (PQFP) 496 (**1**), 542 (**1**)
plastic strain 145 (**1**), 153 (**1**), 214 (**1**), 223 (**1**), 300 (**1**), 317 (**1**), 401 (**1**), 430 (**1**)–432 (**1**), 443 (**1**)–447 (**1**), 552 (**2**), 562 (**2**), 701 (**2**)
plastic wire-bond ball grid array 283 (**1**), 288 (**1**)
plated-through-hole (PTH) 289 (**1**), 302 (**1**), 502 (**1**)
plating material 424 (**1**)
pneumatic piston 605 (**1**)
Poincare map 637 (**1**), 643 (**1**), 644 (**1**)
point of presence (POP) 362 (**2**)
Poisson's ratio 16 (**1**), 23 (**1**), 26 (**1**), 147 (**1**), 159 (**1**), 167 (**1**), 212 (**1**), 398 (**1**), 524 (**1**), 572 (**1**), 635 (**1**), 234 (**2**), 347 (**2**), 409 (**2**), 413 (**2**), 422 (**2**), 451 (**2**), 481 (**2**), 549 (**2**), 653 (**2**), 687 (**2**)
poly-arylethers (PAE) 704 (**2**)
polybutadiene rubber 5 (**1**)
polycarbonate 4 (**1**)
polyethylene 5 (**1**)
poly(ethylene terephthalate) (PET) 403 (**2**), 419 (**2**)
polyimide (PI) 142 (**1**), 145 (**1**), 79 (**2**), 138 (**2**), 313 (**2**), 364 (**2**), 413 (**2**)
polyisoprene rubber 5 (**1**)
polymer 3 (**1**), 65 (**1**), 113 (**1**), 142 (**1**), 157 (**1**), 185 (**1**), 212 (**1**), 263 (**1**), 269 (**1**), 277 (**1**), 585 (**1**), 13 (**2**), 108 (**2**), 131 (**2**), 138 (**2**), 244 (**2**), 370 (**2**), 379 (**2**), 403 (**2**), 529 (**2**), 534 (**2**), 540 (**2**), 574 (**2**), 589 (**2**), 612 (**2**)–614 (**2**)
– fiber Bragg grating (PFBG) 66 (**1**), 92 (**1**), 99 (**1**)
– film 137 (**1**)
– materials 3 (**1**), 65 (**1**)
PFBG–LED illumination 87 (**1**)
PFBG–SM LD illumination 92 (**1**)
polymer adhesion 535 (**2**)
polymer materials 313 (**2**)
polymer matrix 536 (**2**), 614 (**2**), 619 (**2**)
polymer-coated fiber 269 (**1**), 272 (**1**), 11 (**2**), 16 (**2**)
polymethylmethacrylate (PMMA) 84 (**1**)
– FBG 102 (**1**)
– PFBG 84 (**1**)
PolyMUMPS 304 (**2**)
polynomial model 238 (**1**)
polypropylene 5 (**1**)
polysilicon 324 (**2**)
polystyrene 4 (**1**)
popcorn 511 (**1**), 512 (**1**), 524 (**1**), 525 (**1**), 185 (**2**)
– cracking 524 (**1**), 13 (**2**)
– effect 511 (**1**)
positive displacement piston 517 (**2**)
post-screening experiment 256 (**1**)
potato chip effect 145 (**1**)
power cycling 315 (**1**), 389 (**1**), 400 (**1**), 413 (**1**), 520 (**1**), 8 (**2**), 87 (**2**), 96 (**2**), 125 (**2**), 126 (**2**), 192 (**2**), 215 (**2**), 293 (**2**), 698 (**2**)
power spectral density (PSD) 91 (**2**)
predictive modeling 262 (**1**), 557 (**1**), 222 (**2**), 447 (**2**)
pressurized blister test 404 (**2**)
pressurized bulge testing 165 (**1**)
printed circuit board (PCB) 264 (**1**), 352 (**1**), 377 (**1**), 459 (**1**), 482 (**1**), 556 (**1**), 8 (**2**), 244 (**2**), 313 (**2**), 366 (**2**), 380 (**2**), 573 (**2**), 599 (**2**), 604 (**2**), 692 (**2**)
printed wiring board (PWB) 283 (**1**), 295 (**1**), 313 (**1**), 629 (**1**), 631 (**1**), 108 (**2**), 135 (**2**), 257 (**2**), 270 (**2**), 271 (**2**), 276 (**2**), 294 (**2**), 652 (**2**)
pristine fiber 608 (**1**), 612 (**1**)
probabilistic design 46 (**2**), 49 (**2**), 66 (**2**), 230 (**2**)
probabilistic models 15 (**2**)
probability 183 (**1**), 184 (**1**), 217 (**1**), 235 (**1**), 261 (**1**), 319 (**1**), 388 (**1**), 455 (**1**), 571 (**1**), 574 (**1**), 581 (**1**), 589 (**1**)–591 (**1**), 15 (**2**), 35 (**2**), 44 (**2**), 67 (**2**), 87 (**2**), 104 (**2**), 177 (**2**), 184 (**2**), 201 (**2**), 203 (**2**), 209 (**2**), 229 (**2**), 230 (**2**), 253 (**2**), 268 (**2**), 389 (**2**), 443 (**2**), 533 (**2**), 574 (**2**), 578 (**2**), 589 (**2**), 607 (**2**)
– density distribution 217 (**1**), 197 (**2**)
– density function (PDF) 319 (**1**), 455 (**1**), 574 (**1**), 590 (**1**), 180 (**2**), 195 (**2**), 229 (**2**)
probability-based reliability engineering 205 (**2**)

INDEX

product development/verification tests (PDTs)   215 (**2**), 217 (**2**)
product level drop tests   258 (**2**)
product reliability   136 (**1**), 219 (**1**), 313 (**1**)
production-sampling AST   193 (**2**)
Prony series   11 (**1**), 52 (**1**)
pull strength   104 (**2**), 185 (**2**)
pull test   258 (**2**)
pure gold   102 (**2**)
pyramidal structure   639 (**2**)

quad flat pack no-lead (QFN)   270 (**2**), 272 (**2**), 642 (**2**)
quad flat package (QFP)   284 (**1**), 544 (**1**), 573 (**2**)
qualification tests (QTs)   260 (**1**), 483 (**1**), 210 (**2**), 212 (**2**), 214 (**2**), 217 (**2**)
quality assurance (QA)   283 (**1**)
quantum confined (QC)   342 (**2**)
quantum efficiency   348 (**2**)
quantum resistance   184 (**1**)

radial basis function (RBF) model   241 (**1**)
Raman microscopy   677 (**2**)
Raman spectroscopy   346 (**2**)
random vibration tests   215 (**2**)
ray tracing method   25 (**2**)
Rayleigh criterion   514 (**1**)
real-time observation   470 (**1**), 485 (**1**)
receiver   364 (**2**), 368 (**2**), 487 (**2**)
recrystallization   151 (**1**), 335 (**1**)
recrystallized grain boundaries (RGB)   275 (**2**)
reference beam   332 (**2**)
reference grating   510 (**1**)
reference mirror   506 (**1**)
reflected power spectrum   90 (**1**), 97 (**1**)
reflection peak   120 (**1**)
reflectivity spectrum   89 (**1**), 96 (**1**)
reflow process   285 (**1**), 298 (**1**), 329 (**1**), 387 (**1**), 413 (**1**), 460 (**1**), 491 (**1**)
reflow soldering   5 (**1**), 316 (**1**), 330 (**1**), 381 (**1**), 412 (**1**), 8 (**2**), 74 (**2**), 496 (**2**)
refractive index   68 (**1**), 69 (**1**), 84 (**1**), 113 (**1**), 115 (**1**), 118 (**1**), 120 (**1**)–123 (**1**), 128 (**1**), 273 (**1**)–275 (**1**), 27 (**2**), 387 (**2**), 477 (**2**), 491 (**2**), 503 (**2**), 520 (**2**)
refractive losses   27 (**2**)
regular Fabry-Perot etalon   114 (**1**)
regular Fabry-Perot filter   116 (**1**)
regular fiber   587 (**1**)
relative dispersion slope (RDS)   127 (**1**)

relative humidity (RH)   500 (**1**), 576 (**1**), 587 (**1**), 225 (**2**), 254 (**2**), 316 (**2**), 438 (**2**), 615 (**2**), 618 (**2**)
relaxation description   13 (**1**), 18 (**1**)
relaxation modulus functions   8 (**1**)
relaxation shear modulus   27 (**1**)
relaxation time   7 (**1**)
reliability   3 (**1**), 53 (**1**), 144 (**1**), 217 (**1**), 270 (**1**), 283 (**1**), 288 (**1**), 292 (**1**), 315 (**1**), 337 (**1**), 395 (**1**), 454 (**1**), 475 (**1**), 524 (**1**), 557 (**1**), 571 (**1**), 585 (**1**), 591 (**1**), 684 (**1**), 4 (**2**), 15 (**2**), 46 (**2**), 47 (**2**), 83 (**2**), 87 (**2**), 97 (**2**), 104 (**2**), 110 (**2**), 124 (**2**), 129 (**2**), 136 (**2**), 152 (**2**), 177 (**2**), 178 (**2**), 186 (**2**), 187 (**2**), 203 (**2**), 204 (**2**), 208 (**2**), 209 (**2**), 211 (**2**)–213 (**2**), 230 (**2**), 253 (**2**), 267 (**2**), 268 (**2**), 281 (**2**), 293 (**2**), 294 (**2**), 300 (**2**), 301 (**2**), 303 (**2**), 316 (**2**), 348 (**2**), 357 (**2**), 362 (**2**), 363 (**2**), 372 (**2**), 383 (**2**), 384 (**2**), 388 (**2**), 395 (**2**), 397 (**2**), 399 (**2**), 403 (**2**), 433 (**2**), 442 (**2**), 473 (**2**), 488 (**2**), 513 (**2**), 522 (**2**), 554 (**2**), 559 (**2**), 572 (**2**), 602 (**2**), 612 (**2**), 623 (**2**), 630 (**2**), 644 (**2**), 647 (**2**), 652 (**2**), 677 (**2**), 691 (**2**)
– costs money   208 (**2**)
– distribution function (RDF)   571 (**1**), 574 (**1**)
– engineering   270 (**1**), 319 (**1**), 204 (**2**), 205 (**2**), 214 (**2**)
– failures   6 (**1**), 143 (**1**), 205 (**1**), 217 (**1**), 262 (**1**), 270 (**1**), 289 (**1**), 303 (**1**), 314 (**1**), 395 (**1**), 455 (**1**), 4 (**2**), 13 (**2**), 182 (**2**), 210 (**2**), 258 (**2**), 317 (**2**)
– hardware   270 (**1**), 191 (**2**), 294 (**2**)
– lead-free CSPs in thermal cycling   337 (**1**)
– lead-free lead-free CSPs in drop testing   341 (**1**)
– models   389 (**1**)
– software   550 (**1**), 110 (**2**), 188 (**2**)
– standards   135 (**1**), 483 (**1**), 207 (**2**), 208 (**2**)
– test   199 (**1**), 260 (**1**), 454 (**1**)
– testing   305 (**1**), 315 (**1**), 316 (**1**), 335 (**1**), 483 (**1**), 4 (**2**), 255 (**2**), 267 (**2**), 280 (**2**), 294 (**2**), 314 (**2**), 341 (**2**), 546 (**2**), 685 (**2**)
reparability   206 (**2**)
repeatability   126 (**1**), 603 (**1**), 110 (**2**), 206 (**2**), 259 (**2**), 370 (**2**), 399 (**2**)
repeated stress concentration   413 (**1**)
residual resistivity ratio (RRR)   151 (**1**)
residual strain   96 (**2**)
residual stress   4 (**1**), 145 (**1**), 150 (**1**), 156 (**1**), 163 (**1**), 166 (**1**), 170 (**1**), 385 (**1**), 460 (**1**),

267 (**2**), 405 (**2**), 409 (**2**), 413 (**2**), 423 (**2**), 432 (**2**), 700 (**2**)
resin   6 (**1**), 35 (**1**), 39 (**1**)–41 (**1**), 55 (**1**), 336 (**1**), 459 (**1**), 503 (**1**), 109 (**2**), 244 (**2**), 272 (**2**), 441 (**2**), 509 (**2**), 613 (**2**)
resistance heater   465 (**1**)
resistance welding   483 (**1**), 488 (**2**)
response surface modeling (RSM)   206 (**1**), 236 (**1**), 242 (**1**)
– method   226 (**1**)
response time   129 (**1**)
response transition   651 (**1**)
Restriction of Hazardous Substances (ROHS)   429 (**1**)
rework methods   384 (**1**)
rheologically simple   20 (**1**)
rise time   124 (**1**)
risk   283 (**1**), 285 (**1**), 313 (**1**), 319 (**1**), 337 (**1**), 386 (**1**), 81 (**2**), 180 (**2**), 263 (**2**), 267 (**2**), 268 (**2**), 291 (**2**), 318 (**2**), 372 (**2**), 514 (**2**), 547 (**2**)
robust design   206 (**1**), 226 (**1**), 228 (**1**), 251 (**1**), 40 (**2**), 41 (**2**), 190 (**2**)
robust design objective   228 (**1**)
robust scheme   470 (**1**)
robustness   140 (**1**), 205 (**1**), 260 (**1**), 263 (**1**), 23 (**2**), 187 (**2**), 192 (**2**), 199 (**2**), 397 (**2**)
root cause   205 (**1**), 215 (**1**), 337 (**1**), 384 (**1**), 204 (**2**), 255 (**2**), 348 (**2**)
– analysis   215 (**1**), 177 (**2**), 180 (**2**), 294 (**2**)
root of the sum of the squares (RSS)   331 (**2**)
Rouard method   82 (**1**)
rubber   4 (**1**), 5 (**1**), 479 (**1**), 605 (**1**), 392 (**2**)
rupture strain   419 (**2**), 421 (**2**)

safe stress model   598 (**1**)
sample preparation   468 (**1**), 602 (**1**)
sandwich beam shear tool (SBT)   39 (**1**)
scanning acoustic microscopy (SAM)   321 (**1**), 523 (**1**), 539 (**1**), 541 (**1**), 282 (**2**), 353 (**2**), 644 (**2**)
scanning electron microscopy (SEM)   148 (**1**), 602 (**1**), 72 (**2**), 127 (**2**), 234 (**2**), 350 (**2**), 512 (**2**)
scanning force microscopy (SFM)   233 (**2**), 238 (**2**), 239 (**2**), 242 (**2**), 250 (**2**)
scanning probe microscopy   185 (**1**)
scanning transmission electron microscopy-atomic number (STEM-Z)   343 (**2**)
scratch test   404 (**2**)
screening experiment   255 (**1**)

screening objective   227 (**1**)
sea of leads (SoL)   139 (**2**), 140 (**2**)
second-order elastic constants   137 (**1**)
seeding by particle   602 (**1**)
self-assembled monolayer (SAM)   301 (**2**), 316 (**2**), 318 (**2**), 621 (**2**)
self-pumped phase conjugators (SPPC)   112 (**1**)
semi-crystalline polymers   5 (**1**)
semiconductor optical amplifier (SOA)   489 (**2**)
sequential RSM   244 (**1**)
shadow moiré   459 (**1**), 476 (**1**), 509 (**1**), 652 (**2**), 656 (**2**), 661 (**2**)
Shapiro-Wilk test   319 (**1**)
shear creep compliance   24 (**1**)
shear modulus   27 (**1**), 212 (**1**), 60 (**2**), 97 (**2**), 347 (**2**)
shear relaxation modulus   15 (**1**), 19 (**1**)
– curve   33 (**1**)
shear strain   219 (**1**), 481 (**1**)
shear strength   93 (**2**), 102 (**2**), 114 (**2**), 263 (**2**), 451 (**2**), 547 (**2**), 554 (**2**), 555 (**2**), 599 (**2**), 601 (**2**)
shear stress   19 (**1**), 28 (**1**), 41 (**1**), 218 (**1**), 334 (**1**), 527 (**1**), 61 (**2**), 67 (**2**), 136 (**2**), 275 (**2**), 352 (**2**), 394 (**2**), 493 (**2**), 510 (**2**), 511 (**2**), 537 (**2**), 684 (**2**)
shearing strength   390 (**1**)
shearing stress   474 (**2**)
shock load   302 (**1**), 556 (**1**), 261 (**2**)
shock response spectrum (SRS)   259 (**2**)
shock test   556 (**1**), 262 (**2**), 642 (**2**)
short bare fibers   685 (**1**)
short-range thickness variation SR-TV   343 (**2**)
short-term reliability   272 (**1**), 9 (**2**), 187 (**2**), 215 (**2**)
shutter   129 (**1**)
side-contacted nanotubes   197 (**1**)
silane   393 (**2**), 506 (**2**), 508 (**2**)
silica glass   65 (**1**)
– fiber   571 (**1**)
silicon nitride film   142 (**1**)
silicon optical bench (SiOB)   152 (**2**), 162 (**2**)
silicon-on-insulator (SOI)   135 (**1**)
silicon-on-silicon flip-chip assemblies   12 (**2**)
silver-tin   9 (**2**)
simulated annealing   34 (**2**), 35 (**2**)
single mode laser diode (SM LD)   71 (**1**), 95 (**1**)
single walled CNT (SWCNT)   182 (**1**), 183 (**1**)
single-coated fiber   269 (**1**)

INDEX

single-mode fiber (SMF) 27 (**2**), 151 (**2**), 167 (**2**), 330 (**2**), 333 (**2**), 362 (**2**), 365 (**2**), 489 (**2**), 492 (**2**), 513 (**2**)
single-region power-law model 596 (**1**)
sinusoidal vibration tests 215 (**2**)
SiO_2 185 (**1**)
small displacement 558 (**1**)
small form factor pluggable (SFP) 364 (**2**)
Sn-Ag-Bi 353 (**1**), 412 (**1**)
Sn-Ag-Cu 351 (**1**), 353 (**1**), 359 (**1**), 375 (**1**), 378 (**1**), 389 (**1**), 412 (**1**)
Sn-Ag-Cu alloys 378 (**1**)
Sn-Cu 331 (**1**), 353 (**1**), 355 (**1**), 378 (**1**), 412 (**1**)
– alloy 378 (**1**)
Sn-Pb eutectic solder 351 (**1**), 273 (**2**)
Sn-Pb solder 352 (**1**), 384 (**1**), 386 (**1**)–388 (**1**), 394 (**1**), 401 (**1**), 405 (**1**), 411 (**1**), 413 (**1**), 423 (**1**), 424 (**1**), 433 (**1**), 446 (**1**), 273 (**2**)
Sn-Zn-Bi 353 (**1**), 412 (**1**)
SnPb-based solders 313 (**1**)
soft solder 9 (**2**), 129 (**2**), 348 (**2**)
soft-core model 578 (**2**)
solder 294 (**1**)–299 (**1**), 302 (**1**)–306 (**1**), 310 (**1**), 313 (**1**)–318 (**1**), 323 (**1**)–325 (**1**), 330 (**1**)–334 (**1**), 347 (**1**), 361 (**1**), 375 (**1**)–391 (**1**), 393 (**1**)–405 (**1**), 411 (**1**)–425 (**1**), 439 (**1**)–455 (**1**), 491 (**1**)–494 (**1**), 8 (**2**), 77 (**2**), 145 (**2**), 185 (**2**), 258 (**2**), 262 (**2**), 274 (**2**), 284 (**2**), 306 (**2**), 309 (**2**)–311 (**2**), 348 (**2**), 349 (**2**), 377 (**2**), 384 (**2**), 389 (**2**), 450 (**2**), 479 (**2**), 488 (**2**), 497 (**2**)–500 (**2**), 502 (**2**), 522 (**2**), 527 (**2**), 542 (**2**), 546 (**2**), 557 (**2**), 565 (**2**), 601 (**2**), 604 (**2**), 606 (**2**), 611 (**2**), 619 (**2**), 620 (**2**), 625 (**2**), 646 (**2**), 672 (**2**), 685 (**2**), 698 (**2**), 699 (**2**)
– ball 483 (**1**), 489 (**1**), 517 (**1**), 74 (**2**), 136 (**2**), 263 (**2**), 270 (**2**), 274 (**2**), 282 (**2**), 476 (**2**), 700 (**2**)
– bump 143 (**1**), 313 (**1**), 337 (**1**)–339 (**1**), 401 (**1**), 502 (**1**), 74 (**2**), 306 (**2**), 652 (**2**), 653 (**2**)
– column interposer 294 (**1**)
– fatigue failure mode 412 (**1**), 415 (**1**)
– glass 143 (**1**), 358 (**1**), 389 (**1**), 502 (**1**), 12 (**2**), 74 (**2**), 143 (**2**)
– interconnection 283 (**1**), 306 (**1**), 313 (**1**), 328 (**1**), 351 (**1**), 256 (**2**)
– interface 272 (**1**), 298 (**1**), 310 (**1**), 324 (**1**), 330 (**1**)–333 (**1**), 339 (**1**), 347 (**1**), 357 (**1**), 411 (**1**)–413 (**1**), 415 (**1**)–423 (**1**), 525 (**1**), 270 (**2**), 356 (**2**), 493 (**2**), 495 (**2**)
– mask 306 (**1**), 311 (**1**), 74 (**2**), 260 (**2**), 273 (**2**), 450 (**2**), 657 (**2**), 658 (**2**), 674 (**2**)
– – defined (SMD) 260 (**2**)
– materials 272 (**1**)
– pad 297 (**1**), 302 (**1**)–304 (**1**), 325 (**1**), 338 (**1**), 347 (**1**), 398 (**1**), 411 (**1**), 417 (**1**), 491 (**1**), 525 (**1**), 76 (**2**), 110 (**2**), 140 (**2**), 185 (**2**), 312 (**2**)
– paste 296 (**1**), 298 (**1**), 304 (**1**), 316 (**1**), 337 (**1**), 341 (**1**), 354 (**1**), 382 (**1**), 387 (**1**), 77 (**2**), 542 (**2**), 573 (**2**)
– reflow 285 (**1**), 298 (**1**), 316 (**1**), 337 (**1**), 380 (**1**), 386 (**1**), 388 (**1**), 413 (**1**), 429 (**1**), 513 (**1**), 525 (**1**), 75 (**2**), 102 (**2**), 110 (**2**), 142 (**2**), 185 (**2**), 557 (**2**), 621 (**2**)
– reflow process 429 (**1**), 523 (**1**)
– strength 141 (**1**), 273 (**1**), 318 (**1**), 342 (**1**), 391 (**1**), 414 (**1**), 424 (**1**), 262 (**2**), 263 (**2**)
– volume 303 (**1**)
– wetting 330 (**1**), 351 (**1**), 357 (**1**), 358 (**1**), 383 (**1**), 387 (**1**), 143 (**2**)
solder joint 220 (**1**), 225 (**1**), 272 (**1**), 289 (**1**), 297 (**1**), 351 (**1**), 356 (**1**), 380 (**1**), 413 (**1**), 8 (**2**), 75 (**2**), 80 (**2**), 136 (**2**), 142 (**2**), 163 (**2**), 192 (**2**), 225 (**2**), 260 (**2**), 261 (**2**), 268 (**2**), 275 (**2**)–277 (**2**), 279 (**2**), 281 (**2**), 293 (**2**), 309 (**2**), 313 (**2**), 432 (**2**), 498 (**2**), 499 (**2**), 502 (**2**), 552 (**2**), 604 (**2**), 651 (**2**), 685 (**2**), 698 (**2**), 700 (**2**)
– failure 220 (**1**), 225 (**1**), 290 (**1**), 295 (**1**), 306 (**1**), 388 (**1**)–390 (**1**), 411 (**1**)–414 (**1**), 418 (**1**)–425 (**1**), 434 (**1**), 451 (**1**), 281 (**2**)
– fracture 220 (**1**), 225 (**1**), 390 (**1**), 396 (**1**), 413 (**1**), 417 (**1**)–419 (**1**), 424 (**1**), 454 (**1**), 280 (**2**)
– reliability 285 (**1**), 292 (**1**), 297 (**1**), 306 (**1**), 389 (**1**)
solderability 302 (**1**), 325 (**1**), 336 (**1**), 384 (**1**), 185 (**2**), 272 (**2**), 284 (**2**)
soldered assembly 315 (**1**), 309 (**2**)
soldering 253 (**1**), 254 (**1**), 314 (**1**), 316 (**1**), 323 (**1**), 330 (**1**), 337 (**1**), 348 (**1**), 351 (**1**), 74 (**2**), 185 (**2**), 302 (**2**), 488 (**2**), 493 (**2**), 494 (**2**), 511 (**2**), 522 (**2**), 552 (**2**), 563 (**2**), 602 (**2**), 604 (**2**), 606 (**2**), 633 (**2**), 646 (**2**)
solid surface 183 (**1**), 192 (**1**), 360 (**1**)
solidification 324 (**1**), 328 (**1**), 330 (**1**), 347 (**1**), 352 (**1**)
– structure 325 (**1**)
Solomon's low cycle fatigue model 432 (**1**)

730 INDEX

specimen gratings   461 (**1**), 479 (**1**)
spectral dispersion   89 (**1**)
spectral heat generation density   87 (**1**), 93 (**1**)
spectral power density   71 (**1**), 86 (**1**), *see also* power spectral density (PSD)
splitter   506 (**1**), 186 (**2**), 332 (**2**), 384 (**2**), 487 (**2**), 656 (**2**)
spreading   145 (**1**), 357 (**1**)
– rate   360 (**1**)
spring   6 (**1**), 209 (**1**), 558 (**1**), 15 (**2**), 31 (**2**), 52 (**2**), 73 (**2**), 261 (**2**), 326 (**2**)
– constant   209 (**1**), 558 (**1**), 564 (**1**), 15 (**2**)
– element   6 (**1**), 209 (**1**), 561 (**1**), 52 (**2**)
staggered array configuration   289 (**1**)
stain rate   700 (**2**)
static fatigue   610 (**1**), 611 (**1**), 4 (**2**)
– test   600 (**1**), 605 (**1**), 620 (**1**), 623 (**1**)
static stress fatigue   186 (**2**)
static testing   19 (**1**)
– of creep compliances   24 (**1**)
statistical modeling   584 (**1**)
statistical testing   319 (**1**)
steady-state crack growth   223 (**1**)
steady-state creep strain rate   440 (**1**)
steady-state in-plane thermal field   638 (**1**)
steady-state peeling   409 (**2**)
steady-state temperature effect   633 (**1**)
steady-state thermal field   633 (**1**), 634 (**1**), 660 (**1**)
stencil design   302 (**1**)
stencil parameter   304 (**1**)
stiff adherends   473 (**2**)
stiffness coefficients   29 (**1**)
stiffness matrix   210 (**1**)
stochastic modeling   594 (**1**), 24 (**2**)
stochastical model   240 (**1**)
storage bulk modulus   39 (**1**)
storage longitudinal modulus   39 (**1**)
storage modulus curve   33 (**1**)
storage shear modulus   39 (**1**)
strain   7 (**1**), 97 (**1**), 139 (**1**), 143 (**1**), 173 (**1**), 185 (**1**), 210 (**1**), 211 (**1**), 213 (**1**), 262 (**1**), 270 (**1**), 294 (**1**), 313 (**1**), 316 (**1**), 318 (**1**), 333 (**1**), 385 (**1**), 389 (**1**), 419 (**1**), 430 (**1**), 448 (**1**), 460 (**1**), 475 (**1**), 568 (**1**), 572 (**1**), 668 (**1**), 685 (**1**), 687 (**1**), 3 (**2**), 52 (**2**), 54 (**2**), 61 (**2**), 97 (**2**), 138 (**2**), 141 (**2**), 219 (**2**), 233 (**2**), 234 (**2**), 241 (**2**), 257 (**2**)–259 (**2**), 347 (**2**), 373 (**2**), 403 (**2**), 425 (**2**), 426 (**2**), 442 (**2**), 451 (**2**), 456 (**2**), 497 (**2**), 500 (**2**), 552 (**2**), 686 (**2**)
– analysis   430 (**1**), 462 (**1**), 480 (**1**)
– based approach   223 (**1**)
– constant   7 (**1**)
– distribution   218 (**1**), 385 (**1**), 421 (**1**), 574 (**1**), 578 (**1**), 258 (**2**), 418 (**2**), 677 (**2**)
– effective   7 (**1**)
– energy   166 (**1**), 210 (**1**), 213 (**1**)–215 (**1**), 225 (**1**), 404 (**1**), 433 (**1**), 533 (**1**), 673 (**1**), 692 (**1**), 347 (**2**), 410 (**2**), 476 (**2**), 547 (**2**), 686 (**2**)
– energy release rate (SERR)   166 (**1**), 169 (**1**), 529 (**1**), 549 (**1**)
– field   181 (**1**), 301 (**1**), 385 (**1**), 460 (**1**), 475 (**1**), 499 (**1**), 537 (**1**), 631 (**1**), 673 (**1**), 684 (**1**), 242 (**2**)
– gage techniques   460 (**1**)
– hardening coefficient   412 (**2**), 422 (**2**)
– inelastic   215 (**1**), 222 (**1**), 225 (**1**), 393 (**1**), 403 (**1**), 433 (**1**), 448 (**1**), 243 (**2**), 292 (**2**)
– initial   10 (**1**), 692 (**1**)
– plastic   145 (**1**), 153 (**1**), 223 (**1**), 300 (**1**), 317 (**1**)
– rate   166 (**1**), 214 (**1**), 224 (**1**), 334 (**1**), 342 (**1**), 344 (**1**), 393 (**1**), 430 (**1**), 442 (**1**), 491 (**1**), 537 (**1**), 605 (**1**), 52 (**2**), 125 (**2**), 257 (**2**), 262 (**2**), 279 (**2**), 497 (**2**), 537 (**2**)
– tensile   185 (**1**)
– tensor   532 (**1**)
– volumetric   14 (**1**)
strength   141 (**1**), 157 (**1**), 172 (**1**), 198 (**1**), 263 (**1**), 270 (**1**), 556 (**1**), 571 (**1**), 591 (**1**), 613 (**1**), 625 (**1**), 5 (**2**), 8 (**2**), 11 (**2**), 26 (**2**), 46 (**2**), 75 (**2**), 82 (**2**), 114 (**2**), 191 (**2**), 198 (**2**), 227 (**2**), 263 (**2**), 401 (**2**), 405 (**2**), 473 (**2**), 511 (**2**)
– distribution   357 (**1**), 426 (**1**), 576 (**1**), 582 (**1**), 584 (**1**), 585 (**1**), 591 (**1**), 595 (**1**), 606 (**1**), 655 (**1**), 196 (**2**), 229 (**2**), 258 (**2**)
– distribution function   229 (**2**)
– tests   590 (**1**)
stress   6 (**1**), 136 (**1**), 144 (**1**), 145 (**1**), 185 (**1**), 211 (**1**), 222 (**1**), 248 (**1**), 262 (**1**), 270 (**1**), 272 (**1**), 299 (**1**), 313 (**1**), 316 (**1**), 333 (**1**), 347 (**1**), 380 (**1**), 385 (**1**), 389 (**1**), 419 (**1**), 430 (**1**), 448 (**1**), 460 (**1**), 475 (**1**), 524 (**1**), 556 (**1**), 572 (**1**), 575 (**1**), 595 (**1**), 597 (**1**), 657 (**1**), 668 (**1**), 684 (**1**), 687 (**1**), 3 (**2**), 10 (**2**), 15 (**2**), 23 (**2**), 52 (**2**), 54 (**2**), 77 (**2**), 110 (**2**), 125 (**2**), 136 (**2**), 146 (**2**), 152 (**2**), 189 (**2**), 198 (**2**), 204 (**2**), 216 (**2**), 221 (**2**), 241 (**2**), 275 (**2**), 348 (**2**), 350 (**2**), 355 (**2**), 357 (**2**), 358 (**2**), 373 (**2**), 374 (**2**), 376 (**2**), 378 (**2**), 385 (**2**)–387 (**2**), 393 (**2**), 395 (**2**),

INDEX

397 (**2**), 404 (**2**), 409 (**2**), 411 (**2**), 442 (**2**), 473 (**2**), 494 (**2**), 497 (**2**), 502 (**2**), 523 (**2**), 574 (**2**), 681 (**2**), 686 (**2**), 688 (**2**), 690 (**2**), 694 (**2**), 697 (**2**)
– compensation layer (SCL)   138 (**2**)
– compressive   145 (**1**), 166 (**1**), 271 (**1**), 384 (**1**)
– cracking   220 (**1**), 225 (**1**), 253 (**1**), 298 (**1**), 325 (**1**), 603 (**1**), 87 (**2**), 185 (**2**)
– distribution   41 (**1**), 218 (**1**), 295 (**1**), 272 (**1**), 299 (**1**), 389 (**1**), 524 (**1**), 527 (**1**), 654 (**1**), 655 (**1**), 58 (**2**), 162 (**2**), 229 (**2**), 419 (**2**), 677 (**2**), 682 (**2**)
– – function   217 (**1**), 388 (**1**), 571 (**1**), 574 (**1**), 229 (**2**)
– field   157 (**1**), 220 (**1**), 536 (**1**), 654 (**1**), 661 (**1**), 242 (**2**)
– film   145 (**1**)
– generation   145 (**1**), 550 (**2**)
– gradient   385 (**1**), 350 (**2**)
– initial   10 (**1**), 392 (**1**)
– intensity factor (SIF)   524 (**1**), 525 (**1**), 528 (**1**), 226 (**2**), 248 (**2**), 451 (**2**), 466 (**2**)
– intrinsic   146 (**1**), 677 (**2**), 678 (**2**), 681 (**2**)
– rate   166 (**1**), 170 (**1**), 214 (**1**), 224 (**1**), 334 (**1**), 344 (**1**), 393 (**1**), 430 (**1**), 442 (**1**), 598 (**1**), 623 (**1**), 52 (**2**), 125 (**2**), 177 (**2**), 256 (**2**), 536 (**2**)
– relaxation   6 (**1**), 8 (**1**), 19 (**1**), 34 (**1**), 52 (**1**), 145 (**1**), 215 (**1**), 389 (**1**), 419 (**1**), 493 (**1**), 4 (**2**), 228 (**2**), 348 (**2**), 423 (**2**), 604 (**2**)
– residual   148 (**1**), 156 (**1**), 609 (**1**)
– screening   254 (**1**)–257 (**1**), 389 (**1**), 191 (**2**)
– service   613 (**1**)
– state   9 (**1**), 334 (**1**)
– tensile hydrostatic   155 (**1**)
– tensor   672 (**1**)
– testing   19 (**1**)–21 (**1**), 28 (**1**), 157 (**1**), 174 (**1**), 262 (**1**), 272 (**1**), 347 (**1**), 390 (**1**), 454 (**1**), 595 (**1**)–598 (**1**), 613 (**1**), 623 (**1**), 177 (**2**), 191 (**2**), 198 (**2**), 542 (**2**)
– thermal   145 (**1**), 156 (**1**), 222 (**1**), 388 (**1**), 398 (**1**), 401 (**1**)–405 (**1**), 500 (**1**), 532 (**1**), 627 (**1**), 654 (**1**), 662 (**1**), 185 (**2**), 201 (**2**), 217 (**2**), 404 (**2**), 703 (**2**)
– thin film   145 (**1**)
– volumetric   14 (**1**)
stress–strain relationship   10 (**1**), 220 (**1**), 269 (**1**), 421 (**1**), 683 (**1**), 688 (**1**), 694 (**1**), 421 (**2**), 422 (**2**), 424 (**2**), 427 (**2**), 497 (**2**), 500 (**2**)
stress-free nano-wires   563 (**1**)

– wire   559 (**1**), 560 (**1**)
stress/strain analysis   270 (**1**), 476 (**1**)
striction   262 (**1**), 303 (**2**)
structural analysis   209 (**1**), 270 (**1**), 277 (**1**), 3 (**2**), 480 (**2**)
structural nonlinearity   271 (**1**)
structural reliability   271 (**1**), 3 (**2**), 4 (**2**), 210 (**2**), 474 (**2**)
structure function   633 (**2**), 643 (**2**), 645 (**2**), 649 (**2**)
structure function evaluation methodology   630 (**2**), 636 (**2**)
structure zone model   148 (**1**)
stud bump   115 (**2**), 377 (**2**), 528 (**2**)
subpixel analysis   236 (**2**)
substrate   6 (**1**), 65 (**1**), 137 (**1**), 187 (**1**), 218 (**1**), 292 (**1**), 296 (**1**), 323 (**1**), 330 (**1**), 333 (**1**), 481 (**1**), 501 (**1**), 517 (**1**), 557 (**1**), 6 (**2**), 13 (**2**), 14 (**2**), 47 (**2**), 53 (**2**), 74 (**2**), 81 (**2**), 109 (**2**), 135 (**2**), 142 (**2**), 149 (**2**), 256 (**2**), 267 (**2**), 307 (**2**), 311 (**2**), 357 (**2**), 364 (**2**), 387 (**2**), 393 (**2**), 395 (**2**), 400 (**2**), 401 (**2**), 403 (**2**), 405 (**2**), 459 (**2**), 466 (**2**), 494 (**2**), 504 (**2**), 552 (**2**), 564 (**2**), 573 (**2**), 605 (**2**), 648 (**2**), 651 (**2**), 657 (**2**), 661 (**2**), 664 (**2**), 669 (**2**), 672 (**2**), 683 (**2**), 688 (**2**), 693 (**2**)
– dissolution   363 (**1**)
– temperature   149 (**1**)
surface insulation resistance (SIR)   384 (**1**), 264 (**2**), 266 (**2**)
surface laminar circuitry (SLC)   271 (**2**)
surface mount technology (SMT)   285 (**1**), 411 (**1**), 429 (**1**), 135 (**2**), 141 (**2**), 163 (**2**), 368 (**2**), 380 (**2**), 527 (**2**), 595 (**2**), 611 (**2**), 613 (**2**)
surface relief grating   67 (**1**)
surface tension   360 (**1**), 389 (**1**), 401 (**1**), 480 (**1**), 74 (**2**), 83 (**2**), 143 (**2**), 163 (**2**), 169 (**2**), 311 (**2**), 476 (**2**), 486 (**2**), 504 (**2**), 619 (**2**)
surface-mounted device (SMD)   595 (**2**), 602 (**2**), 618 (**2**)
switchable optical add/drop multiplexer (SOADM)   121 (**1**)
system camera (CCD)   331 (**2**)
system decomposition   31 (**2**), 34 (**2**)

Taguchi approach   251 (**1**)
Taguchi method   250 (**1**)
tape automated bonding (TAB)   506 (**1**), 72 (**2**), 75 (**2**), 80 (**2**)

– technology   506 (**1**)
target transmission   130 (**1**)
temperature coefficient of resistance (TCR)   541 (**2**), 554 (**2**)
temperature cycling   6 (**1**), 289 (**1**), 388 (**1**), 389 (**1**), 430 (**1**), 454 (**1**), 460 (**1**), 8 (**2**), 124 (**2**), 129 (**2**), 132 (**2**), 185 (**2**), 192 (**2**), 348 (**2**), 394 (**2**), 397 (**2**), 560 (**2**), 563 (**2**), 604 (**2**), 620 (**2**), 698 (**2**), 703 (**2**)
tensile creep compliance   25 (**1**)
tensile failure   386 (**2**)
tensile properties   435 (**1**)
tensile strain   426 (**2**), 502 (**2**)
tensile strength   156 (**1**), 174 (**1**), 175 (**1**), 185 (**1**), 419 (**1**), 96 (**2**), 386 (**2**), 403 (**2**), 419 (**2**)
tensile stress   271 (**1**), 355 (**2**), 357 (**2**), 425 (**2**), 506 (**2**), 691 (**2**)
tensile test program   434 (**1**)
tensile testing   21 (**1**), 157 (**1**)
termination metallurgy   384 (**1**)
ternary eutectic reaction   355 (**1**)
test assemblies   316 (**1**)
test results   169 (**1**), 306 (**1**)
test vehicle (TV)   298 (**1**)
testability   206 (**2**)
testing time   315 (**1**)
tetraethoxysilane (TEOS)   192 (**1**)
theoretical maximum stress   20 (**1**)
theory-of-elasticity method   5 (**2**)
theory-of-elasticity treatment   5 (**2**)
thermal aging   93 (**2**), 114 (**2**), 262 (**2**), 271 (**2**), 293 (**2**), 445 (**2**), 453 (**2**), 461 (**2**)
thermal capacitance   630 (**2**), 632 (**2**), 639 (**2**), 643 (**2**), 645 (**2**)
thermal coefficient of expansion (TCE)   384 (**2**), 386 (**2**), 388 (**2**), *see also* coefficient of thermal expansion (CTE)
thermal conductivity   84 (**1**), 186 (**1**), 212 (**1**), 519 (**1**), 97 (**2**), 389 (**2**), 392 (**2**), 496 (**2**), 603 (**2**), 623 (**2**), 639 (**2**)
thermal CVD   187 (**1**)
thermal cycle (TC)   293 (**1**), 306 (**1**), 449 (**1**), 454 (**1**), 489 (**1**), 357 (**2**), 691 (**2**), 701 (**2**)
– profiles   292 (**1**)
– range   297 (**1**)
– test   297 (**1**), 305 (**1**)
thermal cycling   4 (**1**), 5 (**1**), 186 (**1**), 286 (**1**), 291 (**1**), 299 (**1**)–302 (**1**), 348 (**1**), 555 (**1**), 87 (**2**), 139 (**2**), 192 (**2**), 215 (**2**), 220 (**2**), 242 (**2**), 255 (**2**), 273 (**2**), 274 (**2**), 386 (**2**), 488 (**2**), 510 (**2**), 519 (**2**), 539 (**2**), 542 (**2**), 557 (**2**), 604 (**2**), 618 (**2**), 698 (**2**), *see also* temperature cycling
– loading   450 (**1**)
– reliability   430 (**1**), 255 (**2**), 261 (**2**), 269 (**2**)
– – test   430 (**1**)
– test   316 (**1**), 431 (**1**), 268 (**2**)
thermal deformation   272 (**1**), 400 (**1**), 406 (**1**), 459 (**1**), 470 (**1**), 486 (**1**), 494 (**1**)
thermal effect   636 (**1**)
thermal expansion   4 (**1**), 34 (**1**), 42 (**1**), 65 (**1**), 90 (**1**), 155 (**1**), 164 (**1**), 186 (**1**), 198 (**1**), 222 (**1**), 270 (**1**), 300 (**1**), 306 (**1**), 481 (**1**), 7 (**2**), 14 (**2**), 59 (**2**), 63 (**2**), 87 (**2**), 473 (**2**), 572 (**2**), 604 (**2**)
– coefficient   82 (**1**), 137 (**1**), 145 (**1**), 155 (**1**), *see also* coefficient of thermal expansion (CTE)
thermal fatigue strength   411 (**1**)
thermal fatigue test   413 (**1**)
thermal field   206 (**1**), 657 (**1**), 661 (**1**)
– assumption   633 (**1**), 638 (**1**)
thermal fluctuations   577 (**1**)
thermal gradient   475 (**1**), 215 (**2**), 393 (**2**), 396 (**2**), 398 (**2**)
thermal load   544 (**1**), 642 (**1**), 658 (**1**), 661 (**1**), 3 (**2**), 242 (**2**), 695 (**2**)
thermal loading   270 (**1**), 389 (**1**), 516 (**1**), 233 (**2**), 651 (**2**), 668 (**2**)
thermal mechanical fatigue life assessment   405 (**1**)
thermal mismatch   6 (**1**), 215 (**1**), 389 (**1**), 398 (**1**), 403 (**1**), 405 (**1**), 686 (**1**), 244 (**2**), 306 (**2**), 410 (**2**)
thermal modeling   321 (**1**)
thermal properties   186 (**1**), 321 (**1**), 398 (**1**), 400 (**1**), 387 (**2**), 503 (**2**), 553 (**2**)
thermal resistance   291 (**1**), 602 (**2**), 604 (**2**), 630 (**2**), 632 (**2**), 637 (**2**), 639 (**2**), 643 (**2**), 644 (**2**), 649 (**2**)
thermal resonance   645 (**1**)
thermal response   643 (**1**), 644 (**1**)
thermal shock (TS)   213 (**1**), 291 (**1**), 299 (**1**), 337 (**1**), 383 (**1**), 449 (**1**), 454 (**1**), 215 (**2**), 396 (**2**), 698 (**2**)
– testing   337 (**1**)
thermal simulation   242 (**1**), 321 (**1**), 637 (**2**)
thermal strain   389 (**1**), 401 (**1**), 405 (**1**), 406 (**1**), 499 (**1**), 281 (**2**), 303 (**2**)
thermal stress   155 (**1**), 198 (**1**), 3 (**2**), 12 (**2**), 14 (**2**), 262 (**2**), 273 (**2**), 303 (**2**), 473 (**2**), 493 (**2**), 680 (**2**), 685 (**2**), 698 (**2**), 701 (**2**), 706 (**2**)

– modeling   7 (**2**)
thermal stress-induced voiding   155 (**1**)
thermal-mismatch strain   8 (**2**)
thermally matched assemblies   11 (**2**)
thermo-compression bonding   506 (**1**), 313 (**2**)
thermo-elastic   215 (**1**)
thermo-mechanical reliability   205 (**1**)
thermo-optic coefficient   65 (**1**)
thermo-optic effect   81 (**1**)
thermo-optical analysis   89 (**1**)
thermo-optical model   66 (**1**), 70 (**1**), 80 (**1**)
thermocompression bonding   75 (**2**), 81 (**2**)
thermodynamic   324 (**1**)
thermoelectric cooler (TEC)   469 (**1**), 483 (**1**)
thermomechanical properties   388 (**2**)
thermomechanical reliability   268 (**2**)
thermomechanical shock test   315 (**1**)
thermomechanical testing   317 (**1**)
thermoplastic polymers   4 (**1**)
thermosetting polymers   4 (**1**), 5 (**1**)
thermosonic bonding   81 (**2**), 85 (**2**)
thick film   74 (**2**), 185 (**2**)
thickness variation (TV)   347 (**2**)
thin film   135 (**1**), 479 (**1**), 7 (**2**), 12 (**2**), 73 (**2**),
    101 (**2**), 127 (**2**), 137 (**2**), 141 (**2**), 149 (**2**),
    309 (**2**), 316 (**2**), 347 (**2**), 404 (**2**), 405 (**2**),
    426 (**2**), 496 (**2**), 677 (**2**), 683 (**2**), 708 (**2**)
– structure   142 (**1**), 156 (**1**), 175 (**1**), 7 (**2**)
thin isotropic laminate   630 (**1**), 632 (**1**)
thin small outline package (TSOP)   285 (**1**),
    14 (**2**), 272 (**2**)
thin-profile fine-pitch BGA (TFBGA)   270 (**2**)
thixotropic index (TI)   386 (**2**), 389 (**2**)
three-part-structure (TPS)   147 (**2**)
– probe   147 (**2**)
time dependent creep strain   430 (**1**)
time pressure dispensing   516 (**2**)
time–temperature superposition principle
    23 (**1**), 32 (**1**)
time-dependent modeling   4 (**1**)
time-dependent thermal field   638 (**1**)
time-to-failure (TTF)   319 (**1**), 587 (**1**), 588 (**1**),
    595 (**1**), 597 (**1**), 600 (**1**), 611 (**1**), 618 (**1**),
    623 (**1**), 224 (**2**), 527 (**2**), *see also*
    cycles-to-failure (CTF)
time-to-market   3 (**1**), 205 (**1**), 315 (**1**), 454 (**1**),
    203 (**2**), 208 (**2**), 253 (**2**), 254 (**2**), 399 (**2**)
tin whiskering   384 (**1**)
tin whiskers   385 (**1**), 386 (**1**), 267 (**2**)
tin-lead solder   352 (**1**), 440 (**1**), 263 (**2**),
    527 (**2**)
tiny thermo-mechanical displacements   6 (**2**)

tire pressure monitoring system (TPMS)
    300 (**2**)
total deflection function   559 (**1**)
total internal reflection (TIR)   367 (**2**)
total life approach   431 (**1**)
total mass loss (TML)   511 (**2**)
total shear strain range fatigue model   432 (**1**)
total stress intensity factor   530 (**1**)
tougheners   392 (**2**)
trace fracture   285 (**2**)
transfer matrix method (TMM)   70 (**1**), 82 (**1**)
transient in-plane thermal field   661 (**1**)
transient moisture distribution   438 (**2**)
transient thermal field   633 (**1**), 660 (**1**)
– effect   638 (**1**)
transient transverse thermal field   642 (**1**)
transistor outline (TO)   489 (**2**)
transition probability distribution   581 (**1**)
transmission dispersion   130 (**1**), 131 (**1**)
transmission electron microscopy (TEM)
    343 (**2**)
transmission isolation   130 (**1**)
transmission loss   271 (**1**), 272 (**1**), 684 (**1**),
    4 (**2**), 15 (**2**), 546 (**2**)
transmission speed   126 (**1**)
transmitter   362 (**2**), 364 (**2**), 368 (**2**), 487 (**2**)
transverse optical (TO)   350 (**2**)
tri-material assembly   58 (**2**), 493 (**2**)
tri-material model   651 (**2**)
true Bragg wavelength shift   98 (**1**)
true interface fatigue mode   415 (**1**)
tunable laser   119 (**1**), 4 (**2**)
twist testing   281 (**2**)
twist-off technique   8 (**2**)
two-region power-law model   598 (**1**)
two-wave mixing (TWM)   112 (**1**)
Twyman-Green interferometry   475 (**1**),
    505 (**1**), 679 (**2**), 694 (**2**)

U-groove   515 (**2**)
U displacement field   466 (**1**), 477 (**1**), 494 (**1**)
ultimate tensile stress (UTS)   156 (**1**), 436 (**1**),
    437 (**1**), 439 (**1**)
ultra large scale integrated (ULSI)   137 (**1**),
    141 (**1**), 143 (**1**), 191 (**1**), 196 (**1**)
ultra short reach (USR)   364 (**2**)
– interconnects   364 (**2**)
– optical interconnects   366 (**2**)
ultra violet (UV)   371 (**2**), 501 (**2**), 521 (**2**),
    565 (**2**)
ultra-low expansion (ULE)   497 (**1**), 501 (**1**)
ultrasonic bonding   83 (**2**)

ultrasonic energy 81 (**2**), 125 (**2**)
uncoupled axial vibrations 673 (**1**)
under bump metalization (UBM) 143 (**1**), 683 (**2**)
underfill 6 (**1**), 34 (**1**), 35 (**1**), 37 (**1**), 220 (**1**), 450 (**1**)–453 (**1**), 481 (**1**), 501 (**1**), 502 (**1**), 516 (**1**), 432 (**2**), 532 (**2**), 621 (**2**), 625 (**2**), 651 (**2**), 662 (**2**), 666 (**2**), 668 (**2**), 671 (**2**), 673 (**2**), 674 (**2**), 685 (**2**), 688 (**2**), 698 (**2**)
– flow modeling 532 (**2**)
– polymers 6 (**1**)
uniform curing 113 (**1**), 118 (**1**)
universal testing machine 434 (**1**), 604 (**1**)
US Military Standards (MIL-STDs) 206 (**2**), 207 (**2**)
UV-ozone cleaning 104 (**2**)

V-groove 152 (**2**), 164 (**2**), 384 (**2**), 394 (**2**), 496 (**2**), 506 (**2**)
V-number 27 (**2**)
vapor deposition techniques 385 (**1**)
variable order boundary element method (VOBEM) 536 (**1**), 546 (**1**), 550 (**1**)
variable strain rate shear test 391 (**1**)
VCR during test 465 (**1**)
V displacement field 466 (**1**), 477 (**1**), 494 (**1**)
vector-loop-equations (VLEs) 328 (**2**)
verification tests 538 (**1**)
vertical cavity surface emitting laser (VCSEL) 139 (**1**), 37 (**2**), 63 (**2**), 67 (**2**), 223 (**2**), 363 (**2**), 370 (**2**), 380 (**2**), 489 (**2**)
– transreceiver 23 (**2**)
vertical geometry 191 (**1**)
very large scale integrated (VLSI) 137 (**1**), 190 (**2**), 703 (**2**)
very short reach (VSR) 364 (**2**)
– interconnects 364 (**2**), 380 (**2**)
very-thin-profile fine-pitch BGA (VFBGA) 270 (**2**)
vibration loading 454 (**1**)
virtual master grating 462 (**1**)
virtual prototyping 242 (**1**)
virtual thermo-mechanical prototyping method 205 (**1**)
visco-plastic phenomena 4 (**2**)
viscoelastic models 52 (**2**)
viscoplastic analysis 430 (**1**), 701 (**2**)
viscoplastic solid 698 (**2**)
viscoplastic strain 430 (**1**), 702 (**2**)
– energy density 703 (**2**)
visual inspection 305 (**1**)

volumetric creep compliance 15 (**1**), 25 (**1**)
von-Karman's equations 13 (**2**)

wafer 140 (**1**), 146 (**1**)–148 (**1**), 150 (**1**), 157 (**1**), 164 (**1**)–167 (**1**), 187 (**1**), 188 (**1**), 192 (**1**), 198 (**1**), 261 (**1**), 287 (**1**), 318 (**1**), 347 (**1**), 385 (**1**), 627 (**1**), 111 (**2**), 135 (**2**), 142 (**2**), 153 (**2**), 294 (**2**), 302 (**2**), 343 (**2**), 363 (**2**), 364 (**2**), 395 (**2**), 413 (**2**), 625 (**2**), 652 (**2**), 677 (**2**), 678 (**2**), 706 (**2**)
– curvature technique 164 (**1**)
– level CSP (WLCSP) 136 (**2**), 145 (**2**)
– level packaging (WLP) 135 (**2**), 149 (**2**)
– level test 145 (**2**), 149 (**2**)
– level underfill 141 (**2**)–144 (**2**)
– on wafer (WOW) 139 (**2**)
wafer-level capping 302 (**2**), 320 (**2**)
wafer-level packaging 299 (**2**), 320 (**2**)
warpage 6 (**1**), 398 (**1**), 476 (**1**), 508 (**1**), 509 (**1**), 514 (**1**), 305 (**2**), 309 (**2**), 652 (**2**), 661 (**2**)–663 (**2**), 666 (**2**), 670 (**2**), 672 (**2**)
– analysis 505 (**1**)
– control 5 (**1**)
wave soldering 316 (**1**), 383 (**1**), 185 (**2**)
– process 381 (**1**), 429 (**1**)
waveguide 65 (**1**), 67 (**1**), 74 (**1**), 76 (**1**), 117 (**1**), 121 (**1**), 122 (**1**), 25 (**2**), 313 (**2**), 367 (**2**), 376 (**2**), 380 (**2**), 521 (**2**)
wavelength division multiplexing (WDM) 68 (**1**), 114 (**1**), 126 (**1**)
– filters 132 (**1**)
– systems 114 (**1**)
wavelength selective 2×2 switch 117 (**1**)
WB-PBGA package 511 (**1**)
Weibull distribution 217 (**1**), 218 (**1**), 319 (**1**), 337 (**1**), 342 (**1**), 388 (**1**), 455 (**1**), 571 (**1**), 577 (**1**), 183 (**2**), 269 (**2**)
Weibull failure rate distribution 183 (**2**)
Weibull life distribution model 455 (**1**)
Weibull modulus 604 (**1**)
Weibull parameter 299 (**1**), 319 (**1**), 597 (**1**)
Weibull plot 337 (**1**), 341 (**1**), 577 (**1**), 578 (**1**), 607 (**1**), 608 (**1**), 610 (**1**), 183 (**2**)
Weibull reliability analysis 318 (**1**)
Weibull shape 319 (**1**), 320 (**1**), 183 (**2**)
wettability 360 (**1**), 361 (**1**), 378 (**1**), 382 (**1**), 384 (**1**), 411 (**1**), 458 (**2**)
wetting angle 360 (**1**), 361 (**1**)
wetting balance curve 362 (**1**)
wetting balance test 361 (**1**)
wetting curve 361 (**1**)
whole-field deformation 680 (**2**)

wide area vertical expansion (WAVE)   139 (**2**)
Wilcoxon Rank-Sum test   319 (**1**)
wire   146 (**1**), 180 (**1**), 183 (**1**), 253 (**1**),
    283 (**1**), 293 (**1**), 387 (**1**), 486 (**1**), 508 (**1**),
    557 (**1**), 564 (**1**), 566 (**1**)
– array (WA)   558 (**1**), 567 (**1**), 474 (**2**),
    475 (**2**), 479 (**2**)
– bond   293 (**1**), 511 (**1**), 121 (**2**), 124 (**2**),
    185 (**2**), 276 (**2**)
– bonding   486 (**1**), 294 (**2**)
– strength   108 (**2**)
wire-bond plastic ball grid array (WB-PBGA)
    486 (**1**)
wirebond failure   94 (**2**), 105 (**2**)
wirebond pull strength   102 (**2**), 104 (**2**),
    185 (**2**)
wirebond reliability   87 (**2**)
wirebond testing   89 (**2**)
wirebonding   72 (**2**), 79 (**2**), 111 (**2**), 116 (**2**),
    139 (**2**), 606 (**2**)
– machine   110 (**2**)

X-ray diffraction (XRD)   354 (**2**)
X-ray inspection   305 (**1**), 321 (**1**), 282 (**2**)
X-ray photoelectron spectroscopy (XPS)
    460 (**2**), 468 (**2**), 541 (**2**)
X-ray reflectometry method   146 (**1**)

yield strain   412 (**2**), 422 (**2**)
yield strength   156 (**1**), 335 (**1**)
yield stress   148 (**1**), 253 (**1**), 333 (**1**), 435 (**1**),
    439 (**1**)
Young's modulus   17 (**1**), 23 (**1**), 147 (**1**),
    156 (**1**), 159 (**1**), 167 (**1**), 185 (**1**), 198 (**1**),
    212 (**1**), 213 (**1**), 253 (**1**), 269 (**1**), 398 (**1**),
    501 (**1**), 524 (**1**), 558 (**1**), 562 (**1**), 635 (**1**),
    7 (**2**), 11 (**2**), 12 (**2**), 14 (**2**), 32 (**2**), 38 (**2**),
    39 (**2**), 241 (**2**), 248 (**2**), 260 (**2**), 374 (**2**),
    409 (**2**), 413 (**2**), 421 (**2**), 422 (**2**), 475 (**2**),
    476 (**2**), 510 (**2**), 547 (**2**), 549 (**2**), 668 (**2**),
    671 (**2**), 687 (**2**)

zero stress condition   394 (**2**)
zero stress temperature   396 (**2**), 402 (**2**)